Table of Formula Weights*

$AgBr$	187.772	$H_2C_2O_4$	90.035	MgO	40.304
$AgCl$	143.321	$H_2C_2O_4\ 2H_2O$	126.066	$Mg_2P_2O_7$	222.553
$Ag_2C_2O_4$	303.756	$HCOOH$	46.026	MnO_2	86.937
Ag_2CrO_4	331.730	HCl	36.461	$NaBr$	102.894
AgI	234.773	$HClO_4$	100.459	$NaCHO_2$	68.007
$AgNO_3$	169.873	HF	20.006	$NaC_2H_3O_2(NaOAc)$	82.034
$AgSCN$	165.95	HNO_2	47.013	$NaCl$	58.443
Al_2O_3	101.961	HNO_3	63.013	$NaCN$	49.007
As_2O_3	197.841	H_2O	18.015	Na_2CO_3	105.989
$BaCl_2$	208.24	H_3PO_4	97.995	$Na_2C_2O_4$	133.999
$BaCO_3$	197.34	$HgCl_2$	271.50	$NaHCO_3$	84.007
$BaCrO_4$	253.32	Hg_2Cl_2	472.09	NaH_2PO_4	119.977
BaO	153.33	$Hg(NO_3)_2$	324.60	Na_2HPO_4	141.959
$Ba(OH)_2$	171.34	H_2S	34.08	$NaNO_2$	68.995
BaS	169.39	H_2SO_4	98.07	$NaNO_3$	84.995
$BaSO_4$	233.39	KBr	119.002	Na_2O	61.979
$CaCO_3$	100.09	$KBrO_3$	167.000	$NaOH$	39.997
CaC_2O_4	128.10	KCl	74.551	Na_2SO_4	142.04
CaF_2	78.08	$KClO_3$	122.550	$Na_2S_2O_3 \cdot 5H_2O$	248.17
CaO	56.08	$KClO_4$	138.549	NH_3	17.030
$Ca(OH)_2$	74.09	KCN	65.116	N_2H_4	32.045
$CaSO_4$	136.14	K_2CrO_4	194.190	NH_4Cl	53.491
$Ce(HSO_4)_4$	528.38	$K_2Cr_2O_7$	294.185	N_2H_5Cl	68.506
$Ce(NO_3)_4 \cdot 2NH_4NO_3$	548.23	$KHC_8H_4O_4\ (KHP)$	204.223	$(NH_4)_3PO_4 \cdot 12MoO_3$	1876.34
$Ce(SO_4)_2$	332.24	KHC_2O_4	128.126	$(NH_4)_2PtCl_6$	443.88
$Ce(SO_4)_2 \cdot 2(NH_4)_2SO_4 \cdot 2H_2O$	632.53	$KHC_2O_4 \cdot H_2C_2O_4$	218.161	$Ni(C_4H_7O_2N_2)_2$	288.93
CO_2	44.010	KI	166.003	$PbCO_3$	267.2
Cr_2O_3	151.990	KIO_3	214.001	$PbCrO_4$	323.2
CuO	79.545	$KMnO_4$	158.034	$Pb(IO_3)_2$	557.0
$Cu(OH)_2$	97.561	K_2O	94.196	PbO	223.2
$CuSCN$	121.62	KOH	56.106	$PbSO_4$	303.3
FeO	71.846	K_2PtCl_6	486.00	P_2O_5	141.945
Fe_2O_3	159.692	$KSCN$	97.18	SO_2	64.059
Fe_3O_4	231.539	K_2SO_4	174.25	SO_3	80.06
FeS_2	119.97	$Mg(C_9H_6ON)_2 \cdot 2H_2O$	348.640	SnO_2	150.69
$FeSO_4$	151.90	$MgCO_3$	84.314	SrC_2O_4	175.64
$HAsO_2$	107.928	MgC_2O_4	112.325	U_3O_8	842.082
$HC_2H_3O_2\ (HOAc)$	60.052	$MgNH_4PO_4 \cdot 6H_2O$	245.406	$Zn_2P_2O_7$	304.70

*No more than three decimals are carried, although in some instances four could be retained according to the rules of significant figures.

QUANTITATIVE ANALYSIS

Sixth
Edition

QUANTITATIVE ANALYSIS

R. A. DAY, JR.
Emory University, Emeritus

A. L. UNDERWOOD
Emory University

PRENTICE HALL, *Englewood Cliffs, New Jersey 07632*

Library of Congress Cataloging-in-Publication Data

DAY, R. A. (REUBEN ALEXANDER), DATE
    Quantitative analysis / R.A. Day, Jr., A.L. Underwood. — 6th ed.
      p.   cm.

    Includes index.
    ISBN 0-13-747155-6 :
    1. Chemistry, Analytic—Quantitative.   I. Underwood, A. L.
(Arthur Louis), date   . II. Title.
QD101.2.D37   1991
545—dc20   90-21670
                                      CIP

Editorial/production supervision and
  interior design: Virginia Huebner
Acquisitions editor: Daniel Joraanstad
Cover design: William Frost Associates
Manufacturing buyer: Lori Bulwin
Prepress buyer: Paula Massenaro
Cover photo: Paul Silverman/Fundamental Photographs

Printed in the United States of America
10 9 8 7 6 5 4 3 2 1

ISBN 0-13-747155-6

Prentice-Hall International (UK) Limited, *London*
Prentice-Hall of Australia Pty. Limited, *Sydney*
Prentice-Hall Canada Inc., *Toronto*
Prentice-Hall Hispanoamericana, S.A., *Mexico*
Prentice-Hall of India Private Limited, *New Delhi*
Prentice-Hall of Japan, Inc., *Tokyo*
Simon & Schuster Asia Pte. Ltd., *Singapore*
Editora Prentice-Hall do Brasil, Ltda., *Rio de Janeiro*

Contents

Preface

"Analytical chemistry continues to be taught in a variety of ways at different schools. Courses are found at every level from the freshman to the senior year. Some are for chemistry majors; others are service courses for students in such fields as agriculture, biology, engineering, and medical technology. No single book can be optimized for all of these, nor can it accommodate the viewpoints of every teacher. Nevertheless we have tried to keep the book useful for several types of courses by avoiding extremes, writing clearly and carefully, making chapters (or small groups of chapters) as independent as possible, and utilizing experience with earlier editions."

This paragraph from the Preface to our 5th Edition is still appropriate. Nevertheless, although old friends will find the new edition recognizably "Day & Underwood," there have been numerous changes. The first eleven chapters have been modified the least, mostly by shortening Chapters 4 and 11, rewriting portions of Chapters 2, 7, and 8 which some students have found troublesome (propagation of errors, metallochromic indicators, and equilibria involving polyprotic acids), and introducing a more systematic approach to the stoichiometric calculations used in constructing titration curves. The concept of the equivalent and the normality concentration system, which many teachers prefer not to use, are in a separate section of Chapter 3 which is easily omitted. Chapter 21 now includes a full discussion of electronic balances, and several new laboratory experiments are found in Chapter 22.

A numbering scheme for sections and subsections of chapters has been introduced to help teachers assign and refer to specific topics. The increased number of topical headings may clarify for the student the organization within each chapter and make more explicit the purpose behind each discussion or exercise. We are grateful to Professor Joseph Topping of Towson State University for his help in this connection.

A new feature of this edition is our use of boxes. In some of these are found discussions of such topics as mass spectrometry, affinity chromatography, and differential pulse polarography which are not covered in many elementary courses but which students with curiosity may find interesting and, at the level written, understandable. Other boxes relate the narrowly analytical applications in the body of a chapter to a broader world of chemistry and sometimes biology. Teachers can easily inform students of their own expectations in regard to these side lights, although the more aggressive learners may find them interesting in any event.

Chapter 19 has been entirely redone to show more effectively what analytical chemists do and its importance to a technological society. The two major themes are the interactions of analytical chemistry with other fields and the importance of quality assurance in analytical work. Lastly, two supplements are available to support the text. The Solutions Manual provides complete solutions to problems in the text. The Laboratory Manual provides experimental procedures in a convenient paperback format.

Professor James Ingle of Oregon State University corrected some nonstandard terminology in Chapter 15 and also suggested changes in emphasis which reflect modern practice in emission spectroscopy. Professor Hubert L. Youmans, Western Carolina University, has again contributed several laboratory experiments. Professors Orest Popovych, Brooklyn College, CUNY, has offered helpful comments. We are also indebted to our other reviewers, Professors R. H. Dinius, Auburn University, M. Dale Hawley, Kansas State University, and Marvin Rowe, Texas A & M University. We thank Dan Joraanstad, the Science Editor at Prentice-Hall, for his enthusiastic encouragement, and we are greatly indebted to our Production Editor, Virginia Huebner, for both her help and her patience.

R. A. DAY, Jr.
Atlanta, Georgia　　　　　　　　　　　　　　　　A. L. UNDERWOOD

ABOUT THE AUTHORS

R. A. Day, Jr. received the Ph.D. in physical chemistry in 1940 from Princeton University. He served on the Emory University chemistry faculty from 1940 to 1981 except for a 2-year interval as a chemist on the Manhattan Project. He has published on ion exchange separations, kinetics of hydrocarbon oxidation, and organic electrochemistry. He is a member of Phi Beta Kappa and ODK, and has served as Chairman of the Georgia Section of the American Chemical Society and as Chairman of the Electroorganic Division of the Electrochemical Society. Professor Day won Emory's Williams Award for outstanding teaching in 1980.

A. L. Underwood received his Ph.D. from the University of Rochester in 1951. After a postdoctoral year in analytical chemistry at the Massachusetts Institute of Technology, he joined the Chemistry Department at Emory University, where he has served continuously. Professor Underwood is a member of Phi Beta Kappa, Sigma Xi, and the American Chemical Society. He is a Fellow of the American Association for the Advancement of Science. He has about 70 publications to his credit, of which the most recent involve surfactant micelles and their uses in analytical chemistry. In his earlier work, electroanalytical techniques were employed in studies of the redox properties of biological molecules.

QUANTITATIVE ANALYSIS

1

Introduction

ANALYTICAL CHEMISTRY

It was once possible to divide chemistry into several clear and well-defined branches—analytical, inorganic, organic, physical, and biological chemistry. Although there was always a certain overlap among these simple categories, it was not difficult to define the branches in terms that were acceptable to most chemists. It was generally clear into which category any particular chemist fitted, and a label such as "organic chemist" usually implied a reasonably clear picture of the sort of things such a chemist did.

Since World War II there has been a general blurring of the defined branches of chemistry. Actually, the boundaries between chemistry itself and other major sciences such as physics and biology are considerably less clear than they used to be. Fields such as chemical physics, biophysical chemistry, physical organic chemistry, geochemistry, and chemical oceanography have achieved recognition, although precise definitions of these fields may be difficult to formulate.

Analytical chemistry is concerned with the theory and practice of methods used to determine the composition of matter. In developing methods of analysis the analytical chemist feels free to draw upon the principles from any field of science—chemistry, physics, biology, engineering, computer science, etc. For example, instruments developed by physicists, such as the mass spectrometer, the x-ray dispersive spectrometer, and the infrared spectrophotometer, have found wide applications in solving analytical problems.

There are today many new techniques available for application to analytical problems, and for this reason the analytical chemist needs to have a good knowledge of a number of scientific fields. The explosion of technological developments in recent years has created analytical problems which demand increasingly sophisticated knowledge and instrumentation for their solution. Typical examples of such problems are determining traces of impurities at the parts-per-billion level in ultrapure semiconductor materials, deducing the sequence of about 20 different amino acids in a giant protein molecule, detecting traces of unusual molecules in the polluted atmosphere of a smog-bound city, determining pesticide residues at the parts-per-billion level in food products, and determining the nature and concentration of complex organic molecules in, say, the nucleus of a single cell.

The solutions to a host of problems such as these have been developed by research workers with the most diverse backgrounds. Workers in many fields are constantly confronted by analytical problems, and in many cases they work out their own solutions. What distinguishes the analytical chemist from other workers is an interest in the methods and techniques in their own right. Workers in other fields often need to develop new analytical methods for their own purposes, but their primary interests do not lie in the method itself. To analytical chemists, developing methods is the challenging part of the research. They are likely to be skeptical of data presented without a full disclosure of experimental details, and they retain a critical attitude toward results which some workers would like to accept so as to get on with other things. Analytical chemists deal with real, practical systems, and much of their effort is expended in an attempt to apply sound theory to actual chemical situations.

1.2

QUALITATIVE AND QUANTITATIVE ANALYSIS

Analytical chemistry can be divided into areas called qualitative analysis and quantitative analysis. *Qualitative* analysis deals with the identification of substances. It is concerned with *what* elements or compounds are present in a sample. The student's first encounter with qualitative analysis is often in the general chemistry course where a number of elements are separated and identified by precipitation with hydrogen sulfide. Organic products synthesized in the laboratory may be identified using such instrumental techniques as infrared and nuclear magnetic resonance spectroscopy.

Quantitative analysis is concerned with the determination of *how much* of a particular substance is present in a sample. The substance determined, often referred to as the *desired constituent* or *analyte,* may constitute a small or large part of the sample analyzed. If the analyte constitutes more than about 1% of the sample, it is considered a *major* constituent. It is considered *minor* if it amounts to 0.01 to 1% of the sample. Finally, a substance present to the extent of less than 0.01% is considered a *trace* constituent.

Another classification of quantitative analysis may be based upon the size of the sample available for analysis. The subdivisions are not clear-cut, but merge imperceptibly into one another and are roughly as follows: When a sample weighing more than 0.1 g is available, the analysis is spoken of as *macro; semimicro* analyses are performed on samples of perhaps 10 to 100 mg; *micro* analyses deal with samples weighing from 1 to 10 mg; and *ultramicro* analyses involve samples on the order of a microgram ($1 \mu g = 10^{-6} g$).

ANALYTICAL METHODOLOGY

In the introductory course in quantitative analysis students deal mainly with major constituents of macro samples. They seldom perform a *complete* quantitative analysis of a sample. A complete analysis actually consists of five main steps: (1) sampling, that is, selecting a representative sample of the material to be analyzed; (2) dissolution of the sample; (3) conversion of the analyte into a form suitable for measurement; (4) measurement; and (5) calculation and interpretation of the measurement. Often beginners carry out only steps 4 and 5, since these are usually the easiest ones.

In addition to the steps mentioned above, there are other operations that may be required. If the sample is a solid, it may be necessary to dry it before performing the analysis. An accurate measurement of the weight of the sample (the volume if it is a gas) must be made, since quantitative results are usually reported in relative terms, for example, the number of grams of analyte per 100 g of sample (percent by weight).

At this time we shall make some general comments on the steps in an analysis. In subsequent chapters these topics will be developed in much greater detail.

1.3a. *Sampling*

Beginning students seldom encounter the problem of sampling, since the samples they are given are usually homogeneous, or nearly so. Nevertheless, they should be aware of the importance of sampling and should know where to find proper directions when they are confronted with an unfamiliar problem. We shall discuss briefly the sampling of solids, liquids, and gases to give a general idea of the nature of the problems involved.

Before carrying out an analysis a chemist attempts to obtain a sample that is representative of all the components and their amounts as contained in the bulk sample. The process involves statistical reasoning, in that conclusions will be drawn about the composition of the bulk sample from the analysis of a very small portion of material.

Solids. Coal is a particularly difficult material to sample, and we shall use it to illustrate methods used for solid materials. The first step in the sampling procedure is to select a large portion of coal, called the *gross* sample, which though not homogeneous itself, represents the average composition of the entire mass. The size of the sample needed depends on such factors as particle size and homogeneity of the particles. In the case of coal the gross sample must be about 1000 lb if the particles are no greater than about 1 in. in any dimension.

There are many techniques used to obtain the gross sample. If the coal is in motion on a conveyor of some type, a definite fraction may be continuously diverted to give the gross sample. If, on the other hand, the coal is being shoveled from a car, every fiftieth shovelful might be placed aside to form the sample.

After the sample has been selected, it is ground or crushed and systematically mixed and reduced in size. One method used for reducing the sample of coal involves piling it in a cone with a shovel, flattening the cone, and dividing it into four equal parts, two of which are discarded. A mechanical device for subdividing the sample is called a *riffle*. The riffle consists of a row of small sloping chutes arranged so that alternate chutes discharge the sample in opposite directions. In this manner the sample is halved automatically.

In the laboratory further grinding of the sample may be done with a mortar and pestle. It is often necessary to grind a sample to pass through a sieve of a certain mesh. One hopes that the final laboratory sample, 1 g or so, is representative of the gross sample. The analytical data obtained cannot be better than the care exercised in the sampling procedure.

Liquids. If the liquid to be analyzed is homogeneous, the sampling procedure is straightforward. The process is much more difficult if the liquid is heterogeneous. In the case of a liquid circulating in, for example, a pipe system, samples are often taken from different points in the system. In a lake or river, samples may be taken at different locations and at different depths. Sometimes the analyst may not wish to have an average sample for the entire liquid system. For example, in testing the natural purification of a river contaminated with sewage, samples may be taken at a number of places downstream from the sewer outlet.

Devices called *grab samplers* may be used to collect samples from large bodies of water at various depths. Such a device consists of a sample bottle inside a metal container sufficiently heavy to force the empty bottle to the desired depth. The sample bottle is closed by a stopper, which has a line attached to it, and held by the person taking the sample. The device is lowered to the desired depth, the stopper is pulled out, and the sample bottle is filled. The sampler may have a ball float which automatically seals the bottle when it is filled.

Gases. There is much interest today in sampling the atmosphere because of efforts directed toward improvement of the quality of the air we breathe. Air is, of course, a complex mixture which contains particulate matter as well as numerous gaseous compounds. Its composition depends on a number of factors, such as location, temperature, wind, and rain.

In the collection of an atmospheric sample for analysis the volume taken and the rate and duration of sampling are important factors. The air is passed through a series of fine filters to isolate particulate matter, and through a column of a solution where a chemical reaction occurs to trap the desired component. After collection on a filter, particulate matter may be determined by chemical analysis or by weighing.

The general requirements for sampling solids, liquids, and gases may be found in general reference works.[1-4] Directions are also available in the publications of various groups, such as the American Society for Testing Materials and the American Oil Chemists' Society.

1.3b. Dissolving the Sample

Many of the samples analyzed in the beginning course in quantitative analysis are soluble in water. Generally speaking, however, naturally occurring materi-

[1] W. W. Walton and J. I. Hoffman, "Principles and Methods of Sampling," Chap. 4, p. 67, of I. M. Kolthoff and P. J. Elving, eds., *Treatise on Analytical Chemistry,* Part 1, Vol. 1, Interscience Publishers, Inc., New York, 1959.

[2] W. W. Anderson in "Standard Methods of Chemical Analysis," F. J. Welcher, ed., 6th ed., Vol. 2, Part A, p. 28, D. Van Nostrand, Princeton, N.J., 1963.

[3] W. V. Cropper, ibid., Vol. 2, Part A, p. 39.

[4] C. W. Wilson, ibid., Vol. 2, Part B, p. 1507.

als, such as ores, and metallic products, such as alloys, must be given special treatments to effect their solution. While each material may present a specific problem, the two most common methods employed in dissolving samples are (1) treatment with hydrochloric, nitric, sulfuric, or perchloric acid, and (2) fusion with an acidic or basic flux followed by treatment with water or an acid.

The solvent action of acids depends upon several factors:

1. The reduction of hydrogen ion by metals more active than hydrogen: for example,

$$Zn(s) + 2H^+ \longrightarrow Zn^{2+} + H_2(g)$$

2. The combination of hydrogen ion with anion of a weak acid: for example,

$$CaCO_3(s) + 2H^+ \longrightarrow Ca^{2+} + H_2O + CO_2(g)$$

3. The oxidizing properties of the anion of the acid: for example,

$$3Cu(s) + 2NO_3^- + 8H^+ \longrightarrow 3Cu^{2+} + 2NO(g) + 4H_2O$$

4. The tendency of the anion of the acid to form soluble complexes with the cation of the substance dissolved: for example,

$$Fe^{3+} + Cl^- \longrightarrow FeCl^{2+}$$

Hydrochloric and nitric acids are most commonly used to dissolve samples. The chloride ion is not an oxidizing agent as is nitrate ion, but it has a strong tendency to form soluble complexes with many elements. A very powerful solvent, *aqua regia,* is obtained by mixing these two acids.

Many substances that are resistant to attack by water or acids are more soluble after fusion with an appropriate flux. Basic fluxes such as sodium carbonate are used to attack acidic materials such as silicates. Acidic fluxes such as potassium hydrogen sulfate are used with basic materials such as iron ores. Oxidizing or reducing substances can also be used in certain cases. Sodium peroxide, for example, is often employed as a flux.

1.3c. *Conversion of the Analyte to a Measurable Form*

Before a physical or chemical measurement can be made to determine the amount of analyte in a solution of the sample, it is usually necessary to solve the problem of "interferences." Suppose, for example, that the analyst wishes to determine the amount of copper in a sample by adding potassium iodide and titrating the liberated iodine with sodium thiosulfate. If the solution also contains iron(III) ion, this ion will interfere, since it also oxidizes iodide to iodine. The interference can be prevented by adding sodium fluoride to the solution, converting iron(III) into the stable complex FeF_6^{3-}. This is an illustration of a general method in which interferences are effectively "immobilized" by alteration of their chemical nature.

A second method involves physical separation of the analyte from the interferences. Suppose that one wishes to determine magnesium in a sample which also contains iron(III) ion and the magnesium is to be precipitated as the oxalate. The iron will interfere, since it also forms a precipitate with oxalate. The iron can be precipitated as the hydroxide using ammonia at a pH of about 6.5. The magnesium is not precipitated at this pH, and hence the interference is removed.

In a *gravimetric* analysis the analyte is physically separated from all other components of the sample as well as from the solvent. For example, the chloride in a sample may be determined by precipitation of silver chloride, which is then filtered, dried, and weighed. Precipitation is one of the more widely used techniques for separating the analyte from interferences. Other important methods include electrolysis, solvent extraction, chromatography, and volatilization.

1.3d. Measurement

The measurement step in an analysis can be carried out by chemical, physical, or biological means. The laboratory technique employed has led to the classification of quantitative methods into the subdivisions *titrimetric* (*volumetric*), *gravimetric,* and *instrumental*. A *titrimetric* analysis involves measurement of the volume of a solution of known concentration which is required to react with the analyte. In a *gravimetric* method the measurement is one of weight; an example was mentioned above in which chloride is determined by precipitating and weighing silver chloride. The term *instrumental* analysis is used rather loosely; it originally referred to the use of a special instrument in the measurement step. Actually, instruments may be used in any or all steps of the analysis, and, strictly speaking, burets and analytical balances are instruments. Spectroscopy, both absorption and emission, is perhaps the most widely used instrumental method and is generally discussed in some detail in introductory texts. Other instrumental methods include potentiometry, polarography, coulometry, conductimetry, polarimetry, refractometry, and mass spectrometry.

1.3e. Calculation and Interpretation of the Measurements

The final step in an analysis is calculation of the percentage of analyte in the sample. The principles involved in such calculations are normally straightforward. For example, titrimetric and gravimetric methods are based on the simple stoichiometric relationships of chemical reactions. In spectrophotometric methods the property measured, absorbance, is directly proportional to the concentration of the analyte in the solution. On the other hand, interpretation of the results obtained by analytical methods is not always simple. Since errors can be made in any measurement, the analytical chemist must consider this possibility in interpreting the results. The methods of statistics are commonly used and are especially useful in expressing the significance of analytical data. We shall devote a full chapter to the topic of errors and the treatment of analytical data.

2

Errors and the Treatment of Analytical Data

In a science such as chemistry much effort is expended in gathering quantitative data. Such data are derived from *experimental* measurements and are therefore subject to error. It is important for students of chemistry to study the errors that attend their measurements in order to minimize them. Students should also develop the ability to evaluate experimental data and to draw justified conclusions while rejecting interpretations that may be unwarranted.

In this chapter we shall examine the methods used by scientists in assessing the significance of experimental results. Most of the techniques we shall consider are based upon statistical concepts, and although we cannot examine in detail the foundations of probability theory, we should be able to use this theory to express the reliability of our data. We shall also see how different sets of data may be compared to learn whether they are *really* different or not, and we shall also learn how errors are propagated through a series of experimental steps and calculations. Finally, we shall learn how to express properly the results of our measurements so that the maximum amount of information is conveyed to other people.

2.1

ERROR

The term *error* as used here refers to the numerical difference between a measured value and the true value. The *true* value of any quantity is really something we never know, although scientists generally accept a value as being true when it is believed that the uncertainty in the value is less than the uncertainty in something else with which it is being compared. For example, the percentage composition of a standard sample certified by the National Institute of Standards and Technology (NIST, formerly the National Bureau of Standards) may be treated as correct in evaluating a new analytical method. Differences between the standard values and the results obtained by the new method are then treated as errors in the latter.

2.1a. Determinate Errors

Errors which can, at least in principle, be ascribed to definite causes are termed *determinate* or *systematic* errors. Determinate errors are generally unidirectional with respect to the true value, in contrast to *indeterminate* errors, discussed below, which lead to both high and low results with equal probability. Determinate errors are often reproducible, and in many cases they can be predicted by a person who thoroughly understands all the aspects of the measurement. Examples of sources of determinate errors are an incorrectly calibrated instrument, such as a buret, balance, or pH meter, an impurity in a reagent, a side reaction in a titration, and heating a sample at too high a temperature.

Determinate errors have been classified as *methodic, operative,* and *instrumental* in accordance with their origin in (a) the method of analysis as it reflects the properties of the chemical systems involved, (b) ineptitude of the experimenter, and (c) failure of measuring devices to perform in accordance with required standards. Frequently, the source of an error may lie in more than one of these categories. For example, some error may always be expected in weighing a hygroscopic substance, but it may be increased if the analyst has poor balance technique; the environment outside the system may influence the error, as, for example, in the effect of humidity upon the error in weighing a hygroscopic substance.

Constant Errors

Determinate errors can also be classified as *constant* or *proportional*. A constant error is independent of the magnitude of the measured quantity and becomes less significant as the magnitude increases. For example, if a constant endpoint error of 0.10 ml is made in a series of titrations, this represents a *relative* error of 1% for a sample requiring 10 ml of titrant, but only 0.2% if 50 ml of titrant is used.

Proportional Errors

The absolute value of this type of error varies with sample size in such a way that the relative error remains constant. A substance that interferes in an analytical method may lead to such an error if present in the sample. For example, in the iodometric determination of an oxidant like chlorate, another oxidizing agent such as bromate would cause high results if its presence were unsuspected and not corrected for. Taking larger samples would increase the absolute error, but the relative error would remain constant provided the sample was homogeneous.

2.1b. Indeterminate Errors

Indeterminate errors, as the name implies, cannot be attributed to any known cause, but they inevitably attend measurements made by human beings.

They are random in nature and lead to both high and low results with equal probability. They cannot be eliminated or corrected and are the ultimate limitation on the measurement. They can be treated by statistics, and repeated measurement of the same variable can have the effect of reducing their importance.

Scientists routinely repeat their measurements several times and refine them to the point where it is mere coincidence if replicates agree to the last recorded digit. Sooner or later, the point is approached where unpredictable and imperceptible factors introduce what appear to be random fluctuations in the measured quantity. In some cases, it may be possible to specify definite variables that are beyond control near the performance limit of an instrument: noise and drift in an electronic circuit, vibrations in a building caused by passing traffic, temperature variations, and the like. Often the inability of the eye to detect slight changes in a readout device may be invoked as a source of error. To be sure, variations which a slipshod person considers random may appear obvious and controllable to a careful onlooker, but nevertheless a point must be reached where anyone, however meticulous, will encounter random errors which cannot be further reduced.

It is tempting at first glance to retreat from indeterminate errors simply by performing coarser measurements. After backing off to the point where scatter in the data ceases to exist, an observer will obtain exactly the same result each time, and superficially this seems as good as recording an additional digit which varies from one time to the next. But this withdrawal from the challenge to push measurements as far as possible is unacceptable to most scientists. More cogent, however, is the fact that the average of a number of fine observations with random scatter is more precise than coarser data which agree perfectly. Data that exhibit random scatter may be subjected to an analysis that does attach significance to the last recorded digit, as we shall see below.

2.1c. Accuracy and Precision

The terms *accuracy* and *precision,* often used synonymously in ordinary discourse, should be carefully distinguished in connection with scientific data.

Accuracy

An accurate result is one that agrees closely with the true value of a measured quantity. The comparison is usually made on the basis of an inverse measure of the accuracy, that is, the error (the smaller the error, the greater the accuracy). The *absolute error* is the difference between the experimental value and the true value. For example, if an analyst finds a value of 20.44% iron in a sample which actually contains 20.34%, the absolute error is

$$20.44 - 20.34 = 0.10\%$$

The error is most frequently expressed relative to the size of the measured quantity, for example, in percent or in parts per thousand. Here the *relative error* is

$$\frac{0.10}{20.34} \times 100 = 0.5\%$$

or

$$\frac{0.10}{20.34} \times 1000 = 5 \text{ ppt}$$

Precision

The term *precision* refers to the agreement among a group of experimental results; it implies nothing about their relation to the true value. Precise values may well be inaccurate, since an error causing deviation from the true value may affect all the measurements equally and hence not impair their precision. A determinate error which leads to inaccuracy may or may not affect precision, depending upon how nearly constant it remains throughout a series of measurements. The precision is commonly stated in terms of the *standard deviation, average deviation,* or *range.* These terms will be defined later. As in the case of error (above), the precision can be expressed on an absolute or a relative basis.

2.2

DISTRIBUTION OF RANDOM ERRORS

After the search for determinate errors has been carried as far as possible and all precautions taken and corrections applied, the remaining fluctuations in the data are found to be random in nature. Results that scatter in a random fashion are best treated by the powerful techniques of statistics. It will now be our goal to show how these techniques are applied and what information they furnish beyond what may be seen by simply inspecting the data.

2.2a. *Frequency Distributions*

Table 2.1 contains some actual data obtained by a person who prepared 60 replicate colored solutions and measured their absorbance values with a spectrophotometer. (Absorbance is discussed in a later chapter, but the nature of the measured quantity need not concern us here.) The data in Table 2.1 have not been treated in any way but are simply listed in the order in which they were obtained. We are here concerned, not with the "correct" result, only with the relationships of the measured values among themselves. It is apparent that the values in Table

TABLE 2.1 Individual Values, Unorganized

1	0.458	21	0.462	41	0.450
2	0.450	22	0.450	42	0.455
3	0.465	23	0.454	43	0.456
4	0.452	24	0.446	44	0.456
5	0.452	25	0.464	45	0.459
6	0.447	26	0.461	46	0.454
7	0.459	27	0.463	47	0.455
8	0.451	28	0.457	48	0.458
9	0.446	29	0.460	49	0.457
10	0.467	30	0.451	50	0.456
11	0.452	31	0.456	51	0.455
12	0.463	32	0.455	52	0.460
13	0.456	33	0.451	53	0.456
14	0.456	34	0.462	54	0.463
15	0.449	35	0.451	55	0.457
16	0.454	36	0.469	56	0.456
17	0.456	37	0.458	57	0.457
18	0.441	38	0.458	58	0.453
19	0.457	39	0.456	59	0.455
20	0.459	40	0.454	60	0.453

2.1 must be treated in some manner before they can be discussed intelligently. A reader with an exceptionally quick eye may notice that the lowest value is 0.441 and the highest 0.469, and perhaps it is apparent that many values are between 0.45 and 0.46, but on the whole the table is relatively uninstructive. Let us now enumerate some steps that will enable us to interpret the data more fully.

1. Arrange the results in order from lowest to highest. This has been done in Table 2.2. This simple operation discloses information not so readily apparent in the raw data, namely, the maximum and minimum values and, by simple counting, the middle or median value. This is still an inadequate presentation of the data, however; the mind does not grasp the meaning of 60 numbers on a piece of paper, regardless of how they are arranged. We need more compactness in order to make practical use of the data.

TABLE 2.2   Individual Values Arranged in Order

1	0.441	21	0.454	41	0.457
2	0.446	22	0.455	42	0.458
3	0.446	23	0.455	43	0.458
4	0.447	24	0.455	44	0.458
5	0.449	25	0.455	45	0.458
6	0.450	26	0.455	46	0.459
7	0.450	27	0.456	47	0.459
8	0.450	28	0.456	48	0.459
9	0.451	29	0.456	49	0.460
10	0.451	30	0.456	50	0.460
11	0.451	31	0.456	51	0.461
12	0.451	32	0.456	52	0.462
13	0.452	33	0.456	53	0.462
14	0.452	34	0.456	54	0.463
15	0.452	35	0.456	55	0.463
16	0.453	36	0.456	56	0.463
17	0.453	37	0.457	57	0.464
18	0.454	38	0.457	58	0.465
19	0.454	39	0.457	59	0.467
20	0.454	40	0.457	60	0.469

2. Condense the data by grouping them into cells. We divide the range from the lowest to the highest value into a convenient number of intervals or *cells* and then count the number of values falling within each cell. Strictly, this process involves some loss of information, but this is more than compensated for by the increased efficiency with which the significance of the condensed data may be perceived. In order to proceed, we must first decide upon the number of cells to be used and choose their boundaries. Usually the range is divided into equal intervals, and sometimes confusion is avoided by choosing cell boundaries halfway between possible observed values. In the present case, the absorbance was recorded to three decimal places, and we choose cell boundaries such as 0.4605 so that none of the values coincides with a boundary. Judgment is required in selecting the number of cells: 13 to 20 are sometimes recommended, but 10 or even fewer may be preferable if the number of values to be grouped is small, say, less than 250. A fairly satisfactory grouping of our data into 8 cells is shown in Table 2.3.

    A glance at Table 2.3 shows that information buried in Tables 2.1 and 2.2 is now obvious. Thus, although the values range from 0.441 to 0.469, we see immediately that very few results are below 0.448 or above 0.464.

TABLE 2.3 Grouping of Individual Values into Cells

CELL MIDPOINT	CELL BOUNDARIES	NUMBER OF VALUES
	0.4405	
0.4425		1
	0.4445	
0.4465		3
	0.4485	
0.4505		11
	0.4525	
0.4545		21
	0.4565	
0.4585		14
	0.4605	
0.4625		7
	0.4645	
0.4665		2
	0.4685	
0.4705		1
	0.4725	

3. Devise a pictorial representation of the frequency distribution. This step is actually unnecessary, and it is rarely performed except for teaching purposes or for popular presentation of what might otherwise be "dry" data to laypeople. Two types of graphs are shown in Fig. 2.1: The *histogram* consists of contiguous columns of heights proportional to the frequencies, erected upon the full widths of the cells; the *frequency polygon* is constructed by plotting frequencies at cell midpoints and connecting the points with straight lines.

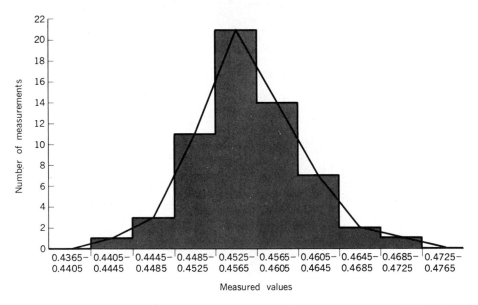

Figure 2.1 Histogram and frequency polygon for absorbance measurements of 60 replicate solutions.

2.2b. *The Normal Error Curve*

The limiting case approached by the frequency polygon as more and more replicate measurements are performed is the *normal* or *Gaussian* distribution curve, shown in Fig. 2.2. This curve is the locus of a mathematical function which is well known, and it is more easily handled than the less ideal and more irregular curves often obtained with a smaller number of observations. Data are often treated as though they were normally distributed in order to simplify their

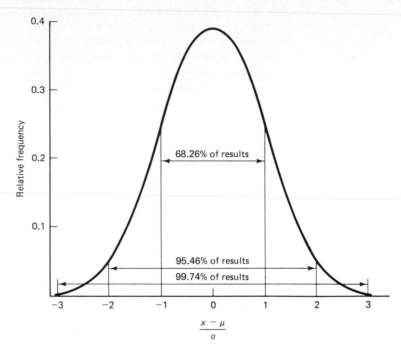

Figure 2.2 Normal distribution curve; relative frequencies of deviations from the mean for a normally distributed infinite population; deviations $(x - \mu)$ are in units of σ.

analysis, and we may look upon the normal error curve as a model which is approximated more or less closely by real data.

Population and Sample

It is supposed that there exists a "universe" of data made up of an infinite number of individual measurements, and it is actually this **infinite population** to which the normal error function pertains. A finite number of replicate measurements is considered by statisticians to be **a sample** drawn in a random fashion from a hypothetical infinite population; thus it is hoped that the sample is a representative one, and fluctuations in its individual values may be considered to be normally distributed so that the terminology and techniques associated with the normal error function may be employed in their analysis.

The equation of the normal curve may be written for our purposes as follows:

$$y = \frac{1}{\sigma\sqrt{2\pi}} e^{-(x-\mu)^2/2\sigma^2}$$

Here y represents the relative frequency with which random sampling of the infinite population will bring to hand a particular value x.

Population Parameters

The quantities μ and σ, called the population parameters, specify the distribution. μ is the *mean* of the infinite population, and since we are not here concerned with determinate errors, we may consider that μ gives the correct magnitude of the measured quantity. It is clearly impractical to determine μ by actually averaging an infinite number of measured values, but we shall see below that a statement can be made from a finite series of measurements regarding the probability that μ lies within a certain interval. To the extent of our confidence in having eliminated determinate errors, such a statement approaches an assessment of the true value of the measured quantity. σ, which is called the *standard deviation,* is the distance from the mean to either of the two inflection points of the distribution curve and may be thought of as a measure of the spread or scatter of the values making up the population; σ thus relates to precision. π has its usual significance, and e is the base of the natural logarithm system. The term $(x - \mu)$ represents simply the extent to which an individual value x deviates from the mean.

Normalization of the Distribution Function

The distribution function may be normalized by setting the area under the curve equal to unity, representing a total probability of 1 for the whole population. Since the curve approaches the abscissa asymptotically on either side of the mean, there is a small but finite probability of encountering enormous deviations from the mean. A person who happened to encounter one of these in performing a series of laboratory observations would be unfortunate indeed; some of us who have faith in never obtaining such a "wild" result in our own work are inclined to the view that the normal distribution as a model for real data breaks down and that only the central region of the distribution curve is pertinent when applied to scientific measurements by competent workers. The area under the curve between any two values of $(x - \mu)$ gives the fraction of the total population having magnitudes between these two values. It may be shown that about two-thirds (actually 68.26%) of all the values in an infinite population fall within the limits $\mu \pm \sigma$, while $\mu \pm 2\sigma$ includes 95.46%, and $\mu \pm 3\sigma$ practically all (99.74%) of the values. Happily, then, small errors are more probable than large ones. Since the normal curve is symmetrical, high and low results are equally probable once determinate errors have been dismissed.

When a worker goes into the laboratory and measures something, we suppose that the result is one of an infinite population of such values that might be obtained in an eternity of such activity; then the chances are roughly 2 : 1 that the measured value will be no farther than σ from the mean of the infinite population, and about 20 : 1 that the result will lie in the range $\mu \pm 2\sigma$. In practice, of course, we can never find σ for an infinite population, but the standard deviation of a finite number of observations may be taken as an estimate of σ. Thus we may predict something about the likelihood of occurrence of an error of a certain magnitude in the work of a particular individual once enough measurements have been made to permit estimation of the characteristics of the particular infinite population.

2.3

STATISTICAL TREATMENT OF FINITE SAMPLES

Although there is no doubt as to its mathematical meaning, the normal distribution of an infinite population is fiction so far as real laboratory work is concerned. We must now turn our attention to techniques for handling scientific data as we obtain them in practice.

2.3a. Measures of Central Tendency and Variability

The *central tendency* of a group of results is simply that value about which the individual results tend to "cluster." For an infinite population, it is μ, the mean of such a sample.

Mean

The *mean* of a finite number of measurements, $x_1, x_2, x_3, \ldots, x_n$, is often designated $\bar{x}$ to distinguish it from μ. Of course, $\bar{x}$ approaches μ as a limit when n, the number of measured values, approaches infinity. Calculation of the mean involves simply averaging the individual results:

$$\bar{x} = \frac{x_1 + x_2 + x_3 + \cdots + x_n}{n} = \frac{\sum\limits_{i=1}^{i=n} x_i}{n}$$

The mean is generally the most useful measure of central tendency. It may be shown that the mean of n results is $\sqrt{n}$ times as reliable as any one of the individual results. Thus there is a diminishing return from accumulating more and more replicate measurements: The mean of 4 results is twice as reliable as 1 result in measuring central tendency; the mean of 9 results is three times as reliable; the mean of 25 results, five times as reliable etc. Thus, generally speaking, it is inefficient for a careful worker who gets good precision to repeat a measurement more than a few times. Of course the need for increased reliability, and the price to be paid for it, must be decided on the basis of the importance of the results and the use to which they are to be put.

Median

The *median* of an odd number of results is simply the middle value when the results are listed in order; for an even number of results, the median is the average of the two middle ones. In a truly symmetrical distribution, the mean and the median are identical. Generally speaking, the median is a less efficient measure of central tendency than is the mean, but in certain instances it may be useful, particularly in dealing with very small samples.

Range

Since two parameters, μ and σ, are required to specify a frequency distribution, it is clear that two populations may have the same central tendency but differ in "spread" or *variability* (or, as some say, *dispersion*), as suggested in Fig. 2.3. For a finite number of values, the simplest measure of variability is the *range,* which is the difference between the largest and smallest values. Like the median, the range is sometimes useful in small sample statistics, but generally speaking it is an inefficient measure of variability. Note, for example, that one "wild" result exerts its full impact upon the range, whereas its effect is diluted by all the other results in the better measures of variability noted below.

Average Deviation

The *average deviation* from the mean is often given in scientific papers as a measure of variability, although strictly it is not very significant from a statistical

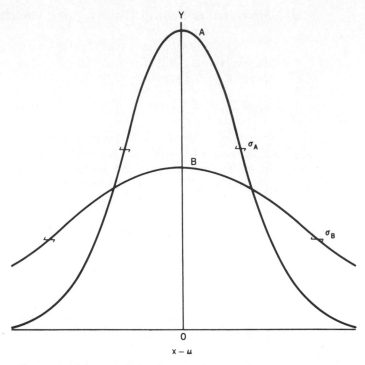

Figure 2.3 Two populations with the same central tendency μ, but different variability.

point of view, particularly for a small number of observations. For a large group of data which are normally distributed, the average deviation approaches 0.8σ. To calculate the average or mean deviation, one simply finds the differences between individual results and the mean, regardless of sign, adds these individual deviations, and divides by the number of results:

$$\text{Average deviation} = \overline{d} = \frac{\sum\limits_{i=1}^{i=n} |x_i - \overline{x}|}{n}$$

Relative Average Deviation

Often the average deviation is expressed relative to the magnitude of the measured quantity, for example, as a percentage:

$$\text{Relative average deviation (\%)} = \frac{\overline{d}}{\overline{x}} \times 100 = \frac{\sum\limits_{i=1}^{i=n} |x_i - \overline{x}|/n}{\overline{x}} \times 100$$

Because analytical results are often expressed as percentages (e.g., percentage of iron in an iron ore sample), it may be confusing to report relative deviations on a percentage basis, and it is preferable to use parts per thousand instead of percent (parts per hundred):

$$\text{Relative average deviation (ppt)} = \frac{\sum\limits_{i=1}^{i=n} |x_i - \overline{x}|/n}{\overline{x}} \times 1000$$

Standard Deviation

The *standard deviation* is much more meaningful statistically than is the average deviation. The symbol s is used for the standard deviation of a finite number of values; σ is reserved for the population parameter. The standard deviation, which may be thought of as a root-mean-square deviation of values from their average, is calculated using the formula

$$s = \sqrt{\frac{\sum\limits_{i=1}^{i=n} |x_i - \bar{x}|^2}{n - 1}}$$

If n is large (say 50 or more), then, of course, it is immaterial whether the term in the denominator is $n - 1$ (which is strictly correct) or n. When the standard deviation is expressed as a percentage of the mean, it is called the *coefficient of variation, v*:

$$v = \frac{s}{\bar{x}} \times 100$$

Variance

The *variance*, which is s^2, is fundamentally more important in statistics than is s itself, but the latter is much more commonly used in treating chemical data.

For the data in Tables 2.1, 2.2, and 2.3 the following measures of central tendency and variability were calculated:

$$\text{Mean:} \quad \bar{x} = 0.456$$

$$\text{Median:} \quad M = 0.456$$

$$\text{Range:} \quad R = 0.028$$

$$\text{Average deviation:} \quad \bar{d} = 0.0038$$

$$\text{Relative average deviation:} \quad \frac{\bar{d}}{\bar{x}} \times 1000 = 8.3 \text{ ppt}$$

$$\text{Standard deviation:} \quad s = 0.0052$$

$$\text{Coefficient of variation:} \quad v = \frac{s}{\bar{x}} \times 100 = 1.1\%$$

The following example illustrates calculation of the foregoing terms in the case of a determination of the normality of a solution.

EXAMPLE 2.1

Calculation of fundamental statistical parameters

The normality of a solution is determined by four separate titrations, the results being 0.2041, 0.2049, 0.2039, and 0.2043. Calculate the mean, median, range, average deviation, relative average deviation, standard deviation, and coefficient of variation.

$$\text{Mean:} \quad \bar{x} = \frac{0.2041 + 0.2049 + 0.2039 + 0.2043}{4}$$

$$\bar{x} = 0.2043$$

Median: $\quad M = \dfrac{0.2041 + 0.2043}{2}$

$\qquad\qquad M = 0.2042$

Range: $\quad R = 0.2049 - 0.2039$

$\qquad\qquad R = 0.0010$

Average deviation: $\quad \bar{d} = \dfrac{(0.0002) + (0.0006) + (0.0004) + (0.0000)}{4}$

$\qquad\qquad\qquad \bar{d} = 0.0003$

Relative average deviation: $\quad \dfrac{\bar{d}}{\bar{x}} \times 1000 = \dfrac{0.0003}{0.2043} \times 1000$

$\qquad\qquad\qquad\qquad\qquad\qquad\qquad = 1.5 \; ppt$

Standard deviation: $\quad s = \sqrt{\dfrac{(0.0002)^2 + (0.0006)^2 + (0.0004)^2 + (0.0000)^2}{4 - 1}}$

$\qquad\qquad\qquad s = 0.0004$

Coefficient of variation: $\quad v = \dfrac{0.0004}{0.2043} \times 100$

$\qquad\qquad\qquad\qquad v = 0.2\%$ $\qquad\qquad\qquad\qquad\qquad\square$

2.3b. Student's t

We have seen that, given μ and σ for the normal distribution of an infinite population, a precise statement can be made regarding the odds of drawing from the population an observation lying outside certain limits. But in practical work, we deal with finite numbers of observations, and we know not μ and σ, but rather $\bar{x}$ and s, which are only estimates of μ and σ. Since these estimates are subject to uncertainty, what we really have is a sort of blurred distribution curve on which to base any predictions we wish to make. This naturally widens the limits corresponding to any given odds that an individual observation will fall outside such limits. An English chemist, W. S. Gosset, writing under the pen name of Student, studied the problem of making predictions based upon a finite sample drawn from an unknown population and published a solution in 1908.[1] The theory of Student's work is beyond the scope of this book, but we may accept it as soundly based and see how it may be used in chemistry. The quantity t (often called *Student's t*) is defined by the expression

$$\pm t = (\bar{x} - \mu)\frac{\sqrt{n}}{s}$$

Tables of t-values relating to various odds or probability levels and for varying degrees of freedom may be found in statistical compilations; a portion of such a table is reproduced here in Table 2.4. *Degrees of freedom* in the present connec-

[1] *Biometrika*, **6**, 1 (1908).

TABLE 2.4 Some Values of Student's *t*

NUMBER OF OBSERVATIONS, n	NUMBER OF DEGREES OF FREEDOM, $n-1$	PROBABILITY LEVELS			
		50%	90%	95%	99%
2	1	1.000	6.314	12.706	63.66
3	2	0.816	2.920	4.303	9.925
4	3	0.765	2.353	3.182	5.841
5	4	0.741	2.132	2.776	4.604
6	5	0.727	2.015	2.571	4.032
7	6	0.718	1.943	2.447	3.707
8	7	0.711	1.895	2.365	3.500
9	8	0.706	1.860	2.306	3.355
10	9	0.703	1.833	2.262	3.250
11	10	0.700	1.812	2.228	3.169
21	20	0.687	1.725	2.086	2.845
∞	∞	0.674	1.645	1.960	2.576

tion are one less than n, the number of observations.[2] The *t*-values are calculated to take into account the fact that $\bar{x}$ will not in general be the same as μ and to compensate for the uncertainty in using s as an estimate of σ. Values of t such as those in Table 2.4 are used in several statistical methods, some of which are outlined below.

2.3c. *Confidence Interval of the Mean*

By rearranging the equation above which defines t, we obtain the *confidence interval* of the mean, or *confidence limits:*

$$\mu = \bar{x} \pm \frac{ts}{\sqrt{n}}$$

We might use this to estimate the probability that the population mean, μ, lies within a certain region centered at $\bar{x}$, the experimental mean of our measurements. It is more usual in treating analytical data, however, to adopt an acceptable probability and then find the limits on either side of $\bar{x}$ to which we must go in order to be assured that we have embraced μ. It may be seen in Table 2.4 that *t*-values increase as n, the number of observations, decreases. This is reasonable, since the smaller n becomes, the less information is available for estimating the population parameters. Increases in t exactly compensate for the lessening information.

The following example illustrates the use of Table 2.4.

[2] Degrees of freedom may be defined as the number of individual observations that could be allowed to vary under the condition that $\bar{x}$ and s, once determined, be held constant. For example, once the mean is obtained and we decide to keep it constant, all but one observation can be varied; the last one is fixed by $\bar{x}$ and all the other x_i values, and the degrees of freedom then equal $n-1$. In general, if s is calculated from the same number of observations as were used to calculate $\bar{x}$ (which would normally be the case in treating analytical data), then the degrees of freedom equal $n-1$.

EXAMPLE 2.2

Calculation of confidence intervals

A chemist determined the percentage of iron in an ore, obtaining the following results: $\bar{x} = 15.30$, $s = 0.10$, $n = 4$.

(a) Calculate the 90% confidence interval of the mean.
From Table 2.4, $t = 2.353$ for $n = 4$. Hence

$$\mu = 15.30 \pm \frac{2.353 \times 0.10}{\sqrt{4}}$$

$$\mu = 15.30 \pm 0.12$$

(b) Calculate the 99% confidence interval of the mean.
From Table 2.4, $t = 5.841$ for $n = 4$. Hence

$$\mu = 15.30 \pm \frac{5.841 \times 0.10}{\sqrt{4}}$$

$$\mu = 15.30 \pm 0.29 \qquad \square$$

The meaning of confidence intervals is sometimes confused by the beginning student. The correct interpretation, using part (a) of the example above, is as follows: Suppose the chemist repeats the analysis ten times, each time performing four determinations and calculating an interval as illustrated above. One would obtain ten intervals, such as 15.30 ± 0.12, 15.28 ± 0.14, 15.33 ± 0.11, etc., and could expect nine of these ten intervals to embrace the population mean, μ. It is a common misconception that 90% of the experimental means would lie within the interval 15.30 ± 0.12. Predicting the interval within which future $\bar{x}$ values will lie is a different statistical problem which can be treated only with another sort of limits which are much wider than the confidence limits discussed here.

In some cases where analyses have been repeated extensively, a chemist may have a reliable estimate of the population standard deviation, σ. There is then no uncertainty in the value of σ, and the confidence interval is given by

$$\mu = \bar{x} \pm \frac{Z\sigma}{\sqrt{n}}$$

where Z is simply the value of t at $n = \infty$ (Table 2.4). Note that in the example above the confidence interval for part (a) would be given by

$$15.30 \pm \frac{1.645 \times 0.10}{\sqrt{4}} = 15.30 \pm 0.08$$

The interval is narrower since the uncertainty in σ has been removed.

It is possible to calculate a confidence interval from the range, R, of a series of measurements, using the relationship

$$\mu = \bar{x} \pm c_n R$$

Values of c_n for various numbers of observations and probability levels have been tabulated; some of these are given in Table 2.5. The values of c_n are based upon estimates of s obtained from the range. It should be emphasized that while it is easy to calculate a confidence interval from the range, an occasional large error will have an undue impact upon the result. The range is normally used in this way only when dealing with a very small number of observations, say, ten or less.

TABLE 2.5 Some Values of c_n for Calculating Confidence Intervals from the Range

NUMBER OF OBSERVATIONS	PROBABILITY LEVELS 95%	99%
2	6.353	31.828
3	1.304	3.008
4	0.717	1.316
5	0.507	0.843
6	0.399	0.628

2.3d. *Testing for Significance*

Comparison of Two Means

Suppose that a sample is analyzed by two different methods, each repeated several times, and that the mean values obtained are different. Statistics, of course, cannot say which value is "right," but there is a prior question in any case, namely, is the difference between the two values significant? It is possible simply by the influence of random fluctuations to get two different values using two methods; but it is likewise possible that one (or even both) of the methods is subject to a determinate error. There is a test, using Student's t, that will tell (with a given probability) whether it is worthwhile to seek an assignable cause for the difference between the two means. It is clear that the greater the scatter in the two sets of data, the less likely it is that differences between the two means are real.

The statistical approach to this problem is to set up the *null hypothesis*. This hypothesis states, in the present example, that the two means are identical. The t-test gives a *yes* or *no* answer to the correctness of the null hypothesis with a certain confidence, such as 95 or 99%. The procedure is as follows: Suppose that a sample has been analyzed by two different methods, yielding means $\bar{x}_1$ and $\bar{x}_2$ and *standard deviations* s_1 and s_2; n_1 and n_2 are the number of individual results obtained by the two methods. The first step is to calculate a t-value using the formula

$$t = \frac{|\bar{x}_1 - \bar{x}_2|}{s} \sqrt{\frac{n_1 n_2}{n_1 + n_2}}$$

(This procedure presupposes that s_1 and s_2 are the same; there is a test for this, noted below.) Second, enter a t-table such as Table 2.4 at a degree of freedom given by $(n_1 + n_2 - 2)$ and at the desired probability level. If the value in the table is greater than the t calculated from the data, the null hypothesis is substantiated (i.e., $\bar{x}_1$ and $\bar{x}_2$ are the same with a certain probability). If the t-value in the table is less than the calculated t, then by this test the null hypothesis is incorrect and it might be profitable to look for a reason to explain the difference between $\bar{x}_1$ and $\bar{x}_2$.

Comparison of Two Standard Deviations

If s_1 and s_2 are really different, a much more complicated procedure, which is not discussed here, must be used. Usually in analytical work involving methods that would by ordinary common sense be considered comparable, s_1 and s_2 are about the same. A test is available for deciding whether a difference between s_1 and s_2 is significant: this is the *variance-ratio test* or *F-test*. The procedure is simple: Find the ratio $F = s_1^2/s_2^2$, placing the larger s-value in the numerator so that $F > 1$; then go to a table of F-values. If the F-value in the table is less than the calculated F-value, then the two standard deviations are significantly different; otherwise, they are not. Some sample F-values are given in Table 2.6 for a probability level of 95%. The F-test may be used to determine the validity of the simple t-test described here, but it may also be of interest in its own right to determine whether two analytical procedures yield significantly different precision.

The following example illustrates this procedure.

TABLE 2.6 *F*-Values at the 95% Probability Level

$n - 1$ FOR SMALLER s^2	$n - 1$ FOR LARGER s^2					
	3	4	5	6	10	20
3	9.28	9.12	9.01	8.94	8.79	8.66
4	6.59	6.39	6.26	6.16	5.96	5.80
5	5.41	5.19	5.05	4.95	4.74	4.56
6	4.76	4.53	4.39	4.28	4.06	3.87
10	3.71	3.48	3.33	3.22	2.98	2.77
20	3.10	2.87	2.71	2.60	2.35	2.12

EXAMPLE 2.3

Comparison of two analytical methods

A sample of soda ash (Na_2CO_3) is analyzed by two different methods, giving the following results for the percentage of Na_2CO_3:

Method 1	Method 2
$\bar{x}_1 = 42.34$	$\bar{x}_2 = 42.44$
$s_1 = 0.10$	$s_2 = 0.12$
$n_1 = 5$	$n_2 = 4$

(a) Are s_1 and s_2 significantly different? Apply the variance-ratio, or F-test:

$$F = \frac{s_2^2}{s_1^2} = 1.44$$

Consult Table 2.6 under column $n - 1 = 3$ (since $s_2 > s_1$) and row $n - 1 = 4$, find $F = 6.59$. Since $6.59 > 1.44$, the standard deviations are not significantly different.

(b) Are the two means significantly different at the 95% probability level? Calculate a t-value (either s_1 or s_2 may be used):

$$t = \frac{|42.34 - 42.44|}{0.10} \sqrt{\frac{5 \times 4}{5 + 4}}$$

$$t = 1.491$$

Consult Table 2.4 at degrees of freedom $n_1 + n_2 - 2 = 7$, finding t for the 95% probability level $= 2.365$. Since $1.491 < 2.365$, the null hypothesis is correct and the difference is not significant. □

Comparison of an Experimental and a True Mean

Sometimes it may be of interest to compare two results, one of which is considered a priori to be highly reliable. An example of this might be a comparison of the mean $\bar{x}$ of several analyses of a sample certified by the National Institute of Standards and Technology. The goal would be to decide whether the method employed gives results that agree with the Institute's. In this case the Institute's value is taken as μ in the equation defining Student's t, and a t-value is calculated using $\bar{x}$, n, and s for the analytical results at hand. If the calculated t-value is greater than that in the t-table for $(n - 1)$ degrees of freedom and the desired probability, then the analytical method in question has given a mean value significantly different from the NIST value; otherwise, differences in the two values would be attributable to chance alone.

The following example is an illustration.

EXAMPLE 2.4

Comparison of an experimental and a true mean

A chemist analyzes a sample of iron ore furnished by the National Institute of Standards and Technology and obtains the following results: $\bar{x} = 10.52$, $s = 0.05$, $n = 10$. The NIST value for this sample is 10.60% Fe. Are the results significantly different at the 95% probability level?

Calculate t from the equation

$$\mu = \bar{x} \pm \frac{ts}{\sqrt{n}}$$

$$10.60 = 10.52 \pm \frac{t \times 0.05}{\sqrt{10}}$$

$$t = 5.06$$

In Table 2.4, at degrees of freedom = 9 and 95% probability level, $t = 2.262$. Since $5.06 > 2.262$, the results are significantly different from the NIST value.

□

2.3e. Criteria for Rejection of an Observation

Sometimes a person performing measurements is faced with one result in a set of replicates which seems to be out of line with the others, and he then must decide whether to exclude this result from further consideration. This problem is encountered in beginning analytical chemistry courses, later in physical chemistry laboratory work, and even in advanced research, although hopefully with lessening frequency as the student progresses. It is a generally accepted rule in scientific work that a measurement is to be automatically rejected when it is known that an error was made; this is a determinate situation with which we are not concerned here. It should be noted that it is incorrect (but all too human) to reject results which were subject to known errors only when they appear to be discordant. The only way to avoid an unconscious introduction of bias into the measurements is to reject every result where an error was known to be made, regardless of its agreement with the others. The problem to which we address ourselves here is a different one: How do we decide whether to throw out a result which appears discordant when there is no known reason to suspect it?

If the number of replicate values is large, the question of rejecting one value is not an important one; first, a single value will have only a small effect upon the mean, and second, statistical considerations give a clear answer regarding the probability that the suspected result is a member of the same population as the others. On the other hand, a real dilemma arises when the number of replicates is small: the divergent result exerts a significant effect upon the mean, while at the same time there are insufficient data to permit a real statistical analysis of the status of the suspected result.

The many different recommendations that have been promulgated by various writers attest to the conclusion that the question of rejecting or retaining one divergent value from a small sample really cannot satisfactorily be answered. Some of the more widely recommended criteria for rejection are considered below, and the student is referred to the excellent discussion by Blaedel et al.,[3] and interesting briefer commentaries by Laitinen[4] and Wilson.[5]

In the first place, it is necessary to decide how large the difference between the suspected result and the other data must be before the result is to be discarded. If the minimum difference is made too small, valid data may be rejected too frequently; this is said to be an "error of the first kind." On the other hand, setting the minimum difference too high leads to "errors of the second kind," that is, too frequent retention of highly erroneous values. The various recommendations for criteria of rejection steer one course or another between the Scylla and Charybdis of these two types of errors, some closer to one and some closer to the other.

Rules Based on the Average Deviation

Two rules used by chemists for years were called the "2.5d" and the "4d" rules. To apply the rules one first calculated the mean and average deviation of the "good" results and determined the deviation of the suspected result from the mean of the good ones. If the deviation of the suspected result was at least 4 times the average deviation of the good results (2.5 times for the 2.5d rule), the suspected result was discarded. Otherwise it was retained. Strictly, the limit for rejection is too low for both rules. Errors of the first kind are made too often (valid data rejected). The degree of confidence often quoted for these rules is based upon large sample statistics extended to small samples without proper compensation.

Rule Based on the Range

The Q-test, described by Dean and Dixon,[6] is statistically correct, and it is very easy to apply. When the Q-test calls for rejection, confidence is high (90%) that the suspected result was indeed subject to some special error. Using the Q-test for rejection, errors of the first kind are highly unlikely. However, when applied to small sets of data (say, three to five results), the Q-test allows rejection only of results that deviate widely, and hence leads frequently to errors of the second kind (retention of erroneous results). Thus, the Q-test provides excellent justification for the rejection of grossly erroneous values, but it does not eliminate

[3] W. J. Blaedel, V. W. Meloche, and J. A. Ramsey, *J. Chem. Ed.,* **28**, 643 (1951).

[4] H. A. Laitinen, *Chemical Analysis,* McGraw-Hill Book Company, New York, 1960, p. 574.

[5] E. B. Wilson, Jr., *An Introduction to Scientific Research,* McGraw-Hill Book Company, New York, 1952, p. 256.

[6] R. B. Dean and W. J. Dixon, *Anal. Chem.,* **23**, 636 (1951).

the dilemma with suspicious but less deviant values. The reason for this, of course, is that with small samples only crude guesses of the real population distribution are possible, and thus sound statistics lends assurance only to the rejection of widely divergent results.

The Q-test is applied as follows:

1. Calculate the range of the results.
2. Find the difference between the suspected result and its nearest neighbor.
3. Divide the difference obtained in step 2 by the range from step 1 to obtain the rejection quotient, Q.
4. Consult a table of Q-values. If the computed value of Q is greater than the value in the table, the result can be discarded with 90% confidence that it was indeed subject to some factor which did not operate on the other results.

Some Q-values are given in Table 2.7.

TABLE 2.7 Values of Rejection Quotient, Q

NUMBER OF OBSERVATIONS	$Q_{0.90}$
3	0.94
4	0.76
5	0.64
6	0.56
7	0.51
8	0.47
9	0.44
10	0.41

The following example illustrates application of the Q-test.

EXAMPLE 2.5

Application of the Q-test

Five determinations of the vitamin C content of a citrus fruit drink gave the following results: 0.218, 0.219, 0.230, 0.215, and 0.220 mg/ml. Apply the Q-test to see if the 0.230 value can be discarded.

The value of Q is

$$Q = \frac{0.230 - 0.220}{0.230 - 0.215}$$

$$Q = 0.67$$

The value in Table 2.7 at $n = 5$ is $Q = 0.64$. Since $0.67 > 0.64$, the rule says that the result can be discarded. ☐

As noted above, the Q-test affirms the rejection of a result at a confidence level of 90%. Willingness to reject a result with less confidence would make possible a Q-test which allowed retention of fewer deviant values (errors of the second kind). While this appears superficially attractive, there are valid reasons for conservatism in rejecting measurements. Actually, low confidence levels (say, 50%) are scarcely meaningful when only a small number of observations are involved. Further, it must be remembered that the collection of data is a meaningful enterprise with a purpose and that a result which has been painstakingly obtained should not be discarded hastily. Rather the measurement should be repeated until the dilemma of the discordant result has evaporated through the operation of two factors: dilution of any one result by all the others will lessen its significance, and as the number of observations increases, statistical evaluation of the suspected result will become more meaningful.

A sort of compromise between outright rejection and the retention of a suspected value is sometimes recommended, which is to report the median of all the results rather than a mean either with or without the deviant value. The median is influenced by the *existence* of one discordant result, but it is not affected by the *extent* to which the result differs from the others. For a sample containing three to five values, Blaedel et al. recommend testing the suspected value with the Q-test and rejecting it if the test allows this; if not, the median is reported rather than the mean. Some writers (e.g., Wilson) recommend that the highest and lowest values both be rejected and the mean of the others reported: "The best procedure to use depends on what is known about the frequency of occurrence of wild values, on the cost of additional observations, and on the penalties for the various types of error. In the absence of special arguments, the use of the interior average . . . would appear to be good practice."[7] It may be noted that this interior average and the median are necessarily identical in the special case where there are just three results.

2.4

CONTROL CHARTS

The control chart method was originally developed as a system for keeping track of quality during large-scale manufacturing operations. Often a production run is too large to permit individual inspection of each item (say, razor blades or ball bearings), and in some cases the quality test is destructive (as in measuring the stress required to break an object) and hence cannot be applied to each specimen produced by a company. In such cases, some sort of spot-checking of a few of the samples coming off the production line is necessary, and judgment is required to decide whether the manufacturing process is under control or whether a costly shutdown is justified in order to seek the cause of a deviation from the specifications in the tested results. The control chart method has also proved useful in keeping track of the performance of analytical methods in busy laboratories where the same types of samples are repeatedly analyzed day after day over long periods of time. The method tends to distinguish with a high degree of efficiency definite trends or periodically recurring anomalies from random fluctuations. The control chart method can be discussed only briefly here; the interested reader is referred to books on the subject[8,9] and several briefer discussions.[10–14]

[7] E. B. Wilson, Jr., loc. cit., p. 257.

[8] E. L. Grant, *Statistical Quality Control*, 2nd ed., McGraw-Hill Book Company, New York, 1952.

[9] W. A. Shewhart, *Economic Control of Quality of Manufactured Product*, D. Van Nostrand Co., New York, 1931.

[10] H. A. Laitinen, loc. cit., p. 560.

[11] E. B. Wilson, Jr., loc. cit., p. 263.

[12] G. Wernimont, *Ind. Eng. Chem., Anal. Ed.*, **18**, 587 (1946).

[13] J. A. Mitchell, *Ibid.*, **19**, 961 (1947).

[14] J. K. Taylor, *Anal. Chem.*, **53**, 1588A (1981).

Let us suppose that a company manufactures some chemical material, and that as part of the quality control program, the analytical laboratory performs each day a certain analysis on samples bled from the plant output, perhaps for the percentage of water in the product. Let us further suppose that the laboratory checks its water determination each day by running a standard sample of known water content through the analytical procedure. We are interested here in how the control chart for the laboratory analysis is set up and used. The plant could also use a control chart method, based upon the laboratory reports, for monitoring the quality of the product, but here we are concerned with the laboratory's checking its own analytical method.

The control chart for the analysis is set up as follows (see Fig. 2.4). The percentage of water in the standard sample is indicated on the chart by a horizontal line. The standard sample is analyzed daily, and the average of five weekly results is plotted, week after week, on the chart. Also placed on the chart are the *control limits*. Analytical results falling outside these limits are considered to result from the operation of some definite factor which is worth investigating and correcting. When results fall within the limits, the method is "under control," and fluctuations are only random and indeterminate. (The analogous conclusion with a production control chart is that when samples test outside the control limits, there is justification for shutting down the process and looking for the trouble.) Clearly, the control limits must be set in an arbitrary manner; one must decide how large must be the probability of an assignable cause for a deviant result before one can state that something is wrong with the analysis. It seems usual in practice to set the control limits at the expected value $\pm 3s$; there is no fundamental aspect of probability theory demanding this, but apparently experience has shown that these are sound limits economically as a basis for action. Sometimes two sets of control limits are placed on the chart, "inner limits" at about $\pm 2s$ to warn of possible trouble, and "outer limits" of $\pm 3s$ demanding a corrective. (Actually the chances are 1 in 20 that an observation subject only to random scatter will lie outside limits of $\pm 1.96\sigma$; 99.7% of a group of results should fall within the $\pm 3\sigma$ limits unless a definite cause is operating on the analysis.) If the analysis is one that has been performed many times, the laboratory may have a value for s which is a good estimate of σ. Otherwise, the control limits can be established temporarily on the basis of an s value obtained from a few results and then adjusted later as more data become available. Parallel control charts for ranges,

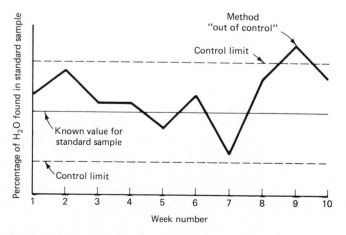

Figure 2.4 Control chart.

standard deviations, etc., may be employed to help the laboratory personnel keep track of the precision of an analytical method.

2.5

PROPAGATION OF ERRORS

Usually, the numerical result of a measurement is not of interest in its own right, but rather is used, sometimes in conjunction with several other measurements, to calculate the quantity which is actually desired. Attention is naturally focused upon the precision and accuracy of the final, computed quantity, but it is instructive to see how errors in the individual measurements are propagated into this result. A rigorous treatment of this problem requires more space than is available and mathematics beyond that presupposed for this book. An interesting elementary approach has been given by Waser,[15] and the interested student may find the elements of a more sophisticated treatment discussed briefly by Wilson.[16] A discussion with particular emphasis on analytical chemistry has been given by Benedetti-Pichler.[17]

2.5a.   *Determinate Errors*

Addition and Subtraction

Consider a computed result, R, based upon the measured quantities A, B, and C. Let α, β, and γ represent the absolute determinate errors in A, B, and C, respectively, and let ρ represent the resulting error in R. If the actual measurements are $A + \alpha$, $B + \beta$, and $C + \gamma$, we can see how the errors are transmitted through addition and subtraction. Suppose $R = A + B - C$. Changing each quantity by the amount of its error, we may write

$$R + \rho = (A + \alpha) + (B + \beta) - (C + \gamma)$$

or

$$R + \rho = (A + B - C) + (\alpha + \beta - \gamma)$$

Subtracting $R = A + B - C$ gives

$$\rho = \alpha + \beta - \gamma$$

Multiplication and Division

Now suppose, on the other hand, that multiplication and division are involved; i.e., let $R = AB/C$. Again the actual measurements are $A + \alpha$, $B + \beta$, and $C + \gamma$. Then

$$R + \rho = \frac{(A + \alpha)(B + \beta)}{C + \gamma} = \frac{AB + \alpha B + \beta A + \alpha \beta}{C + \gamma}$$

[15] J. Waser, *Quantitative Chemistry*, rev. ed., W. A. Benjamin, Inc., New York, 1964, p. 371.

[16] Wilson, loc. cit., p. 272.

[17] A. A. Benedetti-Pichler, *Ind. Eng. Chem., Anal. Ed.*, **8**, 373 (1936).

Let us neglect $\alpha\beta$, since it may be supposed that the errors are very small compared with the measured values. Then subtracting $R = AB/C$ gives

$$\rho = \frac{AB + \alpha B + \beta A}{C + \gamma} - \frac{AB}{C}$$

Placing the right-hand terms over a common denominator gives

$$\rho = \frac{\alpha BC + \beta AC - \gamma AB}{C(C + \gamma)}$$

It is now convenient to consider the relative error, ρ/R, by dividing by $R = AB/C$, which leads, after appropriate cancellation, to

$$\frac{\rho}{R} = \frac{\alpha BC + \beta AC - \gamma AB}{AB(C + \gamma)}$$

Since γ is very small compared with C, this reduces to

$$\frac{\rho}{R} = \frac{\alpha}{A} + \frac{\beta}{B} - \frac{\gamma}{C}$$

Thus it is found that determinate errors are propagated as follows:

1. Where addition or subtraction is involved, the *absolute* determinate errors are transmitted directly into the result.
2. Where multiplication or division is involved, the *relative* determinate errors are transmitted directly into the result.

The following examples illustrate the propagation of determinate errors.

EXAMPLE 2.6

Addition and subtraction

In adding up the molecular weight (MW) of KSCN from the atomic weights, a student accidentally recorded the atomic weight of K as 39.10 rather than 39.01, and that of S as 32.60 rather than 32.06. The other weights were recorded correctly. Calculate the absolute error made in the molecular weight of KSCN.

The determinate errors are $K = -0.09$ and $S = +0.54$. Hence the error in the molecular weight is

$$-0.09 + 0.54 = +0.45$$

Note that the correct molecular weight is 97.18, whereas the student would have obtained 97.63, or an error of $+0.45$. □

EXAMPLE 2.7

Multiplication and division

A student analyzed a sample for Cl by precipitating and weighing AgCl. A 0.8625-g sample gave a precipitate of AgCl weighing 0.7864 g. By mistake the student used the atomic weight (AW) of Cl as 35.345 rather than the correct value of 35.453. Calculate the error in the percentage Cl the student would make.

The correct % Cl is given by the expression

$$\% \text{ Cl} = \frac{0.7864 \times 35.453/143.321}{0.8625} \times 100 = 22.55$$

The determinate error made by the student would be

$$35.345 - 35.453 = -0.108$$

The relative errors are

$$\text{In AW of Cl} = -\frac{0.108}{35.453} = -0.00305$$

$$\text{In MW of AgCl} = -\frac{0.108}{143.321} = -0.00075$$

The relative error in the final result is then

$$\frac{\rho}{R} = -0.00305 - (-0.00075) = -0.0023 \text{ or } -2.3 \text{ ppt}$$

and the absolute error, $\rho = -0.05$. Using the incorrect weights the student would obtain a value of $22.55 - 0.05 = 22.50\%$ Cl. This can be confirmed by using the incorrect weights in the expression for %Cl above. $\qquad\square$

2.5b. Indeterminate Errors

In the case of determinate errors it was reasonable to assume, at least for the purpose of illustration, that each measurement of some quantity A was attended by a definite error α; we were able to work with the errors of individual measurements. Indeterminate errors, on the other hand, are manifested by scatter in the data when a measurement is performed more than once. No sign can be attached to random errors, since it is equally probable that the error will be positive or negative. The uncertainty in a calculated result is given as an interval such as 14.50 ± 0.08, usually expressed as the standard deviation, s, or as a confidence interval (Section 2.3c), $\pm ts/n$.

Addition and Subtraction

In addition or subtraction, statistical theory tells us that the *absolute variances* (squares of the standard deviations) of the measured values are additive in determining the most probable uncertainty in the result. That is, for $R = A + B - C$,

$$s_R^2 = s_A^2 + s_B^2 + s_C^2$$

Multiplication and Division

In multiplication or division, the squares of the *relative variances* are transmitted; that is, for $R = AB/C$,

$$\left(\frac{s_R}{R}\right)^2 = \left(\frac{s_A}{A}\right)^2 + \left(\frac{s_B}{B}\right)^2 + \left(\frac{s_C}{C}\right)^2$$

The following examples illustrate the propagation of indeterminate errors.

EXAMPLE 2.8

Uncertainty in reading a buret

A common operation in the analytical laboratory is the measurement of volume using a 50-mL buret. The smallest graduation is 0.1 mL, and the chemist estimates a reading to 0.01 mL by mentally dividing the smallest division into 10 equal parts, thus obtaining a reading such as 1.42 mL (see Fig. 20.5). Assuming that the uncertainty in such a reading is ± 0.02 mL, let us calculate the uncertainty in a volume read from a buret. Two readings are required; the initial reading is subtracted from the final reading to give the volume delivered.

Since the uncertainty in each reading is 0.02 mL, the uncertainty in the volume is simply

$$\sqrt{(0.02)^2 + (0.02)^2} = 0.03 \text{ mL}$$

It is interesting to note that if both buret readings had been 0.02 mL high, or if both had been 0.02 mL low, no error would occur since the errors would cancel. On the other hand, if one reading had been 0.02 mL high and the other 0.02 mL low, an error of 0.04 mL would have been made. As a general rule, chemists try to use a volume of at least 30 to 40 mL in a titration. This ensures that the relative error in determining the volume delivered from the buret is no greater than about 1 ppt. □

2.5c. Calculation of Analytical Results

Since a final result can be no more accurate than the least accurate measurement involved, the chemist should be familiar with the uncertainties in all the measurements in a multistep operation. Once the weakest link in the operation is determined, then the care taken in other steps can be adjusted so that the result will not be impaired, and at the same time valuable labor and time will not be uselessly expended.

In titrimetric and gravimetric methods, measurements of weight and volume are normally involved. We have seen (Example 8 above) that the chemist tries to use 30 to 40 mL in titrations in order to minimize the uncertainty in the volume measurement. Normally, the weight of sample taken for analysis is at least 0.2 g, since the uncertainty in weighing on an analytical balance may be as much as 0.2 mg. These procedures ensure that the relative errors will be about 1 ppt (0.1%) or less.

Sometimes the factor determining the overall uncertainty in an experimental result is not an instrumental measurement, but some factor beyond control of the analyst. For example, with modern instruments, voltages can be measured to the nearest 0.01 mV. Such an uncertainty leads to an error in determining the activity of an ion of only 0.4%. (See Examples 1 and 2, Chapter 12.) However, because of the presence of a liquid junction potential (Section 10.1d), an error as high as 3.8% can occur. Hence the chemist must be aware of such factors in order to express results properly.

EXAMPLE 2.9

Uncertainty in calculation of results.

A chemist determines the percentage of copper in an ore using the following data: Milliliters of titrant = 30.34 ± 0.03 mL; molarity of titrant = 0.1012 ± 0.0002 mmol/mL; milligrams of sample = 1073.2 ± 0.2; AW of Cu = 63.546 ± 0.003 mg/mmol. The calculation is

$$\frac{30.34 \times 0.1012 \times 63.546}{1073.2} \times 100 = 18.18\%$$

Calculate the uncertainty in the % Cu caused by the propagation of uncertainties in the data.

The relative uncertainties in ppt are

$$\frac{0.03}{30.34} \times 1000 = 1.0 \text{ ppt}$$

$$\frac{0.0002}{0.1012} \times 1000 = 2.0 \text{ ppt}$$

$$\frac{0.003}{63.546} \times 1000 = 0.05 \text{ ppt}$$

$$\frac{0.2}{1073.2} \times 1000 = 0.2 \text{ ppt}$$

Then,

$$\frac{\rho}{18.18} \times 1000 = (1.0)^2 + (2.0)^2 + (0.05)^2 + (0.2)^2$$

$$\rho = 0.04$$

Hence the result can be written $18.18 \pm 0.04\%$ Cu. □

2.6

SIGNIFICANT FIGURES

When a computation is made from experimental data, the uncertainty in the final result can be calculated by the methods described in Section 2.5 above. However, the procedure of calculating the square roots of the sum of the squares of the relative uncertainties is frequently more trouble than it is worth. A simpler procedure that can be employed to estimate the uncertainty involves the use of *significant figures*. This procedure gives only a rough estimate of the uncertainty, but in most situations encountered in analyses, an estimate is all that is needed.

Most scientists define significant figures as follows: *All digits that are certain plus one which contains some uncertainty are said to be significant figures.*[18] For example, in reading the volume on a buret where the smallest graduation is 0.1 mL (see Example 8 above), the figures 1.4 can be read with certainty. The second decimal is estimated by mentally dividing the smallest division into ten equal parts. The final reading of 1.42 mL then contains three significant figures, two certain, and one with some uncertainty.

It is important to use only significant figures in expressing experimental data. The use of too many or too few figures may mislead another person with respect to the precision of the measurement. If a volume is recorded as 12.346 mL, for example, another worker will assume that the volume was measured on a buret with graduations at 0.01-mL intervals and that the third decimal was estimated by reading between the graduations.

2.6a. Recognition of Significant Figures

The digit zero may or may not be a significant figure, depending upon its function in the number. In a buret reading of, say, 10.06 mL, both zeros are measured numbers and are therefore significant; the number contains four significant figures. Suppose the foregoing volume is expressed as liters, that is, .01006 L. The number of significant figures is not increased by changing the unit of volume; the number of significant figures is still 4. The function of the initial zero is to locate the decimal point; hence, *initial zeros are not significant*. Usually most people place a zero before the decimal, as 0.01006, and this zero also is not significant. *Terminal zeros are significant*. For example, a weight of 10.2050 g

---

[18] Note that in terms of this definition it is improper to say that "one should use the appropriate number of significant figures." Rather, one should say that *only significant figures* should be recorded. One can record too many digits, but not too many significant figures.

contains six significant figures. When it is necessary to use terminal zeros merely to locate the decimal properly, powers of ten may be used to avoid confusion with regard to the number of digits that are significant. For example, a weight of 24.0 g expressed as milligrams should not be written as 24,000. There are only three significant figures in the number, and this is indicated by writing 24.0×10^3, or 2.40×10^4.

2.6b. Computation Rules

The following rules are suggested to ensure that a calculated result contains only the number of digits justified by the experimental data.

Addition and Subtraction

In addition and subtraction, keep only as many decimal places as occur in the number which has the *fewest decimals*. The following is an example.

EXAMPLE 2.10

Significant figures in addition.

Add algebraically the following numbers: $14.23 + 8.145 - 3.6750 + 120.4$.

The final answer should be expressed to only one decimal place. It is recommended that the numbers be rounded[19] preliminarily to two decimal places, then added. The final result is then rounded to one decimal place.

$$
\begin{array}{rr}
+14.23 & +14.23 \\
+8.145 & +8.14 \\
-3.6750 & -3.68 \\
+120.4 & +120.4 \\
\hline
 & 139.09
\end{array}
$$

Final rounding gives: 139.1

This procedure tends to avoid the accumulation of rounding errors in the final result. □

Multiplication and Division

In multiplication and division, retain in each term and in the answer a number of significant figures that will indicate a *relative uncertainty* no greater than that of the term with the *greatest* relative uncertainty. Some people refer to this as the "weakest-link" approach: the answer cannot be any more precisely known than the weakest link in the data. Consider the following example.

EXAMPLE 2.11

Significant figures in multiplication and division

The percentage of chromium in an ore is calculated from titrimetric data as follows:

$$
\% \text{ Cr} = \frac{40.36 \text{ mL} \times 0.0999 \text{ mmol/mL} \times (51.9961 \text{ mg/mmol}/3)}{346.6 \text{ mg}} \times 100
$$

$$
\% \text{ Cr} = 20.173826 \ldots
$$

Round off this result to contain only significant figures.

[19] In rounding numbers, drop the last digit if it is less than 5: 4.33 becomes 4.3. If the last digit is greater than 5, increase the preceding digit by one: 4.36 becomes 4.4. If the digit to be dropped is 5, round the preceding digit to the nearest even number: 4.35 becomes 4.4; 4.65 becomes 4.6. This procedure avoids a tendency to round in one direction only.

According to the rule, we need to identify that portion of the data which has the greatest relative uncertainty. Since no uncertainties are specified, we assume at least ±1 in the last digit. (Actually, a chemist may have a good idea as to some of the uncertainties. See Section 2.5c.) If this is true, the relative uncertainties are 1 part in 4036 (about 0.25 ppt), 1 part in 999 (about 1 ppt), 1 part in 519961 (about 0.002 ppt), and 1 part in 3464 (about 0.3 ppt). There is no uncertainty in the numbers 3 and 100; these are sometimes called "counting numbers." Therefore, the term with the greatest uncertainty is the concentration term, 0.0999, known to about 1 ppt.

Before multiplication and division, the other terms may be rounded off, but only in such a manner that none of them comes out with an uncertainty greater than about 1 ppt. However, with the hand calculators used today, not much effort is saved in rounding off numbers before the final calculation.

To indicate a relative uncertainty in the % Cr no greater than 1 ppt, then, the above number should be rounded to 20.18%. It is understood that there is some uncertainty in the final digit, but the actual amount is not specified. Here we could probably estimate it to be ±2, since the uncertainty in the weakest link is 1 ppt. ☐

It should be emphasized that the ultimate criterion in rounding off an answer in multiplication and division is the relative uncertainty, not the number of significant figures in the data. Note that the weakest link in the problem above, 0.0999, contains three significant figures, whereas the final result, 20.18, contains four. If the concentration term had been 0.1001 (four SFs) instead of 0.0999 (three SFs), it would still have been the weakest link (1 ppt), and the answer would have been rounded to 20.20 (four SFs). Some chemists use a rule which says to keep as many significant figures in the final answer as are found in that portion of the data with the least number of significant figures. In most cases this simple rule gives the same result as the rule cited here. Its shortcoming is apparent when the first digit of the value with the least number of significant figures is a 9, as in the example above.

2.6c. Logarithms and Significant Figures

A logarithm is composed of two parts: a whole number, the *characteristic*, plus a decimal fraction, the *mantissa*. The characteristic is a function of the position of the decimal in the number whose logarithm is being determined and therefore is not a significant figure. The mantissa is the same regardless of the position of the decimal, and all the digits are considered significant.

Consider the following illustration.

log (sig fig) = # dec places

EXAMPLE 2.12

(a) Express properly the logarithm of 2.4×10^5.

Using a four-place log table we find that the mantissa is 0.3802. The characteristic is 5. The log is expressed as 5.38, since there are two significant figures in the number 2.4×10^5.

(b) Express the antilog of 2.974.

Using a four-place log table we see that 0.9741 is the mantissa of the number 942. If a calculator is used to obtain the antilog, the result is 941.8896. The answer should contain only three significant figures and hence should be written 942 or 9.42×10^2. ☐

METHOD OF LEAST SQUARES

Many analytical procedures involve instrumental measurements of a physical parameter which is directly proportional to the concentration of the analyte. Common examples are the determination of concentration by measuring the absorbance of a solution using a spectrophotometer, and by measuring peak areas obtained with a gas chromatograph. A series of solutions of known concentrations is prepared, and the instrument response is measured for each standard solution. The response is then plotted vs. concentration to give a *standard curve* or *calibration graph*. In many cases there is a straight-line relation between concentration and instrument response. However, the experimental points seldom fall *exactly* on a straight line because of indeterminate errors in the instrument readings. The problem facing the analyst is to draw the "best" straight line through the points so as to minimize the error in determining the concentration of an unknown using the calibration graph. Deciding where to draw the line could be a subjective process, and different analysts might vary somewhat in their decisions.

Fortunately, statistics provides a mathematical relationship which enables the chemist to calculate objectively the slope and intercept of the "best" straight line. The process is called the *method of least squares*. Not only can the best line be determined, but also the uncertainties in the use of the calibration graph for analysis of an unknown can be specified.

The derivations of the mathematical relationships are beyond the scope of the text. We shall illustrate the method here by applying it to a simple case; the data are shown in Table 2.8 and plotted in Fig. 2.5. In Fig. 2.5 we let the numbers on the x-axis represent the concentrations of the standard solutions. The numbers on the y-axis represent the instrument response. The problem is to draw the best straight line through these points.

Assuming that the relationship is a linear one, we write the equation for a straight line:

$$y = mx + b$$

where m is the slope and b the intercept on the y-axis. It is also assumed that the values of x, the concentrations of the standards, are free of error. The failure of

TABLE 2.8 Method of Least Squares

x	y	xy	x^2	y^2
1.00	2.96	2.96	1.00	8.76
2.00	5.05	10.10	4.00	25.50
3.00	7.03	21.09	9.00	49.42
4.00	8.92	35.68	16.00	79.57
5.00	10.94	54.70	25.00	119.68
$\sum x =$	$\sum y =$	$\sum xy =$	$\sum x^2 =$	$\sum y^2 =$
15.00	34.90	124.53	55.00	282.93

$$C = \sum x^2 - \frac{(\sum x)^2}{n} = 55.00 - \frac{(15.00)^2}{5} = 10.00$$

$$D = \sum y^2 - \frac{(\sum y)^2}{n} = 282.93 - \frac{(34.90)^2}{5} = 39.33$$

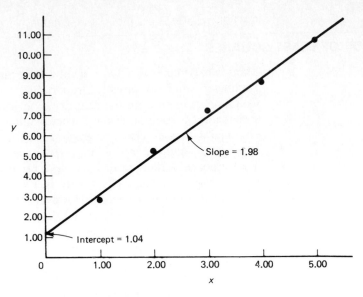

Figure 2.5 Method of least squares.

the data points to fall exactly on the line is assumed to be caused entirely by the indeterminate errors in the instrument readings, y. The sum of the squares of the deviations of the actual instrument readings from the correct values are minimized by adjusting the values of the slope, m, and the intercept, b. If a linear relationship between x and y does exist, this puts the line through the best estimates of the true mean values.

Table 2.8 contains not only values of x and y for the graph in Fig. 2.5, but also values of x^2, y^2, and xy and the sums of all these terms. The quantities C and D are defined for convenience in presenting later formulas. The term n is the number of data points, 5 in this example.

The equations given by statistics for the slope and intercept of the line are as follows:

$$\text{Slope: } m = \frac{\Sigma\, xy - (\Sigma\, x\, \Sigma\, y)/n}{C} = \frac{124.53 - (15.00 \times 34.90)/5}{10.00} = 1.98$$

$$\text{Intercept: } b = \frac{\Sigma\, y - m\, \Sigma\, x}{n} = \frac{34.90 - 1.98 \times 15.00}{5} = 1.04$$

Hence the equation for our line is

$$y = 1.98x + 1.04$$

The standard deviation of the y-values, s_y, is given by

$$s_y = \sqrt{\frac{(D - m^2 C)}{n - 2}} = \sqrt{\frac{39.33 - (1.98)^2(10.00)}{5 - 2}} = 0.20$$

The number of degrees of freedom here is $n - 2$, since two degrees have been "used up" in calculating the values of m and b. The standard deviation of the slope, s_m, is

$$s_m = \frac{s_y}{\sqrt{C}} = \frac{0.20}{\sqrt{10.00}} = 0.06$$

The 90% confidence limit of the slope is

$$CL_{0.90} = 1.98 \pm ts_m$$

The value of $t_{0.90}$ at $(n - 2)$ degrees of freedom (Table 2.4) is 2.353. Hence

$$CL_{0.90} = 1.98 \pm 2.353 \times 0.06$$

$$CL_{0.90} = 1.98 \pm 0.14$$

Suppose the calibration graph is now used to determine the concentration of an unknown. Let us say that three determinations are carried out, giving y-values of 6.25, 6.27, and 6.31, or an average value, $\bar{y}_u$, of 6.28. Then

$$6.28 = 1.98x + 1.04$$

or

$$x = 4.13$$

The standard deviation in this result, s_u, is given by

$$s_u = \frac{s_y}{m} \sqrt{\frac{1}{n_u} + \frac{1}{n} + \frac{(\bar{y}_u - \bar{y})^2}{m^2 C}}$$

where n_u is the number of determinations, here 3, n is the number of points in the calibration graph, here 5, and $\bar{y}$ is the average of the y values in the calibration graph, here $34.90/5 = 6.98$. Hence

$$s_u = \frac{0.20}{1.98} \sqrt{\frac{1}{3} + \frac{1}{5} + \frac{(6.28 - 6.98)^2}{(1.98)^2(10.00)}}$$

$$s_u = 0.07$$

If the analysis had been based on a single determination, say $y = 6.28$, the value of s_u would be

$$s_u = \frac{0.20}{1.98} \sqrt{\frac{1}{1} + \frac{1}{5} + \frac{(6.28 - 6.98)^2}{(1.98)^2(10.00)}}$$

or

$$s_u = 0.11$$

KEY TERMS

Absolute error. The difference between the experimental value and the true value.

Accuracy. A measure of how closely a measured quantity agrees with the true value.

Average deviation. The average of the differences between the individual results and the mean.

Coefficient of variation. The standard deviation expressed as a percentage of the mean.

Confidence interval. The interval within which there is a specified probability that the true value will occur.

Constant error. A determinate error of constant magnitude; its relative value decreases as the size of the sample increases.

Control chart. A chart used for keeping track of quality during large-scale manufacturing operations.

Degrees of freedom. The number of individual observations which can be allowed to vary under conditions that $\bar{x}$ and s, once determined, be held constant.

Determinate error. An error which can be ascribed to a definite cause; it is unidirectional and can normally be discovered and corrected.

F-test. (See variance-ratio test.)

Gaussian (normal) distribution. A mathematical function giving the relative frequency of an error as a function of its magnitude. A graph of the function gives a "bell-shaped" curve.

Indeterminate error. A random, unpredictable error which cannot be eliminated; plus and minus indeterminate errors are equally probable.

Mean. The average of a series of results.

Median. The middle value of a series of results listed in order; for an even number of results, the median is the average of the two middle ones.

Method of least squares. A statistical method for determining the "best" straight line through a series of points.

Precision. A measure of the agreement among a group of results.

Proportional error. A determinate error which varies with sample size in such a way that the relative error remains constant.

Q-test. A statistical test for discarding an experi-

mental result with a specified probability or confidence.

Range. The difference between the largest and smallest values in a set of measurements.

Relative error. The absolute error divided by the magnitude of the measured quantity.

Significant figures. All digits in a number which are certain plus one which contains some uncertainty are said to be significant.

Standard deviation. The distance from the mean to the inflection point of the normal distribution curve.

Student's t. A statistical term defined as $\pm t = (\bar{x} - \mu)\sqrt{n}/s$.

t-test. A statistical test which tells with a specified probability whether the means of two determinations differ significantly.

Variance. The square of the standard deviation.

Variance-ratio test (F-test). A test which uses the ratio of the variances of two sets of results to determine if the standard deviations are significantly different.

QUESTIONS

1. *Terms.* Explain clearly the meaning of the following terms: (a) mean; (b) median; (c) standard deviation; (d) relative average deviation; (e) variance; (f) absolute error; (g) relative error.
2. *Errors.* Explain whether the following errors are determinate or indeterminate. If the error is determinate, tell whether it is methodic, operative, or instrumental, and how it could be eliminated. (a) A 50-mL pipet actually delivers 50.1 mL. (b) An acid-base indicator changes at pH 6 rather than pH 4. (c) A worker does not allow time for a buret to drain before reading the volume delivered. (d) A worker miscalculates the molecular weight of the analyte. (e) A side reaction occurs during a titration.
3. *Confidence intervals.* Explain clearly the meaning of (a) a 95% and (b) a 99% confidence interval of the mean.
4. *Testing for significance.* Explain clearly how

to test two sets of results to determine if they differ significantly.
5. *Rejection of a result.* In rejecting results, what is meant by making an "error of the first kind?" An "error of the second kind?"
6. *Q-test.* What kind of errors are most likely to be made when applying the Q-test? Why do some people recommend using the median rather than the mean in some instances?

Multiple choice. In the following multiple-choice questions, select the *one best* answer.

7. The average of 64 results is how many more times reliable than the average of 4 results? (a) 2; (b) 4; (c) 8; (d) 16.
8. Which of these statements is true? (a) The variance is the square root of the standard deviation. (b) Precise values are always accurate. (c) The number 0.02040 contains only

four significant figures. (d) Two of the above are true.

9. Students performed an analysis obtaining a percentage purity of 18.54% and a relative standard deviation of 4.0 ppt. Later they discovered they had made an error of a factor of 2 and that the result should have been 9.27%. What is the correct relative standard deviation in ppt? __(a) 2.0; (b) 4.0; (c) 8.0; (d) $\sqrt{2.0}$.

10. Titrator A obtains a mean value of 12.96% and a standard deviation of 0.05 for the purity of a sample. Titrator B obtains corresponding values of 13.12% and 0.08. The true percent purity is 13.08. Compared to titrator B, titrator A is (a) less accurate but more precise; (b) more accurate and more precise; (c) less accurate and less precise; (d) more accurate but less precise.

PROBLEMS

(Unless otherwise indicated, it is to be understood that any weight or volume involves the difference of two readings.)

1. *Precision.* Refer to the table of atomic weights on the inside front cover. The uncertainty in the last digit is given in parentheses. (a) Which of the atomic weights is known with the greatest precision, that of hydrogen, oxygen, or iron? (b) Express each relative precision in parts per thousand.

2. *Precision.* Repeat Problem 1 for these elements: fluorine, sodium, and aluminum.

3. *Mean, median, etc.* Analyst A reported the following percentages of FeO in a sample: 16.65, 16.70, 16.68, 16.60, 16.58, and 16.63. For this set of results calculate (a) the mean, median, range, average deviation, relative average deviation (ppt), standard deviation, and coefficient of variation; (b) the 95% confidence interval of the mean, first from the standard deviation and then from the range.

4. *Mean, median, etc.* Analyst B reported the following percentages of FeO in the same sample as in Problem 3: 16.60, 16.74, 16.66, 16.64, 16.82, and 16.50. For this set of results calculate (a) the mean, median, range, average deviation, relative average deviation (ppt), standard deviation, and coefficient of variation; (b) the 95% confidence interval of the mean, first from the standard deviation and then from the range.

5. *Accuracy.* The National Institute of Standards and Technology reported a value of 16.55% FeO in the sample in Problems 3 and 4. (a) Calculate the absolute and relative errors of analysts A and B. (b) What can you say about the work of the two analysts?

6. *Mean and average deviation.* A student standardized a solution of HCl and found the following molarities: 0.1016, 0.1024, 0.1012, and 0.1010. Calculate the mean and relative average deviation of these results in ppt.

7. *Mean and average deviation.* A student standardized a solution of iodine and found the following molarities: 0.0512, 0.0520, 0.0516, and 0.0506. Calculate the mean and relative average deviation of these results in ppt.

8. *Precision.* The uncertainty in each reading on a semimicro balance is ±0.01 mg. How large a sample should be taken for analysis so that the maximum relative uncertainty in the sample weight will be 1.0 ppt?

9. *Precision.* The uncertainty in each reading on a trip balance is ±0.01 g. How large a sample should be taken using this balance so that the maximum relative uncertainty in the weight will be 2.0 ppt?

10. *Precision.* If the uncertainty in each reading on a buret is ±0.02 mL, what is the maximum uncertainty in reading a volume of 10 mL on this buret?

11. *Precision.* (a) A beaker is placed on an analytical balance and weighed (one reading) to the fourth decimal place. If the beaker weighs exactly 100 g, what is the relative uncertainty in the weight? (b) If a relative uncertainty in the weight of the beaker of only 0.1% is desired, to what decimal place should it be weighed (one reading)?

12. *Constant error.* In a certain method of determining silica, SiO_2 is precipitated and

weighed. It is found that the amount of SiO_2 obtained is always 0.5 mg too high regardless of the weight of the sample taken for analysis. Calculate the relative error in ppt in a sample which contains 5.00% SiO_2 if the size of the sample analyzed is (a) 0.200 g; (b) 0.600 g; (c) 1.00 g.

13. *Proportional error*. A sample is to be analyzed for chloride by titration with silver nitrate. The sample actually contains sufficient bromide to cause the amount of chloride to appear 0.20% high. Calculate the error in the number of milligrams of chloride an analyst would find in a sample which contains 25.0% chloride if the size of the sample analyzed is (a) 0.200 g; (b) 0.600 g; (c) 1.00 g.

14. *Relative error*. (a) A relative error of 0.5% is how many ppt? (b) A relative error of 2.0 parts per 500 is what percent error? (c) A relative error of 0.01 ppt is how many ppm? (d) A relative error of 5 parts per 2000 is how many parts per 100?

15. *Relative error*. Assuming an uncertainty of ± 1 in the last digit, what is the relative uncertainty in ppt in the following numbers? (a) 40; (b) 500; (c) 6.02×10^{23}; (d) 9.99; (e) 10.01.

16. *Relative error*. Assuming an uncertainty of ± 1 in the last digit, how should the following numbers be expressed to indicate the given uncertainty? (a) 2000, 0.5%; (b) 1 million, 1 ppt; (c) 50, 0.02%; (d) 25, 0.4%.

17. *Testing for significance*. A blood sample is sent to two different laboratories to be analyzed for cholesterol. The results obtained for the concentration (mg/dL) were:

Laboratory 1	Laboratory 2
$\bar{x}_1 = 243$	$\bar{x}_2 = 258$
$s_1 = 13$	$s_2 = 15$
$n_1 = 11$	$n_2 = 11$

(a) Are the standard deviations significantly different at the 95% level? (b) Are the two means significantly different (i) at the 90% level; (ii) at the 95% level; (iii) at the 99% level? Use 14 as the value of s in calculating t.

18. *Testing for significance*. Repeat Problem 17 for the case where $\bar{x}_1 = 241$ and $\bar{x}_2 = 259$.

19. *Testing for significance*. Repeat Problem 17 for $s_1 = 16$ and $s_2 = 18$. Use 17 as the value of s in calculating t.

20. *Testing for significance*. Repeat Problem 17 for $n_1 = 5$ and $n_2 = 6$.

21. *Testing for significance*. An analyst develops a new method to determine manganese in an ore. He analyzes a sample from the National Institute of Standards and Technology by this method and obtains the following data: mean of four results = 8.82%; standard deviation = 0.10. The NIST value is 8.72%. Are the results significantly different at the 95% level?

22. *Testing for significance*. Suppose the data in Problem 21 were all the same except that the mean was based on nine results. Would the difference be significant at the 95% level?

23. *Testing for significance*. Repeat Problem 22 except for a standard deviation of 0.14. Is the difference significant at the 95% level?

24. *Rejection of a result*. A student obtained the following results for the normality of a solution: 0.1026, 0.1019, 0.1047, 0.1016, and 0.1022. (a) Can any result be rejected by the Q-test? (b) What value should be reported for the normality?

25. *Rejection of a result*. A technician obtained the following results for the concentration (mg/dL) of cholesterol in a blood sample: 240, 265, 230, 238, and 244. (a) Can any result be rejected by the Q-test? (b) What value should be reported for the concentration?

26. *Rejection of a result*. Suppose the technician in Problem 25 ran one additional determination and obtained a value of 236. (a) Can any result now be rejected by the Q-test? (b) What value should be reported for the concentration?

27. *Rejection of a result*. A student obtained the following results for the percentage purity of a sample: 10.30, 10.44, 10.38, and 10.34. What is (a) the highest, and (b) the lowest value a fifth result could be without being discarded by the Q-test?

28. *F-test*. Two students are given the same sample to analyze. Student A makes seven determinations with a standard deviation of 0.09. Student B makes five determinations with a standard deviation of 0.04. Does the difference in standard deviations imply a significant difference in the techniques of the two students?

29. *F-test.* Suppose in Problem 28 the students' standard deviations were the same, but student A made five determinations and student B made seven. Is there a significant difference in the techniques of the two students? Explain the different conclusion from that in Problem 28.

30. *F-test.* Two students are given the same sample to analyze. Student A makes six determinations with a standard deviation of 0.06. Student B also makes six determinations. How large can the standard deviation of the latter student be without suggesting significant difference in the techniques of the two?

31. *Confidence interval of the mean.* A chemist analyzes an iron ore for FeO and obtains a value of 12.35% with a standard deviation of 0.08. Calculate the 95% confidence interval of the mean based on (a) four determinations, and (b) eight determinations.

32. *Confidence interval of the mean.* Suppose the procedure used for determining iron in Problem 31 has been run many times and it is known that the standard deviation for this procedure is 0.08. Calculate the 95% confidence interval of the mean if the result of 12.35 is based on (a) four determinations; (b) nine determinations. (*Hint:* Use the value of t at $n = \infty$ in Table 2.4.)

33. *Confidence interval of the mean.* The standard deviation of a method for determining manganese in steel is known from a large number of determinations to be 0.07. How many determinations must be run by this method if (a) the 90% confidence interval of the mean is to be ±0.06; (b) the 99% confidence interval of the mean is to be ±0.06?

34. *Propagation of errors.* Make the following calculations assuming that the uncertainties are determinate errors:

(a) $(6.48 \pm 0.02) + (10.64 \pm 0.03) - (7.04 \pm 0.04)$

(b) $\dfrac{(42.48 \pm 0.04)(0.1105 \pm 0.0002)(168.6 \pm 0.2)}{38.66 \pm 0.02}$

35. *Propagation of errors.* Make the following calculations. Assume that the uncertainties are standard deviations and calculate the standard deviation in the final result.

(a) $(12.384 \pm 0.002) + (7.82 \pm 0.04) - (5.6 \pm 0.1)$

(b) $\dfrac{(39.84 \pm 0.04)(0.0994 \pm 0.0001)(224.3 \pm 0.2)}{426.4 \pm 0.3}$

36. *Significant figures.* How many significant figures does each of these numbers contains? (a) 0.0380; (b) 6.022×10^{23}; (c) 9.99; (d) 10.00; (e) 96,500; (f) 2.08×10^{-6}.

37. *Relative error.* If the uncertainty in the last digit of each number in Problem 36 is ±1, what is the relative uncertainty in each number in ppt?

38. *Significant figures.* Express the results of the following calculations using only significant figures.

(a) $\dfrac{0.0382 \times 3.65 \times 10^3 \times 2.304}{8.64 \times 10^4}$

(b) $\dfrac{4.25 \times 10^2 \times 30.20 \times 0.0720}{8.64 \times 10^{-3}}$

(c) $24.364 + 5.6 + 1.3420$

39. *Significant figures.* Express the results of the following calculations using only significant figures.

(a) $\dfrac{41.24 \times 0.0994 \times 56.02}{22.267}$

(b) $\dfrac{0.005681 \times 2.463}{22.30 \times 0.304}$

(c) $25.4623 + 0.620 - 8.14302$

40. *Significant figures.* How should the percentage of (a) P in $Mg_2P_2O_7$, and (b) Cr in $K_2Cr_2O_7$ be properly expressed using the atomic weights given in the text and the rules for significant figures?

41. *Significant figures.* Express properly the following: (a) log of 4.82×10^4; (b) log of 2.624×10^{-5}; (c) antilog of 4.350; (d) antilog of 0.0626.

42. *Absolute and relative errors.* A student analyzed a sample for sulfur by precipitating and weighing $BaSO_4$. A 0.6283-g sample gave a precipitate of $BaSO_4$ weighing 0.4816 g. By mistake the student used the atomic weight of sulfur as 32.60 instead of 32.06 both in calculating the molecular weight of $BaSO_4$ and in calculating the gravimetric factor. Calculate the relative error in ppt and the absolute error in the % S the student would make from this mistake.

43. *Absolute and relative errors.* Suppose the analysis given in Example 21, Section 3.3d, is to be carried out and by mistake the analyst reads the buret as 39.90 mL rather than the correct 40.00 mL. Calculate the relative and absolute errors in the percentage $Na_2C_2O_4$ caused by this error.

44. *Sample size.* A sample contains about 10% of the ion B^{2-}. It is determined by precipitating the compound A_2B. A has an atomic weight about twice that of B. If the uncertainty in determining the weight of the precipitate is not to exceed 1 ppt on a balance sensitive to 0.1 mg, what size sample should be taken for analysis?

45. *Sample size.* A sample containing about 10% sulfur is to be analyzed by oxidizing the S to SO_4^{2-}, and precipitating and weighing $BaSO_4$. If the uncertainty in determining the weight of $BaSO_4$ is not to exceed 1 ppt on a balance sensitive to 0.1 mg, how large a sample should be taken for analysis?

46. *Method of least squares.* The concentration of lead in a solution was determined polarographically by measuring the diffusion current obtained with a series of standard solutions. The molarities of the standards were (a) 2.0×10^{-4}; (b) 5.0×10^{-4}; (c) 1.0×10^{-3}; (d) 1.5×10^{-3}; (e) 2.0×10^{-3}. The corresponding diffusion currents in microamperes were (a) 2.8; (b) 3.9; (c) 6.2; (d) 7.8; (e) 10.1. Using the method of least squares, calculate (a) the equation for the best straight line through these points; (b) the standard deviation of the current values; (c) the standard deviation of the slope; (d) the 90% confidence interval of the slope. A solution of

unknown concentration gave a diffusion current of 4.8 μA. Calculate (e) the concentration of the unknown; (f) the standard deviation of this result if the measurement is (i) a single one; (ii) the average of four determinations.

47. *Method of least squares.* The concentration of glucose in aqueous solution can be determined by oxidation with ferricyanide ion, $Fe(CN)_6^{3-}$. The yellow ferricyanide ion is reduced by glucose to colorless ferrocyanide ion, $Fe(CN)_6^{4-}$. The absorbance of the ferricyanide ion, measured at 420 nm, decreases as the concentration of glucose increases. The decrease in absorbance is proportional to the concentration of glucose.

The concentrations of the standard solutions in milligrams of glucose per milliliter in a determination were (a) 0.0050; (b) 0.010; (c) 0.020; (d) 0.030; (e) 0.040. The corresponding absorbances were (a) 0.750; (b) 0.642; (c) 0.498; (d) 0.290; (e) 0.132. Using the method of least squares, calculate (a) the equation of the best straight line through these points; (b) the standard deviation of the absorbance values; (c) the standard deviation of the slope; (d) the 90% confidence interval of the slope. A solution of unknown concentration gave an absorbance of 0.422. Calculate (e) the concentration of the unknown; (f) the standard deviation of this result if the measurement is (i) a single one; (ii) the average of five determinations.

3

Titrimetric Methods of Analysis

We have already mentioned (page 6) that *titrimetric* analysis is one of the major divisions of analytical chemistry and that the calculations involved are based on the simple stoichiometric relationships of chemical reactions. In this chapter we shall discuss the general principles involved in titrimetry and the stoichiometric calculations which are employed.

3.1

GENERAL PRINCIPLES

A titrimetric method of analysis is based on a chemical reaction such as

$$a\text{A} + t\text{T} \longrightarrow \text{products}$$

where a molecules of the analyte, A, react with t molecules of the reagent, T. The reagent T, called the *titrant*, is added incrementally, normally from a buret, in the form of a solution of known concentration. The latter solution is called a *standard* solution, and its concentration is determined by a process called *standardization*. The addition of the titrant is continued until an amount of T chemically equivalent to that of A has been added. It is then said that the *equivalence point* of the titration has been reached. In order to know when to stop the addition of titrant the chemist may use a chemical substance, called an *indicator*, which responds to the appearance of excess titrant by changing color. This color change may or may not occur precisely at the equivalence point. The point in the titration where the indicator changes color is termed the *end point*. It is desirable, of course, that the end point be as close as possible to the equivalence point. Choosing indicators to make the two points coincide (or correcting for the difference between the two) is one of the important aspects of titrimetric analysis. Visual indicators are only one of several methods used to detect the end point of a titration. Other techniques which detect the sudden change in a physical or chemical property of a solution are also available.

The term *titration* refers to the process of measuring the volume of titrant required to reach the equivalence point. For many years the term *volumetric* analysis was used rather than titrimetric. However, from a rigorous standpoint the term *titrimetric* is preferable because volume measurements need not be confined to titrations. In certain analyses, for example, one might measure the volume of a gas.

3.1a. Reaction Used for Titrations

The chemical reactions which may serve as the basis for titrimetric determinations are conveniently grouped into four types:

1. *Acid-base*. There are a large number of acids and bases which can be determined by titrimetry. If HA represents the acid to be determined and B the base, the reactions are

$$HA + OH^- \longrightarrow A^- + H_2O$$

and

$$B + H_3O^+ \longrightarrow BH^+ + H_2O$$

The titrants are generally standard solutions of strong electrolytes, such as sodium hydroxide and hydrochloric acid.

2. *Oxidation-reduction* (*redox*). Chemical reactions involving oxidation-reduction are widely used in titrimetric analyses. For example, iron in the +2 oxidation state can be titrated with a standard solution of cerium(IV) sulfate:

$$Fe^{2+} + Ce^{4+} \longrightarrow Fe^{3+} + Ce^{3+}$$

Another oxidizing agent which is widely used as a titrant is potassium permanganate, $KMnO_4$. Its reaction with iron(II) in acid solution is

$$5Fe^{2+} + MnO_4^- + 8H^+ \longrightarrow 5Fe^{3+} + Mn^{2+} + 4H_2O$$

3. *Precipitation*. The precipitation of silver cation with the halogen anions is a widely used titrimetric procedure. The reaction is

$$Ag^+ + X^- \longrightarrow AgX(s)$$

where X^- can be chloride, bromide, iodide, or thiocyanate (SCN^-) ion.

4. *Complex formation*. An example of a reaction in which a stable complex is formed is that between silver and cyanide ions:

$$Ag^+ + 2CN^- \longrightarrow Ag(CN)_2^-$$

This reaction is the basis of the so-called Liebig method for the determination of cyanide. Certain organic reagents, such as ethylenediaminetetraacetic acid (EDTA), form stable complexes with a number of metal ions and are widely used for the titrimetric determination of these metals.

3.1b. Requirements for Reactions Used in Titrimetric Analysis

Of the host of known chemical reactions, relatively few can be used as the basis for titrations. A reaction must satisfy certain requirements before it can be used:

1. The reaction must proceed according to a definite chemical equation. There should be no side reactions.
2. The reaction should proceed to virtual completion at the equivalence point. Another way of saying this is that the equilibrium constant of the reaction should be very large. If this requirement is met, there will be a large, abrupt change in the concentration of the analyte (or of the titrant) at the equivalence point.
3. Some method must be available for determining when the equivalence point is reached. An indicator should be available, or some instrumental method may be used to tell the analyst when to stop the addition of titrant.
4. It is desirable that the reaction be rapid, so that the titration can be completed in a few minutes.

Consider, as an example of a reaction well suited for titrations, determination of the concentration of a hydrochloric acid solution by titration with standard sodium hydroxide. There is only one reaction,

$$H_3O^+ + OH^- \longrightarrow 2H_2O \qquad K = 1 \times 10^{14}$$

and it is exceedingly fast. The reaction goes to virtual completion. At the equivalence point the pH of the solution changes by several units for a few drops of titrant, and a number of indicators are available which respond to this pH change by changing color.

On the other hand the reaction between boric acid and sodium hydroxide,

$$HBO_2 + OH^- \rightleftharpoons BO_2^- + H_2O \qquad K = 6 \times 10^4$$

is not sufficiently complete to satisfy requirement 2; the equilibrium constant is only about 6×10^4. For this reason, the pH change for a few drops of titrant at the equivalence point is very small, and the volume of titrant needed cannot be determined with good accuracy.

The reaction between ethyl alcohol and acetic acid is also unsuitable for titration. It is too slow for convenience and does not go well to completion. The reaction between tin(II) and potassium permanganate is not satisfactory unless air is excluded. A side reaction can occur because tin is readily oxidized by atmospheric oxygen. The precipitation of certain metal ions by sulfide ion satisfies all the requirements above except number 3; that is, no suitable indicators are available.

3.2

CONCENTRATION SYSTEMS

In this section we shall consider the methods used by the analytical chemist to express the *concentration* of a solution, viz., the relative amounts of solute and solvent. The systems of *molarity* and *normality*[1] are most commonly employed,

---

[1] We shall consider the term *equivalent* and the concentration system *normality* in Section 3.4.

since they are based on the *volume* of solution, the quantity measured in a titration. *Formality* and *analytical concentration* are useful in situations where dissociation or complex formation occurs. The *percent-by-weight* system is commonly employed to express approximate concentrations of laboratory reagents. For very dilute solutions *parts per million* or *parts per billion* units are convenient.

We shall first review the concept of the mole and molecular weight.

3.2a. Molecular and Formula Weights

The student will recall that the *mole* (or *mol*) is defined as the amount of a substance which contains as many entities as there are atoms in 12 g of the carbon-12 isotope, $^{12}_{6}C$. The entities may be atoms, molecules, ions, or electrons. Since 12 g of carbon-12 contains Avogadro's number of atoms, it follows that 1 mol of any substance contains 6.022×10^{23} elementary particles. If the particles are molecules, the weight in grams of a mole of the substance is called the *gram-molecular weight* (usually shortened to *molecular weight*). Hence the molecular weight of H_2 is 2.016 g/mol and contains 6.022×10^{23} H_2 molecules. If the particles are atoms, the weight in grams of 1 mol of the substance is called the *gram-atomic weight* (usually shortened to *atomic weight*). The atomic weight of copper is 63.54 g/mol and contains 6.022×10^{23} Cu atoms.

The term *gram-formula weight* (or *formula weight*) is the summation of the atomic weights of all the atoms in the chemical formula of a substance and is normally the same as the molecular weight. Some chemists use formula weight rather than molecular weight in cases where it would be inappropriate to talk about "molecules" of a substance, particularly ionic compounds. In sodium chloride (NaCl), for example, the smallest units in the solid state are Na^+ and Cl^- ions; a molecule of NaCl does not exist. Since the mole, as defined above, refers to other entities as well as molecules, we shall use the term *molecular weight* as synonymous with *formula weight* in such cases. It is understood that this usage does not imply anything about the structure of the compound. We shall use formula weight in defining the concentration system *formality* (Section 3.2c).

In titrimetric procedures the volume of titrant used is normally in the range of 0.010 to 0.050 liter. For convenience this volume is usually expressed in *milliliters* (mL), where 1 mL = 0.001 liter. The quantity of solute present in this volume is on the order of 0.0010 to 0.0050 mol and may weigh 0.040 to 0.200 g. Again for convenience these quantities are expressed in *millimoles* (mmol), where 1 mmol = 0.001 mol, and *milligrams* (mg), where 1 mg = 0.001 g. Thus volumes of 10 to 50 mL normally contain 1.0 to 5.0 mmol of solute which may weigh 40 to 200 mg. In practice chemists customarily use moles, grams, and liters when dealing with volumes greater than 1 liter, and millimoles, milligrams, and milliliters when the volume is considerably less than 1 liter. Note that the molecular weight of a substance can be expressed in either the large or the small units. For example, the molecular weight of NaOH is 40.00 g/mol or 40.00 mg/mmol.

3.2b. Molarity

This concentration system is based on the volume of *solution* and hence is convenient to use in laboratory procedures where the volume of solution is the quantity measured. It is defined as follows:

$$\text{Molarity} = \text{number of moles per liter of solution}$$

or

$$M = \frac{n}{V}$$

where M is the molarity, n the number of moles of solute, and V the volume of solution in liters. Since

$$n = \frac{g}{\text{MW}}$$

where g is the grams of solute and MW the molecular weight of solute, it follows that

$$M = \frac{g}{\text{MW} \times V}$$

This equation can be solved for grams of solute, giving

$$g = M \times V \times \text{MW}$$

The following examples illustrate the molarity system of concentration.

EXAMPLE 3.1

Calculation of molarity from mass and volume

Calculate the molarity of a solution which contains 6.00 g of NaCl (MW 58.44) in 200 mL of solution.

$$M \text{ (mol/liter)} = \frac{6.00 \text{ g NaCl} \times 1000 \text{ mL/liter}}{58.44 \text{ g/mol NaCl} \times 200 \text{ mL}}$$

$$M = 0.513 \text{ mol/liter} \qquad \square$$

EXAMPLE 3.2

Calculation of moles and mass from concentration and volume

Calculate the number of moles and the number of grams of $KMnO_4$ (MW 158.0) in 3.00 liters of a 0.250 M solution.

$$n = M \times V$$

$$n \text{ (mol)} = 0.250 \text{ mol/liter} \times 3.00 \text{ liter}$$

$$n = 0.750 \text{ mol}$$

$$g \text{ (grams)} = n \times \text{MW}$$

$$g = 0.750 \text{ mol} \times 158.0 \text{ g/mol}$$

$$g = 119 \qquad \square$$

EXAMPLE 3.3

Molarity calculation involving density

Calculate the molarity of a solution of H_2SO_4 which has a density[3] of 1.30 g/mL and contains 32.6% SO_3 by weight. The molecular weight of SO_3 is 80.06 g/mol.
One liter of solution contains

$$\text{g/liter } SO_3 = 1.30 \text{ g/mL} \times 1000 \text{ mL/liter} \times 0.326$$

$$\text{g/liter } SO_3 = 424$$

[3] The *density* of a solution is the mass of a unit volume, usually in kilograms per liter or grams per milliliter. *Specific gravity* is dimensionless; it is the ratio of the mass of a substance to the mass of an equal volume of water at 4°C. Since water has a density of exactly 1.00 g/mL at 4°C, the two terms are interchangeable when the metric system is employed.

The molarity is

$$M \text{ (mol/liter)} = \frac{424 \cancel{g} \text{ SO}_3}{80.06 \cancel{g}/\text{mol SO}_3 \times 1.00 \text{ liter}}$$

$$M = 5.30 \text{ mol/liter}$$

Since 1 mol of SO_3 produces 1 mol of H_2SO_4 in water:

$$SO_3 + H_2O \longrightarrow H_2SO_4$$

there are 5.30 mol/liter of H_2SO_4 in the solution. $\qquad\qquad\qquad\qquad$ □

3.2c. *Formality or Analytical Concentration*

Many compounds undergo dissociation or complex formation when dissolved in certain solutions. For example, the weak electrolyte acetic acid (CH_3COOH, or HOAc) dissociates slightly into ions when dissolved in water:

$$HOAc + H_2O \rightleftharpoons H_3O^+ + OAc^-$$

Now if 0.100 mol of HOAc is dissolved in 0.100 liter of aqueous solution and dissociates to the extent of 1.3%, the solution is not 0.100 M in HOAc molecules. Rather, the solution is 0.0987 M in HOAc molecules and 0.0013 M in OAc^- and H_3O^+ ions. In such cases many chemists use the term *formality* (F) or *analytical concentration*[2] (C_x) to indicate the *total* concentration of species arising from acetic acid. In this example

$$F = C_a = [\text{HOAc}] + [\text{OAc}^-]$$

$$F = C_a = 0.0987 + 0.0013 = 0.100$$

Formality is defined as

$$F = \frac{n_f}{V}$$

where n_f is the number of formula weights of solute and V is the volume of solution in liters. Since

$$n_f = \frac{g}{\text{FW}}$$

where g is the number of grams of solute and FW is the formula weight, then

$$F = \frac{g}{\text{FW} \times V}$$

The following example illustrates this system of concentration.

EXAMPLE 3.4

Calculation of formality from mass and volume

A sample of dichloroacetic acid, $Cl_2CHCOOH$(FW 128.94), weighing 6.447 g is dissolved in 500 mL of solution. At this concentration the acid is about 45% dissociated:

$$Cl_2CHCOOH \rightleftharpoons H^+ + Cl_2CHCOO^-$$

Calculate the formality of the dichloroacetic acid and the molarities of the two species $Cl_2CHCOOH$ and Cl_2CHCOO^-.

---

[2] Some texts call this the *molar analytical concentration*.

$$F = \frac{g}{FW \times V}$$

$$F = \frac{6.447 \ g}{128.94 \ g/FW \times 0.500 \ \text{liter}}$$

$$F = 0.100 \ FW/\text{liter}$$

This is the *total* concentration of the species arising from dichloroacetic acid. The *equilibrium concentrations* of dichloroacetic molecules and dichloroacetate ions are

$$[\text{Cl}_2\text{CHCOO}^-] = 0.100 \times 0.45 = 0.045 \ M$$

$$[\text{Cl}_2\text{CHCOOH}] = 0.100 \times 0.55 = 0.055 \ M$$

Such concentrations are expressed as molarities and are indicated by enclosing the molecule or ion in brackets. Hence

$$F = c_a = [\text{Cl}_2\text{CHCOOH}] + [\text{Cl}_2\text{CHCOO}^-]$$

$$F = c_a = 0.055 + 0.045 = 0.100 \qquad \square$$

In most examples we shall encounter in the text, molarity and formality can be used interchangeably. In the few cases where a distinction is needed it will be noted.

3.2d. Weight Percent

This system of concentration is commonly employed to express approximate concentrations of laboratory reagents. It specifies the number of grams of solute per 100 g of solution. Mathematically this is expressed as follows:

$$P = \frac{w}{w + w_0} \times 100$$

where P is the percent by weight of solute, w the number of grams of solute, and w_0 the number of grams of solvent.

The following examples illustrate the weight percent system of concentration.

EXAMPLE 3.5

Calculation of weight percent from mass of solute and solvent

A sample of NaOH weighing 5.0 g is dissolved in 45 g of water. (1 g of water is approximately 1 mL.) Calculate the weight percent NaOH in the solution.

$$P = \frac{5.0}{5.0 + 45} \times 100$$

$$P = 10\% \qquad \square$$

EXAMPLE 3.6

Calculation of required volume from density and weight percent

Concentrated HCl (MW 36.5) has a density of 1.19 g/mL and is 37% by weight HCl. How many milliliters of the concentrated acid should be diluted to 1.00 liter with water to prepare a 0.100 M solution?

$$\text{grams HCl needed} = 1.00 \ \text{liter} \times 0.100 \ \text{mol/liter} \times 36.5 \ \text{g/mol}$$

$$\text{grams HCl needed} = 3.65$$

$$\text{grams HCl per milliliter} = 1.19 \ \text{g/mL} \times 0.37 = 0.44$$

$$\frac{3.65 \ g}{0.44 \ g/\text{mL}} = 8.3 \ \text{mL} \qquad \square$$

3.2e.   *Parts per Million (ppm)*

This system is convenient for expressing the concentrations of very dilute solutions. It specifies the number of parts of solute in 1 million parts of solution and can be expressed mathematically as

$$\text{ppm} = \frac{w}{w + w_0} \times 10^6$$

where w is the number of grams of solute and w_0 the number of grams of solvent. Since w is usually very small compared to w_0, this becomes

$$\text{ppm} = \frac{w}{w_0} \times 10^6$$

One liter of water at room temperature weighs approximately 10^6 mg. Hence a convenient relationship to remember is that 1 mg of solute in 1 liter of water is a concentration[4] of 1 ppm.

The following example illustrates a calculation involving parts per million.

EXAMPLE 3.7

Calculation of mass required to prepare a solution of given ppm

If drinking water contains 1.5 ppm of NaF, how many liters of water can be fluoridated with 1.0 lb (454 g) of NaF?

Let V = liters of water to be fluoridated. Since 1 ppm = 1 mg/liter of water,

$$\frac{454 \times 10^3 \text{ mg NaF}}{V \text{ (liters)}} = 1.5$$

$$V = 3.0 \times 10^5 \text{ liters}$$

For even more dilute solutions the system *parts per billion* (ppb) is employed, where

$$\text{ppb} = \frac{w}{w_0} \times 10^9$$

3.3

STOICHIOMETRIC CALCULATIONS

After a solution is prepared and its concentration determined accurately, it can be employed as a titrant in the determination of the purity of an unknown sample. The calculations involved, called *stoichiometric*, are based on the mole and mass relations between the elements and compounds as expressed by a chemical equation.

3.3a.   *Standardization of Solutions*

It has been previously mentioned that the process by which the concentration of a solution is accurately determined is called *standardization*. A *standard* solution can sometimes be prepared by dissolving an accurately weighed sample of the desired solute in an accurately measured volume of solution. This method

---

[4] The term *milligram percent* is sometimes employed in clinical chemistry. It is the number of milligrams of solute per 100 mL of solution (i.e., 1 mg percent is 1 part per 100,000).

is not generally applicable, however, since relatively few chemical reagents can be obtained in sufficiently pure form to meet the analyst's demand for accuracy. The few substances which are adequate in this regard are called *primary standards*. More commonly a solution is standardized by a titration in which it reacts with a weighed portion of a primary standard.

The reaction between the titrant and the substance selected as a primary standard should fulfill the requirements for titrimetric analysis (page 45). In addition, the primary standard should have the following characteristics:

1. It should be readily available in a pure form or in a state of known purity at a reasonable cost. In general, the total amount of impurities should not exceed 0.01 to 0.02%, and it should be possible to test for impurities by qualitative tests of known sensitivity.
2. The substance should be stable. It should be easy to dry and should not be so hygroscopic that it takes up water during weighing. It should not lose weight on exposure to air. Salt hydrates are not normally employed as primary standards.
3. It is desirable that the primary standard have a reasonably high equivalent weight in order to minimize the consequences of errors in weighing.

For acid-base titrations it is customary to prepare solutions of an acid and base of approximately the desired concentration and then to standardize one of the solutions against a primary standard. The solution thus standardized can be used as a *secondary standard* to obtain the concentration of the other solution. For highly accurate work, it is preferable to standardize both the acid and base independently against primary standards. A widely used primary standard for base solutions is the compound potassium hydrogen phthalate, $KHC_8H_4O_4$, abbreviated KHP. Sulfamic acid, HSO_3NH_2, and potassium hydrogen iodate, $KH(IO_3)_2$, are both strong acids and are excellent primary standards. Sodium carbonate, Na_2CO_3, and tris(hydroxymethyl) aminomethane, $(CH_2OH)_3CNH_2$, known as TRIS or THAM, are common primary standards for strong acids.

Many primary standards are available for redox reagents. Table 3.1 gives a summary of the primary standards commonly used in the laboratory for the reagents listed. The methods used for balancing redox equations are reviewed in Appendix II.

For precipitation and complex formation titrations, pure salts are usually employed as primary standards. Sodium or potassium chloride can be used to

TABLE 3.1 Primary Standards for Redox Reagents

SOLUTION TO BE STANDARDIZED	PRIMARY STANDARD	REACTION
$KMnO_4$	As_2O_3	$5H_3AsO_3 + 2MnO_4^- + 6H^+ \longrightarrow 2Mn^{2+} + 5H_3AsO_4 + 3H_2O$
$KMnO_4$	$Na_2C_2O_4$	$5C_2O_4^{2-} + 2MnO_4^- + 16H^+ \longrightarrow 2Mn^{2+} + 10CO_2 + 8H_2O$
$KMnO_4$	Fe	$5Fe^{2+} + MnO_4^- + 8H^+ \longrightarrow 5Fe^{3+} + Mn^{2+} + 4H_2O$
$Ce(SO_4)_2$	Fe	$Fe^{2+} + Ce^{4+} \longrightarrow Fe^{3+} + Ce^{3+}$
$K_2Cr_2O_7$	Fe	$6Fe^{2+} + Cr_2O_7^{2-} + 14H^+ \longrightarrow 6Fe^{3+} + 2Cr^{3+} + 7H_2O$
$Na_2S_2O_3$	$K_2Cr_2O_7$	$Cr_2O_7^{2-} + 6I^- + 14H^+ \longrightarrow 2Cr^{3+} + 3I_2 + 7H_2O$
		$I_2 + 2S_2O_3^{2-} \longrightarrow 2I^- + S_4O_6^{2-}$
$Na_2S_2O_3$	Cu	$2Cu^{2+} + 4I^- \longrightarrow 2CuI(s) + I_2$
		$I_2 + 2S_2O_3^{2-} \longrightarrow 2I^- + S_4O_6^{2-}$
I_2	As_2O_3	$HAsO_2 + I_2 + 2H_2O \longrightarrow H_3AsO_4 + 2I^- + 2H^+$

standardize a solution of silver nitrate, the reaction being

$$Ag^+ + Cl^- \longrightarrow AgCl(s)$$

Calcium carbonate, $CaCO_3$, is used as a primary standard for solutions of the complexing agent ethylenediaminetetraacetic acid (EDTA). The reaction is

$$Ca^{2+} + Y^{4-} \longrightarrow CaY^{2-}$$

where Y^{4-} stands for the anion of EDTA.

The following examples illustrate the calculations involved in standardizing a solution. Example 3.9 involves "back-titration"; frequently the analyst "overruns" the end point—i.e., adds too much titrant—and then back-titrates with a second solution. The concentration of this second solution must be known in order to correct for the excess titrant.

EXAMPLE 3.8 *Standardization of an acid solution*	A sample of pure sodium carbonate, Na_2CO_3, weighing 0.3542 g is dissolved in water and titrated with a solution of hydrochloric acid. A volume of 30.23 mL is required to reach the methyl orange end point, the reaction being

$$Na_2CO_3 + 2HCl \longrightarrow 2NaCl + H_2O + CO_2$$

Calculate the molarity of the acid.

At the equivalence point

$$\text{mmol HCl} = 2 \times \text{mmol } Na_2CO_3$$

$$V_{HCl} \times M_{HCl} = 2 \times \frac{\text{mg } Na_2CO_3}{\text{MW } Na_2CO_3}$$

$$30.23 \times M_{HCl} = 2 \times \frac{354.2}{106.0}$$

$$M_{HCl} = 0.2211 \text{ mmol/mL} \qquad \square$$

EXAMPLE 3.9 *Standardization of an oxidant solution*	A sample of pure sodium oxalate, $Na_2C_2O_4$, weighing 0.2856 g is dissolved in water, sulfuric acid is added, and the solution is titrated at 70°C, requiring 45.12 mL of a $KMnO_4$ solution. The end point is overrun, and back-titration is carried out with 1.74 mL of 0.0516 M oxalic acid solution. Calculate the molarity of the $KMnO_4$ solution.

The reaction written ionically is

$$5C_2O_4^{2-} + 2MnO_4^- + 16H^+ \longrightarrow 2Mn^{2+} + 10CO_2 + 8H_2O$$

At the equivalence point:

$$5 \times \text{mmol permanganate} = 2 \times \text{mmol oxalate}$$

$$5 \times \text{mmol } KMnO_4 = 2(\text{mmol } Na_2C_2O_4 + \text{mmol } H_2C_2O_4)$$

$$5 \times V_{KMnO_4} \times M_{KMnO_4} = 2\left(\frac{\text{mg } Na_2C_2O_4}{\text{MW } Na_2C_2O_4} + V_{H_2C_2O_4} \times M_{H_2C_2O_4} \right)$$

$$5 \times 45.12 \times M_{KMnO_4} = 2\left(\frac{285.6}{134.0} + 1.74 \times 0.0516 \right)$$

$$M_{KMnO_4} = 0.01969 \text{ mmol/mL} \qquad \square$$

Sometimes it is convenient or necessary to add excess titrant and to back-titrate the excess with a solution of known concentration. For example, in the

Volhard method (Section 9.2b) for chloride, excess silver nitrate is added to precipitate AgCl:

$$Ag^+ + Cl^- \longrightarrow AgCl(s)$$

The excess silver is titrated with a standard solution of potassium thiocyanate:

$$Ag^+ + SCN^- \longrightarrow AgSCN(s)$$

Iron (III) serves as the indicator.

3.3b. Aliquots

Sometimes the analyst weighs a large sample of the primary standard (or unknown), dissolves it in a volumetric flask, and withdraws a portion of the solution using a pipet. The portion withdrawn with the pipet is called an *aliquot*. An aliquot is a known portion of the whole, usually some simple fraction. This process of dilution to a known volume and removing a portion for titration is called *taking an aliquot*. The following example illustrates this procedure.

EXAMPLE 3.10

Calculation involving an aliquot portion

A sample of pure $CaCO_3$ (MW 100.09) weighing 0.4148 g is dissolved in 1 : 1 hydrochloric acid, and the solution is diluted to 500.0 mL in a volumetric flask. A 50.00-mL aliquot is withdrawn with a pipet and placed in an Erlenmeyer flask. The solution is titrated with 40.34 mL of an EDTA (Section 8.2) solution using Eriochrome Black T indicator. Calculate the molarity of the EDTA solution.

The reaction in the titration is

$$Ca^{2+} + Y^{4-} \longrightarrow CaY^{2-}$$

where Y^{4-} stands for the anion of EDTA. At the equivalence point

$$mmol\ EDTA = mmol\ CaCO_3$$

$$V \times M_{EDTA} = \frac{mg\ CaCO_3}{MW\ CaCO_3}$$

The weight of $CaCO_3$ in the aliquot is one-tenth of 0.4148 g, or 0.04148 g, since 50.00 mL was taken from a volume of 500.0 mL. Hence

$$40.34 \times M_{EDTA} = \frac{41.48\ mg}{100.09\ mg/mmol}$$

$$M_{EDTA} = 0.01027\ mmol/mL \qquad \square$$

3.3c. Dilution

Laboratory procedures in analytical chemistry often call for taking an aliquot of a standard solution and diluting it to a larger volume in a volumetric flask. This technique is especially useful in spectrophotometric procedures to adjust the concentration of solute so that the error in measuring the absorbance of the solution is minimized (Chapter 14).

The calculation involved in a dilution is straightforward. Since no chemical reaction occurs, the number of moles of solute in the original solution must be the same as the number of moles in the final solution. The following example illustrates the calculation.

EXAMPLE 3. 11

Dilution calculation

A 0.0200 *M* solution of $KMnO_4$ is prepared by dissolving a weighed amount of the salt in a 1-liter volumetric flask. A 25-ml aliquot of this solution is placed in a 500-mL volumetric flask, and the flask filled to the mark with water. Calculate the molarity of the solution in the 500-mL flask.

We know that since $mmol_1 = mmol_2$,

$$V_1 \times M_1 = V_2 \times M_2$$

Then

$$25.0 \times 0.0200 = 500 \times M_2$$

$$M_2 = 0.00100 \text{ mmol/mL}$$

Sometimes the student may hear the expression that the solution has undergone a "20-fold dilution." This refers to the fact that the concentration has been reduced by a factor of 25.0/500, or 1/20.

3.3d. Calculation of Percent Purity

To analyze a sample of unknown purity the analyst weighs accurately a portion of the sample, dissolves it appropriately, and titrates it with a standard solution. If the titration reaction is

$$a\text{A} + t\text{T} \longrightarrow \text{products}$$

where *a* molecules of the analyte, A, react with *t* molecules of the titrant, T, then at the equivalence point

$$t \times \text{mmol A} = a \times \text{mmol T}$$

$$\text{mmol A} = \frac{a}{t} \times \text{mmol T}$$

If *V* and *M* represent the volume (mL) and molarity (mmol/mL), respectively, of the titrant, and MW_A is the molecular weight of the analyte, then

$$\text{mmol A} = \frac{a}{t} \times V \times M$$

$$\text{mg A} = \frac{a}{t} \times V \times M \times MW_A$$

The percent by weight of A is

$$\% \text{ A} = \frac{\text{mg analyte}}{\text{mg sample}} \times 100$$

$$\% \text{ A} = \frac{\frac{a}{t} \times V(\text{mL}) \times M(\text{mmol/mL}) \times MW_A(\text{mg/mmol})}{\text{weight of sample (mg)}} \times 100$$

Note the cancellation of units to give percentage, which is dimensionless.

EXAMPLE 3. 12

Calculation of percent purity

A 1.000-g sample containing $Na_2C_2O_4$ (MW 126.0) is titrated with 40.00 mL of 0.0200 *M* $KMnO_4$ in acid solution. The ionic reaction is

$$5C_2O_4^{2-} + 2MnO_4^- + 16H^+ \longrightarrow 10CO_2 + 2Mn^{2+} + 8H_2O$$

Calculate the percentage of $Na_2C_2O_4$ in the sample.

$$\% \ Na_2C_2O_4 = \frac{\frac{5}{2} \times 40.00 \times 0.0200 \times 126.0}{1000} \times 100$$

$$\% \ Na_2C_2O_4 = 25.20 \qquad \square$$

EXAMPLE 3.13

Calculation of percent purity

A sample of iron ore weighing 0.6428 g is dissolved in acid, the iron is reduced to Fe^{2+}, and the solution is titrated with 36.30 mL of 0.01753 M $K_2Cr_2O_7$ solution. The ionic reaction is

$$6Fe^{2+} + Cr_2O_7^{2-} + 14H^+ \longrightarrow 6Fe^{3+} + 2Cr^{3+} + 7H_2O$$

(a) Calculate the percentage of iron (AW 55.847) in the sample. (b) Express the percentage as Fe_2O_3 (MW 159.69) rather than as Fe.

(a) $$\% \ Fe = \frac{6 \times 36.30 \ mL \times 0.01753 \ mmol/mL \times 55.847 \ mg/mmol}{642.8 \ mg} \times 100$$

$$\% \ Fe = 33.17$$

(b) To express the percentage as Fe_2O_3, note that 1 mmol of Fe_2O_3 yields 2 mmol Fe^{2+} which react with $\frac{1}{3}$ mmol of $Cr_2O_7^{2-}$. Then

$$\% \ Fe_2O_3 = \frac{3 \times 36.30 \ mL \times 0.01753 \ mmol/mL \times 159.69 \ mg/mmol}{642.8 \ mg} \times 100$$

$$\% \ Fe_2O_3 = 47.43$$

EXAMPLE 3.14

Calculation of %N using the Kjeldahl method

In the Kjeldahl method for nitrogen, the element is converted into NH_3, which is then distilled into a known volume of standard acid. There is more than enough acid to neutralize the NH_3, and the excess is titrated with standard base.

The ammonia from a 1.325-g sample of fertilizer is distilled into 50.00 mL of 0.1015 M H_2SO_4, and 25.32 mL of 0.1980 M NaOH is required for back-titration. Calculate the percentage of nitrogen (N) in the sample.

Note that 1 mmol of H_2SO_4 reacts with 2 mmol of NaOH and 2 mmol of NH_3. Hence

$$mmol \ NH_3 + mmol \ NaOH = 2 \times mmol \ H_2SO_4$$

$$mmol \ NH_3 + 25.32 \times 0.1980 = 2 \times 50.00 \times 0.1015$$

$$mmol \ NH_3 = 5.137$$

Then

$$\% \ N = \frac{5.137 \ mmol \times 14.007 \ mg/mmol}{1325 \ mg} \times 100$$

$$\% \ N = 5.43 \qquad \square$$

EXAMPLE 3.15

Calculation involving an indirect titration

Copper can be determined titrimetrically by allowing Cu^{2+} ion to react with excess KI, liberating I_2:

$$2Cu^{2+} + 4I^- \longrightarrow 2CuI(s) + I_2$$

The liberated I_2 is then titrated with a standard solution of sodium thiosulfate, $Na_2S_2O_3$:

$$I_2 + 2S_2O_3^{2-} \longrightarrow 2I^- + S_4O_6^{2-}$$

This is sometimes called an *indirect* titration, since the analyte does not react directly with the titrant. The I_2 which reacts with the titrant is, of course, chemically equivalent to the analyte.

A 2.165-g sample of a copper ore is dissolved, and excess KI added to liberate I_2. The I_2 required 31.43 mL of 0.0978 M $Na_2S_2O_3$ for titration. Calculate the percentage of copper in the ore.

We note that

$$2 \text{ mmol } Cu^{2+} = 1 \text{ mmol } I_2 = 2 \text{ mmol } Na_2S_2O_3$$

or

$$1 \text{ mmol } Cu^{2+} = 1 \text{ mmol } Na_2S_2O_3$$

Then

$$\% \text{ Cu} = \frac{31.42 \text{ mL} \times 0.0978 \text{ mmol/mL} \times 63.546 \text{ mg/mmol}}{2165 \text{ mg}} \times 100$$

$$\% \text{ Cu} = 9.02 \qquad \qquad \square$$

3.3e. *Coulometric Analysis*

A substance can be determined by electrolysis, and the procedure can be either gravimetric or titrimetric. In the titrimetric technique, called *coulometry*, the amount of analyte is determined by measuring the quantity of electricity required to react with it completely. The titrant in this case can be considered to be electrons. A mole of electrons is called a *faraday*, and in electrical units the faraday is equal to 96,486 coulombs (C) of charge:

$$6.0221 \times 10^{23} \text{ } e/F \times 1.6022 \times 10^{-19} \text{ } C/e = 96,486 \text{ } C/F$$

For most purposes this number is rounded to 96,500. It should also be noted that the unit of current, the ampere (A), is defined as 1 C/s. By measuring the current and time required for a reaction to go to completion, the chemist can calculate the number of moles of electrons and hence the number of moles of analyte in the sample.

EXAMPLE 3.16

Calculation for a coulometric analysis

A sample of copper ore weighing 2.132 g is dissolved in acid, and the copper is electrolyzed:

$$Cu^{2+} + 2e \longrightarrow Cu$$

If 8.04 min is required for the electrolysis using a constant current of 2.00 A, calculate the percentage of copper in the ore.

First, calculate the number of millimoles of electrons:

$$\text{mmol} = \frac{8.04 \text{ min} \times 60.0 \text{ s/min} \times 2.00 \text{ C/s}}{96.5 \text{ C/mmol}}$$

$$\text{mmol} = 10.0$$

Since 1 mmol of Cu^{2+} reacts with 2 mmol of electrons, the sample must contain 5.00 mmol of Cu. The percentage of Cu is

$$\% \text{ Cu} = \frac{5.00 \text{ mmol} \times 63.55 \text{ mg/mmol}}{2132 \text{ mg}} \times 100$$

$$\% \text{ Cu} = 14.9 \qquad \qquad \square$$

3.4

EQUIVALENT WEIGHTS AND THE NORMALITY SYSTEM OF CONCENTRATION[5]

Many years ago chemists introduced the term *equivalent* in an attempt to simplify stoichiometric calculations in titrimetry. We have seen that a titration involves adding the titrant until an amount *chemically equivalent* to the analyte is reached and that this point is called the *equivalence point* (EPt). At the EPt the number of mols of analyte and titrant may or may not be equal. The equivalent is defined so that *at the EPt the equivalents of analyte and titrant are always equal.*

Consider the following acid-base reactions, all written molecularly:

$$HCl + NaOH \longrightarrow NaCl + H_2O \qquad (1)$$

$$H_2SO_4 + 2NaOH \longrightarrow Na_2SO_4 + 2H_2O \qquad (2)$$

$$2HCl + Ca(OH)_2 \longrightarrow CaSO_4 + 2H_2O \qquad (3)$$

$$H_2SO_4 + Ca(OH)_2 \longrightarrow CaSO_4 + 2H_2O \qquad (4)$$

It can be seen that 1 mol of H_2SO_4 reacts with twice as many mols of NaOH as does 1 mol of HCl, and 1 mol of $Ca(OH)_2$ reacts with twice as many mols of HCl as does 1 mol of NaOH. Then $\frac{1}{2}$ mol of H_2SO_4 and $\frac{1}{2}$ mol of $Ca(OH)_2$ are each *chemically equivalent* to 1 mol of HCl and 1 mol of NaOH, respectively. They are also equivalent to each other, as shown in reaction (4).

The ionic reaction which occurs in strong acid–strong base titrations is (writing H^+ for H_3O^+)

$$H^+ + OH^- \longrightarrow H_2O$$

One mole of HCl furnishes 1 mol of H^+, whereas 1 mol of H_2SO_4 furnishes 2 mol of H^+. Likewise, 1 mol of NaOH furnishes 1 mol of OH^-, whereas 1 mol of $Ca(OH)_2$ furnishes 2 mol of OH^-.

The *gram-equivalent weight* (usually shortened to *equivalent weight*, EW) of an acid or base is then defined as the weight in grams required to furnish or react with 1 mol of H^+ (1.008 g). The EW of a substance is called an *equivalent* (eq), just as the MW is called a mole. A *milliequivalent* (meq) is one-thousandth of an equivalent, or

$$1000 \text{ meq} = 1 \text{ eq}$$

If n is the number of moles of H^+ furnished by 1 mol of an acid, or reacted with by 1 mol of a base, the relation between the molecular and equivalent weights is

$$EW = \frac{MW}{n}$$

For HCl and NaOH, $n = 1$ and MW and EW are the same. For H_2SO_4 and $Ca(OH)_2$, $n = 2$ and EW is one-half MW.

It is obvious from the definition of equivalent weight that one equivalent of any acid reacts with one equivalent of any base. And at the EPt of the titration reaction

$$aA + tT \longrightarrow \text{ products}$$

[5] This section can be omitted if the instructor prefers not to introduce equivalents and normality.

the mathematical relation

$$\text{Equivalents of analyte} = \text{equivalents of titrant}$$

is always true. In terms of moles the mathematical relation at the EPt is always

$$t \times \text{mol analyte} = a \times \text{mol titrant}$$

For reaction (2) above this would be

$$2 \times \text{mol } H_2SO_4 = \text{mol NaOH}$$

For oxidation-reduction reactions the gram-equivalent weight is defined as the weight in grams required to furnish or react with 1 mol of electrons. For precipitation and complex-formation reactions the gram-equivalent weight is defined as the weight in grams of the substance required to furnish or react with 1 mol of a univalent cation, $\frac{1}{2}$ mol of a divalent cation, $\frac{1}{3}$ mol of a trivalent cation[6], etc.

It should be noted that compounds which undergo more than one reaction will have more than one equivalent weight. For example, the reaction of phosphoric acid with a strong base can be stopped when the following reaction is complete:

$$H_3PO_4 + OH^- \longrightarrow H_2PO_4^- + H_2O$$

In this reaction the EW of H_3PO_4 is the same as the MW. But the titration can be carried further until this reaction is complete:

$$H_3PO_4 + 2OH^- \longrightarrow HPO_4^{2-} + 2H_2O$$

In this reaction the EW of H_3PO_4 is MW/2. Of course if moles are used rather than equivalents, the values of a and t in the titration reactions are different. In the first, $a = 1$ and $t = 1$, and in the second, $a = 1$ and $t = 2$.

Following are some examples illustrating the calculation of equivalent weights.

EXAMPLE 3.17

Equivalent weight of an acid

Calculate the EW of weight SO_3 used as an acid in aqueous solution.

SO_3 is the anhydride of sulfuric acid, H_2SO_4. When the latter acid is titrated with a strong base, it furnishes two protons:

$$SO_3 + H_2O \longrightarrow H_2SO_4 \longrightarrow 2H^+ + SO_4^{2-}$$

Hence 1 mol of SO_3 is responsible for furnishing 2 mol of H^+, and

$$EW = \frac{MW}{2} = \frac{80.06}{2}$$

$$EW = 40.03 \text{ g/eq}$$

The EW of H_2SO_4 is also one-half the MW, or $98.07/2 = 49.04$ g/eq. □

---

[6] The cation referred to is the one directly involved in the titration reaction. For example, the equivalent weight of $AgNO_3$ in the reaction

$$Ag^+ + 2KCN \longrightarrow Ag(CN)_2^- + 2K^+$$

is the molecular weight, 169.87 g, since this amount of salt furnishes 1 mol of the univalent cation, Ag^+. The equivalent weight of KCN is twice the molecular weight, since 2 mol of KCN react with 1 mol (1 eq) of Ag^+.

EXAMPLE 3.18

Equivalent weight of an oxidizing agent

The permanganate ion, MnO_4^-, can undergo the following reactions in solutions of varying acidity:

$$MnO_4^- + e \longrightarrow MnO_4^{2-} \tag{1}$$

$$MnO_4^- + 4H^+ + 3e \longrightarrow MnO_2 + 2H_2O \tag{2}$$

$$MnO_4^- + 8H^+ + 4e \longrightarrow Mn^{3+} + 4H_2O \tag{3}$$

$$MnO_4^- + 8H^+ + 5e \longrightarrow Mn^{2+} + 4H_2O \tag{4}$$

Calculate the EW of $KMnO_4$ (MW 158.03) in each reaction.

In reaction (1), $n = 1$ and

$$EW = MW = 158.03 \text{ g/eq}$$

In reaction (2), $n = 3$ and

$$EW = MW/3 = 52.68 \text{ g/eq}$$

In reaction (3), $n = 4$ and

$$EW = MW/4 = 39.51 \text{ g/eq}$$

In reaction (4), $n = 5$ and

$$EW = MW/5 = 31.61 \text{ g/eq} \qquad \square$$

EXAMPLE 3.19

Equivalent weight of a reducing agent

Calculate the EWs of $Na_2C_2O_4$, the reducing agent, and $K_2Cr_2O_7$, the oxidizing agent, in the following reaction:

$$3C_2O_4^{2-} + Cr_2O_7^{2-} + 14H^+ \longrightarrow 2Cr^{3+} + 6CO_2 + 7H_2O$$

The number of electrons gained or lost can be determined from the change in oxidation number or from the half-reactions (Appendix II). The half-reactions are

$$C_2O_4^{2-} \longrightarrow 2CO_2 + 2e$$

$$Cr_2O_7^{2-} + 14H^+ + 6e \longrightarrow 2Cr^{3+} + 7H_2O$$

The oxalate ion furnishes two electrons, and the dichromate ion gains six electrons. Hence the EWs are

$$Na_2C_2O_4: \frac{MW}{2} = \frac{134.0}{2} = 67.00 \text{ g/eq}$$

$$K_2Cr_2O_7: \frac{MW}{6} = \frac{294.2}{6} = 49.03 \text{ g/eq} \qquad \square$$

EXAMPLE 3.20

Equivalent weights in precipitation reactions

Calculate the EWs of $AgNO_3$ and $BaCl_2$ in the reaction

$$2Ag^+ + BaCl_2 \longrightarrow 2AgCl(s) + Ba^{2+}$$

One mole of $AgNO_3$ furnishes 1 mol of the univalent cation, Ag^+; 1 mol of $BaCl_2$ reacts with 2 mol of Ag^+. Hence

$$EW \ AgNO_3 = \frac{MW}{1} = \frac{169.9}{1} = 169.9 \text{ g/eq}$$

$$EW \ BaCl_2 = \frac{MW}{2} = \frac{208.2}{2} = 104.1 \text{ g/eq} \qquad \square$$

3.4a. Normality

Like molarity and formality the normality system of concentration is based on the volume of solution. It is defined as follows:

Normality = number of equivalents per liter of solution

or
$$N = \frac{eq}{V}$$

where N is the normality, eq the number of equivalents, and V the volume of solution in liters. Since

$$eq = \frac{g}{EW}$$

where g is the grams of solute and EW the equivalent weight, it follows that

$$N = \frac{g}{EW \times V}$$

Solving this equation for grams of solute gives

$$g = N \times V \times EW$$

The relation between normality and molarity is

$$N = nM$$

where n is the number of moles of hydrogen ion, electrons, or univalent cation furnished by or combined with the reacting substance (page 57).

The following examples illustrate some calculations of normality and percent purity.

EXAMPLE 3.21

Calculation of normality from mass and volume

Arsenic can be determined by titration with a standard iodine solution. The reaction is

$$HAsO_2 + I_2 + 2H_2O \longrightarrow H_3AsO_4 + 2H^+ + 2I^-$$

Pure As_2O_3 can be used as a primary standard in determining the concentration of the iodine solution.

A sample of pure As_2O_3 weighing 4.0136 g is dissolved in 800.0 mL of solution. Calculate the normality of the solution when it is used in the reaction above. Also calculate the molarity of the solution.

Note that each arsenic atom loses two electrons in being oxidized from $HAsO_2$ to H_3AsO_4. Each As_2O_3 contains two arsenic atoms and hence loses four electrons:

$$As_2O_3 \longrightarrow 2HAsO_2 \longrightarrow 2H_3AsO_4 + 4e$$

The equivalent weight of As_2O_3 is one-fourth of the molecular weight. Hence

$$N = \frac{4.0136 \text{ g}}{197.84/4 \text{ g/eq} \times 0.800 \text{ liter}}$$

$$N = 0.1014 \text{ eq/liter}$$

$$N = 4 \times M$$

$$M = 0.02535 \text{ mol/liter} \qquad \square$$

EXAMPLE 3.22   Calculate the normality of a solution of nickel nitrate made by dissolving 2.00 g of pure nickel metal in nitric acid and diluting the solution to 500 ml. The nickel is to be titrated with KCN, the following reaction occurring:

$$Ni^{2+} + 4CN^- \longrightarrow Ni(CN)_4^{2-}$$

Also calculate the molarity.

The equivalent weight of nickel is one-half the atomic weight, since nickel is a divalent cation. Hence

$$N = \frac{2.00 \text{ g}}{58.70/2 \text{ g/eq} \times 0.500 \text{ liter}}$$

$$N = 0.136 \text{ eq/liter}$$

$$N = 2 \times M$$

$$M = 0.0680 \text{ mol/liter} \qquad \Box$$

EXAMPLE 3.23   A sample of pure sodium carbonate, Na_2CO_3, weighing 0.3542 g is dissolved in water and titrated with a solution of hydrochloric acid. A volume of 30.23 mL is required to reach the methyl orange end point, the reaction being

$$Na_2CO_3 + 2HCl \longrightarrow 2NaCl + H_2O + CO_2$$

Calculate the normality of the acid solution. (This is the same problem as Example 3.8 where it was worked with moles.)

At the equivalence point

$$\text{meq HCl} = \text{meq Na}_2\text{CO}_3$$

Since Na_2CO_3 reacts with two H^+, the EW is one-half the MW, or $106.0/2 = 53.00$ mg/meq. Hence

$$V_{\text{HCl}} \times N_{\text{HCl}} = \frac{\text{mg Na}_2\text{CO}_3}{\text{EW Na}_2\text{CO}_3}$$

$$30.23 \times N_{\text{HCl}} = \frac{354.2}{106.0/2} = 2 \times \frac{354.2}{106.0}$$

$$N_{\text{HCl}} = 0.2211 \text{ meq/mL}$$

Note that the molarity of the acid is the same as the normality since HCl furnishes one H^+. $\qquad \Box$

EXAMPLE 3.24

Calculation of percent purity using normality

A 1.000-g sample containing $Na_2C_2O_4$ (MW 126.0) is titrated with 40.00 mL of 0.0200 M $KMnO_4$ in acid solution. The ionic reaction is

$$5C_2O_4^{2-} + 2MnO_4^- + 16H^+ \longrightarrow 10CO_2 + 2Mn^{2+} + 8H_2O$$

Calculate the percentage of $Na_2C_2O_4$ in the sample. (This is the same problem as Example 3.12 where it was worked with moles.)

First note that in the above reaction MnO_4^- gains five electrons and $C_2O_4^{2-}$ loses two. Hence the normality of the $KMnO_4$ solution is $5 \times 0.0200 = 0.1000$ meq/mL, and the EW of $Na_2C_2O_4$ is $MW/2 = 63.00$ mg/meq. The percentage purity is given by the equation

$$\% = \frac{V \text{ (mL)} \times N \text{ (meq/mL)} \times EW \text{ (mg/meq)}}{\text{weight of sample (mg)}} \times 100$$

Then

$$\% \ Na_2C_2O_4 = \frac{40.00 \ mL \times 0.1000 \ meq/mL \times 63.00 \ mg/meq}{1000 \ mg} \times 100$$

$$\% \ Na_2C_2O_4 = 25.20 \qquad \qquad \square$$

EXAMPLE 3.25   A sample of iron ore weighing 0.6428 g is dissolved in acid, the iron reduced to Fe^{2+}, and the solution titrated with 36.30 mL of a 0.01753 M solution of $K_2Cr_2O_7$. The reaction is

$$6Fe^{2+} + Cr_2O_7^{2-} + 14H^+ \longrightarrow 6Fe^{3+} + 2Cr^{3+} + 7H_2O$$

(a) Calculate the percentage of iron (Fe, Aw = 55.847 mg/mmol) in the sample.
(b) Express the percentage as Fe_2O_3. (This is the same problem as Example 3.13, where it was solved with moles.)

(a) First note that $Cr_2O_7^{2-}$ gains $6e$; hence $N = 6 \times 0.01753 = 0.1052$ meq/mL. Since Fe loses $1e$, AW = EW = 55.847 mg/meq. Then

$$\% \ Fe = \frac{36.30 \ mL \times 0.1052 \ meq/mL \times 55.847 \ mg/meq}{642.8 \ mg} \times 100$$

$$\% \ Fe = 33.18$$

(b) To express the percentage as Fe_2O_3, note that since each Fe atom loses $1e$ and each Fe_2O_3 contains two Fe atoms, the EW of Fe_2O_3 is one-half the MW = 159.69/2 = 79.85 mg/meq. Then

$$\% \ Fe_2O_3 = \frac{36.30 \ mL \times 0.1052 \ meq/mL \times 79.85 \ mg/meq}{642.8 \ mg} \times 100$$

$$\% \ Fe_2O_3 = 47.43 \qquad \qquad \square$$

KEY TERMS

Aliquot.   A portion of the whole, usually a simple fraction. A portion of a sample withdrawn from a volumetric flask with a pipet is called an aliquot.

Analytical concentration.   The total number of moles per liter of a solute regardless of any reactions that might occur when the solute dissolves. Used synonymously with *formality*.

Equivalent.   The amount of a substance which furnishes or reacts with 1 mol of H^+ (acid-base), 1 mol of electrons (redox), or 1 mol of a univalent cation (precipitation and complex formation).

Equivalent weight.   The weight in grams of one equivalent of a substance.

Equivalence point.   The point in a titration where chemically equivalent amounts of analyte and titrant are present.

End point.   The point in a titration where there

is a sudden change in a physical property of the solution, such as the color of an indicator.

Formula weight.   The summation of atomic weights of all the atoms in the chemical formula of a substance.

Formality.   The number of formula weights of solute per liter of solution; synonymous with *analytical concentration*.

Indicator.   A chemical substance which exhibits different colors in the presence of excess analyte or titrant.

Mole.   The number of grams of a substance containing as many entities as there are atoms in 12 g of the carbon-12 isotope.

Molecular weight.   The weight in grams of 1 mol of a substance.

Molarity.   The number of moles of solute per liter of solution.

Normality. The number of equivalents of solute per liter of solution.

Primary standard. A substance available in a pure form or state of known purity which is used in standardizing a solution.

Standardization. The process by which the concentration of a solution is accurately ascertained.

Standard solution. A solution whose concentration has been accurately determined.

Titrant. The reagent (a standard solution) which is added from a buret to react with the analyte.

Titration. The process of measuring the volume of titrant required to reach the equivalence point.

Weight percent. The number of grams of solute per 100 g of solution.

QUESTIONS

1. *Types of reactions.* List and give examples of the four types of chemical reactions which can be used as the basis of titrimetric analyses.

2. *Requirements for reactions.* What requirements must be satisfied for a reaction to be used in a titration?

3. *Errors.* Explain the effect that the following errors would have on the normality calculated for a solution of NaOH being standardized against pure KHP. Would the error cause the calculated normality to be high or low, or would it have no effect? (a) The buret containing NaOH is read too quickly, not allowing sufficient time for drainage. (b) The initial reading of the NaOH buret is recorded as 0.90 mL when it is actually 1.10 mL. (c) The weight of KHP is recorded as 0.7682 g when it is actually 0.7862 g. (d) The sample is dissolved in 150 mL of water although the directions call for 100 mL.

4. *Errors.* Repeat question 3 for the analysis of an impure sample of KHP by titration with a standard solution of NaOH. Explain the effect of the errors on the percentage of KHP calculated in the sample.

Multiple-choice. In the following multiple-choice questions select the *one best* answer.

5. One part per million is the same as (a) 1 mg/kg; (b) 1 μg/g; (c) 1 ng/mg; (d) all of the above.

6. The normality of a 0.10 M solution of $KMnO_4$ is (a) 0.50 (b) 0.30; (c) 0.10; (d) more information is needed to answer this.

7. Suppose the substance A, molecular weight MW, reacts with permanganate as follows:

$$5A + 2MnO_4^- + \cdots \longrightarrow 2Mn^{2+} + \cdots$$

The EW of A in this reaction is MW divided by (a) 2; (b) 3; (c) 4; (d) 5.

8. One of the chief ores of uranium is pitchblende, U_3O_8. When pitchblende is treated with HNO_3, uranyl nitrate, $UO_2(NO_3)_2$, is formed. The EW of U_3O_8 in this reaction is MW divided by (a) 2; (b) 3; (c) $\frac{2}{3}$; (d) $\frac{3}{2}$.

9. A sample is analyzed and reported to contain 21% NaOH. Later it is found that the basic material is KOH rather than NaOH. The percentage NaOH in the sample is (a) 29.4; (b) 37.5; (c) 21; (d) 15.

10. How many coulombs of electricity are required to liberate 11.2 mL (STP) of O_2 gas by the reaction

$$2H_2O \longrightarrow 4H^+ + O_2 + 4e$$

(a) 96.5; (b) 96.5/2; (c) 96.5 × 2; (d) 96.5 × 4

PROBLEMS

1. *Molarity.* Calculate the molarity of each of these solutions: (a) 25.0 g of H_2SO_4 in 400 mL of solution; (b) 20.0 mg of NaOH in 50.0 mL of solution; (c) 5.30 g of Na_2CO_3 in 100 mL of solution.

2. *Molarity.* Calculate the molarity of each of

these solutions: (a) 30.0 g of $NaHCO_3$ in 2.50 liters of solution; (b) 30.0 mg of KOH in 75.0 mL of solution; (c) 2.00 g of $AgNO_3$ in 250 mL of solution.

3. *Molarity*. (a) Calculate the number of grams of NaOH needed to prepare 2.00 liters of a 0.100 *M* solution; (b) In how many milliliters of solution should 79.0 mg of $KMnO_4$ be dissolved to prepare a 0.0200 *M* solution? (c) If a 0.150 *M* solution is produced by dissolving 51.0 g of a solute in 2.00 liters of solution, what is the MW of the solute?

4. *Molarity*. (a) Calculate the number of grams of As_2O_3 needed to prepare 0.500 liter of a 0.100 *M* solution; (b) In how many milliliters of solution should 200 mg of $Na_2S_2O_3 \cdot 5H_2O$ be dissolved to prepare a 0.120 *M* solution?; (c) If a 0.0200 *M* solution is produced by dissolving 42.0 mg of a solute in 25.0 mL of solution, what is the MW of the solute?

5. *Density and molarity*. (a) Calculate the molarity of a solution of $HClO_4$ given that the density is 1.242 g/mL and the percent by weight $HClO_4$ is 34.0; (b) Calculate the number of milliliters of nitric acid, density 1.42 g/mL, 72% HNO_3 by weight, that should be taken to prepare 1.5 liters of a 0.15 *M* solution.

6. *Density and molarity*. (a) Calculate the molarity of concentrated ammonia, given that the density is 0.90 g/mL and the percent by weight NH_3 is 28. (b) Calculate the number of milliliters of sulfuric acid, density 1.30 g/mL, 32.6% SO_3 by weight, that should be taken to prepare 2.00 liters of a 0.0100 *M* solution.

7. *Formality*. A solution is prepared by dissolving 5.80 g of monochloroacetic acid, $CH_2ClCOOH$ (FW 94.5), in 600 mL of solution. At this concentration the acid is about 12% dissociated. Calculate (a) the formality of the monochloroacetic acid; (b) the molarities of the species $CH_2ClCOOH$ and CH_2ClCOO^-.

8. *Formality*. A solution is prepared by dissolving 9.88 g of trichloroacetic acid, Cl_3CCOOH (FW 163.39), in water and diluting to a volume of 500 mL. At this concentration the acid is about 70% dissociated. Calculate (a) the formality of the trichloroacetic acid; (b) the molarities of the species Cl_3CCOOH and Cl_3CCOO^-.

9. *Parts per million*. (a) 5.00 liters of a water sample is found to contain 0.0162 g of $MgCO_3$. Calculate the concentration of $MgCO_3$ in ppm. (b) A water sample is 0.00026 *M* in $CaCO_3$. Calculate the concentration in ppm.

10. *Parts per million*. (a) How many pounds of NaF would be required to fluoridate 2.0 million liters of water to a level of 1.2 ppm? (1 lb = 454 g.). (b) A certain water sample contains 5.0 ppb of $CaCO_3$. Calculate the molarity of the solution.

11. *Dilution of solutions*. (a) 20.0 mL of a 0.240 *M* solution is diluted to 600 mL with water. What is the molarity of the final solution? (b) What volume of water (mL) should be added to 250 mL of a 0.400 *M* solution to make the molarity 0.100? Assume the volumes are additive.

12. *Dilution of solutions*. A 10.0-mL aliquot of a 0.120 *M* solution is diluted to 250 mL in a volumetric flask (solution A). A 25.0-mL aliquot of solution A is diluted to 1.00 liter in another volumetric flask (solution B). Calculate the molarity of solution B. What fold dilution did the original solution undergo?

13. *Plasma volume*. In the determination of plasma volume a small amount of a nontoxic dye is injected intravenously and its concentration determined as soon as equilibrium is reached. 20 mL of an Evans blue solution (0.40 mg/mL) was injected into a patient, and after 5 min a blood sample was withdrawn. The sample was found to contain 0.25 mg% of the dye. Calculate the patient's plasma volume in liters.

14. *Mixing solutions*. How many milliliters of a 0.60 *M* solution of NaOH should be added to 100 mL of a 0.12 *M* NaOH solution to make the molarity 0.20? Assume the volumes are additive.

15. *Mixing solutions*. 50 mL of 0.10 *M* NaOH is mixed with 50 mL of 0.060 *M* H_2SO_4. Is the resulting solution acidic, basic, or neutral? Calculate the molarity of the reagent in excess if that is the case. Assume the volumes are additive.

16. *Mixing solutions*. 50 mL of 0.12 *M* $KMnO_4$ is mixed with 50 mL of 0.080 *M* $Na_2C_2O_4$ in a strongly acidic solution. This reaction occurs:

$$5C_2O_4^{2-} + 2MnO_4^- + 16H^+ \longrightarrow$$
$$10CO_2 + 2Mn^{2+} + 8H_2O$$

Assuming the final volume is 100 mL, calculate (a) the molarity of the reagent in excess, and (b) the molarity of the Mn^{2+} ion.

17. *Volume relations.* It is found that 41.26 mL of NaOH solution is required to titrate 43.48 mL of HCl. (a) Calculate the volume ratio of the two solutions expressed as (i) 1.000 mL HCl = ? mL NaOH, and (ii) 1.000 mL NaOH = ? mL HCl. (b) If the molarity of the HCl is 0.0994, what is the molarity of the NaOH?

18. *Equivalent weights.* Given these unbalanced equations:
    (i) $KOH + H_3PO_4 \longrightarrow K_2HPO_4 + H_2O$
    (ii) $Cr_2O_7^{2-} + C_2O_4^{2-} + H^+ \longrightarrow$
    $$Cr^{3+} + CO_2 + H_2O$$
    (iii) $CaCl_2 + K_2SO_4 \longrightarrow CaSO_4(s) + KCl$
    (iv) $Ag^+ + CN^- \longrightarrow Ag(CN)_2^-$
    Balance each equation and calculate the equivalent weights of these substances when used in the above reactions (a) KOH; (b) H_3PO_4; (c) $K_2Cr_2O_7$; (d) $Na_2C_2O_4$; (e) $CaCl_2$; (f) K_2SO_4; (g) $AgNO_3$; (h) KCN.

19. *Equivalent weights.* Given the unbalanced equations:
    (i) $CaO + HCl \longrightarrow CaCl_2 + H_2O$
    (ii) $MnO_4^- + CN^- + H_2O \longrightarrow$
    $$MnO_2(s) + CNO^- + OH^-$$
    (iii) $Ag^+ + SCN^- \longrightarrow AgSCN(s)$
    (iv) $Hg^{2+} + Cl^- \longrightarrow HgCl_2$
    Balance each equation and calculate the equivalent weights of these substances when used in the above reactions: (a) CaO; (b) HCl; (c) $KMnO_4$; (d) KCN; (e) $AgNO_3$; (f) KSCN; (g) $Hg(NO_3)_2$; (h) KCl used in reaction (iv).

20. *Normality.* Calculate the normality of each of these solutions, assuming that the reagents undergo the reactions listed in Table 3.1. (a) 0.0500 mol of $K_2Cr_2O_7$ in 2.50 liters of solution; (b) 10.0 mmol of $Na_2S_2O_3$ in 200 mL of solution; (c) 0.150 mol of I_2 in 3.00 liters of solution.

21. *Normality.* Calculate the normality of each of these solutions, assuming that the reagents undergo the reactions listed in Table 3.1. (a) 0.0500 mol of $KMnO_4$ in 2.50 liters of solution; (b) 40.0 mmol of $Na_2C_2O_4$ in 1.25

liters of solution; (c) 0.0250 mol of As_2O_3 in 500 mL of solution.

22. *Standardization.* A sodium hydroxide solution is standardized using potassium hydrogen phthalate (KHP, page 51) as a primary standard. A sample of KHP weighing 0.8482 g required 42.30 mL of NaOH for titration. Calculate the molarity of the NaOH.

23. *Standardization.* A solution of hydrochloric acid is standardized using pure Na_2CO_3 as the primary standard. The net ionic reaction is
    $$CO_3^{2-} + 2H^+ \longrightarrow H_2O + CO_2(g)$$
    A sample of Na_2CO_3 weighing 0.2520 g required 38.64 mL of HCl for titration. Calculate the molarity of the HCl solution.

24. *Standardization.* A solution of $KMnO_4$ is standardized using pure As_2O_3 as the primary standard. (See Table 3.1 for the reaction.) A sample of As_2O_3 weighing 0.2248 g requires 44.22 mL of $KMnO_4$ for titration. Calculate the molarity of the $KMnO_4$ solution.

25. *Standardization.* A solution of sodium thiosulfate, $Na_2S_2O_3$, is standardized using pure copper as the primary standard. (See Table 3.1 for reactions.) A sample of copper weighing 0.2624 g is dissolved in acid, excess KI is added, and the liberated I_2 titrated with 42.18 mL of $Na_2S_2O_3$ solution. Calculate the molarity of the $Na_2S_2O_3$ solution.

26. *Normality.* Calculate the normality of each of the solutions in Problems 22, 23, 24, and 25.

27. *Standardization with back-titration.* From the following data calculate the molarities of the NaOH and HCl solutions: weight of pure KHP = 0.8463 g; mL NaOH = 43.48; mL HCl used in back-titration = 1.62; 1.000 mL NaOH = 1.024 mL HCl.

28. *Titrimetric analysis.* A 0.5820-g sample containing chloride ion is titrated with 32.46 mL of 0.1082 *M* $AgNO_3$ solution. (a) Calculate the percentage of Cl^- in the sample. (b) Suppose the chloride is present as $CaCl_2$. Calculate the percentage of $CaCl_2$ in the sample.

29. *Titrimetric analysis with back-titration.* A 1.876-g sample containing oxalic acid ($H_2C_2O_4$) requires 38.84 mL of 0.1032 *M* NaOH for titration:
    $$H_2C_2O_4 + 2NaOH \longrightarrow Na_2C_2O_4 + 2H_2O$$

1.38 mL of 0.0992 M HCl is used in back-titration. (a) Calculate the percentage of $H_2C_2O_4$ in the sample. (b) Calculate the percentage expressed as the dihydrate, $H_2C_2O_4 \cdot 2H_2O$.

30. *Titrimetric analysis—Volhard Method.* A 0.8165-g sample containing chloride is analyzed by the Volhard method (page 53). The sample is dissolved in water, and 50.00 mL of 0.1214 M $AgNO_3$ is added to precipitate chloride ion. The excess $AgNO_3$ is titrated with 11.76 mL of 0.1019 M KSCN. Calculate the percentage of Cl^- in the sample.

31. *Titrimetric analysis—aliquot portions.* A sample containing arsenic and weighing 3.458 g is dissolved and diluted to exactly 250.0 mL in a volumetric flask. A 50-mL aliquot (page 53) is withdrawn with a pipet and titrated with 34.15 mL of a 0.0566 M solution of I_2. (See Table 3.1 for the reaction.) Calculate (a) the percentage As in the sample; (b) the percentage expressed as As_2O_3.

32. *Analysis of pyrolusite.* The ore pyrolusite can be analyzed for MnO_2 (and other higher metal oxides) by adding an excess of pure $Na_2C_2O_4$ and heating with H_2SO_4. This reaction occurs:

$$MnO_2(s) + C_2O_4^{2-} + 4H^+ \longrightarrow$$
$$Mn^{2+} + 2CO_2 + 2H_2O$$

The excess oxalate is titrated with standard $KMnO_4$ is acid solution (Table 3.1).

Given these data: weight of ore = 1.000 g; weight of $Na_2C_2O_4$ = 0.4020 g; volume of 0.0200 M $KMnO_4$ = 20.00 mL, calculate (a) the percentage of MnO_2 in the sample. (b) Since oxides other than MnO_2 can contribute to the oxidation, the purity is sometimes expressed as percentage oxygen. Calculate the percentage oxygen in the sample. *Hint*: Write the equation for O_2 oxidizing $C_2O_4^{2-}$ in acid solution.

33. *Kjeldahl method.* In the Kjeldahl method for nitrogen, the element is converted into NH_3, which is then distilled into a known volume of standard acid. There is more than enough acid to neutralize the NH_3, and the excess is titrated with standard base.

The protein content of spinal fluid can be determined by this method. The percentage protein is obtained by multiplying the percentage N by 6.25. Calculate the percent-

age of protein in a sample of spinal fluid from these data: weight of fluid = 2.00 g; volume of 0.0900 M H_2SO_4 = 5.00 mL; volume of 0.0900 M NaOH = 6.82 mL.

34. *Mercurimetric determination of chloride.* Chloride ion can be determined by titration with mercury(II) ion to form slightly dissociated $HgCl_2$:

$$Hg^{2+} + 2Cl^- \longrightarrow HgCl_2$$

The method is useful for determining Cl^- in serum and in urine samples. Calculate the Cl^- content of a serum sample (mmol/liter) from these data: 2.00 mL of serum after proper treatment required 11.18 mL of 0.0104 M $Hg(NO_3)_2$ for titration to the diphenylcarbazone end point.

35. *Determination of calcium in blood.* Calcium can be determined in blood by precipitating CaC_2O_4, dissolving the precipitate in H_2SO_4, and titrating the oxalate ion with a standard solution of $KMnO_4$. (See Table 3.1 for the reaction.) A 10.0-mL blood sample from a patient is diluted to 50.0 mL in a volumetric flask. A 20.0-mL aliquot from the flask is treated with excess oxalate to precipitate CaC_2O_4. The precipitate is dissolved in acid and titrated with 1.52 mL of 0.00108 M $KMnO_4$. Calculate the number of milligrams of Ca^{2+} per 10.0 mL of blood.

36. *Iron ore mixture.* In the titrimetric determination of iron the sample is dissolved in acid and all the iron reduced to Fe^{2+}. The Fe^{2+} is then oxidized to Fe^{3+} by titrating with a standard solution of an oxidizing agent. A certain iron ore contains 19.7% FeO and 14.3% Fe_2O_3. What volume (mL) of 0.0204 M $KMnO_4$ will be required to titrate the iron in a 0.642-g sample of the ore? The titration is done in acid solution (Table 3.1).

37. *Analysis of an oxalate mixture.* A 0.3000-g sample is a mixture of only $Na_2C_2O_4$ and $H_2C_2O_4$. It required 48.30 mL of a 0.0230 M solution of $KMnO_4$ for titration in acid solution. (See Table 3.1 for reaction.) How many milliliters of a 0.0836 M solution of NaOH will be required to titrate another 0.3000-g sample of the same mixture in an acid-base titration? $H_2C_2O_4 = 2H^+$.

38. *Analysis of an oxalate mixture.* A 1.000-g sample containing KHC_2O_4, $H_2C_2O_4$, and impurities required 46.48 mL of 0.0980 M

NaOH for titration. ($H_2C_2O_4 = 2H^+$.) A duplicate sample required 48.33 mL of 0.0220 M KMnO$_4$ for titration in acid solution. (See Table 3.1 for the reaction.) Calculate the percentages of KHC$_2$O$_4$ and H$_2$C$_2$O$_4$ in the sample.

39. *Titrimetric determination of silver*. The silver in a 1.000-g sample is determined by first precipitating the silver as Ag$_2$CrO$_4$. The precipitate is filtered, dissolved in acid, and treated with excess KI. The chromate is reduced to Cr^{3+}, and the iodide is oxidized to I$_2$. The I$_2$ requires 26.73 mL of 0.1044 M Na$_2$S$_2$O$_3$ for titration. (See Table 3.1 for the reaction.) Calculate the percentage of Ag in the sample.

40. *Factor weight solutions*. It is possible to adjust the concentration of a standard solution and the weight of sample taken for analysis so that the number of milliliters used in a titration equals the percentage of analyte (or a fraction thereof). (a) What weight of sample should be taken for analysis so that the volume of 0.1074 M NaOH used for titration equals the percentage of potassium acid phthalate (KHP, page 51) in the sample? The MW of KHP is 204.22 mg/mmol, and KHP furnishes one H$^+$. *Hint*: Recall that

$$\% \text{ Analyte} = \frac{V \times N \times \text{EW}}{\text{mg sample}} \times 100$$

and here $V = \%$ analyte. (What modification in the above expression is needed if molarity and molecular weight are used?) (b) What should be the molarity of an HCl solution so that the milliliters of titrant equal the % Na$_2$CO$_3$ in a 0.5000-g sample? The reaction is

$$CO_3^{2-} + 2H^+ \longrightarrow H_2O + CO_2$$

41. *Titer*. A system of concentration used by some chemists is called *titer*. The titer of a solution is the weight of a substance that is chemically equivalent to (1 mL) of the solution. For example, if 1.00 mL of a hydrochloric acid solution exactly reacts with 4.00 mg of sodium hydroxide, the HCl solution is said to have an NaOH titer of 4.00 mg/mL. Derive the relation between titer (T) and molarity (M) for the reaction

$$a\text{A} + t\text{T} \longrightarrow \text{products}$$

What is the relation between titer and normality?

42. *Titer*. Calculate (a) the Na$_2$CO$_3$ titer (Problem 41) in mg/mL of a 0.1200 M solution of HCl. The reaction is

$$CO_3^{2-} + 2H^+ \longrightarrow H_2O + CO_2$$

(b) the Fe$_2$O$_3$ titer of a 0.0200 M solution of K$_2$Cr$_2$O$_7$; (c) the N of a solution of KMnO$_4$ which has a Na$_2$C$_2$O$_4$ titer of 8.04 mg/mL. See Table 3.1 for the reactions in (b) and (c).

43. *Sample size*. How many grams of sample should be taken for analysis so that about 40 mL of a 0.10 M NaOH solution will be used if (a) the sample is pure KHP?; (b) the sample contains about 25% KHP?; (c) the sample contains about 25% Na$_2$CO$_3$? (Na$_2$CO$_3$ = 2H$^+$). 0.10 M HCL is the titrant.

44. *Titrant concentration*. A sample containing about 20% KHP is to be titrated with sodium hydroxide. If it is desired to use about 40 mL of titrant, how many grams of sample should be taken for analysis if the titrant concentration is (a) 1.0 M; (b) 0.10 M; (c) 0.010 M?

45. *Degree of hydration*. A 0.4500-g sample of supposedly pure BaCl$_2 \cdot$2H$_2$O is dissolved in water and requires 38.88 mL of 0.1000 M AgNO$_3$ for titration of Cl$^-$ instead of the anticipated volume. If the discrepancy is caused by partial dehydration of the salt to BaCl$_2$, what percentage by weight of the sample is BaCl$_2$?

46. *Titrimetric analysis*. A 1.000-g sample of a substance of unknown equivalent weight requires 40.00 mL of a standard solution for titration. It is found that the percentage purity is exactly 250.0 times the normality of the titrant. What is the EW of the substance?

47. *Coulometric analysis*. A 0.7820-g sample of an alloy containing 94.5% Cu is dissolved, and the copper deposited by electrolysis. If a constant current of 3.50 A is used, how many minutes will be required to complete the electrolysis?

48. *Coulometric analysis*. A 2.38-g sample of an alloy containing silver is dissolved, and the silver deposited by electrolysis. It requires 5.20 min using a constant current of 2.25 A to complete the electrolysis. Calculate the percentage of Ag in the alloy.

4

Gravimetric Methods of Analysis

It has been previously mentioned that *gravimetric* analysis is one of the divisions of analytical chemistry. The measurement step in a gravimetric method is a weighing. The analyte is physically separated from all other components of the sample as well as from the solvent. Precipitation is a widely used technique for separating the analyte from interferences; electrolysis, solvent extraction, chromatography, and volatilization are other important methods of separation.

In this chapter we shall discuss the general principles involved in gravimetric analysis, including stoichiometric calculations. We shall also examine the topic of formation and properties of precipitates as this relates to the use of precipitation in gravimetric analysis. Other methods of separation will be discussed in later chapters.

4.1

GENERAL PRINCIPLES OF GRAVIMETRIC ANALYSIS

A gravimetric method of analysis is usually based on a chemical reaction such as

$$a\text{A} + r\text{R} \longrightarrow \text{A}_a\text{R}_r$$

where a molecules of the analyte, A, react with r molecules of the reagent, R. The product, A_aR_r, is usually a slightly soluble substance which can be weighed as such after drying, or which can be ignited to another compound of known composition and then weighed. For example, calcium can be determined gravimetrically by precipitation of calcium oxalate and ignition of the oxalate to calcium oxide:

$$Ca^{2+} + C_2O_4^{2-} \longrightarrow CaC_2O_4(s)$$

$$CaC_2O_4(s) \longrightarrow CaO(s) + CO_2(g) + CO(g)$$

An excess of reagent R is normally added to repress the solubility of the precipitate.

The following requirements should be met in order that a gravimetric method be successful:

1. The separation process should be sufficiently complete so that the quantity of analyte left unprecipitated is 0.1 mg or less in determining a major constituent of a macro sample.
2. The substance weighed should have a definite composition and should be pure, or very nearly so. Otherwise, erroneous results may be obtained.

The second requirement is the more difficult for the analyst to meet. Errors due to such factors as solubility of the precipitate can generally be minimized and seldom cause significant errors. It is the problem of obtaining pure and filterable precipitates which is of major importance. Much research has been done on the formation and properties of precipitates, and considerable knowledge has been gained which enables the analyst to minimize the problem of contamination of precipitates.

4.2

STOICHIOMETRY OF GRAVIMETRIC REACTIONS

In the usual gravimetric procedure a precipitate is weighed, and from this value the weight of analyte in the sample is calculated. The percentage of analyte, A, is then

$$\% \ A = \frac{\text{weight of A}}{\text{weight of sample}} \times 100$$

4.2a. Utilization of the Gravimetric Factor in Gravimetric Calculations

A common example of gravimetric analysis is the determination of iron in an ore sample. Assume, for example, that a 0.4852-g sample of iron ore is dissolved in acid, and iron is oxidized to the +3 state and then precipitated as the hydrous oxide, $Fe_2O_3 \cdot xH_2O$. The precipitate is filtered, washed, and ignited to Fe_2O_3, which is found to weigh 0.2481 g. Calculate the percentage of iron (Fe) in the sample.

Let g be the grams of Fe in the precipitate. The reactions are

$$2Fe^{3+} \longrightarrow Fe_2O_3 \cdot xH_2O \longrightarrow Fe_2O_3(s)$$

Since 2 mol of Fe^{3+} produce 1 mol of Fe_2O_3,

$$\text{mol Fe} = 2 \times \text{mol Fe}_2O_3$$

$$\frac{g}{55.85 \ g/\text{mol}} = 2 \times \frac{0.2481 \ g}{159.69 \ g/\text{mol}}$$

$$g = 0.2481 \text{ g} \times \frac{2 \times 55.85 \ \cancel{\text{g/mol}}}{159.69 \ \cancel{\text{g/mol}}} = 0.1735 \text{ g}$$

$$\% \text{ Fe} = \frac{0.1735 \cancel{\text{g}}}{0.4582 \cancel{\text{g}}} \times 100$$

$$\% \text{ Fe} = 35.76$$

Note that in calculating the weight of Fe in the precipitate, the weight of precipitate, 0.2481 g, is multiplied by the factor ($2 \times 55.85/159.69$) to give 0.1735 g. This factor is called a *gravimetric factor*. It is simply the number of grams of Fe in 1 g of Fe_2O_3. Hence, multiplying it by the weight of precipitate gives the number of grams of Fe in the precipitate. The calculation can be set up in one step as

$$\% \text{ A} = \frac{\text{weight of precipitate} \times \text{gravimetric factor}}{\text{weight of sample}} \times 100$$

Of course it is not necessary to use the concept of the gravimetric factor in calculating the percentage of analyte in a sample. If the concept is used, two points should be noted. First, the molecular (or atomic) weight of the analyte is in the numerator; that of the substance weighed is in the denominator. Second, the number of molecules or atoms in the numerator and denominator must be chemically equivalent. Thus the gravimetric factor for Fe in Fe_2O_3 is commonly written $2Fe/Fe_2O_3$, where Fe stands for the atomic weight of iron, and Fe_2O_3 stands for the molecular weight of iron(III) oxide. Other examples are: Fe in Fe_3O_4 is $3Fe/Fe_3O_4$, and MgO in $Mg_2P_2O_7$ is $2MgO/Mg_2P_2O_7$. Handbooks of chemistry and physics usually contain rather lengthy lists of the numerical values of these factors.

The following examples illustrate some applications of stoichiometric calculations in gravimetric analysis.

EXAMPLE 1

Calculation of the amount of precipitating reagent required

Calculate the number of milliliters of ammonia, density 0.99 g/mL, 2.3% by weight NH_3, which will be required to precipitate as $Fe(OH)_3$ the iron in a 0.70-g sample that contains 25% Fe_2O_3.

The precipitation reaction is

$$Fe^{3+} + 3NH_3 + 3H_2O \longrightarrow Fe(OH)_3(s) + 3NH_4^+$$

and

$$3 \times \text{mol Fe}^{3+} = \text{mol NH}_3$$

$$\text{mol Fe}_2O_3 \text{ in sample} = \frac{0.70 \times 0.25}{159.69} = 0.0011$$

$$\text{mol Fe}^{3+} = 2 \times \text{mol Fe}_2O_3 = 2 \times 0.0011 = 0.0022$$

$$M_{NH_3} = \frac{0.99 \text{ g/mL} \times 1000 \text{ mL/liter} \times 0.023}{17.03 \text{ g/mol}}$$

$$M_{NH_3} = 1.34 \text{ mol/liter}$$

$$\text{mol NH}_3 = V \times M \quad \text{where} \quad V = \text{liters of NH}_3$$

Hence

$$3 \times 0.0022 = V \times 1.34$$

$$V = 0.0049 \text{ liters, or } 4.9 \text{ mL} \qquad \square$$

EXAMPLE 2

Calculation of the optimum sample size

What size sample containing 12.0% chlorine (Cl) should be taken for analysis if the chemist wishes to obtain a precipitate of AgCl which weighs 0.500 g?

The precipitation reaction is

$$Ag^+ + Cl^- \longrightarrow AgCl(s)$$

and

$$mol\ Cl^- = mol\ AgCl$$

If w = grams of sample, then

$$\frac{w \times 0.120}{35.45} = \frac{0.500}{143.32}$$

$$w = 1.03\ g \qquad \square$$

EXAMPLE 3

Calculation of a determinate error

In the gravimetric determination of sulfur the ignited precipitate of $BaSO_4$ is sometimes partially reduced to BaS (page 87). This causes an error, of course, if the analyst does not realize this and convert the BaS back to $BaSO_4$. Suppose a sample which contains 32.3% SO_3 is analyzed and 20.0% of the final precipitate that is weighed is BaS. (80.0% is $BaSO_4$.) What percentage of SO_3 would the analyst calculate if he assumed the entire precipitate was $BaSO_4$?

Letting f = fraction of SO_3 he would calculate ($100f$ = percent) and w_p = weight of the mixture of $BaSO_4$ and BaS obtained from a 1.000-g sample. Then

$$\frac{w_p \times \dfrac{SO_3}{BaSO_4}}{1.000} \times 100 = 100f$$

and

$$w_p = f \times \frac{BaSO_4}{SO_3} \qquad (1)$$

Since 80.0% of the precipitate is $BaSO_4$ and 20.0% is BaS, the correct percentage of SO_3 is given by

$$\frac{0.800w_p \times \dfrac{SO_3}{BaSO_4} + 0.200w_p \times \dfrac{SO_3}{BaS}}{1.000} \times 100 = 32.3 \qquad (2)$$

Substituting Eq. (1) in Eq. (2) gives

$$\frac{0.800f + 0.200f \times \dfrac{BaSO_4}{BaS}}{1.000} \times 100 = 32.3$$

Inserting molecular weights and solving for f gives $f = 0.300$; that is, the analyst would calculate $100 \times 0.300 = 30.0\%$ SO_3. $\qquad \square$

4.3

FORMATION AND PROPERTIES OF PRECIPITATES[1]

We have previously mentioned that a problem of major importance in gravimetric analysis is the formation of pure and filterable precipitates. Insight into this prob-

[1] An excellent summary of this topic at a more advanced level is given by H. A. Laitinen and W. E. Harris, *Chemical Analysis,* 2nd ed., McGraw-Hill Book Company, New York, 1975, p. 142.

lem can be gained by studying the rate at which particles are built up into solid aggregates sufficiently large to settle from the solution as a precipitate. It is this aspect of precipitation that we shall now discuss.

4.3a. Colloids

Let us consider the precipitation of silver chloride starting with Ag^+ and Cl^- ions in solution. The ions are on the order of a few ångstrom units (10^{-8} cm) in diameter. When the solubility product is surpassed, Ag^+ and Cl^- ions begin clinging together, forming particles called *nuclei* which then grow sufficiently large to be pulled to the bottom of the container by the force of gravity. As a general rule, it is said that a spherical particle must have a diameter greater than about 10^{-4} cm before it will settle from the solution as a precipitate. During the growth process the size of the particle passes through what is called the *colloidal* range. Particles with diameters of about 10^{-7} to 10^{-4} cm (1 to 1000 nm) are said to be *colloids*.

Surface Charge of Colloids

Colloidal particles are electrically charged because of the *adsorption* of ions to their surfaces. When the AgCl particles are of colloidal size, there are a large number of Ag^+ and Cl^- ions on the surface. Small particles have a large surface-to-mass ratio, and the surface ions attract ions of opposite charge from the solution. Here the solution contains Na^+, Cl^-, and NO_3^- ions (as well as H^+ and OH^-). The surface Ag^+ ions attract Cl^- and NO_3^- ions, and the surface Cl^- ions attract Na^+ ions. In general (Paneth-Fajans-Hahn rule), that ion in solution which is most strongly adsorbed is the one common to the lattice, in this case the chloride ion.[2] Thus the surface of the particle acquires a layer of chloride ions and the particles become negatively charged. The process is represented schematically in Fig. 4.1. The Cl^- ions are said to form a *primary* layer; they in turn attract Na^+ ions, forming a *secondary* layer. The secondary layer is held more loosely than the primary layer.

The primary and secondary layers are considered to constitute an *electrical double layer* which imparts a degree of stability to the colloidal dispersion. These layers cause colloidal particles to repel one another, and the particles therefore

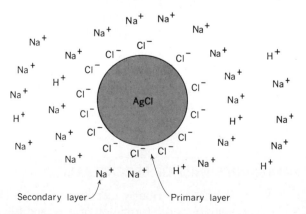

Figure 4.1 Schematic picture of colloidal particle.

[2] If no common ion is present, this rule says that the ion in solution that forms the least soluble compound with one of the lattice ions is the most strongly adsorbed.

resist combination to form larger particles which will settle from the solution. The particles can be made to *coagulate* (or *flocculate*), that is, to cohere and form larger clumps of material that will settle from the solution, by removal of the charge contributed by the primary layer. In the example of silver chloride, coagulation can be achieved by further addition of silver nitrate until equivalent amounts of silver and chloride ions are present. Since silver ions are more strongly attracted to the primary layer of chloride ions than are sodium ions, they replace sodium ions in the secondary layer and then "neutralize" the negative charge contributed by the primary layer. Stripped of their charge, the particles immediately cohere and form clumps of material which are sufficiently large to settle from the solution.

Lyophilic and Lyophobic Precipitates

Some colloids when coagulated, carry down large quantities of water, giving a jelly-like precipitate. Such materials are termed *gels* or *hydrogels* if water is the solvent. The solid material is also referred to as an *emulsoid,* or said to be *lyophilic,* meaning that it has a strong affinity for the solvent. (If water is the solvent, the term *hydrophilic* is used.) Iron(III) and aluminum hydroxides and silicic acid are familiar examples of emulsoids. A colloid that has only a small affinity for water is called a *suspensoid,* or said to be *lyophobic,* and when coagulation takes place, very little solvent is retained. Silver chloride is this type of material, and the small amount of water retained upon coagulation of silver chloride is easily removed by drying above 100°C. The water retained by an emulsoid, such as iron(III) hydroxide, is much more strongly held, and high temperatures are required for complete dehydration.

Peptization

Coagulation of colloidal dispersions can be brought about by ions other than those of the precipitate itself. When coagulation of a colloid occurs, the coagulating ions may be dragged down with the precipitate. If these ions are dissolved when a precipitate is washed, the solid-particles will go back into a colloidal dispersion and pass through the filter. Such a process of dispersing an insoluble material into a liquid as a colloid is termed *peptization* and must be avoided in quantitative procedures.

Minimizing Peptization

When peptization may occur, an electrolyte is dissolved in the wash water to replace the ions which are washed away. For this reason dilute nitric acid is added to the water used to wash a silver chloride precipitate. When the precipitate is dried, any nitric acid retained by the silver chloride is volatilized and does not interfere in the analysis.

4.3b.   *The Precipitation Process and Particle Size*

Nucleation and Particle Growth

When precipitation occurs, the size of the particles of the precipitate is thought to be determined by the relative rates of two processes: (1) the formation of nuclei, called *nucleation,* and (2) the growth of these nuclei to form particles sufficiently large to precipitate. If the rate of nucleation is small compared to the rate of growth of nuclei, fewer particles are finally produced and these particles are relatively large in size. Such a precipitate is more easily filtered and is frequently purer than a precipitate whose particles are relatively small. Hence the analyst tries to adjust conditions during precipitation so that the rate of nucleation is small compared to the rate at which the particles grow in size.

von Weimarn's Theory of Relative Supersaturation

von Weimarn[3] was the first to make a systematic study of the relationship between the size of the particles of a precipitate and the rate of precipitation. He proposed that the initial rate of precipitation is proportional to the *relative supersaturation*, where

$$\text{Relative supersaturation} = \frac{Q - S}{S}$$

Here Q is the total concentration of the substance momentarily produced in the solution by mixing the reagents, and S is the equilibrium solubility. The term $Q - S$ represents the degree of supersaturation at the moment precipitation begins. The larger this term, the greater the number of nuclei and the smaller the particles of precipitate. The term S in the denominator represents the force resisting precipitation or causing the precipitate to redissolve. The greater the value of S, the smaller the ratio and hence the smaller the number of nuclei formed. Since the analyst is interested in obtaining large particles, conditions should be adjusted to make the ratio $(Q - S)/S$ as small as possible.

Selecting or Predicting Optimal Experimental Conditions from von Weimarn's Theory

Actually the von Weimarn expression has only qualitative significance, but it serves as an excellent guide to the selection of conditions of precipitation.[4] One does find experimentally that relatively large particles of precipitate are obtained if the degree of supersaturation is kept low. The value of $(Q - S)/S$ can be decreased by decreasing Q or increasing S. In practice one routinely brings about a moderate decrease in Q by (1) using reasonably dilute solutions, and (2) adding the precipitating agent slowly. Frequently it is possible to increase the value of S markedly and thus effect a large decrease in the ratio. This can be done by taking advantage of the factors that may increase solubility: temperature, pH, or the use of complexing agents. Precipitations are quite commonly carried out at elevated temperatures for this reason. Salts of weak acids, such as CaC_2O_4 and ZnS, are better precipitated in weakly acidic rather than alkaline solution. Barium sulfate is better precipitated in 0.01 to 0.05 M hydrochloric acid solution, since the solubility is increased by the formation of bisulfate ion. A compound such as $Fe(OH)_3$ is so insoluble that even in acid solution the value of $(Q - S)/S$ is still so large that a gelatinous precipitate results. However, a dense precipitate of iron as the basic formate can be obtained by homogeneous precipitation (Section 4.3.d).

[3] P. P. von Weimarn, *Chem. Rev.*, **2**, 217 (1925); *Kolloid-Beihefte*, **18**, 44 (1923).

[4] Extensive studies have been made of the precipitation process since the work of von Weimarn, and some theories are better in explaining certain details of the precipitation process. See A. E. Nielsen, *Acta Chem. Scand.*, **14**, 1654 (1960); R. A. Johnson and J. D. O'Rourke, *J. Amer. Chem. Soc.*, **76**, 2124 (1954); R. Becker and W. Doring, *Ann. Phys. (Leipzig)*, **24**, 719 (1935).

Digestion of Precipitates

In addition to controlling conditions during the actual precipitation process, the analyst has one other recourse after the precipitate is formed. This is to *digest,* or *age,* the precipitate, that is, to allow it to stand in contact with the mother liquor, frequently at elevated temperature, for some time before filtration. The small particles of a crystalline substance, such as barium sulfate, being more soluble than the larger ones, dissolve more readily, making the solution supersaturated with respect to the larger particles. In order to establish equilibrium with respect to the larger particles, additional material must leave the solution and enter the solid phase. The ions then deposit on the larger particles, causing these particles to grow even larger. Thus the larger particles grow at the expense of the smaller ones. This process, sometimes called *Ostwald ripening,* is useful for increasing the particle size of crystalline precipitates such as barium sulfate and calcium oxalate. Curdy precipitates, such as silver chloride, and gelatinous precipitates, such as iron(III) hydroxide, are not digested. Either the particles of the latter compound are so insoluble, or the small particles do not differ sufficiently in solubility from the larger ones, that no appreciable growth in size occurs. Even with crystalline precipitates it is necessary to employ conditions which increase solubility if a beneficial effect is to be attained in a reasonable time. This is the reason for the frequent use of elevated temperatures during digestion.

Thus to obtain a precipitate of large particle size, precipitation is carried out by slow mixing of dilute solutions under conditions of increased solubility of the precipitate. Crystalline precipitates are normally digested at elevated temperature before filtration to further increase particle size.

4.3c. *Coprecipitation and Purity of Precipitates*

One of the most difficult problems that faces the analyst in employing precipitation as a means of separation and gravimetric determination is obtaining the precipitate in a high degree of purity. Substances that are normally soluble can be carried down during precipitation of the desired substance by a process called *coprecipitation.* For example, when sulfuric acid is added to a solution of barium chloride containing a small amount of nitrate ions, the precipitate of barium sulfate is found to contain barium nitrate. It is said that the nitrate is *coprecipitated* with the sulfate.

Occlusion

A crystalline precipitate, such as $BaSO_4$, sometimes adsorbs impurities when the particles are small. As the particles grow in size, the impurity may become enclosed in the crystal. This type of contamination is called *occlusion* to distinguish it from the case where the solid does not grow around the impurity. Occluded impurities cannot be removed by washing the precipitate, but the quality of the precipitate can often be improved by digestion.

Surface Adsorption

Curdy precipitates, such as AgCl, also adsorb impurities on the primary particles of the substance. However, such particles do not grow beyond colloidal dimensions, and they finally precipitate as a coagulated colloid. The resulting curd is still made up of fine particles that have not grown together to form an extensive lattice structure. Thus curdy precipitates do not enclose, or occlude, foreign ions as do crystalline precipitates. The impurities on the surfaces of the tiny particles can normally be washed off, since the particles are not firmly bound to one another and the wash liquid can penetrate to all parts of the curd. Peptization of the precipitate must be avoided, and hence the wash liquid must contain a volatile electrolyte.

Box 4.1   Coprecipitation and the Production of Plutonium

Coprecipitation is a phenomenon the analytical chemist ordinarily tries to avoid. However, the fact that precipitates tend to adsorb foreign substances is not always detrimental; coprecipitation has been widely used to isolate traces of radioactive isotopes. When these isotopes are produced in nuclear reactions, the quantity formed may be extremely small, and ordinary precipitation procedures generally fail at minute concentrations.

One technique used by chemists to precipitate and sometimes isolate elements at trace or ultratrace levels is to add a macro quantity of an element which, when precipitated, will "carry" the radioactive isotope by coprecipitation. For example, when CuS is precipitated, it will carry traces of ions such as Hg^{2+}, Bi^{3+}, and Pb^{2+} which also form acid-insoluble sulfides. A radioactive isotope can even be "purified" by repeated precipitation with fresh portions of carrier.

In World War II, American scientists produced the transuranium element plutonium (Pu) from uranium (U) in nuclear reactors (called "piles" at that time). Chemists were faced with the problem of separating the plutonium from uranium and the fission products produced by the splitting of U-235 atoms. A number of separation methods were considered: volatilization, ion exchange, solvent extraction, and precipitation. Precipitation, using a carrier, was considered the most promising in regard to future "scale-up."

The separation process was first worked out with invisible tracer quantities of Pu and later scaled up. By September 1942, visible quantities of a pure Pu compound, free of carrier material, had been prepared. The separation procedure involved precipitating Pu(IV) as the phosphate and using $BiPO_4$ as the carrier, leaving U(VI) in solution. The precipitate was dissolved, the plutonium was oxidized to the VI state, and the carrier was reprecipitated, along with some fission products, while Pu(VI) remained in solution. The Pu(VI) was then precipitated as the fluoride, using lanthanum fluoride as the carrier. Successive cycles were carried out until elimination of the fission products was achieved.

The actual large-scale process for bomb material at the Hanford Works was much more complicated than the description above would indicate. There were about 30 major chemical reactions involving hundreds of operations before the plutonium was finally separated. The separation plant consisted of a series of compartments with heavy concrete walls, almost completely buried in the ground. The various stages of precipitation, dissolution, oxidation, and reduction occurred in these compartments, with all the equipment operated remotely. In spite of the complexity of the process, it was operated successfully for several years. It was later replaced by a solvent extraction procedure (Box 16.1).

Adsorption on Gelatinous Precipitates

The primary particles of a gelatinous precipitate are much larger in number and have much smaller dimensions than those of crystalline or curdy precipitates. The surface area exposed to the solution by such a precipitate is extremely large. A large quantity of water is adsorbed, rendering the precipitate gelatinous. The adsorption of foreign ions can also be quite extensive. Since the flocculated primary particles do not grow into larger crystals, the impurities are not occluded, as with $BaSO_4$, but are held by adsorption on the surface of the tiny particles. The electric charge on the primary particles of substances such as $Fe(OH)_3$ and $Al(OH)_3$ is primarily a function of the pH of the solution, since H^+ and OH^- ions are readily adsorbed. The precipitates tend to be positively charged at low pH and negatively charged at high pH. Thus anions tend to be coprecipitated at low pH, and cations at high pH.

Minimizing Coprecipitation

The following procedures are routinely used to minimize coprecipitation.

1. *Method of addition of the two reagents.* If it is known that either the sample or the precipitant contains a contaminating ion, the solution containing this ion can be added to the other solution. In this way the concentration of the contaminant is kept at a minimum during the early stages of precipitation. In the case of hydrous oxides the charge carried by the primary particles can be controlled.
2. *Washing.* Adsorbed impurities can be removed by washing unless they are occluded. With curdy and gelatinous precipitates one must have an electrolyte in the wash solution to avoid peptization.
3. *Digestion.* This technique is of considerable benefit to crystalline precipitates, of some benefit to curdy precipitates, but not used for gelatinous precipitates.
4. *Reprecipitation.* If the substance can be readily redissolved (as salts of weak acids in stronger acids), it can be filtered, redissolved, and reprecipitated. The contaminating ion will be present in a lower concentration during the second precipitation, and consequently a smaller amount will be coprecipitated.
5. *Separation.* The impurity may be separated or its chemical nature changed by some reaction before the precipitate is formed.

4.3d. *Precipitation from Homogeneous Solution*

When a precipitant is added to a solution, even when the solution is dilute and well stirred, there will always be some local regions of high concentration. However, by using a procedure in which the precipitant is produced as a result of the reaction that *takes place in the solution,* such local effects can be avoided. This technique is usually called *precipitation from homogeneous solution,* and it can lead to both large and pure particles of a precipitate. The best-known example of this method is use of the hydrolysis of urea to increase the pH and precipitate hydrous oxides, or salts of weak acids. Urea hydrolyzes according to the equation

$$CO(NH_2)_2 + H_2O \longrightarrow CO_2 + 2NH_3$$

The hydrolysis is slow at room temperature but is fairly rapid at 100°C. Thus the pH can be well controlled in effecting separation by controlling the temperature and duration of heating. Also, the carbon dioxide liberated as bubbles prevents "bumping." Precipitation is usually complete in 1 to 2 h. During this slow growth the particles have time to attain a large size without the occurrence of imperfections in the lattice structure, and therefore the amount of occluded impurity is minimized.

Examples of Precipitation from Homogeneous Solution

Calcium oxalate has been precipitated by neutralizing an acid solution of calcium containing excess oxalate by the hydrolysis of urea:

$$Ca^{2+} + H_2C_2O_4 + CO(NH_2)_2 + H_2O \longrightarrow CaC_2O_4(s) + CO_2 + 2NH_4^+$$

Barium sulfate has been precipitated in this manner by hydrolyzing sulfamic acid or dimethyl sulfate. The hydrolysis reactions are

$$NH_2SO_3H + 2H_2O \longrightarrow NH_4^+ + SO_4^{2-} + H_3O^+$$

$$(CH_3O)_2SO_2 + 4H_2O \longrightarrow 2CH_3OH + SO_4^{2-} + 2H_3O^+$$

Less calcium is coprecipitated with $BaSO_4$ when the latter is precipitated homogeneously. Other ions that have been generated homogeneously include phosphate, from trimethyl phosphate, and oxalate, from ethyl oxalate. The hydrolysis equations are

$$(CH_3O)_3PO + 3H_2O \longrightarrow 3CH_3OH + H_3PO_4$$

$$(C_2H_5)_2C_2O_4 + 2H_2O \longrightarrow 2C_2H_5OH + H_2C_2O_4$$

Hydrous oxides are gelatinous whether formed under ordinary analytical conditions or homogeneously. However, Willard and his co-workers[5] have obtained dense precipitates of iron and aluminum by precipitation with urea in the presence of certain anions. The succinate ion is best for aluminum, and formate is best for iron. The precipitates are of indefinite composition but contain basic salts of aluminum and succinate or of iron and formate. Coprecipitation of foreign ions is less than when the hydrous oxides are precipitated by addition of ammonia.

4.3e. *Postprecipitation*

The process by which an impurity is deposited *after* precipitation of the desired substance is termed *postprecipitation*. This process differs from coprecipitation principally in the fact that the amount of contamination increases the longer the desired precipitate is left in contact with the mother liquor. When there is a possibility that postprecipitation may occur, directions call for filtration to be made shortly after the desired precipitate is formed.

Examples of Postprecipitation

Postprecipitation occurs when the solution is supersaturated with a foreign substance that precipitates very slowly. For example, zinc sulfide does not readily precipitate from solutions containing zinc ion, hydrogen ion (0.1 to 0.2 *M*), and those that are saturated with hydrogen sulfide. However, if mercury(II) sulfide is precipitated under the same conditions in the presence of zinc, over 90% of the zinc comes down as the sulfide within 20 min. Apparently, zinc sulfide forms very stable supersaturated solutions. When mercury(II) sulfide is present, sulfide ions are strongly adsorbed at the interface of the solid and solution. The solubility

[5] H. H. Willard, *Anal. Chem.,* **22,** 1372 (1950).

product constant of zinc sulfide is exceeded to an even greater extent at the interface than in the bulk of the solution, and the rate of precipitation is increased.

Magnesium oxalate forms stable, supersaturated solutions, and unless precautions are taken, postprecipitates on calcium oxalate when calcium and magnesium are separated by precipitation of the latter compound. Postprecipitation can be avoided by using as high acidity as possible and filtering off the calcium precipitate within 1 or 2 h after precipitation.

4.4

DRYING AND IGNITION OF PRECIPITATES

In any gravimetric procedure involving precipitation, one must finally convert the separated substance into a form suitable for weighing. It is necessary that the substance weighed be pure, stable, and of definite composition for the results of the analysis to be accurate. Even if coprecipitation has been minimized, there still remains the problem of complete removal of water and of any electrolytes added to the wash water. Some precipitates are weighed in the same chemical form as that in which they precipitate. Others undergo chemical changes during ignition, and these reactions must go to completion for correct results. The procedure used in this final step depends both upon the chemical properties of the precipitate and upon the tenacity with which water is held by the solid.

4.4a. Air Drying (Ambient Temperature)

Some precipitates can be dried sufficiently for analytical determination without resort to high temperatures. For example, $MgNH_4PO_4 \cdot 6H_2O$ is sometimes dried by washing with a mixture of alcohol and ether and drawing air over the precipitate for a few minutes. However, this procedure is not usually recommended because of the danger of incomplete removal of water by washing.

4.4b. Air Drying (Low Temperature)

Some precipitates lose water readily in an oven at temperatures of 100 to 130°C. Silver chloride does not adsorb water strongly and is normally dried in this manner for ordinary analytical work. In the determination of atomic weights, however, it has been found necessary to fuse AgCl in order to remove the last traces of water.

4.4c. Ignition of Precipitates

Ignition at high temperature is required for complete removal of water that is occluded or very strongly adsorbed, and for complete conversion of some precipitates to the desired compound. Gelatinous precipitates such as the hydrous oxides adsorb water quite strongly and must be heated to very high temperatures to remove water completely. The ignition of CaC_2O_4 to CaO is an example of a chemical change that requires a high temperature for complete reaction.

4.4d. Errors During Ignition

Errors other than incomplete removal of water or volatile electrolytes can occur during ignition. One of the most serious is reduction of the precipitate by carbon when filter paper is employed. Substances such as AgCl that are very easily reduced are never filtered on paper; filtering crucibles are always employed. Precipitates also can be overignited, leading to decomposition to substances of indefinite composition. Errors can result from reabsorption of water or carbon dioxide by an ignited precipitate upon cooling. Crucibles should be properly covered and kept in a desiccator as they cool.

4.4e. *Determination of Optimum Drying and Ignition Temperatures*

Thermogravimetric Analysis

Studies on the ignition temperatures required for different precipitates can be made using the technique of *thermogravimetric analysis*. This method involves the use of a *thermobalance,* which allows a sample to be weighed while it is in a furnace. The data are recorded in the form of a graph of weight of the precipitate vs. temperature, and the graph is called a *pyrolysis* curve. The pyrolysis curves for a few substances are shown in Fig. 4.2. The curves for calcium and magnesium oxalates are particularly interesting. The monohydrate $CaC_2O_4 \cdot H_2O$ is stable at 100°C and then loses water up to about 226°C. Up to 398°C the form CaC_2O_4 is stable, and then the oxalate loses CO abruptly to form $CaCO_3$. The carbonate is stable in the range of about 420 to 600°C, and then the dissociation to CaO begins. The weight finally becomes constant at about 850°C. Magnesium oxalate differs in its behavior in that it loses CO and CO_2, simultaneously forming MgO with no intermediate $MgCO_3$.

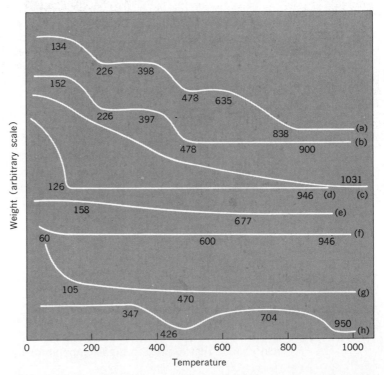

Figure 4.2 Pyrolysis curves: (a) CaC_2O_4, (b) MgC_2O_4, (c) Al_2O_3, precipitated by aqueous NH_3. (d) Al_2O_3, precipitated by urea, (e) $BaSO_4$, (f) AgCl, (g) Fe_2O_3, (h) CuSCN. (Taken by permission from C. Duval. *Inorganic Thermogravimetric Analysis.* 2nd ed., Elsevier Publishing Co., New York, 1963).

Instrumentation for Thermogravimetric Analysis

Commercial instruments are available today which are fully automated and programmable. For example, there is a commercial multisample thermogravimetric analyzer which can handle up to 19 samples unattended. The sample carousel and balance weighing pan are located inside the furnace, and all weighing is performed automatically. The instrument can be programmed to perform as many as five weight loss steps per program. Temperature, rate of increase in temperature, and the atmospheric gas (oxygen or nitrogen) are selectable for each step. Applications of this instrument include measurement of loss on ignition of cement; moisture and ash determination of food and agricultural products; and compositional analysis of polymers, rubber, and ceramics.

4.5

ORGANIC PRECIPITANTS

Many inorganic ions can be precipitated with certain organic reagents called *organic precipitants*. A number of these reagents are useful not only for separations by precipitation, but also by solvent extraction. The latter will be discussed in Chapter 16.

Most of the organic precipitants on which our discussion will be centered combine with cations to form *chelate* rings. We shall be concerned here with neutral metal chelate compounds which are insoluble in water. In Chapter 8 we will discuss reagents which form soluble 1 : 1 complex ions with various cations and which can be employed as titrants for metals. In this chapter we shall also mention a few examples of organic precipitants which form saltlike precipitates with metal ions.

4.5a. *Reagents Forming Chelate Compounds*

Generally speaking, most of the better-known organic precipitants which form chelate compounds with cations contain both an acidic and a basic functional group. The metal, interacting with both of these groups, itself becomes one member of a heterocyclic ring. 8-Hydroxyquinoline (often called 8-quinolinol, or oxine) forms insoluble compounds with a number of metal ions, aluminum for one. The formation of this compound may be formulated as follows:

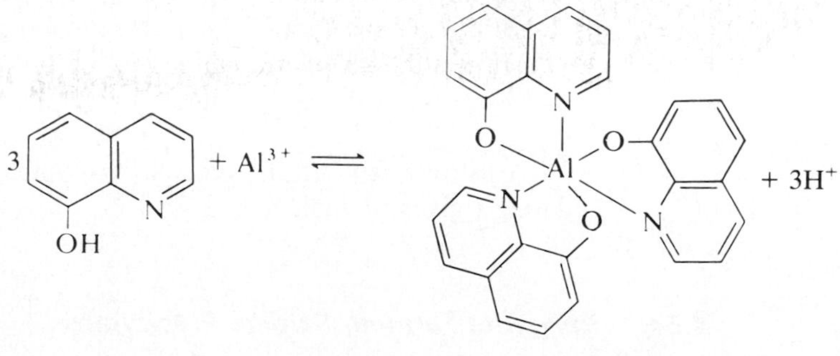

Aluminum replaces the acidic hydrogen of the hydroxyl group. At the same time, the unshared pair of electrons on the nitrogen is donated to the aluminum, thereby forming a five-membered ring. From the strain theory of organic chemistry it is expected that rings of this type would be mainly five- and six-membered. Hence the acidic and basic functional groups in the organic molecule must be situated in positions with respect to each other which permit the closure of such rings.

A neutral chelate compound of the type described is essentially organic in nature. The metal ion becomes simply one of the members of an organic ring structure, and its usual properties and reactions are no longer readily demonstrable. With the reservation in mind that exceptions can be found, we may state generally that such chelate compounds are insoluble in water but soluble in less polar solvents such as chloroform and carbon tetrachloride. We shall see in Chapter 16 that this differential solubility may be utilized in effecting separations by extraction processes, and in Chapter 14 we shall mention briefly the use of chelates in colorimetric analysis. At this point, we wish to consider only the precipitation of metal ions by these organic reagents.

4.5b. Advantages and Disadvantages of Organic Precipitants

Let us consider first the advantages offered by organic precipitants.

1. Many of the chelate compounds are very insoluble in water, as noted above, so that metal ions may be quantitatively precipitated.
2. The organic precipitant often has a high molecular weight. Thus a small amount of metal may yield a large weight of precipitate, minimizing weighing errors.
3. Some of the organic reagents are fairly selective, yielding precipitates with only a limited number of cations. By controlling such factors as pH and the concentration of masking reagents, the selectivity of an organic reagent can often be greatly enhanced.
4. The precipitates obtained with organic reagents are often coarse and bulky and hence easily handled.
5. In some cases a metal can be precipitated with an organic reagent, the precipitate collected and dissolved, and the organic molecule titrated, furnishing an indirect titrimetric method for the metal.

There are some disadvantages in the use of organic precipitants.

1. Many of the chelate compounds do not have good weighing forms and are used only for separations, not determinations.
2. There is a danger of contaminating the precipitate with the chelating agent itself because of the latter's limited solubility in water.

A list of a few organic precipitants which have found use in analytical procedures is given in Table 4.1.

4.5c. Reagents Forming Saltlike Precipitates

Some organic precipitants form salts rather than chelate complexes with inorganic ions. Oxalic acid is well known in analytical processes for its use in the

TABLE 4.1 Some Common Organic Precipitants

Compound	Chelate with metal of valence n	Comments
CH₃—C=N—OH / CH₃—C=N—OH Dimethylglyoxime	(chelate structure)	Principally used for determination of nickel
8-Hydroxyquinoline	(chelate structure)	Precipitates many elements but can be used for group separations by controlling pH
α-Nitroso-β-naphthol	(chelate structure)	Principally used for precipitation of cobalt in presence of large amounts of nickel
Cupferron	(chelate structure)	Mainly used for separations, such as iron and titanium from aluminum
α-Benzoin oxime	(chelate structure)	Good reagent for copper; also precipitates bismuth and zinc
Thionalide	(chelate structure)	Used for precipitation and determination of elements of H₂S group
Quinaldic acid	(chelate structure)	Used for determination of cadmium, copper, and zinc

precipitation of calcium; calcium oxalate is a typical insoluble salt. There are a number of such organic compounds which form precipitates with both cations and anions. A few of these are listed in Table 4.2.

TABLE 4.2 Reagents Forming Saltlike Precipitates

Reagent	Comments
$Na^+B(C_6H_5)_4^-$ Sodium tetraphenylboron	Used principally for K^+; in 0.1 M HCl only NH_4^+, Hg^{2+}, Rb^+, and Cs^+ interfere
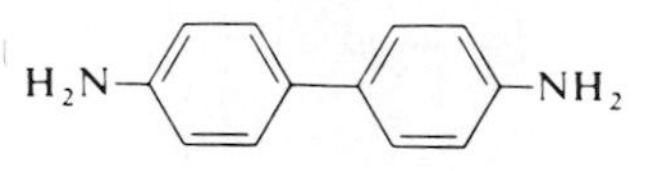 H_2N-⟨benzidine structure⟩$-NH_2$ Benzidine	Used principally for SO_4^{2-}
$(C_6H_5)_4As^+Cl^-$ Tetraphenylarsonium chloride	Used for Tl^+; other metals which form precipitates include tin, gold, zinc, mercury, and cadmium
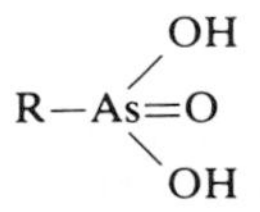 $R-As=O$ with OH, OH groups Arsonic acids; R = phenyl, n-propyl, etc.	Acids precipitate quadrivalent metal ions such as tin, thorium, and zirconium from acid media

4.6

APPLICATIONS OF GRAVIMETRIC ANALYSIS

4.6a.   *Quantitation of Inorganic and Organic Compounds*

A gravimetric method for almost every element in the periodic table has been reported in the chemical literature.[6] We have listed in Table 4.3 some examples where precipitation reactions have been used for the elements in various groups of the periodic table. The substances precipitated and the form which is finally weighed are given in the table. In addition to inorganic substances, organic compounds have been analyzed by gravimetric techniques. Two examples are the determination of cholesterol in cereals and of lactose in milk products. Cholesterol, a steroid alcohol, can be precipitated by an organic saponin called digitonin. Digitonin, a compound of high molecular weight (1214), forms a 1 : 1 insoluble complex with cholesterol. A number of pharmaceutical compounds are determined gravimetrically either by isolating the pure form of the medicinal agent without a chemical reaction or by conversion of the sodium salt to the acid form.

4.6b.   *Elemental Analysis*

The elemental analysis of organic compounds can be done gravimetrically. For example, carbon and hydrogen can be determined by burning the sample in a stream of oxygen and absorbing the CO_2 and H_2O produced on appropriate ab-

[6] See, for example, C. L. Wilson and D. W. Wilson, *Comprehensive Analytical Chemistry,* Vol. IC, Elsevier Publishing Co., New York, 1962.

TABLE 4.3 Gravimetric Determinations of Various Elements

GROUP IN PERIODIC TABLE	ELEMENT	PRECIPITATE	SUBSTANCE WEIGHED
IA	Potassium	$KClO_4$	$KClO_4$
IIA	Calcium	CaC_2O_4	CaO
IIIA	Aluminum	$Al_2O_3 \cdot xH_2O$	Al_2O_3
IVA	Silicon	$SiO_2 \cdot xH_2O$	SiO_2
VA	Phosphorus	$MgNH_4PO_4 \cdot 6H_2O$	$Mg_2P_2O_7$
VIA	Sulfur	$BaSO_4$	$BaSO_4$
VIIA	Chlorine	$AgCl$	$AgCl$
IB	Silver	$AgCl$	$AgCl$
IIB	Zinc	$ZnNH_4PO_4$	$Zn_2P_2O_7$
IIIB	Scandium	Scandium oxinate	Scandium oxinate
IVB	Titanium	Titanium cupferrate	TiO_2
VB	Vanadium	$HgVO_3$	V_2O_5
VIB	Chromium	$Cr_2O_3 \cdot xH_2O$	Cr_2O_3
VIIB	Manganese	MnO_2	Mn_2O_3
VIII	Iron	$Fe_2O_3 \cdot xH_2O$	Fe_2O_3
	Cobalt	CoS	$CoSO_4$
	Nickel	Nickel dimethylglyoxime	Nickel dimethylglyoxime

sorbents. The absorption tubes are weighed before and after the combustion to obtain the weights of CO_2 and H_2O produced. In many laboratories, the weighing of absorption tubes has been replaced by quantitation of the combustion gases in a carrier gas stream flowing through a thermal conductivity cell (see Chapter 17 for a discussion of such thermal conductivity measurements).

4.6c. The Role of Gravimetric Analysis in Modern Analytical Chemistry

The student may well have heard the argument that modern instrumental methods have taken the place of classical gravimetric techniques. Although it is true that the gravimetric technique has been displaced in its routine aspects by instrumental methods, gravimetric analysis is still of great importance in the field of analytical chemistry. There are still many instances where it represents the best choice for solving a particular analytical problem. In general, where only a few determinations are required, a gravimetric procedure may actually be faster and more accurate than an instrumental method which requires extensive calibration or standardization. Instruments generally furnish only relative measurements and must be calibrated on the basis of a classical gravimetric or titrimetric method. In providing the standards required to check the performance of an instrumental method, the gravimetric technique offers a direct and comparatively simple approach.

4.6d. Accuracy and Sensitivity of Gravimetric Methods

Gravimetric methods compare favorably with other analytical techniques in terms of the accuracy attainable. If the analyte is a major constituent ($> 1\%$ of the sample), accuracy of a few parts per thousand can be expected if the sample is not too complex. If the analyte is present in minor or trace amounts (less than 1%), a gravimetric method is generally not employed.

4.6e. Specificity of Gravimetric Methods

In general, gravimetric methods are not very *specific*. Some reagents are *selective* in that they form precipitates with only a certain group of ions. The selectivity of precipitating agents can often be improved by controlling such factors as *p*H and the concentration of certain masking agents.

As illustrations of gravimetric procedures we shall discuss briefly the determinations commonly used as student exercises. These involve the precipitation of silver chloride, barium sulfate, and iron(III) hydroxide.

4.6f. Examples of Common Gravimetric Procedures

1. Silver Chloride

Silver chloride precipitates in curds or lumps resulting from the coagulation of colloidal material (Section 4.3a). The precipitate is easily filtered and then washed with water containing a little nitric acid. The acid prevents peptization of the precipitate and is volatilized when the precipitate is dried. Silver chloride is normally filtered through a sintered glass or porous porcelain crucible and dried at 110 to 130°C.

Errors

The precipitation of silver chloride generally gives excellent analytical results. The principal error arises from decomposition of the precipitate by sunlight:

$$2AgCl(s) \longrightarrow 2Ag(s) + Cl_2(g)$$

The extent of this reaction is negligible unless the precipitate is exposed to direct sunlight.

The solubility of silver chloride in water is slight, and losses due to solubility are negligible. However, alkali and ammonium salts, as well as large concentrations of acids, should be avoided since they increase the solubility.

Other Applications

In addition to the determination of silver and chloride, the precipitation of silver chloride can be used to determine chlorine in oxidation states other than -1. Hypochlorites (ClO^-), chlorites (ClO_2^-), and chlorates (ClO_3^-) may be determined by first reducing these ions to chloride and then precipitating silver chloride. Chlorine in organic compounds is often determined by this precipitation after the organic chlorine is converted to sodium chloride by fusion with sodium peroxide.

Bromide and iodide may be determined by precipitation of their silver salts. Also, oxygen-containing anions, such as hypobromite (BrO^-), bromate (BrO_3^-), hypoiodite (IO^-), iodate (IO_3^-), and periodate (IO_4^-), can first be reduced to bromide or iodide and then precipitated as the silver salts.

2. Barium Sulfate

Barium sulfate is a crystalline precipitate. It is only slightly soluble in water, and losses due to solubility are small. The precipitation is carried out in about 0.01 *M* hydrochloric acid for the purpose of obtaining larger particles, and a purer precipitate (page 77) and to prevent the precipitation of such salts as $BaCO_3$.

Errors

Coprecipitation of foreign substances with barium sulfate is very pronounced. The anions most strongly coprecipitated are nitrate and chlorite. Cations, particularly divalent and trivalent ones that form slightly soluble sulfates, are strongly coprecipitated, iron(III) being one of the most prominent examples. The procedures summarized on page 77 are employed where practical to minimize coprecipitation. The digestion process employed to increase particle size also leads to some purification. Reprecipitation is not employed, since a suitable solvent is not available.

Barium sulfate is normally filtered with filter paper and washed with hot water. The filter paper must be burned off carefully with a plentiful supply of air. The sulfate is reduced rather readily by the carbon of the paper:

$$BaSO_4(s) + 4C(s) \longrightarrow BaS(s) + 4CO(g)$$

If this reduction occurs, the results are low and usually the precision is poor. The precipitate can be reconverted to the sulfate by moistening it with sulfuric acid and reigniting. A porous porcelain crucible can be employed instead of paper.

Other Applications

The sulfur in sulfides, sulfites, thiosulfates, and tetrathionates can be determined by oxidizing the sulfur to sulfate and then precipitating barium sulfate. Permanganate is often used to effect the oxidation. Sulfur in organic compounds is determined by oxidizing the element to sulfate with sodium peroxide. Ores of sulfur, such as pyrites, FeS_2, and chalcopyrite, $CuFeS_2$, may be fused with sodium peroxide to oxidize the sulfur to sulfate.

Other cations often precipitated as sulfates are lead and strontium. Both of these sulfates are more soluble than barium sulfate. Alcohol is added in the determination of strontium to decrease the solubility of the sulfate. The determination of lead in brass can be done by the precipitation of lead sulfate.

3. Iron(III) Hydroxide

The gravimetric determination of iron involves the precipitation of iron(III) hydroxide (actually $Fe_2O_3 \cdot xH_2O$, called a *hydrous oxide*), followed by ignition at high temperature to Fe_2O_3. The method is used in rock analyses, where iron is separated from elements such as calcium and magnesium by the precipitation. Iron ores are usually dissolved in hydrochloric acid, and nitric acid or bromine is used to oxidize the iron to the $+3$ oxidation state.

The hydrous oxide of iron is a gelatinous precipitate which is very insoluble in water ($K_{sp} = 1 \times 10^{-36}$). Coagulation of the colloidal material is aided by precipitation from a hot solution. The precipitate is washed with water containing a small amount of ammonium nitrate to prevent peptization. Filtration is carried out using rapid filter paper. The paper is burned off and the precipitate ignited at fairly high temperatures to remove all the water.

Errors

Coprecipitation by adsorption of foreign ions during precipitation can cause serious errors. By using some of the procedures described on page 77, however, coprecipitation can be held to a minimum. Precipitation is generally made from acid solution so that the colloidal particles are positively charged and cations are less strongly adsorbed. Since the oxide can be readily dissolved in acids, reprecipitation is used to advantage to rid the precipitate of adsorbed impurities.

Iron(III) oxide, Fe_2O_3, is fairly easily reduced to either Fe_3O_4 or Fe by the carbon of the filter paper. The ignited precipitate can be treated with concentrated nitric acid and reignited to form Fe_2O_3 again.

Other Applications

Several other metals are precipitated as hydrous oxides by ammonia. Among these are aluminum, chromium(III), titanium(IV), and manganese(IV). The hydroxides of aluminum and chromium are amphoteric, and a large excess of ammonia must be avoided because these substances redissolve. Manganese(II) is incompletely precipitated, but if an oxidizing agent, such as bromine, is added, hydrous manganese dioxide is precipitated. All these hydrous oxides can be ignited to the oxides Al_2O_3, Cr_2O_3, TiO_2, and Mn_3O_4. Thus the precipitation with ammonia can be used for determining each element provided the others are absent. The method is seldom used for manganese, since the conversion of MnO_2 to Mn_3O_4 is not quantitative.

Precipitation with ammonia is used for the quantitative separation of iron and the foregoing elements from alkali and alkaline earth cations in the analysis of rocks, such as limestone.

KEY TERMS

Absorption. The holding of one substance *inside* another.

Adsorption. The holding of one substance *on the surface* of another.

Chelate ring. A heterocyclic ring formed by a cation and an organic precipitating agent.

Coagulation. The coherence of particles of colloidal size to form larger particles which precipitate.

Colloids. Particles with diameters of about 10^{-4} to 10^{-7} cm which are electrically charged, repel each other, and resist coagulation.

Coprecipitation. The process by which a normally soluble substance is carried down during precipitation of the desired substance.

Digestion. The process in which a precipitate stands in contact with the mother liquor to promote crystal growth.

Electrical double layer. The region comprising the electrically charged surface of a colloid plus the surrounding layer of opposite charge.

Emulsoid. A colloid which has a strong affinity for the solvent.

Gel (hydrogel). A coagulated colloid which carries down large quantities of solvent (water) to produce a jellylike precipitate.

Gravimetric factor. The number of grams of analyte in 1 g of precipitate.

Hydrophilic. "Water-loving"—having a great affinity for water.

Hydrophobic. "Water-hating"—having little affinity for water.

Lyophilic. "Solvent-loving"—having a great affinity for the solvent.

Lyophobic. "Solvent-hating"—having little affinity for the solvent.

Nucleation. The formation of small particles (nuclei) when the solubility product of a substance is exceeded; further growth of the nuclei leads to formation of the precipitate.

Occlusion. The process by which an impurity is

held within a crystal by growth of the crystal around it.

Peptization. The process of dispersing an insoluble substance into a liquid as a colloid.

Postprecipitation. The process by which an impurity is deposited after precipitation of the analyte.

Precipitation from homogeneous solution. Precipitation in which the precipitating agent is produced homogeneously throughout the solution.

Primary layer. The layer of ions held by adsorption on the surface of a colloid.

Relative supersaturation. The ratio of the excess concentration of substance to its equilibrium solubility.

Reprecipitation. The dissolving of a precipitate and precipitation a second time to minimize coprecipitation.

Secondary layer. The layer of ions loosely held by an electrically charged colloidal particle.

Suspensoid. A colloid which has only a slight affinity for water.

Thermobalance. A balance which allows a sample to be weighed while it is in a furnace.

QUESTIONS

1. *Requirements.* What two requirements should be met for a reaction to serve as the basis of a gravimetric method?
2. *Gravimetric factor.* What is meant by a gravimetric factor? Show how it arises in solving a stoichiometric problem by the mole method.
3. *Terms.* Explain the meaning of the following terms: (a) gelatinous precipitate; (b) primary layer; (c) lyophobic colloid; (d) emulsoid; (e) peptization; (f) nucleation.
4. *Homogeneous precipitation.* Explain what is meant by precipitation from homogeneous solution. How does this procedure lead to both larger and purer particles of precipitate?
5. *Properties of precipitates.* Explain (a) why gelatinous precipitates are not digested; (b) why curdy precipitates are not digested.
6. *Occlusion.* Explain why occluded impurities are not removed by washing. What procedure is helpful in removing occluded impurities?
7. *Coprecipitation.* Why is $Fe(OH)_3$ usually precipitated from acidic solutions?
8. *Organic precipitants.* Why are chelate compounds generally insoluble in water and soluble in organic solvents?
9. *Ignition of precipitates.* Point out the possible errors in the final step of igniting and weighing a precipitate. Why is AgCl never filtered on paper?

10. *Particle size.* What conditions should be used to precipitate a sulfide, such as ZnS, in order to obtain as large particles as possible?

Multiple-choice. In the following multiple-choice questions, select the *one best* answer.

11. A precipitate of $Fe(OH)_3$ is contaminated with $Mg(OH)_2$. The best way to get rid of the impurity is (a) washing; (b) digestion; (c) ignition; (d) reprecipitation.
12. Colloids which carry down only a small quantity of water when coagulated are said to be (a) lyophobic; (b) suspensoids; (c) emulsoids; (d) two of the above; (e) none of the above.
13. The process of dispersing an insoluble material into a liquid as a colloid is called (a) occlusion; (b) nucleation; (c) peptization; (d) coagulation.
14. Which of the following *does not* promote the formation of large crystals of CaC_2O_4? $H_2C_2O_4$ is a weak acid. (a) Slow mixing of dilute solutions; (b) decreasing $(Q - S)/S$; (c) digestion; (d) precipitation at high pH rather than low pH.
15. AgCl is precipitated by adding $AgNO_3$ to an aqueous NaCl solution. The ion most strongly adsorbed to the surface of the colloidal particles before the EPt is (a) Na^+; (b) Cl^-; (c) Ag^+; (d) H_3O^+; (e) OH^-.

PROBLEMS

1. *Percent purity.* A 0.7203-g sample containing chloride is dissolved, and the chloride precipitated as AgCl. The precipitate is washed, dried, and found to weigh 0.4026 g. Calculate (a) the percentage of chloride (Cl) in the sample; (b) the percentage expressed as NaCl.

2. *Percent purity.* The sulfur in a 0.8245-g sample is converted to sulfate, and the sulfate precipitated as $BaSO_4$. The precipitate is washed, ignited, and found to weigh 0.3084 g. Calculate (a) the percentage of sulfur (S) in the sample; (b) the percentage expressed as SO_3.

3. *Determination of calcium.* The calcium in a 0.8432-g sample is precipitated as CaC_2O_4. The precipitate is washed, ignited to $CaCO_3$, and found to weigh 0.3462 g. Calculate the percentage of CaO in the sample.

4. *Determination of aluminum.* A 1.000-g sample of an alum is dissolved, and the sulfate precipitated as $BaSO_4$. The precipitate weighed 0.3486 g after it was washed and ignited. Assuming all the sulfate and aluminum in the sample were in the form $K_2Al_2(SO_4)_3 \cdot 24H_2O$, calculate the percentage of Al_2O_3 in the sample.

5. *Determination of chloride.* The chloride in a sample is to be determined gravimetrically by precipitating and weighing AgCl. What weight in grams of sample should be taken so that the percentage of Cl is obtained by simply multiplying the weight of the AgCl precipitate by 10?

6. *Gravimetric factors.* Calculate the gravimetric factors for the following. The substance weighed is given first, then the substance sought. (a) K_2PtCl_6, K; (b) $CaCO_3$, CaO; (c) Ag_2CrO_4, Cr_2O_3; (d) $(NH_4)_2PtCl_6$, NH_3.

7. *Gravimetric factors.* Calculate the gravimetric factors for the following. The substance weighed is given first, then the substance sought. (a) $Zn_2P_2O_7$, ZnO; (b) $BaSO_4$, FeS_2; (c) U_3O_8, UO_2; (d) $KClO_4$, K_2O.

8. *Precipitation of iron.* Calculate the number of milliliters of ammonia, density of 0.99 g/mL, 2.3% by weight NH_3, which will be required to precipitate the iron as $Fe(OH)_3$ in a 0.74-g sample that contains 22% FeO.

9. *Precipitation of barium sulfate.* Calculate the number of milliliters of a solution which contains 16 g of $BaCl_2$/liter which will be required to precipitate the sulfur as $BaSO_4$ in a 0.60-g sample that contains 12% S.

10. *Sample size.* What size sample containing 14.4% Cl should be taken for analysis to obtain a precipitate of AgCl which weighs 0.440 g?

11. *Sample size.* What size sample which contains 18.0% Fe_3O_4 should be taken for analysis in order to obtain a precipitate of Fe_2O_3 which weighs 0.400 g?

12. *Sample size.* Calcium is determined in a rock sample by precipitating $CaCO_3$ and igniting the precipitate to CaO. What weight of sample should be taken for analysis so that each milligram of precipitate represents 0.200% CaO in the sample?

13. *Organic analysis.* A 1.000-g sample of a pure organic compound containing chlorine is fused with Na_2O_2 to convert the chlorine to NaCl. The sample is then dissolved in water, and the chloride precipitated with $AgNO_3$, giving 1.950 g of AgCl. If the MW of the organic compound is about 147, how many chlorine atoms does each molecule contain?

14. *Organic precipitant.* The nickel in a 0.4036 g sample of an ore is precipitated with dimethylglyoxime, giving a precipitate which weighs 0.2894 g. Calculate the percentage of nickel in the ore.

15. *Atomic weight.* A sample of pure NaCl weighing 0.65310 g is dissolved in water, and the chloride precipitated as AgCl. If the AgCl weighs 1.6010 g, calculate the atomic weight of sodium. Assume that the atomic weights of Cl and Ag are 35.453 and 107.868, respectively.

16. *Mixture.* A 0.6000-g sample consisting of only CaC_2O_4 and MgC_2O_4 is heated at 500°C, converting the two salts to $CaCO_3$ and $MgCO_3$. The sample then weighs 0.4650 g. If the sample had been heated at 900°C, where the products are CaO and MgO, what would the mixture of oxides have weighed?

17. *Mixture.* A 0.4531-g sample containing only $CaCO_3$ and $MgCO_3$ is treated with excess HCl, and 112 mL (STP) of CO_2 is produced. Assuming that the reactions went to completion, what percentage of the sample is $MgCO_3$?

18. *Determination of sodium and potassium.* The sodium and potassium in a sample weighing 0.9250 g are converted into NaCl and KCl salts. The chloride mixture weighs 0.6065 g. The chlorides are then converted into Na_2SO_4 and K_2SO_4, the mixture weighing 0.7190 g. Calculate the percentages of Na_2O and K_2O in the original sample.

19. *Molecular weight.* A 1.000-g sample containing 75.0% K_2SO_4 and 25.0% of another metal sulfate, MSO_4, is dissolved and the sulfate precipitated as $BaSO_4$. If the $BaSO_4$ precipitate weighs 1.490 g, what is the MW of MSO_4?

20. *Errors.* A sample which actually contains 32.3% SO_3 is analyzed by precipitating, igniting, and weighing $BaSO_4$. During ignition, 20.0% of the $BaSO_4$ precipitate is reduced to BaS. What percentage of SO_3 would the analyst calculate, assuming that it was not known that the error occurred?

21. *Indirect analysis.* A 0.6200-g sample that contains NaCl, NaBr, and impurities gives a precipitate of AgCl and AgBr that weighs 0.5120 g. Another 0.6200-g sample is titrated with 0.1070 M $AgNO_3$, requiring 28.10 mL. Calculate the percentages of NaCl and NaBr in the sample.

22. *Indirect analysis.* A 0.4800-g sample that contains KCl, KI, and inert material gives a precipitate of AgCl and AgI that weighs 0.2720 g. A 0.7200-g sample of the same material is titrated with 0.1020 M $AgNO_3$, requiring 25.80 mL. Calculate the percentages of KCl and KI in the sample.

23. *Gravimetric analysis.* A certain sample of phosphate rock contains 26.26% P_2O_5. A 0.5428-g sample is analyzed by precipitating $MgNH_4PO_4 \cdot 6H_2O$ and igniting the precipitate to $Mg_2P_2O_7$. Calculate the weights obtained of (a) $MgNH_4PO_4 \cdot 6H_2O$, and (b) $Mg_2P_2O_7$.

24. *Stoichiometry.* All of the oxygen in a 0.5434-g sample of a pure oxide of iron is removed by reduction in a stream of H_2. The loss in weight is 0.1210 g. What is the formula of the iron oxide?

5

Review of Chemical Equilibrium

One of the requirements for a reaction used in a titration is that it go essentially to completion at the equivalence point (page 45). Similarly, in a gravimetric analysis the reaction used for the separation process must be so complete that an insignificant amount of analyte is left behind. Stoichiometric calculations, treated in Chapters 3 and 4, tell us nothing about the completeness of a reaction. We need to know the magnitude of the *equilibrium constant* of a reaction in order to calculate the degree to which a reaction goes to completion under a given set of conditions.

Equilibrium calculations are introduced in all standard texts on general chemistry. Here we review how equilibrium constants are formulated and used for acid-base, precipitation, complex formation, and oxidation-reduction reactions. In subsequent chapters we shall apply the principles of equilibrium to determine the feasibility of specific titrations and to decide whether certain separation processes can be made complete.

5.1

THE EQUILIBRIUM CONSTANT

Chemical reactions, such as the formation of hydrogen iodide from hydrogen and iodine in the gaseous phase,

$$H_2(g) + I_2(g) \rightleftharpoons 2HI(g)$$

are generally *reversible*, and when the rates of the forward and reverse reactions are equal, the concentrations of the reactants and products remain constant with time. We say that the reaction has reached a state of *equilibrium*.

It is found experimentally that the extent to which reactions are complete when equilibrium is reached varies tremendously. In some cases the concentrations of products are much larger than those of reactants; in other cases the reverse is true. The equilibrium concentrations reflect the intrinsic tendencies of the atoms to exist as molecules of reactant or product. Although there may be an infinite number of concentrations which satisfy the equilibrium condition, there is only one general expression which is found to be constant at a given temperature for a reaction at equilibrium. For the general reaction in aqueous solution,

$$A(aq + B(aq) \rightleftharpoons C(aq) + D(aq)$$

this expression is

$$\frac{[C]_e[D]_e}{[A]_e[B]_e} = K$$

and is called the *equilibrium constant*. Here the brackets designate concentrations in moles per liter (molarity) at equilibrium. The fraction is often called the *mass action* expression, from the work of Guldberg and Waage (1864), who used the term *active mass* to denote the amount of chemical reactivity. These workers were not clear about the meaning of active masses. Chemists today use the term *activity* as the best meaning of active mass, and *concentration* as an approximation when activity is not known. The concentrations employed are usually molality or molarity for reactions in aqueous solution. For gases, molarity or partial pressure is employed. We shall discuss activity in more detail later.

For the completely general reaction

$$aA(aq) + bB(aq) \rightleftharpoons cC(aq) + dD(aq)$$

the equilibrium constant is

$$\frac{[C]_e^c[D]_e^d}{[A]_e^a[B]_e^b} = K$$

The exponents in this expression are the coefficients of the reactants and products in the balanced equation for the reaction. Table 5.1 contains the equilibrium ex-

TABLE 5.1 Equilibrium Constants for Various Reactions

REACTION	EQUILIBRIUM EXPRESSION	NUMERICAL VALUE AT 25°C
$HOAc(aq) + H_2O \rightleftharpoons H_3O^+ + OAc_c^-$	$\dfrac{[H_3O^+][OAc^-]}{[HOAc]}$	1.8×10^{-5}
$Al(OH)_3(s) \rightleftharpoons Al^{3+} + 3OH^-$	$[A]^{3+}[OH^-]^3$	5×10^{-33}
$Ag^+ + 2NH_3(aq) \rightleftharpoons Ag(NH_3)_2^+$	$\dfrac{[Ag(NH_3)_2^+]}{[Ag^+][NH_3]^2}$	1.4×10^7
$Sn^{2+} + 2Ce^{4+} \rightleftharpoons Sn^{4+} + 2Ce^{3+}$	$\dfrac{[Sn^{4+}][Ce^{3+}]^2}{[Sn^{2+}][Ce^{4+}]^2}$	5.4×10^{43}
$HOAc + OH^- \rightleftharpoons OAc^- + H_2O$	$\dfrac{[OAc^-]}{[HOAc][OH^-]}$	1.8×10^9
$Ag^+ + Cl^- \rightleftharpoons AgCl(s)$	$\dfrac{1}{[Ag^+][Cl^-]}$	1×10^{10}

pressions for several reactions of the type we will encounter later in our discussions. Note that if a reactant is a solid or liquid, its concentration does not appear in the equilibrium expression. The reason is that the concentration of a solid or liquid is constant. Increasing the amount of a solid or liquid in a reacting system does not change its concentration. The number of moles increases, but the volume also increases, and the number of moles per liter is unchanged.

5.2

AQUEOUS SOLUTIONS

Water is one of the most plentiful compounds in nature and is essential to life processes. It dissolves many substances and serves as the medium in which a wide variety of chemical reactions take place. Most of the analytical processes we will discuss involve reactions in aqueous solutions.

5.2a. *Weak and Strong Electrolytes*

Aqueous solutions of certain compounds are good conductors of an electric current because of the presence of positive and negative ions. Such compounds are called *electrolytes,* whereas compounds whose aqueous solutions do not conduct current are called *nonelectrolytes*. Sodium chloride, NaCl, is completely dissociated into Na^+ and Cl^- ions in aqueous solution and is an electrolyte. Ethylene glycol, CH_2OH—CH_2OH, a common antifreeze, is not dissociated in aqueous solution and is an example of a nonelectrolyte. Most *ionic* compounds are completely dissociated in water and are called *strong* electrolytes. Many *covalent* compounds dissociate to only a slight degree when dissolved in water and are called *weak* electrolytes. Acetic acid (CH_3COOH, abbreviated HOAc), ammonia, and water are examples of weak electrolytes. Table 5.2 gives a short list of the most common weak and strong acids and bases encountered in the introductory course in quantitative analysis.

TABLE 5.2 Strong and Weak Acids and Bases in Aqueous Solution

ACIDS		BASES	
Strong	Weak	Strong	Weak
HCl	HOAc	NaOH	NH_3
HNO_3	HF	KOH	CH_3NH_2
$HClO_4$	H_2CO_3	$Ba(OH)_2$	$C_6H_5NH_2$ (aniline)
H_2SO_4	H_2S	$(CH_3)_4NOH$	

We write equilibrium constants for the dissociation of weak electrolytes into ions and call them *dissociation* constants. For example, the dissociation of water is written

$$2H_2O \rightleftharpoons H_3O^+ + OH^- \qquad K_w = [H_3O^+][OH^-]$$

For acetic acid and ammonia we write

$$HOAc + H_2O \rightleftharpoons H_3O^+ + OAc^- \qquad K_a = \frac{[H_3O^+][OAc^-]}{[HOAc]}$$

$$NH_3 + H_2O \rightleftharpoons NH_4^+ + OH^- \qquad K_b = \frac{[NH_4^+][OH^-]}{[NH_3]}$$

The subscripts are used for convenience. K_a is used as the symbol for the dissociation constant of a weak acid; K_b is used for a weak base. K_w refers to the dissociation of water and is sometimes called the *autoprotolysis* constant of water.

Salts

The term *salt* deserves special comment. A salt is the product other than water formed when an acid reacts with a base. For example, when hydrochloric acid and sodium hydroxide react, the products are a salt (sodium chloride) and water. Written molecularly, this is

$$HCl + NaOH \longrightarrow NaCl + H_2O$$

Written in ionic form, this is

$$H^+ + Cl^- + Na^+ + OH^- \longrightarrow Na^+ + Cl^- + H_2O$$

The *net* ionic reaction is (H^+ is hydrated in water)

$$H_3O^+ + OH^- \longrightarrow 2H_2O$$

Note that the salt, $NaCl$, is not a molecule, but is a strong electrolyte, completely dissociated into Na^+ and Cl^- ions. Most salts are strong electrolytes, although sometimes we write their formulas molecularly, as Na_2SO_4, KNO_3, $CaCl_2$, etc. A few salts are weak electrolytes; one we will encounter later is mercury(II) chloride, $HgCl_2$. The reaction

$$Hg^{2+} + 2Cl^- \longrightarrow HgCl_2$$

is sufficiently complete at the equivalence point that it can be used for the titration of chloride ions.

Reactions of Salts of Weak Electrolytes

Salts of weak electrolytes react with water to produce either hydrogen or hydroxide ions. For example, an aqueous solution of the salt sodium acetate, NaOAc, is basic because the acetate ion reacts with water to produce hydroxide ions (hydrolysis):

$$OAc^- + H_2O \rightleftharpoons HOAc + OH^-$$

The reason that this reaction occurs to an appreciable extent is that HOAc is a weak acid and prefers to remain as molecules. The Na^+ ion does not react with water to produce NaOH molecules and H_3O^+ ions, since NaOH is a strong base and prefers to remain as Na^+ and OH^- ions. The salt ammonium chloride, NH_4Cl, is acidic, since ammonium ions react with water to produce H_3O^+ ions:

$$NH_4^+ + H_2O \rightleftharpoons NH_3 + H_3O^+$$

The Cl^- ion does not react with water to produce HCl and OH^- ions, since HCl is a strong acid.

The equilibrium constant for the reaction of acetate ion with water is

$$\frac{[HOAc][OH^-]}{[OAc^-]} = K_b$$

The symbol K_b is used to indicate that the acetate ion is a base; it is said that the acetate ion is the *conjugate* base of acetic acid. (See discussion of Brønsted theory, page 129.) There is a simple relation between the K_a of an acid and the K_b of its conjugate base. This can be seen by multiplying the two expressions:

$$\frac{[H_3O^+][\cancel{OAc^-}]}{[\cancel{HOAc}]} \times \frac{[\cancel{HOAc}][OH^-]}{[\cancel{OAc^-}]} = [H_3O^+][OH^-] = K_w$$

or

$$K_a \times K_b = K_w$$

The same relation holds for the weak base NH_3 and its conjugate acid, NH_4^+, or for any such conjugate pair.

In discussing solutions containing salts such as sodium acetate or ammonium chloride, chemists sometimes refer to the ion OAc^- or NH_4^+ as the salt. It is understood that the salt is NaOAc or NH_4Cl. Since the acetate or ammonium ions undergo reaction with water molecules, they are the species of interest in equilibrium calculations of the hydrogen ion concentration.

5.3

ACTIVITY AND ACTIVITY COEFFICIENTS

It has been previously mentioned that the equilibrium expression is not strictly a constant if concentrations, such as molarity or molality, are used as the measure of "active masses" of the reacting species. Strong electrolytes, such as NaCl, though completely dissociated into Na^+ and Cl^- ions in the solid state, do not behave as though the ions are independent particles in aqueous solutions. Physical properties of salt solutions, such as conductivity and freezing point, suggest that the ions may be "clustered" together, with positive ions having more negative than positive ions in their immediate neighborhood, and negative ions in turn having an excess of positive ions around them. Under such conditions, the effectiveness of ions in determining the rate of chemical reactions, and also in altering physical properties of the solvent, is less than it would be were each ion capable of acting independently. Only in very dilute solutions are ions sufficiently free of the influence of neighboring ions to act as independent particles.

In order to obtain agreement between experimental and theoretical equilibrium calculations, the chemist multiplies *actual concentrations* (molarities, for example) by certain numbers, called *activity coefficients,* to obtain *effective concentrations,* called *activities.* The activity of a species A is defined as follows:

$$a_A = f_A[A]$$

where a_A is the activity, f_A the activity coefficient, and [A] the molarity of species A. For example, the activity of the hydronium ion is

$$a_{H_3O^+} = f_{H_3O^+}[H_3O^+]$$

and that of the hydroxide ion is

$$a_{OH^-} = f_{OH^-}[OH^-]$$

The true constant for the dissociation of water, K_w, is

$$K_w = a_{H_3O^+} \times a_{OH^-}$$

Either the activity or the activity coefficient could be made dimensionless, but generally it is the activity coefficient which is so treated. Hence activity is expressed in the same units as concentration. In aqueous solutions this is usually moles per liter. From the definition it can be seen that the more ideally a solution behaves, i.e., the closer activity is to concentration, the closer the activity coefficient is to unity. At infinite dilution $f_A = 1$ and $a_A = [A]$.

There is an enormous amount of experimental evidence to indicate that electrostatic interaction between ions and between ions and solvent molecules can cause large differences between activity and concentration. Such interactions are general, not just between ions undergoing chemical reaction. For example, many compounds show increased solubility in the presence of electrolytes containing no ions which react chemically with ions of the precipitate. The solubility product constant, K_{sp}, of $BaSO_4$,

$$K_{sp} = [Ba^{2+}][SO_4^{2-}]$$

increases from 1.0×10^{-10} in water to 2.9×10^{-10} in 0.010 M aqueous KNO_3 at 25°C. The effect is caused by electrostatic interactions of the ions. Clustering of NO_3^- ions about Ba^{2+} and of K^+ ions about SO_4^{2-} tends to shield Ba^{2+} and SO_4^{2-} ions from each other and to hamper their effectiveness in forming $BaSO_4$.

The ion product constant of water, $[H_3O^+][OH^-]$, for the dissociation

$$2H_2O \rightleftharpoons H_3O^+ + OH^-$$

increases from 1.01×10^{-14} in pure water to 1.65×10^{-14} in 0.10 M NaCl at 25°C. Again the effect is electrostatic, Na^+ and Cl^- ions reducing the effectiveness of H_3O^+ and OH^- ions in reacting to form H_2O.

In addition to the concentration of the added electrolyte, the magnitudes of the charges on the ions determine the degree of electrostatic interaction with other ions in solution. For example, an ion of charge $+2$ will reduce the activity of a substance to a larger degree than will an equal concentration of an ion of charge $+1$. The term now used as a measure of the magnitude of the electrostatic environment of a solute is *ionic strength*, defined by the equation

$$\mu = \frac{1}{2} \sum_i C_i Z_i^2$$

where C_i is the molar concentration and Z_i the charge of each ionic species in the solution. Thus doubly charged ions, such as Mg^{2+} and SO_4^{2-}, have $(2)^2$ or 4 times as much effect as singly charged ions such as Na^+ and Cl^- in lowering the activity of a reacting species.

We may now summarize the facts known about nonideal behavior of electrolytes by listing the properties of activity coefficients.

1. In electrolyte solutions activity coefficients first decrease as the ionic strength increases from infinite dilution, then pass through a fairly flat minimum in the region of 0.4 to 1.0 ionic strength, and then increase to values often greater than 1 at high ionic strengths. Apparently, as the electrolyte concentration is increased to large values, more and more water molecules are bound to the solute ions, leaving fewer free or unbound water molecules to act as the solvent. The activity may then become greater than the concentration, and the activity coefficient greater than 1. In examples we shall be considering in the text the ionic strength will be sufficiently small so that activity coefficients can be expected to be less than 1.
2. It is the ionic strength, not the nature of the electrolyte in the solution, that determines the value of the activity coefficient.
3. The activity coefficient is dependent on the magnitude of the electrical charge

on the ion, but not on its sign. At a given ionic strength the activity coefficient of a +2 ion will be farther from unity than that for a +1 ion.

4. Activity coefficients depend upon the size of the hydrated ions. Generally, smaller ions show greater departure from ideal behavior than do larger ones at the same ionic strength.

5. In general the presence of ions will have a lesser effect upon the activity of a neutral molecule than upon that of another electrolyte. Ions do influence molecules to some degree by interacting with existing dipoles, or even inducing them. But it is reasonable to take the activity coefficient of a neutral molecule as unity under normal solution conditions.

5.3a. The Debye-Hückel Limiting Law

In 1923, Peter Debye and Erich Hückel presented a theory of interionic-attraction effects which enables one to calculate activity coefficients theoretically. They assumed that ions were point charges at relatively large distances from each other, that the dielectric constant of the electrolyte solution was independent of the concentration of the solute, and that the dielectric constant of water could be used in all calculations. From their treatment the equation

$$-\log f_i = AZ_i^2\sqrt{\mu}$$

is derived, where f_i is the activity coefficient of a single ionic species, such as Na^+ or Cl^-, Z_i the charge of the ion, μ the ionic strength of the solution, and A a constant. For water at 25°C, A, a collection of constants including temperature, the dielectric constant, and Avogadro's number, is 0.512. The equation has been of great value in providing a law for extrapolation to zero concentration and is commonly referred to as the *Debye-Hückel limiting law*.

It is not possible to test the above equation experimentally, since it is impossible to prepare a solution which contains only a single ionic species. The *mean activity coefficient* for a 1:1 type electrolyte is defined by the equation

$$f_\pm = \sqrt{f_+f_-}$$

This quantity can be measured by various physicochemical techniques and has been found to agree satisfactorily with values calculated by the Debye-Hückel equation. For example, the activity coefficient calculated for a 0.01 M solution of HCl is 0.89, and the mean value measured experimentally is 0.90.

For a binary electrolyte A_mB_n the Debye-Hückel equation is written

$$-\log f_\pm = 0.512Z_A Z_B\sqrt{\mu}$$

where Z_A and Z_B are the charges on the cation and anion taken without regard to sign.

5.3b. The Extended Debye-Hückel Equation

The simple Debye-Hückel equation, or limiting law, has been found to give useful results only in very dilute solutions. As the ionic strength increases, mean activity coefficients calculated from the equation are significantly smaller than the experimental values. For example, the calculated value for the mean activity coefficient in 0.10 M HCl is 0.69, whereas the experimental value is 0.80.

Debye and Hückel later extended their theoretical treatment, assuming that ions are not point charges, but have significant size, and that they are present in solutions of greater ionic strength than originally assumed. They derived the so-called *extended* Debye-Hückel equation,

$$-\log f_{\pm} = \frac{0.512 Z_A Z_B \sqrt{\mu}}{1 + aB\sqrt{\mu}}$$

where a is the diameter of the ion in ångstrom units ($1\ \text{Å} = 10^{-10}$ m), and B a term which depends on the absolute temperature and the dielectric constant of the solvent. Its value at 298°K using the dielectric constant for pure water (78.5) is 0.328.

The term a, sometimes called the *ion-size parameter*, is said to correspond to the distance of closest approach of the cation and anion. There is considerable uncertainty about the magnitude of this term, but it seems to be about 3 Å for most singly charged ions and as large as 11 Å for ions of higher charge. Table 5.3 gives some activity coefficients calculated from the extended Debye-Hückel equation and includes values of the "effective" diameter used in the calculations. To calculate the mean activity coefficient for a binary electrolyte, such as HCl, the values of a are averaged. If a value of a of 6 Å ($a = 9$ for H_3O^+ and 3 for Cl^-) is used, the calculated value of $f_{\pm}$ for HCl by the extended equation is 0.79, in good agreement with the experimental value of 0.80.

TABLE 5.3 Activity Coefficients Calculated from the Extended Debye-Hückel Equation

ION	SIZE a(Å)	IONIC STRENGTH				
		0.0010	0.0050	0.010	0.050	0.10
H_3O^+	9	0.967	0.933	0.913	0.853	0.825
Li^+	6	0.966	0.929	0.906	0.833	0.795
Na^+, OAc^-, IO_3^-, HCO_3^-	4	0.965	0.927	0.901	0.816	0.768
NH_4^+, K^+, Rb^+, Ag^+						
F^-, Cl^-, Br^-, I^-, OH^-, NO_3^-, CN^-, ClO_4^-, ClO_3^-, SCN^-, HS^-, HCO_2^-	3	0.964	0.925	0.898	0.806	0.753
Mg^{2+}	8	0.871	0.755	0.688	0.513	0.443
Ca^{2+}, Cu^{2+}, Sn^{2+}, Zn^{2+}, Fe^{2+}	6	0.869	0.746	0.674	0.481	0.399
Sr^{2+}, Ba^{2+}, Cd^{2+}, Hg^{2+}, Pb^{2+}, CO_3^{2-}, SO_3^{2-}	5	0.868	0.742	0.667	0.462	0.375
Hg_2^{2+}, SO_4^{2-}, CrO_4^{2-}, $S_2O_3^{2-}$	4	0.867	0.737	0.659	0.443	0.349
Al^{3+}, Cr^{3+}, Fe^{3+}, Ce^{3+}, La^{3+},	9	0.736	0.538	0.441	0.240	0.176
$Cr(NH_3)_6^{3+}$, PO_4^{3-}, $Fe(CN)_6^{3-}$	4	0.725	0.503	0.391	0.160	0.093
Ce^{4+}, Sn^{4+}, Th^{4+}	11	0.585	0.346	0.250	0.097	0.062
$Fe(CN)_6^{4-}$	5	0.567	0.303	0.198	0.046	0.020

Activity coefficients calculated by the extended Debye-Hückel equation using the values of a given in Table 5.3 are in good agreement with experimentally measured mean values up to ionic strengths of about 0.10. However, the equation does not predict the observed minimum in experimental values in the region of 0.4 to 1.0 ionic strength. For solutions of higher ionic strength, a number of modifications of the Debye-Hückel equation have been proposed. Most of these have an empirical correction term added to or subtracted from the equation. The one proposed by Davies,[1]

[1] C. W. Davies, *Ion Association*, Butterworths, London, 1962.

$$-\log f_{\pm} = 0.512 Z_A Z_B \left(\frac{\sqrt{\mu}}{1 + \sqrt{\mu}} - 0.2\mu \right)$$

is useful for obtaining approximate values of activity coefficients up to ionic strengths of about 0.60.

5.3c. The Thermodynamic Equilibrium Constant

In thermodynamic terms, the true equilibrium constant (designated $K°$) for the general reaction

$$a\mathrm{A} + b\mathrm{B} \rightleftharpoons c\mathrm{C} + d\mathrm{D}$$

is expressed in terms of activities,

$$K° = \frac{(a_C)^c (a_D)^d}{(a_A)^a (a_B)^b}$$

Since we have defined the activity of a species by the expression

$$a_A = f_A[\mathrm{A}]$$

we can write

$$K° = \frac{(f_C[\mathrm{C}])^c (f_D[\mathrm{D}])^d}{(f_A[\mathrm{A}])^a (f_B[\mathrm{B}])^b} = \frac{[\mathrm{C}]^c [\mathrm{D}]^d}{[\mathrm{A}]^a [\mathrm{B}]^b} \cdot \frac{f_C^c f_D^d}{f_A^a f_B^b}$$

or

$$K° = K \cdot \frac{f_C^c f_D^d}{f_A^a f_B^b}$$

Here K represents the common equilibrium constant in terms of concentrations, and it is often referred to as the "concentration quotient." Since $K°$ is a true constant and since the activity coefficients of reactants and products change with the ionic strength of a solution, K is not strictly constant. However, at infinite dilution the activity coefficients approach unity, and K does become equal to $K°$.

In cases where activity coefficients are known, equilibrium calculations should be made using the thermodynamic value of the equilibrium constant. However, it is rare that activity coefficients are known in the complex, concentrated solutions frequently encountered in analytical chemistry. Therefore, in most calculations in the text we shall use molarities as an approximation of activities and shall assume that K is constant. Fortunately, many of the answers we shall be seeking regarding, for example, the feasibility of a titration, involve *relative* changes in equilibrium concentrations and are not greatly affected by neglecting activities. We shall make a practice, however, of reminding the student that activities should be used in equilibrium calculations, and we shall point out examples where activity effects may appreciably affect the answer we are seeking.

5.3d. Standard States

So far our discussion has been confined to reactants which are in aqueous solution, and we have decided to use molarity as an approximation of the activity of such substances. Many of the reactions we encounter in analytical chemistry involve reactants which are gases, liquids, or solids, and we shall need to make some provision for the activities of such substances. In thermodynamics the relation between activity and free energy is given by the equation

$$\Delta G = 2.3RT \log \frac{a_2}{a_1}$$

where ΔG is the free energy change for the transfer of 1 mol of a substance from a state of activity a_1 to one of activity a_2. The free energy of a chemical species depends not only upon the nature and quantity of the substance but also upon temperature and pressure. Hence it is customary to adopt an arbitrary reference or *standard state* and to assign to it an activity of *unity;* then values for the absolute free energy of an element or the free energy of formation of a compound can be calculated and tabulated.

The customary choices for standard states, all at 25°C, are as follows:

1. The activity of a solute is the same as its molality in very dilute solution, where ideal behavior may be assumed. That is,

$$\frac{a}{m} = 1 \qquad \text{when } m \longrightarrow 0$$

where m is the molality. Since molality and molarity are very nearly the same in dilute aqueous solutions, we shall use molarity rather than molality as an approximation of activity. The standard state is a hypothetical one in which the solute is at 1 M concentration, but the environment about the solute is the same as that in an ideal solution.

2. For a perfect gas the standard state is 1 atm, and the activity is then the same as the pressure of the gas. For a real gas the standard state is that in which the so-called fugacity[2] is unity. Since at low pressures a real gas approaches ideal behavior, making fugacity and pressure approximately equal, we will take the pressure of a gas as its activity. Thus

$$\frac{a}{P} = 1 \qquad \text{when } P \longrightarrow 0$$

3. The activity of a pure liquid or solid (in its most stable crystalline state) acting as a solvent for other substances is unity. That is, the standard state is a mole fraction of unity, where X is the mole fraction of the solvent.

$$\frac{a}{X} = 1 \qquad \text{when } X \longrightarrow 1$$

If the activity of the liquid or solid is changed by dissolving in it a solute, the activity of the solvent is still given by the mole fraction. In most examples that we shall encounter, it will still be acceptable to take a value of unity as the activity of the solvent. For example, a liter of a 0.1 M aqueous solution of a solute contains 0.1 mol of that solute and about 55.3 mol of water. The mole fraction of water is thus about $55.3/55.4 \cong 1$. The possible effect of the solute upon the activity of water will be ignored in our calculations.

The following example illustrates an equilibrium calculation using the conventions described above.

[2] The fugacity is the same as the vapor pressure when the vapor is a perfect gas, and it may be regarded as an "ideal" or "corrected" vapor pressure.

EXAMPLE 5.1

Use of activities to calculate [H⁺]

The equilibrium constant of the reaction

$$2AgI(s) + H_2(g) \rightleftharpoons 2Ag(s) + 2H^+_{aq} + 2I^-_{aq}$$

is 8.2×10^{-6} at 25°C.

Calculate the hydrogen ion concentration in a system at equilibrium if the pressure of H_2 gas is 0.50 atm and the concentration of I^- is 0.10 M.

The equilibrium expression is

$$K° = \frac{(a_{Ag})^2(a_{H^+})^2(a_{I^-})^2}{(a_{AgI})^2(a_{H_2})} = 8.2 \times 10^{-6}$$

According to our conventions,

$a_{Ag} = 1$, since silver is a pure solid

$a_{H^+} = [H^+]$, since this is a soluble electrolyte

$a_{I^-} = 0.10$, since this is a soluble electrolyte

$a_{AgI} = 1$, since silver iodide is a pure solid

$a_{H_2} = 0.50$ atm, since this is the pressure of H_2 gas

Substituting,

$$\frac{(1)^2[H^+]^2(0.10)^2}{(1)^2(0.50)} = 8.2 \times 10^{-6}$$

$$[H^+] = 0.020 \ M \qquad \square$$

The following examples illustrate calculations of ionic strength, activity coefficients from the Debye-Hückel and Davies equations, and the equilibrium constant using activity coefficients.

EXAMPLE 5.2

Ionic strength

Calculate the value of the ionic strength of these solutions: (a) 0.10 M NaCl; (b) 0.10 M Na_2SO_4; (c) 0.10 M $MgSO_4$; (d) 0.10 M NaCl + 0.10 M $MgSO_4$.

By definition

$$\mu = \frac{1}{2} \sum_i C_i Z_i^2$$

(a) For NaCl, $[Na^+] = 0.10 \ M$ and $[Cl^-] = 0.10 \ M$. Hence

$$\mu = \tfrac{1}{2}(C_{Na^+} \cdot Z_{Na^+}^2 + C_{Cl^-} \cdot Z_{Cl^-}^2)$$

$$\mu = \tfrac{1}{2}[0.10(1)^2 + 0.10(-1)^2]$$

$$\mu = 0.10$$

(b) For Na_2SO_4, $[Na^+] = 0.20 \ M$ and $[SO_4^{2-}] = 0.10 \ M$. Hence

$$\mu = \tfrac{1}{2}[0.20(1)^2 + 0.10(-2)^2]$$

$$\mu = 0.30$$

(c) For $MgSO_4$, $[Mg^{2+}] = 0.10 \ M$ and $[SO_4^{2-}] = 0.10 \ M$. Hence

$$\mu = \tfrac{1}{2}[0.10(2)^2 + 0.10(-2)^2]$$

$$\mu = 0.40$$

(d) In solutions containing more than one salt where no chemical reaction occurs, the ionic strength is the sum of the values calculated individually; here $0.10 + 0.40 = 0.50$. Alternatively, the value can be calculated from the total concentration and charges of all ions. $\qquad \square$

EXAMPLE 5.3

Activity coefficients

Calculate the mean activity coefficient for $CaCl_2$ in a 0.300 M solution of the salt using (a) the Debye-Hückel limiting law, (b) the extended Debye-Hückel equation, and (c) the Davies equation. The experimental value is 0.455.

$[Ca^{2+}] = 0.300$ M and $[Cl^-] = 0.600$ M. The ionic strength is

$$\mu = \tfrac{1}{2}[0.300(2)^2 + 0.600(-1)^2]$$

$$\mu = 0.900$$

(a) Using the Debye-Hückel limiting law, we have

$$-\log f_{\pm} = 0.512 Z_{Ca^{2+}} \cdot Z_{Cl^-} \sqrt{\mu}$$

$$-\log f_{\pm} = 0.512(2)(1)\sqrt{0.90}$$

$$f_{\pm} = 0.107$$

(b) Using the extended Debye-Hückel equation, we have

$$-\log f_{\pm} = \frac{0.512 Z_A Z_B \sqrt{\mu}}{1 + a \cdot 0.328\sqrt{\mu}}$$

and taking a as 4.5 Å (the average of 6 Å for Ca^{2+} and 3 Å for Cl^- from Table 5.3), we have

$$-\log f_{\pm} = \frac{0.512(2)(1)\sqrt{0.90}}{1 + (4.5)(0.328)(\sqrt{0.90})}$$

$$f_{\pm} = 0.394$$

(c) Using the Davies equation, we have

$$-\log f_{\pm} = 0.512 Z_A Z_B \left\{ \frac{\sqrt{\mu}}{1 + \sqrt{\mu}} - 0.2\mu \right\}$$

$$-\log f_{\pm} = 0.512(2)(1) \left\{ \frac{\sqrt{0.90}}{1 + \sqrt{0.90}} - 0.2 \cdot 0.90 \right\}$$

$$f_{\pm} = 0.485 \qquad \qquad \square$$

EXAMPLE 5.4

Equilibrium constant

Calculate the value of K_w for water in 0.100 M NaCl solution. The experimental value is 1.65×10^{-14}.

The thermodynamic value of the equilibrium constant for the dissociation of water

$$2H_2O \rightleftharpoons H_3O^+ + OH^-$$

is

$$K_w = \frac{a_{H_3O^+} \cdot a_{OH^-}}{(a_{H_2O})^2} = 1.01 \times 10^{-14}$$

Water is a liquid acting as a solvent, and its activity is its mole fraction, here $55.3/55.4 \cong 1$ (page 101). Then

$$K_w^{\circ} = [H_3O^+][OH^-] \cdot f_{H_3O^+} \cdot f_{OH^-}$$

$$K_w^{\circ} = K_w \cdot f_{H_3O^+} \cdot f_{OH^-}^{\circ}$$

where K_w is the constant in terms of concentrations.

We can calculate the individual activity coefficients from the extended Debye-Hückel equation and the data in Table 5.3, giving $f_{H_3O^+} = 0.825$ and

$f_{OH^-} = 0.753$. Using these values, we obtain

$$K_w = \frac{1.01 \times 10^{-14}}{0.825 \times 0.753}$$

$$K_w = 1.61 \times 10^{-14} \qquad \square$$

5.4

EQUILIBRIUM CALCULATIONS

We shall now review briefly the formulation of equilibrium constants for the four types of chemical reactions used in titrimetric analyses. We shall also examine some typical equilibrium calculations which are of interest in analytical chemistry.

5.4a. Acid-Base Equilibria

We have already mentioned that water is a weak electrolyte, dissociating into H_3O^+ and OH^- ions:

$$2H_2O \rightleftharpoons H_3O^+ + OH^-$$

The extent of dissociation of water has been measured experimentally, and at 25°C the hydronium and hydroxide ion concentrations have been found to be 1.0×10^{-7} M. This means that the value of K_w, the autoprotolysis constant of water, is 1.0×10^{-14} at 25°C:

$$K_w = [H_3O^+][OH^-]$$

$$K_w = (1.0 \times 10^{-7})(1.0 \times 10^{-7})$$

$$K_w = 1.0 \times 10^{-14}$$

The value of K_w at several other temperatures is shown in Table 5.4.

TABLE 5.4 Ion Product Constant, K_w, of Water at Different Temperatures

TEMPERATURE, °C	K_w	pK_w
0	1.14×10^{-15}	14.943
15	4.51×10^{-15}	14.346
25	1.01×10^{-14}	13.996
35	2.09×10^{-14}	13.680
50	5.47×10^{-14}	13.262
100	$4.9 \ \times 10^{-13}$	12.31

The term pH is convenient for expressing hydrogen ion concentrations, since the latter values are very small and may vary over many orders of magnitude during titrations. Sorensen, in 1909, defined pH as the negative logarithm of the hydrogen ion concentration. It was later realized that the electromotive force (emf) of a galvanic cell used to measure pH (Chapter 12) was dependent more upon the *activity* of hydrogen ion than upon the *concentration*. Hence the proper definition of pH is

$$pH = -\log a_{H_3O^+} = -\log f_{H_3O^+}[H_3O^+]$$

For simplicity we will use the approximation

$$pH = -\log [H_3O^+] = \log \frac{1}{[H_3O^+]}$$

Some texts label this approximation p_CH. In Chapter 12 the potentiometric determination of pH is discussed in detail.

Note that pH is defined in such a way as to convert a negative power of ten into a small positive number. Thus a hydrogen ion concentration of 1.0×10^{-1} corresponds to a pH value of 1.00, and a value of 1.0×10^{-13} becomes $pH = 13.00$. Such numbers, ranging from, say, 0 or 1 up to perhaps 13 or 14, are conveniently plotted on titration curves. We shall consistently express pH values to two decimal places for reasons discussed later.

The following examples illustrate the conversion of hydrogen ion concentration to pH, and vice versa.

EXAMPLE 5.5

Calculation of pH from hydrogen ion concentration

The hydrogen ion concentration of a solution is 5.0×10^{-7}. Calculate the pH.

$$pH = -\log (5.0 \times 10^{-7})$$
$$pH = -(\log 5.0 + \log 10^{-7})$$
$$pH = -(0.70 - 7.00)$$
$$pH = 7.00 - 0.70$$
$$pH = 6.30 \qquad \square$$

EXAMPLE 5.6

Calculation of hydrogen ion concentration from pH

The pH of a solution is 10.70. Calculate the hydrogen ion concentration.

$$pH = 10.70 = -\log[H_3O^+]$$
$$\log[H_3O^+] = -10.70 = 0.30 - 11.00$$
$$[H_3O^+] = \text{antilog } 0.30 \times \text{antilog}(-11)$$
$$[H_3O^+] = 2.0 \times 10^{-11} \qquad \square$$

It is often convenient to use other p-functions analogous to pH. For example, the following functions are frequently used:

$$pOH = -\log[OH^-]$$
$$pK_a = -\log K_a$$
$$pK_b = -\log K_b$$
$$pK_w = -\log K_w$$

Note that since $[H_3O^+][OH^-] = K_w = 1.0 \times 10^{-14}$ at 25°C, then

$$pH + pOH = pK_w = 14.00$$

Also note that in pure water (neutral), $[H_3O^+] = [OH^-]$ and

$$pH = pOH = \frac{pK_w}{2}$$

At 25°C the values of the pH and pH are 7.00 in neutral solutions. In acidic solutions $[H_3O^+] > 10.0 \times 10^{-7}$ and $pH < 7.00$. In basic solutions $[H_3O^+] < 1.0 \times 10^{-7}$ and $pH > 7.00$.

The following examples illustrate calculations involving acid-base equilibria.

EXAMPLE 5.7

Calculation of K_a

In 0.100 M solution at 25°C acetic acid is found to be 1.34% dissociated. Calculate the value of K_a for HOAc.

The dissociation reaction is

$$\text{HOAc} + \text{H}_2\text{O} \rightleftharpoons \text{H}_3\text{O}^+ + \text{OAc}^-$$

The concentrations of hydrogen and acetate ions are approximately equal, since these ions are formed in equimolar amounts. (We neglect the small amount of hydrogen ion formed by the dissociation of water.) Hence

$$[\text{H}_3\text{O}^+] \cong [\text{OAc}^-] = 0.100 \text{ mmol/mL} \times 0.0134$$

$$[\text{H}_3\text{O}^+] \cong [\text{OAc}^-] = 0.00134 \text{ mmol/mL}$$

The concentration of undissociated acetic acid molecules is the original concentration minus the amount dissociated:

$$[\text{HOAc}] = 0.100 - 0.00134$$

$$[\text{HOAc}] \cong 0.0987 \cong 0.10 \text{ mmol/mL}$$

Hence

$$K_a = \frac{[\text{H}_3\text{O}^+][\text{OAc}^-]}{[\text{HOAc}]}$$

$$K_a = \frac{(0.00134)(0.00134)}{0.10}$$

$$K_a = 1.8 \times 10^{-5} \qquad \qquad \square$$

EXAMPLE 5.8

Calculation of pH and pOH from concentration

Calculate the pH and pOH of a 0.050 M solution of HCl. HCl is a strong acid, completely dissociated. Hence

$$[\text{H}_3\text{O}^+] = 0.050 \quad \text{or} \quad 5.0 \times 10^{-2}$$

$$pH = 2.00 - \log 5.0$$

$$pH = 1.30$$

Since $pH + pOH = 14.00$, $pOH = 12.70$. $\qquad \qquad \square$

EXAMPLE 5.9

Calculation of pH of a weak acid; use of the quadratic equation

Calculate the pH of a 0.050 M solution of acetic acid. The K_a of HOAc is 1.8×10^{-5}.

This is a weak acid and is only slightly dissociated. We can calculate the hydrogen ion concentration using the dissociation constant.

$$\text{HOAc} + \text{H}_2\text{O} \rightleftharpoons \text{H}_3\text{O}^+ + \text{OAc}^-$$

$$[\text{H}_3\text{O}^+] \cong [\text{OAc}^-]$$

$$[\text{HOAc}] = 0.050 - [\text{H}_3\text{O}^+]$$

Substituting in the equilibrium expression,

$$\frac{[\text{H}_3\text{O}^+][\text{OAc}^-]}{[\text{HOAc}]} = K_a$$

$$\frac{[\text{H}_3\text{O}^+]^2}{0.050 - [\text{H}_3\text{O}^+]} = 1.8 \times 10^{-5}$$

This is a quadratic equation which can be written in the form

$$[H_3O^+]^2 + 1.8 \times 10^{-5}[H_3O^+] - 9.0 \times 10^{-7} = 0$$

It can be solved using the "quadratic formula." For the equation

$$ax^2 + bx + c = 0$$

the solution for x is given by the formula

$$x = \frac{-b \pm \sqrt{b^2 - 4ac}}{2a}$$

Here $x = [H_3O^+]$, $a = 1$, $b = 1.8 \times 10^{-5}$, and $c = -9.0 \times 10^{-7}$. Substituting in the above formula gives

$$[H_3O^+] = 9.4 \times 10^{-4}\ M$$

$$pH = 3.03$$

The solution can be simplified by recognizing that the acid is weak and the amount of H_3O^+ produced is small with respect to the total amount of HOAc present. We assume

$$[HOAc] = 0.050 - [H_3O^+] \cong 0.050$$

Then the equation becomes

$$\frac{[H_3O^+]^2}{0.050} = 1.8 \times 10^{-5}$$

$$[H_3O^+] = 9.5 \times 10^{-4}\ M$$

$$pH = 3.02$$

The percentage error in this approximation is

$$\frac{(9.5 - 9.4) \times 10^{-4}}{9.4 \times 10^{-4}} \times 100 = 1.1\%$$

Instead of the quadratic formula the method of *successive approximations* can be used. In using this method one first drops the $[H_3O]^+$ term in the denominator just as we did above, giving as a *first* approximation the value 9.5×10^{-4}. This value is then inserted in the denominator of the original equation, and the equation is solved to give a *second* approximation,

$$\frac{[H_3O^+]^2}{0.050 - 0.00095} = 1.8 \times 10^{-5}$$

$$[H_3O^+]^2 = 9.4 \times 10^{-4}$$

The procedure is repeated to give a *third* approximation,

$$\frac{[H_3O^+]^2}{0.050 - 0.00094} = 1.8 \times 10^{-5}$$

$$[H_3O^+] = 9.4 \times 10^{-4}$$

When successive approximations agree, the procedure is complete. □

Table 5.5 lists the error made by simply dropping the $[H_3O^+]$ term relative to the concentration of the weak acid. It is calculated for different values of K_a and concentrations of the acid. Note that the error increases as the value of K_a increases and as the concentration of acid decreases. We will rarely deal with con-

TABLE 5.5 Error Introduced by Approximation in Calculation of Hydrogen Ion Concentration[†]

VALUE OF K_a	ANALYTICAL CONCENTRATION, c_a	APPROXIMATE $[H_3O^+]$	EXACT $[H_3O^+]$[‡]	PERCENT ERROR
1.00×10^{-6}	0.100	3.16×10^{-4}	3.16×10^{-4}	0.0
	0.0500	2.24×10^{-4}	2.23×10^{-4}	0.4
	0.0100	1.00×10^{-4}	0.995×10^{-4}	0.5
1.00×10^{-5}	0.100	1.00×10^{-3}	0.995×10^{-3}	0.5
	0.0500	7.07×10^{-4}	7.02×10^{-4}	0.7
	0.0100	3.16×10^{-4}	3.11×10^{-4}	1.6
1.00×10^{-4}	0.100	3.16×10^{-3}	3.11×10^{-3}	1.6
	0.0500	2.24×10^{-3}	2.19×10^{-3}	2.3
	0.0100	1.00×10^{-3}	0.95×10^{-3}	5.3
1.00×10^{-3}	0.100	1.00×10^{-2}	0.95×10^{-2}	5.3
	0.0500	7.07×10^{-3}	6.55×10^{-3}	7.9
	0.0100	3.16×10^{-3}	2.70×10^{-3}	17.0

[†] *Note:* The approximation made is $c_a - [H_3O^+] \cong c_a$.

[‡] From the equation $\dfrac{[H_3O^+]^2}{c_a - [H_3O^+]} = K_a$.

centrations below $0.0100\ M$, but we do occasionally encounter an acid with a K_a as large as 10^{-3} or even 10^{-2}. A solution by the quadratic formula or by the method of successive approximations is advised when the weak electrolyte has such a large dissociation constant.

EXAMPLE 5.10

The common-ion effect

Calculate the pH of a solution which is $0.050\ M$ in acetic acid and $0.10\ M$ in sodium acetate.

The dissociation of acetic acid

$$HOAc + H_2O \rightleftharpoons H_3O^+ + OAc^-$$

is decreased by the presence of the common ion, acetate, from the completely dissociated salt, sodium acetate:

$$NaOAc \longrightarrow Na^+ + OAc^-$$

According to the *principle of LeChâtelier*, increasing the acetate ion concentration by adding sodium acetate to the acetic acid solution shifts the equilibrium to the left, thereby reducing the hydrogen ion concentration. The concentrations are as follows:

$$[HOAc] = 0.050 - [H_3O^+]$$

$$[OAc^-] = 0.10 + [H_3O^+]$$

Note that the concentrations of H_3O^+ and OAc^- are not equal as in Example 9, because of the additional acetate ion from the sodium acetate.

Noting again that the dissociation of acetic acid is small, we can make the approximations

$$[HOAc] \cong 0.050\ M$$

$$[OAc^-] \cong 0.10\ M$$

Then, substituting in the equilibrium expression,

$$\frac{[H_3O^+](0.10)}{0.050} = 1.8 \times 10^{-5}$$

$$[H_3O^+] = 9.0 \times 10^{-6}$$

$$pH = 5.05$$

The calculation just given is sometimes set up in a manner that seems different at first glance but really amounts to the same thing. Solve the K_a expression for $[H_3O^+]$, take logarithms of both sides of the equation, and multiply by -1:

$$\frac{[H_3O^+][OAc^-]}{[HOAc]} = K_a$$

$$[H_3O^+] = K_a \times \frac{[HOAc]}{[OAc^-]}$$

$$-\log [H_3O^+] = -\log K_a - \log \frac{[HOAc]}{[OAc^-]}$$

$$pH = pK_a - \log \frac{[HOAc]}{[OAc^-]}$$

The equation in this form shows that the pH is a function of pK_a and the ratio of acid concentration to that of the salt, or conjugate base. Since HOAc and OAc^- are in the same solution, the volume cancels and the ratio of millimoles is the same as the ratio of molar concentrations. This logarithmic form of the dissociation constant frequently appears in biochemistry or physiology textbooks under the designation *Henderson-Hasselbalch equation*. In these texts the log term is sometimes inverted and the equation written

$$pH = pK_a + \log \frac{[OAc^-]}{[HOAc]}$$

For a weak base B and its conjugate acid BH^+, the equation is

$$pOH = pK_b + \log \frac{[BH^+]}{[B]} \qquad \square$$

EXAMPLE 5.11

The common-ion effect

Calculate the pH of a solution made by mixing 50 mL of 0.10 M NH_3 and 50 mL 0.040 M HCl. The K_b of NH_3 is 1.8×10^{-5}.

Here the common ion, NH_4^+, is formed by reaction of the acid and base. We start with

$$50 \text{ mL} \times 0.10 \text{ mmol/mL} = 5.00 \text{ mmol of } NH_3$$

and

$$50 \text{ mL} \times 0.040 \text{ mmol/mL} = 2.0 \text{ mmol of HCl}$$

The reaction is

	NH_3	+	H_3O^+	$\longrightarrow$	NH_4^+	+	H_2O
Original (mmol)	5.0		2.0		—		
Change (mmol)	-2.0		-2.0		$+2.0$		
Equilibrium (mmol)	3.0		—		2.0		

The equilibrium involved is

$$NH_3 + H_2O \rightleftharpoons NH_4^+ + OH^-$$

and the concentrations are

$$[NH_3] = \frac{3.0 \text{ mmol}}{100 \text{ mL}} - [OH^-] \cong 0.030 \; M$$

$$[NH_4^+] = \frac{2.0 \text{ mmol}}{100 \text{ mL}} + [OH^-] \cong 0.020 \; M$$

Using the logarithmic expression from Example 10,

$$pOH = pK_b + \log \frac{[NH_4^+]}{[NH_3]}$$

$$pOH = 4.74 + \log \frac{0.020}{0.030}$$

$$pOH = 4.56$$

$$pH = 9.44 \hspace{4cm} \square$$

EXAMPLE 5.12

Hydrolysis of a salt

Calculate the *p*H of a 0.10 *M* solution of sodium acetate.

We have previously noted that the acetate ion is a base, reacting with water to produce hydroxide ions:

$$OAc^- + H_2O \; \rightleftharpoons \; HOAc + OH^-$$

The equilibrium expression is

$$\frac{[HOAc][OH^-]}{[OAc^-]} = K_b$$

Note that since $K_a \times K_b = K_w$ and $K_a = 1.8 \times 10^{-5}$, $K_b = 5.6 \times 10^{-10}$. The concentrations are

$$[HOAc] \cong [OH^-]$$

$$[OAc^-]$$

$$[HOAc] = 0.10 - [OH^-]$$

Substituting in the expression for K_b, we have[3]

$$\frac{[OH^-]^2}{0.10 - [OH^-]} = 5.6 \times 10^{-10}$$

We make the approximation

$$0.10 - [OH^-] \cong 0.10$$

since OAc^- is a weak base. Hence

$$\frac{[OH^-]^2}{0.10} = 5.6 \times 10^{-10}$$

$$[OH^-] = 7.5 \times 10^{-6}$$

$$pOH = 5.12$$

$$pH = 8.88 \hspace{4cm} \square$$

---

[3] Some students get the impression that in these calculations the numerator always contains a squared term. This is the case when the solution contains *only* a *weak acid* or a *weak base,* conjugate or not. When the solution contains a weak acid or base *plus* the salt, as in Examples 10 and 11, the concentration terms in the numerator are not equal, and hence there is no squared term.

5.4b. Solubility Equilibria

The equilibrium constant expressing the solubility of a precipitate in water is the familiar *solubility product constant*. For a precipitate of silver chloride, the equilibrium constant of the reaction

$$AgCl(s) \rightleftharpoons Ag^+(aq) + Cl^-(aq)$$

is

$$K° = \frac{a_{Ag^+} a_{Cl^-}}{a_{AgCl}}$$

The activity of solid AgCl is constant, and by convention we take it to be unity (page 101). The solid is only slightly soluble; hence the concentrations of Ag^+ and Cl^- ions are small and, unless large concentrations of other ions are present, activities can be aproximated by molarities, giving

$$K_{sp} = [Ag^+][Cl^-]$$

The constant K_{sp} is called the solubility product constant.

The proper equilibrium expressions for a few other salts are given below:

Salt	K_{sp}
$BaSO_4$	$[Ba^{2+}][SO_4^{2-}]$
Ag_2CrO_4	$[Ag^+]^2[CrO_4^{2-}]$
CaF_2	$[Ca^{2+}][F^-]^2$
$Al(OH)_3$	$[Al^{3+}][OH^-]^3$

A general expression for the salt A_xB_y dissociating as follows:

$$A_xB_y = xA^{y+} + yB^{x-}$$

is

$$K_{sp} = [A^{y+}]^x[B^{x-}]^y$$

The numerical value of a solubility product constant can be readily calculated from the solubility of the compound. The calculation can be reversed, of course, and the solubility calculated from the K_{sp}. If the ions of the precipitate undergo a reaction such as hydrolysis or complex formation, the calculations are more complicated. Such cases will be considered in a later chapter. Typical computations are illustrated in the following examples:

EXAMPLE 5.13

Calculation of K_{sp}

The solubility of barium sulfate (MW 233) at 25°C is 0.0023 mg/mL of solution. Calculate the value of K_{sp}.

The molarity is

$$\frac{0.0023 \text{ mg BaSO}_4}{1.0 \text{ mL}} \times \frac{1 \text{ mmol BaSO}_4}{233 \text{ mg BaSO}_4} = 1.0 \times 10^{-5} \text{ mmol/mL BaSO}_4$$

Since each millimole of $BaSO_4$ yields 1 mmol of Ba^{2+} and 1 mmol of SO_4^{2-},

$$[Ba^{2+}] = [SO_4^{2-}] = 1.0 \times 10^{-5}$$

$$K_{sp} = [Ba^{2+}][SO_4^{2-}] = [1.0 \times 10^{-5}]^2 = 1.0 \times 10^{-10} \qquad \square$$

EXAMPLE 5.14

Calculation of K_{sp}

The solubility of silver chromate (MW 332) is 0.0279 g/liter at 25°C. Calculate K_{sp} neglecting hydrolysis of the chromate ion.

The molarity is

$$\frac{0.0279 \text{ g Ag}_2\text{CrO}_4}{1.0 \text{ liter}} \times \frac{1 \text{ mol Ag}_2\text{CrO}_4}{332 \text{ g Ag}_2\text{CrO}_4} = 8.4 \times 10^{-5} \text{ mol/liter Ag}_2\text{CrO}_4$$

Since each Ag_2CrO_4 yields two Ag^+ ions and one CrO_4^{2-} ion,

$$[Ag^+] = 2 \times 8.4 \times 10^{-5} = 1.7 \times 10^{-4}$$

$$[CrO_4^{2-}] = 8.4 \times 10^{-5}$$

Therefore,

$$K_{sp} = [Ag^+]^2[CrO_4^{2-}] = (1.7 \times 10^{-4})^2(8.4 \times 10^{-5})$$

$$K_{sp} = 2.4 \times 10^{-12} \qquad \square$$

It should be noted that one can judge on inspection the relative molar solubilities of two compounds from their solubility product constants only if they are the same type of compound, i.e., both type AB or AB$_2$, etc. The solubility product constants of both AgCl and BaSO$_4$ are about 1×10^{-10}, and hence both are soluble to the extent of 1×10^{-5} mol/liter. However, a compound of the type AB$_2$ with the same molar solubility would have a smaller solubility product constant, 4×10^{-15}.

EXAMPLE 5.15

Calculation of precipitating reagent concentration

The silver in a solution is precipitated by the addition of chloride ion. The final volume of the solution is 500 mL. What should be the concentration of Cl^- if no more than 0.10 mg of Ag^+ remains unprecipitated?

The concentration of Ag^+ is

$$[Ag^+] = \frac{0.10 \text{ mg}}{107.9 \text{ mg/mmol} \times 500 \text{mL}}$$

$$[Ag^+] = 1.9 \times 10^{-6} \ M$$

$$[Ag^+][Cl^-] = K_{sp} = 1.0 \times 10^{-10}$$

$$(1.9 \times 10^{-6})(Cl^-) = 1.0 \times 10^{-10}$$

$$[Cl^-] = 5.3 \times 10^{-5} \qquad \square$$

EXAMPLE 5.16

Predicting precipitate formation

Calcium fluoride, CaF_2, has a K_{sp} of 4×10^{-11}. Predict whether a precipitate forms or not when the following solutions are mixed: (a) 100 mL of $2.0 \times 10^{-4} \ M$ Ca^{2+} plus 100 mL of $2.0 \times 10^{-4} \ M$ F^-; (b) 100 mL of $2.0 \times 10^{-2} \ M$ Ca^{2+} plus 100 mL of $6.0 \times 10^{-3} \ M$ F^-.

If the product $[Ca^{2+}][F^{-2}]^2$ exceeds K_{sp}, precipitation will occur; if not, precipitation will not occur. Note that the concentrations are halved when the solutions are mixed.

(a) $[Ca^{2+}] = 1.0 \times 10^{-4}$ and $[F^-] = 1.0 \times 10^{-4}$

$$[Ca^{2+}][F^-]^2 = (1.0 \times 10^{-4})(1.0 \times 10^{-4})^2 = 1.0 \times 10^{-12}.$$

Since $1.0 \times 10^{-12} < 4 \times 10^{-11}$, precipitation does not occur.

(b) $[Ca^{2+}] = 1.0 \times 10^{-2}$ and $[F^-] = 3.0 \times 10^{-3}$

$$[Ca^{2+}][F^-]^2 = (1.0 \times 10^{-2})(3.0 \times 10^{-3})^2 = 9.0 \times 10^{-8}.$$

Since $9.0 \times 10^{-8} > 4 \times 10^{-11}$, precipitation does occur. $\qquad \square$

5.4c. Complex-Formation Equilibria

Reactions involving complex formation are utilized by the chemist in both titrimetric and gravimetric procedures. We shall discuss the topic in detail in a later chapter. At this point we shall use the complex formed by silver and ammonia to illustrate formulation of the equilibrium constant and its use in a calculation.

Students of chemistry learn early in their careers that solid silver chloride will dissolve in a solution of ammonia. The equation can be written molecularly as

$$AgCl(s) + 2NH_3(aq) \longrightarrow Ag(NH_3)_2Cl$$

The compound $Ag(NH_3)_2Cl$ is called a *complex*. Actually, the compound is ionic, dissociating into $Ag(NH_3)_2^+$ and Cl^- ions, and the species $Ag(NH_3)_2^+$ is called a *complex ion*.

The silver-ammonia complex ion is formed in steps by the addition of molecules of ammonia, called the *ligand*, to silver ion, called the *central metal ion*:

$$Ag^+ + NH_3 \; \rightleftharpoons \; AgNH_3^+ \tag{1}$$

$$AgNH_3^+ + NH_3 \; \rightleftharpoons \; Ag(NH_3)_2^+ \tag{2}$$

The equilibrium constants for the two reactions are

$$\frac{[AgNH_3^+]}{[Ag^+][NH_3]} = K_1$$

and

$$\frac{[Ag(NH_3)_2^+]}{[AgNH_3^+][NH_3]} = K_2$$

They are called the *stability* or *formation* constants of the complexes. Sometimes the reactions are written as the dissociation of the complex:

$$AgNH_3^+ \; \rightleftharpoons \; Ag^+ + NH_3$$

and the equilibrium constant is called the *instability* or *dissociation* constant. The stability and instability constants are simply reciprocals of one another. We shall write only stability constants in our discussion of this topic.

Note that if the two "stepwise" constants are multiplied together, we obtain the formation constant for the complex $Ag(NH_3)_2^+$:

$$\frac{[AgNH_3^+]}{[Ag^+][NH_3]} \times \frac{[Ag(NH_3)_2^+]}{[AgNH_3^+][NH_3]} = K_1 \times K_2$$

$$\frac{[Ag(NH_3)_2^+]}{[AgNH_3^+][NH_3]} = \beta_2 \qquad \frac{[Ag(NH_3)_2^+]}{[Ag^+][NH_3]^2}$$

The *overall formation constant* for the general reaction

$$M + nL \; \rightleftharpoons \; ML_n$$

is usually given the symbol β_n. Here M is the central metal ion and L the ligand. The relations between stepwise and overall formation constants are $\beta_1 = K_1$, $\beta_2 = K_1K_2$, $\beta_3 = K_1K_2K_3$, $\beta_4 = K_1K_2K_3K_4$, and so on. Here β_2 for $Ag(NH_3)_2^+$ is $2.3 \times 10^3 \times 6.0 \times 10^3 = 1.4 \times 10^7$. The symbol β is also commonly used to represent the fraction of a metal ion in the uncomplexed form (page 209). To avoid confusion, we shall use the symbol K rather than β for the overall formation constant, and we shall indicate which stepwise constants are involved.

Example 17 illustrates the use of the stability constant in a calculation.

EXAMPLE 5.17

Use of the stability constant to calculate concentration

The ion Hg^{2+} reacts with Cl^- ion as follows:

$$Hg^{2+} + Cl^- \rightleftharpoons HgCl^+ \qquad K_1 = 5.5 \times 10^6$$

$$HgCl^+ + Cl^- \rightleftharpoons HgCl_2 \qquad K_2 = 3.0 \times 10^6$$

For the reaction

$$Hg^{2+} + 2Cl^- \rightleftharpoons HgCl_2 \qquad K = K_1 \times K_2 = 1.65 \times 10^{13}$$

calculate the concentration of Hg^{2+} at the EPt in the titration of 2.0 mmol of Hg^{2+} with Cl^- according to the last equation. The final volume of solution is 100 mL.

At the EPt

$$[Cl^-] = 2[Hg^{2+}]$$

$$[HgCl_2] = 0.020 - [Hg^{2+}]$$

Since the stability constant for $HgCl_2$ is large (1.65×10^{13}) the value of $[Hg^{2+}]$ should be small, and it is reasonable to assume that it can be neglected when subtracted from 0.020. Hence we assume

$$[HgCl_2] = 0.020 - [Hg^{2+}] \cong 0.020$$

Substituting in the equilibrium expression,

$$\frac{[HgCl_2]}{[Hg^{2+}][Cl^-]^2} = 1.65 \times 10^{13}$$

$$\frac{0.020}{[Hg^{2+}](2[Hg^{2+}])^2} = 1.65 \times 10^{13}$$

$$[Hg^{2+}] = 6.7 \times 10^{-6} \, M \qquad \square$$

5.4d. *Oxidation-Reduction Equilibria*

The equilibrium constant of an oxidation-reduction reaction is obtained from the potential of an appropriate galvanic cell. We shall consider this topic in detail in Chapter 10. The equilibrium expressions for such reactions are formulated in the usual way. For example, when iron(II) is titrated with cerium(IV), the reaction is

$$Fe^{2+} + Ce^{4+} \rightleftharpoons Fe^{3+} + Ce^{3+}$$

The equilibrium constant for this reaction is

$$\frac{[Fe^{3+}][Ce^{3+}]}{[Fe^{2+}][Ce^{4+}]} = K$$

The following example illustrates a calculation using this equilibrium constant.

EXAMPLE 5.18

Use of the equilibrium constant for a redox reaction

5.00 mmol of Fe^{2+} is titrated with Ce^{4+} in sulfuric acid solution. Calculate the concentration of Fe^{2+} at the equivalence point. The volume of solution is 100 mL and the K for the reaction is 7.6×10^{12}.

Since the K is large, we shall assume that the reaction goes essentially to completion. Then at the equivalence point

$$[Fe^{2+}] = [Ce^{4+}]$$

and

$$[Fe^{3+}] = [Ce^{3+}] = 0.050 - [Fe^{2+}]$$

Since $[Fe^{2+}]$ is small,

$$[Fe^{3+}] = [Ce^{3+}] \cong 0.050$$

Hence

$$\frac{(0.050)(0.050)}{[Fe^{2+}]^2} = 7.6 \times 10^{12}$$

$$[Fe^{2+}] = 1.8 \times 10^{-8}\,M \qquad \square$$

5.4c. *Simultaneous Equilibria*

We have now reviewed the formulation of equilibrium constants for the four types of reactions used in titrimetric analyses. It is quite common for two or more of these equilibria to be established in the same solution. If any one ion is a participant in more than one of the equilibria, the reactions are said to "interact," or it is said that the equilibria are established simultaneously. The principle of LeChâtelier can be used to make qualitative predictions of the result of such interactions. Quantitative calculations can be fairly complex.

As an example, consider the effect of pH on the solubility of the salt of a weak acid, say calcium carbonate, $CaCO_3$. The solubility equilibrium is

$$CaCO_3(s) \rightleftharpoons Ca^{2+} + CO_3^{2-} \qquad (1)$$

The carbonate ion is a base and reacts with hydrogen ions in two steps:

$$CO_3^{2-} + H_3O^+ \rightleftharpoons HCO_3^- + H_2O \qquad (2)$$

$$HCO_3^- + H_3O^+ \rightleftharpoons H_2CO_3 + H_2O \qquad (3)$$

According to LeChâtelier's principle, increasing the concentration of H_3O^+ (lowering the pH) shifts equilibrium (2) to the right, lowering the CO_3^{2-} concentration. This in turn shifts equilibrium (1) to the right, causing more solid $CaCO_3$ to dissolve. It is a general principle that salts of weak acids, such as oxalates, sulfides, carbonates, and hydroxides, are more soluble in acidic than basic solutions.

The dissolving of AgCl in aqueous ammonia is an example of the interaction of solubility and complex formation equilibria. The solubility reaction is

$$AgCl(s) \rightleftharpoons Ag^+ + Cl^- \qquad (1)$$

and the reactions forming complexes are

$$Ag^+ + NH_3 \rightleftharpoons AgNH_3^+ \qquad (2)$$

$$AgNH_3^+ + NH_3 \rightleftharpoons Ag(NH_3)_2^+ \qquad (3)$$

The addition of ammonia to a saturated solution of silver chloride lowers the concentration of Ag^+ by shifting equilibria (2) and (3) to the right. This, in turn, shifts equilibrium (1) to the right, causing more AgCl to dissolve.

SYSTEMATIC EQUILIBRIUM CALCULATIONS

In our study of analytical chemistry we often need to calculate the concentration of some species in an aqueous solution in which there are several interacting equilibria. The mathematical calculations may be quite complex. However, there is a procedure, called a *systematic treatment of equilibrium,* which can be used in such cases, and often reasonable assumptions can be made which greatly simplify the mathematical operations.

The general approach is to identify all the chemical species in the solution and to find a set of equations sufficient in principle to permit calculation of the concentration of each species. This means we need as many *independent* mathematical equations as we have chemical species. If we have five unknowns and five equations, we can in principle solve for all unknowns. When there are more equations than can be solved simultaneously with convenience, we can sometimes invoke our chemical knowledge to simplify the mathematical problem.

The procedure is as follows:

1. Identify all the species in the solution. Water molecules are not counted since their concentration in dilute solutions is essentially constant at about 55.3 mol/liter.
2. Write all the equilibrium constants involving the unknown species. These constants must be independent equations. For example, if one uses the dissociation constant of a weak acid, K_a, and the autoprotolysis constant of water, K_w, one could not also use the dissociation constant of the anion of the weak acid, K_b, since $K_b = K_w/K_a$. No additional information on the relations between concentrations is given by the constant K_b.
3. Write the *mass-balance* (or material balance) equation. This equation is simply a statement of the conservation of matter. For example, in a $0.10\ F$ solution of acetic acid, all of the acid ends up as HOAc molecules or OAc$^-$ ions. The mass-balance equation on acetate is

$$[HOAc] + [OAc^-] = 0.10$$

More than one mass-balance equation may be needed, as we shall see in a subsequent example.
4. Write the *charge-balance* equation, which is based on what is called the *electroneutrality condition: The total concentration of positive charge must equal the total concentration of negative charge.*
5. Using chemical principles, make approximations to simplify the mathematical problem. Frequently, for example, we are able to drop certain terms because our knowledge of chemistry tells us that they are negligible.
6. Solve the equations for the unknown concentrations. Then substitute them into the original equations to determine the validity of the approximations which were made.

We shall illustrate the above procedure with two examples.

EXAMPLE 5.19   Calculate the concentrations of all the species in a $0.10\ F$ solution of HOAc.

Simultaneous equilibria involving a weak acid

1. There are four species in the solution in addition to the H$_2$O molecule: H$_3$O$^+$, OH$^-$, HOAc, and OAc$^-$.

2. The equilibria and equilibrium constants are

$$HOAc + H_2O \rightleftharpoons H_3O^+ + OAc^- \qquad K_a = \frac{[H_3O^+][OAc^-]}{[HOAc]} \qquad (1)$$

$$2H_2O \rightleftharpoons H_3O^+ + OH^- \qquad K_w = [H_3O^+][OH^-] \qquad (2)$$

3. The mass-balance equation on acetate is

$$[HOAc] + [OAc^-] = 0.10 \qquad (3)$$

4. The charge-balance equation is

$$[H_3O^+] = [OH^-] + [OAc^-] \qquad (4)$$

5. We now have four equations and four unknowns. In seeking a solution for these equations we may first decide to neglect $[OH^-]$ as compared with $[OAc^-]$, since acetic acid, although weak, is appreciably acidic. Equation (4) becomes

$$[H_3O^+] = [OH^-] + [OAc^-] \cong [OAc^-]$$

Also, since acetic acid is weak, $[OAc^-]$ is small compared with $[HOAc]$, and Eq. (3) becomes

$$[HOAc] + [OAc^-] \cong [HOAc] = 0.10$$

Substitution of the modified Eqs. (3) and (4) into Eq. (1) gives

$$\frac{[H_3O^+]^2}{0.10} = 1.8 \times 10^{-5}$$

whence $[H_3O^+] = 1.34 \times 10^{-3}$ M, and from Eq. (2), $[OH^-] = 7.5 \times 10^{-12}$ M. From the modified Eq. (1) we see that $[OAc^-] = 1.34 \times 10^{-3}$ M, and since we already know that $[HOAc] = 0.10$ M, we now have our four concentrations.

We may now check the assumptions we made in neglecting $[OH^-]$ in Eq. 4 and $[OAc^-]$ in Eq. (3). Equation (4) becomes

$$1.34 \times 10^{-3} \cong 7.5 \times 10^{-12} + 1.34 \times 10^{-3}$$

This is obviously a good approximation. Equation (3) becomes

$$0.10 + 0.00134 \cong 0.10$$

The relative error incurred here is

$$\frac{0.00134}{0.10} \times 100 = 1.3\% \qquad \square$$

EXAMPLE 5.20

Calculation of solubility in the presence of a complexing ion

Calculate the molar solubility of AgCl in 0.010 M NH$_3$. (This is the concentration of the free, or uncomplexed, NH$_3$ molecules in the solution.) Also calculate the equilibrium concentrations of the various species in the solution. The concentrations of H_3O^+ and OH^- can be disregarded, since they are not involved in the solubility process.

1. There are five species in the solution: Ag^+, Cl^-, $AgNH_3^+$, $Ag(NH_3)_2^+$, and NH$_3$. We are given the concentration of NH$_3$ molecules, but we do not know how much AgCl dissolves. Our fifth unknown is s, the molar solubility.

2. The equilibria and equilibrium constants are

$$\text{AgCl(s)} \rightleftharpoons \text{Ag}^+ + \text{Cl}^- \qquad [\text{Ag}^+][\text{Cl}^-] = 1.0 \times 10^{-10} \qquad (1)$$

$$\text{Ag}^+ + \text{NH}_3 \rightleftharpoons \text{AgNH}_3^+ \qquad \frac{[\text{AgNH}_3^+]}{[\text{Ag}^+][\text{NH}_3]} = 2.3 \times 10^3 \qquad (2)$$

$$\text{AgNH}_3^+ + \text{NH}_3 \rightleftharpoons \text{Ag(NH}_3)_2^+ \qquad \frac{[\text{Ag(NH}_3)_2^+]}{[\text{AgNH}_3^+][\text{NH}_3]} = 6.0 \times 10^3 \qquad (3)$$

3. The mass balance on chloride ion is

$$[\text{Cl}^-] = s$$

The mass balance on silver is

$$[\text{Ag}^+] + [\text{AgNH}_3^+] + [\text{Ag(NH}_3)_2^+] = s \qquad (5)$$

Note that this is also the charge-balance equation, since s = [Cl⁻].

4. Since both of the stability constants are large, we may logically assume that most of the silver is in the form $\text{Ag(NH}_3)_2^+$ and write Eq. (5) as

$$[\text{Ag(NH}_3)_2^+] \cong s \qquad (6)$$

Substituting Eq. (4) in Eq. (1) gives

$$[\text{Ag}^+] = \frac{1.0 \times 10^{-10}}{s} \qquad (7)$$

When Eqs. (2) and (3) are multiplied, one obtains

$$\frac{[\text{Ag(NH}_3)_2^+]}{[\text{Ag}^+][\text{NH}_3]^2} = 1.4 \times 10^7 \qquad (8)$$

Substituting Eqs. (6) and (7) in Eq. (8) gives

$$\frac{s}{\dfrac{1.0 \times 10^{-10}}{s}(0.010)^2} = 1.4 \times 10^7$$

$$s = 3.7 \times 10^{-4}$$

Then $[\text{Cl}^-] = s = 3.7 \times 10^{-4}$,

$$[\text{Ag}^+] = \frac{1.0 \times 10^{-10}}{3.7 \times 10^{-10}} = 2.7 \times 10^{-7}$$

and $[\text{Ag(NH}_3)_2^+ = s = 3.7 \times 10^{-4}$.
Substituting in Eq. (3) gives

$$\frac{3.7 \times 10^{-4}}{[\text{AgNH}_3^+][0.010]} = 6.0 \times 10^3$$

$$[\text{AgNH}_3^+] = 6.2 \times 10^{-6}$$

The approximation made in Eq. (5) can be checked:

$$2.7 \times 10^{-7} + 6.2 \times 10^{-6} + 3.7 \times 10^{-4} \cong 3.7 \times 10^{-4}$$

The left-hand term differs from the right-hand term by about 1.8%.

See Example 11, Section 9.3.g, for the same calculation using the "effective" equilibrium constant approach. □

5.6

COMPLETENESS OF REACTIONS USED FOR ANALYSIS

We can now examine the question as to exactly how large equilibrium constants should be for the reactions used in titrimetric and gravimetric procedures to go to "completion." First we must decide just what we mean by a reaction's being complete. No reaction goes absolutely to completion, of course. What the chemist desires is that the reaction be so complete that the amount of unreacted analyte is analytically undetectable (page 45). In the determination of a major constituent in a macro sample, the quantity of analyte is normally 0.1 g or greater and the analytical balance can detect as little as 0.1 mg. Hence the chemist usually regards a reaction as being complete if the amount of unreacted analyte is 0.1 mg or less, or if all but 0.1% of the analyte has reacted.

In a separation process using a precipitation reaction, the chemist can take advantage of the common-ion effect (page 108) if the equilibrium constant is not very large. *Excess* precipitating agent can be added, forcing the reaction to completion. But this cannot be done in titrimetric procedures. Titrimetry, by its very nature, generally precludes forcing a reaction to completion by using a large excess of reactant. The reaction must be complete *at or near the equivalence point*.

Let us now calculate just how large an equilibrium constant should be for a reaction to be complete at the equivalence point of a titration. For simplicity we shall consider a precipitation titration:

$$A(aq) + B(aq) \longrightarrow AB(s)$$

where the equilibrium constant is

$$K = \frac{1}{[A][B]}$$

Suppose that we titrate 5.000 mmol of A with 5.000 mmol of B and the volume of solution at the EPt is 100 mL. Let us calculate the value of K if 99.9% of A has reacted at the EPt; i.e., 0.1% (0.005 mmol) is unreacted. The concentrations of A and B are

$$[A] = [B] = \frac{0.005 \text{ mmol}}{100 \text{ mL}} = 5.0 \times 10^{-5} M$$

and

$$K = \frac{1}{(5.0 \times 10^{-5})(5.0 \times 10^{-5})} = 4.0 \times 10^{8}$$

Had we asked that 99.99% be reacted at the EPt, the value of K would be 4.0×10^{10}. Table 5.6 lists examples of the four types of reactions used in titrations with the values of K required for 99.9% and 99.99% completion. We shall discuss the question of feasibility of titrations in detail in later chapters.

5.6a. End Point Detection in Titrations

If the titration reaction goes well to completion, there will be a large, abrupt change in the concentration of the analyte (and of the titrant) near the EPt (page 45). The analyst can convert this abrupt change into a signal to stop the addition of titrant—the EPt has been reached.

TABLE 5.6 Values of Equilibrium Constants for Titrations

REACTION	K	VALUE REQUIRED FOR [†]	
		99.9%	99.99%
Acid-base:			
$HOAc + OH^- \rightleftharpoons OAc^- + H_2O$	$\dfrac{[OAc^-]}{[HOAc][OH^-]}$	2×10^7	2×10^9
Complex formation:			
$Ca^{2+} + EDTA^{4-} \rightleftharpoons CaEDTA^{2-}$	$\dfrac{[CaEDTA^{2-}]}{[Ca^{2+}][EDTA^{4-}]}$	2×10^7	2×10^9
Precipitation:			
$Ag^+ + Cl^- \rightleftharpoons AgCl(s)$	$\dfrac{1}{[Ag^+][Cl^-]}$	4×10^8	4×10^{10}
Oxidation-Reduction:			
$Fe^{2+} + Ce^{4+} \rightleftharpoons Fe^{3+} + Ce^{3+}$	$\dfrac{[Fe^{3+}][Ce^{3+}]}{[Fe^{2+}][Ce^{4+}]}$	1×10^6	1×10^8

[†] Calculated for the titration of 5.000 mmol of analyte, 100 ml volume at the equivalence point.

Table 5.7 contains data for the titration of 50 mL of $0.10\ M$ A with 50 mL of $0.10\ M$ B for the precipitation reaction

$$A(aq) + B(aq) \rightleftharpoons AB(s)$$

From the data in this table it can be seen that the *relative change* in [A] is small at the start of the titration and that the rate of change rapidly increases as the EPt is approached. If the titration is followed graphically by plotting [A] vs. milliliters of B (Fig. 5.1), this rapid rate of change is obscured because the linear concentration scale is not sufficiently large to reveal the enormous change that occurs. Since the concentration extends over several orders of magnitude, it is best displayed graphically by plotting a logarithmic function of the concentration, $pA = -\log A$, against the volume of titrant (Fig. 5.1). In such a plot, called a *titration curve*, the value of pA rises slowly at the start of the titration, rapidly increases as the EPt is approached, and increases slowly after the EPt. Values of pA are also shown in Table 5.7, where the change in pA per change in volume, $\Delta pA/\Delta V$, can be seen to increase rapidly at the EPt.

TABLE 5.7 Change in Concentration of Analyte during Titration of 50 ml of 0.10 M A With 50 ml of 0.10 M B[†]

ML B	% A PPTD.	% A UNPPTD.	[A]	pA	$\Delta pA/\Delta V$
0	0	100	0.10	1.00	0.017
10	20	80	0.067	1.17	0.020
20	40	60	0.043	1.37	0.023
30	60	40	0.025	1.60	0.036
40	80	20	0.011	1.96	0.114
49.0	98	2	0.0010	3.00	1.11
49.9	99.8	0.2	1×10^{-4}	4.00	11.1
49.99	99.9	0.1	1×10^{-5}	5.00	30.0
50.00	99.99	0.01	5×10^{-6}	5.30	

[†] $A(aq) + B(aq) \rightleftharpoons AB(s)$, $K = 4.0 \times 10^{10}$.

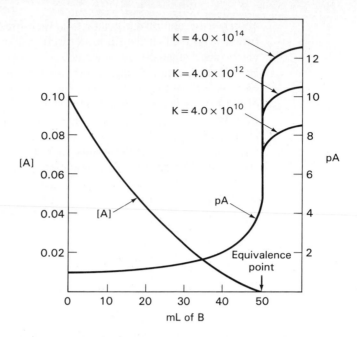

Figure 5.1 Plots of [A] and pA in titration of 50 mL of 0.10M A with 0.10M B for three values of K.

The effect of the magnitude of K on $\Delta pA/\Delta V$ at the EPt is shown in Table 5.8. In Figure 5.1 the effect of the value of K on the titration curve can be seen. Note that the curve is not affected before the EPt, but after the EPt the larger K results in larger values of pA, and hence a larger change for a given volume of titrant.

We shall discuss in detail later the methods the chemist uses to convert the large change in analyte concentration into a signal that the EPt has been reached. It should be pointed out that while the entire titration curve is of interest, actually only points near the EPt are needed to detect the end of the titration.

TABLE 5.8 Effect of Value of Equilibrium Constant on Concentration on Analyte at Equivalence Point[†]

K	% A PPTD.	% A UNPPTD.	[A]	pA	$\Delta pA/0.1$ ML
4.0×10^{8}	99.9	0.1	5.0×10^{-5}	4.30	3.0
4.0×10^{10}	99.99	0.01	5.0×10^{-6}	5.30	130
4.0×10^{12}	99.999	0.001	5.0×10^{-7}	6.30	230
4.0×10^{14}	99.9999	0.0001	5.0×10^{-8}	7.30	330

[†] Titration of 50 ml of 0.10 M A with 50 ml of 0.10 M B. A(aq) + B(aq) $\rightleftharpoons$ AB(s).

5.6b. *Use of Linear Plots in Titrations*

Sometimes the analyst does employ linear plots to depict a titration curve. Examples are photometric (Chapter 14) and amperometric titrations (Chapter 13). In such titrations a physical property which is proportional to the concentration of the analyte (or titrant) is measured. For example, in a photometric titration light absorbed by one of the reactants is measured and plotted against the volume of titrant (Fig. 5.2). If the titration reaction is essentially complete, the end point is easily determined by the intersection of two straight lines on either side of the EPt. If the reaction is appreciably incomplete, the curve will be rounded near the EPt, as seen in Fig. 5.2. In this case the end point is located by the intersection of the *extrapolated straight lines* drawn through points taken well before and after the EPt. If these measurements are made well away from equivalence, the excess analyte or titrant will be sufficient to force the reaction to completion by the common-ion effect. Then the points should fall on a straight line, and extrapolation should yield an accurate determination of the EPt volume.

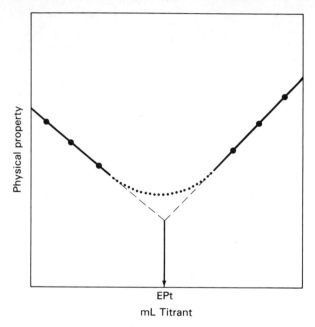

Figure 5.2 Linear titration curve. Extrapolation of straight lines to intersection gives EPt.

KEY TERMS

Activity. The measure of the "active mass" of a substance which gives a "true" constant when used in the equilibrium expression.

Activity coefficient. The number by which concentration is multiplied to give activity.

Autoprotolysis constant. The equilibrium constant for a reaction in which one solvent molecule loses a proton to another, as $2H_2O \rightleftharpoons H_3O^+ + OH^-$.

Charge-balance equation. The equation expressing the electroneutrality principle; i.e., the total concentration of positive charge must equal the total concentration of negative charge.

Common-ion effect. The effect produced by an ion, say from a salt, which is the same ion produced by the dissociation of a weak electrolyte. The "common" ion shifts the dissociation equilibrium in accordance with LeChâtelier's principle.

Central metal atom. A cation which accepts electrons from a ligand to form a complex ion.

Debye-Hückel equations. Theoretical equations which enable one to calculate the activity coefficient as a function of the ionic strength of a solution.

Electrolyte. A compound which produces positive and negative ions in solution. *Strong* electrolytes are completely dissociated, whereas *weak* electrolytes are only partially dissociated.

Electroneutrality principle. The principle which says that a solution must be electrically neutral; i.e., the total concentration of positive charge must equal the total concentration of negative charge.

Equilibrium constant. A general expression which is found to be constant for a reaction at equilibrium. For the reaction $aA + bB \rightleftharpoons cC + dD$ the thermodynamic constant is

$$K^\circ = \frac{(a_C)^c(a_D)^d}{(a_A)^a(a_B)^b}$$

Formation constant. The equilibrium constant for a reaction in which a complex is formed. Also called a *stability constant*.

Henderson-Hasselbalch equation. The dissociation constant of a weak acid solved for pH; $pH = pK_a + \log [\text{base}]/[\text{acid}]$.

Hydrolysis. An acid-base reaction of a cation or anion with water.

Ionic strength. A measure of the magnitude of the electrostatic environment of a solute, defined as one-half the sum of the concentration of each ion multiplied by the square of its electrical charge.

LeChâtelier's principle. A principle which states that if a stress is applied to a system at equilibrium, the equilibrium shifts in a direction that tends to reduce the stress.

Ligand. An anion or neutral molecule which forms a complex ion with a cation by donating one or more pairs of electrons.

Mass balance. The equation expressing the sum of the concentrations of all species arising from one substance through dissociation or association reactions.

Nonelectrolyte. A substance which does not dissociate into ions in solution.

pH. The negative logarithm of the hydrogen ion concentration.

pK. The negative logarithm of an equilibrium constant.

Salt. The product other than water which is formed when an acid reacts with a base.

Simultaneous equilibria. Equilibria established in the same solution in which one molecule or ion is a participant in more than one of the equilibria.

Solubility product constant, K_{sp}. The constant for the equilibrium established between a slightly soluble salt and its ions in solution.

Stability constant. The equilibrium constant for a reaction in which a complex is formed. Also called a *formation constant*.

Standard state. The state of a substance assigned an activity of unity at a given temperature and pressure.

Systematic treatment of equilibria. A method of solving for the concentrations of the species involved in solution equilibria.

Titration curve. A plot of some function of the concentration of the analyte, or of a property proportional to this concentration, vs. milliliters of titrant.

QUESTIONS

1. *Equilibrium constants.* Write the equilibrium constant expressions for each of these reactions:
   (a) $Fe(OH)_3(s) \rightleftharpoons Fe^{3+} + 3OH^-$
   (b) $H_3PO_4 + 2H_2O \rightleftharpoons 2H_3O^+ + HPO_4^{2-}$
   (c) $2Cu^{2+} + 4I^- \rightleftharpoons 2CuI(s) + I_2$
   (d) $Zn^{2+} + 4NH_3 \rightleftharpoons Zn(NH_3)_4^{2+}$
   (e) $MnO_2(s) + HAsO_2 + 2H^+ \rightleftharpoons Mn^{2+} + H_3AsO_4$

2. *Electrolytes.* Identify each of these substances as (a) weak or strong electrolyte, and (b) acid, base, or salt:
   (a) $NaHSO_4$; (b) $Ca(OH)_2$; (c) H_2CO_3;
   (d) $MgBr_2$; (e) H_2SO_4; (f) $HgCl_2$;
   (g) $K_2Cr_2O_7$; (h) NH_3; (i) NH_4Cl;
   (j) K_2SO_3; (k) HBO_2.

3. *Hydrolysis.* Write the net ionic reaction, if any, which occurs when these substances are dissolved in water: (a) CaF_2; (b) $KOAc$;

(c) $NaNO_3$; (d) NH_3; (e) CO_2; (f) NH_4OAc.

4. *Principle of Le Châtelier.* In which direction will this equilibrium be shifted by making the indicated changes?

$$Fe(OH)_3(s) \rightleftharpoons Fe^{3+} + 3OH^-$$

(a) Increase $[H_3O^+]$. (b) Increase $[Fe^{3+}]$. (c) Increase $[OH^-]$. (d) Add more solid $Fe(OH)_3$. (e) Add more water.

5. *Activity coefficients.* Explain how each of these factors affects the magnitude of the activity coefficient of an ion: (a) Magnitude of the ionic charge; (b) size of the hydrated ion; (c) ionic strength; (d) sign of the ionic charge.

6. *Charge-balance equation.* Write the charge-balance equation for these solutions: (a) 0.10 M HCl; (b) 0.10 M NaOAc; (c) 0.10 M NH_3.

Multiple-choice. In the following multiple-choice questions select the *one best* answer.

7. An aqueous solution is made by mixing equal volumes of 0.10 M HOAc and 0.10 M NaOH. Which equation is correct for this solution?

(a) $[H_3O^+] + [OH^-] = [Na^+] + [OAc^-]$;
(b) $[HOAc] + [OAc^-] = 0.10$;
(c) $[H_3O^+] + [Na^+] = [OH^-] + [OAc^-]$;
(d) More than one of the above.

8. For a 0.10 M solution of Na_2SO_4, which equation is correct?
(a) $[H_3O^+] + [Na^+] = [OH^-] + [HSO_4^-] + 2[SO_4^{2-}]$; (b) $[HSO_4^-] + [SO_4^{2-}] = 0.10$;
(c) $[Na^+] = 0.20$; (d) all of the above.

9. Which of these compounds should increase in solubility in acid solution? (a) KCl; (b) CaF_2; (c) $Ba(OH)_2$; (d) $NaClO_4$; (e) two of the above.

10. Which of the following is the most basic solution?
(a) $pH = 10$; (b) $pH = 5$; (c) $[OH^-] = 10^{-12}$; (d) $[H_3O^+] = 10^{-12}$.

11. Which statement is true? (a) Addition of NaOAc to a solution of HOAc increases the pH. (b) Addition of NH_3 to a saturated solution of AgCl causes more AgCl to dissolve. (c) Addition of NH_4Cl to a solution of NH_3 lowers the pH. (d) All of the above are true. (e) None of the above is true.

12. In a solution in which $[H_3O^+] = 10$, (a) $pH = 1$; (b) $pH = -1$; (c) $pOH = 15$; (d) more than one of the above.

PROBLEMS

1. *Ionic strength.* Calculate the ionic strength of these solutions: (a) 0.16 M $NaClO_4$; (b) 0.20 M $CuSO_4$; (c) 0.05 M Na_3PO_4.

2. *Ionic strength.* Calculate the ionic strength of these solutions: (a) 0.15 M $NaNO_3$; (b) 0.12 M Na_2SO_4; (c) 0.05 M $MgSO_4$ + 0.10 M $Al(NO_3)_3$.

3. *Activity coefficients.* Calculate the mean activity coefficient for $CaCl_2$ in a 0.400 M solution of the salt using (a) the Debye-Hückel limiting law; (b) the extended Debye-Hückel equation; (c) the Davies equation. The experimental value is 0.448.

4. *Activity coefficients.* Calculate the mean activity coefficient for $CaCl_2$ in a 0.100 M solution of the salt using (a) the Debye-Hückel limiting law; (b) the extended Debye-Hückel equation; (c) the Davies equation. The experimental value is 0.518.

5. *Equilibrium constants.* Calculate the value of the equilibrium constant in terms of con-

centration for these reactions at the indicated ionic strength:
(a) $HCO_3^- + H_2O \rightleftharpoons H_2CO_3 + OH^-$
$K_b = 2.2 \times 10^{-8}$; $\mu = 0.10°$
(b) $BaF_2(s) \rightleftharpoons Ba^{2+} + 2F^-$
$K_{sp} = 3 \times 10^{-6}$; $\mu = 0.05°$

6. *Equilibrium constants.* Calculate the value of the equilibrium constant in terms of concentration for these reactions at the indicated ionic strength:
(a) $CO_3^{2-} + H_2O \rightleftharpoons HCO_3^- + OH^-$
$K_b = 2.3 \times 10^{-4}$; $\mu = 0.10$
(b) $CaCO_3(s) \rightleftharpoons Ca^{2+} + CO_3^{2-}$
$K_{sp} = 5 \times 10^{-9}$; $\mu = 0.05$

7. *pH and pOH.* Convert the following hydrogen ion concentrations to pH, and hydroxide ion concentrations to pOH:
(a) $[H_3O^+] = 0.080$; (b) $[OH^-] = 0.40$;
(c) $[H_3O^+] = 5.0$; (d) $[OH^-] = 4.0 \times 10^{-10}$.

8. *pH and pOH.* Convert the following hydrogen ion concentrations to *pH*, and hydroxide ion concentrations to *pOH*:
   (a) $[H_3O^+] = 3.6 \times 10^{-8}$; (b) $[OH^-] = 2.0$;
   (c) $[H_3O^+] = 3.0 \times 10^{-15}$; (d) $[OH^-] = 5.0 \times 10^{-5}$.

9. *pH and pOH.* Convert the following *pH* values to hydrogen ion concentrations, and *pOH* values to hydroxide ion concentrations:
   (a) *pH* = -0.25; (b) *pOH* = 14.40;
   (c) *pH* = 9.30; (d) *pOH* = 5.74.

10. *pH and pOH.* Convert the following *pH* values to hydrogen ion concentrations, and *pOH* values to hydroxide ion concentrations:
    (a) *pH* = 9.26; (b) *pOH* = 0.40;
    (c) *pH* = 1.25; (d) *pOH* = -0.40.

11. *p-Functions.* Convert the following to the corresponding *p*-functions: (a) $K_{sp} = 4 \times 10^{-12}$; (b) $[Cl^-] = 0.025$ *M*; (c) $K_b = 5.0 \times 10^{-6}$; (d) $K_a = 1.8 \times 10^{-5}$.

12. *p-Functions.* Make the following conversions:
    (a) *pMg* = 0.50 to $[Mg^{2+}]$; (b) $pK_b = 9.74$ to K_b; (c) $pK_{sp} = 15.30$ to K_{sp}; (d) $pK_a = 3.30$ to K_a.

13. *Dissociation constant.* A weak acid, HX, is 1.0% dissociated in a 0.050 *M* solution.
    (a) Calculate the value of K_a for the acid.
    (b) Calculate the percentage dissociation in a 0.10 *M* solution. (c) At what concentration is the acid 0.50% dissociated?

14. *Dissociation constant.* A weak base, B (MW 60.0 g/mol), dissociates as follows:

$$B + H_2O \rightleftharpoons BH^+ + OH^-$$

A solution is prepared by dissolving 2.55 g of B in 250 mL of solution. The *pH* is found to be 11.30. Calculate the dissociation constant of the base.

15. *pH calculations.* Calculate the *pH* of these solutions:
    (a) 0.25 g of $Ba(OH)_2$ in 1.25 liters of solution.
    (b) 2.00 g of $HClO_4$ in 3.50 liters of solution.
    (c) 0.500 g of HCOOH in 250 mL of solution.
    (d) 0.60 g of NH_3 in 300 mL of solution.

16. *pH calculations.* Calculate the *pH* of these solutions:
    (a) 10 g of KOH in 300 mL of solution.
    (b) 1.50 g of HCl in 4.00 liters of solution.
    (c) 500 mg of HOCl in 250 mL of solution.

17. *Common-ion effect.* Calculate the *pH* of a solution which is (a) 0.20 *M* in HF and 0.10 *M* in NaF; (b) 0.12 *M* in NH_3 and 0.24 *M* in NH_4Cl; (c) 0.10 *M* in KOH and 0.20 *M* in KCl.

18. *Common-ion effect.* Calculate the number of grams of NH_4Cl which should be added to 400 mL of 0.10 *M* NH_3 to make the *pH* 10.70.

19. *Salt solutions.* Calculate the *pH* of these aqueous solutions: (a) 0.30 *M* $NaHCO_3$; (b) 0.20 *M* NH_4Cl; (c) 0.10 *M* $NaNO_3$.

20. *Mixtures of solutions.* Calculate the *pH* of the solutions made as indicated. Assume that the volumes are additive. (a) 60 mL of 0.10 *M* HOAc + 40 mL of 0.15 *M* NaOH; (b) 80 mL of 0.10 *M* NH_3 + 20 mL of 0.20 *M* HCl; (c) 100 mL of 0.10 *M* HCl + 100 mL of 0.08 *M* $Ba(OH)_2$.

21. *Mixtures of solutions.* Calculate the *pH* of the solutions made as indicated. Assume the volumes are additive. (a) 50 mL of 0.10 *M* HOCl + 50 mL of 0.08 *M* NaOH; (b) 40 mL of 0.20 *M* NH_3 + 60 mL of 0.10 *M* HCl; (c) 40 mL of 0.15 *M* HF + 60 mL of 0.10 *M* KOH.

22. *Mixtures of solutions.* Calculate the *pH* of the solutions made as indicated. Assume the volumes are additive. (a) 60 mL of 0.12 M HCN + 40 mL of 0.20 M NaOH; (b) 50 mL of 0.06 M NH_3 + 60 mL of 0.05 M HCl; (c) 40 mL of 0.15 M $Ba(OH)_2$ + 100 mL of 0.12 M HCl.

23. *Dissociation constant.* The *pH* of a 0.10 M solution of the salt NaX (HX is a weak acid) is 8.30. Calculate the dissociation constant of HX.

24. *Dissociation constant.* A weak base, B, has a molecular weight of 80.0 g/mol. A solution prepared by dissolving 2.00 g of B in 200 mL of solution has a *pH* of 9.70. Calculate the dissociation constant of B.

25. *Mixtures of solutions.* Calculate the *pH* of the solutions which result from mixing equal volumes of these solutions of strong acids and bases: (a) *pH* 1.00 + *pH* 3.00; (b) *pH* 2.00 + *pH* 7.00; (c) *pH* 2.00 + *pH* 12.00; (d) *pH* 2.00 + *pH* 13.00; (e) *pH* 4.00 + *pH* 10.00.

26. *Mixtures of solutions.* A chemist wishes to prepare 200 mL of a solution of *pH* 12.60 from solutions of HCl, *pH* = 0.70, and NaOH, *pOH* = 0.40. How many milliliters

of each solution should be mixed to prepare the desired solution? Assume the volumes are additive.

27. *Approximations.* Calculate the hydrogen ion concentration of the following solutions in two ways: (1) Neglect the anion concentration as was done on page 107; (2) Solve using the quadratic formula or method of successive approximations. (a) 0.050 M dichloroacetic acid; (b) 0.050 M $NaHSO_4$; (c) 0.050 M chloroacetic acid; (d) 0.050 M formic acid; (e) 0.050 M acetic acid. Calculate the percent error in the hydrogen ion concentration in each case.

28. *Systematic calculations.* A solution is prepared by mixing 40 mL of 0.15 M HF and 60 mL of 0.10 M NaOH. The volumes are additive. Calculate the concentrations of all species in the solution using the systematic equilibrium method. Check your approximations.

29. *Systematic calculations.* Calculate the molar solubility of AgBr in 4.0 M NH_3. (This is the concentration of the free, or uncomplexed, NH_3 molecules.) Also calculate the equilibrium concentrations of the various species in the solutions and check your approximations.

30. *Solubility product constant.* Calculate the solubility product constants for the given solubilities: (a) AgI, 0.000235 mg/100 mL; (b) $Mg(OH)_2$, 0.00793 g/liter; (c) Ag_2CrO_4, 0.0262 mg/mL.

31. *Solubility.* From the solubility product constants listed in Table 3, Appendix I, calculate the solubility of the salt in water. Neglect such effects as hydrolysis. (a) $PbSO_4$ in mg/mL; (b) CaF_2 in g/liter; (c) $Cu(IO_3)_2$ in mol/liter.

32. *Solubility, common-ion effect.* Calculate the following molar solubilities: (a) $BaSO_4$ in 0.02 M K_2SO_4; (b) MgF_2 in 0.10 M NaF; (c) Ag_2CO_3 in 0.02 M $AgNO_3$.

33. *Precipitation.* Calcium fluoride, CaF_2, has a K_{sp} of 4×10^{-11}. Predict whether a precipitate forms or not when the following solutions are mixed:
(a) 100 mL of 2.0×10^{-4} M $Ca(NO_3)_2$ + 100 mL of 2.0×10^{-4} M NaF;
(b) 100 mL of 2.0×10^{-3} M $Ca(NO_3)_2$ + 100 mL of 6.0×10^{-4} M NaF.

34. *Solubility.* 100 mL of 0.050 M $Pb(NO_3)_2$ is mixed with 100 mL of 0.060 M KIO_3, pre-

cipitating $Pb(IO_3)_2$. (a) Calculate the molar solubility of $Pb(IO_3)_2$ in the solution. (b) Calculate the concentrations of these ions: Pb^{2+}, NO_3^-, K^+, and IO_3^-.

35. *Solubility.* To 100 mL of 0.10 M $AgNO_3$ is added 100 mL of 0.12 M NaCl. (a) Calculate the number of milligrams of Ag^+ not precipitated. (b) If the precipitate of AgCl is washed with 200 mL of water at 25°C, what is the maximum number of milligrams of AgCl that could be lost by solubility in the wash water?

36. *Precipitation.* To 100 mL of a solution containing 2.0 g of $Ba(NO_3)_2$ is added 2.0 g of NaF. (a) Calculate the number of grams of BaF_2 formed. (b) How many milligrams of barium remain unprecipitated?

37. *Precipitation.* The barium ion in a solution is to be precipitated by adding KIO_3. What must be the final $[IO_3^-]$ if all but 0.10 mg of Ba^{2+} is precipitated? The final volume of the solution is 150 mL.

38. *Oxidation-reduction equilibrium.* The equilibrium constant of the reaction

$$Sn^{2+} + 2Ce^{4+} \rightleftharpoons Sn^{4+} + 2Ce^{3+}$$

is 5.4×10^{43} in sulfuric acid solution at 25°C. 20.0 mL of a 0.10 M solution of Sn^{2+} is titrated with 0.10 M Ce^{4+} in sulfuric acid. Calculate the concentration of Sn^{2+} after the addition of (a) 20 mL, (b) 40 mL, and (c) 60 mL of the Ce^{4+} solution.

39. *Oxidation-reduction equilibrium.* The equilibrium constant of the reaction

$$2Ag^+ + Cu(s) \rightleftharpoons 2Ag(s) + Cu^{2+}$$

is 2.1×10^{15} at 25°C. An excess of copper turnings is added to a 0.10 M solution of Ag^+. Calculate the concentration of Ag^+ after equilibrium is reached.

40. *Oxidation-reduction equilibrium.* The equilibrium constant of the reaction

$$H^+ + Ag(s) \rightleftharpoons \tfrac{1}{2}H_2(g) + Ag^+$$

is 2.8×10^{-14} at 25°C. Suppose some pure silver metal is placed in 100 mL of 0.10 M HNO_3 which contains no Ag^+ ions. Assuming that the above reaction actually reaches equilibrium and that the pressure of H_2 is 1.0 atm, calculate the number of grams of silver that dissolve.

41. *Equilibrium constants.* From the values of the two dissociation constants of H_2CO_3 (Table 1, Appendix I) calculate the values of the equilibrium constants of these reactions:

(a) $H_2CO_3 + 2H_2O \rightleftharpoons 2H_3O^+ + CO_3^{2-}$

(b) $2HCO_3^- \rightleftharpoons H_2CO_3 + CO_3^{2-}$

42. *Completeness of reactions.* For the reaction

$$HOAc + OH^- \rightleftharpoons OAc^- + H_2O$$

calculate the value of the equilibrium constant for the following percentages of conversion of HOAc into OAc$^-$ at the EPt: (a) 67%; (b) 95%; (c) 99%

43. *Completeness of reactions.* Given the all-solution reaction

$$A + B \rightleftharpoons C + D$$

50.00 mL of 0.10 M A is titrated with 0.10 M B. Calculate the value of $\Delta pA/\Delta V$ from 49.95 mL to 50.00 mL of titrant for these values of the equilibrium constant: (a) 1.0×10^{10}; (b) 1.0×10^{12}.

6

Acid-Base Titrations

Acid-base equilibrium is an extremely important topic throughout chemistry and in other fields which utilize chemistry, such as biology, medicine, and agriculture. Titrations involving acids and bases are widely employed in the analytical control of many products of commerce, and the dissociation of acids and bases exerts an important influence upon metabolic processes in the living cell. Acid-base equilibrium, as it is taught in analytical chemistry courses, offers inexperienced students the opportunity to broaden their understanding of chemical equilibrium and to gain confidence in applying this understanding to a wide variety of problems.

In this chapter we shall calculate titration curves for strong and weak acids, learn how to select the proper indicator for a given titration, and discuss the feasibility of acid-base titrations. We shall also discuss such topics as the Brønstead theory, buffer solutions, titrations in nonaqueous solutions, and applications of acid-base titrations.

6.1

BRØNSTED TREATMENT OF ACIDS AND BASES

Although substances with acidic and basic properties had been known for hundreds of years, the quantitative treatment of acid-base equilibria became possible after 1887, when Arrhenius presented his theory of electrolytic dissociation. In water solution, according to Arrhenius, acids dissociate into hydrogen ions and anions, and bases dissociate into hydroxide ions and cations:

$$\text{Acid:} \quad HX \rightleftharpoons H^+ + X^-$$

$$\text{Base:} \quad BOH \rightleftharpoons OH^- + B^+$$

By applying to these dissociations the principles of chemical equilibrium which had been well systematized before the turn of the century, the behavior of acids and bases in aqueous solution could be quantitatively described, at least approximately. The Debye-Hückel theory (1923) permitted a refined treatment that was even better.

6.1a. The Brønsted Theory

In 1923, Brønsted presented a new view of acid-base behavior which retained the soundness of the Arrhenius equilibrium treatment but which was conceptually broader and facilitated the correlation of a much larger body of information.[1] In Brønsted terms, an acid is any substance that can give up a proton, and a base is a substance that can accept a proton. The hydroxide ion, to be sure, is such a proton acceptor and hence a Brønsted base, but it is not unique; it is one of many species that can exhibit basic behavior. When an acid yields a proton, the deficient species must have some proton affinity, and hence it is a base. Thus in the Brønsted treatment we encounter *conjugate* acid-base pairs:

$$\underset{\text{Acid}}{HB} \rightleftharpoons H^+ + \underset{\text{Base}}{B}$$

The acid HB may be electrically neutral, anionic, or cationic (e.g., HCL, HSO_4^-, NH_4^+), and thus we have not specified the charge on either HB or B.

As the elemental unit of positive charge, the proton possesses a charge density which makes its independent existence in a solution extremely unlikely. Thus, in order to transform HB into B, a proton acceptor (i.e., another base) must be present. Often, as in the dissociation of acetic acid in water, this base may be the solvent itself:

$$HOAc \rightleftharpoons H^+ + OAc^-$$
$$\underline{H_2O + H^+ \rightleftharpoons H_3O^+}$$
$$\underset{\text{Acid}_1 \quad \text{Base}_2}{HOAc + H_2O} \rightleftharpoons \underset{\text{Acid}_2 \quad \text{Base}_1}{H_3O^+ + OAc^-}$$

The interaction of the two conjugate acid-base pairs (designated by subscripts 1 and 2) leads to an equilibrium in which some of the acetic acid molecules have transferred their protons to water. The protonated water molecule or hydrated proton, H_3O^+, may be called a *hydronium ion,* but it is usually designated simply *hydrogen ion* and often written H^+.[2]

---

[1] The same ideas were proposed independently by Lowry in 1924; some writers speak of the Brønsted-Lowry theory.

[2] The proton in aqueous solution may actually be more heavily hydrated than H_3O^+. For example, a species $H_9O_4^+$ ($H_3O^+ \cdot 3H_2O$ or $H^+ \cdot 4H_2O$) has been postulated on the basis of the infrared spectra of strong acid solutions, studies of the extraction of strong acids from water into certain organic solvents, and other experimental evidence. For an interesting review, see H. L. Clever, *J. Chem. Ed.,* **40,** 637 (1963).

Water is not the only solvent to which acids can transfer their protons, and we may write a general dissociation equation, where S is any solvent capable of accepting a proton:

$$HB + S \rightleftharpoons HS^+ + B$$

The species HS^+ is the solvated proton (H_3O^+ in water solution, H_2OAc^+ in glacial acetic acid, $H_3SO_4^+$ in sulfuric acid, $C_2H_5OH_2^+$ in ethanol, etc.). One of the important contributions of Brønsted theory is its emphasis on the role of the solvent in the dissociation of acids and bases. We may suppose that an acid has a certain intrinsic "acidity" if we wish, but the Brønsted treatment makes clear that the extent to which such an acid is dissociated in solution depends importantly upon the basicity of the solvent. Thus perchloric acid, $HClO_4$, is a strong acid, completely dissociated in water solution, but it is only slightly dissociated in non-aqueous sulfuric acid.

6.1b.   *The Leveling Effect*

If HB is inherently a stronger acid than HS^+, it will transfer its proton to the solvent; in other words, the position of the equilibrium in the reaction $HB + S \rightleftharpoons HS^+ + B$ will lie toward the right. If HB is very much stronger than HS^+, the equilibrium will lie far to the right, and HB will be essentially 100% dissociated. A series of different acids, all of which are very much stronger than the solvated proton, will dissociate completely; such solutions will be brought to a level of acidity governed by the acid strength of HS^+. This is known as the *leveling effect*. Thus in aqueous solution the acids perchloric, nitric, and hydrochloric are equally strong, whereas in the less basic solvent, glacial acetic acid, the three acids are not leveled, and perchloric is stronger than the other two.

In Brønsted terms, the dissociation of bases is treated in a similar fashion, except that here the process is promoted by the *acidity* of the solvent. Again, the general case may be formulated as the interaction of two conjugate pairs:

$$SH \rightleftharpoons S^- + H^+$$
$$B + H^+ \rightleftharpoons BH^+$$
$$\overline{B + SH \rightleftharpoons BH^+ + S^-}$$

$$\text{Base}_1 \quad \text{Acid}_2 \qquad \text{Acid}_1 \quad \text{Base}_2$$

An example is $NH_3 + H_2O \rightleftharpoons NH_4^+ + OH^-$. As with acids, bases may be of any charge type (neutral, cationic, or anionic). The charges have been placed in the foregoing equations simply to show that the base and its conjugate acid differ by 1. If the solvent is sufficiently acidic, we may again encounter a leveling effect in which a series of bases are brought to a level of basicity in solution determined by the species S^-. In water, for example, so-called basic anhydrides like CaO yield OH^- by a process which may be written

$$O^{2-} + H_2O \rightleftharpoons 2OH^-$$

In anhydrous sulfuric acid sulfates are analogous to the basic anhydrides in the aqueous system:

$$SO_4^{2-} + H_2SO_4 \rightleftharpoons 2HSO_4^-$$

Neutralization reactions involving strong acids and bases in the various solvents become, in Brønsted terms, simply reactions between the cation and the anion of the solvent because of the leveling effect. Water, for example, dissociates as follows:

$$2H_2O \rightleftharpoons H_3O^+ + OH^-$$

One of the two water molecules in the equation acts as an acid, the other as a base, which is to say that water is *amphoteric*. Neutralization of strong acids and bases is simply the reverse of this self-dissociation or autoprotolysis reaction:

$$H_3O^+ + OH^- \rightleftharpoons 2H_2O$$

Likewise, in liquid ammonia solution, strong acids and bases are leveled to NH_4^+ and NH_2^-, respectively, and neutralization may be written

$$NH_4^+ + NH_2^- \rightleftharpoons 2NH_3$$

In sulfuric acid as a solvent, the reaction becomes

$$H_3SO_4^+ + HSO_4^- \rightleftharpoons 2H_2SO_4$$

The Brønsted treatment offers the conceptual advantage of unifying a number of acid-base processes which, in other terms, many appear different. Hydrolysis, for example, need no longer be distinguished as a special process. The hydrolysis of a salt like sodium acetate is simply the dissociation reaction of the acetate ion as a base:

$$OAc^- + H_2O \rightleftharpoons HOAc + OH^-$$

It may be seen that it will be a property of a conjugate acid-base pair that a strong acid has a weak conjugate base. Thus chloride ion, the conjugate base of the strong acid hydrochloric, is too weak a base to abstract protons from water, and hydrolysis of the chloride ion is negligible.

6.2

TITRATION CURVES

In examining a reaction to determine whether or not it can be used for a titration, it is of interest to construct a *titration curve* (page 120). For acid-base reactions a titration curve consists of a plot of pH or pOH vs. milliliters of titrant. Such curves are helpful in judging the feasibility of a titration and in selecting the proper indicator. We shall examine two cases, the titration of a strong acid with a strong base and the titration of a weak acid with a strong base.

In all calculations in this section we shall make the approximations used in Examples 7 through 11, pages 106–109. In Section 6.3 we shall use the systematic equilibrium approach (page 116) to check these approximations.

6.2a. *Strong Acid–Strong Base Titration*

Strong acids and bases are completely dissociated in aqueous solution (page 94). Hence the pH at various points during a tiration can be calculated directly from the stoichiometric amounts of acid and base that have been allowed to react. At the equivalence point the pH is determined by the extent to which water dissociates. At 25°C the pH of pure water is 7.00.

The following example illustrates the calculations involved in constructing a titration curve.

EXAMPLE 6.1

Strong acid–strong base

50.0 mL of 0.100 M HCl is titrated with 0.100 M NaOH. Calculate the pH at the start of the titration and after the addition of 10.0, 50.0, and 60.0 mL of titrant.

(a) *Initial pH.* HCl is a strong acid and is completely dissociated. Hence

$$[H_3O^+] = 0.100$$

$$pH = 1.00$$

(b) *pH after addition of 10.0 mL of base.* We started with 50.0 mL × 0.100 mmol/mL = 5.00 mmol of HCl and have added 10.0 mL × 0.100 mmol/mL = 1.00 mmol of NaOH. The reaction is

mmol	H_3O^+ +	OH^-	$\longrightarrow$	$2H_2O$
Initial:	5.00	1.00		
Change:	-1.00	-1.00		
Equilibrium:	4.00	—		

The reaction goes well to completion, since the equilibrium constant, K, is $1/K_w$ or 1.0×10^{14}. The concentration of H_3O^+ is

$$[H_3O^+] = \frac{4.00 \text{ mmol}}{60.0 \text{ mL}} = 6.67 \times 10^{-2} \text{ mmol/mL}$$

$$pH = 2 - \log 6.67 = 1.18$$

The pH values for other volumes of titrant can be calculated in a similar fashion.

(c) *pH at the equivalence point.* We started with 50.0 mL × 0.100 mmol/mL = 5.00 mmol HCl and have added 50.0 mL × 0.100 mmol/mL = 5.00 mmol of NaOH. The reaction is

mmol	H_3O^+ +	OH^-	$\longrightarrow$	$2H_2O$
Initial:	5.00	5.00		
Change:	-5.00	-5.00		
Equilibrium:	—	—		

The equilibrium is

$$2H_2O \rightleftharpoons H_3O^+ + OH^-$$

and

$$[H_3O^+][OH^-] = K_w = 1.0 \times 10^{-14}$$

Since

$$[H_3O^+] = [OH^-]$$

$$[H_3O^+]^2 = 1.0 \times 10^{-14}$$

$$[H_3O^+] = 1.0 \times 10^{-7}$$

$$pH = 7.00$$

(d) *pH after addition of 60.0 mL of base.* We started with 50.0 mL × 0.100 mmol/mL = 5.00 mmol of HCl and have added 60.0 mL × 0.100 mmol/mL = 5.00 mmol of NaOH. The reaction is

$$
\begin{array}{llll}
\text{mmol} & \text{H}_3\text{O}^+ + \text{OH}^- & \longrightarrow & 2\text{H}_2\text{O} \\
\text{Initial:} & 5.00 \quad 6.00 & & \\
\text{Change:} & \underline{-5.00 \quad -5.00} & & \\
\text{Equilibrium:} & \text{---} \quad\;\; 1.00 & &
\end{array}
$$

The OH^- ion concentration is

$$[OH^-] = \frac{1.00 \text{ mmol}}{110 \text{ mL}} = 9.1 \times 10^{-3} \, M$$

$$pOH = 3 - \log 9.1 = 2.04$$

$$pH = 14.00 - 2.04 = 11.96 \qquad\qquad \square$$

Box 6.1 Bases Derived from Ammonia

In Table A-1, Appendix I, it may be seen that all of the neutral molecules listed as weak bases contain nitrogen and can be considered derivatives of ammonia. (The other bases in the table are anions of weak acids such as HCO_3^-, F^-, and OCl^-.) These compounds owe their basicity to the unshared pair of electrons on nitrogen, which can accept protons from donors such as water:

$$:\text{NH}_3 + \text{H}_2\text{O} \rightleftharpoons \text{NH}_4^+ + \text{OH}^-$$

Amines are organic compounds in which hydrogen atoms of NH_3 have been replaced by alkyl or aryl groups:

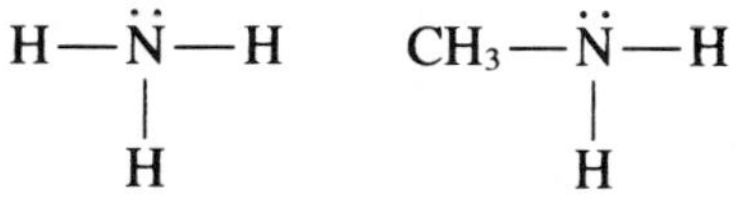

$$
\begin{array}{cc}
\text{H}-\ddot{\text{N}}-\text{H} & \text{CH}_3-\ddot{\text{N}}-\text{H} \\
\quad\;\; | & \qquad\;\; | \\
\quad\;\; \text{H} & \qquad\;\; \text{H} \\
\text{Ammonia} & \text{Methylamine}
\end{array}
$$

$$
\begin{array}{ccc}
\text{CH}_3-\ddot{\text{N}}-\text{CH}_3 & \text{CH}_3-\ddot{\text{N}}-\text{CH}_3 & \text{C}_6\text{H}_5-\ddot{\text{N}}-\text{H} \\
\qquad\; | & \qquad\;\; | & \qquad\;\; | \\
\qquad\; \text{H} & \qquad\;\; \text{CH}_3 & \qquad\;\; \text{H} \\
\text{Dimethylamine} & \text{Trimethylamine} & \text{Aniline}
\end{array}
$$

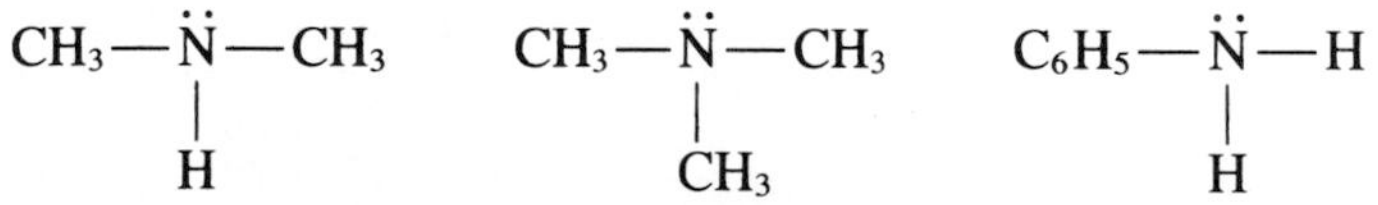

In aqueous solution, the unshared pair is involved in hydrogen bonding with a water molecule (the hydrogen bond is dotted):

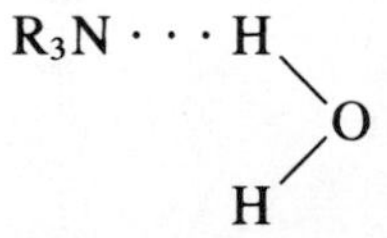

$$
\text{R}_3\text{N} \cdots \text{H}\diagdown \\
\qquad\qquad\qquad \text{O} \\
\qquad\quad \text{H}\diagup
$$

In addition, in primary and secondary amines, $\diagdown\text{N}-\text{H}$ can form a hydrogen bond with another water molecule, leading to clusters:

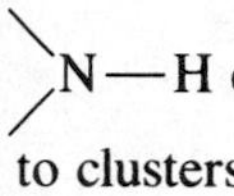

The high solubility of low-molecular-weight amines in water is partly attributable to the extensive hydrogen bonding between amine and solvent.

Alkylation of ammonia increases the basicity by the plus inductive effect of the alkyl group (cf. NH_3, $K_b = 1.8 \times 10^{-5}$; $C_2H_5NH_2$, $K_b = 5.6 \times 10^{-4}$; $(C_2H_5)_2NH$, $K_b = 1.3 \times 10^{-3}$). An aryl group has the opposite effect upon the basicity (e.g., K_b for aniline is 4.6×10^{-10}) because the nitrogen lone pair is partly shared with the aromatic ring (resonance effect).

The fourth alkylation produces a quaternary ammonium salt: $R_3N + R\!-\!X \rightarrow R_4N^+ X^-$. If X^- is exchanged for OH^-, the compound is a strong base, like NaOH. Salts such as tetra-n-butylammonium chloride, $(C_4H_9)_4N^+Cl^-$, are sometimes used in electroanalytical chemistry (e.g., in polarography, Chapter 13) when an electrolyte is desired which will not itself reduce at the cathode. Cathode potentials more negative by about 0.7 V can be attained when a salt such as NaCl or KCl is replaced by a quaternary ammonium salt. This permits the observation of electron-transfer reactions of analytes more difficult to reduce.

The pH values at other points in the titration are given in Table 6.1, and the titration curve is shown in Fig. 6.1. Note that initially the pH rises only gradually as the titrant is added, rises more rapidly as the equivalence point is approached, and then increases by about 5.40 units for the addition of only 0.10 ml of base at

TABLE 6.1 Titration of a Strong Acid and a Weak Acid with NaOH (50.0 mL of 0.100 M Acid Titrated with 0.100 M NaOH)

NaOH, mL	VOLUME OF SOLUTION	pH, HCl	pH, HB††
0.00	50.0	1.00	3.00
10.00	60.0	1.18	4.40
20.00	70.0	1.37	4.82
25.00	75.0	1.48	5.00
30.00	80.0	1.60	5.18
40.00	90.0	1.95	5.60
49.00	99.0	3.00	6.69
49.90	99.9	4.00	7.70
†49.95	99.95	4.30	8.00
50.00	100.0	7.00	8.85
†50.05	100.05	9.70	9.70
50.10	100.10	10.00	10.00
51.00	101.0	11.00	11.00
60.00	110.0	11.96	11.96
70.00	120.0	12.23	12.23

† Assuming 20 drops per milliliter, these values are 1 drop before and 1 drop after the equivalence point.

†† $K_a = 1.0 \times 10^{-5}$.

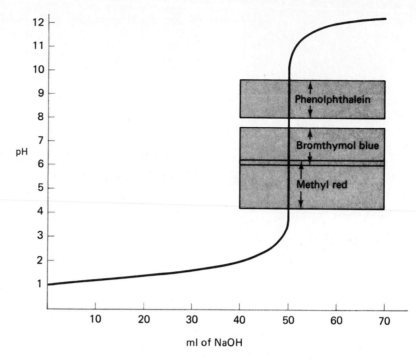

Figure 6.1 Strong acid-strong base titration curve: 50 mL of 0.10 M HCl titrated with 0.10 M NaOH.

the equivalence point. Beyond the equivalence point the pH again increases only slowly as titrant is added.

The shaded areas in Fig. 6.1 are the ranges over which three visual indicators (see Section 6.6 for discussion of indicators) change color. It is apparent that as the strong acid is titrated, the large increase in pH at the equivalence point is sufficient to span the ranges of all three indicators. Hence any one of these indicators would change color within one or two drops of the equivalence point.

The titration curve for a strong base titrated with a strong acid, say NaOH with HCl, would be indentical to the curve in Fig. 6.1 if pOH were plotted vs. volume of HCl. If pH is plotted, the curve in Fig. 6.1 is simply inverted, starting at a high value and dropping to a low pH after the equivalence point.

6.2b. *Weak Acid–Strong Base Titration*

Let us now consider the titration of a weak acid with a strong base. We reviewed the principles involved in the dissociation of weak acids in Examples 9 through 12, pages 106–110. The following example illustrates the calculations needed to construct a titration curve.

EXAMPLE 6.2

Weak acid–strong base

50.00 mL of a 0.100 M solution[3] of the weak acid, HB, $K_a = 1.0 \times 10^{-5}$, is titrated with 0.100 M NaOH. Calculate the pH at the start of the titration and after the addition of 10.0, 50.0, and 60.0 mL of titrant.

[3] Since HB is slightly dissociated, the solution is really 0.100 F in HB (Section 3.2.c). As the calculation shows, the solution is 0.099 M in HB and 0.0010 M in B⁻. Also, since sodium hydroxide is completely dissociated, the titrant is actually 0.100 F in NaOH, 0.100 M in Na⁺, and 0.100 M in OH⁻. However, most chemists seem to prefer to use the molar designation in this situation. Since the calculations of pH are not affected, we shall also designate the concentrations as molarities.

(a) *Initial pH*. Since HB is weakly dissociated, producing one B^- and one H_3O^+,

$$HB + H_2O \rightleftharpoons H_3O^+ + B^-$$

we assume that

$$[H_3O^+] \cong [B^-]$$

and

$$[HB] = 0.100 - [H_3O^+] \cong 0.100$$

Substituting in the expression for K_a, we have

$$\frac{[H_3O^+][B^-]}{[HB]} = K_a$$

$$\frac{[H_3O]^2}{0.100} = 1.0 \times 10^{-5}$$

$$[H_3O^+] = 1.0 \times 10^{-3}$$

$$pH = 3.00$$

(b) *pH after addition of 10.0 mL of base*. We started with 50.0 mL × 0.100 mmol/mL = 5.00 mmol of HB and have added 10.0 mL × 0.100 mmol/mL = 1.00 mmol of OH^-. The reaction which occurs is

mmol	HB	+	OH	$\longrightarrow$	B^-	+ H_2O
Initial:	5.00		1.00		—	
Change:	−1.00		−1.00		+1.00	
Equilibrium:	4.00		—		1.00	

The dissociation reaction and equilibrium concentrations are

$$HB + H_2O \rightleftharpoons H_3O^+ + B^-$$

$$\frac{4.00}{60.0} - [H_3O^+] \qquad\qquad [H_3O^+] \qquad \frac{1.00}{60.0} + [H_3O^+]$$

Since $[H_3O^+]$ is small,

$$[HB] \cong 4.00/60.0 \qquad \text{and} \qquad [B^-] \cong 1.00/60.0$$

The expression for K_a is

$$\frac{[H_3O^+][B^-]}{[HB]} = K_a$$

Then

$$\frac{[H_3O^+](1.00/60.0)}{4.00/60.0} = 1.0 \times 10^{-5}$$

Note that the volume cancels. Hence

$$[H_3O^+] = 4.0 \times 10^{-5}$$

$$pH = 5.0 - \log 4.0 = 4.40$$

Alternatively, we could substitute in the Henderson-Hasselbalch equation (page 109):

$$pH = pK_a + \log \frac{[B^-]}{[HB]}$$

$$pH = 5.00 + \log \frac{1.00/6.00}{4.00/60.0}$$

$$pH = 4.40$$

(c) *pH at the equivalence point*. We started with 5.00 mmol of HB and have added 50.0 mL $\times$ 0.100 mmol/mL = 5.00 mmol of OH^-. The reaction which occurs is

mmol	HB	+	OH^-	$\longrightarrow$	B^-	+ H_2O
Initial:	5.00		5.00		—	
Change:	−5.00		−5.00		+5.00	
Equilibrium:	—		—		5.00	

B^- is a base. The dissociation reaction and equilibrium concentrations are

$$B^- + H_2O \rightleftharpoons HB + OH^-$$

$$\frac{5.00}{100} - [OH^-] \qquad [HB] \quad [OH^-]$$

The expression for K_b is

$$\frac{[HB][OH^-]}{[B^-]} = K_b = \frac{K_w}{K_a} = \frac{1.0 \times 10^{-14}}{1.0 \times 10^{-5}} = 1.0 \times 10^{-9}$$

Since B^- is a weak base, we assume that $[OH^-]$ is small and

$$[B^-] = \frac{5.00}{100} - [OH^-] \cong 0.0500$$

Since the dissociation produces one HB and one OH^-, we assume that

$$[HB] \cong [OH^-]$$

Then

$$\frac{[OH^-]^2}{0.0500} = 1.0 \times 10^{-9}$$

$$[OH^-] = 7.1 \times 10^{-6}$$

$$pOH = 5.15 \quad \text{and} \quad pH = 8.85$$

(d) *pH after the addition of 60.0 mL of base*. We started with 5.00 mL of HB and have added 60.0 mL $\times$ 0.100 mmol/mL = 6.00 mmol of OH^-. The reaction which occurs is

mmol	HB	+	OH^-	$\longrightarrow$	B^-	+ H_2O
Initial:	5.00		6.00		—	
Change:	−5.00		−5.00		+5.00	
Equilibrium:	—		1.00		5.00	

There is now 1.00 mmol excess of OH^-. There is also a small amount of OH^- produced by the base B^- (reverse of the above reaction).

$$B^- + H_2O \rightleftharpoons HB + OH^-$$

This can be neglected, however, since excess OH^- shifts the equilibrium to the left. Hence

$$[OH^-] = \frac{1.00 \text{ mmol}}{110 \text{ mL}} = 9.1 \times 10^{-3} \text{ mmol/mL}$$

$$pOH = 2.04$$

$$pH = 11.96$$

The pH values at other points in the titration are given in Table 6.1, and the titration curve is shown in Fig. 6.2. The curve for the titration of a strong acid is included in this figure for comparison. Note that the curve for a weak acid begins to rise rapidly as base is first added; the rate of increase slows down as the concentration of B^- increases. The solution is said to be *buffered* (see page 150) in this region where the rate of increase in pH is slow. Note that when half of the acid is neutralized, $[HB] \cong [B^-] = 2.5$ mmol/125 mL. Since

$$pH = pK_a + \log \frac{[B^-]}{[HB]}$$

$$pH \cong pK_a = 5.00$$

After the halfway point, the pH slowly increases again until the large change occurs at the equivalence point.

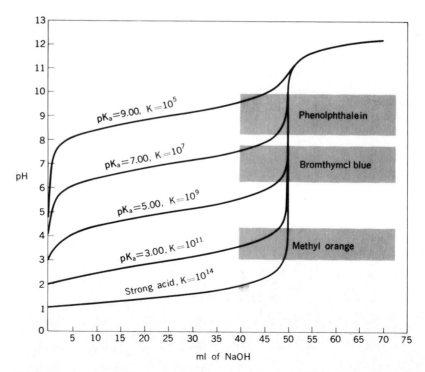

Figure 6.2 Typical acid-base titration curves: 50.0 mL of 0.100 M monoprotic acid titrated with 0.100 M NaOH; pK_a values of the acids are shown on the curves.

Titration curves for weak acids of pK_a values 3.00, 5.00, 7.00, and 9.00 are included in Fig. 6.2. The value for the K of the reaction occurring during titration,

$$HB + OH^- \rightleftharpoons B^- + H_2O$$

is K_a/K_w, or 10^{11}, 10^9, 10^7, and 10^5, respectively, for acids with the above pK_a values. It is obvious that the weaker the acid, the smaller the value of K and the smaller the change in pH at the equivalence point.

It should also be noted that the pH at the equivalence point for an acid of $K_a = 1.0 \times 10^{-5}$ is 8.85. For the addition of 0.10 ml of titrant at the equivalence point the pH changes from 8.00 to 9.70. The pH range over which the indicator phenolphthalein changes color is about 8.0 to 9.6, and this indicator will change color at the equivalence point of this titration.

6.3

SYSTEMATIC EQUILIBRIUM CALCULATIONS

In this section we shall check the approximations we made in the preceding calculations (Examples 1 and 2) by using the systematic equilibrium approach (page 116). It should be recalled that in this approach we identify all the chemical species in the solution and find a set of equations sufficient in principle to permit calculation of the concentration of each species.

6.3a. *Strong Acid-Strong Base Titration*

(a) *Initial pH*. In Example 1 (page 132) we said that $[H_3O^+]$ was 0.100 M and the pH 1.00. The assumption we are making is that the $[H_3O^+]$ contributed by the dissociation of H_2O molecules,

$$2H_2O \rightleftharpoons H_3O^+ + OH^-$$

is negligible. In this solution we have three chemical species: H_3O^+, OH^-, and Cl^-. To solve for these three concentrations we need three independent equations. These are the autoprotolysis constant of water:

$$[H_3O^+][OH^-] = K_w = 1.0 \times 10^{-14} \tag{1}$$

the charge-balance equation:

$$[H_3O^+] = [OH^-] + [Cl^-] \tag{2}$$

and the mass balance on chloride:

$$[Cl^-] = 0.10 \tag{3}$$

We recognize that water is a very weak acid compared with HCl, and thus $[OH^-]$ is very small, smaller, indeed, than the 10^{-7} M found in pure water, because the dissociation of water is repressed by the H_3O^+ from the strong acid HCl. Thus we neglect $[OH^-]$ as compared with $[Cl^-] = 0.10$ M and obtain

$$[H_3O^+] = [OH^-] + 0.10 \cong 0.10$$

From Eq. (1) it is now readily obtained that

$$[OH^-] = \frac{K_w}{[H_3O^+]} = \frac{1.0 \times 10^{-14}}{1.0 \times 10^{-1}} = 1.0 \times 10^{-13}$$

Thus the pH of the solution is 1.00, and its pOH is 13.00. Note that the approximation $[H_3O^+] \cong [Cl^-]$ is obviously alright in fairly concentrated HCl solutions. However, in very dilute HCl, say $1.0 \times 10^{-7}\,M$, a large error would result if the H_3O^+ produced by water were neglected. In such a case one would need to substitute for $[OH^-]$ in Eq. (2), giving

$$[H_3O^+] = \frac{K_w}{[H_3O^+]} + [Cl^-] = \frac{1.0 \times 10^{-14}}{[H_3O^+]} + 10^{-7}$$

a quadratic equation which can be solved for $[H_3O^+]$.

(b) *After addition of 10.0 mL of base.* Our assumption again is that the $[H_3O^+]$ contributed by the dissociation of H_2O molecules is negligible. We have four species in the solution: H_3O^+, OH^-, Cl^-, and Na^+. The four equations are

$$[H_3O^+][OH^-] = 1.0 \times 10^{-14} \qquad (1)$$

$$[Na^+] + [H_3O^+] = [OH^-] + [Cl^-] \qquad (2)$$

$$[Cl^-] = 5.00 \text{ mmol}/60.0 \text{ mL} = 0.0833 \qquad (3)$$

$$[Na^+] = 1.00 \text{ mmol}/60.0 \text{ mL} = 0.0167 \qquad (4)$$

Since the solution is acidic, $[OH^-]$ can be dropped in Eq. (2), giving

$$0.0167 + [H_3O^+] = 0.0833$$

$$[H_3O^+] = 0.0666\,M$$

From Eq. (1)

$$[OH^-] = 1.50 \times 10^{-13}$$

It should be obvious that the error made in dropping $[OH^-]$ in Eq. (2) is negligible. Since the $[H_3O^+]$ contributed by the dissociation of H_2O molecules is the same as the $[OH^-]$, our original assumption that it could be neglected is a good one.

(c) *After addition of 60.0 mL of base.* Here we neglected the $[OH^-]$ contributed by the dissociation of H_2O molecules. We still have four species: H_3O^+, OH^-, Cl^-, and Na^+. Our four equations are

$$[H_3O^+][OH^-] = 1.0 \times 10^{-14} \qquad (1)$$

$$[Na^+] + [H_3O^+] = [OH^-] + [Cl^-] \qquad (2)$$

$$[Cl^-] = 5.00 \text{ mmol}/110 \text{ mL} = 0.0455 \qquad (3)$$

$$[Na^+] = 60.0 \text{ mmol}/110 \text{ mL} = 0.0545 \qquad (4)$$

Since the solution is basic we can drop $[H_3O^+]$ in Eq. (2), giving

$$0.0545 = [OH^-] + 0.0455$$

$$[OH^-] = 0.0090$$

From Eq. (1)

$$[H_3O^+] = 1.11 \times 10^{-12}$$

Again it is obvious that the error made in dropping $[H_3O^+]$ in Eq. (2) is negligible. And since the $[OH^-]$ contributed by the dissociation of H_2O molecules is also 1.11×10^{-12}, our assumption that it could be neglected was quite valid.

6.3b. Weak Acid-Strong Base Titration

(a) *Initial pH*. In Example 2 (page 135) we started with 50.0 mL of 0.100 M HB. In this solution we have four species: H_3O^+, OH^-, HB, and B^-. Our four equations are

$$[H_3O^+][OH^-] = 1.0 \times 10^{-14} \tag{1}$$

$$\frac{[H_3O^+][B^-]}{[HB]} = 1.0 \times 10^{-5} \tag{2}$$

$$[H_3O^+] = [B^-] + [OH^-] \tag{3}$$

$$[HB] + [B^-] = 0.10 \tag{4}$$

The third equation is the charge-balance expression, and the fourth the mass balance on B.

Since the solution is acidic, we assume that $[OH^-]$ can be neglected in Eq. (3), giving

$$[H_3O^+] \cong [B^-]$$

Also, since HB is weak, $[B^-]$ is small compared to $[HB]$. Equation (4) becomes

$$[HB] \cong 0.10$$

Substituting into Eq. (2) and solving gives

$$[H_3O^+] = 1.0 \times 10^{-3} \quad \text{and} \quad [OH^-] = 1.0 \times 10^{-11}$$

Note that the assumption that $[OH^-]$ can be neglected in Eq. (3) is a good one, since

$$1.0 \times 10^{-3} \cong 1.0 \times 10^{-3} + 1.0 \times 10^{-11}$$

Checking the second assumption in Eq. (4) gives

$$0.10 + 0.001 \cong 0.10$$

Here the error is larger. The relative error (disregarding significant figures) is

$$\frac{0.001}{0.10} \times 100 = 1.0\%$$

(b) *After addition of* 10.0 *mL of base*. We now have five species in the solution: H_3O^+, OH^-, HB, B^-, and Na^+. Our five equations are K_a, K_w, the charge-balance equation,

$$[H_3O^+] + [Na^+] = [B^-] + [OH^-] \tag{1}$$

and the mass-balance equations,

$$[HB] + [B^-] = \frac{5.00 \text{ mmol}}{60.0 \text{ mL}} = 0.0833 \tag{2}$$

$$[Na^+] = \frac{1.00 \text{ mmol}}{60.0 \text{ mL}} = 0.0167 \tag{3}$$

Since the solution is acidic, we assume that $[OH^-]$ is small; Eq. (1) becomes

$$[H_3O^+] + 0.0167 \cong [B^-]$$

But since HB is a weak acid, $[H_3O^+]$ is probably much smaller than 0.0167. Then

$$[B^-] \cong 0.0167$$

Substituting this into Eq. (2) gives

$$[HB] + 0.0167 = 0.0833$$

$$[HB] = 0.0666$$

These values of [HB] and $[B^-]$ can be substituted into the expression for K_a, giving

$$\frac{[H_3O^+] \times 0.0167}{0.0666} = 1.0 \times 10^{-5}$$

$$[H_3O^+] = 4.0 \times 10^{-5} \quad \text{and} \quad [OH^-] = 2.5 \times 10^{-10}$$

Checking the assumption in Eq. (1) gives

$$4.0 \times 10^{-5} + 0.0167 \cong 0.0167 + 2.5 \times 10^{-10}$$

The left-hand term differs from the right-hand term by only about 0.2%.

(c) *Equivalence point.* The solution contains the same five species as in part (b). The five equations are K_b, K_w, the charge-balance equation,

$$[H_3O^+] + [Na^+] = [B^-] + [OH^-] \tag{1}$$

and the mass-balance equations,

$$[HB] + [B^-] = \frac{5.00 \text{ mmol}}{100 \text{ mL}} = 0.050 \tag{2}$$

$$[Na^+] = 0.050 \tag{3}$$

In this situation $[H_3O^+]$, $[OH^-]$, and [HB] are all small terms, and in order to obtain a useful expression involving them, Eqs. (1) and (2) should be added, giving

$$[H_3O^+] + [Na^+] + [HB] = [OH^-] + 0.050$$

Since $[Na^+] = 0.050$, this becomes

$$[H_3O^+] + [HB] = [OH^-] \tag{4}$$

Since the solution is basic, $[H_3O^+]$ is small and we can assume it is negligible, giving

$$[HB] \cong [OH^-]$$

If [HB] is neglected in Eq. (2), we have

$$[B^-] \cong 0.050$$

We can substitute in the expression for K_b, giving

$$\frac{[OH^-]^2}{0.050} = 1.0 \times 10^{-9}$$

$$[OH^-] = [HB] = 7.1 \times 10^{-6} \quad \text{and} \quad [H_3O^+] = 1.4 \times 10^{-9}$$

Checking the assumption in Eq. (4),

$$1.4 \times 10^{-9} + 7.1 \times 10^{-6} = 7.1 \times 10^{-6}$$

The left-hand term differs from the right-hand term by only about 0.02%. Checking Eq. (2),

$$7.1 \times 10^{-6} + 0.050 = 0.050$$

The two terms differ by only about 0.01%.

(d) *After addition of* 60.0 *mL of base*. The solution contains the same five species as in part (b). The five equations are: K_w, K_a (or K_b), the charge-balance equation,

$$[H_3O^+] + [Na^+] = [OH^-] + [B^-] \tag{1}$$

and the mass-balance equations,

$$[HB] + [B^-] = \frac{5.0 \text{ mmol}}{110 \text{ mL}} = 0.045 \tag{2}$$

$$[Na^+] = \frac{6.0 \text{ mmol}}{110 \text{ mL}} = 0.055 \tag{3}$$

We assume that $[H_3O^+]$ and $[HB]$ are negligible, giving

$$[B^-] = 0.045$$

and

$$0.055 = [OH^-] + 0.045$$

$$[OH^-] = 1.0 \times 10^{-2} \quad \text{and} \quad [H_3O^+] = 1.0 \times 10^{-12}$$

$[HB]$ can be obtained from K_a:

$$\frac{(1.0 \times 10^{-12})(0.045)}{[HB]} = 1.0 \times 10^{-5}$$

$$[HB] = 4.5 \times 10^{-9}$$

Checking Eq. (2) gives

$$4.5 \times 10^{-9} + 0.045 = 0.045$$

The left-hand side differs from the right-hand side by about $1 \times 10^{-5}\%$. Substitution in Eq. (1) gives

$$1.0 \times 10^{-12} + 0.055 = 0.010 + 0.045$$

The two sides differ by only about $2 \times 10^{-9}\%$.

6.4

ACID-BASE INDICATORS

6.4a. Theory of Indicator Behavior

The analyst takes advantage of the large change in pH that occurs in titrations in order to determine when the equivalence point is reached. There are many weak organic acids and bases in which the undissociated and ionic forms show different colors. Such molecules may be used to determine when sufficient titrant has been added and are termed *visual indicators*. A simple example is *p*-nitrophenol, which is a weak acid, dissociating as follows:

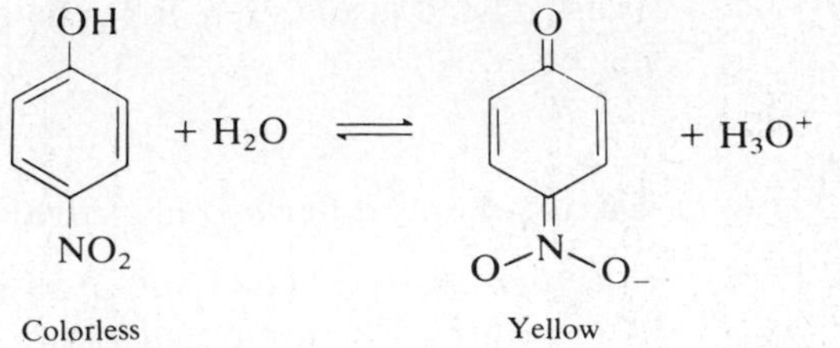

OH
+ H$_2$O $\rightleftharpoons$ + H$_3$O$^+$

NO$_2$
O$-$N$-$O$^-$

Colorless   Yellow

The undissociated form is colorless, but the anion, which has a system of alternating single and double bonds (a conjugated system), is yellow. Molecules or ions having such conjugated systems absorb light of longer wavelengths than comparable molecules in which no conjugated system exists. The light absorbed is often in the visible portion of the spectrum, and hence the molecule or ion is colored (See Chapter 14).

The well-known indicator phenolphthalein (below) is a diprotic acid and is colorless. It dissociates first to a colorless form and then, on losing the second proton, to an ion with a conjugated system; a red color results.[4] Methyl orange,

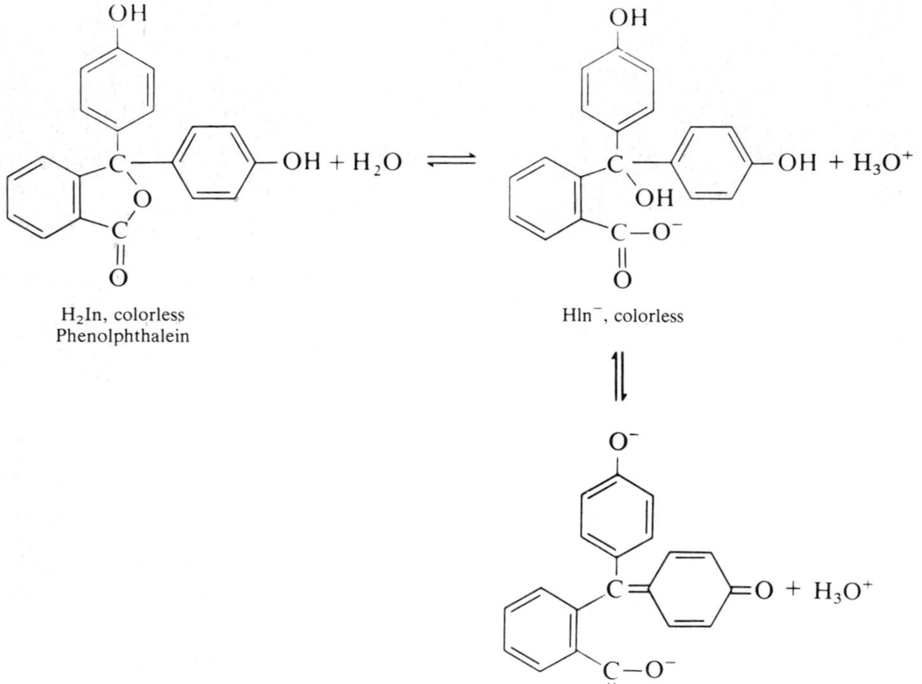

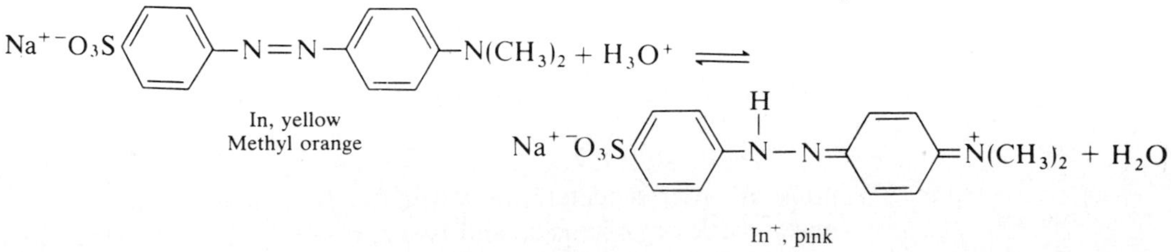

H$_2$In, colorless
Phenolphthalein

Hln$^-$, colorless

In^{2-}, red

Na^{+}$^-$O$_3$S$-$N$=$N$-$N(CH$_3$)$_2$ + H$_3$O$^+$ $\rightleftharpoons$

In, yellow
Methyl orange

Na^{+}$^-$O$_3$S$-$N$-$N$=$$=N^+$(CH$_3$)$_2$ + H$_2$O

In$^+$, pink

[4] In 65 to 98% H$_2$SO$_4$ the colorless form of phenolphthalein adds a proton and turns an orange-red color. Above pH 11, the In^{2-} ion adds an OH$^-$ ion and turns colorless. See G. Wittke, *J. Chem. Ed.*, **60**, 239 (1983).

another widely used indicator, is a base and is yellow in the molecular form. Addition of a proton gives a cation which is pink in color.

6.4b. Determining the Color-Change Range of an Indicator

For simplicity, let us designate an acid indicator as HIn, and a basic indicator as In. The dissociation expressions are

$$HIn + H_2O \rightleftharpoons H_3O^+ + In^-$$

$$In + H_2O \rightleftharpoons InH^+ + OH^-$$

The dissociation constant of the acid is

$$K_a = \frac{[H_3O^+][In^-]}{[HIn]}$$

In the logarithmic form, this becomes

$$pH = pK_a - \log \frac{[HIn]}{[In^-]}$$

Let us for illustration assume that the molecule HIn is red in color and the ion In$^-$ is yellow. Both forms are present, of course, in a solution of the indicator, their relative concentrations depending upon the pH. The color that the human eye detects depends upon the relative amounts of the two forms. Obviously, in solutions of low pH, the acid HIn predominates and we would expect to see only a red color. In solutions of high pH, In$^-$ should predominate and the color should be yellow. At intermediate pH values, where the two forms are in about equal concentrations, the color might be orange.

Suppose that the pK_a of HIn is 5.00 and that a few drops of HIn are added to a solution of a strong acid which is being titrated with a strong base. The quantity of HIn added is so small that the amount of titrant used by HIn can be considered negligible. Now let us follow the ratio of the two colored forms as the pH changes during the titration. This is shown in Table 6.2. Let us also assume that the solution appears red to the eye when the [HIn]/[In$^-$] ratio is as large as 10 : 1, and yellow when this ratio is 1 : 10 or less. In such a case, the minimum change in pH, designated ΔpH, required to cause a color change from red to yellow is 2 units:

$$\begin{array}{ll} \text{Yellow:} & pH_y = pK_a + \log 10/1 = 5 + 1 \\ \text{Red:} & pH_r = pK_a + \log 1/10 = 5 - 1 \\ \hline & \Delta pH = pH_y - pH_r = 6 - 4 = 2 \end{array}$$

TABLE 6.2 Ratio of Colored Forms of Indicator at Various pH Values

pH SOLUTION	[HIn]/[In$^-$] RATIO	COLOR	
1	10,000 : 1	Red	
2	1000 : 1	Red	
3	100 : 1	Red	
4	10 : 1	Red	⎫
5	1 : 1	Orange	⎬ Range
6	1 : 10	Yellow	⎭
7	1 : 100	Yellow	
8	1 : 1000	Yellow	

This minimum change in pH required for a color change is referred to as the *indicator range*. In our example, the range is 4 to 6. At intermediate pH values, the color shown by the indicator is not red or yellow but some shade of orange. At pH 5, the pK_a of HIn, the two colored forms are in equal concentrations; that is, HIn is half-neutralized. Frequently, one hears a statement such as, "An indicator which changed color at pH 5 was employed." This means that the pK_a of the indicator is 5, and the range is approximately pH 4 to 6.

Table 6.3 lists some acid-base indicators together with their approximate ranges. Note that the ranges are roughly 1 to 2 pH units, in general agreement with the assumption we made above. Actually, the range may not be symmetrical about the pK of the indicator, since a higher ratio may be required for the observer to see one form than is required to see the other. It should also be noted that various indicators change color at widely different pH values. It is necessary for the analyst to select the proper indicator for the titration.

TABLE 6.3 Some Acid-Base Indicators

INDICATOR	COLOR CHANGE WITH INCREASING pH	pH RANGE
Picric acid	Colorless to yellow	0.1–0.8
Thymol blue	Red to yellow	1.2–2.8
2,6-Dinitrophenol	Colorless to yellow	2.0–4.0
Methyl yellow	Red to yellow	2.9–4.0
Bromphenol blue	Yellow to blue	3.0–4.6
Methyl orange	Red to yellow	3.1–4.4
Bromcresol green	Yellow to blue	3.8–5.4
Methyl red	Red to yellow	4.2–6.2
Litmus	Red to blue	5.0–8.0
Methyl purple	Purple to green	4.8–5.4
p-Nitrophenol	Colorless to yellow	5.6–7.6
Bromcresol purple	Yellow to purple	5.2–6.8
Bromthymol blue	Yellow to blue	6.0–7.6
Neutral red	Red to yellow	6.8–8.0
Phenol red	Yellow to blue	6.8–8.4
p-α-Naphtholphthalein	Yellow to blue	7.0–9.0
Phenolphthalein	Colorless to red	8.0–9.6
Thymolphthalein	Colorless to blue	9.3–10.6
Alizarin yellow R	Yellow to violet	10.1–12.0
1,3,5-Trinitrobenzene	Colorless to orange	12.0–14.0

6.4c. Selection of Proper Indicator

In Fig. 6.2 the shaded areas are the indicator ranges of methyl orange (3.1 to 4.4), bromthymol blue (6.0 to 7.6), and phenolphthalein (8.0 to 9.6). It is apparent that as a strong acid is titrated, the large change in pH at the equivalence point is sufficient to span the ranges of all three indicators. Hence any one of these indicators would change color within one or two drops of the equivalence point, as would any other indicator changing color between pH 4 and 10.

In the titration of weaker acids, the choice of indicators is much more limited. For an acid of pK_a 5, approximately that of acetic acid, the pH is higher than 7 at the equivalence point and the change in pH is relatively small. Phenolphthalein changes color at approximately the equivalence point and is a suitable indicator.

In the case of a very weak acid, for example, $pK_a = 9$, no large change in

pH occurs in the vicinity of the equivalence point. Hence a large volume of base would be required to change the color of an indicator, and the equivalence point could not be detected with the usually desired precision.

As a general rule, then, one should select an indicator which changes color at approximately the pH at the equivalence point of the titration. For weak acids, the pH at the equivalence point is above 7, and phenolphthalein is the usual choice. For weak bases, where the pH is below 7, methyl red (4.2 to 6.2) or methyl orange is widely used. For strong acids and strong bases, methyl red, bromthymol blue, and phenolphthalein are suitable.

6.4d. Indicator Errors

There are at least two sources of error in determination of the end point of a titration using visual indicators. One occurs when the indicator employed does not change color at the proper pH. This is a determinate error and can be corrected by the determination of an *indicator blank*. The latter is simply the volume of acid or base required to change the pH from that at the equivalence point to the pH at which the indicator changes color. The indicator blank is usually determined experimentally.

A second error occurs in the case of very weak acids (or bases) where the slope of the titration curve is not great and hence the color change at the end point is not sharp. Even if the proper indicator is employed, an indeterminate error occurs and is reflected in a lack of precision in deciding exactly when the color change occurs. The use of a nonaqueous solvent (page 161) may improve the sharpness of the end point in such cases.

In order to sharpen the color change shown by some indicators, mixtures of two indicators, or of an indicator and an indifferent dye, are sometimes used. The familiar "modified methyl orange" for carbonate titrations is a mixture of methyl orange and the dye xylene cyanole FF. The dye absorbs some of the wavelengths of light that are transmitted by both colored forms, thus cutting down on the overlapping of the two colors. At an intermediate pH, the methyl orange assumes a color which is almost complementary to that of xylene cyanole FF, and the solution thus appears gray. This color change is more easily detected than the gradual change of methyl orange from yellow to red through a number of shades of orange. Many mixtures of two indicators have been recommended for improved color changes.

6.5

FEASIBILITY OF ACID-BASE TITRATIONS

We have previously mentioned that for a chemical reaction to be suitable for use in a titration, the reaction must be complete at the equivalence point. The degree of completeness of the reaction determines the size and sharpness of the vertical portion of the titration curve. The larger the equilibrium constant, the more complete the reaction, the larger the change in pH near the equivalence point, and the easier it is to locate the equivalence point with good precision. The completeness of the reaction is related to the practical feasibility of the titration. Theoretically, it may be possible to locate the equivalence point of a reaction which does not go well to completion, but practically, it may be a difficult problem.

The equilibrium constant for the titration of a strong acid with a strong base is quite large:

$$H_3O^+ + OH^- \rightleftharpoons 2H_2O \qquad K = \frac{1}{K_w} = 1.0 \times 10^{14}$$

We have noted the large ΔpH which occurs at the equivalence point, 5.40 units for $\Delta V = 0.10$ mL, and have pointed out that because of this large change, several indicators could be used to determine the equivalence volume with a precision of a few parts per thousand. Hence we say that the titration is *feasible*.

6.5a. *Magnitude of the Equilibrium Constant*

How large must the equilibrium constant be for a titration to be feasible? It is difficult to give an unequivocal answer to this question. The concentrations of the substance titrated and the titrant influence the magnitude of ΔpH, and under certain circumstances an analyst might be satisfied with less precision than we specified above. However, if we are given a specific set of conditions to be met, we can make a rather simple calculation to determine the magnitude of K. It is generally desired that essentially all of the substance titrated be converted into product at or near the equivalence point. In Chapter 5 (page 119) we calculated values of K for 99.9% and 99.99% conversion of the analyte into product at the equivalence point. It is also desirable that the pH change by 1 or 2 units on the addition of a few drops of titrant at the equivalence point if a visual indicator is to be employed. The following example illustrates a calculation of K_a for a weak acid and K for the titration reaction for a specific statement as to feasibility requirements.

EXAMPLE 6.3

Calculation of minimum K for a feasible titration

50 mL of 0.10 M HA is titrated with 0.10 M strong base. (a) Calculate the minimum value of K so that when 49.95 mL of titrant has been added, the reaction between HA and OH^- is essentially complete and the pH changes by 2.00 units on the addition of two more drops (0.10 mL) of titrant. (b) Repeat the calculation for $\Delta pH = 1.00$ unit.

(a) The pH 0.05 mL beyond the equivalence point can be calculated as follows:

$$[OH^-] = \frac{0.05 \times 0.10}{100.05} = 5 \times 10^{-5}\ M$$

$$pOH = 4.30$$

$$pH = 9.70$$

If ΔpH is to be 2.00 units, the pH 0.05 mL before the equivalence point must be 7.70. At this point, if the reaction is complete, we have only 0.005 mmol of HA unreacted. Hence

$$pH = pK_a + \log \frac{[A^-]}{[HA]}$$

$$7.70 = pK_a + \log \frac{4.995}{0.005}$$

$$pK_a = 4.70$$

$$K_a = 2.0 \times 10^{-5}$$

$$K = \frac{K_a}{K_w} = \frac{2.0 \times 10^{-5}}{1.0 \times 10^{-14}} = 2.0 \times 10^9$$

(b) If $\Delta p\text{H} = 1.00$, then

$$8.70 = pK_a + \log \frac{4.995}{0.005}$$

$$pK_a = 5.70$$

$$K_a = 2.0 \times 10^{-6}$$

$$K = 2.0 \times 10^{8} \qquad \square$$

6.5b. *Effect of Concentration*

The magnitude of $\Delta p\text{H}$ at the equivalence point also depends upon the concentrations of the analyte and the titrant. The effect of concentration on the change in $p\text{H}$ for the strong acid–strong base titration is shown is Fig. 6.3. The $\Delta p\text{H}$ decreases as the concentrations of analyte and titrant decrease.

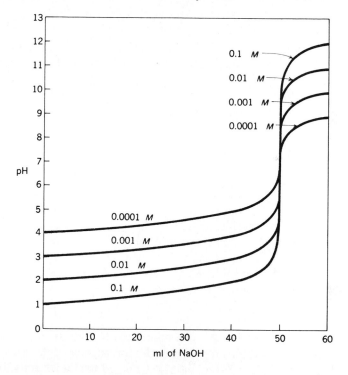

Figure 6.3 Effect of concentration on titration curves of strong acids with strong bases. 50 mL acid titrated with base of same molarity as that of the acid.

For weak acids the effect of concentration as well as the magnitude of K_a on $\Delta p\text{H}$ is shown in Table 6.4. The following conclusions can be drawn from the table:

1. The smaller the value of K_a, the higher the $p\text{H}$ at the the equivalence point and the smaller $\Delta p\text{H}$.
2. (a) Increasing the amount of HA titrated in the same initial volume decreases $\Delta p\text{H}$. However, this increases the volume of titrant required, rendering a given error in determining the end point a smaller relative error. (b) If the same amount of HA is titrated but the initial volume is decreased, $\Delta p\text{H}$ is in-

TABLE 6.4 ΔpH for Titration of a Weak Acid, HA, with 0.1 M Strong Base

K_a OF HA	mmol HA TITRATED	INITIAL VOLUME, mL	pH 0.05 mL BEFORE EQ. PT.	pH AT EQ. PT.	pH 0.05 mL AFTER EQ. PT.	ΔpH FOR 0.10 mL
	2.5	75	7.70	8.70	9.70	2.00
	2.5	50	7.70	8.76	9.82	2.12
1×10^{-5}	2.5	25	7.70	8.85	10.00	2.30
	5.0	75	8.00	8.80	9.60	1.60
	5.0	50	8.00	8.85	9.70	1.70
	5.0	25	8.00	8.91	9.82	1.82
5×10^{-6}	2.5	75	8.00	8.85	9.70	1.70
1×10^{-6}	2.5	75	8.70	9.20	9.70	1.00
5×10^{-7}	2.5	75	9.00	9.35	9.70	0.70
$1 \times 10^{-7\dagger}$	2.5	75	9.70	9.70	9.70	0.00

† Usual approximate calculations.

creased. This is caused primarily by the fact that excess titrant is in a smaller volume. (See below.)

3. Increasing the concentration of titrant increases ΔpH. This decreases the volume of titrant required, thus making a given error a larger relative error.

For the titration of a given amount of a certain weak acid, the procedure recommended to increase ΔpH is as described in 2(b). Starting with a smaller volume, that is, an increased concentration of HA, will increase ΔpH at the equivalence point while using the same volume of titrant.

As a general rule it can be said that a precision of a few parts per thousand can be obtained in the titration of a 0.05 M solution of a weak acid or base of dissociation constant as low as 1×10^{-6}, using a 0.10 M titrant. This corresponds to a value of K of about 1×10^{8}. Still weaker acids or bases may be titrated with sacrifice of precision in determining the end point.

A salt of a weak acid, that is, a Brønsted base, can be titrated feasibly with a strong acid if the acid itself is too weak for feasible titration. For example, an acid HA, $K_a = 1 \times 10^{-9}$, is too weak for feasible titration. The dissociation constant of the conjugate base A$^-$ is 1×10^{-5}, since

$$K_a \times K_b = 1 \times 10^{-14}$$

Hence A$^-$ can be titrated feasibly with a strong acid. Similar conclusions can be drawn for the titration of salts of weak bases.

It might also be noted that in aqueous solutions the most complete acid-base reaction is that between H$_3$O$^+$ and OH$^-$ and the value of K is 1×10^{14}. The large change in pH which occurs in this case is roughly 6 units, from about pH 4 to 10. One can conclude that if the pH at the equivalence point of a titration falls below 4 or above 10, the magnitude of ΔpH/ΔV will not be very large. It is doubtful that the titration will be considered feasible. One might also note that the closer the equivalence point pH is to 4 or 10, the smaller will be the value of ΔpH/ΔV.

6.6

BUFFER SOLUTIONS

A solution which resists large changes in pH when hydrogen or hydroxide ions are added, or when the solution is diluted, is called a *buffer solution*. We have encountered such solutions in the titration of a weak acid with a strong base. Note the flat portion of the titration curve for the acid $pK_a = 5.00$ in Fig. 6.2. The pH increases only about 2 units during the addition of about 45 mL of NaOH. Then at the equivalence point, where the solution is no longer buffered, the pH increases nearly 2 units for the addition of only 0.10 mL of NaOH.

Many chemical and biological processes are very sensitive to changes in the pH of a solution, and it is extremely important to maintain as constant a pH as possible. Buffer solutions are therefore of considerable interest in the chemical and biological sciences.

6.6a. *Behavior of Buffer Solutions*

In general, buffer solutions contain a conjugate acid-base pair, such as HOAc–OAc$^-$ or NH$_3$–NH$_4^+$. (We have discussed the calculation of the pH of such solutions as examples of the *common-ion effect* in Example 10; Section 5.4a.) These components react with any hydrogen or hydroxide ions entering the solution. For example, if the buffer solution contains HOAc and NaOAc, any hydrogen ion entering the solution is consumed by reaction with acetate ion:

$$H_3O^+ + OAc^- \longrightarrow HOAc + H_2O$$

and hydroxide ion reacts with acetic acid molecules:

$$OH^- + HOAc \longrightarrow OAc^- + H_2O$$

The pH does not change appreciably because it is necessary to change the conjugate acid-base ratio by a factor of ten to change the pH by 1 unit:

$$pH = pK_a - \log \frac{[HOAc]}{[OAc^-]}$$

or

$$pH = pK_a + \log \frac{[OAc^-]}{[HOAc]}$$

The following example illustrates the difference between a buffered and an unbuffered solution in resisting a change in pH when hydroxide ion is added.

EXAMPLE 6.4

Comparison of buffered and unbuffered solutions

There are two flasks containing (a) 100 mL of pure water, $pH = 7.00$, and (b) 100 mL of a solution containing 10 mmol of the acid HA, $pK_a = 7.00$, and 10 mmol of the conjugate base A$^-$. The pH of this solution is also 7.00. To each flask is added 1.0 mmol of solid NaOH. Calculate the pH of the resulting solutions and the change in pH.

(a) In pure water the hydroxide ion concentration becomes

$$[OH^-] = 1.00 \text{ mmol}/100 \text{ mL} = 0.0100 \ M$$

$$pOH = 2.00$$

$$pH = 12.00$$

The change in pH is from 7.00 to 12.00, or 5.00 units.

(b) In this solution the 1.0 mmol of OH$^-$ reacts with 1.0 mmol of HA, decreasing the mmol of HA to 9.0 and increasing the mmol of A$^-$ to 11.0. Hence,

$$pH = 7.00 + \log \frac{11}{9}$$

$$pH = 7.09$$

Here the change in pH is only from 7.00 to 7.09, or 0.09 units. □

6.6b. *Buffer Effectiveness*

The effectiveness of a buffer solution in resisting change in pH per unit of strong acid or base added is greatest when the ratio of buffer acid to salt is unity. In the titration of a weak acid this point of maximum effectiveness is reached when the acid is half-neutralized, or $pH = pK_a$. This can be seen from the following calculation:

EXAMPLE 6.5

Calculation of the point of maximum effectiveness of buffered solution

Calculate the slope of the titration curve for the weak acid HA titrated with OH$^-$ and find its minimum value. Let a = original mmol of HA and b = mmol of OH$^-$ added. Then

$$[HA] = \frac{(a - b)}{v}$$

$$[A^-] = \frac{b}{v}$$

where v is the volume of solution.

$$pH = pK_a + \log \frac{b}{a - b} = pK_a - \log(a - b) + \log b$$

Differentiating, the slope is

$$\frac{dpH}{db} = \frac{0.43}{a - b} + \frac{0.43}{b} = \frac{0.43a}{b(a - b)}$$

To find the minimum value of the slope, differentiate the preceding expression and equate to zero,

$$\frac{d^2pH}{db^2} = -\frac{0.43a(a - 2b)}{b^2(a - b)^2} = 0$$

$$b = \frac{a}{2}$$

That is, $[HA] = [A^-]$, and at this point $pH = pK_a$. $\square$

It may be instructive to examine Table 6.5, which shows the change in pH produced during the titration of two different amounts of acetic acid at intervals

TABLE 6.5 Change in pH during Titration of Acetic Acid

mmol OH$^-$ ADDED TO 10 mmol HOAc	pH[†]	ΔpH	mmol OH$^-$ ADDED TO 20 mmol HOAc	pH[†]	ΔpH
0	2.87	—	0	2.72	—
1	3.79	0.92	1	3.46	0.74
2	4.14	0.35	2	3.79	0.33
3	4.37	0.23	3	3.99	0.20
4	4.56	0.19	9	4.65	—
5	4.74	0.18	10	4.74	0.09
6	4.92	0.18	11	4.83	0.09
7	5.11	0.19	12	4.92	0.09
8	5.34	0.23	18	5.69	—
9	5.69	0.35	19	6.02	0.33
10	8.87	3.18	20	9.02	3.00

[†] Calculated for 100 mL volume assuming no change in volume as base added.

of 1 mmol of added base. It is apparent that at the start of the titration the solution is not well buffered, and the pH rises rapidly as base is added. This explains the initial rapid rise in the titration curves of weak acids shown in Fig. 6.2. The rate of rise in pH decreases, passes through a minimum at $pH = pK_a$, and then slowly increases again. At the equivalence point, a large change occurs, since the acid is exhausted and the solution is no longer buffered.

6.6c. *Buffer Capacity*

The capacity of a buffer is a measure of its effectiveness in resisting changes in pH upon the addition of acid or base. The greater the concentrations of the acid and conjugate base, the greater the capacity of the buffer. This is evident from Table 6.5 in that twice as much base is required to increase the pH of the more concentrated solution from 3.79 to 4.74 than is needed for the more dilute solution. Buffer capacity can be defined more quantitatively as the number of moles of strong base required to change the pH of 1 liter of solution by 1 pH unit. The term *range* of a buffer is ill-defined, but it is evident from Table 6.5 that little buffering action is obtained if the acid-salt ratio is greater than 9 : 1 or less than 1 : 9.

In preparing a buffer of a desired pH, the analyst should select an acid-salt (or base-salt) system in which the pK_a of the acid is as close as possible to the desired pH. By this selection, the ratio of acid to salt is near unity, and maximal effectiveness against an increase or decrease in pH is obtained. The actual concentrations of acid and salt employed depend upon the desired resistance to change in pH. These points are illustrated in the following example.

EXAMPLE 6.6   (a) It is desired to prepare 100 mL of a buffer of pH 5.00. Acetic, benzoic, and formic acids and their salts are available for use. Which acid should be used for maximum effectiveness against increase or decrease in pH? What acid-salt ratio should be used?

The pK_a values of these acids are: acetic. 4.74; benzoic, 4.18; and formic, 3.68. The pK_a of acetic is closest to the desired pH, and this acid and its salt should be used. Then

$$pH = pK_a + \log \frac{[OAc^-]}{[HOAc]}$$

$$5.00 = 4.74 + \log \frac{[OAc^-]}{[HOAc]}$$

Taking antilogs, $[OAc^-]/[HOAC] = 1.8/1$; i.e., there should be 1.8 times as much salt as acid. This is the molar ratio, not the ratio of grams.

(b) If it is desired that the change in pH of the buffer be no more than 0.10 unit for the addition of 1 mmol of either acid or base, what minimum concentrations of the acid and salt should be used?

Since there is less acid present than salt, a greater change in pH will result if base is added. Hence if we calculate on the basis that base is added, the condition will be more than satisfied for the addition of acid. Thus, if

$$x = \text{mmol acid originally present}$$

$$1.8x = \text{mmol salt originally present}$$

If 1 mmol OH$^-$ is added, then

$$x - 1 = \text{mmol acid remaining}$$

$$1.8x + 1 = \text{mmol acid salt}$$

Then

$$5.10 = 4.74 + \log \frac{1.8x + 1}{x - 1}$$

Solving gives $x = 6.6$ mmol, and $1.8x = 11.9$ mmol. The molar concentrations are then $[\text{HOAc}] = 0.066$ mmol/mL, and $[\text{OAc}^-] = 0.119$ mmol/mL. □

6.6d. Buffers and Solutions of Strong Acids and Bases

Concentrated solutions of strong acids and bases resist large changes in pH, and the titration curves are flat over a fairly wide range of pH (Fig. 6.1). Such solutions may be used to maintain constant pH at fairly low or high values. For example, a $0.100\ M$ solution of HCl can be used as a buffer of pH 1.00. 100 mL of this solution contains 10.0 mmol of H_3O^+, and the addition of 1.00 mmol of OH^- to the solution changes the pH by only about 0.05 unit. However, a 0.0100 M solution of HCl used as a buffer of pH 2.00 has a very low capacity for removing hydroxide ion. 100 mL of this solution contains only 1.00 mmol of H_3O^+, and the addition of 1.00 mmol of OH^- will change the pH from 2.00 to 7.00.

Solutions of strong acids and bases also undergo an appreciable change in pH upon dilution. If 100 mL of water is added to 100 mL of $0.100\ M$ HCl, the $[H_3O^+]$ becomes 0.050 and the pH changes from 1.00 to 1.30. In contrast, the pH of a buffer made with a conjugate acid-base pair is theoretically independent of the volume of the solution, since the pH is dependent on the *ratio* of the base to acid as expressed by the Henderson-Hasselbalch equation:

$$p\text{H} = pK_a + \log \frac{[\text{base}]}{[\text{acid}]}$$

An appreciable change in pH upon dilution will occur if the concentrations of the buffering components are below about $10^{-4}\ M$. At such low concentrations the dissociation of water molecules can become an important factor. However, one does not normally employ a buffer using such low concentrations, since the buffering capacity is very low.

6.6e. Preparation of Buffer Solutions

We have seen how to calculate the concentrations of a conjugate acid-base pair required to prepare a buffer of a certain pH and capacity (page 153). If we do prepare such a solution and then measure the pH in the laboratory, we will probably find that the measured value differs somewhat from the calculated value. There are at least three reasons for such differences: (1) uncertainties in the values of the dissociation constants of the weak acids and bases, (2) errors caused by the approximations used in our calculations, (3) activity effects. Usually the ionic strength of a buffer is sufficiently high to cause activity coefficients to deviate considerably from unity.

Experimental values of pH are normally measured in the laboratory by a potentiometric method, using a glass electrode and a so-called pH meter. This topic is discussed in some detail in Chapter 12. When one prepares a buffer for use in the laboratory one can measure the pH using a pH meter which has been calibrated with a buffer recommended by the NIST.

Acid-base systems commonly employed to prepare buffers in the laboratory include (1) phthalic acid–potassium hydrogen phthalate, potassium dihydrogen phosphate–dipotassium hydrogen phosphate, and boric acid–sodium borate, pH 2 to 10, known as Clark and Lubs buffers; (2) citric acid–disodium hydrogen phosphate, pH 2 to 8, known as McIlvaine's buffers; (3) sodium carbonate–sodium bicarbonate, pH 9.6 to 11; (4) disodium hydrogen phosphate–sodium hydroxide, pH 10.9 to 12. Directions for preparing these buffers can be found in chemical handbooks. The compositions of some buffers recommended by the NIST are given in Chapter 12. For many practical purposes a solution saturated with potassium hydrogen tartrate is a very convenient buffer.[5] This solution has a pH of 3.57 at 25°C and a temperature coefficient of -0.0014 deg^{-1}.

6.6f. Physiological Buffers

It is of interest to point out that the principles of acid-base chemistry discussed in this chapter are of direct significance in such fields as biochemistry and physiology. The great physiologist Claude Bernard was the first to emphasize that the fluids of the body provide an "internal environment" in which the body cells live and perform their many functions protected from the inconstancy of the external environment. Living tissues are extremely sensitive to changes in the composition of the fluids that bathe them, and the regulatory mechanisms within the body which maintain the constancy of the internal environment comprise one of the most important phases in the study of the biological sciences.

A very important aspect of this regulation is the maintenance of a nearly constant pH in the blood and other fluids of the body. Substances that are acidic or alkaline in character are ingested in the diet and are formed continually by metabolic reactions, but the pH of the blood normally remains constant within about 0.1 pH unit (7.35 to 7.45).

The two principal routes for the elimination of acids from the body are the lungs and kidneys. It is estimated that in one day the normal human adult eliminates the equivalent of about 30 liters of 1 M acid by way of the lungs, and about 100 mL of 1 M acid through the kidneys.[6] To handle such large amounts of acid, the normal adult has enough buffers in approximately 5 liters of blood to absorb about 150 mL of 1 M acid. The proton acceptors found in tissues, such as the muscles, can handle about five times as much acid as the blood buffers.

The principal buffers in the blood are proteins, bicarbonate, phosphates, hemoglobin (HHb), and oxyhemoglobin (HHbO$_2$). Carbon dioxide is formed metabolically in the tissues and is carried away by the blood primarily as bicarbonate ion. A typical reaction is

$$H_2O + CO_2(aq) + Hb^-(aq) \rightleftharpoons HHb(aq) + HCO_3^-$$

$$\underset{\text{Base}}{} \quad \underset{\text{Acid}}{} \quad \underset{\rightarrow \text{To lungs}}{}$$

Note that H$_2$CO$_3$ is a stronger acid ($pK_{a_1} = 6.1$ under conditions in the blood) than is hemoglobin ($pK_a = 7.93$); hence the reaction above tends to go to the

[5] J. J. Lingane, *Electroanalytical Chemistry*, 2nd ed., Interscience Publishers, Inc., New York, 1958, p. 77.

[6] W. R. Frisell, *Acid-Base Chemistry in Medicine*, Macmillan Publishing Co., Inc., New York, 1968.

right. In the blood, at pH 7.4, the ratio of bicarbonate to free CO_2 can be calculated from the equation

$$7.4 = 6.1 + \log \frac{[HCO_3^-]}{[CO_2]}$$

The ratio $[HCO_3^-]/[CO_2]$ is about 20 : 1, showing that the predominant form in the blood is bicarbonate ion.

In the lungs carbon dioxide is released by the reaction

$$HCO_3^-(aq) + HHbO_2(aq) \rightleftharpoons HbO_2^-(aq) + H_2O + CO_2(g)$$

$\llcorner\!\!\rightarrow$ To tissues $\qquad$ $\llcorner\!\!\rightarrow$ Exhaled

When blood is oxygenated in the lungs, hemoglobin is converted into oxyhemoglobin. Since oxyhemoglobin is a stronger acid ($pK_a = 6.68$) than hemoglobin, this facilitates the conversion of HCO_3^- to CO_2 by the reaction above.

The phosphate buffer system is found mostly in the red cells. Its reaction is

$$H_2PO_4^- + H_2O \rightleftharpoons HPO_4^{2-} + H_3O^+$$

The pK_a of $H_2PO_4^-$ is about 7.2; hence this system exhibits its maximal effectiveness very close to physiological pH.

Disturbances in the pH of the blood are seen clinically in certain diseases. For example, untreated diabetes sometimes give rise to an acidosis which may be fatal. Kidney failure, or chronic nephritis, leads to retention of $H_2PO_4^-$ and an increase in the amount of carbon dioxide in the blood:

$$H_2PO_4^- + HCO_3^- \rightleftharpoons HPO_4^{2-} + H_2O + CO_2$$

6.7

APPLICATIONS OF ACID-BASE TITRATIONS

Acid-base titrations are widely used for chemical analyses. In most applications water is the solvent, and we shall restrict our discussion at this point to aqueous solutions. In the next section we will discuss the use of nonaqueous solvents.

6.7a. Acid-Base Reagents

In laboratory practice it is customary to prepare and standardize one solution of an acid and one of a base. These two solutions can then be used to analyze unknown samples of acids and bases. Since acid solutions are more easily preserved than basic solutions, an acid is normally chosen as a permanent reference standard in preference to a base.

Acids

In choosing an acid to use in a standard solution, the following factors should be considered. (1) The acid should be strong, that is, highly dissociated. (2) The acid should not be volatile. (3) A solution of the acid should be stable. (4) Salts of the acid should be soluble. (5) The acid should not be a sufficiently strong oxidizing agent to destroy organic compounds used as indicators.

Hydrochloric and sulfuric acids are most widely employed for standard solutions, although neither satisfies all the foregoing requirements. The chloride salts of silver, lead, and mercury(I) ion are insoluble, as are the sulfates of the alkaline earth metals and lead. This does not normally lead to trouble, however, in most applications of acid-base titrations. Hydrogen chloride is a gas, but is not appreciably volatile from solutions in the concentration range normally employed because it is so highly dissociated in aqueous solution. A solution as concentrated as 0.5 N can be boiled for some time without losing hydrogen chloride if the solution is not allowed to concentrate by evaporation. Nitric acid is seldom used, because it is a strong oxidizing agent, and its solutions decompose when heated or exposed to light. Perchloric is a strong acid, nonvolatile and stable toward reduction in dilute solutions. The potassium and ammonium salts may precipitate from concentrated solutions when formed during a titration. Perchloric acid is commonly preferred for nonaqueous titrations. It is inherently a stronger acid than hydrochloric acid and is more strongly dissociated in an acidic solvent, such as glacial acetic acid.

Bases and the Carbonate Error

Sodium hydroxide is the most commonly used base. Potassium hydroxide offers no advantage over sodium hydroxide and is more expensive. Sodium hydroxide is always contaminated by small amounts of impurities, the most serious of which is sodium carbonate. When CO_2 is absorbed by a solution of NaOH, the following reaction occurs:

$$CO_2 + 2OH^- \longrightarrow CO_3^{2-} + H_2O$$

Carbonate ion is a base, but it combines with hydrogen ion in two steps:

$$CO_3^{2-} + H_3O^+ \longrightarrow HCO_3^- + H_2O \qquad (1)$$

$$HCO_3^- + H_3O^+ \longrightarrow H_2CO_3 + H_2O \qquad (2)$$

If phenolphthalein is employed as the indicator, the color change occurs when reaction (1) is complete; that is, the carbonate ion has reacted with only one H_3O^+ ion. This results in an error, since two OH^- ions were used in the formation of one CO_3^{2-}. If methyl orange is used as the indicator, the color change occurs when reaction (2) is complete and no error occurs, since each CO_3^{2-} ion combines with two H_3O^+ ions. However, in the titration of weak acids phenolphthalein is the proper indicator to use, and if CO_2 has been absorbed by the titrant, an error will occur.

There are several ways to minimize the "carbonate error." Barium hydroxide can be used as the titrant. If CO_2 is absorbed by a solution of this base, a precipitate of barium carbonate is apparent:

$$Ba^{2+} + 2OH^- + CO_2 \longrightarrow BaCO_3(s) + H_2O$$

Since barium hydroxide is of limited solubility in water, solutions cannot be more concentrated than about 0.05 N.

The most common method used to avoid the carbonate error is to prepare carbonate-free sodium hydroxide and then protect the solution from the uptake of CO_2 from the air. Carbonate-free sodium hydroxide can be readily prepared from a concentrated solution of the base, one which is about 50% by weight NaOH. Sodium carbonate is insoluble in the concentrated NaOH solution and settles to the bottom of the container. The solution is decanted from the solid Na_2CO_3 and diluted to the desired concentration. It is then stored in a bottle equipped with a tube containing a solid material (soda lime or Ascarite) which absorbs CO_2 from any air that enters.

The acid-base solutions used in the laboratory are usually in the concentration range of about 0.05 to 0.5 N, most often about 0.1 N. Solutions of such concentrations require reasonable volumes (30 to 50 mL) for titration of samples which are of convenient size to weigh on the analytical balance. For example, 0.6000 g of a pure substance of equivalent weight 200 will require 30 mL of a 0.1 N solution for titration.

6.7b. Primary Standards

In laboratory practice it is customary to prepare solutions of an acid and a base of approximately the desired concentration and then to standardize the solutions against a primary standard. It is possible to prepare a standard solution of hydrochloric acid by direct weighing of a portion of constant-boiling HCl of known density, followed by dilution in a volumetric flask. More frequently, however, solutions of this acid are standardized in the customary manner against a primary standard.

Requirements for a Primary Standard Material

The reaction between the substance selected as a primary standard and the acid or base should obviously fulfill the requirements for titrimetric analysis. In addition, the primary standard should have the following characteristics:

1. It should be readily available in a pure form or in a state of known purity. In general, the total amount of impurities should not exceed 0.01 to 0.02%, and it should be possible to test for impurities by qualitative tests of known sensitivity.
2. The substance should be easy to dry and should not be so hygroscopic that it takes up water during weighing. It should not lose weight on exposure to air. Salt hydrates are not normally employed as primary standards.
3. It is desirable that the primary standard have a high equivalent weight in order to minimize the consequences of errors in weighing.
4. It is preferable that the acid or base be strong, that is, highly dissociated. However, a weak acid or base may be employed as a primary standard with no great disadvantages, especially when the standard solution is to be used to analyze samples of weak acids or bases.

Examples of Primary Standard Materials

The compound potassium hydrogen phthalate (page 51), $KHC_8H_4O_4$ (abbreviated KHP), is an excellent primary standard for base solutions. It is readily available in purity of 99.95% or better from the National Bureau of Standards and from chemical supply houses. It is stable on drying, is nonhygroscopic, and has a high equivalent weight, 204.2 g/eq. It is a weak, monoprotic acid, but since base solutions are frequently used to determine weak acids, this is no disadvantage. Phenolphthalein indicator is employed in the titration, and the base solution should be carbonate-free.

Sulfamic acid, HSO_3NH_2, is a strong monoprotic acid, and either phenolphthalein or methyl red indicator can be employed in the titration with a strong base. It is readily available, inexpensive, and easily purified by recrystallization from water. It is a white crystalline solid, nonhygroscopic, and stable at temperatures up to 130°C. Its equivalent weight is 97.09, considerably less than that of

KHP. However, the weight which would normally be employed to standardize solutions 0.1 N or greater is sufficiently large to keep weighing errors small. Sulfamic acid is readily soluble in water, and most of its salts are soluble.

The compound potassium hydrogen iodate, $KH(IO_3)_2$, a strong monoprotic acid, is also an excellent primary standard for base solutions. It is readily available in a form sufficiently pure for use as a primary standard. It is a white, crystalline, nonhygroscopic solid, and it has a high equivalent weight, 389.91. It is sufficiently stable to be dried at 110°C.

Sulfosalicylic acid, which has the formula

forms a double salt with potassium which we may represent as $KHSa \cdot K_2Sa$, where Sa represents the doubly charged anion. This salt has been proposed by Butler and Bates[7] as a primary standard for base solutions. It has a molecular weight of 550.655 and a pK_a of 2.85, and it can be prepared in very pure form. Since it is a stronger acid than KHP (pK_a of $HP^- = 5.41$) it will give a larger change in pH at the equivalence point than the latter acid.

Sodium carbonate, Na_2CO_3, is widely used as a primary standard for solutions of strong acids. It is readily available in a very pure state, except for small amounts of sodium bicarbonate, $NaHCO_3$. The bicarbonate can be converted completely into carbonate by heating the substance to constant weight at 270 to 300°C. Sodium carbonate is somewhat hygroscopic but can be weighed without great difficulty. The carbonate can be titrated to sodium bicarbonate, using phenolphthalein indicator, and the equivalent weight is the molecular weight, 106.0. More commonly it is titrated to carbonic acid using methyl orange indicator. The equivalent weight in this case is one-half the molecular weight, 53.00.

The organic base tris(hydroxymethyl)aminomethane, $(CH_2OH)_3CNH_2$, also called TRIS or THAM, is an excellent primary standard for acid solutions. It is available commercially in purity of 99.95% and is readily dried and weighed. Its reaction with hydrochloric acid is

$$(CH_2OH)_3CNH_2 + H_3O^+ \longrightarrow (CH_2OH)_3CNH_3^+ + H_2O$$

and its equivalent weight is 121.14 g/eq.

6.7c. *Analyses Using Acid-Base Titrations*

A wide variety of acidic and basic substances, both inorganic and organic, can be determined by an acid-base titration. There are also many examples in which the analyte can be converted chemically into an acid or base and then determined by titration. We shall discuss briefly a few examples.

Nitrogen

The determination of nitrogen by titration of ammonia with strong acid is an important application of acid-base titrations. The procedure depends upon the

[7] R. Butler and R. G. Bates, *Anal. Chem,* **48,** 1669 (1976).

oxidation state of nitrogen in the compound to be analyzed. If nitrogen is present as the ammonium salt, oxidation state -3, ammonia can be liberated by the addition of strong base:

$$NH_4^+ + OH^- \longrightarrow NH_3(g) + H_2O$$

The sample is heated in a distilling flask with excess base, and the evolved ammonia is caught in excess standard sulfuric or hydrochloric acid. The excess acid is then titrated with standard base.

If, on the other hand, the nitrogen is attached to carbon, as in many organic compounds (proteins, etc.), ammonia is not readily evolved when the compound is heated with strong base. A more drastic treatment is required to break the carbon-nitrogen bond. Kjeldahl, in 1883, suggested a preliminary treatment of the nitrogen compound with hot concentrated sulfuric acid. The organic material is dehydrated, the carbon oxidized to CO_2, and the nitrogen converted to ammonium sulfate. Addition of strong alkali then liberates ammonia which can be absorbed and titrated as mentioned above.

The Kjeldahl method has been widely studied, and various modifications have been proposed. Amines, amides, nitriles, cyanates, and isocyanates are particularly suited to the method. If a preliminary reduction step is employed, the procedure will also handle compounds containing nitro, nitroso, and azo groups. The method is the standard procedure for determination of the protein content of certain grains, meat, and animal food. Old as it is, this method is still widely used in biochemistry, nutrition, and agriculture.

Sulfur

This element can be determined in organic substances by burning the sample in a stream of oxygen, converting sulfur into SO_2 and SO_3. The gas formed by the reaction is passed through an aqueous solution of H_2O_2 to oxidize SO_2 to SO_3. The sulfuric acid is titrated with standard base:

$$H_2SO_4 + 2OH^- \longrightarrow SO_4^{2-} + 2H_2O$$

Boron

This element can be determined in organic compounds by combustion in a nickel bomb to convert boron to boric acid. Boric acid is too weak to titrate feasibly, but the addition of mannitol forms a strong acid which can be titrated with aqueous NaOH.

Carbonate Mixtures

We have previously mentioned that carbonate ion is titrated in two steps:

$$CO_3^{2-} + H_3O^+ \longrightarrow HCO_3^- + H_2O \qquad \text{(phenolphthalein)}$$
$$HCO_3^- + H_3O^+ \longrightarrow H_2CO_3 + H_2O \qquad \text{(methyl orange)}$$

Phenolphthalein serves as the indicator for the first step in the titration, and methyl orange for the second. The titration of NaOH is complete at the phenolphthalein end point, and only a drop or two of additional titrant is required to reach the methyl orange end point. (See Fig. 6.2.)

Sodium hydroxide is commonly contaminated with sodium carbonate; sodium carbonate and sodium bicarbonate often occur together. It is possible to analyze mixtures of these compounds by titration with standard acid, using the two indicators mentioned above. We shall discuss details of the calculations in Chapter 7.

Organic Functional Groups

A number of organic functional groups can be determined by acid-base titration. Carboxylic acids, $R-CCOH$, generally have pK_a values of about 4 to 6 and are readily titrated. Sulfonic acids, $R-SO_3H$, are generally strong and readily soluble in water. They can be titrated with standard base.

Alcohols can be determined by the addition of excess acetic anhydride:

$$(CH_3CO)_2O + ROH \longrightarrow CH_3COOR + CH_3COOH$$

$\qquad$ Acetic anhydride $\quad$ Alcohol $\qquad\qquad$ Ester $\qquad$ Acetic acid

The excess anhydride is hydrolyzed to acetic acid,

$$(CH_3CO)_2O + H_2O \longrightarrow 2CH_3COOH$$

and the total acid produced by the two reactions is titrated with standard base.

Aliphatic amines, such as CH_3NH_2, generally have pK_b values of about 5 and can be titrated directly with standard acid. Aromatic amines, such as aniline, $C_6H_5NH_2$, have pK_b values of about 10 and are too weak to titrate in aqueous solution.

Esters can be determined by first hydrolyzing the compound with excess base:

$$R_1COOR_2 + OH^- \longrightarrow R_1COO^- + R_2OH$$

The excess base is then titrated with standard acid.

6.8

NONAQUEOUS TITRATIONS

6.8a. The Role of the Solvent in Acid-Base Reactions

Consider an acid, HB, which we wish to titrate with base, say NaOH. We have discussed the feasibility of this titration in terms of the strength of HB, using its dissociation constant, K_a, as a measure. But, as pointed out earlier, in terms of the Brønsted theory, K_a is really a measure of the tendency of HB to transfer a proton to the solvent, water:

$$HB + H_2O \rightleftharpoons H_3O^+ + B$$

That is, K_a is not a measure of the intrinsic acid strength of HB, because the basicity of water is also involved in this reaction. The same acid might dissociate to a much greater degree in a more basic solvent, say on organic amine:

$$HB + RNH_2 \rightleftharpoons RNH_3^+ + B$$

That is, there will be a greater concentration of solvated protons in the latter solvent. Thus it might appear that if HB is too weak an acid to be titrated feasibly in aqueous solution, we could enhance its "acidity" and hence its "titratability" by choosing a solvent more basic than water.

Actually, in a practical sense, this is often the case, but the above discussion is misleading as it stands. In fact, dissociation is not at all necessary for successful acid-base titrations. Excellent titrations have been performed in nonpolar solvents like benzene and chloroform which do not promote dissociation to any appreciable extent. Indeed, it is *not* the greater basicity of the organic amine that makes it a better solvent than water for titration of the very weak acid HB. It is a better solvent for this titration because it is a *weaker acid* than water. In the aqueous system the titration reaction is

$$HB + OH^- \rightleftharpoons H_2O + B^- \qquad K = \frac{K_a}{K_w} \qquad (1)$$

Water is a product of the titration reaction, and furthermore it is present in large excess. Thus, to the extent that water is acidic, it competes against the acid we wish to titrate and prevents the titration reaction from going to completion unless HB is itself sufficiently strong. This can be seen from the constant K; the constant is larger the larger K_a, and the smaller the autoprotolysis constant of the solvent. In general terms, we wish the following reaction to go to completion:

$$HB + S^- \rightleftharpoons HS + B^- \qquad K = \frac{K_a}{K_{auto}} \qquad (2)$$

Here HS is the solvent, S^- the conjugate base, and K_{auto} the autoprotolysis constant of the solvent. If HS is a weaker acid than water, K for reaction (2) will be larger than K for reaction (1). It often happens that the solvent is also more basic than water, but it is not correct to fixate upon this latter aspect.

In any case, we find that many titrations of weak acids and bases which are not feasible in water solution can be performed in other solvents. A variety of solvents have now been studied, and various methods of end-point detection are available. Much of the work is empirical because we do not have acidity scales in all these solvents as we have for water. But even on this basis, the field of nonaqueous titrations has become important in analytical chemistry.

Solvent Systems

Several classifications of solvents have been proposed. Laitinen[8] considers four types. *Amphiprotic* solvents possess both acidic and basic properties as does water. They undergo autoprotolysis, and, as we noted above, the degree to which the titration reaction goes to completion is a function of this reaction. Some, such as methanol and ethanol, have acid-base properties comparable to water and, along with water, are called *neutral* solvents. Others, called *acid* solvents, such as acetic acid, formic acid, and sulfuric acid, are much stronger acids and weaker bases than water. *Basic* solvents such as liquid ammonia and ethylenediamine have greater basicity and weaker acidity than water.

Aprotic, or inert, solvents are neither appreciably acidic nor basic and hence show little or no tendency to undergo autoprotolysis reactions. Examples are benzene, carbon tetrachloride, and chloroform.

Another group of solvents, called basic solvents, have a strong affinity for protons but are not appreciably acidic. Examples are ether, pyridine, and various ketones. Pyridine, for example, can accept a proton from an acid such as water:

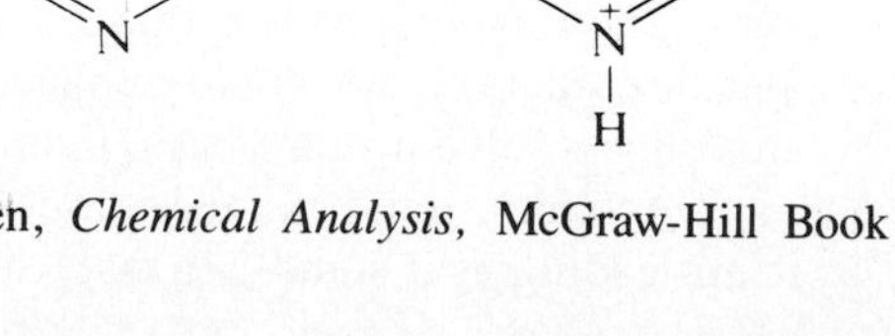

[8] H. A. Laitinen, *Chemical Analysis*, McGraw-Hill Book Company, New York, 1960, p. 60.

On the other hand, pyridine has no tendency to furnish a proton. Consequently, no autoprotolysis reaction can be written.

A fourth class of solvents would be those with acidic but no basic properties. No examples of such solvents are known.

Differentiating Ability of a Solvent

We have previously pointed out that water levels the mineral acids perchloric, hydrochloric, and nitric (page 130). That is, in aqueous solution these acids appear equally strong. However, in an acidic solvent such as acetic acid, the greater strength of perchloric acid over, say, hydrochloric acid, allows it to be titrated in a separate step from the latter acid. Of the two equilibria,

$$HClO_4 + HOAc \rightleftharpoons H_2OAc^+ + ClO_4^- \tag{1}$$

$$HCl + HOAc \rightleftharpoons H_2OAc^+ + Cl^- \tag{2}$$

the first goes much further to the right than the second. Hence in a titration of a mixture of the two acids in acetic acid solvent, two breaks in the titration curve are found, and the acids are said to be *differentiated*.

There are two properties of the solvent which determine its leveling or differentiating ability. One is the intrinsic acid-base character of the solvent, and the second is the autoprotolysis constant. For example, water is sufficiently strong a base to level HCl and $HClO_4$, but not HCl and HOAc. The latter two acids are differentiated in aqueous solution. Acetic acid is a weaker base than water and differentiates $HClO_4$ and HCl. Ammonia, however, is a stronger base than water and levels not only HCl and $HClO_4$ but also HCl and HOAc. An inert solvent, having no appreciable acidic or basic properties, exerts no leveling effect and hence is very suitable for differentiating mixtures of compounds of varying acidity.

We have previously pointed out that the neutralization of a strong acid with a strong base in aqueous solution is simply the reverse of the autoprotolysis reaction of the solvent:

$$H_3O^+ + OH^- \rightleftharpoons 2H_2O$$

The K for this reaction is $1/K_w = 1 \times 10^{14}$. The magnitude of this constant determines the size of the break at the equivalence point or the useful range over which breaks in titration curves can be detected. For the titration of $0.1\ N$ reagents, the steep break is about 6 pH units, from pH 4 to 10. Below pH 4 and above pH 10 water levels acids and bases dissolved in it. These two pH extremes correspond to concentrations of H_3O^+ and OH^- of $10^{-4}\ M$ each.

Consider ethanol, C_2H_5OH ($pK_{auto} = 19.5$) as a solvent. The autoprotolysis constant is 3×10^{-20}, and hence K for the reaction of strong acid with strong base,

$$C_2H_5OH_2^+ + OC_2H_5^- \rightleftharpoons 2C_2H_5OH$$

is $1/K_{auto}$, or 3×10^{19}. Reasoning as in the case of water above, we see that the large break in the titration curve would occur between the limits of $10^{-4}\ M$ strong acid and strong base. That is, the useful pH range in ethanol is roughly $19.5 - 2 \times 4 = 11.5$ units, almost twice the useful range of water. In general, we can conclude that the useful pH range for a solvent in differentiating acids and bases increases as the autoprotolysis constant becomes smaller.

Dielectric Constant

Another property of a solvent which is of importance in nonaqueous titrations is the dielectric constant. In amphiprotic solvents the dissociation of a weak acid into separate ions is thought to occur as follows:

$$HB + HS \underset{}{\overset{1}{\rightleftharpoons}} \{H_2S^+B^-\} \underset{}{\overset{2}{\rightleftharpoons}} H_2S^+ + B^-$$

$$\text{Ion pair} \qquad\qquad \text{Separate ions}$$

The first step is called *ionization*, and the product is called an *ion pair*. In the second step, complete separation of the ion occurs.[9] Solvents with high dielectric constants encourage complete dissociation into ions by lessening the energy required for the process. In solvents of low dielectric constant, considerable ion pairing occurs.

The acidity of an ion such as NH_4^+ is not greatly affected by the dielectric constant of the solvent, since no ion-pair production occurs:

$$NH_4^+ + H_2O \rightleftharpoons \{NH_3H_3O^+\} \rightleftharpoons NH_3 + H_3O^+$$

On the other hand, the autoprotolysis constant of the solvent is increased, the larger the dielectric constant:

$$HS + HS \rightleftharpoons \{H_2S^+S^-\} \rightleftharpoons H_2S^+ + S^-$$

since charge separation does occur.

Generally, a high dielectric constant is desirable for amphiprotic solvents. A factor of prime importance is solubility; a high dielectric constant generally favors the solubility of polar reagents and samples. Water is a unique solvent in having a very high dielectric constant and a relatively small autoprotolysis constant.

6.8b. Completeness of the Titration Reaction

In general we represent the titration of a weak acid, HX, with the solvent anion (base), S^-, as follows:

$$HX + S^- \rightleftharpoons HS + X^-$$

In solvents of low dielectric constant where ion-pair formation may occur, we can represent the reaction as

$$H^+X^- + M^+S^- \rightleftharpoons HS + M^+X^-$$

From LeChâtelier's principle we can conclude that the reaction goes farther to completion the more highly dissociated the ion pairs H^+X^- and M^+S^- and the more highly associated the ion pairs of the salt M^+X^-. In addition, the lower the autoprotolysis constant of the solvent, the larger the K for the titration reaction.

---

[9] This process explains the fact that HF is a weak acid in water, whereas HCl, HBr, and HI are all completely dissociated. HF does lose its proton completely to water, but the resulting ion pair is only slightly dissociated because of strong hydrogen bond formation:

$$HF + H_2O \longrightarrow \left\{ H\text{—}O\text{—}H^+ \cdots F^- \atop \quad\ H \right\} \rightleftharpoons H_3O^+ + F^-$$

This also explains the fact that, whereas ammonia, methylamine, dimethylamine, and trimethylamine are all weak bases, tetramethylammonium hydroxide, $(CH_3)_4NOH$, is a strong base. In the latter molecule there is no hydrogen available for forming a hydrogen bond with the oxygen of the hydroxide ion.

Concentration factors (page 149) also are to be considered in nonaqueous systems.

6.8c. Titrants

Perchloric acid is by far the most widely used acid for the titration of weak bases, because it is a very strong acid which is readily available. It is normally obtained commercially as 72% $HClO_4$ by weight, the remainder being water; this is an azeotrope of $HClO_4$ and H_2O, and it represents approximately the composition $HClO_4 \cdot H_2O$, which some writers formulate as hydronium perchlorate $H_3O^+ClO_4^-$. Weak bases are titrated most often in glacial acetic acid solution. In such cases, the titrant is perchloric acid, say $0.1\ M$, in the same solvent. Because the presence of water may be deleterious(see above), the desired quantity of 72% $HClO_4$ is mixed with acetic acid, and then acetic anhydride is added in approximately the correct amount to react with the water estimated to be present. The product of this reaction is, of course, acetic acid.

A somewhat larger variety of strong bases are used, including alkali hydroxides, tetraalkylammonium hydroxides, and sodium or potassium methoxide or ethoxide. Common solvents for these bases are lower alcohols and mixtures of benzene with methanol or ethanol.

Normally the effect of temperature upon measured titrant volumes can be ignored with aqueous solutions under ordinary room temperature variations. Organic solvents such as acetic acid, benzene, and methanol, on the other hand, have fairly large coefficients of thermal expansion, and the volume changes may not be negligible if the titrant is at a different temperature from that at which it ws standardized. Correction for the effect of a temperature change upon the volume of titrant may be made by means of an equation of the type

$$V_T = V_0(1 + \alpha T + \beta T^2 + \gamma T^3)$$

where V_0 is the volume at $0°C$ and V_T the volume at $T\,°C$. Values of α, β, and γ for various liquids may be found in handbooks. Practically, β and γ are usually small enough so that βT^2 and γT^3 may be ignored. Suppose the titrant were at $30°C$ when an unknown was titrated, whereas it had been standardized at $25°C$. Neglecting the higher-order terms in the above equation and eliminating V_0 between the two temperatures involved gives

$$V_{25} = V_{30} \times \frac{1 + 25\alpha}{1 + 30\alpha}$$

Using a handbook value for α, we can readily calculate the volume of titrant that would have been consumed had the titration been performed at $25°C$. For a mixed solvent such as benzene-methanol, a value for α may be used, weighted according to the volume fractions or the mole fractions of the two solvents in the mixture (if the mixture is nonideal, an exact, theoretically valid value of α cannot be calculated from the information that is normally available, but an adequate value can be obtained with weighted means).

6.8d. End-Point Detection

A number of visual indicators are available, generally under trivial names such as cresol red, methyl red, azo violet, and crystal violet. The rationale of indicator selection does not have a good theoretical base, and the choice is often best made on the basis of experience, trial and error, or reference to analogous cases found in the literature.

Potentiometric end-point methods (Chapter 12) are frequently employed, although, in general, electrode behavior in nonaqueous solvents is not well understood. Again, the safest approach is to see what other workers have used in similar situations. Other instrumental end points such as conductometric and photometric (Chapter 14) have been used successfully.

6.8e. *Applications*

The number of compounds which have been titrated in nonaqueous media is much too large for listing here. Very weak acids, such as phenols, have been titrated in ethylenediamine. Because carboxylic acids are sufficiently strong, only moderately basic solvents such as methanol and ethanol need be employed. Nonaqueous titrations have become important in the pharmaceutical industry. For example, most of the well-known sulfa drug group can be determined by titration as acids (the acidity is conferred by the sulfonamide group, $-SO_2-NH-$) with alkali methoxide in benzene-methanol or dimethylformamide solution.

Weak bases, such as amines, amino acids, and anions of weak acids, have been titrated in glacial acetic acid solution using perchloric acid. Alkaloids have very weak basic properties and can be titrated in acidic or inert solvents.

The solvent methyl isobutyl ketone, a basic but not acidic solvent, has been used for the titration of a wide range of acids and bases. It has been used to differentiate a five-component acid mixture: perchloric, hydrochloric, salicyclic, acetic, and phenol. (See Fig. 6.4.) This mixture ranges from the strongest mineral acid, perchloric, to phenol, a very weak acid. The titrant for acids is a solution of tetrabutylammonium hydroxide in isopropanol; for the titration of bases, perchloric acid dissolved in dioxane is usually employed.

KEY TERMS

Amphiprotic solvent. A solvent which possesses both acidic and basic properties.

Amphoteric substance. A molecule which can act both as an acid and as a base.

Aprotic solvent. A solvent which is neither appreciably acidic or basic.

Brønsted acid. A substance which furnishes a proton.

Brønsted base. A substance which accepts a proton.

Buffer capacity. A measure of the effectiveness of a buffer in resisting changes in pH; the capacity is greater the greater the concentrations of the conjugate acid-base pair.

Buffer effectiveness. The change in pH produced per unit of acid or base added to a buffer.

For the same concentrations a buffer is more effective the closer to unity the ratio of buffer acid to base.

Buffer solution. A solution which contains a conjugate acid-base pair. Such a solution resists large changes in pH when H_3O^+ or OH^- ions are added and when the solution is diluted.

Carbonate error. The error caused when CO_2 is absorbed by a standard base solution and the solution is used to titrate a weak acid with phenolphthalein indicator.

Conjugate acid-base pair. An acid-base pair whose members differ only by a proton, as HCl and Cl^-.

Differentiating ability of a solvent. If a mixture of two acids gives two separate titration curves in a

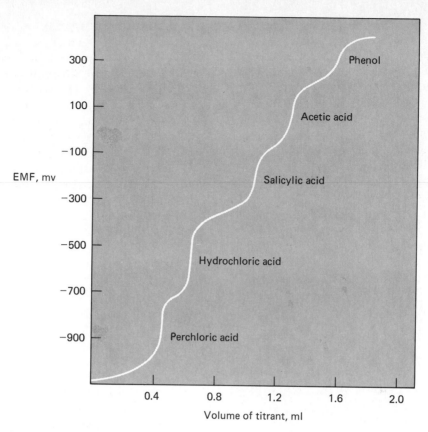

Figure 6.4 Titration of a five-component acid mixture in methyl isobutyl ketone using 0.2 *N* tetrabutylammonium hydroxide titrant, glass-platinum electrodes. (Courtesy of D. B. Bruss and G. E. A. Wyld, *Anal. Chem. 29*, 232 (1957).)

solvent, the solvent is said to have *differentiated* the two acids. If only a single curve is obtained, the solvent is said to have *leveled* the acids (page 130).

Feasible titration. A titration (acid-base) which gives a *p*H change of 1 or 2 units for the addition of a few drops of titrant at the equivalence point.

Indicator. A visual acid-base indicator is a weak organic acid or base which shows different colors in the molecular and ionic forms.

Indicator blank. The volume of acid or base required to change the *p*H from that at the equiva-lence point to the *p*H at which the indicator changes color.

Leveling effect. The effect in which a series of acids, all of which are much stronger than the sol-vated proton, dissociate completely and appear to be equally strong.

Range of an indicator. That portion of the *p*H scale over which an indicator changes color, roughly the *p*K of the indicator ±1 unit.

Titration curve. For acid-base reactions a titra-tion curve is a plot of *p*H (or *p*OH) vs. milliliters of titrant.

QUESTIONS

1. *Standard solutions.* Point out the factors to be considered in choosing an acid to use as a standard solution in the laboratory.

2. *Primary standards.* What characteristics should a substance have to be used as a pri-mary standard for acid or base solutions?

Name several commonly used primary standards.

3. *Carbonate error*. Explain why the "carbonate error" occurs and how it can be minimized in the laboratory.

4. *Physiological buffers*. (a) What are the principal buffers in the blood? (b) How are acids eliminated from the body? (c) How does oxygenating the blood in the lungs facilitate the liberation of CO_2?

5. *Feasibility*. Explain what is meant by a feasible acid-base titration. What factors determine the size of the ΔpH at the equivalence point?

Multiple-choice. In the following multiple-choice questions select the *one best* answer. Given the following acids, bases, and indicators, answer the questions below.

Acid	pK_a	Base	pK_b	Indicator
HA	Strong	MOH	Strong	I (base), $pK_b = 9$
HB	4.40	NOH	8.80	II (acid), $pK_a = 5$
HC	9.10	ROH	4.10	III (base), $pK_b = 5$

6. Which titration/s is/are feasible (0.10 M solution)? (a) HC with NaOH; (b) ROH with

HCl; (c) HA with NaOH; (d) NOH with HCl; (e) two of the above.

7. Which titration/s is/are feasible (0.10 M solution)? (a) NCl with NaOH; (b) NaC with NaOH; (c) HA with NaOH; (d) all of the above; (e) none of the above.

8. HB is titrated with 0.10 M NaOH. Which is the proper indicator? (a) I; (b) II; (c) III; (d) I and II.

9. Which is the strongest conjugate base? (a) OAc^-; (b) F^-; (c) NO_2^-; (d) OCl^-.

10. Which of these salts can be titrated feasibly? (a) $NaNO_2$ with HCl; (b) CH_3NH_3Cl with NaOH; (c) NaCN with HCl; (d) NH_4Cl with NaOH; (e) two of the above.

11. Which titration (0.10 M solution) will give the largest change in pH at the EPt? (a) Benzoic acid with NaOH; (b) formic acid with NaOH; (c) pyridine with HCl; (d) monochloroacetic acid with NaOH.

12. Which of these statements is true? (a) An aprotic solvent has acidic properties. (b) The titration reaction is more complete the smaller the autoprotolysis constant. (c) Dissociation into ions is necessary for successful acid-base titrations. (d) A low dielectric constant is desirable for amphiprotic solvents.

PROBLEMS

1. *Titration curve*. 40.00 mL of 0.1100 M HCl is diluted to 100 mL with water and titrated with 0.1000 M NaOH. Calculate the pH after addition of the following volumes (mL) of titrant: (a) 0.00; (b) 10.00; (c) 22.00; (d) 40.00; (e) 43.95; (f) 44.00; (g) 44.05; (h) 50.00. Plot the titration curve. Select an indicator from Table 6.3.

2. *Titration curve*. Repeat Problem 1 for the titration of 40.00 mL of 0.1100 M benzoic acid with 0.1000 M NaOH.

3. *Titration curve*. Repeat Problem 1 for the titration of 40.00 mL of 0.1100 M HCN with 0.1000 M NaOH.

4. *Titration curve*. 40.00 mL of 0.0900 M NaOH is diluted to 100 mL and titrated with 0.1000 M HCl. Calculate the pH after addition of the following volumes (mL) of titrant: (a) 0.00; (b) 10.00; (c) 18.00; (d) 30.00; (e) 35.95; (f) 36.00; (g) 36.05; (h) 40.00. Plot the titration curve and select an indicator.

5. *Titration curve*. Repeat Problem 4 for the titration of 40.00 mL of 0.0900 M NH_3 with 0.1000 M HCl.

6. *Titration curve*. 40.00 mL of 0.1100 M NaCN is diluted to 100 mL and titrated with 0.1000 M HCl. Calculate the pH after addition of the following volumes (mL) of titrant: (a) 0.00; (b) 10.00; (c) 22.00; (d) 40.00; (e) 43.95; (f) 44.00; (g) 44.05; (h) 50.00. Plot the titration curve. Select an indicator from Table 6.3.

7. *Fraction neutralized*. A solution of formic acid is titrated with NaOH. Calculate the pH when the following percentages of acid have been neutralized: (a) 25%; (b) 33%; (c) 50%; (d) 75%; (e) 99%; (f) 99.9%; (g) 99.99%.

8. *Fraction neutralized*. A solution of ethylamine is titrated with HCl. Calculate the percentage of the base neutralized at these pH

values: (a) 11.23; (b) 11.05; (c) 10.75; (d) 10.15; (e) 7.25.

9. *Fraction neutralized.* 25 mL of 0.10 M HCl is diluted to 100 mL with water and titrated with 0.10 M NaOH. Calculate the fraction of the acid neutralized at these pH values: (a) 2.00; (b) 3.00; (c) 4.00; (d) 5.00; (e) 6.00; (f) 7.00.

10. *Derivations.* Derive the following expressions for calculating the pH of a solution at the specified conditions: (a) Equivalence point in the titration of a weak acid HA with NaOH:

$$pH = \tfrac{1}{2}pK_w + \tfrac{1}{2}pK_a + \tfrac{1}{2}\log[A^-]$$

(b) Equivalence point in the titration of a weak base BOH with HCl:

$$pH = \tfrac{1}{2}pK_w - \tfrac{1}{2}pK_b - \tfrac{1}{2}\log[B^+]$$

(c) Solution of a weak acid HA:

$$pH = \tfrac{1}{2}pK_a - \tfrac{1}{2}\log[HA]$$

11. *pH calculations.* Calculate the pH of these solutions: (a) 3.0 g of formic acid + 4.0 g of sodium formate in 500 mL of solution; (b) 0.90 g of formic acid dissolved in 100 mL of 0.080 M NaOH; (c) 2.0 g of formic acid + 8.0 g of sodium formate + 0.40 g of HCl in 300 mL of solution.

12. *pH calculations.* Methylamine, CH_3NH_2, has a MW of 31.06 g/mol; methylammonium chloride, CH_3NH_3Cl, has a MW of 67.52. Calculate the pH of these solutions: (a) 0.60 g of CH_3NH_2 + 0.80 g of CH_3NH_3Cl in 400 mL of solution; (b) 0.50 g of CH_3NH_2 dissolved in 100 mL of 0.090 M HCl; (c) 0.40 g of CH_3NH_2 + 0.50 g of CH_3NH_3Cl + 0.15 g of HCl in 200 mL of solution.

13. *Weak acids and bases.* Cocaine is a weak base, $K_b = 2.6 \times 10^{-6}$. Representing the base as B and the hydrochloride salt BHCl, calculate the pH of a solution made by mixing 40 mL of 0.075 M BHCl with 60 mL of 0.050 M NaOH.

14. *Weak acids and bases.* The pH of a 0.10 M solution of the sodium salt of hydroxyacetic acid is 8.41. Calculate the K_a of the acid.

15. *Titration of a weak base.* A sample of the weak base hydroxylamine is titrated with 0.0900 M HCl. 30.0 mL is required to reach the EPt. After the addition of 10.0 mL of ti-

trant the pH is 6.26. Calculate the pK_b of hydroxylamine.

16. *Approximations.* Calculate the $[H_3O^+]$ in these dilute HCl solutions: (a) 1.0×10^{-6} M; (b) 2.0×10^{-7} M; (c) 1.0×10^{-7} M. First neglect the H_3O^+ produced by the dissociation of water and then include this in the calculation. What is the percentage error in each case?

17. *Approximations.* In Section 6.3b we calculated the $[H_3O^+]$ and the error in our assumptions after 10.0 mL of titrant had been added. Repeat this calculation for these values for the dissociation constant of HB: (a) 1.0×10^{-4}; (b) 1.0×10^{-3}.

18. *Approximations.* Using the usual approximations, calculate the pH of a 1.0×10^{-6} M solution of the weak acid HA, $K_a = 1.0 \times 10^{-10}$. What is the fallacy in this calculation? What is the correct pH?

19. *Hydrolysis.* Calculate the percentage hydrolysis of the following ions in 0.10 M solution: (a) NH_4^+; (b) CN^-; (c) S^{2-} to HS^-.

20. *Indicator range.* Suppose an indicator, HIn, shows a red color when at least 80% or more of it is in the HIn form and a blue color when at least 80% or more is in the In^- form. The pK_a of the indicator is 5.40. What is the range of the indicator?

21. *Indicator error.* A sample of dimethylamine is titrated with HCl. By mistake the wrong indicator is used, and the titration is stopped at a pH of 10.12. If the sample actually contains 24.40% dimethylamine, what percentage would be found?

22. *Carbonate error.* 100 mL of 0.200 M NaOH absorbs some CO_2 from the air. When the solution is titrated with standard HCl using phenolphthalein indicator, it is found to be 0.180 M. How many millimoles of CO_2 were absorbed?

23. *Buffer solution.* A student wishes to prepare 400 mL of a buffer of pH 8.40 using the base TRIS (Section 6.7.b) and its hydrochloride salt. (a) What should be the molar ratio of base to salt for the buffer? (b) How many grams of the salt should be added to 400 mL of a 0.12 M solution of the base to get the desired pH? The MW of the salt is 157.60 g/mol. The pK_b of TRIS is 5.93.

24. *Buffer solution.* Suppose you wish to prepare 500 mL of an acetate buffer of pH 5.04

in which the concentration of HOAc is 0.10 M. You have available some glacial acetic acid (17 M) and some 0.25 M NaOH. How many milliliters of these two solutions should be mixed and diluted to 500 mL to give such a buffer?

25. *Buffer solution.* Suppose you wish to prepare 100 mL of a buffer of pH 9.50 using 0.30 M NH$_3$ and 0.10 M HCl. How many milliliters of each solution should be taken and mixed to prepare this buffer? Assume the volumes are additive.

26. *Buffer capacity.* It is desired to prepare 100 mL of a buffer of pH 8.80. Three bases, ammonia, hydrazine, and ethylamine, and their salts are available for use. (a) Which base should be chosen for maximum effectiveness against an increase or decrease in pH? (b) What should the molar ratio of base to salt be to obtain the desired pH? (c) If it is desired that the change in pH be no greater than 0.12 pH unit for the addition of 1.0 mmol of either acid or base, what minimum concentrations of base and salt should be used?

27. *Buffer capacity.* Two buffer solutions of pH 5.30 are prepared from the weak acid HA, $pK_a = 5.00$, and its salt NaA. Buffer I is 0.40 M in NaA and 0.20 M in HA; buffer II is 0.20 M in NaA and 0.10 M in HA. Calculate the change in pH which occurs on the addition of (a) 5.0 mmol of acid to 100 mL of each buffer, and (b) 5.0 mmol of base to 100 mL of each buffer.

28. *Buffer solution.* A buffer solution is prepared by dissolving 9.20 g of formic acid and 6.80 g of sodium formate in 1.00 liter of solution. (a) Calculate the pH of the buffer. (b) Calculate the pH of the solution which results when the following are added to separate 100-mL portions of the buffer: (i) 5.00 mmol of NaOH; (ii) 5.00 mmol of HCl; (iii) 800 mg of NaOH; (iv) 365 mg of HCl; (v) 1.00 g of NaOH. (c) Note the ΔpH in each case in part (b).

29. *Buffer effectiveness.* Three buffers are prepared from the weak acid HB, $pK_a = 5.40$, and is salt. Buffer I is 0.40 M in B$^-$ and 0.20 M in HB. Buffer II is 0.35 M in B$^-$ and 0.25 M in HB. Buffer III is 0.30 M in B$^-$ and 0.30 M in HB. To 100 mL of each buffer

is added 10 mL of 1.0 M NaOH. Calculate the ΔpH for each buffer.

30. *Buffer solution.* A buffer solution prepared from formic acid and sodium formate has a pH of 4.04. When 30 mL of 0.20 M HCl is added to 100 mL of the buffer, the pH changes to 3.74. Calculate the molarities of the acid and its salt in the original buffer.

31. *Dilution of a buffer.* Given two solutions: A has a pH of 9.26 and is 0.20 M in NH$_3$ and 0.20 M in NH$_4$Cl, and B is a solution of NaOH, $pH = 9.26$. 10 mL of each solution is diluted to 40 mL. Calculate the pH of the resulting solutions.

32. *Buffer solution.* 5.0 mmol of HCl is added to 100 mL of each solution A and B in Problem 31. Calculate the pH values of the resulting solutions.

33. *Buffer solution.* The electrolytic reduction of an organic nitro compound was carried out in a solution buffered by acetic acid and sodium acetate. The reaction was

$$RNO_2 + 4H_3O^+ + 4e \longrightarrow$$

$$RNHOH + 5H_2O$$

200 mL of a 0.0100 M solution of RNO$_2$ buffered initially at pH 4.44 was reduced, with the reaction above going virtually to completion. The total acetate concentration, [HOAc] + [OAc$^-$], was 0.480 M. Calculate the pH of the solution after the reduction is complete.

34. *Feasibility of a titration.* 30.00 mL of a 0.1200 M solution of a weak base, B, is diluted to 100 mL and titrated with 0.0900 M HCl. Calculate the value of the equilibrium constant for the reaction

$$B + H_3O^+ \rightleftharpoons BH^+ + H_2O$$

so that one drop (0.05 mL) before the EPt the reaction is essentially complete and the pH changes by 1.80 units on the addition of two more drops (0.10 mL) of titrant.

35. *Feasibility of a titration.* 2.5 mmol of a weak acid, HA, $pK_a = 7.00$, is dissolved in 75 mL of water and titrated with 0.10 M NaOH. Using the usual approximations, calculate the pH (a) 0.05 mL before the EPt; (b) at the EPt; (c) 0.05 mL after the EPt. Comment on the feasibility of the titration and the validity of the approximations.

36. *Effect of concentration.* 2.5 mmol of a weak acid, HY, $K_a = 1.0 \times 10^{-5}$, is titrated with 0.10 M NaOH. Calculate the pH 0.05 mL before the EPt, at the EPt, and 0.05 mL beyond the EPt for the cases where the acid was dissolved initially in a volume of (a) 75 mL; (b) 50 mL; (c) 25 mL of solution. What is the value of ΔpH in each case?

37. Repeat Problem 36 but starting with 5.0 mmol of the same weak acid.

38. *Nonfeasible titration.* 30 mL of a 0.10 M solution of NaOAc is diluted to 70 mL with water and titrated with 0.10 M HCl. (a) What is the value of the equilibrium constant for the titration reaction? (b) Calculate the pH at the equivalence point; (c) Calculate the pH two drops (0.10 mL) after the equivalence point. Comment on the feasibility of the titration.

39. *Ion trapping.* A model used to explain the absorption of a drug such as aspirin (a weak acid, designated HAsp), is as follows:

Membrane

Blood plasma   Stomach

$pH = 7.4$ $pH = 1.0$

$H^+ + Asp^- \rightleftharpoons HAsp \rightleftharpoons HAsp \rightleftharpoons H^+ + Asp^-$

It is assumed that ions such as H^+ and Asp^- do not penetrate the membrane, but that the undissociated form, HAsp, equilibrates freely across the membrane. At equilibrium the concentration of HAsp is the same on both sides of the membrane, but there is more *total drug* on the side where the degree of dissociation is greater. The mechanism is known as *ion trapping*.

Aspirin is a weak acid with a pk_a of 3.5. Calculate the ratio of total drug, [HAsp] + [Asp$^-$], in the blood plasma to total drug in the stomach, assuming that the model above is correct.

40. A 2.00-g sample containing 25.0% of the weak acid HX is dissolved in 60.0 mL of solution and titrated with 0.100 M NaOH. When half the acid is neutralized, the pH is 6.00. At the EPt the pH is 9.30. Calculate the MW of HX.

41. *Carbonic acid.* The value normally given in tables for the first dissociation constant, K_{a_1}, of carbonic acid

$$H_2CO_3 + H_2O \rightleftharpoons H_3O^+ + HCO_3^-$$

is 4.6×10^{-7}. This value is 10^2 to 10^4 times smaller than the K_a values for other carboxylic acids (CH$_3$COOH, 1.8×10^{-5}; HCOOH, 1.8×10^{-4}). When CO_2 dissolves in water, only about 0.2% is converted into H_2CO_3 molecules:

$$CO_2 + H_2O \rightleftharpoons H_2CO_3$$

$$K = 1.3 \times 10^{-3}$$

If the K_{a_1} value of 4.6×10^{-7} applies to the expression

$$\frac{[H_3O^+][HCO_3^-]}{[CO_2(aq) + H_2CO_3]}$$

and a saturated solution of CO_2 is about 0.03 M at 1 atm, calculate the true of K_{a_1}, viz.,

$$\frac{[H_3O^+][HCO_3^-]}{[H_2CO_3]}$$

7

Acid-Base Equilibria in Complex Systems

In the previous chapter we confined our attention to acids and bases which furnish or react with a single hydrogen ion. An acid which furnishes only one proton is called a *monoprotic* acid. Carbonic acid, H_2CO_3, furnishes two hydrogen ions and is called a *diprotic* acid; H_3PO_4 is a *triprotic* acid, etc. In general, acids which furnish two or more protons are called *polyprotic* acids. Phosphoric acid and some amino acids are important polyprotic acids. Phosphates are involved in buffers in the body fluids of living systems, and amino acids are the units from which proteins are built.

As might be expected, the equilibrium calculations involving polyprotic acids are more complex than those for monoprotic acids. However, reasonable assumptions can be made in some cases, which enable the chemist to make good approximations of the pH values of solutions of such acids and their salts. The purpose of this chapter is to examine a few of the more important calculations which involve equilibria of polyprotic acids.

7.1

POLYPROTIC ACIDS

A solution of the hypothetical acid H_2B actually contains two acids, H_2B and HB^-. The dissociation reactions and equilibrium constants are

$$H_2B + H_2O \rightleftharpoons H_3O^+ + HB^- \qquad K_{a_1} = \frac{[H_3O^+][HB^-]}{[H_2B]}$$

$$HB^- + H_2O \rightleftharpoons H_3O^+ + B^{2-} \qquad K_{a_2} = \frac{[H_3O^+][B^{2-}]}{[HB^-]}$$

The dissociation reactions and equilibrium constants of the conjugate bases B^{2-} and HB^- are

$$B^{2-} + H_2O \rightleftharpoons HB^- + OH^- \qquad K_{b_1} = \frac{K_w}{K_{a_2}} = \frac{[HB^-][OH^-]}{[B^{2-}]}$$

$$HB^- + H_2O \rightleftharpoons H_2B + OH^- \qquad K_{b_2} = \frac{K_w}{K_{a_1}} = \frac{[H_2B][OH^-]}{[HB^-]}$$

Note the relations between the acid and base constants:

$$K_{a_1} \times K_{b_2} = K_w$$

and

$$K_{a_2} \times K_{b_1} = K_w$$

In a solution of the diprotic acid all three species, H_2B, HB^-, and B^{2-}, are present to some extent. In the following sections we shall calculate the equilibrium concentrations of these species and the pH of solutions of the diprotic acid and its two salts, NaHB and Na_2B.

7.1a. Solution of H_2B

Suppose we have a 0.10 F solution of the diprotic acid, H_2B, where $K_{a_1} = 1.0 \times 10^{-3}$ and $K_{a_2} = 1.0 \times 10^{-7}$. Let us calculate the pH of the solution and the concentrations of H_2B, HB^-, and B^{2-}. Our problem is more complicated than in previous cases in that there are two acids furnishing protons:

$$H_2B + H_2O \rightleftharpoons H_3O^+ + HB^- \qquad K_{a_1} = 1.0 \times 10^{-3}$$
$$HB^- + H_2O \rightleftharpoons H_3O^+ + B^{2-} \qquad K_{a_2} = 1.0 \times 10^{-7}$$

However, since H_2B is a much stronger acid than HB^-, it is reasonable to assume that we can neglect the H_3O^+ furnished by HB^-. The problem then reduces to that of a monoprotic acid, which we have considered before. We assume that

$$[H_3O^+] \cong [HB^-]$$

and

$$[H_2B] = 0.10 - [H_3O^+]$$

Substituting in the expression for K_{a_1},

$$\frac{[H_3O^+]^2}{0.10 - [H_3O^+]} = 1.0 \times 10^{-3}$$

Since K_{a_1} is relatively large, it is best to solve the complete quadratic, or use the method of successive approximations (Section 5.4.a), obtaining

$$[H_3O^+] = 0.0095 \qquad \text{and} \qquad pH = 2.02$$

Then

$$[HB^-] = 0.0095$$

and

$$[H_2B] = 0.10 - 0.0095 = 0.09$$

To obtain the concentration of B^{2-} we use K_{a_2}:

$$\frac{[H_3O^+][B^{2-}]}{[HB^-]} = 1.0 \times 10^{-7}$$

Since $[H_3O^+] = [HB^-]$,

$$[B^{2-}] = K_{a_2} = 1.0 \times 10^{-7}$$

Note that the H_3O^+ contributed by dissociation of the acid HB^- would also be 1.0×10^{-7}. Hence our assumption that the dissociation of HB^- could be neglected is a good one.

In this example the ratio K_{a_1}/K_{a_2} is 10^4. We may ask how valid our calculation would have been had the ratio been smaller, say 10^2, or even 10. In such a case, HB^- would be much stronger compared to H_2B and would furnish a greater share of the protons to the solution. The error made in treating the solution as that of a monoprotic acid would be greater, of course, but calculations show that only a low percentage error would result even when the ratio is as small as 10.

7.1b. Solution of Na₂B

Let us now calculate the pH of a 0.10 F solution of Na_2B and the concentrations of H_2B, HB^-, and B^{2-}. The principal species, B^{2-}, is a base, dissociating (hydrolyzing) in two steps:

$$B^{2-} + H_2O \;\rightleftharpoons\; HB^- + OH^- \qquad K_{b_1} = \frac{K_w}{K_{a_2}} = 1.0 \times 10^{-7}$$

$$HB^- + H_2O \;\rightleftharpoons\; H_2B + OH^- \qquad K_{b_2} = \frac{K_w}{K_{a_1}} = 1.0 \times 10^{-11}$$

Here B^{2-} is a much stronger base than HB^-, and we assume that all the OH^- ions come from the first step in the hydrolysis. Then our concentrations are

$$[OH^-] \cong [HB^-]$$

and

$$[B^{2-}] = 0.10 - [OH^-] \cong 0.10$$

Note that since K_{b_1} is small, little error is made in assuming $[B^{2-}]$ to be 0.10 (Table 5.5, Section 5.4.a). Substituting in the expression for K_{b_1},

$$\frac{[OH^-]^2}{0.10} = 1.0 \times 10^{-7}$$

$$[OH^-] = 1.0 \times 10^{-4}$$

$$pOH = 4.00 \qquad \text{and} \qquad pH = 10.00$$

Then

$$[HB^-] = 1.0 \times 10^{-4}$$

and

$$[B^{2-}] = 0.10$$

To obtain the concentration of H_2B, we substitute in the expression for K_{b_2}:

$$\frac{[H_2B][OH^-]}{[HB^-]} = 1.0 \times 10^{-11}$$

Since $[OH^-] = [HB^-]$,

$$[H_2B] = 1.0 \times 10^{-11}$$

Here again our assumption that we could neglect the second step in the hydrolysis is quite valid.

7.1c.   *Solution of NaHB*

The problem of calculating the pH of a solution of the salt NaHB is complicated by the fact that the principal species, HB^-, is both an acid and a base. Its reaction as a base is

$$HB^- + H_2O \;\rightleftharpoons\; H_2B + OH^- \qquad K_{b_2} = \frac{K_w}{K_{a_1}} = 1.0 \times 10^{-11}$$

and as an acid,

$$HB^- + H_2O \;\rightleftharpoons\; H_3O^+ + B^{2-} \qquad K_{a_2} = 1.0 \times 10^{-7}$$

Note that HB^- is a *weak* acid and a *weak* base, but it is a stronger acid than it is a base. Hence, we can expect the solution to be slightly acidic.

In this complex situation it is convenient to use the systematic method (Section 5.5) in treating the equilibria involved. Our problem is to calculate the pH of a $0.10\,F$ solution of NaHB and the concentrations of the various species in the solution. There are six of these species: H_3O^+, OH^-, Na^+, H_2B, HB^-, and B^{2-}. To solve for the concentrations of these species we need six independent equations. These are K_{a_1}, K_{a_2}, K_w, the charge-balance equation,

$$[Na^+] + [H_3O^+] = [OH^-] + [HB^-] + 2[B^{2-}] \tag{1}$$

the mass-balance equation on B,

$$[H_2B] + [HB^-] + [B^{2-}] = 0.10 \tag{2}$$

and the mass-balance equation on sodium,

$$[Na^+] = 0.10 \tag{3}$$

We add Eqs. (1) and (2), noting that $[Na^+] = 0.10$ gives

$$[H_2B] + [H_3O^+] = [B^{2-}] + [OH^-] \tag{4}$$

If values for $[H_2B]$, $[B^{2-}]$, and $[OH^-]$ are obtained from K_{a_1}, K_{a_2}, and K_w and substituted in Eq. (4), we obtain

$$\frac{[H_3O^+][HB^-]}{K_{a_1}} + [H_3O^+] = \frac{[HB^-]K_{a_2}}{[H_3O^+]} + \frac{K_w}{[H_3O^+]} \tag{5}$$

Multiplying Eq. (5) by $[H_3O^+]$ gives

$$\frac{[H_3O^+]^2[HB^-]}{K_{a_1}} + [H_3O^+]^2 = [HB^-]K_{a_2} + K_w \tag{6}$$

Factoring out $[H_3O^+]^2$ gives

$$[H_3O^+]^2\left\{\frac{[HB^-]}{K_{a_1}} + 1\right\} = [HB^-]K_{a_2} + K_w \tag{7}$$

or

$$[H_3O^+]^2 \left\{ \frac{[HB^-] + K_{a_1}}{K_{a_1}} \right\} = K_{a_2}[HB^-] + K_w \tag{8}$$

Then

$$[H_3O^+]^2 = \frac{K_{a_2}[HB^-] + K_w}{\dfrac{[HB^-] + K_{a_1}}{K_{a_1}}} \tag{9}$$

or

$$[H_3O^+]^2 = \frac{K_{a_1}K_{a_2}[HB^-] + K_{a_1}K_w}{[HB^-] + K_{a_1}} \tag{10}$$

Taking the square root,

$$[H_3O^+] = \sqrt{\frac{K_{a_1}K_{a_2}[HB^-] + K_{a_1}K_w}{[HB^-] + K_{a_1}}} \tag{11}$$

Since HB^- is the principal species in the solution, we may reasonably assume that $[H_2B]$ and $[B^{2-}]$ can be dropped in Eq. (2), giving

$$[HB^-] \cong 0.10$$

Substituting this value and the values of the equilibrium constants in Eq. (11) gives

$$[H_3O^+] = 9.95 \times 10^{-6}$$

In terms of significant figures this should be rounded to 1.0×10^{-5}. Hence we obtain the pH of the solution to be 5.00.

The concentrations of the other species can be obtained from the appropriate equilibrium constant. From K_w,

$$[OH^-] = \frac{1.0 \times 10^{-14}}{1.0 \times 10^{-5}} = 1.0 \times 10^{-9}$$

From K_{a_2},

$$\frac{(1.0 \times 10^{-5})[B^{2-}]}{0.10} = 1.0 \times 10^{-11}$$

$$[B^{2-}] = 1.0 \times 10^{-7}$$

From K_{a_1},

$$\frac{(1.0 \times 10^{-5})(0.10)}{[H_2B]} = 1.0 \times 10^{-3}$$

$$[H_2B] = 1.0 \times 10^{-3}$$

The approximation made in Eq. (2) is seen to be a reasonable one:

$$1.0 \times 10^{-3} + 0.10 + 1.0 \times 10^{-7} \cong 0.10$$

In general this approximation is good if K_{a_1} and K_{a_2} are small and if the concentration of the salt is not too low. It should also be noted that under conditions where $[HB^-] \gg K_{a_1}$ and $K_{a_1}K_{a_2}[HB^-] \gg K_{a_1}K_w$, Eq. (11) reduces to

$$[H_3O^+] = \sqrt{K_{a_1}K_{a_2}}$$

and

$$pH = \tfrac{1}{2}(pK_{a_1} + pK_{a_2}) \tag{12}$$

In the example above using Eq. (12) our calculation of pH is simply

$$pH = \tfrac{1}{2}(3.00 + 7.00) = 5.00$$

Some examples illustrating the use of Eqs. (11) and (12) and showing the effect of concentration of the salt are given in Problem 1 at the end of the chapter.

7.1d. *Titration Curves for Polyprotic Acids*

We can now calculate the titration curve for a polyprotic acid. The following example is an illustration.

EXAMPLE 7.1

Titration of a diprotic acid

50.0 mL of 0.100 M H_2B is titrated with 0.100 M NaOH. The dissociation constants are $K_{a_1} = 1.0 \times 10^{-3}$ and $K_{a_2} = 1.0 \times 10^{-7}$. Calculate the pH at the start of the titration and after the addition of 10.0, 50.0, 60.0, and 100.0 mL of base.

(a) *Initial pH.* As seen in Section 7.1.a we can treat H_2B as a monoprotic acid:

$$H_2B + H_2O \rightleftharpoons H_3O^+ + HB^-$$

Then using the usual approximations,

$$\frac{[H_3O^+]^2}{0.10} = 1.0 \times 10^{-3}$$

$$[H_3O^+] = 1.0 \times 10^{-2}$$

$$pH = 2.00$$

In Section 7.1.a we noted that because K_{a_1} is relatively large, the error in taking $[H_2B] = 0.10$ is appreciable. If the complete quadratic is solved, the value of $[H_3O^+]$ is found to be 9.5×10^{-3}.

(b) *pH after addition of 10.0 mL of base.* We started with 50.0 mL $\times$ 0.100 mmol/mL = 5.00 mmol of H_2B and have added 10.0 mL $\times$ 0.100 mmol/mL = 1.00 mmol of OH^-. The reaction is

mmol	H_2B	+	OH^-	$\longrightarrow$	HB^-	+	H_2O
Initial:	5.00		1.00		—		
Change:	-1.00		-1.00		$+1.00$		
Equilibrium:	4.00		—		1.00		

The dissociation reaction and equilibrium concentrations are

$$H_2B + H_2O \longrightarrow H_3O^+ + HB^-$$

$$\frac{4.00}{60.0} - [H_3O^+] \qquad [H_3O^+] \qquad \frac{1.00}{60.0} + [H_3O^+]$$

Since $[H_3O^+]$ is small, $[H_2B] \cong 4.00/60.0$ and $[HB^-] \cong 1.00/60.0$. Substituting in the Henderson-Hasselbalch equation and noting that the volume cancels,

$$pH = pK_{a_1} + \log \frac{[HB^-]}{[H_2B]}$$

$$pH = 3.00 + \log \frac{1.00}{4.00}$$

$$pH = 2.40$$

Here again the error made by our usual approximation is large because of the size of K_{a_1}. The more exact solution gives a pH of 2.52. The pH at other points up to the first equivalence point can be calculated in the same manner.

(c) *pH at the first equivalence point*. We started with 50.0 mL × 0.100 mmol/mL of H_2B and have added 50.0 mL × 0.100 mmol/mL = 5.00 mmol of OH^-. The reaction is

mmol	H_2B	+	OH^-	$\longrightarrow$	HB^-	+	H_2O
Initial:	5.00		5.00				
Change:	−5.00		−5.00		+5.00		
Equilibrium:	—		—		5.00		

Since HB^- is the predominant species, the pH can be approximated (Section 7.1.c) by the expression

$$pH = \tfrac{1}{2}(pK_{a_1} + pK_{a_2})$$

Hence

$$pH = \tfrac{1}{2}(3.00 + 7.00)$$

$$pH = 5.00$$

(d) *pH during the titration of HB^-: 60.0 mL of base added*. 50.0 mL of OH^- is used to form 5.00 mmol of HB^-. The additional 10.0 mL × 0.100 mmol/mL = 1.00 mmol of OH^- reacts with HB^-:

mmol	HB^-	+	OH^-	$\longrightarrow$	H_2O	+	B^{2-}
Initial:	5.00		1.00				—
Change:	−1.00		−1.00				+1.00
Equilibrium:	4.00		—				1.00

The dissociation reaction and equilibrium concentrations are

$$HB^- + H_2O \longrightarrow H_3O^+ + B^{2-}$$

$$\frac{4.00}{110} - [H_3O^+] \qquad [H_3O^+] \qquad \frac{1.00}{110} + [H_3O^+]$$

Since $[H_3O^+]$ is small, $[HB^-] \cong 4.00/110$ and $[B^{2-}] \cong 1.00/110$. Substituting in the Henderson-Hasselbalch equation and noting that the volume cancels,

$$pH = pK_{a_2} + \log \frac{[B^{2-}]}{[HB^-]}$$

$$pH = 7.00 + \log \frac{1.00}{4.00}$$

$$pH = 6.40$$

The pH at other points up to the second equivalence point can be calculated in the same manner.

(e) *pH at the second equivalence point.* We started with 50.0 mL × 0.100 mmol/mL = 5.00 mmol of H_2B and have added 100 mL × 0.100 mmol/mL = 10.0 mmol of OH^-. The reaction is

mmol	H_2B	+	$2OH^-$	$\longrightarrow$	B^{2-}	+	$2H_2O$
Initial:	5.00		10.0		—		
Change:	−5.00		−10.0		+5.00		
Equilibrium:	—		—		5.00		

The *p*H is determined by the first step in the hydrolysis of B^{2-}:

$$B^{2-} + H_2O \longrightarrow HB^- + OH^-$$

The equilibrium expression is

$$\frac{[HB^-][OH^-]}{[B^{2-}]} = \frac{K_w}{K_{a_2}}$$

Since $[HB^-] \cong [OH^-]$ and $[B^{2-}] \cong 5.00/150 = 0.0333$

$$\frac{[OH^-]^2}{0.0333} = \frac{1.0 \times 10^{-14}}{1.0 \times 10^{-7}}$$

$$[OH^-] = 5.8 \times 10^{-5}$$

$$p\text{OH} = 4.24$$

$$p\text{H} = 9.76$$

Values beyond the second equivalence point are calculated from the amount of excess base.

$\square$

The titration curve is shown in Fig. 7.1, curve A, where the two titration steps are reasonably distinct. B in Fig. 7.1 is the curve for a diprotic acid in which the ratio of K_{a_1} to K_{a_2} is 10^2. In this case there is only a slight indication of two separate steps in titration, a slightly more rapid rise in *p*H occurring at the first equivalence point. In curve C, for H_2SO_4, where both H_2SO_4 and HSO_4^- are strongly dissociated ($K_{a_2} = 0.012$), the shape is essentially the same as that for a strong monoprotic acid.

In general, one can conclude that in order for the steps in the titration of a polyprotic acid to be distinct, the successive constants must differ by a factor of at least 10^4, or the pK_a values must differ by 4 units. Maleic acid has pK_a values differing by about 4.3 units and hence titrates in two distinct steps. In the case of oxalic acid, the two pK_a values differ by only 3.0 units and the two steps are not sharply separated. Phosphoric is a triprotic acid with pK_a values of 2.12, 7.21, and 12.32. At the first equivalence point, about *p*H 4.62, the value of $\Delta p\text{H}/\Delta V$ is fairly large, and methyl red is a suitable indicator for this step. At the second equivalence point, about *p*H 9.72, $\Delta p\text{H}/\Delta V$ is not as large because $H_2PO_4^-$ is a weaker acid than H_3PO_4. Phenolphthalein can be used to detect this equivalence point. The third acid, HPO_4^{2-}, is too weak for feasible titration (page 148). The value of K for the reaction

$$HPO_4^{2-} + OH^- \rightleftharpoons PO_4^{3-} + H_2O$$

is only $4.8 \times 10^{-13}/1.0 \times 10^{-14} = 48$.

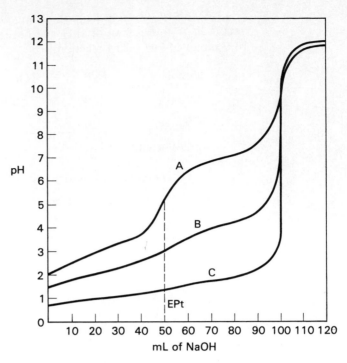

Figure 7.1 Titration curves for diprotic acids. A, $K_{a_1}/K_{a_2} = 10^4$; B, $K_{a_1}/K_{a_2} = 10^2$; C, H_2SO_4, K_{a_1} large, $K_{a_2} = 0.012$.

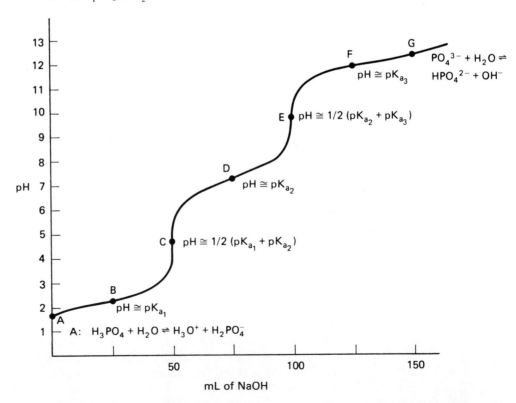

Figure 7.2 Titration curve of phosphoric acid: 50 mL of 0.10 M acid titrated with 0.10 M base. A, start of titration; B, halfway to first E.Pt.; C, first E.Pt.; D, halfway to second E.Pt.; E, second E.Pt.; F, halfway to third E.Pt.; G, third E.Pt.

TABLE 7.1 Approximate Relations between the Concentrations of Species in Phosphate Solutions

PLACE ON TITRATION CURVE[†]	APPROXIMATE RELATION OF PRINCIPAL SPECIES
A	$[H_2PO_4^-] \cong [H_3O^+]$
B	$[H_3PO_4] \cong [H_2PO_4^-]$
C	$[H_3PO_4] \cong [HPO_4^{2-}]$
D	$[H_2PO_4^-] \cong [HPO_4^{2-}]$
E	$[H_2PO_4^-] \cong [PO_4^{3-}]$
F	$[HPO_4^{2-}] \cong [PO_4^{3-}]$
G	$[HPO_4^{2-}] \cong [OH^-]$

[†] Letters correspond to position on titration curve in Fig. 7.2.

Figure 7.2 shows the titration curve of phosphoric acid. The equations or reactions needed to calculate the pH at various stages in the titration are shown in the figure. Table 7.1 contains the approximate relations between the concentrations of the various species at the start of the titration of each species, halfway to the equivalence point, and at the equivalence point. The approximations are poorest at the two ends of the curve because H_3PO_4 has a rather large K_{a_1}, 7.5×10^{-3}, and PO_4^{3-} is a rather strong base, $K_{b_1} = 2.1 \times 10^{-2}$.

7.2

AMINO ACIDS

Amino acids are important biological molecules which serve as building blocks for peptides and proteins. They have the general structure

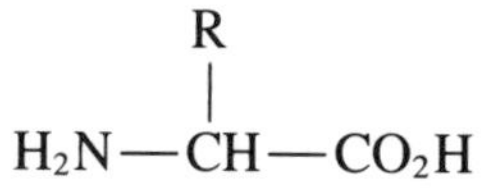

$$\begin{array}{c} R \\ | \\ H_2N-CH-CO_2H \end{array}$$

where R is an organic group which is different in each amino acid. Note that the NH_2 group is attached to the carbon atom adjacent to the CO_2H group. For this reason these molecules are called *alpha* amino acids. There are 20 different amino acids that have been identified as units in the most important plant and animal proteins.

Since amino acids contain both an acidic and a basic group, they are amphoteric and tend to undergo internal proton transfer from the CO_2H group to the NH_2 group:

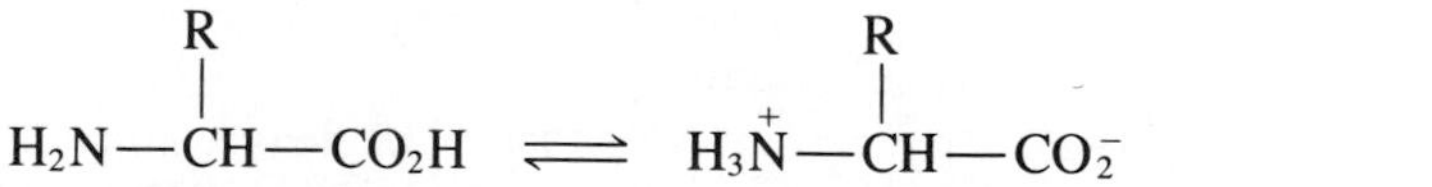

$$\begin{array}{ccc} R & & R \\ | & & | \\ H_2N-CH-CO_2H & \rightleftharpoons & H_3\overset{+}{N}-CH-CO_2^- \\ \text{Neutral molecule} & & \text{Zwitterion} \end{array}$$

The equilibrium greatly favors the dipolar ion, called a *zwitterion*.

In strongly acidic solutions (low pH), the amino acid is protonated and the molecule is positively charged. In strongly basic solutions (high pH), the molecule loses a proton and is negatively charged. At some intermediate pH, called the *isoelectric point*, the molecule is uncharged.

The acid-base equilibria of amino acids are indicated below for the simplest acid, glycine (R = H):

Box 7.1 **Alkaloids**

Many complex organic compounds which are weak bases containing nitrogen occur in nature. Some of these molecules exhibit marked biological activity in humans, and many have proved to be powerful pharmaceutical agents. Some are addictive drugs which have caused severe problems for society. These compounds, sometimes referred to as *alkaloids,* usually contain at least one heterocyclic ring. The structures of a few of the more interesting compounds are shown below:

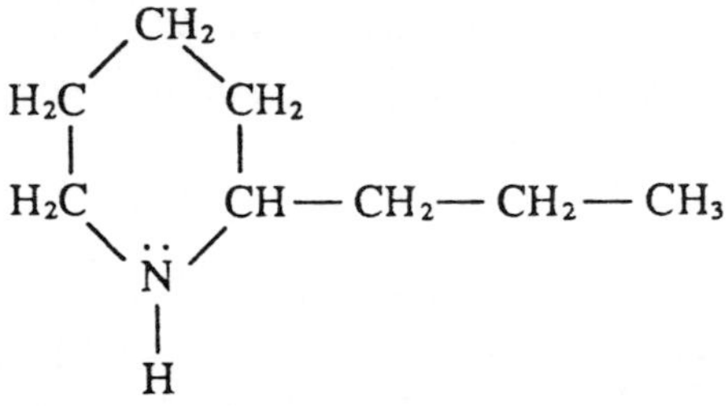

Coniine

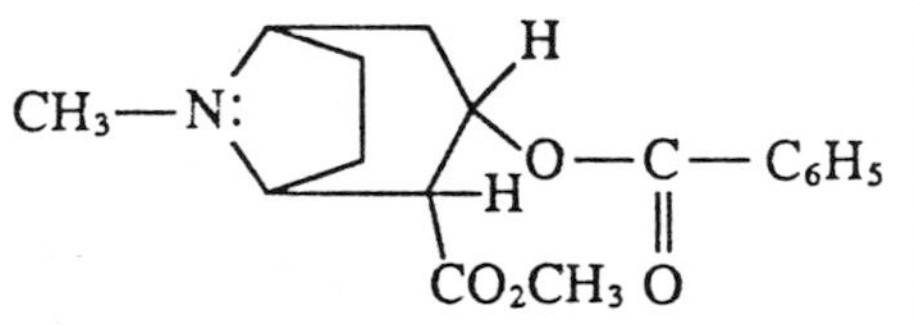

Cocaine

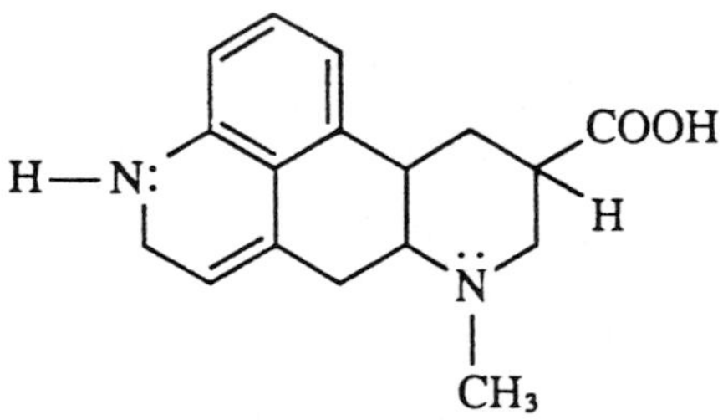

Lysergic acid

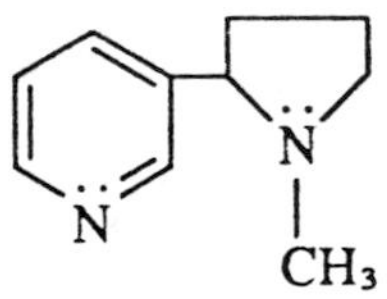

Nicotine

Coniine occurs in the poisonous hemlock, the extract of which is reported to have killed Socrates. Nicotine represents about 75% of the alkaloid content in tobacco. Cocaine is obtained from the leaves of the coca bush found in South America. It induces dependence and is a major problem in society today. The alkaloids found in the fungus ergot are derivatives of lysergic acid. The diethylamide of lysergic acid, usually called LSD, is a powerful hallucinogen, producing psychotic episodes resembling schizophrenia.

Note that all of these compounds contain at least one nitrogen with an unshared pair of electrons. Cocaine has a K_b of 2.6×10^{-6}. After it is extracted from coca leaves it is isolated by adding hydrochloric acid to form a crystalline powder, cocaine hydrochloride. It is in this form that it is shipped to this country from South America.

Although simple acid-base titrations orginated in ancient times, they are still used to assay the purities of many drugs—drugs of abuse as well as legitimate pharmaceutical preparations. Old methods are not always abandoned as new ones appear, and acid-base titrations, when appropriate for the sample, are highly cost-effective.

$$\overset{+}{\text{NH}}_3\text{—CH}_2\text{CO}_2\text{H} \underset{\text{H}^+}{\overset{\text{OH}^-}{\rightleftharpoons}} \overset{+}{\text{NH}}_3\text{—CH}_2\text{CO}_2^- \underset{\text{H}^+}{\overset{\text{OH}^-}{\rightleftharpoons}} \text{NH}_2\text{—CH}_2\text{CO}_2^-$$

Conjugate acid, H_2A^+ Zwitterion, HA   Conjugate base, A^-

The conjugate acid, H_2A^+, is a diprotic acid, dissociating as follows:

$$H_2A^+ + H_2O \rightleftharpoons H_3O^+ + HA \qquad K_{a_1} = \frac{[H_3O^+][HA]}{[H_2A^+]} \qquad (1)$$

$$HA + H_2O \rightleftharpoons H_3O^+ + A^- \qquad K_{a_2} = \frac{[H_3O^+][A^-]}{[HA]} \qquad (2)$$

The concentrations of the species H_2A^+, HA, and A^- can be calculated by the methods discussed in Sections 7.1a, 7.1b, and 7.1c. H_2A^+ can be titrated just as any other diprotic acid, and the calculations of pH are the same as those discussed in Section 7.1d. Generally, pK_{a_1} values of the conjugate acids, H_2A^+, are in the range of about 2 to 5, and a fairly large change in pH occurs at the first equivalence point. Values of pK_{a_2} are usually in the range of 8 to 11, and the change in pH at the second equivalence point is not large. We would normally say that this titration step is not feasible for the purposes of analysis.

Glycine itself is a monoprotic acid with a pK_a of 9.87. The titration curve of this acid is shown in Fig. 7.3, where it can be seen that the end point is not sharp. Alanine (R = CH$_3$) is also a monoprotic acid (pK_a = 9.9), and its titration curve would be of the same form as that of glycine. The salt of alanine, alanine hydrochloride,

$$\overset{+}{\text{NH}}_3\text{—CHCO}_2\text{H} \quad \text{Cl}^-$$
$$|$$
$$\text{CH}_3$$

(which we called the conjugate acid, H_2A^+, above), is a diprotic acid with pK_{a_1} = 2.3 and pK_{a_2} = 9.9. Its titration curve is shown in Fig. 7.4, where it can be seen that the first end point is sharp and the second one is not. Some amino acids have

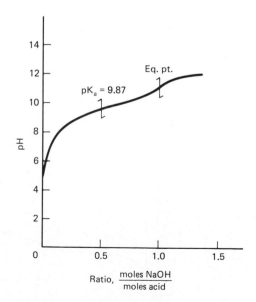

Figure 7.3 Titration curve of an amino acid such as glycine. The equivalence point, which occurs at a mole ratio of 1.0, is not sharp.

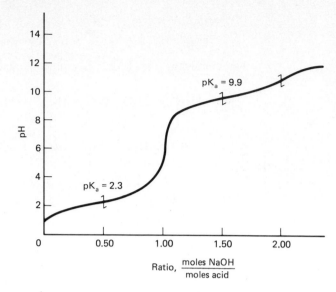

Figure 7.4 Titration curve of an amino acid such as alanine hydrochloride, H_2A^+. The acid titrates in two steps. The first endpoint is sharp; the second is not.

a CO_2H or NH_2 group as part of the R group. For example, aspartic acid hydrochloride is a triprotic acid with the formula

$$\overset{+}{N}H_3 - \underset{|}{CH}CO_2H$$
$$\begin{array}{c} CH_2CO_2H \\ | \end{array}$$

The three pK_a values of this acid are 2.0, 3.9, and 10.0, respectively. This molecule gives a titration curve with three end points similar to that shown in Fig. 7.5. The separation of the first and second steps in the titration is not sharp, since the first two pK_a values differ by only 1.9 units. There is a large change in pH at the second equivalence point.

We mentioned above that pH at which there is an exact balance of positive and negative charges on an amino acid or protein is called the *isoelectric point*.

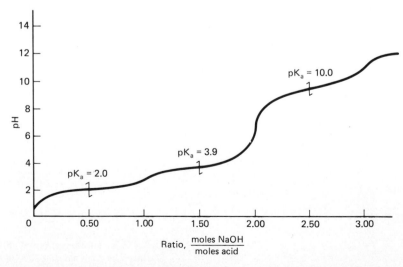

Figure 7.5 Titration curve of an amino acid such as aspartic hydrochloride.

At this pH the zwitterion predominates, and the amino acid will not migrate to either electrode when it is placed between opposite poles of an electric field. At a pH above the isoelectric point the molecule is negatively charged and migrates to the anode; at a pH below this value the molecule migrates to the cathode.

The pH at the isoelectric point of a diprotic acid is given by Eq. (12), page 177. This can be shown as follows. The product of the two dissociation constants, Eqs. (1) and (2), is

$$K_{a_1} \times K_{a_2} = \frac{[H_3O^+]^2[A^-]}{[H_2A^+]} \tag{3}$$

Since there is an exact balance of charge at the isoelectric point,

$$[H_2A^+] = [A^-] \tag{4}$$

Substituting Eq. (4) in Eq. (3) gives

$$[H_3O^+]^2 = K_{a_1} \times K_{a_2}$$

$$[H_3O^+] = \sqrt{K_{a_1}K_{a_2}}$$

or

$$pH = \tfrac{1}{2}(pK_{a_1} + pK_{a_2})$$

For alanine hydrochloride the pH at the isoelectric point is $\tfrac{1}{2}(2.3 + 9.9) = 6.1$.

Biochemists also use the term *isoionic point*. This is the pH obtained when the pure neutral amino acid (zwitterion) is dissolved in water. The isoionic pH is calculated from Eq. (11), page 176, and for a $0.10\ F$ solution of alanine the value obtained is also 6.1. Generally the isoelectric and isoionic pH values are very nearly the same. The isoelectric point is independent of concentration, whereas the isoionic point is slightly concentration dependent.

7.3

TITRATION OF CARBONATES

It was pointed out in Chapter 6 that when CO_2 is absorbed by a standard solution of NaOH, the normality of the solution will be affected if phenolphthalein indicator is employed. It was also mentioned that mixtures of carbonate and hydroxide, or carbonate and bicarbonate, can be determined by titration using phenolphthalein and methyl orange indicators. We would like to examine this topic in more detail now that we have discussed polyprotic acids.

The first pK_a of carbonic acid is 6.34, and the second 10.36, making the difference 4.02 units. We might expect a fair break between the two curves in this case, but K_{a_1} is so small that the break at the first equivalence point is poor. Usually, the carbonate ion is titrated as a base with a strong acid titrant, in which case two fair breaks are obtained, as shown in Fig. 7.6, corresponding to the reactions

$$CO_3^{2-} + H_3O^+ \rightleftharpoons HCO_3^- + H_2O$$

$$HCO_3^- + H_3O^+ \rightleftharpoons H_2CO_3 + H_2O$$

Phenolphthalein, pH range 8.0 to 9.6, is a suitable indicator for the first end point, since the pH of a solution of NaHCO$_3$ is $\tfrac{1}{2}(pK_{a_1} + pK_{a_2})$ or 8.35. Methyl orange, pH range 3.1 to 4.4, is suitable for the second end point. A saturated solution of CO_2 has a pH of about 3.9. Neither end point is very sharp, but the second one can be greatly improved by removal of CO_2. Usually, samples containing only sodium carbonate (soda ash) are neutralized to the methyl orange point, and

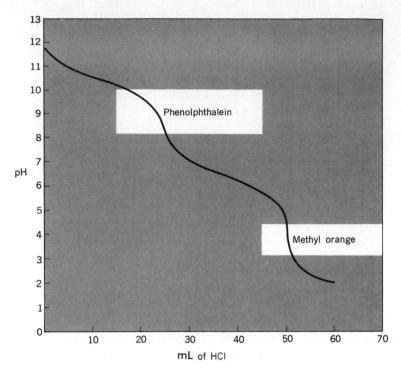

Figure 7.6 Titration curve of Na_2CO_3; 2.5 mmol Na_2CO_3 titrated with 0.10 M HCl.

excess acid is added. Carbon dioxide is removed by boiling the solution, and the excess acid is titrated with standard base.

Mixtures of carbonate and bicarbonate, or of carbonate and hydroxide, can be titrated with standard HCl to the two end points mentioned above. As noted in Fig. 7.7, at the phenolphthalein end point NaOH is completely neutralized, Na_2CO_3 is half-neutralized, and HCO_3^- has not yet reacted. From the phenolphthalein to the methyl orange end point, bicarbonate is being neutralized. Only a few drops of titrant would be required by the NaOH to go from a pH of 8 to 4, and this can be corrected by an indicator blank. In Fig. 7.7 and Table 7.2, v_1 is the volume of acid in milliliters used from the start of the titration to the phenolphthalein end point, and v_2 is the volume from the phenolphthalein end point to the methyl orange.

In Table 7.2 are listed the relations between the volumes of acid used to the two end points for single components and mixtures. The molarity of the HCl is designated by M. The student should be able to verify these relationships, recalling that NaOH reacts completely in the first step, $NaHCO_3$ reacts only in the second step, and Na_2CO_3 reacts in both steps using equal volumes of titrant in the two steps. The mixture of NaOH and $NaHCO_3$ is not considered, since these two compounds react:

$$HCO_3^- + OH^- \rightleftharpoons CO_3^{2-} + H_2O$$

The resulting product is a mixture of CO_3^{2-} and OH^-, a mixture of HCO_3^- and CO_3^{2-}, or CO_3^{2-} alone, depending upon the relative amounts of the two compounds in the sample.

The following examples illustrate the use of the two-indicator method and the effect of CO_2 absorption on the normality of a sodium hydroxide solution.

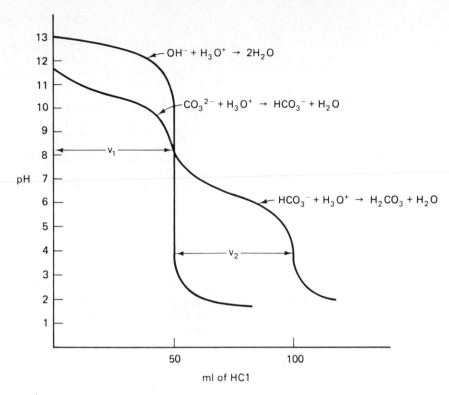

Figure 7.7 Titration curves of NaOH and Na_2CO_3: 50 ml of 0.10 M solution titrated with 0.10 M HCl.

TABLE 7.2 Volume Relations in Carbonate Titrations

SUBSTANCE	RELATION FOR QUALITATIVE IDENTFICATION	MILLIMOLES OF SUBSTANCE PRESENT	
NaOH	$v_2 = 0$		$M \times v_1$
Na_2CO_3	$v_1 = v_2$		$M \times v_1$
$NaHCO_3$	$v_1 = 0$		$M \times v_2$
NaOH + Na_2CO_3	$v_1 > v_2$	NaOH:	$M(v_1 - v_2)$
		Na_2CO_3:	$M \times v_2$
$NaHCO_3$ + Na_2CO_3	$v_1 < v_2$	$NaHCO_3$:	$M(v_2 - v_1)$
		Na_2CO_3:	$M \times v_1$

EXAMPLE 7.2

Use of the "two-indicator" method

A 0.6234-g sample that might contain NaOH, Na_2CO_3, $NaHCO_3$, or a mixture of NaOH + Na_2CO_3 or Na_2CO_3 + $NaHCO_3$ is titrated with 0.1062 M HCl by the two-indicator method. It is found that 40.38 mL of the acid is required to reach the phenolphthalein end point. Methyl orange is then added to the solution, and the titration continued using an additional 12.83 mL of the acid. (a) Identify the base or mixture of bases in the sample. (b) Calculate the percentage of each in the sample.

(a) Since 40.38 mL > 12.83 mL, the sample must contain NaOH and Na_2CO_3.

(b) The volume of titrant used by Na_2CO_3 in the second step is 12.83 mL. An equal volume must have been used in the first step. Hence the volume used by the NaOH is 40.38 − 12.83 = 27.55 mL.

Then

$$\% \text{ Na}_2\text{CO}_3 = \frac{12.83 \times 0.1062 \times 106.0}{623.4} \times 100 = 23.17$$

and

$$\% \text{ NaOH} = \frac{27.55 \times 0.1062 \times 40.00}{623.4} \times 100 = 18.77 \qquad \square$$

EXAMPLE 7.3

The "carbonate error"

A bottle which contains 200 mL of 0.100 M NaOH absorbs 1.00 mmol of CO_2 from the air. If the solution is then titrated with standard acid using phenolphthalein indicator, what normality will be found?

The solution contains

$$200 \text{ mL} \times 0.100 \text{ mmol/mL} = 20.0 \text{ mmol NaOH}$$

The 1.00 mmol of CO_2 reacts with 2.00 mmol of NaOH:

$$2\text{NaOH} + CO_2 \longrightarrow \text{Na}_2\text{CO}_3 + H_2O$$

The resulting solution contains 18.0 mmol of NaOH and 1.00 mmol of Na_2CO_3. On titration to the phenolphthalein end point, the NaOH will use 18.0 mmol of H_3O^+ and the Na_2CO_3 will use 1.00 mmol. Hence the normality found will be

$$\frac{(18.0 + 1.00) \text{ meq}}{200 \text{ mL}} = 0.095 \ N$$

Note the carbonate error. Had methyl orange indicator been employed, the Na_2CO_3 would have used 2.00 mmol of acid and the normality would have been found to be 0.100. $\qquad \square$

7.4

TITRATION OF A MIXTURE OF TWO ACIDS

The conclusions we drew in the previous section concerning titration of the acid H_2B in two steps apply in a similar manner to the titration of a mixture of two weak acids, HX and HY, provided that the initial concentrations of the two acids are the same. If HX, K_{a_1}, is the stronger acid, and HY, K_{a_2}, is the weaker, $pK_{a_1} - pK_{a_2}$ must be at least 4 units for the two titration steps to be reasonably distinct. If the difference in pK_a values is less than this, the two steps are not as distinct, as indicated in Fig. 7.1. The pH at the first equivalence point is $\frac{1}{2}(pK_{a_1} + pK_{a_2})$ if the initial concentrations of HX and HY are the same. If these are not equal, the expression for calculating the pH at the first equivalence point can be obtained as follows.

The charge-balance equation at this point is

$$[\text{Na}^+] + [H_3O^+] = [\text{OH}^-] + [\text{X}^-] + [\text{Y}^-] \qquad (1)$$

The $[\text{Na}^+]$ is the same as the formal concentration of the acid HX; i.e.,

$$[\text{Na}^+] = [\text{HX}] + [\text{X}^-] \qquad (2)$$

Combining Eqs. (1) and (2) we have

$$[H_3O^+] = [\text{OH}^-] + [\text{Y}^-] - [\text{HX}] \qquad (3)$$

Substituting for $[\text{OH}^-]$, $[\text{Y}^-]$, and $[\text{HX}]$ from K_w, K_{a_1}, and K_{a_2}, we have

$$[H_3O^+] = \frac{K_w}{[H_3O^+]} + \frac{K_{a_2}[HY]}{[H_3O^+]} - \frac{[H_3O^+][X^-]}{K_{a_1}} \tag{4}$$

Solving for $[H_3O^+]$ gives

$$[H_3O^+] = \sqrt{\frac{K_{a_1}K_w + K_{a_1}K_{a_2}[HY]}{K_{a_1} + [X^-]}} \tag{5}$$

Assuming $K_{a_2}[HY] \gg K_w$ and $[X^-] \gg K_{a_1}$,

$$[H_3O^+] = \sqrt{\frac{K_{a_1}K_{a_2}[HY]}{[X^-]}} \tag{6}$$

or

$$pH = \tfrac{1}{2}(pK_{a_1} + pK_{a_2}) - \tfrac{1}{2}\log\frac{[HY]}{[X^-]} \tag{7}$$

The principal application of this type of titration is in titrating mixtures of a strong acid and a weak acid, such as hydrochloric and acetic acids. The HCl is titrated first, and if one calculates the pH during the titration, it is logical to disregard the H_3O^+ contributed by the weak HOAc. This follows from LeChâtelier's principle, excess H_3O^+ repressing dissociation of the weak acid. This assumption becomes less valid as the first equivalence point is approached, since the excess H_3O^+ is decreasing in concentration. At the first equivalence point, the HCl is essentially used up and the pH is determined by the dissociation of HOAc. Beyond the first equivalence point the titration curve is that of acetic acid.

Figure 7.8 shows the curve for the titration of 50 mL of a solution which is 0.10 M in HCl and 0.10 M in HOAc. It can be seen that the first equivalence point is a poor one, the value of $\Delta pH/\Delta V$ not being very large. The pH of a

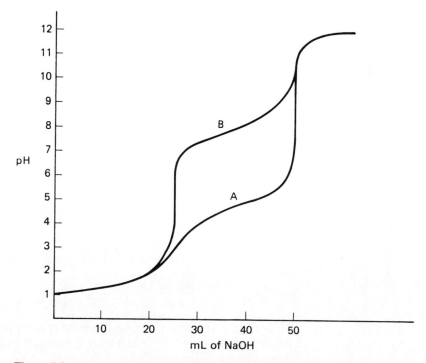

Figure 7.8 Titration curves: *A*, 50 mL of 0.10 *M* HCl and 0.10 *M* HOAc; *B*, 50 mL of 0.10 *M* HCl and 0.10 *M* HX, $K_a = 1 \times 10^{-8}$, titrated with 0.20 *M* NaOH.

0.067 M solution of HOAc is approximately 3. Since this is below pH 4, it is not likely to be considered a feasible titration if a visual indicator is used. (See page 150). The second step in the titration involves the reaction of HOAc with strong base and is a feasible titration.

Figure 7.8 also includes a curve for the titration of a mixture of HCl and an acid HX, $K_a = 1.0 \times 10^{-8}$. It can be seen that the value of ΔpH$/\Delta V$ for the HCl is much larger in the presence of the weaker acid, and this titration step is definitely feasible. The ΔpH$/\Delta V$ is not as large in the second step, since HX is such a weak acid.

7.5

DISTRIBUTION OF ACID-BASE SPECIES AS A FUNCTION OF pH

It is convenient for various purposes to be able to see at a glance the status of dissociation status of common acid-base species as a function of pH. Graphs which show this enable us to determine which of several possible species predominate at a given pH, and they aid in selecting the regions of buffer effectiveness for mixtures of acids or bases and their salts. For example, the pH of blood plasma is held at about 7; it migh be of interest to know whether plasma phosphate exists as H_3PO_4, $H_2PO_4^-$, HPO_4^{2-}, PO_4^{3-}, or some mixture of these species at physiological pH. The type of graph we discuss below can provide answers to such questions almost instantly. It will also be useful in later chapters to be able to calculate the concentration of a particular species in the solution at a given pH. The following examples show the derivation of expressions for these fractions in the cases of a monoprotic and a diprotic acid.

EXAMPLE 7.4

Species distribution— acetic acid

In a solution of acetic acid, calculate the fraction present as HOAc molecules and as OAc$^-$ ions at various pH values. Draw an appropriate graph.

Let c_a represent the *analytical concentration* (page 48). This is the total concentration of all species arising from acetic acid and is simply a mass balance as used previously:

$$c_a = [HOAc] + [OAc^-]$$

From the dissociation constant expression for HOAc, we obtain

$$[OAc^-] = \frac{[HOAc]K_a}{[H_3O^+]}$$

Substitution into the expression for c_a gives

$$c_a = [HOAc] + \frac{[HOAc]K_a}{[H_3O^+]}$$

$$c_a = [HOAc]\left\{1 + \frac{K_a}{[H_3O^+]}\right\}$$

$$\frac{[HOAc]}{c_a} = \frac{1}{1 + (K_a/[H_3O^+])} = \frac{[H_3O^+]}{[H_3O^+] + K_a}$$

$[HOAc]/c_a$ is the fraction of total acetate present in the undissociated form. By a similar approach, it may be shown that the fraction of the acetic acid in the dissociated form is given by

$$\frac{[\text{OAc}^-]}{c_a} = \frac{K_a}{[\text{H}_3\text{O}^+] + K_a}$$

Graphs of these fractions vs. pH are shown in Fig. 7.9. Notice that at a pH roughly 2 units below pK_a, practically all the acetate (about 99%) is in the undissociated form, HOAc, and that the acid is almost completely dissociated at a pH of $(pK_a + 2)$. At the intersection of the two curves, $[\text{OAc}^-]/c_a = [\text{HOAc}]/c_a = 0.5$ and pH $= pK_a$ or $[\text{H}_3\text{O}^+] = K_a$. ☐

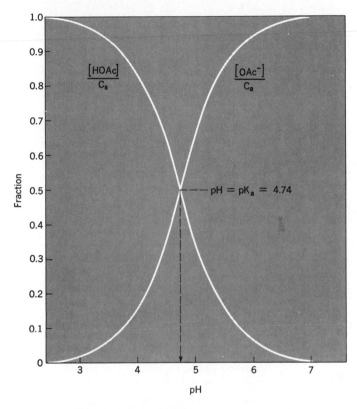

Figure 7.9 Distribution of acetate species as a function of pH.

EXAMPLE 7.5

Species distribution—oxalic acid.

In a solution of the diprotic acid oxalic (H_2Ox), calculate the fractions present as H_2Ox molecules and as HOx^- and Ox^{2-} ions as a function of pH. Draw an appropriate graph.

Here the analytical concentration is given by

$$c_a = [\text{H}_2\text{Ox}] + [\text{HOx}^-] + [\text{Ox}^{2-}]$$

We also have the two dissociation expressions:

$$K_{a_1} = \frac{[\text{H}_3\text{O}^+][\text{HOx}^-]}{[\text{H}_2\text{Ox}]}$$

$$K_{a_2} = \frac{[\text{H}_3\text{O}^+][\text{Ox}^{2-}]}{[\text{HOx}^-]}$$

Rearrangement of the two K_a expressions gives

$$[\text{HOx}^-] = \frac{[\text{H}_2\text{Ox}]K_{a_1}}{[\text{H}_3\text{O}^+]}$$

$$[\text{Ox}^{2-}] = \frac{[\text{HOx}^-]K_{a_2}}{[\text{H}_3\text{O}^+]} = \frac{[\text{H}_2\text{Ox}]K_{a_1}K_{a_2}}{[\text{H}_3\text{O}^+]^2}$$

Substitution into the expression for the analytical concentration yields

$$c_a = [\text{H}_2\text{Ox}] + \frac{[\text{H}_2\text{Ox}]K_{a_1}}{[\text{H}_3\text{O}^+]} + \frac{[\text{H}_2\text{Ox}]K_{a_1}K_{a_2}}{[\text{H}_3\text{O}^+]^2}$$

whence

$$c_a = [\text{H}_2\text{Ox}]\left\{1 + \frac{K_{a_1}}{[\text{H}_3\text{O}^+]} + \frac{K_{a_1}K_{a_2}}{[\text{H}_3\text{O}^+]^2}\right\}$$

$$\frac{[\text{H}_2\text{Ox}]}{c_a} = \frac{1}{1 + \dfrac{K_{a_1}}{[\text{H}_3\text{O}^+]} + \dfrac{K_{a_1}K_{a_2}}{[\text{H}_3\text{O}^+]^2}}$$

$$\frac{[\text{H}_2\text{Ox}]}{c_a} = \frac{[\text{H}_3\text{O}^+]^2}{[\text{H}_3\text{O}^+]^2 + [\text{H}_3\text{O}^+]K_{a_1} + K_{a_1}K_{a_2}}$$

With no more difficulty, the expressions for the fractions present as HOx^- and Ox^{2-} can be derived.

$$\frac{[\text{HOx}^-]}{c_a} = \frac{[\text{H}_3\text{O}^+]K_{a_1}}{[\text{H}_3\text{O}^+]^2 + [\text{H}_3\text{O}^+]K_{a_1} + K_{a_1}K_{a_2}}$$

$$\frac{[\text{Ox}^{2-}]}{c_a} = \frac{K_{a_1}K_{a_2}}{[\text{H}_3\text{O}^+]^2 + [\text{H}_3\text{O}^+]K_{a_1} + K_{a_1}K_{a_2}}$$

Fractions of total oxalate present as H_2Ox, HOx^-, and Ox^{2-} are shown as functions of pH in Fig. 7.10. □

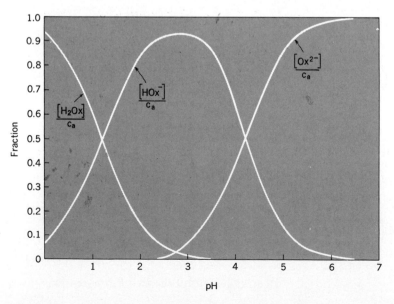

Figure 7.10 Distribution of oxalate species as a function of pH.

Derivation of similar equations for a tri- or even tetraprotic acid, H_3B or H_4B, is more tedious but no more difficult than the above. Figure 7.11 shows the distribution of phosphoric acid species as a function of pH. It may be seen that below pH 5 or so, the only species present in significant concentration are H_3PO_4 and its first dissociation product, $H_2PO_4^-$. Thus the pH of an H_3PO_4 solution can be safely calculated on the basis of the first dissociation constant, as though the acid were monoprotic. As a matter of fact, at no pH are more than two species present in an appreciable amount. In the case of oxalic acid, the two pK_a values are closer than are any pair of H_3PO_4 values; however, only in the pH range 2.5 to 3.0 are all three species discernible in Fig. 7.10, and even here one of the three is predominant.

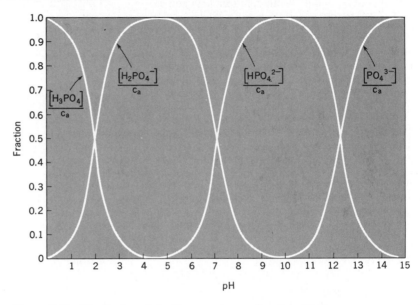

Figure 7.11 Distribution of phosphate species as a function of pH.

KEY TERMS

Amino acid. A compound which contains an amino group on the carbon atom alpha to a carboxyl group: $RCHNH_2CO_2H$.

Diprotic acid. An acid which furnishes two protons.

Isoelectric point. The pH at which there is an exact balance of positive and negative charges on an amino acid.

Isoionic point. The pH obtained when a pure neutral amino acid (zwitterion) is dissolved in water.

Polyprotic acid. An acid which furnishes two or more protons.

Zwitterion. A dipolar ion formed by the internal transfer of a proton in an amino acid: $RCH\overset{+}{N}H_2CO_2^-$.

QUESTIONS

1. *Disproportionation.* The dissociation constants of citric acid (H_3C) are K_{a_1}, K_{a_2}, and K_{a_3}. Derive an expression for the equilibrium constants of these reactions in terms of these three constants:

   (a) $H_2C^- + H_2C^- \rightleftharpoons H_3C + HC^{2-}$
   (b) $HC^{2-} + HC^{2-} \rightleftharpoons H_2C^- + C^{3-}$

2. *Species distribution.* Examine Fig. 7.11 and answer these questions: (a) Which phosphate species is in highest concentration at these pH values? (i) 3.5; (ii) 6.0; (iii) 8.0; (iv) 12.32. (b) Between what pH values is $H_2PO_4^-$ the principal species?

3. *Carbonate mixtures.* If v_1 and v_2 are the milliliters of acid used in the titration of carbonate mixtures as indicated in Fig. 7.7, answer the following: (a) if the sample contains an equal number of millimoles of NaOH and Na_2CO_3, what is the relation between v_1 and v_2? (b) If the sample contains twice as many millimoles of Na_2CO_3 as $NaHCO_3$, what is the relation between v_1 and v_2? (c) If $v_1 = 10$ mL and $v_2 = 20$ mL, what can you conclude about the composition of the sample?

Multiple-choice. In the following multiple-choice questions select the *one best* answer:

4. To 50 mL of 0.10 M H_3PO_4 is added 50 mL of 0.15 M NaOH. In the resulting solution
   (a) $[H_2PO_4^-] \cong [H_3O^+]$;
   (b) $[H_3PO_4] \cong [H_2PO_4^-]$;
   (c) $[H_2PO_4^-] \cong [HPO_4^{2-}]$;
   (d) $[H_3PO_4] \cong [HPO_4^{2-}]$.

5. To 50 mL of 0.12 M H_3PO_4 is added 50 mL of 0.30 M NaOH. The pH of the resulting solution is approximately (a) 12.32; (b) 9.77; (c) 7.21; (d) 11.50.

6. To 50 mL of 0.08 M Na_3PO_4 is added 40 mL of 0.20 M HCl. The pH of the resulting solution is approximately (a) 9.77; (b) 7.21; (c) 4.67; (d) 2.12.

7. What is the value of the equilibrium constant of the reaction

$$CO_3^{2-} + 2H_2O \rightleftharpoons H_2CO_3 + 2OH^-$$

where K_{b_1} and K_{b_2} are the hydrolysis constants of CO_3^{2-} and HCO_3^-, respectively?
   (a) $K_{b_1} \times K_{b_2}$; (b) $\dfrac{K_w}{K_{b_1} \times K_{b_2}}$; (c) K_{b_1}/K_{b_2};
   (d) K_{b_2}/K_w

8. What is the value of the equilibrium constant of the reaction

$$H_3PO_4 + 2OH^- \rightleftharpoons HPO_4^{2-} + 2H_2O$$

   (a) $K_{a_1} K_{a_2}/K_w$; (b) $K_{a_1} K_w/K_{a_2}$;
   (c) $K_{a_1} K_{a_2}/K_w^2$; (d) K_w/K_{a_2}.

9. NaOH is added to a solution of leucine (H_2L^+, $pK_{a_1} = 2.33$; HL, $pK_{a_2} = 9.75$) until the pH reaches a value of 6.04. At this pH the species in highest concentration is (a) HL; (b) L^-; (c) H_2L^+; (d) $[HL] \cong [L^-]$.

10. For the acid H_2A, $pK_{a_1} = 7.00$ and $pK_{a_2} = 11.00$, which of the following indicators is most suitable for the titration (0.10 M solution)

$$A^{2-} + 2H_3O^+ \longrightarrow H_2A + 2H_2O$$

   (a) phenolphthalein; (b) bromthymol blue; (c) methyl red; (d) methyl orange.

PROBLEMS

1. *Approximations.* Calculate the pH of solutions of a diprotic acid, H_2B, for different values of K_{a_1}, K_{a_2}, and $[HB^-]$. Use Eqs. (11) and (12), Section 7.1c.

	K_{a_1}	K_{a_2}	$[HB^-]$
(a)	1.0×10^{-3}	1.0×10^{-7}	0.10
(b)	1.0×10^{-3}	1.0×10^{-7}	0.0010
(c)	1.0×10^{-3}	1.0×10^{-11}	0.10
(d)	1.0×10^{-3}	1.0×10^{-11}	0.0010
(e)	1.0×10^{-3}	1.0×10^{-14}	0.10
(f)	1.0×10^{-3}	1.0×10^{-14}	0.0010

2. *Approximations.* Calculate the pH of the following solutions using Eqs. (11) and (12), Section 7.1c. (a) 0.10 M $NaHSO_3$;
   (b) 0.0010 M $NaHSO_3$;
   (c) 0.010 M Na_2HPO_4;
   (d) 0.010 M $NaHC_2O_4$;
   (e) 0.10 M NaHS.

3. *Species concentrations.* Leucine is an amino acid, $R = -CH_2CH(CH_3)_2$. The acidic form, H_2L^+, has $pK_a = 2.33$, and the pK_a of HL is 9.75. Calculate the concentrations of the three species, H_2L^+, HL, and L^- in (a) 0.050 M H_2L^+ and (b) 0.050 M NaL.

4. *Species concentration.* Calculate the concentrations of H_3O^+, OH^-, Na^+, H_2L^+, HL, and L^- in a 0.050 M solution of the amino

acid leucine, HL. The pK_a values are given in Problem 3.

5. *Mixtures of solutions.* Calculate the pH of the following solutions. Use Eq. (12), Section 7.1c, where appropriate. (a) 40 mL of 0.15 M H_3PO_4 + 30 mL of 0.30 M NaOH; (b) 50 mL of 0.05 M H_3PO_4 + 50 mL of 0.10 M NaOH; (c) 60 mL of 0.12 M Na_2HPO_4 + 30 mL of 0.12 M HCl; (d) 40 mL of 0.10 M H_3PO_4 + 40 mL of 0.10 M Na_3PO_4.

6. *Mixtures of solutions.* Calculate the pH of the following solutions. Use Eq. (12), Section 7.1c, where appropriate. (a) 50 mL of 0.08 M Na_3PO_4 + 40 mL of 0.10 M HCl; (b) 50 mL of 0.08 M Na_3PO_4 + 60 mL of 0.20 M HCl; (c) 30 mL of 0.04 M Na_2HPO_4 + 10 mL of 0.06 M HCl; (d) 60 mL of 0.15 M NaH_2PO_4 + 60 mL of 0.20 M NaOH.

7. Calculate the pH of the following solutions. Use Eq. (12), Section 7.1c, where appropriate. (a) 50 mL of 0.020 M phthalic acid +50 mL of 0.030 M NaOH; (b) 40 mL of 0.08 M H_3PO_4 + 40 mL of 0.18 M NaOH; (c) 60 mL of 0.10 M Na_2HPO_4 + 40 mL of 0.15 M HCl; (d) 50 mL of 0.040 M Na_2CO_3 + 40 mL of 0.050 M HCl.

8. *Equilibrium constants.* Calculate numerical values of the equilibrium constants of these reactions. H_3C is citric acid.
   (a) $2HC_2O_4^- \rightleftharpoons H_2C_2O_4 + C_2O_4^{2-}$
   (b) $H_2C^- + HC^{2-} \rightleftharpoons H_3C + C^{3-}$
   (c) $PO_4^{3-} + 3H_2O \rightleftharpoons H_3PO_4 + 3OH^-$
   (d) $HPO_4^{2-} + H_2O \rightleftharpoons H_2PO_4^- + OH^-$

9. *Isoelectric point.* Calculate the pH at the isoelectric point and at the isoionic point for 0.010 M solutions of these amino acids: (a) serine, $pK_{a_1} = 2.19$, $pK_{a_2} = 9.21$; (b) tryptophan, $pK_{a_1} = 2.35$, $pK_{a_2} = 9.33$.

10. *Titration curve.* 50.0 mL of 0.100 M H_3PO_4 is titrated with 0.100 M NaOH. (a) Calculate the pH after the addition of these volumes (mL) of titrant: 0.00, 10.0, 25.0, 50.0, 65.0, 75.0, 100, and 110. Note that since K_{a_1} is relatively large, the complete quadratic should be solved (or the method of successive approximations used) for the first three volumes. (b) Plot the titration curve. (c) Select suitable indicators for the first two steps in the titration.

11. *Phthalate solution.* How many milliliters of 0.080 M NaOH should be added to 100 mL of 0.060 M phthalic acid to make the $pH = 5.71$?

12. *Phosphate in the blood.* If the pH of blood is 7.40, what are the principal phosphate species at this pH? What is the ratio of the concentrations of the two principal species?

13. *Phosphate solutions.* How many milliliters of 0.20 M NaH_2PO_4 should be added to 40 mL of 0.10 M Na_3PO_4 to make the $pH = 7.21$?

14. *Phosphate solutions.* The pH of a phosphate solution is 7.51. The analytical concentration of phsophate is 0.12 M.
   (a) What are the two principal species in the solution?
   (b) Calculate the concentration of each species.

15. *Phosphate buffer.* A phosphate buffer is prepared by dissolving 8.52 g of Na_2HPO_4 and 10.8 g of NaH_2PO_4 in 500 mL of solution. (a) Calculate the pH of the buffer. (b) 20 mL of 0.30 M NaOH is added to 100 mL of the buffer. Calculate the pH of the resulting solution.

16. *Phosphate buffer.* A phosphate buffer is prepared which is 0.040 M in H_3PO_4 and 0.20 M in NaH_2PO_4. A chemical reaction which produces 8.0 mmol of OH^- is carried out in 100 mL of the buffer. Calculate the pH of the solution at the completion of the reaction.

17. *Phosphate buffer.* Suppose a sample of blood is buffered at pH 7.40. (a) Calculate the $[HPO_4^{2-}]$ in the sample if the $[H_2PO_4^-]$ is 0.012 M. (b) Calculate the pH of the solution resulting from the addition of 6.0 mL of 0.05 M HCl to 100 mL of the blood.

18. *Phosphate buffer.* A buffer solution, 0.12 M in Na_2HPO_4 and 0.020 M in NaH_2PO_4, is prepared. The electrolytic oxidation of 1.50 mmol of the organic compound RNHOH is carried out in 200 mL of the buffer. The reaction is

$$RNHOH + H_2O \longrightarrow RNO_2 + 4H^+ + 4e$$

Calculate the change in pH which occurs during the electrolysis.

19. *Species distribution.* Citric acid, $H_3C_6H_5O_7$, is a triprotic acid. Calculate the fraction of the acid in the molecular and various ionic forms at integral pH values of 1 to 8. Plot a curve similar to Fig. 7.11 and select the best buffering ranges.

20. *Species distribution.* Calculate the fraction of each of the three species H_2Ox, HOx^-, and Ox^{2-} in a solution of oxalic acid at pH 4.40.

21. *Carbonate mixtures.* The samples listed below contain the indicated number of millimoles of Na_2CO_3, $NaOH$, and $NaHCO_3$.

Sample	Na_2CO_3, mmol	NaOH, mmol	$NaHCO_3$, mmol
A	1.5	2.5	0
B	0	2.0	3.0
C	3.0	0	1.0
D	1.8	0	0
E	0	3.0	0
F	0	0	2.5

If these samples are titrated with 0.10 M HCl using the two-indicator method, calculate the values of v_1, the number of milliliters required to reach the phenolphthalein end point, and v_2, the number of milliliters required from the phenolphthalein to the methyl orange end point, for each sample.

22. *Carbonate mixtures.* A sample that may be a mixture of Na_2CO_3 and $NaHCO_3$ or NaOH and Na_2CO_3 is titrated by the two-indicator method. A 0.8034-g sample requires 14.42 mL of 0.1052 M HCl to reach the phenolphthalein end point and an additional 26.58 mL to reach the methyl orange end point. Identify the mixture and calculate the percentage of each component.

23. *Carbonate mixtures.* A 0.7468-g sample of a carbonate mixture required 30.24 mL of 0.1080 M HCl for titration to the phenolphthalein end point and an additional 12.76 mL to reach the methyl orange end point. Identify the mixture and calculate the percentage of each component.

24. *Carbonate error.* A student prepared and standardized a solution of NaOH using phenolphthalein indicator. The normality was found to be 0.0986. Exactly 1.000 liters of the solution was left unprotected and absorbed 0.1100 g of CO_2 from the air. If the solution is again standardized using phenolphthalein, what normality will be found?

25. *Carbonate error.* (a) How many milliliters of 0.1100 M HCl would be required to titrate 50.00 mL of 0.0900 M NaOH to the phenolphthalein end point? (b) Repeat the calculation assuming that the base solution absorbed 0.10 mmol of CO_2 from the air before the titration.

26. *Titration of a mixture of two acids.* 50 mL of a solution which is 0.080 M in the acid HA, $pK_a = 4.20$, and 0.12 M in HB, $pK_a = 8.40$, is titrated with 0.20 M NaOH. Calculate the pH (a) at the first equivalence point, and (b) at the second equivalence point.

27. *Titration of a mixture of two acids.* 50.0 mL of a solution which is 0.100 M in HCl and 0.100 M in the weak acid HX is titrated with 0.200 M NaOH. (a) Calculate the pH at which 99.9% of the HCl has reacted. (b) Calculate the percentage of HX which has reacted in part (a) if the pK_a of HX is (i) 4.00; (ii) 5.00; (iii) 6.00; (iv) 7.00; (v) 8.00.

8

Complex Formation Titrations

One of the types of chemical reactions which may serve as the basis of a titrimetric determination involves the formation of a soluble but slightly dissociated *complex* or *complex ion*. An example is the reaction of silver ion with cyanide ion to form the very stable $Ag(CN)_2^-$ complex ion:

$$Ag^+ + 2CN^- \rightleftharpoons Ag(CN)_2^-$$

The complexes we wish to consider in this chapter are formed by the reaction of a metal ion, a cation, with an anion or neutral molecule. The metal ion in the complex is called the *central atom,* and the group attached to the central atom is called a *ligand*. The number of bonds formed by the central metal atom is called the *coordination number* of the metal. In the complex above, silver is the central metal atom with a coordination number of two, and cyanide is the ligand.

The reaction by which a complex is formed can be regarded as a *Lewis acid-base* reaction. The student will recall that a Lewis acid is an *electron acceptor,* and a Lewis base is an *electron donor*. In forming the complex $Ag(CN)_2^-$ the CN^- ligand acts as the base, donating a pair of electrons to Ag^+, which is the acid. The bond formed between the central metal ion and the ligand is often covalent, but in some cases the interaction may be one of coulombic attraction. Some complexes undergo substitution reactions very rapidly, and the complex is said to be *labile*. An example is

$$Cu(H_2O)_4^{2+} + 4NH_3 \rightleftharpoons Cu(NH_3)_4^{2+} + 4H_2O$$
Light blue   Dark blue

The reaction goes readily to the right with the addition of ammonia to the aquo-complex; addition of a strong acid, which neutralizes the ammonia, shifts the equilibrium rapidly back to the aquo-complex. Some complexes undergo substitution reactions only very slowly and are said to be *nonlabile* or *inert*. Almost all complexes formed by cobalt and chromium in the +3 oxidation state are inert, whereas most of the other complexes of the first series of transition metals are labile. Some examples of typical complexes along with some of their properties are listed in Table 8.1.

TABLE 8.1 Some Typical Complexes

METAL	LIGAND	COMPLEX	COORDINATION NUMBER OF METAL	GEOMETRY	REACTIVITY
Ag^+	NH_3	$Ag(NH_3)_2^+$	2	Linear	Labile
Hg^{2+}	Cl^-	$HgCl_2$	2	Linear	Labile
Cu^{2+}	NH_3	$Cu(NH_3)_4^{2+}$	4	Tetrahedral	Labile
Ni^{2+}	CN^-	$Ni(CN)_4^{2-}$	4	Square planar	Labile
Co^{2+}	H_2O	$Co(H_2O)_6^{2+}$	6	Octahedral	Labile
Co^{3+}	NH_3	$Co(NH_3)_6^{3+}$	6	Octahedral	Inert
Cr^{3+}	CN^-	$Cr(CN)_6^{3-}$	6	Octahedral	Inert
Fe^{3+}	CN^-	$Fe(CN)_6^{3-}$	6	Octahedral	Inert

Molecules or ions which act as ligands generally contain an electronegative atom, such as nitrogen, oxygen, or one of the halogens. Ligands which have only one unshared pair of electrons, for example $:NH_3$, are said to be *unidentate*. Ligands which have two groups capable of forming two bonds with the central atom are said to be *bidentate*. An example is ethylenediamine, $NH_2CH_2CH_2NH_2$, where both nitrogen atoms have unshared electron pairs. Copper(II) ion forms a complex with two molecules of ethylenediamine as follows:

$$Cu^{2+} + 2NH_2CH_2CH_2NH_2 \rightleftharpoons \left[\begin{array}{c} CH_2 \diagup^{NH_2} \diagdown_{} \diagup^{NH_2} \diagdown CH_2 \\ | \quad\quad Cu \quad\quad | \\ CH_2 \diagdown_{NH_2} \diagup^{} \diagdown_{NH_2} \diagup CH_2 \end{array} \right]^{2+}$$

Heterocyclic rings formed by the interaction of a metal ion with two or more functional groups in the same ligand are called *chelate rings;* the organic molecule is a *chelating* agent, and the complexes are called *chelates* or *chelate compounds*. Analytical applications based on the use of chelating agents as titrants for metal ions have shown remarkable growth in recent years.

8.1

STABILITY OF COMPLEXES

We saw in Chapter 5 how the equilibrium constant was formulated for complex formation reactions using the silver-ammonia complex ion as an example. Most of our discussion in this chapter will center on reactions of metal ions with chelating agents. These are generally 1 : 1 reactions in which a soluble complex is formed. We can represent such a reaction in a general manner as

$$M + L \rightleftharpoons ML$$

where M is the central metal cation, L the ligand, and ML the complex. The *sta-*

bility constant of the complex is

$$K = \frac{[ML]}{[M][L]}$$

As mentioned in Section 5.4c, we shall always write the *stability* or *formation* constant of the complex rather than its reciprocal, the *instability* or *dissociation* constant.

We also noted in Section 5.6 (Table 5.6) that the form of the stability constant is the same as that for the titration of a weak acid with a strong base:

$$HOAc + OH^- \rightleftharpoons OAc^- + H_2O \qquad K = \frac{[OAc^-]}{[HOAc][OH^-]}$$

We have seen (Section 6.5) that such a reaction with $K \cong 10^8$ is sufficiently complete at the equivalence point for a feasible titration. We can predict that a reaction yielding a complex of the form ML with a stability constant of the same order of magnitude should give a feasible titration under comparable conditions of concentrations.

8.1a. *Stepwise Formation Constants*

As noted in Section 5.6, the reaction of cations with ligands such as ammonia usually proceeds stepwise. For example, the formation of the complex $Cu(NH_3)_4^{2+}$ proceeds in four steps:

$$Cu^{2+} + NH_3 \rightleftharpoons CuNH_3^{2+} \qquad K_1 = 1.9 \times 10^4$$

$$CuNH_3^{2+} + NH_3 \rightleftharpoons Cu(NH_3)_2^{2+} \qquad K_2 = 3.6 \times 10^3$$

$$Cu(NH_3)_2^{2+} + NH_3 \rightleftharpoons Cu(NH_3)_3^{2+} \qquad K_3 = 7.9 \times 10^2$$

$$Cu(NH_3)_3^{2+} + NH_3 \rightleftharpoons Cu(NH_3)_4^{2+} \qquad K_4 = 1.5 \times 10^2$$

Considering the overall reaction,

$$Cu^{2+} + 4NH_3 \rightleftharpoons Cu(NH_3)_4^{2+}$$

$$K = \frac{[Cu(NH_3)_4^{2+}]}{[Cu^{2+}][NH_3]^4} = K_1 K_2 K_3 K_4 = 8.1 \times 10^{12}$$

the equilibrium constant seems large enough for a feasible titration. The titration of a strong acid with ammonia, $H_3O^+ + NH_3 \rightleftharpoons NH_4^+ + H_2O$, where $K = 1.8 \times 10^9$, is feasible. However, as shown in Fig. 8.1, the titration of strong acid with ammonia gives a large increase in pH at the equivalence point, whereas the titration of Cu^{2+} with ammonia does not.

It is generally true that unless one of the intermediate complexes is extremely stable, there will be no extended range of concentration of complexing agent over which a single species is predominant (except for the last, or highest complex). It may be seen from Fig. 8.1 that pCu rises gradually as ammonia is added, and no sharp break occurs when sufficient titrant has been added to convert all the cation into $Cu(NH_3)_4^{2+}$. The reason for this lies in the fact that not all of the added ammonia is used in one step to form the $Cu(NH_3)_4^{2+}$ complex. Rather, the lower complex species $CuNH_3^{2+}$, $Cu(NH_3)_2^{2+}$, and $Cu(NH_3)_3^{2+}$ are still present in appreciable concentration, i.e., not converted into $Cu(NH_3)_4^{2+}$. Such behavior is predictable from the formation constants of the individual steps given above. It is seen, for example, that there is less tendency for $CuNH_3^{2+}$ to add a

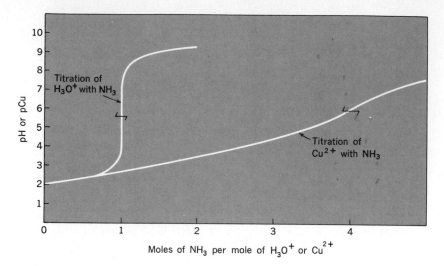

Figure 8.1 Titration of strong acid and of copper (II) ion with ammonia calculated for 10^{-2} M H_3O^+ and Cu^{2+}, assuming no volume change.

second ammonia than for free Cu^{2+} to bind the first one. Actually the tendency to add ammonia molecules decreases at each step of the process.

8.1b. *Chelating Agents*

The difficulty arising from lower complexes can be averted by the use of chelating agents as titrants. Consider, for example, the compound triethylene-tetramine, a quadridentate ligand, often abbreviated "trien." Here, four nitrogen atoms are linked by ethylene bridges in a single molecule which can satisfy copper's normal coordination number of 4 in one step:

$$\left[\begin{array}{c} H_2N \underset{\displaystyle Cu}{\overset{\displaystyle \overset{CH_2-CH_2}{}}{}} NH-CH_2 \\ H_2N \qquad NH-CH_2 \\ CH_2-CH_2 \end{array} \right]^{2+}$$

It may be supposed that formation of the first nitrogen-copper bond brings the other nitrogens of the trien molecule into such proximity that the formation of additional bonds involving these nitrogens is much more probable than the formation of bonds between the copper and other trien molecules. Similarly, it is unlikely that one trien molecule will coordinate with more than one copper. Thus, under ordinary conditions, the stoichiometry of complex formation in this system is 1 Cu^{2+} : 1 trien. The resulting five-membered rings shown in the structural formula are relatively free of strain. The complex is very stable, as shown by its formation constant:

$$Cu^{2+} + trien \rightleftharpoons Cu(trien)^{2+} \qquad K = \frac{[Cu(trien)^{2+}]}{[Cu^{2+}][trien]} = 2.5 \times 10^{20}$$

Thus trien is a good titrant for copper: The ligand and the complex ion are both soluble in water, only a 1 : 1 complex is formed, the equilibrium constant for the titration reaction is large, and the reaction proceeds rapidly.

Only a few metal ions such as copper, cobalt, nickel, zinc, cadmium, and mercury(II) form stable complexes with nitrogen ligands such as ammonia and trien. Certain other metal ions (e.g., aluminum, lead, and bismuth) are better complexed with ligands containing oxygen atoms as electron donors. Certain chelating agents which contain both oxygen and nitrogen are particularly effective in forming stable complexes with a wide variety of metals. Of these, the best known is ethylenediaminetetraacetic acid, sometimes designated (ethylenedinitrilo)tetraacetic acid and often abbreviated EDTA:

$$\begin{array}{ccc}
\text{HOOCCH}_2 & & \text{CH}_2\text{COOH} \\
& \diagdown\ \text{NCH}_2\text{CH}_2\text{N}\ \diagup & \\
\text{HOOCCH}_2 & & \text{CH}_2\text{COOH}
\end{array}$$

The term *chelon* (pronounced "key-loan") was proposed years ago for the entire class of reagents, including polyamines such as trien, polyaminocarboxylic acids such as EDTA, and related compounds that form stable, water-soluble, 1 : 1 complexes with metal ions and hence may be employed as titrants for metals. However, the name has not been widely adopted, and we shall refer to these compounds as *complexing agents*. Titrations involving them will be called *complexometric titrations*.

8.2

COMPLEXOMETRIC TITRATIONS

The suitability of complexing agents such as EDTA as titrants for metal ions was mentioned above. We wish here to examine some of the equilibria involved in these titrations, consider end-point techniques, and show some representative applications. Our discussion will be limited largely to EDTA.

EDTA is potentially a sexidentate ligand which may coordinate with a metal ion through its two nitrogens and four carboxyl groups. It is known from infrared spectra and other measurements that this is the case, for example, with the cobalt(II) ion, which forms an octahedral EDTA complex whose structure is somewhat as shown below:

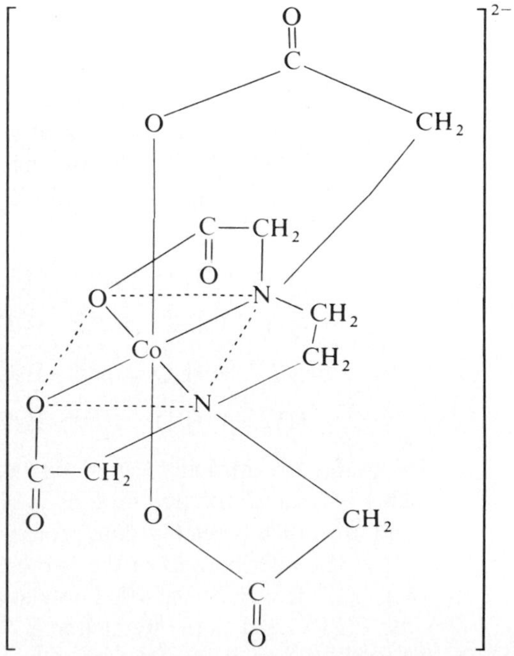

In other cases, EDTA may behave as a quinquedentate or quadridentate ligand having one or two of its carboxyl groups free of strong interaction with the metal.

For convenience, the free acid form of EDTA is often abbreviated H_4Y. The cobalt complex shown above is then written CoY^{2-}, and other complexes become CuY^{2-}, FeY^-, CaY^{2-}, etc. In solutions which are fairly acidic, partial protonation of EDTA without complete rupture of the metal complex may occur, leading to species such as $CuHY^-$; but under the usual conditions all four hydrogens are lost when the ligand is coordinated with a metal ion. At very high pH values, hydroxide ion may penetrate the coordination sphere of the metal, and complexes such as $Cu(OH)Y^{3-}$ may exist.

8.2a. *Equilibria Involved in EDTA Titrations*

We may consider a metal ion such as Cu^{2+}, which is seeking electrons in its reactions, to be analogous to an acid like H_3O^+, and the EDTA anion Y^{4-}, which is an electron donor, to be a base. Then the reaction $Cu^{2+} + Y^{4-} \rightleftharpoons CuY^{2-}$ is analogous to an ordinary neutralization reaction, and it should be a simple matter to calculate pCu values under various conditions, calculate titration curves, discuss feasibility, etc. As a matter of fact, however, the situation is more complicated than this because of the intrusion of other equilibria into the titration situation. We shall discuss some of these in the sections below.

The Absolute Stability or Formation Constant

It is customary to tabulate for various metal ions and chelating agents such as EDTA, values of the equilibrium constants for reactions formulated as follows:

$$M^{n+} + Y^{4-} \rightleftharpoons MY^{-(4-n)} \qquad K_{abs} = \frac{[MY^{-(4-n)}]}{[M^{n+}][Y^{4-}]}$$

K_{abs} is called the *absolute stability constant* or the *absolute formation constant*. Values of some of these constants may be found in Table 2, Appendix I.

The pH Effect

Since the EDTA molecule contains six basic sites—four carboxylate oxygens and two nitrogens—six acid species can exist: H_6Y^{2+}, H_5Y^+, H_4Y, H_3Y^-, H_2Y^{2-}; and HY^{3-}. The first two are relatively strong acids and are not normally of importance in equilibrium calculations. The four dissociation constants of H_4Y are as follows:

$$H_4Y + H_2O \rightleftharpoons H_3O^+ + H_3Y^- \qquad K_{a_1} = 1.02 \times 10^{-2}$$

$$H_3Y^- + H_2O \rightleftharpoons H_3O^+ + H_2Y^{2-} \qquad K_{a_2} = 2.14 \times 10^{-3}$$

$$H_2Y^{2-} + H_2O \rightleftharpoons H_3O^+ + HY^{3-} \qquad K_{a_3} = 6.92 \times 10^{-7}$$

$$HY^{3-} + H_2O \rightleftharpoons H_3O^+ + Y^{4-} \qquad K_{a_4} = 5.50 \times 10^{-11}$$

Note that the third and fourth ionization steps are much weaker than the first two. This is because the two protons in H_2Y^{2-} are attached to the two nitrogen atoms and are lost less readily than protons attached to oxygen.

The distributions of the five EDTA species as functions of pH are shown in Fig. 8.2. It may be seen that only at pH values greater than about 12 does most of the EDTA exist as the tetraanion Y^{4-}. At lower pH values, the protonated species

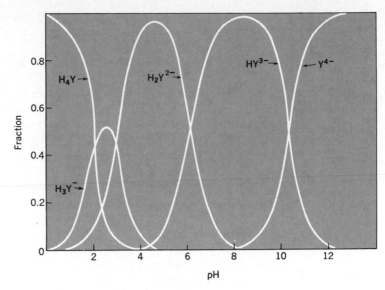

Figure 8.2 Distribution of EDTA species as a function of pH.

HY^{3-}, etc., predominate. We may consider that H_3O^+, then, competes with a metal ion for EDTA, and it is clear that the real tendency to form the metal chelate at any particular pH value is not discernible directly from K_{abs}. For example, at pH 4 the predominant EDTA species is H_2Y^{2-}, and the reaction with a metal such as copper may be written

$$Cu^{2+} + H_2Y^{2-} \rightleftharpoons CuY^{2-} + 2H^+$$

Obviously, as the pH goes down, the equilibrium is shifted away from the formation of CuY^{2-}, and we may expect that there will be a pH value below which the titration of copper with EDTA will not be feasible. We wish to be able to estimate what this value is. Clearly a calculation will involve K_{abs} and the appropriate K_a values of EDTA. Actually, as shown below, it is possible to estimate very easily the minimal pH for a feasible metal ion titration from the K_{abs} value and a simple graph.

Estimation of pH for a Complexometric Titration

The expression for the fraction of EDTA in the Y^{4-} form can be obtained in the same manner as was done for oxalic acid (page 191). Let c_Y represent the total concentration of the uncomplexed EDTA:

$$c_Y = [Y^{4-}] + [HY^{3-}] + [H_2Y^{2-}] + [H_3Y^-] + [H_4Y]$$

Substituting for the concentrations of the various species in terms of the dissociation constants and solving for the fraction in the Y^{4-} form gives

$$\frac{[Y^{4-}]}{c_Y} =$$

$$\frac{K_{a_1}K_{a_2}K_{a_3}K_{a_4}}{[H_3O^+]^4 + [H_3O^+]^3K_{a_1} + [H_3O^+]^2K_{a_1}K_{a_2} + [H_3O^+]K_{a_1}K_{a_2}K_{a_3} + K_{a_1}K_{a_2}K_{a_3}K_{a_4}}$$

Giving the fraction of EDTA in the Y^{4-} form the symbol α_4, we may write

$$\frac{[Y^{4-}]}{c_Y} = \alpha_4$$

or

$$[Y^{4-}] = \alpha_4 c_Y$$

The value of α_4 may obviously be calculated at any desired pH for any chelon whose dissociation constants are known. Shortcuts may be taken in the calculation; for example, it is obvious that at very high pH values, the term containing $[H_3O^+]^4$ will be negligible. In any case, the work has already been done, and graphs or tables showing α-values as functions of pH for a number of chelons may be found in the literature.[1] Because the values extend over a wide range of magnitudes, $-\log \alpha_4$ is usually plotted vs. pH. Such a graph for EDTA is shown in Fig. 8.3. Some values are also given in Table 8.2.

Substitution of $\alpha_4 c_Y$ in the absolute stability constant expression given above yields

$$K_{abs} = \frac{[MY^{-(4-n)}]}{[M^{n+}]\alpha_4 c_Y}$$

or

$$K_{abs}\alpha_4 = \frac{[MY^{-(4-n)}]}{[M^{n+}]c_Y} = K_{eff}$$

K_{eff} is called the *effective* or *conditional stability constant*. Unlike K_{abs}, K_{eff} varies with pH because of the pH dependence of α_4. In certain regards K_{eff} is more immediately useful than K_{abs} because it shows the actual tendency to form the metal complex at the pH value in question. Although K_{eff} values are not customarily tabulated, it is apparent that they may be estimated readily from values of K_{abs}, which are found in tables of constants, and α_4-values obtained from tables such as Table 8.2.

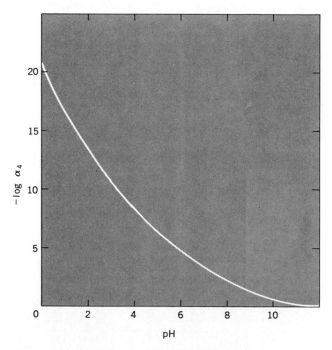

Figure 8.3 Variation of $-\log \alpha_4$ with pH for EDTA.

[1] For example, see C. N. Reilley, R. W. Schmid, and F. S. Sadek, *J. Chem. Ed.*, **36,** 555 (1959).

TABLE 8.2 Values of α_4 for EDTA

pH	α_4	$-\log \alpha_4$
2.0	3.7×10^{-14}	13.44
2.5	1.4×10^{-12}	11.86
3.0	2.5×10^{-11}	10.60
4.0	3.6×10^{-9}	8.44
5.0	3.5×10^{-7}	6.45
6.0	2.2×10^{-5}	4.66
7.0	4.8×10^{-4}	3.33
8.0	5.4×10^{-3}	2.27
9.0	5.2×10^{-2}	1.28
10.0	0.35	0.46
11.0	0.85	0.07
12.0	0.98	0.00

It may be noted that, as the pH goes down, α_4 becomes smaller, and hence K_{eff} becomes smaller. Remember that α_4 is the fraction of EDTA in the Y^{4-} form. Thus at pH values above 12 or so, where EDTA is essentially completely dissociated, α_4 approaches unity ($-\log \alpha_4$ approaches zero), and K_{eff} approaches K_{abs}.

Normally, the solutions of metal ions to be titrated with EDTA are buffered so that the pH will remain constant despite the release of H_3O^+ as the complexes are formed. Thus there is usually a definite basis for estimating K_{eff}, and with this value at hand, it is easy to calculate the titration curve, from which a judgment of feasibility may be made just as with acid-base titrations. The pH is often adjusted to as low a value as is consistent with feasibility. At high pH many metal ions tend to hydrolyze and even precipitate as hydroxides. In most titrations the concentration of the cation is kept as low as 0.010 to 0.0010 M to decrease the chances of precipitation.

8.2b. Complexometric Titration Curves

Titration curves for complexometric titrations can be constructed and are analogous to those for acid-base titrations. Such curves consist of a plot of the negative logarithm of the metal ion concentration (pM) vs. milliliters of titrant. As with acid-base titrations, these curves are helpful in judging the feasibility of a titration and in selecting the proper indicator. The following example shows the calculations involved in deriving a titration curve for Ca^{2+} titrated with EDTA at pH 10.

EXAMPLE 8.1

Calculation of the shape of a complexometric titration curve

50.0 mL of a solution which is 0.0100 M in Ca^{2+} and buffered at pH 10.0 is titrated with 0.0100 M EDTA solution. Calculate values of pCa at various stages of the titration and plot the titration curve.

K_{abs} for CaY^{2-} is 5.0×10^{10}. From Table 8.2, α_4 at pH 10.0 is 0.35. Hence K_{eff} is $5.0 \times 10^{10} \times 0.35 = 1.8 \times 10^{10}$.

(a) *Start of titration.*

$$[Ca^{2+}] = 0.0100 \text{ mmol/mL}$$

$$p\text{Ca} = -\log [Ca^{2+}] = 2.00$$

(b) *After addition of 10.0 mL of titrant.* We started with 50.0 mL $\times$ 0.0100 mmol/mL = 0.500 mmol of Ca^{2+} and have added 10.0 mL $\times$ 0.0100 mmol/mL = 0.100 mmol EDTA. The reaction is

mmol	Ca^{2+}	$+$	Y^{4-}	$\longrightarrow$	CaY^{2-}
Initial:	0.500		0.100		—
Change:	-0.100		-0.100		$+0.100$
Equilibrium:	0.400		—		0.100

There is a large excess of Ca^{2+} at this point, and with a K value on the order of 10^{10} we may assume that the reaction goes to completion. Thus

$$[Ca^{2+}] = \frac{0.400 \text{ mmol}}{60.0 \text{ mL}} = 0.0067 \, M$$

$$pCa = 2.17$$

Similar calculations can be made at various intervals before the equivalence point. In the vicinity of the equivalence point more accurate calculations can be made by not assuming complete reaction, that is, by taking into account Ca^{2+} ions produced by the dissociation of CaY^{2-} and solving the complete quadratic equation. The data in Table 8.3 were calculated by the approximate method.

TABLE 8.3 Titration of 50.0 mL of 0.0100 M Ca^{2+} with 0.0100 M EDTA at pH 10

EDTA, mL	$[Ca^{2+}]$	pCa	% Ca^{2+} REACTED
0.00	0.0100	2.00	0.0
10.0	0.0067	2.17	20.0
20.0	0.0043	2.37	40.0
30.0	0.0025	2.60	60.0
40.0	0.0011	2.96	80.0
49.0	1.0×10^{-4}	4.00	98.0
49.9	1.0×10^{-5}	5.00	99.8
50.0	5.2×10^{-7}	6.28	100.0
50.1	2.8×10^{-8}	7.55	100.0
60.0	2.8×10^{-10}	9.55	100.0

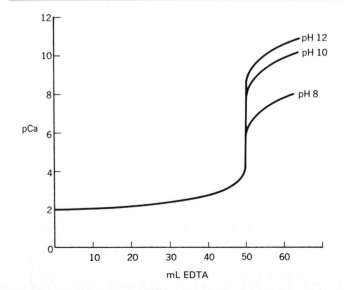

Figure 8.4 Titration curves: 50 mL 0.0100 M Ca^{2+} titrated with 0.0100 M EDTA at pH 8, 10 and 12.

(c) *Equivalence point*. We started with 50.0 mL $\times$ 0.0100 mmol/mL = 0.500 mmol Ca^{2+} and have added 50.0 mL $\times$ 0.0100 mmol/mL = 0.500 mmol of EDTA. The reaction is

mmol	Ca^{2+}	+	Y^{4-}	$\longrightarrow$	CaY^{2-}
Initial:	0.500		0.500		—
Change:	−0.500		−0.500		+0.500
Equilibrium:	—		—		0.500

At this point the concentrations are

$$[Ca^{2+}] = c_Y$$

$$[CaY^{2-}] \cong \frac{0.500 \text{ mmol}}{100 \text{ mL}} = 5.0 \times 10^{-3} \, M$$

The equilibrium expression is

$$\frac{[CaY^{2-}]}{[Ca^{2+}]c_Y} = K_{eff}$$

$$\frac{5.0 \times 10^{-3}}{[Ca^{2+}]^2} = 1.8 \times 10^{10}$$

$$[Ca^{2+}] = 5.2 \times 10^{-7}$$

$$pCa = 6.28$$

(d) *After addition of 60.0 mL of titrant*. We started with 50.0 mL $\times$ 0.0100 mmol/mL = 0.500 mmol of Ca^{2+} and have added 60.0 mL $\times$ 0.0100 mmol/mL = 0.600 mmol of EDTA. The reaction is

mmol	Ca^{2+}	+	Y^{4-}	$\longrightarrow$	CaY^{2-}
Initial:	0.500		0.600		—
Change:	−0.500		−0.500		+0.500
Equilibrium:	—		0.100		0.500

The concentrations are

$$c_Y = \frac{0.100 \text{ mmol}}{100 \text{ mL}} = 9.1 \times 10^{-4} \, M$$

$$[CaY^{2-}] = \frac{0.500 \text{ mmol}}{110 \text{ mL}} = 4.55 \times 10^{-3} \, M$$

The equilibrium expression is

$$\frac{[CaY^{2-}]}{[Ca^{2+}]c_Y} = K_{eff}$$

$$\frac{4.55 \times 10^{-3}}{[Ca^{2+}]9.1 \times 10^{-4}} = 1.8 \times 10^{10}$$

$$[Ca^{2+}] = 2.8 \times 10^{-10}$$

$$pCa = 9.55 \qquad \square$$

The data for this titration are given in Table 8.3, and the titration curve is plotted in Fig. 8.4. The titration curve is of a familiar shape, with a sharp increase in the value of pCa at the equivalence point. Also shown in the figure are curves for the titration done in solutions of pH 8 and pH 12. In these solutions the values of K_{eff} (same as K for the titration) are 2.6×10^8 and 4.9×10^{10}, respectively. Note that the curves are the same up to the equivalence point. The larger increase in pCa is obtained at the higher pH, since K_{eff} is larger in solutions of low hydrogen ion concentration. At low pH, K_{eff} becomes so small that the titration is not feasible.

Feasibility of Complexometric Titrations

The magnitude of K_{eff} or K required for a feasible titration can be calculated as was done for an acid-base titration (Section 6.5). The following example is an illustration.

EXAMPLE 8.2

Calculation of K_{eff} for a feasible titration

50 mL of 0.010 M M^{2+} is titrated with 0.010 M EDTA. Calculate the value of K_{eff} so that when 49.95 mL of titrant has been added, the reaction is essentially complete, and the pM changes by 2.00 units on the addition of two more drops (0.10 mL) of titrant.

One drop before the equivalence point, 0.4995 mmol of EDTA has been added. We started with $50 \times 0.010 = 0.50$ mmol of M^{2+}. There must remain 0.00050 mmol. Hence

$$[M^{2+}] = \frac{0.00050 \text{ mmol}}{99.95 \text{ mL}} = 5 \times 10^{-6} \ M$$

$$pM = 5.30$$

If $\Delta pM = 2.00$ units, then $pM = 7.30$ and $[M^{2+}] = 5 \times 10^{-8} \ M$ when 50.05 mL of titrant is added. At this point

$$c_Y = \frac{0.05 \times 0.010}{100.05} \cong 5 \times 10^{-6}$$

$$[MY^{2-}] \cong \frac{0.5}{100} \cong 5 \times 10^{-3}$$

Hence

$$K_{eff} = \frac{5 \times 10^{-3}}{(5 \times 10^{-8})(5 \times 10^{-6})}$$

$$K_{eff} = K = 2 \times 10^{10}$$

The student should confirm that for $\Delta pM = 1.00$, K_{eff} should be 2×10^9. □

Once a value of K is selected for feasibility, it is easy to determine the lowest pH at which the titration can be carried out. For example, suppose one wishes that in a titration of Zn^{2+} with EDTA, log K_{eff} be at least 8.00. From Table 2, Appendix I, we find that log K_{abs} for ZnY^{2-} is 16.50. Since

$$\log K_{eff} = \log K_{abs} + \log \alpha_4$$

we can calculate values of log K_{eff} at different pH values using the data in Table 8.2. We find that log $K_{eff} = 8.06$ at pH 4 and 10.05 at pH 5. Hence the titration can be carried out at pH 4 with the desired feasibility.

8.2c. *Effect of Other Complexing Agents on EDTA Titrations*

Substances other than the titrant which may be present in the metal ion solution may form complexes with the metal and thus compete against the desired titration reaction. Actually, such complexing is sometimes used deliberately to overcome interferences, in which case the effect of the complexer is called *masking*. For example, nickel forms a very stable complex ion with cyanide, $Ni(CN)_4^{2-}$, whereas lead does not. Thus, in the presence of cyanide, lead can be titrated with EDTA without interference from nickel, despite the fact that the stability constants for NiY^{2-} and PbY^{2-} are nearly the same (log K_{abs} values are 18.6 and 18.0, respectively).

With certain metal ions that hydrolyze readily, it may be necessary to add complexing ligands in order to prevent precipitation of the metal hydroxide. As mentioned above, the solutions are frequently buffered, and buffer anions or neutral molecules such as acetate or ammonia may form complex ions with the metal. Just as the interaction of hydrogen ions with Y^{4-} lowers K_{eff}, so is it lowered by ligands which complex the metal ion. If the stability constants for all the complexes are known, then the effect of the complexers upon the EDTA titration reaction can be calculated. For example, Zn^{2+} forms four complexes with ammonia:

$$Zn^{2+} + NH_3 \rightleftharpoons Zn(NH_3)^{2+} \qquad K_1 = 190$$

$$Zn(NH_3)^{2+} + NH_3 \rightleftharpoons Zn(NH_3)_2^{2+} \qquad K_2 = 210$$

$$Zn(NH_3)_2^{2+} + NH_3 \rightleftharpoons Zn(NH_3)_3^{2+} \qquad K_3 = 250$$

$$Zn(NH_3)_3^{2+} + NH_3 \rightleftharpoons Zn(NH_3)_4^{2+} \qquad K_4 = 110$$

These constants are for an ionic strength of zero. If we designate the total or analytical concentration of all species containing zinc as c_{Zn}, then

$$c_{Zn} = [Zn^{2+}] + [Zn(NH_3)^{2+}] + [Zn(NH_3)_2^{2+}] + [Zn(NH_3)_3^{2+}] + [Zn(NH_3)_4^{2+}]$$

$$c_{Zn} = [Zn^{2+}]\{1 + K_1[NH_3] + K_1K_2[NH_3]^2 + K_1K_2K_3[NH_3]^3 + K_1K_2K_3K_4[NH_3]^4\}$$

Let us designate the fraction of zinc in the uncomplexed form as β_4:

$$\frac{[Zn^{2+}]}{c_{Zn}} = \beta_4$$

or

$$[Zn^{2+}] = \beta_4 c_{Zn}$$

The term β_4 is simply the reciprocal of the terms in the brace of the equation for the total concentration of all species containing zinc. It can be evaluated from the various equilibrium constants and the concentration of NH_3.

For the reaction of Zn^{2+} with EDTA in the presence of ammonia,

$$Zn^{2+} + Y^4 \rightleftharpoons ZnY^{2-}$$

$$K_{abs} = \frac{[ZnY^{2-}]}{[Zn^{2+}][Y^{4-}]} = \frac{[ZnY^2]}{\beta_4 c_{Zn} \alpha_4 c_Y}$$

$$K_{abs}\alpha_4\beta_4 = K_{eff} = \frac{[ZnY^{2-}]}{c_{Zn} c_Y}$$

The following is an example illustrating the calculation of K_{eff} in a solution which contains ammonia.

EXAMPLE 8.3

Calculation of K_{eff}

Given the four constants, K_1, K_2, K_3, and K_4, for the reaction of Zn^{2+} with NH_3 (above) and that K_{abs} for the reaction of Zn^{2+} with EDTA is 3.2×10^{16}, calculate the value of K_{eff} for the reaction of Zn^{2+} with EDTA in a buffer of pH 9.0. Assume that the concentration of free NH_3 in the buffer is 0.10 M.

The value of β_4 is given by

$$\beta_4 = \frac{1}{\begin{array}{c} 1 + 190 \times 0.10 + 190 \times 210 \times (0.10)^2 + 190 \times 210 \times 250 \times (0.10)^3 \\ + 190 \times 210 \times 250 \times 110 \times (0.10)^4 \end{array}}$$

$$\beta_4 = 8.3 \times 10^{-6}$$

At pH 9.0, α_4 is 5.1×10^{-2}. Hence

$$K_{eff} = K_{abs} \times \alpha_4 \times \beta_4$$

$$K_{eff} = 3.2 \times 10^{16} \times 0.052 \times 8.3 \times 10^{-6}$$

$$K_{eff} = 1.4 \times 10^{10} \qquad \qquad \square$$

The effect of the concentration of ammonia on the titration curve on Zn^{2+} with EDTA at pH 9.0 is shown in Fig. 8.5. It can be seen that the break at the equivalence point is smaller the higher the concentration of ammonia. It may be noted that the addition of too much buffer is a common error in EDTA titrations, the resulting complexing action often worsening the end point unnecessarily.

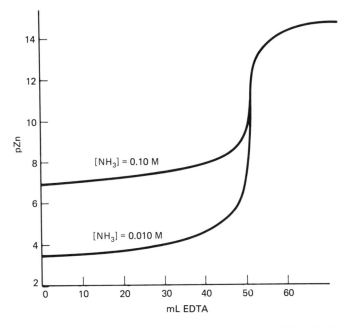

Figure 8.5 Effect of NH_3 concentration on the titration of 50 mL of 0.010 M Zn^{2+} with 0.010 M EDTA at pH 9.

8.2d. *The Hydrolysis Effect*

Hydrolysis of metal ions may compete with the complexometric titration process. Raising the pH makes this effect worse by shifting toward the right equilibria of the type

$$M^{2+} + H_2O \rightleftharpoons M(OH)^+ + H^+$$

Extensive hydrolysis may lead to the precipitation of hydroxides which react only slowly with EDTA even when equilibrium considerations favor formation of the metal complex. Frequently, the appropriate hydrolysis constants for metal ions are not at hand, and hence these effects often cannot be calculated accurately. But of course there is much empirical information which serves experienced persons in deciding how high the pH may be for EDTA titrations of various metal ions. Solubility product constants may sometimes be used to predict where precipitation may occur, although these constants are often quite inaccurate in the case of metal hydroxides.

Sometimes precipitation is actually utilized as a sort of masking in order to circumvent a particular interference. For example, at pH 10, both calcium and magnesium are titrated together with EDTA, only the sum of the two being obtainable. But, if strong base is added to raise the pH above 12 or so, $Mg(OH)_2$ precipitates and calcium alone can be titrated.

Box 8.1

Chelation in Chemistry and Biology

Water-soluble chelating agents such as EDTA provide titrimetric methods for metal ions in aqueous media and may have other uses as well, e.g., as "masking agents," i.e., agents that "hide" metal ions from other reagents with which their interaction is undesired. EDTA is added to mayonnaise, for example, to chelate trace metals such as Fe^{3+} which otherwise catalyze the air oxidation of fatty acids or esters to the odorous products associated with rancidity. Other, less water-soluble, chelating agents are used as precipitants for metal ions (page 81) or to extract metals into organic solvents (page 466). Chelation, however, is not confined to analytical chemistry. Many interesting metal chelate compounds are found in biological systems.

Although there is plenty of iron is most soils (you can see it in Georgia red clay), under oxidizing conditions it is in a very insoluble form [cf. $K_{sp} = 10^{-36}$ for $Fe(OH)_3$ and 10^{-16} for $Fe(OH)_2$ in Table A-3, Appendix I.] Some fertilizers contain iron (II) chelates to provide absorbable iron for plants. Many microorganisms (bacteria and fungi) solve their iron problem by excreting into the surroundings chelating agents called *siderophores* (Greek for "iron bearers"). These compounds, of which one is shown below, form very stable iron (III) chelates

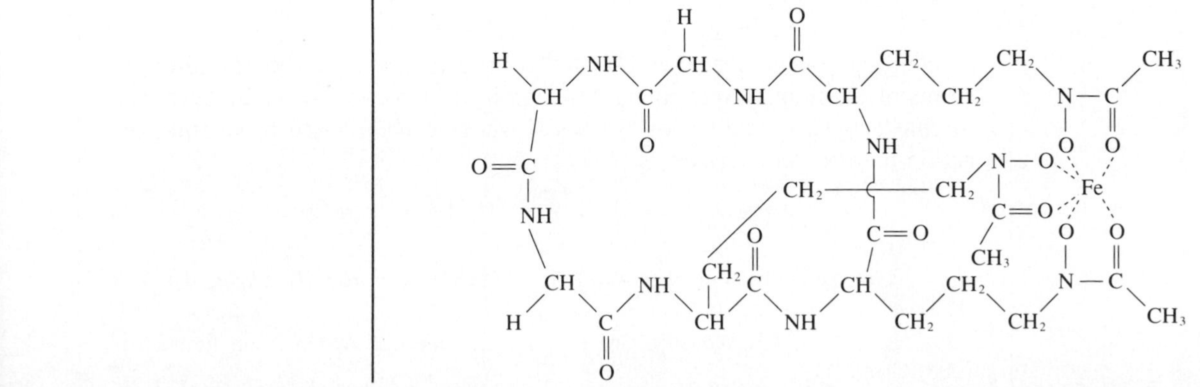

(K_{abs} values range from perhaps 10^{23} up to 10^{52}); they also have an affinity for surface receptors in the cell membrane, where binding initiates the transport of iron into the cell. The iron is released to cellular systems which bind Fe(III) much more weakly than do the siderophores either by reduction to Fe(II) (which greatly lowers K_{abs}) or by metabolic destruction of the siderophores. The compound pictured above is called "ferrichrome"; it is seen to be a cyclic hexapeptide with three side chains bearing chelating hydroxamate groups:

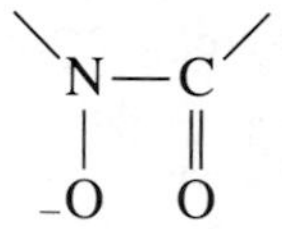

$$
\begin{array}{c}
\diagdown \quad \diagup \\
N - C \\
| \quad \parallel \\
{}_{-}O \quad O
\end{array}
$$

The Fe(III) ion is coordinated octahedrally by the six oxygen electron donor atoms. Reviews of siderophore biochemistry may be found.[†]

Anyone who has seen a green plant has seen chlorophyll. Plants capture solar energy and utilize it for the energetically uphill biosynthesis of complex organic molecules from simple precursors like CO_2 and H_2O. The mechanisms are complex in detail.[‡] The photoreceptor molecule that initiates the process is the magnesium chelate of a tetrapyrrole, which is called chlorophyll a (shown below), where a magnesium ion

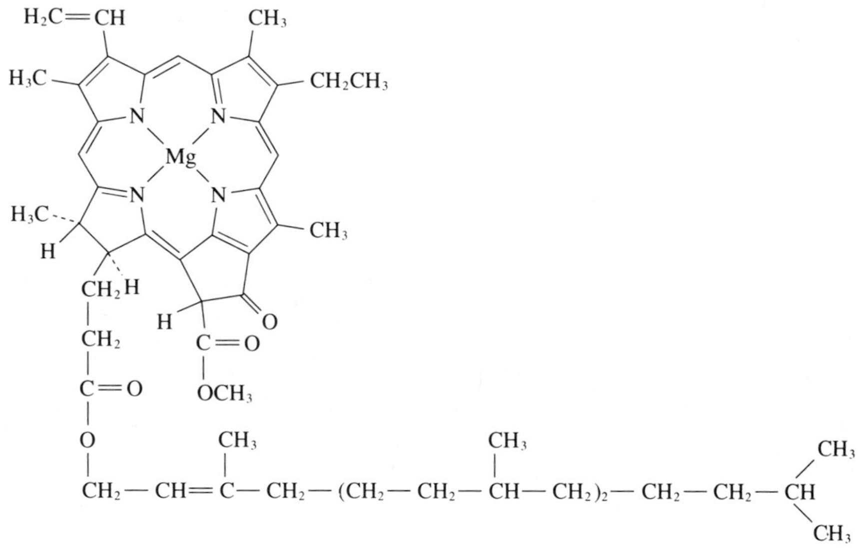

of coordination number 4 is chelated by the four electron donor nitrogen atoms of the four pyrrole rings. All aerobic life on earth owes its existence to chlorophyll, because photosynthesis, whose carbohydrate production reaction may be summarized as

$$n\,H_2O + n\,CO_2 \xrightarrow{\text{light}} (CH_2O)_n + n\,O_2$$

[†] J. B. Neilands, *Ann. Rev. Nutr.*, **1,** 27 (1981); *Ann. Rev. Biochem.*, **50,** 715 (1981).

[‡] L. Stryer, *Biochemistry*, 3rd ed., W. H. Freeman & Co., San Francisco, 1988, Chap. 22.

is our source of oxygen as well as that of the food we eat. An unsophisticated student's guess as to why nature selected a *magnesium* chelate is as good as anybody else's, but we do know at least one reason why a green plant needs magnesium from the soil.

Everybody knows that oxygen is carried throughout our body tissues by hemoglobin which is packaged in our red blood cells. Hemoglobin is a tetrameric protein of four polypeptide chains, each subunit of which binds a nonprotein entity called heme. Heme is the ferrous chelate of a tetrapyrrole called protoporphyrin IX; the chelate is shown below:

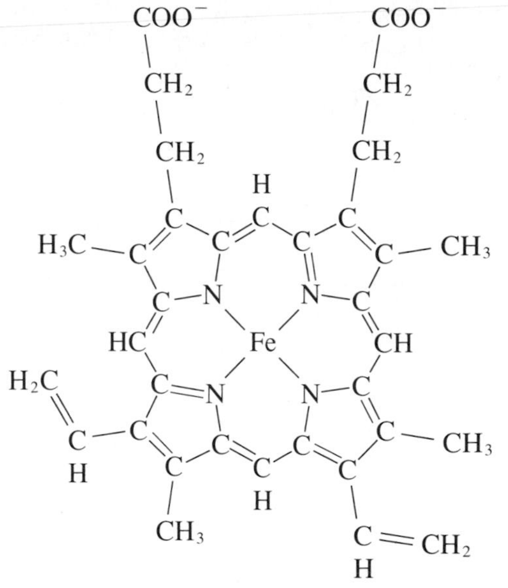

Each heme fits neatly into a pocket in one of the protein moieties. Four of the six coordination positions around the octahedral Fe^{2+} involve the pyrrole nitrogen donors; a fifth is a nitrogen donor in the side chain of a histidyl amino acid residue of the protein, and the sixth is either H_2O or O_2, depending upon the state of oxygenation of the heme. Unlike many synthetic compounds that have been tested as models, hemoglobin binds O_2 reversibly at physiological temperature, with the steepest portion of the binding curve lying between the levels of oxygenation in the lungs and deoxygenation in the body tissues—a perfect delivery system for air-breathing organisms whose tissues require oxygen for the energy-yielding biooxidations that power biological work.

As a last example, pernicious anemia is related to a problem in the absorption of vitamin B_{12} into the body from the gut. This vitamin, a deficiency of which was first countered by feeding raw liver to pernicious anemia patients, is a Co(III) chelate; four coordination positions about the central metal ion are occupied by nitrogens of a tetrapyrrole ring system, similar to what we saw in chlorophyll and heme, while the other two positions involve other ligands which depend upon the method of isolation of the compound and are unclear in the current literature. The vitamin is a reactant in certain enzymic systems. The cobalt ion is *very* tightly bound, too tight to measure; it cannot be displaced from the chelate without practically destroying the ligand.

8.2e. Chelating Agents Other than EDTA

Many other chelating agents have been synthesized. A few of these offer advantages over EDTA in particular situations, although none is so frequently used. The all-nitrogen chelating agents such as triethylenetetramine, mentioned in the introduction to this chapter, are more selective than EDTA. For example, copper can be titrated with trien in the presence of nickel, zinc, and cadmium, whereas with EDTA these metals interfere.

Ethylene glycol-bis-(β-aminoethyl ether)-N,N'-tetraacetic acid (EGTA, below) forms a much more stable chelate with calcium than with magnesium (log K_{abs} = 11.0 vs. 5.4), whereas with EDTA, as noted above, the stabilities are much more nearly the same (log K_{abs} = 10.7 vs. 8.7). Thus

$$\begin{array}{c} \text{HOOCCH}_2 \\ \text{HOOCCH}_2 \end{array} \text{NCH}_2\text{CH}_2-\text{O}-\text{CH}_2\text{CH}_2-\text{O}-\text{CH}_2\text{CH}_2\text{N} \begin{array}{c} \text{CH}_2\text{COOH} \\ \text{CH}_2\text{COOH} \end{array}$$

Ethylene glycol-bis-(β-aminoethyl ether)-N,N'-tetraacetic acid (EGTA)

calcium can be titrated selectively with EGTA in the presence of magnesium, whereas only the sum of the two can be obtained with EDTA unless the magnesium is precipitated.

8.2f. Indicators for EDTA Titrations

When EDTA was first introduced as a titrant, there was a dearth of good visual indicators, and various instrumental end-point techniques were frequently employed. The latter are still valuable in certain situations, but a wide variety of good visual indicators are now available, and usually the visual titrations are the more convenient. We have seen above, using calcium as an example, that there is a large, abrupt break in pM in the vicinity of the equivalence point in a feasible complexometric titration. We wish to convert this into a color change just as acid-base indicators respond to pH changes by changing color. A variety of chemical substances, often called *metallochromic indicators,* are now available for this purpose. Whereas all pH indicators need respond only to hydrogen ion, for complexometric titrations we need a series of substances responsive to pMg, pCa, pCu, etc., although often one indicator may be useful with more than one metal ion.

Basically, metallochromic indicators are colored organic compounds which themselves form chelates with metal ions. The chelate must have a different color from the free indicator, of course, and if large indicator blanks are to be avoided and sharp end points obtained, the indicator must release the metal ion to the EDTA titrant at a pM value very close to that of the equivalence point. This may be considered analogous to the action of an indicator acid in releasing hydrogen ion to hydroxide ion in the titration of an acid. A complete treatment of the equilibria involved is somewhat more complicated than the analogous discussion of acid-base indicators, however, because the common metallochromic indicators also have acid-base properties and respond as pH indicators as well as indicators for pM. Thus, in order to specify the color that a metallochromic indicator will assume in a certain solution, we generally must know both the pH value and the pM value for the particular metal ion which is present. A thorough discussion of the equilibria involved in the action of metallochromic indicators has been given by Reilley and Schmid.[2] We shall present here a somewhat simplified discussion of the indicators Eriochrome Black T and calmagite and then simply note that a number of others are available.

Eriochrome Black T

The structure of Eriochrome Black T is as follows:

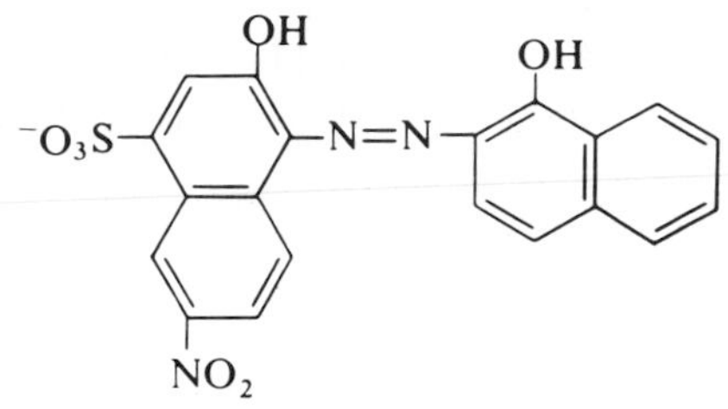

^-O_3S—(ring, OH)—N=N—(ring, OH), NO$_2$

Metal chelates are formed with this molecule by loss of hydrogen ions from the phenolic —OH groups and the formation of bonds between the metal ions and the oxygen atoms as well as the azo group. The molecule is usually represented in abbreviated form as a triprotic acid, H_3In. The sulfonic acid group is shown in the figure as ionized; this is a strong acid group which is dissociated in aqueous solution regardless of pH, and thus the structure shown is that of the ion H_2In^-. This form of the indicator is red. The pK_a value for the dissociation of H_2In^- to HIn^{2-} is 6.3. The latter species is blue. The pK_a value for the ionization of HIn^{2-} to form In^{3-} is 11.6; the latter ion is a yellowish-orange color. The indicator forms stable 1 : 1 complexes, which are wine red in color, with a number of cations, such as Mg^{2+}, Ca^{2+}, Zn^{2+}, and Ni^{2+}. Many EDTA titrations are performed in buffers of pH 8 to 10, the range in which the predominant form of Eriochrome Black T is the blue HIn^{2-} form.

Calmagite

Eriochrome Black T is, unfortunately, unstable in solution, and solutions must be freshly prepared in order to obtain the proper color change. It is still widely used, but another indicator of similar structure, called calmagite, has been developed. Its structure is as follows:

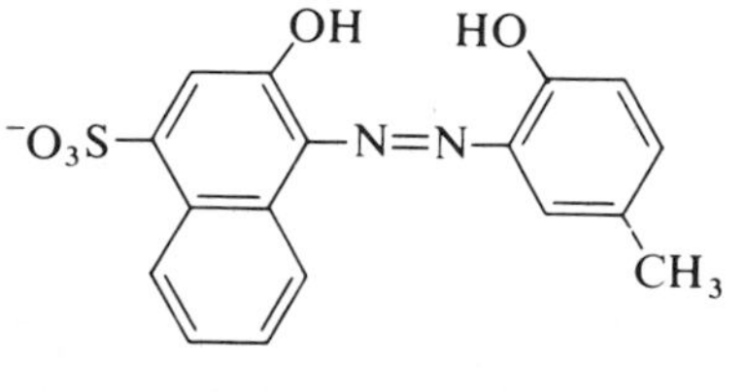

^-O_3S—(ring, OH)—N=N—(ring, HO), CH$_3$

Calmagite is stable in aqueous solution and may be substituted for Eriochrome Black T in procedures which call for the latter indicator. It is also a triprotic acid, H_3In, and the sulfonic acid group is highly dissociated in aqueous solution. The pK_a value for H_2In^- is 8.1, and for HIn^{2-} it is 12.4. The colors are H_2In^- red, HIn^{2-} blue, and In^{3-} reddish orange.

Because both pH and pM determine the color of these metallochromic indicators, it is instructive to examine a graph plotted in terms of these two variables

[2]C. N. Reilley and R. W. Schmid, *Anal. Chem.*, **31,** 887 (1959).

for calmagite, as shown in Fig. 8.6. The metal used in the figure is magnesium. The vertical line separating region II from region III is drawn at the pK_a value of the HIn^{2-} species; in other words, at points along this line there will be equal concentrations of the two species HIn^{2-} and In^{3-}, with the former predominating on the left and the latter predominating in the region to the right. Similarly, regions III and IV are separated by a vertical line drawn at the pK_a value of the species H_2In^-. The curved line separating regions II, III, and IV from region I represents values of pH and pMg where half of the indicator is in the form $MgIn^-$. In other words, as this line is crossed from region II into region I, there will be a color change from reddish orange to red:

$$Mg^{2+} + In^{3-} \rightleftharpoons MgIn^-$$
$$\underset{\text{Red orange}}{\phantom{Mg^{2+} + In^{3-}}} \qquad \underset{\text{Red}}{}$$

Crossing from region III into region I gives a color change from blue to red:

$$Mg^{2+} + HIn^{2-} \rightleftharpoons MgIn^- + H^+$$
$$\underset{\text{blue}}{\phantom{Mg^{2+} + HIn^{2-}}} \qquad \underset{\text{red}}{}$$

Visually, the change from blue to red is much more easily seen than that from orange to red, and hence the indicator is much more attractive to the analyst if the titration can be performed at a pH value below 11 or so.

We may now see why calmagite is a useful visual indicator for the titration of magnesium with EDTA. A calculation of pMg at the equivalence point for the titration of 50.0 mL of 0.0100 M Mg^{2+} with 0.0100 M EDTA at pH 10 (as for Ca^{2+}, page 207) gives a value of 5.26. Referring to Fig. 8.6 we see that at pH 10 the indicator will change from red to blue over a pMg interval of about 4.7 to 6.7. We can calculate this interval in the following manner.

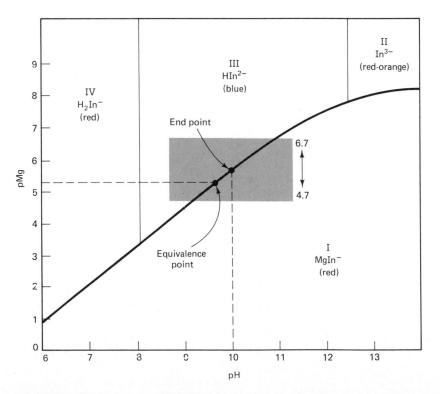

Figure 8.6 Effect of pH and pMg on the color of Calmagite.

The equilibrium constant for the reaction

$$Mg^{2+} + HIn^{2-} \rightleftharpoons MgIn^- + H^+$$

is 5.2×10^{-5}. That is,

$$\frac{[MgIn^-][H^+]}{[Mg^{2+}][HIn^{2-}]} = 5.2 \times 10^{-5}$$

Solving for $[Mg^{2+}]$,

$$[Mg^{2+}] = \frac{[MgIn^-][H^+]}{[HIn^{2-}] \times 5.2 \times 10^{-5}}$$

At pH 10 when the ratio of the two colored forms, $[MgIn^-]/[HIn^{2-}]$, is unity,

$$[Mg^{2+}] = \frac{1.0 \times 10^{-10}}{5.2 \times 10^{-5}}$$

$$[Mg^{2+}] = 1.9 \times 10^{-6}$$

$$pMg = 5.7$$

The interval 5.7 ± 1 is shown in Fig. 8.6, where it can be seen that it easily spans the equivalence point value of $pMg = 5.26$.

The complex formed between Ca^{2+} and calmagite is too weak for a proper color change to occur at pH 10. The equilibrium constant for the reaction

$$Ca^{2+} + HIn^{2-} \rightleftharpoons CaIn^- + H^+$$

is 5.2×10^{-7}. A calculation similar to that above for magnesium gives a value of $pCa = 3.7$ at pH 10 when the ratio of the two colored forms of the indicator is unity. The value of pCa at the equivalence point of a titration (page 207) is 6.28. Hence the color change from red to blue occurs well before the equivalence point.

A number of other indicators are known which can be used for various cations. These are discussed in the article by Reilley and Schmid to which we referred in footnote 2.

8.2g. Applications of EDTA Titrations

EDTA titrations have been carried out successfully on nearly all common cations. These titrations have virtually replaced the former tedious gravimetric analysis for many metals in a variety of samples. There are several procedures which are employed.

Direct Titration

Direct titrations with EDTA can be carried out on at least 25 cations using metallochromic indicators.[3] Complexing agents, such as citrate and tartrate, are often added to prevent precipitation of metal hydroxides. An NH_3–NH_4Cl buffer of pH 9 to 10 is often used for metals which form complexes with ammonia.

The total hardness of water, calcium plus magnesium, can be determined by direct titration with EDTA using Eriochrome Black T or calmagite indicator. As mentioned earlier, the complex between Ca^{2+} and the indicator is too weak for a

---

[3] F. J. Welcher, *The Analytical Use of Ethylenediaminetetraacetic Acid*, D. Van Nostrand Co., Inc., New York, 1958, Chapter 3. See Also G. Schwarzenbach and H. Flaschka, *Complexometric Titrations*, 5th ed., trans. by H. M. N. H. Irving, Methuen & Co. Ltd., London, 1969.

proper color change to occur. However, magnesium forms a stronger complex with the indicator than does calcium, and a proper end point is obtained in an ammonia buffer of pH 10. If the sample titrated does not contain magnesium, a magnesium salt can be added to the EDTA before this solution is standardized. The titrant then (pH 10) is a mixture of MgY^{2-} and Y^{4-}. As this is added to the solution containing Ca^{2+}, the more stable CaY^{2-} is formed, liberating Mg^{2+} to react with the indicator and form the red $MgIn^-$. After the calcium has been used up, additional titrant converts $MgIn^-$ to MgY^{2-} and the indicator reverts to the blue HIn^{2-} form.

Back-titration

Back-titrations are used when the reaction between the cation and EDTA is slow or when a suitable indicator is not available. Excess EDTA is added, and the excess titrated with a standard solution of magnesium using calmagite as the indicator. The magnesium-EDTA complex is of relatively low stability, and the cation being determined is not displaced by magnesium. This method can also be used to determine metals in precipitates, such as lead in lead sulfate and calcium in calcium oxalate.

Replacement Titration

Replacement titrations are useful when no suitable indicator is available for the metal ion being determined. An excess of a solution containing magnesium-EDTA complex is added, and the metal ion, say M^{2+}, displaces the magnesium from the relatively weak EDTA complex:

$$M^{2+} + MgY^{2-} \rightleftharpoons MY^{2-} + Mg^{2+}$$

The displaced Mg^{2+} is then titrated with a standard EDTA solution using calmagite as the indicator.

Indirect Determinations

Indirect determinations of several types have been reported. Sulfate has been determined by adding excess barium ion to precipitate $BaSO_4$. The excess Ba^{2+} is then titrated with EDTA. Phosphate has been determined by titration of the Mg^{2+} in equilibrium with the moderately soluble $MgNH_4PO_4$.

Since metal ions differ in the stability of their EDTA complexes, it is occasionally possible to obtain consecutive end points for more than one metal in a single titration. This situation is somewhat analogous to the titration with base of acids with different dissociation constants. It is necessary, of course, that the stability constants of the metal complexes differ sufficiently; also, the indicator problem is exceptionally critical in such cases. Iron(III) and copper(II) have been determined in a single EDTA titration using photometric detection of the end points,[4] as have lead(II) and bismuth(III).[5]

Selectivity in EDTA Titration

Sometimes by adjusting the pH of a solution it is possible to obtain some degree of selectivity in titrations with EDTA. It is possible, for example, to titrate

[4] A. L. Underwood, *Anal. Chem.* **25,** 1910 (1953).

[5] R. N. Wilhite and A. L. Underwood, *Anal. Chem.,* **27,** 1334 (1955).

in solutions of low pH ions which form very stable complexes. At such low pH values ions which form less stable complexes will not interfere. The log K_{eff} of FeY$^-$ at pH 2 is 11.7, quite large enough for a feasible titration. Iron(III) can be titrated at this pH in the presence of such divalent cations as calcium, barium, and magnesium. The chelates of the latter cations are not formed to any appreciable extent, since the effective stability constants are so low. Nickel(II) can also be titrated in the presence of the alkaline earth cations at pH 3.5, whereas interference will occur at pH 10 in an ammonia buffer. Nickel forms stable complexes with ammonia, while the alkaline earths do not. For this reason the effective stability constant of NiY^{2-} is about the same size as the constants of the alkaline earth complexes in the ammonia buffer, and interference occurs.

Potentiometric EDTA titrations with a mercury indicator electrode are explained in Chapter 12, and photometric titrations of metal ions with EDTA are discussed briefly in Chapter 14.

8.3

METAL-ION BUFFERS

We saw in Chapter 6 that certain systems containing Brønsted conjugate acid-base pairs resist pH changes upon the addition of strong acids or bases. Such systems are said to be buffered. An analogous buffering action with respect to changes in pM is established in solutions containing a metal complex and excess complexing agent.

Consider the equilibrium involving a metal ion M, a ligand L, and a complex ML, where the charges are omitted for convenience:

$$M + L \rightleftharpoons ML$$

$$K_{eff} = \frac{[ML]}{[M]c_L}$$

Here c_L represents the analytical concentration of the ligand. Solving for [M] and then taking logs, we obtain

$$[M] = \frac{1}{K_{eff}} \times \frac{[ML]}{c_L}$$

$$\log [M] = \log \frac{1}{K_{eff}} + \log \frac{[ML]}{c_L}$$

$$p\text{M} = \log K_{eff} - \log \frac{[ML]}{c_L}$$

or

$$p\text{M} = \log K_{eff} + \log \frac{c_L}{[ML]}$$

Compare this with the Henderson-Hasselbalch equation (Section 5.4a).

It is seen that the pM of such a solution is fixed by the value of K_{eff} and the molar ratio of metal complex to ligand. Introduction of additional metal ion to the solution will lead to formation of more ML; in other words, the solution resists the lowering of pM that would otherwise occur if L were absent. Similarly, removal of metal ion will be resisted by the dissociation of ML, and metal ion can be drained by some other reaction without a large rise in pM so long as the capacity of ML to furnish metal ion is not exhausted.

Metal ion buffers have found application in biology and biochemistry in studies on enzyme systems whose catalytic activity exhibits a metal ion dependence. Just as pK_a must be considered in the analogous case of pH buffers, a metal ion buffer will be most efficient if $\log K_{eff}$ is nearly the same as the desired pM value.

8.4

TITRATIONS INVOLVING UNIDENTATE LIGANDS

Because of the stepwise formation of successive complexes as noted above, unidentate ligands are only rarely suitable for the titration of metal ions. However, there are a few examples of important titrations based upon such ligands, and we shall consider briefly the two best-known cases.

8.4a. *Titration of Chloride with Mercury(II)*

The mercury(II) ion-chloride system is unusual in that the last two of the successive complexes in the formation of $HgCl_4^{2-}$ are of much less stability than the first two, as shown by the following successive formation constants:[6]

$$Hg^{2+} + Cl^- \rightleftharpoons HgCl^+ \qquad K_1 = \frac{[HgCl^+]}{[Hg^{2+}][Cl^-]} = 5.5 \times 10^6$$

$$HgCl^+ + Cl^- \rightleftharpoons HgCl_2 \qquad K_2 = \frac{[HgCl_2]}{[HgCl^+][Cl^-]} = 3.0 \times 10^6$$

$$HgCl_2 + Cl^- \rightleftharpoons HgCl_3^- \qquad K_3 = \frac{[HgCl_3^-]}{[HgCl_2][Cl^-]} = 7.1$$

$$HgCl_3^- + Cl^- \rightleftharpoons HgCl_4^{2-} \qquad K_4 = \frac{[HgCl_4^{2-}]}{[HgCl_3^-][Cl^-]} = 10$$

Thus in the titration of a chloride solution with an ionized mercury salt such as mercury(II) nitrate or perchlorate, there is a sudden drop in pHg ($pHg = -\log [Hg^{2+}]$) when the formation of $HgCl_2$ is essentially complete.

One of the common indicators for this titration is sodium nitroprusside, $Na_2Fe(CN)_5NO$. This compound forms a white precipitate of mercury(II) nitroprusside, and the end point is taken as the appearance of a white turbidity in the formerly homogeneous solution. The pHg at the equivalence point of the titration is not so low as might otherwise be expected because of the consumption of mercury(II) ion in the following reaction:

$$Hg^{2+} + HgCl_2 \rightleftharpoons 2HgCl^+ \qquad K = \frac{K_1}{K_2} \cong 1.8$$

Actually, the mercury(II) nitroprusside precipitate is first seen somewhat after the equivalence point, and a correction must be applied in order to obtain the best results. The correction is not really the same as an indicator blank run with distilled water because the reaction shown above does not then take place appreciably. The correction depends upon the final concentration of mercury(II) chloride, and hence varies with the quantity of sample and the final volume. The acidity of the

[6] A. Johnson, I. Quarfort, and L. G. Sillen, *Acta Chem. Scand.*, **1**, 461, 473 (1947).

solution also affects the correction, and there is further variation which depends upon the rate at which the titration is performed and the manner in which the individual analyst views the turbid solution. Typical correction values are given by Kolthoff and Stenger;[7] for example, where the final solution is 100 mL of 0.025 M $HgCl_2$, the correction is roughly 0.2 mL. An advantage of this particular method lies in the fact that the titration may be performed in solutions which are quite acidic, and it works well even in fairly dilute solution, for example, at levels of chloride (10 mg/liter) which frequently occur in natural waters.

Certain organic compounds which form colored complexes with mercury(II) ion have also been employed as indicators for the mercurimetric titration of chloride. The best known are diphenylcarbazide (colorless) and diphenylcarbazone (orange), which form intense-violet mercury(II) complexes. With these indicators, it has been found important to control the pH of the solution being titrated. According to Roberts,[8] diphenylcarbazide performs best at pH 1.5 to 2.0, while Clark[9] found that diphenylcarbazone is best employed at pH 3.2 to 3.3.

It should be pointed out that bromide, thiocyanate, and cyanide may be determined by mercurimetric titration, although there is no advantage over the usual titrations of these ions with silver nitrate. Nitroprusside cannot be used as an indicator in the thiocyanate titration because the appearance of the mercury(II) nitroprusside precipitate is obscured by the slightly soluble thiocyanate. In this case, the usual indicator is iron(III) ion, which acts by the formation of red complexes with thiocyanate, such as $FeSCN^{2+}$. The titration of iodide with mercury is largely unsatisfactory. The complex HgI_4^{2-} forms during the titration; later, a red precipitate of HgI_2 appears through the reaction

$$Hg^{2+} + HgI_4^{2-} \rightleftharpoons 2HgI_2$$

The appearance of this precipitate has been used as an end point, but actually it occurs much too early.

8.4b. *Titration of Cyanide with Silver Ion*

Another titration of some practical importance involving a unidentate ligand and a metal ion is the so-called Liebig titration of cyanide with silver nitrate. The basis of the method is the formation of the very stable complex ion $Ag(CN)_2^-$:

$$2CN^- + Ag^+ \rightleftharpoons Ag(CN)_2^-$$

The equilibrium constant for this reaction as written is about 10^{21}, and this is the only silver-cyanide complex of appreciable stability. Originally, the end point was based upon the appearance of turbidity due to the precipitation of silver cyanide, which may be written as

$$Ag^+ + Ag(CN)_2^- \rightleftharpoons 2AgCN$$

or

$$Ag^+ + Ag(CN)_2^- \rightleftharpoons Ag[Ag(CN)_2]$$

This precipitation occurs after $[CN^-]$ has dropped to a low value, although a calculation based upon the appropriate equilibria shows that it actually comes a little

[7] I. M. Kolthoff and V. A. Stenger, *Volumetric Analysis,* 2nd ed., Vol. II, Interscience Publishers, Inc., New York, 1947, p. 332.

[8] I. Roberts, *Ind. Eng. Chem., Anal. Ed.,* **8,** 365 (1936).

[9] F. E. Clark, *Anal. Chem.,* **22,** 553 (1950).

too early, corresponding to an end-point error on the order of 0.2 ppt. This error is small enough to be accepted, but there is an additional problem: Silver cyanide precipitated locally is slow to redissolve as the solution is stirred, and it is time consuming to perform the titration carefully. Also, there is some difficulty in seeing the silver cyanide precipitate.

In the Deniges modification of Liebig's method, iodide ion is added as the indicator. Precipitated silver iodide is bulky and easy to see, and it is less soluble than silver cyanide and hence precipitates in place of the latter at the end point. This end point occurs, however, too early in the titration. For this reason, ammonia is added, which by forming the soluble species $Ag(NH_3)_2^+$, retards the precipitation of silver iodide until a more propitious time; ammonia does not prevent formation of the much more stable $Ag(CN)_2^-$ and hence does not interfere with the titration reaction.

KEY TERMS

Bidentate ligand. A ligand which has two groups capable of forming two bonds with the central atom.

Central atom. The metal ion in a complex which acts as a Lewis acid in accepting electrons from the ligand.

Chelate ring. A heterocyclic ring formed by the interaction of a metal ion with one or more functional groups in the same ligand.

Chelating agent. The organic molecule involved in forming a chelate ring.

Complex. A molecule or ion formed by the interaction of a metal ion, acting as an electron acceptor, and a ligand, acting as an electron donor.

Complex effect. The effect produced by a substance which forms a complex with a metal ion, thereby competing with the desired titration reaction.

Complexometric titration. A titration which involves the formation of a soluble but slightly dissociated complex or complex ion.

Coordination number. The number of bonds formed by a central metal atom in a complex.

Deniges method. A modification of Liebig's method for titrating cyanide ion with silver nitrate. Iodide ion and ammonia are added, and AgI is precipitated at the end point.

Dissociation constant. The equilibrium constant for the dissociation reaction of a complex; also called the *instability* constant.

EDTA. Ethylenediaminetetraacetic acid—the most widely used chelating agent for the titration of metal ions.

Formation constant. The equilibrium constant for the reaction in which a complex is formed; also called the *stability* constant.

Inert complex. A complex which undergoes substitution of ligand groups only very slowly. It is also said to be *nonlabile*.

Instability constant. Same as *dissociation* constant.

Labile complex. A complex which undergoes substitution of ligand groups very rapidly.

Lewis acid. A substance which accepts a pair of electrons to form a bond, as the central metal ion in a complex.

Lewis base. A substance which furnishes electrons to form a bond, as the ligand groups in a complex.

Liebig method. The method used to determine cyanide ion by titration with silver nitrate to form the stable complex $Ag(CN)_2^-$.

Ligand. The substance which acts as a Lewis base in furnishing electrons to the central metal ion to form a complex.

Masking. Use of a reagent to form a stable complex with an ion which otherwise would interfere with the desired titration reaction.

Metal-ion buffer. A solution which contains a metal complex and excess complexing agent. The solution resists large changes in pM when the metal ion concentration is increased or decreased.

pH effect. The effect of hydrogen ion concentration in competing with a metal ion for the chelating agent in a titration.

Replacement titration. A titration in which a metal ion for which no suitable indicator is available displaces a metal such as magnesium from its relatively weak EDTA complex. The magnesium is then titrated with standard EDTA solution.

Stability constant—absolute. The equilibrium constant for the formation of a complex from metal ion and ligand; also called *formation* constant.

Stability constant—conditional or effective. The equilibrium constant for the formation of a complex from metal ion and ligand which includes the effect of competing equilibria. Its value may depend on such factors as *p*H and concentration of other complexing agents.

Unidentate ligand. A ligand which has only one group capable of forming a bond with the central atom.

QUESTIONS

1. *Terms.* Explain the meaning of the following terms: (a) inert complex; (b) metal chelate; (c) masking; (d) sexidentate ligand; (e) coordination number.

2. *Stability constant.* What is meant by the absolute stability constant of a complex ion? How is it affected by the *p*H of a solution?

3. *Titration curve.* What effect does the concentration of the analyte have on a titration curve such as that shown in Fig. 8.4? What is the effect of the concentration of titrant?

4. *Titration of cyanide ion.* Explain why it is helpful to add iodide ion in the titration of cyanide ion with silver ion. Why is ammonia also added?

5. *Derivation.* Given that $M + L \rightleftharpoons ML$ and K_{abs} is the stability constant, show that the following relation holds:

$$\log K_{abs} = pM + pL - pML$$

Multiple-choice. In the following multiple-choice questions select the *one best* answer:

6. Which statement concerning α_4, the fraction of EDTA in the Y^{4-} form, is correct? (a) α_4 increases as the *p*H increases; (b) $-\log \alpha_4$

increases as the $[H^+]$ increases; (c) $+\log \alpha_4$ increases as the *p*H decreases; (d) all of the above; (e) only two of the above.

7. At *p*H 7 the predominant form of EDTA is (a) H_3Y^-; (b) H_2Y^{2-}; (c) HY^{3-}; (d) equal amounts of H_2Y^{2-} and HY^{3-}.

8. Consider the titration of Zn^{2+} with EDTA. Which statement below is true? (a) The reaction is more complete the lower the *p*H. (b) The titration is more feasible at high $[H^+]$ that at low $[H^+]$. (c) The titration is more feasible in 0.010 M NH_3 than in 0.10 M NH_3. (d) All of the above are true. (e) None of the above are true.

9. 40 mL of 0.025 M Cu^{2+} is titrated with 0.040 M EDTA at *p*H 9. When 50 mL of titrant is added, (a) $\log [Cu^{2+}] = -\log K_{eff}$; (b) $\log [Cu^{2+}] = pK_{eff}$; (c) $pCu = -pK_{eff}$; (d) all of the above; (e) none of the above.

10. Which of these statements is true? (a) The effective stability constant depends upon the *p*H. (b) A Lewis acid is an electron donor. (c) A ligand is a Lewis base. (d) Two of the above are true. (e) All of the above are true.

PROBLEMS

1. *Standardization.* A sample of pure $CaCO_3$ weighing 0.3677 g is dissolved in hydrochloric acid, and the solution diluted to 250.0 mL in a volumetric flask. A 25.00-mL aliquot requires 30.26 mL of an EDTA solution for titration. Calculate (a) the molarity of the

EDTA solution; (b) the number of grams of $Na_2H_2Y \cdot 2H_2O$ (MW 372.2) required to prepare 500.0 mL of the solution.

2. *Hardness of water.* A 200.0-mL sample of well water required 17.94 mL of the EDTA solution in Problem 1 for titration. Calculate the hardness of the water in ppm of $CaCO_3$, although cations other than Ca^{2+} may be present. Recall that 1 ppm is 1 mg/liter.

3. *Hardness of water.* A 50.0-mL sample of water containing both Ca^{2+} and Mg^{2+} is titrated with 16.54 mL of 0.01104 *M* EDTA in an ammonia buffer of *p*H 10. Another 50.0-mL sample is treated with NaOH to precipitate $Mg(OH)_2$ and then titrated at *p*H 13 with 9.26 mL of the same EDTA solution. Calculate the ppm each of $CaCO_3$ and $MgCO_3$ in the sample.

4. *Indirect determination.* Sulfate can be determined indirectly by precipitating $BaSO_4$ with excess Ba^{2+} and titrating the excess with standard EDTA. A 0.3260-g sample containing sulfate is dissolved, and the sulfate precipitated by adding 25.00 mL of 0.03120 *M* $BaCl_2$. After removal of the $BaSO_4$ by filtration the excess Ba^{2+} is titrated with 26.38 mL of 0.01822 *M* EDTA. Calculate the percentage of SO_3 in the sample.

5. *Back-titration.* A 0.2420-g sample containing calcium is dissolved, and the metal precipitated as CaC_2O_4. The precipitate is filtered, washed, and redissolved in acid. The *p*H is adjusted, 25.00 mL of 0.0400 *M* EDTA added, and the excess EDTA titrated with 33.28 mL of 0.01202 *M* Mg^{2+}. Calculate the percentage of Ca in the sample.

6. *Masking.* 50.00 mL of a solution containing both Cd^{2+} and Pb^{2+} requires 40.54 mL of a 0.01934 *M* solution of EDTA for titration of both metals. A second 50.00-mL sample is treated with KCN to mask the Cd^{2+} and then titrated with 22.47 mL of the same EDTA solution. Calculate the molarities of the Cd^{2+} and Pb^{2+} ions.

7. *Analysis of an alloy.* A 0.5832-g sample of an alloy which contains 57.45% bismuth and 31.76% lead is dissolved in nitric acid, and the solution diluted to 250.0 mL in a volumetric flask. A 50.00-mL aliquot is withdrawn, the *p*H adjusted to 1.5, and the bismuth titrated with 0.01104 *M* EDTA. The *p*H of the solution is then increased to 5.0, and the lead titrated with the same EDTA solution. Calculate the volume (mL) of titrant required for each of the two metals.

8. *Liebig method.* A 0.6834-g sample containing 25.63% NaCN is dissolved in water, and then concentrated ammonia and KI solution are added. How many milliliters of 0.0984 *M* $AgNO_3$ solution are required for titration?

9. *Liebig method.* A 0.5012-g sample contains 34.14% KCN and 65.18% NaCN. How many milliliters of 0.1102 *M* $AgNO_3$ will be required for titration of this sample?

10. *Determination of nickel.* Nickel can be determined by addition of excess cyanide ion:

$$Ni^{2+} + 4CN^- \rightleftharpoons Ni(CN)_4^{2-}$$

and back-titration with silver ion:

$$Ag^+ + 2CN^- \rightleftharpoons Ag(CN)_2^-$$

A 0.4000-g sample of a nickel ore is treated to dissolve the metal. The solution is made basic with ammonia, and 50.00 mL of 0.0900 *M* KCN solution is added. Titration of the excess cyanide ion requires 20.34 mL of 0.0500 *M* $AgNO_3$, using KI indicator. Calculate the percentage of nickel in the ore.

11. *Value of α.* Verify the values of α_4 for EDTA given in Table 8.2 at *p*H 2.5, 5.0, and 10.0.

12. *Equilibrium constants.* Calculate the equilibrium constants for the following reactions:
    (a) $Zn^{2+} + 4NH_3 \rightleftharpoons Zn(NH_3)_4^{2+}$
    (b) $Hg^{2+} + 2CN^- \rightleftharpoons Hg(CN)_2$

13. *Equilibrium constants.* Given the pK_a values for calmagite (Section 8.2e) and that K_{abs} for the complex formed between calmagite and Mg^{2+} ($MgIn^-$) is 1.3×10^8, calculate the *K* for the reaction

$$Mg^{2+} + HIn^2 \rightleftharpoons MgIn^- + H^+$$

14. *Effective stability constant.* Calculate the value of K_{eff} for the reaction of Zn^{2+} with EDTA in a buffer of *p*H 10.0. Assume the concentration of free NH_3 in the buffer is 0.020 *M*. Use the constants given in Section 8.2c for the zinc-ammonia complexes.

15. *Eriochrome Black T.* The value of K_{abs} for the complex between Zn^{2+} and Eriochrome Black T is 8.8×10^{12}:

$$Zn^{2+} + In^{3-} \rightleftharpoons ZnIn^-$$

Calculate the *p*H at which the ratio of the two

colored forms of the indicator is unity and the value of pZn is 11.00.

16. *EDTA.* Using the appropriate equations from Chapter 7, calculate the pH of these solutions of salts of EDTA: (a) 0.010 M NaH_3Y; (b) 0.010 M Na_2H_2Y; (c) 0.010 M Na_3HY; (d) 0.0010 M Na_3HY.

17. *Mixtures of solutions.* Calculate the value of pZn in the following solutions, all at pH 9.0, with the concentration of free NH_3 being 0.10 M: (a) 50 mL of 0.010 M Zn^{2+} + 20 mL of 0.020 M EDTA; (b) 50 mL of 0.015 M Zn^{2+} + 30 mL of 0.025 M EDTA; (c) 50 mL of 0.025 M Zn^{2+} + 50 mL of 0.050 M EDTA.

18. *Equivalence point.* Calculate the value of pM at the EPt for the titration of 2.00 mmol of each of these metals with EDTA at pH 4.00. The volume in each case is 100 mL. (a) Hg^{2+}; (b) Ni^{2+}; (c) Fe^{2+}.

19. *Titration curve.* 50.0 mL of a solution which is 0.0200 M in a metal M^{2+} and buffered at pH 9.00 is titrated with 0.0200 M EDTA. The value of $\log K_{abs}$ for MY^{2-} is 14.30. Calculate the value of pM when the following amounts of titrant (mL) are added: (a) 0.00; (b) 25.0; (c) 49.9; (d) 50.0; (e) 50.1; (f) 55.0. Plot the titration curve.

20. *Feasibility of titration.* Repeat Problem 19 at pH 7 and 12. Compare the values of $\Delta pM/\Delta V$ at the EPt for the three different pH values.

21. *Feasibility of titration.* (a) 50.0 mL of a 0.0200 M solution of a cation N^{2+} is titrated with 0.0200 M EDTA. Calculate the value of K_{eff} for the formation of NY^{2-} so that when 49.95 mL of titrant is added the reaction is complete, and that pN changes by 2.00 units on the addition of 0.10 mL of additional titrant. (b) Repeat the calculation for $\Delta pN = 1.0$ unit for 0.10 mL of titrant.

22. *pH-feasibility.* Suppose a chemist wishes the K_{eff} for the reaction

$$M^{2+} + Y^{4-} \rightleftharpoons MY^{2-}$$

to be 4.0×10^{10} for the titration of several cations. Calculate the pH values at which the titration should be carried out for these cations: (a) Fe^{3+}; (b) Mn^{2+}; (c) Ba^{2+}; (d) Ag^+.

23. *pH-fraction complexed.* For the reaction of Mn^{2+} with EDTA,

$$Mn^{2+} + Y^{4-} \rightleftharpoons MnY^{2-}$$

calculate the fraction of manganese in the MnY^{2-} form at these pH values: (a) 2.5; (b) 3.0; (c) 4.0; (d) 6.0. Assume 2.00 mmol of EDTA has been added to 2.00 mmol of Mn^{2+} and that the volume of solution is 100 mL.

24. *Metal-ion buffer.* A metal-ion buffer is prepared by dissolving 1.00 mmol of Mg^{2+} and 1.50 mmol of EDTA in 250 mL of solution at pH 10.0 (a) Calculate the pMg of the buffer. (b) To 25.0 mL of the buffer is added 0.0200 mmol of Mg^{2+}. Calculate the pMg of the resulting solution and the value of ΔpMg.

25. *Complex effect.* 50 mL of a 0.010 M solution of Zn^{2+} is titrated with 0.010 M EDTA at pH 9.00. Calculate the values of pZn for the following volumes of titrant (mL) assuming the concentration of free NH_3 is (a) 0.10 M, and (b) 0.010 M: (i) 0.00; (ii) 20; (iii) 50; (iv) 60. See Example 3, Section 8.2.c, for values of equilibrium constants.

The table below gives values of $\log K_{abs}$ for the metals M, N, Q, R, and S reacting with a complexer L to form complexes. Values of $-\log \alpha$ for L are also given. These data may be needed in the next three problems.

| | METAL | | LIGAND, L | |
Complex	$\log K_{abs}$		pH	$-\log \alpha$
ML	22.60		2.0	16.00
NL	19.30		4.0	10.30
QL	14.20		6.0	5.20
RL	9.70		8.0	3.60
SL	2.80		10.0	1.00

26. *Feasibility.* Which metals (M, N, Q, R, S) should give feasible titrations at pH (a) 2.0; (b) 4.0; (c) 6.0?

27. *Effective stability constant.* Calculate the values of K_{eff} for (a) QL at pH 2.0; (b) ML at pH 8.0; (c) SL at pH 10.0.

28. *Effect of pH.* For the following solutions calculate the value of pQ at pH 6.0 and 8.0: (a) 50 mL of 0.020 MQ + 25 mL of 0.020 M L; (b) 50 mL of 0.020 MQ + 50 mL of 0.020 M L; (c) 50 mL of 0.020 MQ + 75 mL of 0.020 M L.

9

Solubility Equilibria and Precipitation Titrations

Precipitation reactions have been widely used in analytical chemistry in titrations, in gravimetric determinations, and in separating samples into their component parts. Gravimetric methods are no longer widely employed, and the use of precipitation for separations has been largely (but by no means entirely) replaced by the methods discussed in Chapters 16, 17, and 18. Precipitation is still a fundamental technique that is of importance in many analytical procedures.

Formulation of the equilibrium constant expressing the solubility of a precipitate in water is covered in Section 5.4b. Gravimetric methods of analysis are covered in Chapter 4, where the topic of formation and properties of precipitates is discussed. In this chapter we shall first discuss precipitation titrations and then the use of precipitation as a separation technique.

9.1

PRECIPITATION TITRATIONS

Titrations involving precipitation reactions are not nearly so numerous in titrimetric analysis as those involving redox or acid-base reactions. In fact, in a beginning course, examples of such titrations are usually limited to those involving the precipitation of silver ion with anions such as the halogens or thiocyanate. One of the reasons for the limited use of such reactions is the lack of suitable indicators. In some cases, particularly in the titration of dilute solutions, the rate of reaction is too slow for convenience of titration. As the equivalence point is approached and the titrant is added slowly, a high degree of supersaturation does not exist and the rate of precipitation may be very slow. Another difficulty is that the composition of the precipitate is frequently not known because of coprecipitation effects. Although the latter can be minimized or partially corrected for by processes such as aging the precipitate, this is often not possible in a direct titration.

We shall limit our discussion to precipitation titrations involving silver salts, with particular emphasis on the indicators which have been successfully employed in such titrations.

9.1a. Titration Curves for Precipitation Titrations

Titration curves for precipitation titrations can be constructed and are entirely analogous to those for acid-base and complex-formation titrations. Equilibrium calculations based on the solubility product constant are required at the equivalence point, and this topic is reviewed in Section 5.4b. The following example illustrates the calculations involved in precipitation titrations.

EXAMPLE 9.1

Titration of chloride ion with silver ion

50.0 mL of 0.100 M NaCl solution is titrated with 0.100 M AgNO$_3$. Calculate the chloride ion concentration at intervals during the titration and plot pCl vs. milliliters of AgNO$_3$. $pCl = -\log [Cl^-]$, and K_{sp} for AgCl $= 1 \times 10^{-10}$.

(a) *Start of titration.* Since

$$[Cl^-] = 0.100 \text{ mmol/mL}$$

$$pCl = 1.00$$

(b) *After addition of 10.0 mL of AgNO$_3$.* We started with 50.0 mL $\times$ 0.100 mmol/mL $= 5.00$ mmol of Cl$^-$ and have added 10.0 mL $\times$ 0.100 mmol/mL $= 1.00$ mmol of Ag$^+$. The reaction is

mmol	Ag$^+$	+	Cl$^-$	$\longrightarrow$	AgCl(s)
Initial:	1.00		5.00		
Change:	-1.00		-1.00		
Equilibrium:	—		4.00		

Since the reaction goes well to completion, the chloride ion concentration is

$$[Cl^-] \cong \frac{4.00 \text{ mmol}}{60.0 \text{ mL}} = 0.067 \ M$$

$$pCl = 1.17$$

(c) *After addition of 49.9 mL of AgNO$_3$.* We started with 50.0 mL $\times$ 0.100 mmol/mL $= 5.00$ mmol of Cl$^-$ and have added 49.9 mL $\times$ 0.100 mmol/mL $= 4.99$ mmol of Ag$^+$. The reaction is

mmol	Ag$^+$	+	Cl$^-$	$\longrightarrow$	AgCl(s)
Initial:	4.99		5.00		
Change:	-4.99		-4.99		
Equilibrium:	—		0.01		

Assuming complete reaction, the chloride concentration is

$$[Cl^-] \cong \frac{0.01 \text{ mmol}}{99.9 \text{ mL}} = 1.0 \times 10^{-4} \, M$$

$$pCl = 4.00$$

In these calculations we have disregarded the contribution of chloride ions to the solution from the solubility of the AgCl precipitate. This approximation is valid except within one or two drops of the equivalence point.

(d) *Equivalence point.* We started with 50.0 mL × 0.100 mmol/mL = 5.00 mmol of Cl^- and have added 50.0 mL × 0.100 mmol/mL = 5.00 mmol of Ag^+. The reaction is

mmol	Ag^+	+	Cl^-	⟶	AgCl(s)
Initial:	5.00		5.00		
Change:	−5.00		−5.00		
Equilibrium:	—		—		

There is neither excess chloride nor silver ion, and the concentration of each is given by the square root of K_{sp}.

$$AgCl(s) \longrightarrow Ag^+ + Cl^-$$

$$[Ag^+][Cl^-] = K_{sp}$$

$$[Ag^+] = [Cl^-]$$

$$[Cl^-]^2 = 1.0 \times 10^{-10}$$

$$[Cl^-] = 1.0 \times 10^{-5}$$

$$pCl = 5.00$$

(e) *After addition of 60.0 mL of AgNO$_3$.* We started with 50.0 mL × 0.100 mmol/mL = 5.00 mmol of Cl^- and have added 60.0 mL × 0.100 mmol/mL = 6.00 mmol of Ag^+. The reaction is

mmol	Ag^+	+	Cl^-	⟶	AgCl(s)
Initial:	6.00		5.00		
Change:	−5.00		−5.00		
Equilibrium:	1.00		—		

The concentration of excess Ag^+ is

$$[Ag^+] = \frac{1.00 \text{ mmol}}{110 \text{ mL}} = 9.1 \times 10^{-3}$$

$$pAg = 2.04$$

Since

$$pAg + pCl = 10.00$$

$$pCl = 7.96 \qquad \qquad \square$$

The data for this titration are given in Table 9.1, and the titration curve is plotted in Fig. 9.1. The curves for the titration of iodide and bromide ions with silver are also plotted in this figure. Note that the increase in pX (X = Cl, Br, or I) at the equivalence point is greatest for the titration of iodide, since silver iodide

TABLE 9.1 Titration of 50 mL of 0.10 M NaCl with 0.10 M AgNO$_3$

AgNO$_3$, mL	[Cl$^-$]	% Cl$^-$ Pptd.	pCl
0.0	0.10	0.0	1.00
10.0	0.067	20.0	1.17
20.0	0.043	40.0	1.37
30.0	0.025	60.0	1.60
40.0	0.011	80.0	1.96
49.0	0.0010	98.0	3.00
49.9	1.0×10^{-4}	99.8	4.00
50.0	1.0×10^{-5}	100	5.00
50.1	1.0×10^{-6}	100	6.00
51.0	1.0×10^{-7}	100	7.00
60.0	1.1×10^{-8}	100	7.96

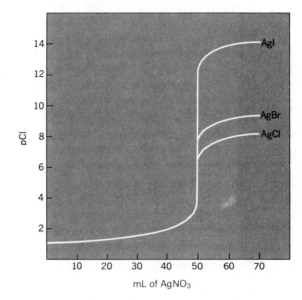

Figure 9.1 Titration curves of NaCl, NaBr, and NaI.
50 mL of 0.1 M salt titrated with 0.1 M AgNO$_3$.

is the least soluble of the three salts. Note also that the value of K for the titration reaction

$$Ag^+ + X^- \rightleftharpoons AgX(s)$$

is

$$K = \frac{1}{[Ag^+][X^-]} = \frac{1}{K_{sp}}$$

Hence the smaller K_{sp}, the larger the K for the titration reaction. For the three salts shown in Fig. 9.1 the values of K are: AgCl, 1×10^{10}; AgBr, 2×10^{12}; and AgI, 1×10^{16}.

9.1b. Feasibility of Precipitation Titrations

The magnitude of K required for a feasible precipitation titration can be calculated as was done previously for acid-base (Section 6.5) and complex-formation titrations (Section 8.2b). The following example illustrates this calculation for a salt of the type MX.

EXAMPLE 9.2

Calculation of K and K_{sp}
for a feasible
precipitation titration

50 mL of 0.10 M NaX is titrated with 50 mL of 0.10 M AgNO$_3$. Calculate the value of K and that of K_{sp} for AgX so that when 49.95 mL of titrant has been added the reaction is complete, and the pX changes by 2.00 units on the addition of two more drops (0.10 mL) of titrant. NaX is a completely dissociated salt, and the titration reaction is

$$Ag^+ + X^- \rightleftharpoons AgX(s) \qquad K = 1/K_{sp}$$

One drop before the equivalence point, 4.995 mmol of Ag$^+$ has been added. We started with $50 \times 0.10 = 5.0$ mmol of X$^-$. Hence 0.0050 mmol remains, and

$$[X^-] = \frac{0.0050 \text{ mmol}}{99.95 \text{ mL}} \cong 5 \times 10^{-5} M$$

$$pX = 4.30$$

If $\Delta pX = 2.00$, then $pX = 6.30$ and $[X^-] = 5 \times 10^{-7} M$ when the volume of titrant is 50.05 mL. Since

$$[Ag^+] = \frac{0.05 \times 0.10}{100.05} \cong 5 \times 10^{-5}$$

$$K = \frac{1}{(5 \times 10^{-5})(5 \times 10^{-7})}$$

$$K = 4 \times 10^{10} \quad \text{and} \quad K_{sp} = 2.5 \times 10^{-11}$$

The student should confirm that for $\Delta pX = 1.00$, K should be 4×10^9. $\square$

The magnitude of ΔpX at the equivalence point in the titration of X$^-$ with Ag$^+$ depends upon the concentrations of the analyte and the titrant. The effect is exactly the same as that for the titration of a strong acid with a strong base (Fig. 6.3, page 149). The lower the concentration of X$^-$, the higher the values of pX before the equivalence point and the smaller the value of ΔpX at the equivalence point. As the concentration of titrant is lowered, the branch of the curve after the equivalence point is lowered, again resulting in a lower value of ΔpX at the equivalence point.

In the titration of Cl$^-$ with Ag$^+$ the concentrations of both reactants should not be much less than 0.10 M if a reasonably good end point is to be obtained.

9.2

INDICATORS FOR PRECIPITATION TITRATIONS INVOLVING SILVER

It has been pointed out that one of the problems associated with precipitation titrations is finding a suitable indicator. In titrations involving silver salts there are three indicators which have been successfully employed for many years. The Mohr method uses chromate ion, CrO_4^{2-}, to precipitate brown Ag$_2$CrO$_4$. The Volhard method uses Fe^{3+} ion to form a colored complex with thiocyanate ion, SCN$^-$. And Fajans' method utilizes adsorption indicators. We shall discuss these three methods briefly.

9.2a. Formation of a Colored Precipitate: The Mohr Method

Just as an acid-base system can be used as an indicator for an acid-base titration, the formation of another precipitate can be used to indicate the completion of a precipitation titration. The best-known example of such a case is the so-called Mohr titration of chloride with silver ion, in which chromate ion is used as the indicator. The first permanent appearance of the reddish silver chromate precipitate is taken as the end point of the titration.

It is necessary, of course, that precipitation of the indicator occur at or near the equivalence point of the titration. Silver chromate is more soluble (about 8.4×10^{-5} mol/liter) than silver chloride (about 1×10^{-5} mol/liter). If silver ions are added to a solution containing a large concentration of chloride ions and a small concentration of chromate ions, silver chloride will precipitate first; silver chromate will not form until the silver ion concentration increases to a large enough value to exceed the K_{sp} of silver chromate. One can readily calculate the concentration of chromate that will lead to precipitation of silver chromate at the equivalence point, where $pAg = pCl = 5.00$. Since the K_{sp} of Ag_2CrO_4 is 2×10^{-12}, and $[Ag^+] = 1 \times 10^{-5}$ at the equivalence point, then

$$[Ag^+]^2[CrO_4^{2-}] = 2 \times 10^{-12}$$

$$[CrO_4^{2-}] = \frac{2 \times 10^{-12}}{(1 \times 10^{-5})^2} = 0.02 \ M$$

Such a high concentration cannot be used in practice, however, since the yellow color of chromate ion makes it difficult to observe formation of the colored precipitate. Normally, a concentration of 0.005 to 0.01 M chromate is employed. The error caused by using such a concentration is quite small. It can be corrected by running an indicator blank or by standardizing the silver nitrate against a pure chloride salt under conditions identical with those used in the analysis.

The Mohr titration is limited to solutions with pH values from about 6 to 10. In more alkaline solutions silver oxide precipitates. In acid solutions the chromate concentration is greatly decreased, since $HCrO_4^-$ is only slightly ionized. Furthermore, hydrogen chromate is in equilibrium with dichromate:

$$2H^+ + 2CrO_4^{2-} \rightleftharpoons 2HCrO_4^- \rightleftharpoons Cr_2O_7^{2-} + H_2O$$

A decrease in chromate ion concentration makes it necessary to add a large excess of silver ions to bring about precipitation of silver chromate and thus leads to large errors. Dichromates are, in general, fairly soluble.

The Mohr method can also be applied to the titration of bromide ion with silver, and also cyanide ion in slightly alkaline solutions. Adsorption effects make the titration of iodide and thiocyanate ions not feasible. Silver cannot be titrated directly with chloride using chromate indicator. The silver chromate precipitate, present initially, redissolves only slowly near the equivalence point. However, one can add excess standard chloride solution and then back-titrate using the chromate indicator.

9.2b. Formation of a Colored Complex: The Volhard Method

The Volhard method is based on the precipitation of silver thiocyanate in nitric acid solution, with iron(III) ion employed to detect excess thiocyanate ion:

$$Ag^+ + SCN^- \rightleftharpoons AgSCN(s)$$

$$Fe^{3+} + SCN^- \rightleftharpoons FeSCN^{2+} \text{ (red)}$$

The method can be used for the direct titration of silver with standard thiocyanate solution or for the indirect titration of chloride, bromide, and iodide ions. In the indirect titration, an excess of standard silver nitrate is added and the excess is titrated with standard thiocyanate.

The Volhard method is widely used for silver and chloride because the titration can be done in acid solution. In fact, it is desirable to employ an acid medium to prevent hydrolysis of the iron(III)-ion indicator. Other common methods for silver and chloride require a nearly neutral solution for successful titration. Many cations precipitate under such conditions and hence interfere in these methods.

In the analysis of chloride an error can occur if the AgCl precipitate is allowed to react with thiocyanate ion:

$$AgCl(s) + SCN^- \rightleftharpoons AgSCN(s) + Cl^-$$

Since AgSCN is less soluble than AgCl, this reaction tends to proceed from left to right and will cause low results in a chloride analysis. The reaction can be prevented by filtering off the AgCl or adding nitrobenzene before titration with thiocyanate. The nitrobenzene apparently forms an oily coating on the AgCl surface, preventing the reaction with thiocyanate.

In the determination of bromide and iodide by the indirect Volhard method, the reaction with thiocyanate does not cause trouble because AgBr has about the same solubility as AgSCN, and AgI is considerably less soluble than AgSCN.

9.2c. The Use of Adsorption Indicators: The Fajans Method

Adsorption of a colored organic compound on the surface of a precipitate may induce electronic shifts in the molecule that alter its color. This phenomenon can be used to detect the end point of precipitation titrations of silver salts. The organic compounds thus employed are referred to as *adsorption indicators*.

The mechanism by which such indicators work was explained by Fajans[1] as follows. In the titration of Cl^- with Ag^+, before the EPt the colloidal particles of AgCl are negatively charged because of adsorption of Cl^- ions from the solution:

$$(AgCl) \cdot Cl^- \mid M^+$$

Primary	Secondary	Excess
layer	layer	chloride

The adsorbed Cl^- ions form the primary layer, causing the colloidal particles to be negatively charged. These particles attract positive ions from the solution to form a more loosely held secondary layer.

Beyond the EPt, excess Ag^+ ions displace Cl^- ions from the primary layer[2] and the particles become positively charged:

$$(AgCl) \cdot Ag^+ \mid X^-$$

Primary	Secondary	Excess
layer	layer	silver

[1] K. Fajans and O. Hassel, *Z. Elektrochem.*, **29**, 495 (1923); see also I. M. Kolthoff, *Chem. Rev.*, **16**, 87 (1935), and K. Fajans, Chapter 7 of W. Bottger, ed., *Newer Methods of Volumetric Analysis,* D. Van Nostrand and Co., New York, 1938.

[2] A precipitate tends to adsorb most readily those ions that form an insoluble compound with one of the ions in the lattice. Thus silver or chloride ions will be more readily adsorbed by a silver chloride precipitate than will, say, sodium or nitrate ions.

Anions in the solution are attracted to form the secondary layer.

Fluorescein is a weak organic acid, which we may represent as HFl. When fluorescein is added to the titration flask, the anion, Fl^-, is not adsorbed by colloidal silver chloride as long as chloride ions are in excess. However, when silver ions are in excess, the Fl^- ions can be attracted to the surface of the positively charged particles:

$$(AgCl) \cdot Ag^+ \mid Fl^-$$

The resulting aggregate is pink, and the color is sufficiently intense to serve as a visual indicator.

A number of factors must be considered in choosing a proper adsorption indicator for a precipitation titration. These are summarized below.

1. The AgCl should not be allowed to coagulate into large particles at the EPt, since this will greatly decrease the surface available for adsorption of the indicator. A protective colloid, such as dextrin, should be added to keep the precipitate highly dispersed. In the presence of dextrin the color change is reversible, and if the end point is overrun, one can back-titrate with a standard chloride solution.
2. The adsorption of the indicator should start just before the EPt and increase rapidly at the EPt. Some unsuitable indicators are so strongly adsorbed that they actually displace the primarily adsorbed ion well before the EPt is reached.
3. The pH of the titration medium must be controlled to ensure a sufficient concentration of the ion of the weak acid or base indicator. Fluorescein, for example, has a K_a of about 10^{-7}, and in solutions more acidic than pH 7 the concentration of Fl^- ions is so small that no color change is observed. Fluorescein can be used only in the pH range of about 7 to 10. Dichlorofluorescein has a K_a of about 10^{-4} and can be used in the pH range 4 to 10.
4. It is preferable that the indicator ion be opposite in charge to the ion added as the titrant. Adsorption of the indicator will then not occur until excess titrant is present. For the titration of silver with chloride, methyl violet, the chloride salt of an organic base, can be employed. The cation is not adsorbed until excess chloride ions are present and the colloid is negatively charged. It is possible to use dichlorofluorescein in this case, but the indicator should not be added until just before the equivalence point.

A list of some adsorption indicators is given in Table 9.2, and Table 9.3 lists some applications of precipitation titrations.

TABLE 9.2 Some Adsorption Indicators

INDICATOR	ION TITRATED	TITRANT	CONDITIONS
Dichlorofluorescein	Cl^-	Ag^+	pH 4
Fluorescein	Cl^-	Ag^+	pH 7–8
Eosin	Br^-, I^-, SCN^-	Ag^+	pH 2
Thorin	SO_4^{2-}	Ba^{2+}	pH 1.5–3.5
Bromcresol green	SCN^-	Ag^+	pH 4–5
Methyl violet	Ag^+	Cl^-	Acid solution
Rhodamine 6 G	Ag^+	Br^-	Sharp in the presence of HNO_3 up to 0.3 M
Orthochrome T	Pb^{2+}	CrO_4^{2-}	Neutral 0.02 M solution
Bromphenol blue	Hg_2^{2+}	Cl^-	0.1 M solution

TABLE 9.3 Determinations by Precipitation Titrations

SPECIES DETERMINED	TITRANT	INDICATOR	COMMENTS
Cl^-, Br^-	$AgNO_3$	K_2CrO_4	Mohr method
Cl^-, Br^-, I^-, SCN^-	$AgNO_3$	Adsorption	Fajans method
Br^-, I^-, SCN^-, AsO_4^{3-}	$AgNO_3$ + $KSCN$	Fe(III)	Volhard method; precipitate need not be filtered.
Cl^-, CN^-, CO_3^{2-}, S^{2-}, $C_2O_4^{2-}$, CrO_4^{2-}	$AgNO_3$ + $KSCN$	Fe(III)	Volhard method; precipitate must be filtered.
F^-	Th(IV)	Alizarin red S	Fajans method
SO_4^{2-}	$BaCl_2$	Tetrahydroxyquinone	Fajans method
PO_4^{3-}	$Pb(OAc)_2$	Dibromofluorescein	Fajans method
$C_2O_4^{2-}$	$Pb(OAc)_2$	Fluorescein	Fajans method
Ag^+	KSCN	Fe(III)	Volhard method
Zn^{2+}	$K_4Fe(CN)_6$	Diphenylamine	Fajans method
Hg_2^{2+}	NaCl	Bromphenol blue	Fajans method

9.3

FACTORS AFFECTING SOLUBILITY

We use the term *solubility* to refer to the concentration of a *saturated* solution of a solute (here crystalline solids) in a given solvent at a certain temperature. In a saturated solution equilibrium exists between the solid and its ions in solution, as for barium sulfate:

$$BaSO_4(s) \rightleftharpoons Ba^{2+} + SO_4^{2-}$$

The equilibrium constant for this process is the familiar *solubility product constant* (Section 5.4b):

$$K_{sp} = [Ba^{2+}][SO_4^{2-}]$$

A saturated solution can be produced by continuing the addition of solute until no further solute dissolves, or by increasing the concentration of ions until precipitation occurs. Precipitation results in the analyte's being physically separated from other substances in the solution, as well as from the solvent itself. Until recent years precipitation was the analyst's most widely used method of separating a sample into its component parts.

The important factors affecting solubility of crystalline solids are temperature, nature of the solvent, and the presence of other ions in the solution. Included in the latter category are ions that may or may not be common to ions in the solid, as well as ions or molecules which form slightly dissociated molecules or complex ions with ions of the solid. We shall discuss these factors in the following sections.

9.3a. Temperature

Most of the inorganic salts in which we are interested increase in solubility as the temperature is increased. It is usually advantageous to carry out the operations of precipitation, filtration, and washing with hot solutions. Particles of large size may result, filtration is faster, and impurities are dissolved more readily. Therefore, directions frequently call for employing hot solutions in cases where the solubility of the precipitate is still negligible at the higher temperature. However, in the case of a fairly soluble compound, such as magnesium ammonium phosphate, the solution must be cooled in ice water before filtration. An appreciable amount of this compound would be lost if the solution were filtered while hot.

The student may recall that lead chloride is separated from silver and mercury(I) chlorides in the qualitative analysis scheme by treatment with hot water. The lead salt dissolves at elevated temperature, leaving the other two salts in the precipitate.

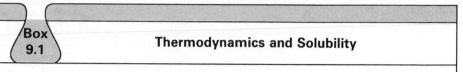

Box 9.1   Thermodynamics and Solubility

An ionic solid such as sodium chloride has a highly ordered structure in which the sodium and chloride ions are arranged in a repeating, regular, geometric array as suggested below. When the solid is placed in water, ions enter the liquid phase, where their arrangement is much more random (but not totally so because of electrostatic constraints):

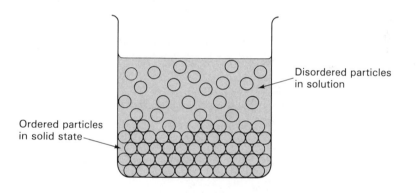

Disordered particles in solution

Ordered particles in solid state

Figure 9.1.1   Dissolving of a solid in a liquid.

Since the dissolution of the solid occurs spontaneously, the process is attended by a free-energy decrease (ΔG is negative). Both heat and entropy contribute to the free-energy change ($\Delta G = \Delta H - T\Delta S$).

Because the solution is less ordered than the solid, ΔS is positive, but ΔH may be either positive (endothermic process) or negative (exothermic process). Energy is required to overcome electrostatic attractions of unlike charges in the crystal and to overcome attractions among water molecules to make space for the ions entering the solution. On the other hand, there are favorable electrostatic interactions of the dipolar water molecules with charged ions; we say the ions become hydrated, a process which releases energy.

Depending upon the relative enthalpies of all of these processes, heat may be, overall, absorbed or released. For most inorganic salts, dissolution is endothermic, which means that the solubility increases with an increase in temperature (Le Châtelier's principle).

9.3b. Choice of Solvent

Most inorganic salts are more soluble in water than in organic solvents. Water has a large dipole moment and is attracted to both cations and anions to form hydrated ions. We have already noted, for example, that the hydrogen ion in water is completely hydrated, forming the H_3O^+ ion. All ions are undoubtedly hydrated to some extent in water solutions, and the energy released by interaction of the ions and the solvent helps overcome the attractive forces tending to hold the ions in the solid lattice. The ions in a crystal do not have so large an attraction for organic solvents, and hence the solubilities are usually less than in water. The analyst can frequently utilize the decreased solubility in organic solvents to separate two substances which are quite soluble in water. For example, a dried nitrate mixture of calcium and strontium nitrates can be separated by treatment with a mixture of alcohol and ether. Calcium nitrate dissolves, leaving strontium nitrate. Potassium can be separated from sodium by precipitating K_2PtCl_6 from an alcohol-water mixed solvent.

9.3c. Common-ion Effect

A precipitate is generally more soluble in pure water than in a solution which contains one of the ions of the precipitate (common-ion effect, page 108). In a solution of silver chloride, for example, the product of the concentrations of silver and chloride ions cannot exceed the value of the solubility product constant, 1×10^{-10}. In pure water, each ion has a concentration of $1 \times 10^{-5}\,M$, but if sufficient silver nitrate is added to make the silver ion concentration $1 \times 10^{-4}\,M$, the chloride ion concentration must decrease to a value of $1 \times 10^{-6}\,M$. The reaction

$$Ag^+ + Cl^- \rightleftharpoons AgCl(s)$$

is forced to the right by excess silver ion, resulting in the precipitation of additional salt and decreasing the quantity of chloride remaining in the solution.

The importance of the common-ion effect in bringing about complete precipitation in quantitative analyses is readily apparent. In carrying out precipitations, the analyst always adds some excess of the precipitating agent to ensure complete precipitation. In washing a precipitate where solubility losses may be appreciable, a common ion may be used in the wash liquid to diminish solubility. The ion should be that of the precipitating agent, of course, not the ion sought. Likewise, the salt used in the wash water should be such that any excess is removed by volatilization when the precipitate is finally heated to constant weight.

In the presence of a large excess of common ion, the solubility of a precipitate may be considerably greater than the value predicted by the solubility product constant. This effect will be discussed later. In general, directions call for adding about 10% excess precipitating agent.

The effect of a common ion on the solubility of a precipitate is illustrated in the following calculations.

EXAMPLE 9.3

The common-ion effect and solubility

Calculate the molar solubility of CaF_2 in (a) water, (b) 0.010 M $CaCl_2$, (c) 0.010 M NaF solution, given the K_{sp} as 4×10^{-11} and neglecting hydrolysis of the fluoride ion.

(a) The equilibrium is

$$CaF_2(s) \rightleftharpoons Ca^{2+} + 2F^-$$

Let s = molar solubility of CaF_2. The mass balances are

$$[Ca^{2+}] = s$$

$$[F^-] = 2s$$

Since

$$[Ca^{2+}][F^-]^2 = K_{sp}$$

$$(s)(2s)^2 = 4 \times 10^{-11}$$

and

$$s = 2.1 \times 10^{-4} \text{ mol/liter}$$

(b) In $0.010\ M$ $CaCl_2$ the mass balances are

$$[Ca^{2+}] = 0.010 + s$$

$$[F^-] = 2s$$

Hence

$$(0.01 + s)(2s)^2 = 4 \times 10^{-11}$$

Since $s \ll 0.01$, this becomes

$$4s^2 = 4 \times 10^{-9}$$

$$s = 3.2 \times 10^{-5} \text{ mol/liter}$$

(c) The mass balances are

$$[Ca^{2+}] = s$$

$$[F^-] = 0.01 + 2s$$

Hence

$$(s)(0.01 + 2s)^2 = 4 \times 10^{-11}$$

Since $2s \ll 0.01$, this becomes

$$s = 4 \times 10^{-7} \text{ mol/liter}$$

Note the extensive reduction in solubility brought about by the common ion. It should also be noted that excess F^- has a greater effect than excess Ca^{2+}. ☐

9.3d. Activity Effect

It has been found that many precipitates show an increased solubility in solutions containing ions which do not react chemically with ions of the precipitate. The effect is referred to by various names, such as *diverse-ion, neutral-salt,* or *activity effect* (Section 5.3). The data in Table 9.4 illustrate the magnitude of the

TABLE 9.4 Solubility of AgCl and $BaSO_4$ in KNO_3 Solutions

MOLARITY KNO_3	MOLARITY AgCl $\times 10^5$	MOLARITY $BaSO_4 \times 10^5$
0.000	1.00	1.00
0.001	1.04	1.21
0.005	1.08	1.48
0.010	1.12	1.70

Source: From data of S. Popoff and E. W. Neuman, *J. Phys. Chem.,* **34,** 1853 (1930); and E. W. Neuman, *J. Am. Chem. Soc.,* **44,** 879 (1933). The K_{sp} of each salt was taken as 1×10^{-10}.

increased solubility of silver chloride and barium sulfate in potassium nitrate solution. It is seen that in 0.010 M KNO$_3$ the solubility of AgCl is increased from the value in water by about 12%, and that of BaSO$_4$ by about 70%.

It was pointed out earlier (page 96) that we are justified in substituting molarity for activity only in very dilute solutions, where activity coefficients are approximately unity. In more concentrated solutions of electrolytes, activity coefficients decrease rapidly because of greater attraction between oppositely charged ions. The effectiveness of the ions in maintaining equilibrium conditions is thus decreased, and additional precipitate must dissolve to restore this activity. The solubility product expression for AgCl is

$$a_{Ag^+} \times a_{Cl^-} = K_{sp}^\circ$$

where K_{sp}° is the equilibrium constant in terms of activities. In terms of concentrations this becomes

$$f_{Ag^+}[Ag^+] \times f_{Cl^-}[Cl^-] = K_{sp}^\circ$$

or

$$[Ag^+][Cl^-] = \frac{K_{sp}^\circ}{f_{Ag^+}f_{Cl^-}} = K_{sp}$$

It is apparent that the smaller the activity coefficients of the two ions, the larger the product of the molar concentrations of the two ions (K_{sp}). This increase in solubility is greater for BaSO$_4$ than for AgCl because activity coefficients of bivalent ions are smaller than those of univalent ions (page 97). In very dilute solutions activity coefficients approach unity, and K_{sp} is approximately the same as K_{sp}°.

The following example illustrates a calculation of solubility using activity coefficients.

EXAMPLE 9.4

Effect of activity considerations in solubility calculations

Calculate the molar solubility of BaSO$_4$ in a 0.010 M solution of KNO$_3$ using activity coefficients calculated by the extended Debye-Hückel equation (Table 5.3, page 99).

Since KNO$_3$ is a 1 : 1 electrolyte the ionic strength is the same as the molarity, 0.010. At this ionic strength the activity coefficient of Ba^{2+} is 0.667, and that of SO$_4^{2-}$ 0.659. Hence

$$K_{sp} = \frac{1.00 \times 10^{-10}}{(0.667)(0.659)} = 2.27 \times 10^{-10}$$

If s is the molar solubility,

$$s^2 = 2.27 \times 10^{-10}$$

and

$$s = 1.51 \times 10^{-5} \text{ mol/liter} \qquad \square$$

The activity effect does not cause serious problems for the analyst since conditions are normally chosen so as to make loss from solubility negligibly small. It is rarely necessary to make a precipitation from a salt solution of very high concentration, and in such cases an estimate of the increased solubility can be made as illustrated above. Errors from other sources are normally more important.

9.3e. *Effect of pH*

The solubility of the salt of a weak acid depends upon the pH of the solution. Some of the more important examples of such salts in analytical chemistry are oxalates, sulfides, hydroxides, carbonates, and phosphates. Hydrogen ion combines with the anion of the salt to form the weak acid, thereby enhancing the solubility of the salt. We shall limit our discussion in this section to solutions which are fairly acidic, so the hydrogen ion concentration is not changed appreciably as the salt dissolves.

Let us consider first the simplest case, that of a salt MA of the weak acid HA. The equilibria to be considered are

$$MA(s) \;\rightleftharpoons\; M^+ + A^-$$

$$HA + H_2O \;\rightleftharpoons\; H_3O^+ + A^-$$

As we did previously (Section 7.5), let us designate c_a the total (analytical) concentration of all species related to the acid HA.

$$c_a = [A^-] + [HA]$$

$$c_a = [A^-]\left\{\frac{[H_3O^+] + K_a}{K_a}\right\}$$

The fraction in the A^- form is

$$\frac{[A^-]}{c_a} = \frac{K_a}{[H_3O^+] + K_a} = \alpha_1$$

Hence

$$[A^-] = \alpha_1 c_a$$

The latter expression can be substituted in the K_{sp}, giving

$$K_{sp} = [M^+][A^-] = [M^+]\alpha_1 c_a$$

or

$$\frac{K_{sp}}{\alpha_1} = K_{\text{eff}} = [M^+]c_a$$

We have designated K_{eff} as the effective solubility product constant, in agreement with the terminology used in Section 8.2.a for the effective stability constant of complexes. The value of K_{eff} varies with pH because of the pH dependence of α_1.

The student should be able to show that for a salt MA_2, the relation is

$$K_{\text{eff}} = \frac{K_{sp}}{\alpha_1^2} = [M^{2+}]c_a^2$$

and that for a diprotic acid, H_2A, the concentration of A^{2-} is given by $\alpha_2 c_a$, where

$$\alpha_2 = \frac{K_{a_1}K_{a_2}}{[H_3O^+]^2 + [H_3O^+]K_{a_1} + K_{a_1}K_{a_2}}$$

$$K_{\text{eff}} = \frac{K_{sp}}{\alpha_2} = [M^{2+}]c_a$$

The following examples illustrate calculations based on the relations just described.

EXAMPLE 9.5

Effect of pH in solubility calculations

Calculate the molar solubility of CaF_2 in an HCl solution, $pH = 3.00$, given that K_{sp} for $CaF_2 = 4 \times 10^{-11}$, and K_a for HF $= 6 \times 10^{-4}$.

First evaluate α_1:

$$\alpha_1 = \frac{6 \times 10^{-4}}{6 \times 10^{-4} + 1 \times 10^{-3}} = 0.38$$

$$\alpha_1^2 = 0.14$$

Hence

$$K_{eff} = \frac{4 \times 10^{-11}}{0.14} = 2.9 \times 10^{-10}$$

Let s = molar solubility of CaF_2. The mass balances are

$$[Ca^{2+}] = s$$

$$c_F = [HF] + [F^-] = 2s$$

and

$$(s)(2s)^2 = 2.9 \times 10^{-10}$$

$$s = 4.2 \times 10^{-4} \text{ mol/liter} \qquad \square$$

EXAMPLE 9.6

Effect of pH in solubility calculations

Calculate the solubility of CaC_2O_4 in an HCl solution of pH 3.00, given $K_{sp} = 2 \times 10^{-9}$, $K_{a_1} = 6.5 \times 10^{-2}$, $K_{a_2} = 6.1 \times 10^{-5}$.

$$\alpha_2 = \frac{6.5 \times 10^{-2} \times 6.1 \times 10^{-5}}{6.5 \times 10^{-2} \times 6.1 \times 10^{-5} + 6.5 \times 10^{-2} \times 10^{-3} + (10^{-3})^2}$$

$$\alpha_2 = 0.057$$

Hence

$$K_{eff} = \frac{2 \times 10^{-9}}{0.057} = 3.5 \times 10^{-8}$$

The mass balances are

$$[Ca^{2+}] = s$$

$$c_{ox} = s$$

Then

$$s^2 = 3.5 \times 10^{-8}$$

$$s = 1.9 \times 10^{-4} \text{ mol/liter} \qquad \square$$

The separation of metal sulfides based upon the control of pH has been used for many years in the qualitative analysis scheme. The metals which form the less soluble sulfides (group II) are precipitated by H_2S in about 0.10 M HCl. Then the pH is raised to precipitate the metals of group III. Hydrogen sulfide is a diprotic acid, and the expression for α_2 (above) is applicable. However, since the two acid constants are so small ($K_{a_1} = 1 \times 10^{-7}$, $K_{a_2} = 1 \times 10^{-15}$), the two terms in the denominator containing the acid constants are negligible compared to the square of the hydrogen ion concentration. The expression becomes (approximately)

$$\alpha_2 \cong \frac{K_{a_1} K_{a_2}}{[H_3O^+]^2}$$

Also, in strongly acidic solution, the analytical concentration of hydrogen sulfide is approximately

$$c_s = [H_2S] + [HS^-] + [S^{2-}] \cong [H_2S]$$

Hence the sulfide ion concentration, $\alpha_2 c_s$, becomes

$$[S^{2-}] = \frac{[H_2S]K_{a_1}K_{a_2}}{[H_3O^+]^2}$$

Since a saturated solution of H_2S is about 0.10 M, this gives

$$[S^{2-}] = \frac{1 \times 10^{-23}}{[H_3O^+]^2}$$

This is the usual expression employed to show how the sulfide ion concentration can be varied by changing the hydrogen ion concentration. The following example illustrates the separation of two metals by employing this principle.

EXAMPLE 9.7

Selective precipitation

100 mL of a solution that is 0.10 M in both Cu^{2+} and Mn^{2+} and 0.20 M in H_3O^+ is saturated with H_2S.

(a) Show which metal sulfide precipitates. K_{sp} for CuS is 4×10^{-38}, for MnS 1×10^{-16}.

The sulfide concentration is given by

$$[S^{2-}] = \frac{1 \times 10^{-23}}{(0.20)^2} = 2.5 \times 10^{-22}$$

The K_{sp} of CuS is greatly exceeded, but that of MnS is not:

$$(0.10)(2.5 \times 10^{-32}) = 2.5 \times 10^{-23} \gg 4 \times 10^{-38}$$
$$= 2.5 \times 10^{-23} \ll 1 \times 10^{-16}$$

Hence CuS precipitates, but MnS does not.

(b) What must be the hydrogen ion concentration for MnS to start to precipitate?

The sulfide ion concentration needed in order for $[Mn^{2+}][S^{2-}]$ to equal the K_{sp} for MnS is

$$(0.10)[S^{2-}] = 1 \times 10^{-16}$$
$$[S^{2-}] = 1 \times 10^{-15}$$

Hence

$$1 \times 10^{-15} = \frac{1 \times 10^{-23}}{[H_3O^+]^2}$$
$$[H_3O^+] = 1 \times 10^{-4} \ M \qquad \square$$

The following example illustrates the separation of two metal hydroxides by control of pH.

EXAMPLE 9.8

Calculation of conditions for selective precipitation

Calculate the pH at which the following hydroxides begin to precipitate if the solution is 0.1 M in each cation: Fe(OH)$_3$, $K_{sp} = 1 \times 10^{-36}$; and Mg(OH)$_2$, $K_{sp} = 1 \times 10^{-11}$.

Iron(III) hydroxide:

Box 9.2

Why Is Silver Fluoride Soluble?

Most students know that silver chloride, AgCl, is insoluble in water. It precipitates in the first group of the qualitative analysis scheme and is the salt used to demonstrate a precipitation titration in the quantitative analysis laboratory. They are not surprised to learn that AgBr and AgI are also insoluble, since bromine and iodine are in the same family of elements as chlorine. But what about fluorine? You may be surprised to learn that AgF is soluble in water to the extent of about 182 g per 100 ml at 15.5°C. A rationalization of this striking difference in behavior has been given in terms of the theory of "hard" and "soft" acids and bases.[†]

This theory classifies Lewis bases with the small atoms N, O, and F (first row of the periodic table) as hard. Lewis bases with P, S, Br, and I are soft; Cl is borderline. *Hard* and *soft* refer to the fact that in the heavier elements the valence shell is more deformable, polarizable, or soft. The valence electrons of the small, very electronegative F and O atoms are more tightly bound, and these bases, which are not so easily polarized, are said to be hard. Metals in low oxidation states with a number of *d* electrons (such as Ag^+) are soft Lewis acids. They are the most deformable and hence the softest metal ions. As the charge on the metal ion increases, its electrons become more firmly bound and it becomes harder. Hard acids include metal ions with noble-gas electron configurations and transition metals in their higher oxidation states, such as Ca^{2+} and Cr^{3+}.

Hard-soft acid-base (HSAB) theory states that hard acids prefer to react with hard bases, and soft acids with soft bases. Ag^+ is a soft acid which binds strongly to the soft bases Cl^-, Br^-, and I^- (Cl^- being the least of the three). In aqueous solution there is competition between the bases H_2O and X^- for the soft Ag^+. The hydrated silver ion, $Ag(H_2O)_x^+$, is stable with respect to the hard base F^- and remains in solution. It is unstable with respect to the softer bases Cl^-, Br^-, and I^-, and the precipitates AgCl, AgBr, and AgI are formed.

[†]R. G. Pearson, *J. Chem. Educ.* **45,** 581, 643 (1968).

$$[Fe^{3+}][OH^-]^3 = 1 \times 10^{-36}$$
$$(0.1)[(OH^-]^3 = 1 \times 10^{-36}$$
$$[OH^-]^3 = 1 \times 10^{-35}$$
$$3pOH = 35$$
$$pOH = 11.7$$
$$pH = 2.3$$

Magnesium hydroxide:

$$[Mg^{2+}][OH^-]^2 = 1 \times 10^{-11}$$
$$(0.1)[OH^-]^2 = 1 \times 10^{-11}$$
$$[OH^-]^2 = 1 \times 10^{-10}$$
$$2pOH = 10.0$$

$$pOH = 5.0$$

$$pH = 9.0 \qquad \square$$

Thus, if an acidic solution containing these two ions is slowly neutralized with base, iron(III) hydroxide will precipitate first. This precipitate can be separated by filtration before the pH is sufficiently high to precipitate magnesium hydroxide. In actual practice, however, the iron(III) hydroxide precipitate is likely to be contaminated by magnesium hydroxide. This arises from the fact that in the region where the two solutions mix, the solubility product constant of magnesium hydroxide may be temporarily exceeded. The magnesium hydroxide may not redissolve as the solution is stirred, and the separation is then not a clean one. Usually, a buffer solution of intermediate pH is employed to diminish the local increase in hydroxide ion concentration. Better still, the pH can be gradually increased by the hydrolysis of a substance such as urea.

9.3f. *Effect of Hydrolysis*

In the previous section we limited our discussion to solutions of fairly high acidity, such that the anion of the weak acid did not change the pH appreciably. Let us now consider the case in which the salt of a weak acid is dissolved not in strong acid but in water. The problem is more complex than the previous one, since the change in hydrogen ion concentration may be of considerable magnitude.

For simplification let us consider that whatever the amount of salt MA dissolves, the anion is completely hydrolyzed:

$$A^- + H_2O \; \rightleftharpoons \; HA + OH^-$$

This is a good approximation if HA is very weak and if MA is not very soluble (i.e., if both K_a and K_{sp} are small). It should be noted that the lower the concentration of A^-, the more complete the hydrolysis reaction.

Let us further consider two extremes, depending upon the magnitude of K_{sp}:

1. The solubility is so low that the pH of water is not changed appreciably by the hydrolysis.
2. The solubility is sufficiently large that the hydroxide ion contribution of water can be neglected.

These cases are illustrated in the following example.

EXAMPLE 9.9

Effect of hydrolysis on solubility

Calculate the molar solubilities in water of (a) CuS, $K_{sp} = 4 \times 10^{-38}$, and (b) MnS, $K_{sp} = 1 \times 10^{-16}$. Consider the hydrolysis reaction

$$S^{2-} + H_2O \; \rightleftharpoons \; HS^- + OH^-$$

(a) Since the solubility of CuS is so low, we shall neglect the OH^- produced by hydrolysis, taking $[OH^-] = 1 \times 10^{-7}$. Hence

$$\alpha_2 = \frac{1 \times 10^{-22}}{(1 \times 10^{-7})^2 + (1 \times 10^{-7})(1 \times 10^{-7}) + 1 \times 10^{-22}}$$

$$\alpha_2 = 5 \times 10^{-9}$$

$$K_{\text{eff}} = \frac{4 \times 10^{-38}}{5 \times 10^{-9}} = 8 \times 10^{-30}$$

Letting s = solubility, the mass balances are

$$[Cu^{2+}] = s$$

$$c_s = s$$

Hence

$$s^2 = 8 \times 10^{-30}$$

$$s = 3 \times 10^{-15}$$

(b) Since the hydrolysis is complete, we can write the reaction as

$$MnS(s) + H_2O \rightleftharpoons Mn^{2+} + HS^- + OH^-$$

the equilibrium constant for which is given by

$$K = \frac{K_{sp}K_w}{K_{a_2}} = \frac{1 \times 10^{-16} \times 1 \times 10^{-14}}{1 \times 10^{-15}}$$

$$K = 1 \times 10^{-15}$$

Letting s = molar solubility, then

$$s = [Mn^{2+}] = [HS^-] = [OH^-]$$

Hence

$$s^3 = 1 \times 10^{-15}$$

$$s = 1 \times 10^{-5} \qquad \square$$

It should be noted that the effect of hydrolysis on the pH of water becomes appreciable for sulfides of the type MS whose K_{sp} values are around 10^{-22}. For a K_{sp} of this value the $[OH^-]$ produced by the hydrolysis reaction is 1×10^{-7} M, the same as that produced by the dissociation of water. For such sulfides with larger solubilities, the $[OH^-]$ produced by hydrolysis is greater than that produced by water. Similar calculations can be made for other types of sulfides, as well as for insoluble salts of other weak acids.

The cation of a salt can undergo hydrolysis just as the anion can, and this will also increase the solubility. Typical hydrolytic reactions of iron(III) ion are

$$Fe^{3+} + HOH \rightleftharpoons FeOH^{2+} + H^+$$

$$FeOH^{2+} + HOH \rightleftharpoons Fe(OH)_2^+ + H^+$$

Many metals have been found to form ionic species containing more than one metal atom, as, for example,

$$2Fe^{3+} + 2H_2O \rightleftharpoons Fe_2(OH)_2^{4+} + 2H^+$$

In the case of aluminum, species such as $Al_6(OH)_{15}^{3+}$ have been postulated to explain certain experimental data.

Because of the complexity of these processes, we shall not consider the topic further here.

9.3g. *Metal Hydroxides*

When a metal hydroxide dissolves in water, the situation is analagous to that considered in Section 9.3f in that the pH may be changed appreciably. Let us consider a solution made by dissolving the hydroxide $M(OH)_2$ in water. There are three species in the solution, M^{2+}, H_3O^+, and OH^-. We are interested in calculating the molar solubility, $[M^{2+}]$, which we shall call s. The three equations relating the three species are

$$[M^{2+}][OH^-]^2 = K_{sp} \qquad (1)$$

$$[H_3O^+][OH^-] = K_w \qquad (2)$$

and the charge-balance equation,

$$2[M^{2+}] + [H_3O^+] = [OH^-] \qquad (3)$$

These equations can be manipulated to give a rigorous solution for the molar solubility. The calculation can be simplified, however, by noting that when $M(OH)_2$ dissolves, the hydroxide ion concentration is increased:

$$M(OH)_2(s) \longrightarrow M^{2+} + 2OH^-$$

This increase in $[OH^-]$ shifts the equilibrium

$$2H_2O \rightleftharpoons H_3O^+ + OH^-$$

to the left, decreasing the hydrogen ion concentration. As we did in Section 9.3f, let us consider two extremes, depending upon the solubility of the hydroxide:

1. The solubility is so slight that the pH of water is not changed appreciably. Then

$$[H_3O^+] = [OH^-] = 1.0 \times 10^{-7}$$

and since

$$[M^{2+}][OH^-]^2 = K_{sp}$$

$$[M^{2+}] = s = \frac{K_{sp}}{(1.0 \times 10^{-7})^2} \qquad (4)$$

2. The solubility is great enough to cause an appreciable increase in $[OH^-]$, thereby decreasing $[H_3O^+]$ to such a small value that it can be neglected. The charge-balance equation, Eq. (3) above, then becomes

$$2[Mg^{2+}] = [OH^-]$$

or

$$[OH^-] = 2s$$

Substituting in the K_{sp} expression,

$$[M^{2+}][OH^-]^2 = K_{sp}$$

$$(s)(2s)^2 = K_{sp}$$

$$s = \sqrt[3]{\frac{K_{sp}}{4}} \qquad (5)$$

$\square$

Example 10 illustrates these calculations.

EXAMPLE 9.10

Solubility of metal hydroxides in water

Calculate the molar solubilities in water of (a) $Al(OH)_3$, $K_{sp} = 5 \times 10^{-33}$, and (b) $Mg(OH)_2$, $K_{sp} = 1 \times 10^{-11}$.

(a) Here the solubility is very low, and we use Eq. (4), noting that the hydroxide is of the type $M(OH)_3$.

$$[Al^{3+}] = s = \frac{5 \times 10^{-33}}{(1.0 \times 10^{-7})^3}$$

$$s = 5 \times 10^{-12} \text{ mmol/mL}$$

Note that if this is the value of s, the $[OH^-]$ produced by the solution process is $3s$ or $3 \times 5 \times 10^{-12} = 1.5 \times 10^{-11}$ mmol/mL. The assumption that the $[OH^-]$ in water is not increased appreciably is obviously valid.

(b) Here the solubility is appreciable, and we use Eq. (5):

$$s = \sqrt[3]{\frac{1 \times 10^{-11}}{4}}$$

$$s = 1.4 \times 10^{-4} \text{ mmol/mL}$$

Note that if this is the value of s, the $[OH^-]$ produced by the solution process is $2s$ or $2 \times 1.4 \times 10^{-4} = 2.8 \times 10^{-4}$ mmol/mL, and

$$[H_3O^+] = \frac{1.0 \times 10^{-14}}{2.8 \times 10^{-4}} = 3.6 \times 10^{-11} \text{ mmol/mL}$$

Our assumption that this concentration could be neglected in the charge-balance equation is thus seen to be valid. □

There are, of course, hydroxides with solubilities that do not fall clearly into one of the two categories above, and calculations using Eqs. (4) and (5) can lead to appreciable errors. As a general rule, one can say that if the K_{sp} of the hydroxide $M(OH)_2$ is 1×10^{-18} or *larger*, the error made using Eq. (5) will be negligible. For K_{sp} values of 1×10^{-24} or *smaller*, the error made using Eq. (4) will be negligible. For intermediate values errors as large as 10 to 20% may be encountered. When Eq. (5) is used, the error increases as the K_{sp} value gets smaller. Errors made using Eq. (4) increase as the K_{sp} value gets larger.

9.3h. Effect of Complex Formation

The solubility of a slightly soluble salt is also dependent upon the concentration of substances which form complexes with the cation of the salt. The effect of hydrolysis, mentioned above, is an example in which the complexing agent is hydroxide ion. The complexing agents normally considered under a heading such as this are neutral molecules and anions, both foreign and common to the precipitate.

One of the best-known examples in analytical chemistry is the effect of ammonia on the solubility of the silver halides, especially silver chloride. Silver chloride can be dissolved in ammonia, and this fact is utilized in separating silver from mercury in the first group of the traditional qualitative analysis scheme. Silver ion forms two complexes with ammonia,

$$Ag^+ + NH_3 \rightleftharpoons Ag(NH_3)^+ \qquad K_1 = 2.3 \times 10^3$$

$$Ag(NH_3)^+ + NH_3 \rightleftharpoons Ag(NH_3)_2^+ \qquad K_2 = 6.0 \times 10^3$$

Designating the fraction of silver in the uncomplexed form β_2 as we did for zinc (Section 8.2.c):

$$\beta_2 = \frac{1}{1 + K_1[NH_3] + K_1 K_2[NH_3]^2} = \frac{[Ag^+]}{c_{Ag}}$$

where c_{Ag} is the analytical concentration of silver. Since

$$K_{sp} = [Ag^+][Cl^-]$$

$$K_{sp} = \beta_2 c_{Ag}[Cl^-]$$

or

$$\frac{K_{sp}}{\beta_2} = K_{eff} = c_{Ag}[Cl^-]$$

The following example illustrates the effect of complexes on the solubility of AgCl. Note that this is the same problem as that in Example 20, Section 5.5. There the calculation was done using the systematic equilibrium approach.

EXAMPLE 9.11

Effect of a complexing agent on solubility

Calculate the molar solubility of AgCl in 0.010 M NH$_3$. (This is the final concentration of free NH$_3$ molecules in the solution.) Given are K_{sp} of AgCl = 1.0×10^{-10} and stability constants $K_1 = 2.3 \times 10^3$ and $K_2 = 6.0 \times 10^3$.

Evaluating β_2,

$$\beta_2 = \frac{1}{1 + 2.3 \times 10^3(10^{-2}) + 1.4 \times 10^7(10^{-2})^2}$$

$$\beta_2 = 7.1 \times 10^{-4}$$

$$K_{eff} = \frac{1.0 \times 10^{-10}}{7.1 \times 10^{-4}} = 1.4 \times 10^{-7}$$

Letting s = molar solubility,

$$s = c_{Ag} = [Cl^-]$$

Hence

$$s^2 = 1.4 \times 10^{-7}$$

$$s = 3.7 \times 10^{-4} \text{ mol/liter} \qquad \square$$

Many precipitates form soluble complexes with the ion of the precipitating agent itself. In such cases, the solubility first decreases because of the common-ion effect, passes through a minimum, and then increases as complex formation becomes appreciable. Silver chloride forms complexes with both silver and chloride ions, such as

$$AgCl + Cl^- \;\rightleftharpoons\; AgCl_2^-$$

$$AgCl_2^- + Cl^- \;\rightleftharpoons\; AgCl_3^{2-}$$

and

$$AgCl + Ag^+ \;\rightleftharpoons\; Ag_2Cl^+$$

In addition, there are a certain number of undissociated AgCl molecules in solution. Figure 9.2 shows the solubility of AgCl in NaCl and AgNO$_3$ solutions. It is interesting to note that AgCl is actually more soluble in 0.1 M AgNO$_3$ and in 1 M NaCl than it is in water. It is because of such effects that only a reasonable excess (usually about 10%) of precipitating agent is used in quantitative precipitations.

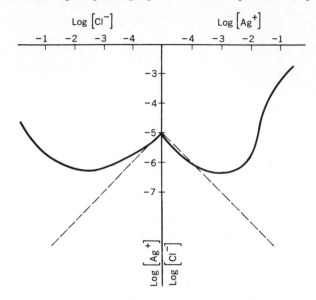

Figure 9.2 Solubility of AgCl in solutions of NaCl and AgNO$_3$. (From H. F. Walton, Principles and Methods of Chemical Analysis, 2nd ed., Prentice-Hall Englewood Cliffs, N.J. Used by permission of the author and publisher.)

KEY TERMS

Activity effect. An increase in the solubility of a compound in the presence of a neutral salt—one which contains no ions which react chemically with the precipitate.

Adsorption indicator. An organic compound which is adsorbed on the surface of a precipitate causing a color change.

Common-ion effect. The decrease in solubility of a compound brought about by the presence in the solution of an ion common to the precipitate.

Complex effect. An increase in solubility of a compound in the presence of a substance which forms soluble complexes with the cation of the salt.

Effective solubility product constant. The solubility product expression of a salt of a weak acid

which includes the effect of pH; its value is dependent on the pH of the solution.

Fajans method. A precipitation titration involving silver salts and using an adsorption indicator to detect the end point.

Mohr method. A precipitation titration using silver ion as the titrant and chromate ion as the indicator.

pH effect. The term used to describe the effect of hydrogen ion concentration on the solubility of the salt of a weak acid.

Volhard method. A precipitation titration in which AgSCN is precipitated with KSCN and Fe(III) indicator is used to detect excess thiocyanate.

QUESTIONS

1. *Solubility*. Explain why AgCl is more soluble in 1 M NaCl and in 0.10 M AgNO$_3$ than it is in water.

2. *Adsorption indicators*. Explain clearly how

an adsorption indicator works. What is the function of dextrin? Why must the pH be controlled?

3. *Mohr method.* By mistake a chloride determination by the Mohr method is carried out in a solution of pH 2.0. Will the results be high or low or will the error not affect the result?

Multiple-choice. In the following multiple-choice questions, select the *one best* answer.

4. The salt whose relative solubility is *most* increased by increasing the ionic strength of the solution is (a) AgI; (b) CuCl; (c) CaSO$_4$; (d) AlPO$_4$.

5. An example of a back-titration is (a) the Volhard determination of Ag$^+$; (b) the Volhard determination of Cl$^-$; (c) the Mohr determination of Cl$^-$; (d) the determination of Cl$^-$ using an adsorption indicator.

6. The salt with the *smallest* molar solubility is (a) AgBr; (b) Zn(OH)$_2$; (c) Ag$_2$C$_2$O$_4$; (d) Ag$_3$AsO$_4$.

7. Which of the following salts of lead should show the greatest percentage increase in solubility for a tenfold increase in [H$^+$]? (a) PbF$_2$; (b) PbCl$_2$; (c) PbBr$_2$; (d) PbI$_2$.

8. In which solution is Ag$_2$CrO$_4$ least soluble? (a) 0.001 M K$_2$CrO$_4$; (b) 0.01 M K$_2$CrO$_4$; (c) 0.001 M AgNO$_3$.

9. Which compound is more soluble at pH 2 than at pH 3? (a) BaF$_2$; (b) CdCO$_3$; (c) Fe(OH)$_2$; (d) all of the above; (e) none of the above.

10. 50 mL of 0.10 M NaCl is titrated with 0.10 M AgNO$_3$. In a second titration 50 mL of 0.10 M NaBr is titrated with 0.10 M AgNO$_3$. Which of the following statements is false? (a) At the start of the titrations, pCl = pBr. (b) When 25 mL of titrant is added in each titration, pCl = pBr. (c) At the EPts, pCl > pBr. (d) When 60 mL of titrant is added in each titration, pCl < pBr.

11. The K_{sp} for Ag$_2$S is 4×10^{-48}; the K_{sp} for NiS is 1×10^{-25}. If 1 mmol of Ni^{2+} is added to 100 mL of a solution containing solid Ag$_2$S in equilibrium with Ag$^+$ and S^{2-} ions, what will happen? (a) NiS will precipitate. (b) More Ag$_2$S will dissolve. (c) More Ag$_2$S will precipitate. (d) Both (a) and (b). (e) Both (a) and (c).

12. At the equivalence point in the titration of CrO$_4^{2-}$ with Ag$^+$,

$$2Ag^+ + CrO_4^{2-} \rightleftharpoons Ag_2CrO_4(s)$$

(a) pAg + 2pCrO$_4$ = pK_{sp}; (b) 3pAg + 2 log 2 = pK_{sp}; (c) 3pCrO$_4$ = pK_{sp} + 2 log 2; (d) all of the above.

PROBLEMS

1. *Solubility product constant.* Calculate the solubility product constants of the following compounds from the given solubilities: (a) AgBr, 0.0118 mg/100 mL; (b) Cu(OH)$_2$, 3.59×10^{-5} g/liter; (c) Ag$_2$CrO$_4$, 0.0263 mg/mL.

2. *Solubility.* From the solubility product constants listed in Table 3, Appendix I, calculate the following solubilities, neglecting such effects as hydrolysis; (a) AgI in mg/liter; (b) Ag$_2$C$_2$O$_4$ in g/100 mL; (c) Pb(IO$_3$)$_2$ in mg/100 mL.

3. *Common-ion effect.* Calculate the molar solubilities of the following, neglecting such effects as hydrolysis: (a) BaSO$_4$ in 0.015 M K$_2$SO$_4$; (b) MgF$_2$ in 0.060 M NaF; (c) Ag$_2$CrO$_4$ in 0.020 M K$_2$CrO$_4$.

4. *Effect of pH.* Calculate the molar solubilites of the following: (a) Fe(OH)$_2$ at pH 11.30; (b) CaF$_2$ in HCl, pH 1.60; (c) Ag$_2$C$_2$O$_4$ in HNO$_3$, pH 2.70.

5. *Effect of pH.* Calculate the molar solubilities of the following: (a) CuS at pH 2.30, solution saturated with H$_2$S; (b) HgS at pH

0.70, solution saturated with H_2S;
(c) $Mg(OH)_2$ at pH 12.70.

6. *Activity effect.* Calculate the molar solubilities of the following salts, neglecting such effects as hydrolysis. (Use activity coefficients from Table 5.3, page 99.) (a) AgCl in 0.0050 M KNO_3; (b) Ag_2CrO_4 in 0.050 M KNO_3; (c) $CaCO_3$ in 0.10 M $NaClO_4$.

7. *Effect of pH.* Calculate the minimum hydrogen ion concentration required to dissolve 0.0020 mol of MgF_2 in 1.0 liter of solution.

8. *Complex effect.* Calculate the molar solubilities of the following: (a) AgCl in 0.20 M NH_3; (b) AgBr in 2.0 M NH_3; (c) AgI in 8.0 M NH_3. The ammonia concentrations are of the uncomplexed NH_3 molecules at equilibrium. See Problem 29, Chapter 5, and compare the systematic equilibrium method of calculation with that using β_2.

9. *Complex effect.* Calculate the minimum concentration of uncomplexed NH_3 molecules required to dissolve 0.020 mol of AgBr in 5.0 liters of solution.

10. *Solubility.* 400 mL of a 0.20 M solution of $Al_2(SO_4)_3$ is added to 400 mL of a 0.40 M solution of $BaCl_2$. Calculate the following: (a) the molar solubility of $BaSO_4$ in the solution; the molarities of Ba^{2+}, SO_4^{2-}, Al^{3+}, and Cl^-. Assume the volumes are additive.

11. *Hydrolysis.* Calculate the molar solubilities of the following, taking into account the hydrolysis of the anion: (a) $CaCO_3$; (b) HgS; (c) Cu_2S, $K_{sp} = 2 \times 10^{-47}$; (d) $MgCO_3$.

12. *Calculation of pX.* Calculate the following; (a) pBr of 0.010 M $MgBr_2$; (b) pCl of 0.020 M NaCl; (c) pCl of 3.0×10^{-4} M $AlCl_3$.

13. *Mixtures.* Calculate the pCl and pAg of the solutions made by mixing (a) 50 mL of 0.080 M $AgNO_3$ + 50 mL of 0.10 M NaCl; (b) 40 mL of 0.20 M $AgNO_3$ + 60 mL of 0.080 M NaCl. Assume the volumes are additive.

14. *Mixtures.* 50 mL of 0.12 M $CaCl_2$ is mixed with 50 mL of 0.10 M $AgNO_3$. Calculate the molarity of Ca^{2+}, Cl^-, Ag^+, and NO_3^- in the resulting solution at equilibrium. Assume the volumes are additive.

15. *Titration curve.* 50 mL of 0.100 M NaI is ti-

trated with 0.100 M $AgNO_3$. Calculate the pI when the following volumes of titrant have been added: (a) 0.00; (b) 25.0; (c) 49.9; (d) 50.0; (e) 50.1; (f) 60.0. Plot the titration curve.

16. *Titration curve.* 50 mL of 0.0500 M K_2CrO_4 is titrated with 0.100 M $AgNO_3$. Calculate the $pCrO_4$ when the following volumes of titrant have been added: (a) 0.00; (b) 25.0; (c) 49.9; (d) 50.0; (e) 50.1; (f) 60.0. Plot the titration curve.

17. *Approximations.* A 50-mL sample of a 0.050 M solution of the salt NaX is titrated with 0.050 M $AgNO_3$, forming the precipitate AgX. Calculate the pX after the addition of 49.9 mL of titrant in two ways: (a) neglecting the $[Ag^+]$ produced by the solubility of AgX, and (b) not neglecting this $[Ag^+]$. Use the following values of K_{sp} for AgX: (a) 1×10^{-10}; (b) 1×10^{-8}; (c) 1×10^{-6}.

18. *Feasibility of titration.* 50 mL of 0.10 M NaX is titrated with 0.10 M $AgNO_3$. Calculate the value of the equilibrium constant for the reaction

$$Ag^+ + X^- \rightleftharpoons AgX(s)$$

so that when 49.9 mL of titrant is added, the reaction is complete and the value of pX changes by 2.00 units for the addition of 0.20 mL more titrant. Also calculate the K_{sp} of AgX.

19. *Feasibility of titration.* 50 mL of 0.10 M Na_2Y is titrated with 0.20 M $AgNO_3$. Calculate the equilibrium constant of the reaction

$$2Ag^+ + Y^{2-} \rightleftharpoons Ag_2Y(s)$$

so that when 49.9 mL of titrant is added, the reaction is complete and the value of pY changes by 2.00 units for the addition of 0.20 mL more titrant. Also calculate the K_{sp} for Ag_2Y.

20. *Volhard method.* Calculate the equilibrium constants for these reactions:
(a) $AgCl(s) + SCN^-(aq) \rightleftharpoons$
$$AgSCN(s) + Cl^-(aq)$$
(b) $AgBr(s) + SCN^-(aq) \rightleftharpoons$
$$AgSCN(s) + Br^-(aq)$$
(c) $AgI(s) + SCN^-(aq) \rightleftharpoons$
$$AgSCN(s) + I^-(aq)$$

Why does this reaction cause trouble in the direct Volhard method for chloride but not for bromide or iodide?

21. *Fraction precipitated.* (a) A base is added to a 0.050 M solution of $MnCl_2$, raising the pH gradually. Calculate the pH of the solution when 50, 90, 99.9, and 99.99% of the manganese has precipitated. (b) Repeat the calculation for a 0.050 M solution of Pb^{2+}.

22. *Solubility.* To 60 mL of 0.10 M NaCl is added 40 mL of 0.16 M $AgNO_3$. (a) Calculate the number of milligrams of Cl^- not precipitated. (b) The precipitate is washed with 100 mL of water at room temperature. Assuming solubility equilibrium is reached, how many milligrams of AgCl dissolve in the wash water?

23. *Sulfide precipitation.* (a) Calculate the pH required to just prevent the precipitation of CdS from a solution which is 0.050 M in Cd^{2+} and saturated with H_2S ($[H_2S] = 0.10$ M); (b) What should the pH be if it is desired to lower the concentration of Cd^{2+} to 10^{-6} M by precipitating CdS?

24. *Sulfide precipitation.* 100 mL of a solution which is 0.10 M in Ni^{2+} and 0.10 M in H_3O^+ is saturated with H_2S ($[H_2S] = 0.10$ M). Calculate the milligrams of Ni^{2+} left in solution, taking into account the fact that the precipitation of NiS produces hydrogen ions.

25. *Separations.* It is desired to separate Pb^{2+} and Mn^{2+} by precipitation with H_2S while controlling the pH. After starting with a strongly acidic solution which is 0.010 M in each cation, the pH is gradually raised by the addition of base. (Assume that $[H_2S]$ remains 0.10 M throughout.) Make the following calculations: (a) At what pH does PbS begin to precipitate? (b) At what pH does $[Pb^{2+}] = 1 \times 10^{-6}$ M? (c) At what pH does MnS begin to precipitate? (d) At what pH does $[Mn^{2+}] = 1 \times 10^{-6}$ M? If the separation is considered feasible when the concentration of one ion can be reduced to 1×10^{-6} M before the other starts to precipitate, is this separation feasible?

26. *Separations.* Repeat Problem 25 for 0.010 M solutions of (a) Pb^{2+} and Zn^{2+}; (b) Zn^{2+} and Mn^{2+}.

27. *Separations.* A solution which is 0.010 M in two hypothetical cations, M^{2+} and N^{2+}, and 0.10 M in H_3O^+ is saturated with H_2S. (a) What is the minimum ratio of the K_{sp} for MS to the K_{sp} for NS so that the concentration of N^{2+} can be reduced to 1×10^{-6} M without precipitating MS? (b) Repeat part (a) where the sulfides are M_2S and N_2S.

28. *Separations.* A solution of pH 1.0 is 0.02 M in each of the following ions: Al^{3+}, Fe^{2+}, and Mg^{2+}. The pH is gradually raised by the addition of base. Calculate the pH values at which each metal hydroxide begins to precipitate and the pH at which the cation concentration reaches 1×10^{-6} M. Is the separation of these three cations by hydroxide precipitation theoretically feasible?

29. *Precipitation of carbonates.* A solution which is 0.060 M in Sr^{2+} and 0.080 M in Ca^{2+} is treated with solid Na_2CO_3, first precipitating $SrCO_3$. What percentage of the Sr^{2+} is precipitated when $CaCO_3$ begins to precipitate?

30. *Precipitation of lead.* 50 mL of a solution which is 0.10 M in Pb^{2+} is mixed with 50 mL of 0.50 M HCl, precipitating $PbCl_2$. (a) Calculate the Pb^{2+} concentration in the resulting solution, assuming precipitation is complete. (b) Calculate the $[H_3O^+]$ in the solution. (c) With the solution now saturated with H_2S (0.10 M), show whether or not PbS will precipitate.

31. *Complexes.* The successive stepwise formation constants for Cd^{2+} and NH_3 are $k_1 = 550$, $k_2 = 162$, $k_3 = 23.5$, and $k_4 = 13.5$. Calculate the molar solubility of CdS in a 0.050 M solution of NH_3 (uncomplexed), calculating first α_2 for the anion, then β_4 for the cation, and then the solubility.

32. *Complexes.* 1.0 mmol of AgCl is dissolved in 750 mL of ammonia, the final concentration of uncomplexed NH_3 molecules being 0.40 M. Calculate the concentration of uncomplexed Ag^+ in the solution.

33. *Buffer solution.* Lead hydroxide is precipitated from an NH_3–NH_4Cl buffer. What must the concentration of NH_3 in the buffer be if all but 0.1 mg of lead is precipitated from 100 mL of a solution which is 0.10 M in NH_4Cl?

34. *Buffer solution.* A hypothetical metal M^{2+} forms an insoluble sulfide MS, $K_{sp} = 1 \times$

10^{-20}. It is desired to precipitate MS and reduce the concentration of M^{2+} to $1 \times 10^{-6}\,M$. Since hydrogen ions are generated by the reaction

$$M^{2+} + H_2S \longrightarrow MS(s) + 2H^+$$

the solution is to be buffered with HOAc and NaOAc. If 100 mL of $0.010\,M$ M^{2+} containing HOAc and NaOAc is saturated with H_2S ($0.10\,M$), calculate the following: (a) the pH when $[M^{2+}] = 1 \times 10^{-6}\,M$; (b) the $[OAc^-]$ so that the pH will not drop below the value in part (a), the original [HOAc] being $0.10\,M$; (c) the pH of the buffer before MS is precipitated.

10

Oxidation-Reduction Equilibria

Chemical reactions which involve oxidation-reduction are more widely used in titrimetric analysis than acid-base, complex-formation, or precipitation reactions. The ions of many elements exist in different oxidation states, resulting in the possibility of a very large number of oxidation-reduction (redox) reactions. Many of these reactions satisfy the requirements for use in titrimetric analysis, and applications are numerous.

Let us recall that *oxidation* is defined as the *loss* of one or more electrons by an atom, molecule, or ion, while *reduction* is *electron gain*. There are no free electrons in ordinary chemical systems, and electron loss by one chemical species is always accompanied by electron gain on the part of another. The term *electron-transfer reaction* is sometimes used for redox reactions.

In this chapter we shall learn how to evaluate equilibrium constants of redox reactions from oxidation potentials (Section 5.4d). We shall calculate titration curves, see how indicators are selected, and learn how to determine the feasibility of a titration. We will find that the electroanalytical techniques discussed in later chapters (e.g., potentiometry, electrolysis, and polarography) require an understanding of the principles discussed here. Finally, students whose futures lie in the biomedical area will see a direct relation between the substance of this chapter and the important topic of biological oxidations.

GALVANIC CELLS

Redox equilibria are conveniently treated in terms of the electromotive force of galvanic cells. A *galvanic* cell is one in which chemical reaction occurs spontaneously, releasing electrical energy which is available for performing useful work. The electromotive force (emf) is measured in units of volts and is referred to as the *voltage* or *potential* of the cell. One volt is the emf required to impart one joule (J) of energy to an electrical charge of one coulomb (C):

$$1 \text{ V} = 1 \text{ J/C}$$

It should be recalled (page 56) that the charge on an electron is 1.60×10^{-19} C and that 1 C/s is one *ampere* (A), the unit of current. Also, the charge on 1 mol of electrons, called the *faraday,* is 96,500 C.

In this section we shall see how the emf of galvanic cells can be used to evaluate the equilibrium constant (and hence the degree of completion) of a redox reaction, and how to select the proper indicator for a titration using emf data.

Box 10.1	**Acid-Base and Redox Reactions**

It is interesting to compare oxidation-reduction with acid-base reactions for similarities and differences between electron transfer and proton transfer. We might think of Fe^{2+} and Fe^{3+} and of Ce^{3+} and Ce^{4+}, for example, as conjugate pairs by analogy with conjugate Brønsted acids and bases (page 129):

$$\underset{\text{Reductant}_1}{Fe^{2+}} + \underset{\text{Oxidant}_2}{Ce^{4+}} \longrightarrow \underset{\text{Oxidant}_1}{Fe^{3+}} + \underset{\text{Reductant}_2}{Ce^{3+}}$$

$$\underset{\text{Acid}_1}{HOAc} + \underset{\text{Base}_2}{NH_3} \longrightarrow \underset{\text{Base}_1}{OAc^-} + \underset{\text{Acid}_2}{NH_4^+}$$

There are important differences, though, between electron transfer and proton transfer. For instance, electrons can travel through wires, while protons cannot. Thus for proton transfer, acid$_1$ and base$_2$ must have a direct encounter in the same solution. Electron transfer can occur directly, but donor and acceptor can be in separate solutions if we wish, electrons then travelling from one to the other through a conductor.

Second, redox reactions are often much slower than acid-base reactions. The latter are extremely fast, so fast, in fact, that they were called "instantaneous" for many years; only in the modern era have techniques been developed for measuring their rates. Rate constants for simple acid-base reactions in aqueous solution are on the order of 10^{10} to 10^{11} $M^{-1}s^{-1}$; the rates are probably diffusion-controlled, which is to say that it takes longer for proton donor and acceptor to find each other than to complete the exchange once the encounter occurs. On the other hand, analytical procedures based upon redox reactions may require elevated temperatures, catalysts, or reagent excess. Slow rates reflect greater complexity; frequently electron transfer is only one part of a multistep process which may involve forming or breaking covalent bonds or various sorts of rearrangements. Compare, for example, $B + H^+ \rightarrow BH^+$ with even these "simple" reactions:

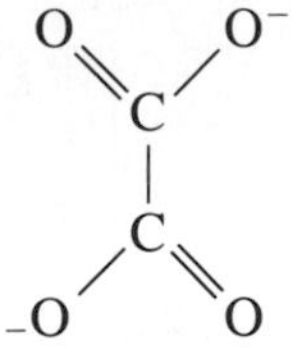

$$\underset{\substack{\text{Oxalate}\\\text{ion}}}{\overset{\text{O}\quad\text{O}^-}{\underset{\text{O}^-\quad\text{O}}{\overset{\text{C}}{\underset{\text{C}}{\Big|}}}} \longrightarrow \underset{\substack{\text{Carbon}\\\text{dioxide}}}{2\text{O}=\text{C}=\text{O}} + 2e$$

$$\underset{\substack{\text{Chromic}\\\text{ion}}}{2Cr^{3+} + 7H_2O} \longrightarrow \underset{\substack{\text{Dichromate}\\\text{ion}}}{Cr_2O_7^{2-} + 14H^+ + 6e}$$

In $Cr_2O_7^{2-}$, each Cr is bonded to four oxygens, one of which forms a bridge between the two chromiums.

Third (and not unrelated to the above), we note the difference between the acid-base and redox properties of our common solvent, water. Protons are rapidly transferred to H_2O to form H_3O^+, and they are just as easily passed along to some other base to regenerate H_2O. Likewise, removal of a proton from water to form OH^- is readily reversed when protons are supplied by another acid. Further, all the species H_3O^+, H_2O, and OH^- are highly soluble. In contrast, the addition of electrons to water generates hydrogen, a slightly soluble gas:

$$2H_2O + 2e \longrightarrow 2OH^- + H_2$$

This product escapes from the solution, and, moreover, even if it remained, many of its reactions are so slow under reasonable conditions that it would seldom be a suitable titrant. Similar considerations arise if water is oxidized:

$$2H_2O \longrightarrow O_2 + 4H^+ + 4e$$

Thus, while strong acids and bases that undergo proton exchanges with water are good titrants, reagents that oxidize or reduce water are usually avoided; the strongest oxidants and reductants are impractical titrants. We do not encounter, then, a levelling effect such as we saw in acid-base reactions. To be sure, reducing agents stronger than H_2 would be levelled to the power of that reagent, and oxidants stronger than O_2 would generate the latter, but we avoid such reagents in most cases. Thus each reagent typically exerts its own characteristic reactivity without mediation by the solvent. It should be noted, however, that some reagents that are strong enough to oxidize or reduce water in fact do so only very slowly; hence their aqueous solutions are sometimes stable enough to serve as titrants if we don't keep them too long.

10.1a. Single-Electrode Potentials

Suppose we place a strip of a metal, say, zinc, in contact with a solution containing zinc ions, as suggested in Fig. 10.1. We may think of the metal itself as comprising zinc ions and electrons. In general, the activities of zinc ions in the metal and solution phases will be different, providing a driving force for the loss or gain of Zn^{2+} by the metal strip. Suppose atoms of zinc enter the solution as indicated in the figure. Electrons are left on the metal surface, and the solution acquires a positive charge. Zinc ions also tend to leave the solution, depositing as

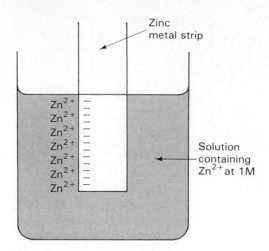

Figure 10.1 Schematic depiction of a single electrode potential.

atoms on the metal strip, and eventually equilibrium is established. At this point a charge separation has been developed, and an electrostatic force, or difference in potential (electrical double layer) has been set up at the interface of the solid and liquid phases. (The term *potential* is used as defined in physics, in terms of the work required to move a charge from an infinite distance or from some other reference point to the site in question.)

The equilibrium process we have described may be written

$$Zn^{2+} + 2e \rightleftharpoons Zn$$

and in the discussion above we assumed that it proceeded somewhat from right to left. The potential established at the metal-solution interface is a measure of the tendency of the Zn–Zn^{2+} couple to lose electrons and is referred to as a *single-electrode potential*. If the metal had been a less active one than zinc, say, copper, the process

$$Cu^{2+} + 2e \rightleftharpoons Cu$$

would not have proceeded so far to the left as in the zinc case. The single-electrode potential of the Cu–Cu^{2+} couple would be different.

A single-electrode potential between a metal and its ions has never been measured absolutely. In order to determine the potential difference between two points, we normally employ some kind of voltmeter, and this requires wires connecting the two points with the meter. But if we attempt to measure the potential difference between the strip of zinc and the solution by attaching one of the meter leads to the metal and inserting the other one into the solution, we introduce by that very action a second electrode. A potential difference will be established between the meter lead and the solution, and we shall be unable to correct for its contribution to the measured voltage. The single-electrode potential between the zinc metal and the solution, then, cannot be measured.[1]

---

[1] It used to be stated that single-electrode potentials were unmeasurable in principle, but a hypothetical experiment has been proposed which theoretically would enable one to calculate a single-electrode potential from measurements of the radiation emitted by the oscillating charges if the electrode were caused to vibrate. See I. Oppenheim, *J. Phys. Chem.*, **68**, 2959 (1964).

10.1b. *Galvanic Cells*

To obtain a useful system upon which meaningful measurements can be performed, we must combine two single electrodes to form a cell. An example is shown in Fig. 10.2, where the two half-cells mentioned in the preceding section have been paired. Wires from the metal strips lead to an ammeter (or voltmeter). The circuit is completed by connecting the two solutions with a so-called salt bridge. This may be an inverted U-tube containing a solution of a salt, such as potassium chloride, with an agar plug at each end.

Now there is more than a hypothetical tendency for electron transfer to occur. The excess electrons remaining on the zinc electrode when Zn^{2+} enters the solution find a place to go. They can travel through the external wiring to the copper electrode where they can be consumed in the reduction of Cu^{2+}. The flow of electrons through the wire is, of course, an electric current that can be measured with the ammeter. The two solutions must remain electrically neutral, and here the salt bridge comes into play, completing the electrical circuit inside the cell. Chloride ions, Cl^-, diffuse into the electrode on the left (Zn), and K^+ into the one on the right (Cu). A current flows through the whole system, carried by electrons in the metallic conductors and by ionic migration through the solution.

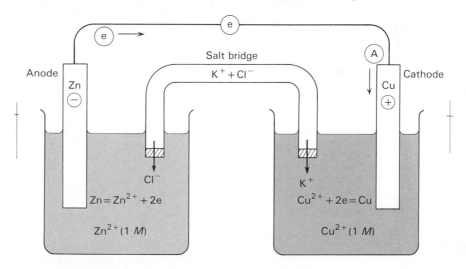

Figure 10.2 A galvanic cell.

This two-electrode system is an electrochemical cell and is an example of a *galvanic* cell. As previously stated, a galvanic cell is one in which some of the energy released spontaneously in a chemical reaction is converted into electrical energy and thereby made available for performing work. In contrast, an electrochemical cell in which the chemical reaction is forced to proceed in the nonspontaneous direction is called an *electrolytic* cell. We shall consider this type of cell in a later chapter.

Normally in analytical chemistry we are not interested in actually doing electrical work using galvanic cells. Rather, we are interested in the maximum voltage, which can be observed only if we do not allow the cell to run down while we are measuring the voltage. Thus, in practice, the ammeter in Fig. 10.2 will be replaced with a voltage-measuring device of very high electrical resistance so that the current drain, and the extent to which the redox reaction proceeds in the cell, are negligible.

10.1c. Schematic Representation of Galvanic Cells

Instead of pictures such as Fig. 10.2, schematic representations of galvanic cells are normally employed. For example, the cell in Fig. 10.2 is written as follows:

$$\underset{E_l}{Zn}\left|\underset{}{Zn^{2+}(1\ M)}\right|\ \underset{E_{j_1}\ E_{j_2}}{}\ \left|\underset{}{Cu^{2+}(1\ M)}\right|\underset{E_r}{Cu}$$

A vertical line (a "slant" when made with a typewriter) indicates a phase boundary across which we suppose a potential exists. A double line indicates a salt bridge used to make electrical contact between the two solutions. Presumably there is a *liquid junction potential* (Section 10.1d) at each end of the salt bridge (E_{j_1} and E_{j_2}), but these are minimized by using KCl as the electrolyte. When we place a voltage-measuring device between the zinc and copper strips, we can measure the potential of the galvanic cell, the difference between the the single-electrode potentials of the two metal electrodes. The subtraction is normally made left from right:

$$E_{cell} = E_r - E_l$$

We shall see the rationale for this in Section 10.1e.

The potential of each single electrode, and hence of the cell, depends upon the concentrations (activities) of the zinc and copper ions. It is customary to define *standard* electrode potentials which are measured with the reacting species in standard states (Sections 5.3d and 10.1f). Here, since both electrolytes are in their standard states (1 M), the expression would be written

$$E^\circ_{cell} = E^\circ_r - E^\circ_l$$

The superscript zero is used to indicate that the reactants are in their standard states.

10.1d. Liquid Junction Potentials

Suppose we could prepare a quiet interface between pure water and a solution of hydrochloric acid, perhaps by very gently sliding out an ultrathin partition separating the two liquids. Immediately, hydrogen and chloride ions would move from the HCl solution into the water, and if we waited long enough, a uniform HCl concentration would prevail throughout. Stirring, of course, would speed the mixing, but we are not going to do that. Now, although both cations and anions migrate across the boundary, a potential will develop at the interface because hydrogen ions move much more rapidly than do chloride ions. Thus there is a slight tendency for a charge separation, as depicted in Fig. 10.3, with the water side of the interface positive and an excess of negative charge on the HCl side. This potential, once established, will itself act upon the migrating ions, decelerating the

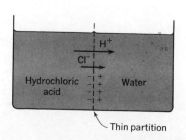

Figure 10.3 Development of a liquid junction potential.

faster ones and speeding up the slower ones. Thus we may visualize a "steady-state" situation, at least for a certain length of time, involving a constant charge separation.

A liquid junction potential may be expected to develop at any interface between two ionic solutions of different compositions as a result of the tendencies of different ions to diffuse at different rates across the boundary. It is possible to calculate the liquid junction potential for a simple case, but in complex electrolyte mixtures the value is generally unknown. Junction potentials may vary over a considerable range and can be quite appreciable in some cases, but typically they are on the order of a few millivolts (thousandths of a volt). The junction potential is minimized by using a salt bridge in which the electrolyte is quite concentrated, and the cation and the anion have comparable mobilities. Potassium and chloride ions represent such a case, and salt bridges of saturated aqueous potassium chloride are widely used.

It is necessary to prevent the KCl solution from running out of the salt bridge, and for this purpose the ends are plugged with a porous material that soaks up solution and permits migration of ions but acts as a barrier to the heavy flow of a liquid stream. Small wads of cotton were used at one time; currently, thin disks of porous glass or small plugs of agar gel are often employed.

In our emf calculations, we shall neglect liquid junction potentials. This is done primarily because we usually do not know what they are, but in fact the errors we incur in this manner are negligible for our purpose here. (We shall see in Chapter 12 some situations where junction potentials must be considered.)

10.1e. *Conventions for Galvanic Cells*

Suppose the galvanic cell represented in Fig. 10.2 is actually constructed in the laboratory. There are several experimental facts that we can determine. First, we can employ a voltmeter and measure the emf of the cell. In making the measurement we will find which electrode is positive and which is negative, i.e., the *polarities* of the electrodes. This information will tell us in which direction electrons flow when the cell discharges spontaneously. In other words, we can now write the *cell reaction,* the redox reaction which occurs as we draw current from the cell.

In the present example our voltmeter would give us a reading of 1.10 V, and we would find that the zinc electrode is negative, the copper positive. If we allow the cell to discharge, electrons will flow through the meter *from the negative to the positive pole,* and hence the spontaneous reaction will be

$$Zn + Cu^{2+} \longrightarrow Zn^{2+} + Cu$$

Note that in this reaction zinc is oxidized, and copper reduced. The electrode at which oxidation occurs is called the *anode;* the one at which reduction occurs is called the *cathode*. In a galvanic cell the anode is negative, the cathode positive.

The above experimental facts are invariant; they do not depend upon the manner in which the schematic diagram of the cell is written on paper. However, it is customary to formulate the cell, as we have done above, so that E°_{cell} is a positive quantity. Since

$$\Delta G^\circ = -nFE^\circ_{cell}$$

if E°_{cell} is positive, ΔG°_{cell} will be negative, and the cell reaction will be spontaneous from left to right by thermodynamic conventions. This result will always be obtained if the cell is formulated with the *anode on the left and the cathode on the right*. Electrons will then flow from left to right as the cell discharges, E°_{cell} will be positive, and the cell reaction will proceed to the right.

Suppose, however, we encounter a galvanic cell in which the direction of spontaneous reaction is not obvious. This is often the case when the concentrations of the reacting species are varied (Section 10.2). There is a simple procedure we can follow which will enable us to easily determine the direction of the cell reaction and the polarities of the electrodes. Let us formulate a cell made up of silver and cadmium:

$$Ag|\,Ag^+(1\ M)\,||\,Cd^{2+}(1\ M)|Cd$$

The steps are as follows:

1. Write the half-reaction for the right-hand electrode with electrons on the left. Record the standard potential (Table A-4, Appendix I).

$$Cd^{2+} + 2e \;\rightleftharpoons\; Cd \qquad E^\circ = -0.40\ V$$

2. Write the half-reaction and standard potential for the left-hand electrode in the same manner:

$$Ag^+ + e \;\rightleftharpoons\; Ag \qquad E^\circ = +0.80\ V$$

3. If necessary, multiply one or both equations by proper integers so that the number of electrons is the same in both equations. Here we need to multiply the silver half-reaction by 2:

$$2Ag^+ + 2e \;\rightleftharpoons\; 2Ag \qquad E^\circ = +0.80\ V$$

Do not double the standard potential. This is an experimental value which does not depend upon how we balance the half-reaction.

4. Subtract the left-hand half-reaction from the right. Also subtract the standard potentials:

$$
\begin{array}{ll}
Cd^{2+} + 2e \;\rightleftharpoons\; Cd & E^\circ = -0.40\ V \\
\underline{2Ag^+ + 2e \;\rightleftharpoons\; 2Ag} & \underline{E^\circ = +0.80\ V} \\
Cd^{2+} + 2Ag \;\rightleftharpoons\; Cd + 2Ag^+ & E^\circ_{cell} = -1.20\ V
\end{array}
$$

5. The sign of E°_{cell} is the same as the polarity of the *right-hand* electrode. Here cadmium is negative, and silver positive.
6. The sign of E°_{cell} also tells the direction of spontaneous reaction. If it is positive (ΔG° negative), the direction is to the right. If it is negative (ΔG° positive), the direction is to the left. Here the direction is leftward; as the cell discharges, silver ion is reduced to metallic silver and cadmium is oxidized to Cd^{2+}.

Suppose the cell diagram were "turned around" and written

$$Cd|\,Cd^{2+}(1\ M)\,||\,Ag^+(1\ M)|Ag$$

If the above procedure is followed, we would write

$$
\begin{array}{ll}
2Ag^+ + 2e \;\rightleftharpoons\; 2Ag & E^\circ = +0.80\ V \\
\underline{Cd^{2+} + 2e \;\rightleftharpoons\; Cd} & \underline{E^\circ = -0.40\ V} \\
Cd + 2Ag^+ \;\rightleftharpoons\; Cd^{2+} + 2Ag & E^\circ_{cell} = +1.20\ V
\end{array}
$$

Note that the cell reaction is also turned around and that the sign of E°_{cell} is now

+1.20 V. This means that the cell reaction is spontaneous to the right, silver being reduced to metallic silver and cadmium oxidized to Cd^{2+}. The electrode polarities are the same as those found above: Cadmium is negative, silver positive.

Box 10.2   Confused Over Signs?

"Whether one calls a quantity 'plus' or 'minus' would seem to be a rather trivial matter, and yet conventions for the signs of the emf of cells and half-reactions, and for the potentials of actual electrodes, have been the cause of much debate."[†] This is a quote from a prominent electroanalytical chemist in 1958. As Lingane pointed out, the confusion arose because chemists failed to recognize that the potential of an electrode and the emf of a half-reaction are distinctly different concepts.

The sign of the potential of an actual physical electrode, say, of metallic zinc in contact with a solution of zinc ions, with respect to the sign of the potential of another electrode of a cell, is *fixed* or *invariant*. In a cell made up of a Zn–Zn^{2+}(1 M) electrode and a standard hydrogen electrode, the zinc electrode is negative with respect to hydrogen regardless of whether we write the electrode on the left or right in a schematic diagram. On the other hand the sign of the potential or emf of the half-reaction which occurs at the zinc electrode depends on the direction in which the reaction is written

$$Zn^{2+} + 2e \rightleftharpoons Zn \qquad E° = -0.76 \text{ V} \qquad (1)$$

$$Zn \rightleftharpoons Zn^{2+} + 2e \qquad E° = +0.76 \text{ V} \qquad (2)$$

The potential of a half-reaction is a *defined* quantity—a thermodynamic concept. The logic of changing its sign (as well as that of ΔG, ΔH, and ΔS) when the direction of the reaction is reversed is obvious. Thermodynamic calculations would be chaotic if a consistent and directional sign convention were not followed.

Further confusion has arisen over the use of the terms *oxidation* and *reduction* potential. In the early part of this century most American physical chemists called the $E°$ of reaction (2) above an *oxidation potential* since the reaction left to right is an oxidation. The authoritative compilation of standard potentials by Latimer[‡] is written in this convention. At that time European chemists commonly used *reduction potentials,* and people spoke of American and European conventions.

In 1953 the International Union of Pure and Applied Chemistry (IUPAC), meeting in Stockholm, recommended that the electrode potential of the zinc electrode be assigned a negative value and that the half-reaction be written as reduction. The name *oxidation potential* is not used. Most texts today use the term *electrode potential* and write the half-reaction as reduction.

[†] J. J. Lingane, *Electroanalytical Chamistry*, 2nd ed., Interscience Publishers, Inc., New York, 1958.

[‡] W. M. Latimer, *Oxidation Potentials*, 2nd ed., Prentice-Hall, Inc., Englewood Cliffs, N.J., 1952.

10.1f. The Standard Hydrogen Electrode

The student will have noticed that after carefully explaining why single-electrode potentials could not be measured, we inserted actual numbers for E_l and E_r in Section 10.1e. We must now explain what we mean by these numbers.

In Section 10.1e it was pointed out that when we measure the potential of the galvanic cell in Fig. 10.2, the quantity we measure is the *difference* in potential between the zinc and copper redox systems, here 1.10 V. We do not know the absolute value of either single-electrode potential; in order to measure the tendency of one redox couple to lose or gain electrons, we must introduce another couple for the purpose of comparison. Since definite potential values are convenient to use, the chemist arbitrarily assigns a value to a certain single electrode. All other single electrodes are then referred to this standard by measuring the emf values of galvanic cells in which one electrode is the standard and the other is the electrode in question.

The redox couple selected as the reference is H_2–H^+:

$$2H^+ + 2e \rightleftharpoons H_2(g)$$

This couple is assigned a potential of exactly zero volts at all temperatures when the pressure of hydrogen gas is 1 atm and the hydrogen ion activity is unity. Physically, the electrode is somewhat as suggested schematically in Fig. 10.4. A platinum surface, rough so as to have a large area, provides an electrical connection to the external circuit and serves as a catalyst for the combination of H atoms formed in the electron-transfer step.

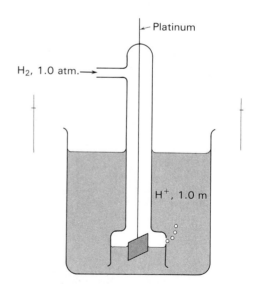

Figure 10.4 The standard hydrogen electrode, $2H^+ + 2e \rightleftharpoons H_2$.

If we set up a cell in which one electrode in Zn–Zn^{2+} (1 *M*) and the other is the standard hydrogen electrode, we find that the potential at 25°C is 0.76 V. We also can determine that the polarity of the zinc electrode is negative, hydrogen positive. If we set up another cell consisting of the Cu–Cu^{2+} (1 *M*) electrode and the standard hydrogen, we find that the potential is 0.34 V, with copper positive and hydrogen negative. We say that the standard potential of the Zn–Zn^{2+} couple is -0.76 V, and that of the Cu–Cu^{2+} couple is 0.34 V referred to the hydrogen electrode. By international agreement the electrode reactions are written as reduction (left to right):

$$Zn^{2+} + 2e \rightleftharpoons Zn \qquad E° = -0.76 \text{ V}$$

$$2H^+ + 2e \rightleftharpoons H_2 \qquad E° = 0.00 \text{ V}$$

$$Cu^{2+} + 2e \rightleftharpoons Cu \qquad E° = +0.34 \text{ V}$$

The voltages are given positive or negative signs in accordance with experimentally determined polarities with respect to hydrogen. Zinc is more negative than hydrogen; this means that the reaction

$$Zn + 2H^+ \longrightarrow Zn^{2+} + H_2$$

will occur spontaneously rightward if both H^+ and Zn^{2+} are at unit activity; zinc loses electrons to hydrogen in this reaction. On the other hand, hydrogen is more negative than copper, the reaction

$$H_2 + Cu^{2+} \longrightarrow 2H^+ + Cu$$

occuring spontaneously rightward if both ions are at unit activity and the pressure of H_2 is 1 atm.

A list of standard potentials of electrodes, all referred to the standard hydrogen electrodes, is given in Table A-4, Appendix I. It should be kept in mind that these values are those in which each reactant is in its standard state; i.e., it is at unit activity. The conventions used to define the activities of pure and impure liquids and solids, gases, and solutes are given in Section 5.3d.

10.1g. Voltage Measurements

As noted in Section 10.1b, we usually desire in analytical chemistry to measure the voltage of a galvanic cell under conditions where the cell actually does no electrical work. That is, we wish to measure the *tendency* of the cell reaction to proceed without allowing it in fact to proceed appreciably. In other words, we do not want the measurement process to lower the voltage we are trying to measure. This generally precludes the use of simple voltmeters, where the galvanic cell itself provides the power to move the galvanometer coil and pointer. A simple device called a *potentiometer* used to be employed to measure the voltages of galvanic cells. Essentially, the cell voltage was opposed by an external, variable voltage provided by a battery, and a balance point was determined in a manner that drew very little current from the test cell. Potentiometers worked very well in many cases but were unsuitable for cells involving very high internal resistances, such as are encountered in pH measurements with glass electrodes (Chapter 12). Electronic voltmeters with very high input resistances were developed after the science of electronics came of age. Not only did these instruments work with cells of high resistance, but they also became so inexpensive and so accurate that they have replaced the potentiometer even for applications where the latter served well. The circuitary of these meters is beyond the scope of this textbook; in principle, the voltage to be measured *controls* but does not *power* the readout device through a circuit that draws currents on the order of 10^{-12} A or even less from the test cell. Electronic voltmeters with solid-state components that provide direct digital readout are now very common.

THE NERNST EQUATION

The potential of a galvanic cell depends upon the activities of the various species which undergo reaction in the cell. The equation which expresses this relationship is called the Nernst equation, after the physical chemist, Nernst, who in 1889 first used the equation to express the relation between the potential of a metal–metal ion electrode and the concentration of the ion in solution.

10.2a. *Derivation of the Nernst Equation*

In a chemical reaction such as

$$aA + bB \rightleftharpoons cC + dD$$

the change in free energy is given by the equation

$$\Delta G = \Delta G^\circ + 2.3RT \log \frac{a_C^c \times a_D^d}{a_A^a \times a_B^b}$$

where ΔG° is the free energy change when all the reactants and products are in their standard states (unit activity), R is the gas constant, 8.314 J/deg-mol, and T is the absolute temperature.

The free-energy change, or work done, by driving Avogadro's number of electrons through a voltage E is $(Ne)E$, where N is Avogadro's number and e is the charge on the electron. The product Ne is 96,500 C, called 1 faraday, or F. Hence

$$\Delta G = -nFE$$

where n is the number of moles of electrons involved in the reaction. If all reactants and products are in their standard states, this becomes

$$\Delta G^\circ = -nFE^\circ$$

Hence

$$-nFE = -nFE^\circ + 2.3RT \log \frac{[C]^c[D]^d}{[A]^a[B]^b}$$

where concentrations are substituted for activities. This can be written as

$$E = E^\circ - \frac{2.3RT}{nF} \log \frac{[C]^c[D]^d}{[A]^a[B]^b}$$

and at 298° K the equation becomes

$$E = E^\circ - \frac{0.059}{n} \log \frac{[C]^c[D]^d}{[A]^a[B]^b}$$

This is the form in which we commonly use the Nernst equation. Note that at equilibrium, $E = 0$, $\Delta G = 0$, and the logarithmic term is the equilibrium constant. Hence

$$\Delta G^\circ = -2.3RT \log K$$

or

$$E^\circ = \frac{0.059}{n} \log K$$

10.2b. Applications of the Nernst Equation

If we know the standard potentials of two redox couples, we can calculate the equilibrium constant for the reaction between the couples using the preceding equation. We can then judge whether the reaction goes sufficiently to completion to be useful in a titrimetric procedure. Before considering titrations, we will illustrate the application of the Nernst equation, as well as other points covered thus far.

EXAMPLE 10.1

Calculation of K using the Nernst equation

A cell is set up as follows:

$$Fe|Fe^{2+}(a = 0.1)\|Cd^{2+}(a = 0.001)|Cd$$

(a) Write the cell reaction. (b) Calculate the voltage of the cell, the polarity of the electrodes, and the direction of spontaneous reaction. (c) Calculate the equilibrium constant of the cell reaction.

(a) The electrode reactions and standard potentials are

$$Cd^{2+} + 2e \rightleftharpoons Cd \qquad E_r^\circ = -0.40 \text{ V}$$

$$Fe^{2+} + 2e \rightleftharpoons Fe \qquad E_l^\circ = -0.44 \text{ V}$$

Substracting, $Fe + Cd^{2+} \rightleftharpoons Fe^{2+} + Cd \qquad E_{cell}^\circ = +0.04 \text{ V}$

(b) The cell potential can be calculated from the single electrode potentials:

$$E_r = -0.40 - \frac{0.059}{2} \log \frac{1}{0.001} = -0.49 \text{ V}$$

$$E_l = -0.44 - \frac{0.059}{2} \log \frac{1}{0.1} = -0.47 \text{ V}$$

Thus

$$E_r - E_l = -0.02 \text{ V}$$

Alternatively, the cell potential can be evaluated from the expression

$$E_{cell} = E_{cell}^\circ - \frac{0.059}{2} \log \frac{a_{Fe}^{2+}}{a_{Cd}^{2+}}$$

$$E_{cell} = +0.04 - \frac{0.059}{2} \log \frac{0.1}{0.001}$$

$$E_{cell} = +0.04 - 0.06 = -0.02 \text{ V}$$

Therefore, the cell reaction, as written above, tends to occur spontaneously from right to left at the given activities. The cadmium electrode is negative, the iron positive. Note that if both ions are at unit activity, $E_{cell}^\circ = +0.04$ V, and the direction is from left to right. The polarities of the electrodes are reversed also.

(c) The equilibrium constant is given by

$$E_{cell}^\circ = \frac{0.059}{n} \log K$$

$$+0.04 = \frac{0.059}{2} \log K$$

$$\log K = 1.36$$

$$K = 23 \qquad \square$$

As a further illustration, consider the following example.

EXAMPLE 10.2

Calculation of K using the Nernst equation

Calculate the potential of the following cell, giving the polarities of the electrodes and the direction of spontaneous reaction. Calculate the equilibrium constant of the cell reaction.

$$Pt, H_2(0.9 \text{ atm})| H^+(0.1 \; M) \| KCl(0.1 \; M), AgCl|Ag$$

The electrode reactions are

$$2AgCl + 2e \rightleftharpoons 2Ag + 2Cl^- \qquad\qquad E_r^\circ = 0.22 \text{ V}$$

$$\underline{2H^+ + 2e \rightleftharpoons H_2 \qquad\qquad\qquad\quad E_l^\circ = 0.00 \text{ V}}$$

$$2AgCl + H_2 \rightleftharpoons 2H^+ + 2Ag + 2Cl^- \qquad E_{cell}^\circ = +0.22 \text{ V}$$

$$E_{cell} = +0.22 - \frac{0.059}{2} \log \frac{(a_{H^+})^2(a_{Ag})^2(a_{Cl^-})^2}{(a_{AgCl})^2(a_{H_2})}$$

In accordance with our conventions regarding activities (page 101)

$a_{AgCl} = 1$, since AgCl is a pure solid

$a_{H_2} = 0.9$, since this is the partial pressure of the gas in atmospheres

$a_{H^+} = 0.1$, since this is a soluble electrolyte

$a_{Ag} = 1$, since silver is a pure solid

$a_{Cl^-} = 0.1$, since this is a soluble electrolyte

Substituting these values and solving gives

$$E_{cell} = +0.34 \text{ V}$$

Hence the silver–silver chloride electrode is positive, and the reaction is spontaneous left to right.

The equilibrium constant is given by

$$0.22 = \frac{0.059}{2} \log K$$

$$\log K = 7.46$$

$$K = 2.9 \times 10^7 \qquad\qquad \square$$

Sometimes one is simply given a reaction and asked to calculate the equilibrium constant from standard potentials. The problem then is to find two half-reactions which, when subtracted, yield the desired equation. Finding the proper two half-reactions may be considered "trial and error," although one should become more adept at it with practice. Consider the following examples.

EXAMPLE 10.3

Calculation of K_w

Calculate K_w, the ion-product constant of water, from data in Table A-4, Appendix I.

We seek two half-reactions which, when subtracted, yield

$$H_2O \rightleftharpoons H^+ + OH^-$$

After searching the table (and perhaps making a false start or two if inexperienced), we come up with:

$$H_2O + e \rightleftharpoons \tfrac{1}{2}H_2 + OH^- \qquad E° = -0.83 \text{ V}$$

$$\underline{H^+ + e \rightleftharpoons \tfrac{1}{2}H_2 \qquad\qquad\qquad E° = 0.00}$$

$$H_2O \rightleftharpoons H^+ + OH^- \qquad E°_{cell} = -0.83 \text{ V}$$

$$E = -0.83 - 0.059 \log [H^+][OH^-]$$

At equilibrium, $E = 0$ and $[H^+][OH^-] = K_w$. Thus

$$\log K_w = \frac{-0.83}{0.059} = -14.0$$

$$K_w = 1 \times 10^{-14} \qquad\qquad\qquad \square$$

EXAMPLE 10.4

Calculation of K_{sp}

Calculate the K_{sp} of $Mn(OH)_2$ from data in Table A-4 Appendix I.

We need two half-reactions which, when subtracted, yield the equation

$$Mn(OH)_2(s) \rightleftharpoons Mn^{2+} + 2OH^-$$

These are

$$Mn(OH)_2(s) + 2e \rightleftharpoons Mn(s) + 2OH^- \qquad E° = -1.59 \text{ V}$$

$$\underline{Mn^{2+} + 2e \rightleftharpoons Mn(s) \qquad\qquad\qquad E° = -1.18 \text{ V}}$$

$$Mn(OH)_2(s) \rightleftharpoons Mn^{2+} + 2OH^- \qquad E°_{cell} = -0.41 \text{ V}$$

Then

$$E = -0.41 - \frac{0.059}{2} \log [Mn^{2+}][OH^-]^2$$

At equilibrium, $E = 0$ and $[Mn^{2+}][OH^-]^2 = K_{sp}$. Thus

$$\log K_{sp} = -\frac{0.41 \times 2}{0.059} = -13.90$$

$$K_{sp} = 1.3 \times 10^{-14} \qquad\qquad\qquad \square$$

Any number of examples could be given, of course, but the principle is already clear: Having a table of standard potentials is tantamount to possessing a great many equilibrium constants.

10.3

TYPES OF ELECTRODES

A metal electrode whose potential responds to the activity of its metal ion, such as the zinc or copper electrodes discussed earlier, is sometimes called an *electrode of the first kind*. This is not, however, the only type. In fact, not all electrodes involve electroactive metals. Several other types are described in Chapter 12. At this point it is useful to discuss *electrodes of the second kind* because they provide the commonest laboratory reference electrodes.

10.3a. Laboratory Reference Electrodes

The hydrogen electrode is inconvenient for routine, practical measurements in the laboratory. It requires a tank of compressed gas, which is heavy and awkward, explosive mixtures of hydrogen and air may be formed, and the catalytic platinum surface is easily poisoned, that is, contaminated with adsorbed substances that inhibit catalytic activity. Thus, in the laboratory, more convenient reference electrodes are commonly employed for measuring the potentials of other half-cells. The potentials of these reference electrodes have themselves been measured against the standard hydrogen electrode. Regardless of what reference electrode is actually employed, it is customary to report any potential as though it had been measured against the standard hydrogen electrode. The commonest reference electrodes are calomel and silver–silver chloride electrodes.

The Calomel Electrode

One form of the calomel electrode is shown in Fig. 10.5. External contact is made through a wire from a pool of mercury. The mercury is in contact with a moist paste prepared by intimately mixing mercury, mercury(I) chloride (calomel), and aqueous KCl solution. As with an electrode of the first kind, the potential is established by the simple redox couple

$$Hg_2^{2+} + 2e \rightleftharpoons 2Hg$$

[It may be recalled that the mercury(I) ion is a dimeric species; i.e., we are dealing with Hg(I), but the ion in soluion is Hg_2^{2+}.] According to the Nernst equation, the potential is given by

$$E = E° - \frac{0.059}{2} \log \frac{1}{[Hg_2^{2+}]}$$

But there is more to the story, because Hg_2^{2+} is involved in another equilibrium: Hg_2Cl_2 is a slightly soluble salt, for which we may write a solubility product constant, K_{sp} (Chapter 9).

$$K_{sp} = [Hg_2^{2+}][Cl^-]^2$$

Rearranging, we obtain

$$[Hg_2^{2+}] = \frac{K_{sp}}{[Cl^-]^2}$$

and substitution into the Nernst equation above yields

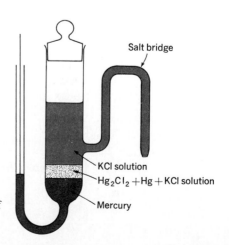

Figure 10.5 Calomel electrode. A diagram of a commercial electrode is shown in Figure 12.1.

Salt bridge

KCl solution
$Hg_2Cl_2 + Hg + KCl$ solution
Mercury

$$E = E° - \frac{0.059}{2} \log \frac{[Cl^-]^2}{K_{sp}}$$

Since $E°$ and K_{sp} are constants, it is seen that the potential reflects the concentration of Cl^-.

Most commonly, the solution is saturated with KCl; the electrode is then called the *saturated calomel electrode* or SCE, and its single-electrode potential is +0.2458 V at 25°C, as determined against a standard hydrogen electrode. If the KCl solution is exactly 1 M, the potential is +0.2847 V; this is sometimes called a *standard* or *normal calomel electrode*.

The student should note that we may write the Nernst equation as we did above, in which case $E°$ is a value for the Hg–Hg_2^{2+} couple, +0.79 V. Alternatively, we can formulate the half-reaction as

$$Hg_2Cl_2 + 2e \rightleftharpoons 2Hg + 2Cl^-$$

whereupon the $E°$ value becomes +0.2847 V. Changing this value to +0.2458 V (paragraph above) allows for the chloride concentration in a saturated KCl solution.

EXAMPLE 10.5

Calculations involving the calomel electrode

Using the potentials given in the preceding discussion, calculate the solubility product constant for mercury(I) chloride and the molar solubility of KCl.

$$E = E° - \frac{0.059}{2} \log \frac{1}{[Hg_2^{2+}]}$$

If we choose E = +0.28 V, then $[Cl^-]$ = 1 M. The mercury(I) ion concentration which obtains in this solution is found as follows:

$$+0.28 = +0.79 - \frac{0.059}{2} \log \frac{1}{[Hg_2^{2+}]}$$

$$[Hg_2^{2+}] = 5.15 \times 10^{-18}$$

$$K_{sp} = [Hg_2^{2+}][Cl^-]^2 = 5.15 \times 10^{-18} \times 1^2 = 5.15 \times 10^{-18}$$

For the SCE:

$$+0.25 = +0.79 - \frac{0.059}{2} \log \frac{[Cl^-]^2}{5.15 \times 10^{-18}}$$

$$[Cl^-] = 3.2 \ M = \text{solubility} \qquad \square$$

Calomel electrodes may be fabricated in a variety of sizes and shapes. Some, of the sort shown in Fig. 10.5, are quite large, while some are small enough to be inserted into blood vessels of experimental animals through syringe needles. It should be emphasized that the potential of an electrode does not depend upon its size. Likewise, the liquid junction between the test solution and the KCl solution in the electrode may be physically established in various ways. Sometimes it involves an agar plug or a sintered-glass frit, sometimes a pinhole, sometimes a small, wet fiber; the point is to permit the migration of ions without allowing excessive solution to pour across the interface. Although contamination from calomel electrodes is inconsequential for most purposes, in critical cases it must be remembered that a small amount of KCl may leak into the test solution. Similarly, a test solution with a large hydrostatic head may contaminate the contents of a calomel electrode.

The Silver–Silver Chloride Electrode

This electrode, also widely used, is similar in principle to the calomel electrode. Usually, a silver wire is coated with a retentive layer of silver chloride by anodizing the silver in a chloride solution. In the finished electrode, the coated wire dips into a potassium chloride solution. As in the calomel case, the potential is basically established by a metal–metal ion electron-transfer reaction

$$Ag^+ + e \rightleftharpoons Ag$$

but the concentration of silver ion is, in turn, governed by the chloride concentration via the K_{sp} for AgCl. The overall half-reaction may thus be written

$$AgCl + e \rightleftharpoons Ag + Cl^-$$

For chloride ion at unit activity, the standard potential is +0.2221 V at 25°C. The following example illustrates the use of a reference electrode.

EXAMPLE 10.6

Calculating an unknown electrode potential using a reference electrode

(a) The potential of a cell made up of an electrode of unknown potential and a standard calomel electrode is 1.04 V, with the calomel electrode positive. Calculate the potential of the unknown electrode referred to the hydrogen electrode.

If we consider our cell as having the unknown electrode on the left, calomel on the right, we must give a positive sign to the potential. That is,

$$E_{cell} = E_r - E_l$$

$$+1.04 = +0.28 - E_l$$

$$E_l = -0.76 \text{ V}$$

(b) In a second measurement with another unknown electrode, the potential of the cell is found to be 0.06 V, with the calomel negative. Calculate the potential of the unknown electrode referred to hydrogen.

In this case, if the calomel is the right-hand electrode as before, the potential of the cell must be written −0.06 V. Hence

$$-0.06 = 0.28 - E_1$$

$$E_l = +0.34 \text{ V}$$

The relation between these potentials is shown schematically in Fig. 10.6. □

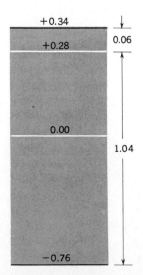

Figure 10.6 Potential relations.

10.3b. Inert Electrodes

So far most of our examples of redox systems have consisted of metals in equilibrium with their ions. There are many important examples of redox systems involving only different oxidation states of ions in solutions. A familiar example is the Fe^{3+}–Fe^{2+} system,

$$Fe^{3+} + e \rightleftharpoons Fe^{2+}$$

A solution of Fe^{3+} and Fe^{2+} ions is a possible source of electrons, but some metal must be inserted into the solution to act as a conductor. A metal such as platinum is normally used because it is not easily attacked by most solutions. The platinum is said to be *inert* electrode, since it does not enter into the reaction.

A cell made up of the Fe^{3+}–Fe^{2+} and Ce^{4+}–Ce^{3+} systems is written as

$$Pt|\,Fe^{3+}(x\,M) + Fe^{2+}(y\,M)\,\|\,Ce^{4+}(a\,M) + Ce^{3+}(b\,M)\,|Pt$$

The student should confirm that the standard potential of this cell as written is +0.84 V, and that Ce^{4+} ion oxidizes Fe^{2+} ion spontaneously if the concentration of each reactant is 1 M.

10.4

FORMAL POTENTIALS

Consider an oxidant and a reductant that constitute a redox couple, say, Fe^{3+} and Fe^{2+}:

$$Fe^{3+} + e \rightleftharpoons Fe^{2+}$$

For the single-electrode potential we have

$$E = E^\circ - 0.059 \log \frac{a_{Fe^{2+}}}{a_{Fe^{3+}}}$$

When $a_{Fe^{2+}} = a_{Fe^{3+}}$, then $E = E^\circ$; that is, the potential is the standard potential, and presumably we could measure this against a reference electrode. But although this appears to be simple enough, in fact it is not easy experimentally to prepare a solution in which we know that the two activities are equal, and determining a true standard potential is often difficult.

In our example here, hydroxides of the metal ions are very insoluble, and to prevent the hydrolysis of Fe^{3+} and Fe^{2+} it is necessary to acidify the solution. But we then introduce the possibility of complex formation between the metal ions and the anion of the acid. If, for example, we use HCl, complexes such as $FeCl^{2+}$, $FeCl_2^+$, etc. (up to $FeCl_6^{3-}$) may form in varying degrees depending upon the chloride concentration. These iron(III) complexes are stronger than the corresponding iron(II) complexes, $FeCl^+$, etc. Thus complexing with Cl^- tends to stabilize the Fe^{3+} state, in effect making Fe^{3+} a weaker oxidant or Fe^{2+} a stronger reductant than would be the case in the absence of the complexing agent.

In the kinds of solutions encountered in analytical chemistry, there is often complexing such as that described above, and electrolyte concentrations are often such that activity coefficients can scarcely be guessed at. Fortunately, however, what we really want to know about redox reagents is how potent they are in these very solutions, not what they might be in some hypothetical state, perhaps at infinite dilution. Thus it is convenient to write our Nernst equation as follows:

$$E = E^{\circ\prime} - 0.059 \log \frac{C_{Fe^{2+}}}{C_{Fe^{3+}}}$$

where the C terms are analytical concentrations and $E^{\circ\prime}$ is called a *formal potential*. For example, if we started with an iron(III) solution in 0.1 M HCl and reduced exactly half of the iron to Fe^{2+}, then the potential E would be equal to the formal potential, $E^{\circ\prime}$, for this redox couple in this particular solution. Changing the HCl concentration, or switching from HCl to, say, sulfuric acid, would change $E^{\circ\prime}$.

$E^{\circ\prime}$ values are used like standard potentials in calculations involving the Nernst equation. Calculated and experimental E values will agree better if $E^{\circ\prime}$ is used instead of E°, provided that the formal potential is the correct one for the particular conditions that obtain.

Formal potentials are often employed in connection with organic compounds, especially by biochemists. Frequently, hydrogen ion participates in the overall redox reaction:

$$Ox + mH^+ + ne \rightleftharpoons Red$$

In such a case, H^+ appears in the Nernst equation:

$$E = E^{\circ} - \frac{0.059}{n} \log \frac{a_{Red}}{a_{Ox} \times a_{H+}^m}$$

Now, in order that $E = E^{\circ}$, not only must the activities of the oxidant and the reductant be the same, but the hydrogen ion activity must be unity. This represents a very acidic solution, on the order of 1 M in strong acid. Because many compounds of biological importance decompose under such conditions, it is often impossible to measure standard potentials directly. Moreover, the potential at physiological pH is of far more interest anyway. Thus biochemistry books often list $E^{\circ\prime}$ values for redox couples at pH 7, and although the term *formal potentials* may not be used, this is what they are.

Box 10.3   Oxidation-Reduction in Living Systems

Redox reactions are paramount among the thousands of processes that occur in living organisms. Dietary carbohydrates, fats, and proteins are "burned" in our cells, releasing energy which drives the "uphill" performance of biological work. As a result of catalysis, this combustion occurs under remarkably mild conditions of temperature, pressure, pH, etc. As we all know from watching fires, the burning of flammable substances is spontaneous once the systems become hot enough to overcome energy barriers, but fuels such as sugar and fat do not oxidize very rapidly at physiological temperature (98.6°F or 37°C). Biocatalysts called *enzymes* lower energies of activation and yield enormous rate enhancements.

Consider the oxidation of glucose, the principal sugar in the blood and a prototypal energy source in biology. The overall half-reaction for its oxidation to carbon dioxide is

$$C_6H_{12}O_6 + 6H_2O \longrightarrow 6CO_2 + 24H^+ + 24e \qquad (1)$$

The ultimate oxidant is oxygen, which is the strongest electron acceptor in the body; the formal potential of +0.83 V at pH 7 for $O_2 + 4H^+ + 4e = 2H_2O$ is the most positive one in biochemistry. Multiplying by 6 to balance electrons,

$$6O_2 + 24H^+ + 24e \longrightarrow 12H_2O \qquad (2)$$

and adding Eqs. (1) and (2) give the overall reaction:

$$C_6H_{12}O_6 + 6O_2 \longrightarrow 6CO_2 + 6H_2O \qquad (3)$$

This reaction, which is "downhill" on a free-energy scale and also exothermic (ΔG and ΔH are nearly the same), occurs in both fires and living cells, but by entirely different pathways.

To make a long story short, the first step in glucose oxidation is conversion into pyruvate:

$$C_6H_{12}O_6 \longrightarrow 2CH_3-\overset{\overset{\displaystyle O}{\|}}{C}-COO^- + 6H^+ + 4e \qquad (4)$$

This process, called glycolysis, actually occurs in ten steps, each catalyzed by its own enzyme. The electron acceptor is NAD^+, and the product NADH. (For the structures of NAD^+ and NADH, as well as the formal potential and other commentary, see Problem 17, Chapter 14.) Doubling the NAD^+–NADH half-reaction to balance electrons gives

$$2NAD^+ + 2H^+ + 4e \longrightarrow 2NADH \qquad (5)$$

and adding Eqs. (4) and (5), we obtain for glycolysis:

$$C_6H_{12}O_6 + 2NAD^+ \longrightarrow 2CH_3-\overset{\overset{\displaystyle O}{\|}}{C}-COO^- + 2NADH + 4H^+ \qquad (6)$$

(Equations must balance in biochemistry as well as anywhere else.)

In the second step of glucose oxidation, the two pyruvates of Eqs. (4) and (6) are decarboxylated, releasing the first two CO_2 molecules for Eq. (3); the other carbon atoms appear as acetylcoenzyme A, a labile thioester which we abbreviate $CH_3-\overset{\overset{\displaystyle O}{\|}}{C}-S-CoA$. This intermediate enters the Krebs cycle, a complex series of oxidative decarboxylations that release the rest of the CO_2. The oxidants are NAD^+ and another redox coenzyme called flavin-adenine dinucleotide (FAD).

Although the glucose has now gone completely to CO_2, we are not finished; the reduced forms of the coenzymes that served as glucose oxidants must be restored to their oxidized condition; the electrons they bear must be delivered to the terminal oxidant, O_2. Now O_2 is a very strong oxidant, and NADH a very strong reductant. Thermodynamically, the direct reaction is highly favorable:

$$
\begin{array}{lr}
O_2 + 4H^+ + 4e \rightleftharpoons 2H_2O & E^{\circ\prime} = +0.83 \text{ V} \\
\underline{2NAD^+ + 2H^+ + 4e \rightleftharpoons 2NADH} & \underline{-0.32 \text{ V}} \\
O_2 + 2NADH + 2H^+ \rightleftharpoons 2H_2O + 2NAD^+ & +1.15 \text{ V}
\end{array}
$$

But this is not the way it is done. Direct oxidations by O_2 often have high energies of activation and are very slow at physiological temperatures. Always remember that *tendency* to react (thermodynamics) and *rate* (kinetics) are two different things. Nature, by developing the right catalysts, directs electron flow through a series of redox couples of increasingly positive potential called the *terminal oxidation chain,* only the last of which (cytochrome oxidase) reacts rapidly with molecular oxygen.

Overall, ΔG for glucose oxidation to CO_2 is the same no matter the pathway, but the biochemical route releases energy a little at a time, in intimate coupling with a storage system that retains nearly 40% of it in the form of reactive compounds that power all the uphill processes we associate with being alive. We broke the process into three parts for convenience, but of course all the reactions are going on all of the time. There is a fairly stable potential within the cell, reflecting steady-state levels of such couples as NAD^+–NADH, of all the other intermediates, and of O_2 itself.

Chemists who deigned to look upon what were, at the time, less quantitative disciplines used to make fun of biologists who poked electrodes into living tissue: "They don't know what they are measuring," it was said. Truth be told, as electrons flow constantly, gently, and rapidly through biological redox systems, no redox couple is ever very far from equilibrium with any other. The system is probably closer to reversible behavior than most physical chemists will ever see in their lifetimes. The oxygen level is probably fairly constant in a normal, unstressed cell, and in fact one potential may tell it all.

Because it is often difficult to measure standard potentials accurately, many of the values in the literature called standard potentials are, in fact, formal potentials. A critical evaluation of the original data is often required in order to determine exactly what the investigators actually measured, regardless of what they chose to call it. Table A-4, Appendix I, doubtless contains both standard and formal potentials. For many purposes in analytical chemistry, calculations based upon these values are sufficiently accurate; for example, in this chapter, uncertainty regarding the exact significance of the potentials employed would not affect conclusions about the feasibility of titrations except for cases which were borderline in any event. A short list of formal potentials is given in Table A-5, Appendix I.

10.5

APPROXIMATIONS USING THE NERNST EQUATION

The mechanism of many redox reactions is complex, and we do not always know the exact nature of the reaction which determines the potential at an electrode surface. For example, the overall reaction for the reduction of dichromate ion to chromium(III) ion is

$$Cr_2O_7^{2-} + 14H^+ + 6e \rightleftharpoons 2Cr^{3+} + 7H_2O$$

The standard potential, $+1.33$ V, was obtained indirectly, not from galvanic cell measurements. The reaction above simply represents the correct stoichiometry. It is thought that the reaction proceeds in steps via an unstable intermediate which is converted into the products. The Nernst expression

$$E = 1.33 - \frac{0.059}{6} \log \frac{[Cr^{3+}]^2}{[Cr_2O_7^{2-}][H^+]^{14}}$$

is not followed. Actually, the potential is practically independent of the concentration of chromium(III) ions.

Similar behavior is found with other complex redox systems, the permanganate–manganese(II) system being one important example. With reactions which involve a large number of hydrogen ions, it is normally found that the potential is strongly dependent upon the hydrogen ion concentration but that the dependence cannot be predicted from the coefficients in the balanced equations. Hence calculations of the potentials of such systems on the basis of these equations give incorrect results. In many cases, however, the error is small enough that conclusions regarding the feasibility of titrations may still be valid.

10.6

TITRATION CURVES

10.6a. Titration of Iron(II) with Cerium(IV) Ion

We wish now to derive a titration curve for a redox reaction, as we have previously done for acid-base, complex-formation, and precipitation reactions. Our example will be the titration of iron(II) with cerium(IV) in sulfuric acid medium. We could calculate the concentration of the analyte, Fe^{2+}, during the titration and plot the value of pFe^{2+} vs. milliliters of titrant, as we have done for other types of reactions. However, it is customary in redox titrations to plot the potential of the $Fe^{2+}-Fe^{3+}$ system, rather than pFe^{2+}, vs. milliliters of titrant. The potential of the redox system and the concentration of Fe^{2+} are related, of course, through the Nernst expression

$$E = E_{Fe}^{\circ} - 0.059 \log \frac{[Fe^{2+}]}{[Fe^{3+}]}$$

Although we will not be measuring this potential experimentally during the titration (this technique, called *potentiometric titration,* is discussed in Chapter 12), it may be helpful to consider how it could be determined if we wished to do so. Let us suppose that the titration is carried out in the setup shown in Fig. 10.7.

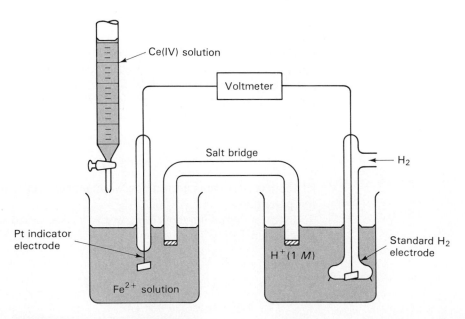

Figure 10.7 Titration of Fe(II) with Ce(IV).

The potential of the $Fe^{2+}-Fe^{3+}$ system in the titration vessel could be measured at any point by making this vessel one half of a galvanic cell. The other half is the standard H_2 electrode. An inert platinum electrode is used in the titration vessel to serve as an *indicator* electrode.

It should be noted that as we perform the titration we are placing the two redox systems in the same container and allowing them to reach equilibrium:

$$Fe^{2+} + Ce^{4+} \rightleftharpoons Fe^{3+} + Ce^{3+}$$

This means that the $Fe^{2+}-Fe^{3+}$ and $Ce^{3+}-Ce^{4+}$ redox systems *have the same potentials throughout the titration*. At any point, the solution has only one potential given by either

$$E = E°_{Fe} - 0.059 \log \frac{[Fe^{2+}]}{[Fe^{3+}]}$$

or

$$E = E°_{Ce} - 0.059 \log \frac{[Ce^{3+}]}{[Ce^{4+}]}$$

Hence we can calculate the potential of the solution in the titration vessel from the Nernst expression for either redox system, and we shall choose the one which is more convenient for our purposes. The values we calculate will be those which would be measured experimentally in the cell shown in Fig. 10.7.

Let us now consider the calculations.

EXAMPLE 10.7

Titration curve for Fe(II)/Ce(IV) system

5.0 mmol of an iron(II) salt is dissolved in 100 mL of sulfuric acid solution and titrated with 0.10 M cerium(IV) sulfate. Calculate the potential of an inert electrode in the solution at various intervals in the titration and plot a titration curve. Use 0.68 V as the formal potential of the $Fe^{2+}-Fe^{3+}$ system in sulfuric acid and 1.44 V for the $Ce^{3+}-Ce^{4+}$ system.

(a) *Start of titration.* The potential is determined by the $Fe^{2+}-Fe^{3+}$ ratio, that is,

$$E = 0.68 - 0.059 \log \frac{[Fe^{2+}]}{[Fe^{3+}]}$$

However, we do not know the iron(III) ion concentration, this being dependent upon how the iron(II) salt was prepared, how much has been oxidized by air, etc. Let us assume that no more than 0.1% of the iron remains in the +3 state, that is, the iron(II)-to-iron(III) ion ratio is 1000 : 1. For such a condition the potential can be calculated:

$$E = 0.68 - 0.059 \log 1000$$

$$E = 0.50 \text{ V}$$

[If the iron(III) ion concentration were actually zero, what would be the value of the potential?]

(b) *After addition of 10 mL of Ce^{4+}.* We started with 5.0 mmol of Fe^{2+} and have added 10 mL × 0.10 mmol/mL = 1.0 mmol of Ce^{4+}. The reaction is

mmol	Fe^{2+}	$+ \, Ce^{4+}$	$\longrightarrow$	Fe^{3+}	$+ \, Ce^{3+}$
Initial:	5.0	1.0		—	—
Change:	−1.0	−1.0		+1.0	+1.0
Equilibrium:	4.0	—		1.0	1.0

We can calculate the potentials from the expression for either redox system; that is,

$$E = 0.68 - 0.059 \log \frac{[Fe^{2+}]}{[Fe^{3+}]}$$

$$E = 1.44 - 0.059 \log \frac{[Ce^{3+}]}{[Ce^{4+}]}$$

The system is at equilibrium; that is, each redox couple has the same potential. It is simpler at this stage of the titration to use the expression for the iron(II)–iron(III) system, since we can estimate the concentrations of these two ions more readily than that of the Ce(IV) ion as seen below:

$$[Fe^{2+}] = \left\{ \frac{4.0}{110} + x \right\} \text{mmol/mL}$$

$$[Fe^{3+}] = \left\{ \frac{1.0}{110} - x \right\} \text{mmol/mL}$$

$$[Ce^{3+}] = \left\{ \frac{1.0}{110} - x \right\} \text{mmol/mL}$$

$$[Ce^{4+}] = x \text{ mmol/mL}$$

The value of x can be calculated, of course, from the equilibrium constant of the reaction. Assuming that the reaction goes well to completion, x is small and may be disregarded in estimating the iron(III) and iron(II) ion concentrations. Hence

$$E = 0.68 - 0.059 \log \frac{4.0/110}{1.0/110}$$

or

$$E = 0.64 \text{ V}$$

(Note that the volume term cancels; that is, the potential is independent of volume.)

Values for the potential of all other points before the equivalence point are calculated in the same manner. In Table 10.1 is a list of such values, and these are plotted in Fig. 10.8. Note that the potential rises slowly in the earlier stages of the titration and begins to increase more rapidly as the equivalence point is approached.

(c) *Equivalence point.* We started with 5.0 mmol of Fe^{2+} and have added 50 mL $\times$ 0.10 mmol/mL = 5.0 mmol of Ce^{4+}. The reaction is

mmol	Fe^{2+} +	Ce^{4+} $\longrightarrow$	Fe^{3+} +	Ce^{3+}
Initial:	5.0	5.0	—	—
Change:	−5.0	−5.0	+5.0	+5.0
Equilibrium:	—	—	5.0	5.0

The concentrations of the reactants and products are then

$$[Fe^{2+}] = [Ce^{4+}] = x$$

$$[Fe^{3+}] = [Ce^{3+}] = \frac{5.0}{150} - x$$

TABLE 10.1 Redox Potential during Titration of 5.0 mmol of Fe^{2+} with 0.10 M Ce^{4+}

Ce^{4+}, mL	Fe^{2+} UNOXIDIZED, mmol	Fe^{2+} OXIDIZED, %	E, V
0.00	5.0	0	—
10.0	4.0	20	+0.64
20.0	3.0	40	0.67
30.0	2.0	60	0.69
40.0	1.0	80	0.72
45.0	0.50	90	0.74
49.50	0.05	99	0.80
49.95	0.005	99.9	0.86
50.0	—	100	1.06
	Ce^{4+} excess, mmol		
50.05	0.005		+1.26
50.50	0.05		1.32
51.0	0.10		1.34
55.0	0.50		1.38
60.0	1.0		1.40

When either of the following expressions is employed, it is necessary to evaluate x, using the equilibrium constant, in order to calculate the potential:

$$E = 0.68 - 0.059 \log \frac{[Fe^{2+}]}{[Fe^{3+}]}$$

$$E = 1.44 - 0.059 \log \frac{[Ce^{3+}]}{[Ce^{4+}]}$$

Notice, however, that if the two equations are added, giving

$$2E = 2.12 - 0.059 \log \frac{[Fe^{2+}][Ce^{3+}]}{[Fe^{3+}][Ce^{4+}]}$$

the logarithmic term is zero, since at the equivalence point

$$[Fe^{2+}] = [Ce^{4+}] \quad \text{and} \quad [Fe^{3+}] = [Ce^{3+}]$$

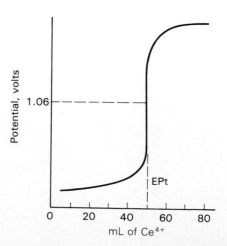

Figure 10.8 Titration of iron (II) with cerium (IV) ion.

Hence

$$2E = 2.12 \quad \text{or} \quad E = 1.06 \text{ V}$$

For any reaction in which the number of electrons lost by the reductant is the same as the number gained by the oxidant, the potential at the equivalence point is simply the arithmetic mean of the two standard potentials:

$$E_{\text{eq pt}} = \frac{E_1^\circ + E_2^\circ}{2}$$

(d) *After addition of 60 mL of* Ce^{4+}. We started with 5.0 mmol of Fe^{2+} and have added 60 mL $\times$ 0.10 mmol/mL = 6.0 mmol of Ce^{4+}. The reaction is

mmol	Fe^{2+} +	Ce^{4+} $\longrightarrow$	Fe^{3+} +	Ce^{3+}
Initial:	5.0	6.0	—	—
Change:	−5.0	−5.0	+5.0	+5.0
Equilibrium:	—	1.0	5.0	5.0

The concentrations are

$$[Fe^{2+}] = x$$

$$[Fe^{3+}] = \left\{ \frac{5.0}{160} - x \right\} \text{ mmol/mL}$$

$$[Ce^{3+}] = \left\{ \frac{5.0}{160} - x \right\} \text{ mmol/mL}$$

$$[Ce^{4+}] = \left\{ \frac{1.0}{160} + x \right\} \text{ mmol/mL}$$

It is now the iron(II) ion concentration that must be evaluated from the equilibrium constant. Hence it is more convenient to employ the expression for the cerium(IV)–cerium(III) system:

$$E = 1.44 - 0.059 \log \frac{[Ce^{3+}]}{[Ce^{4+}]}$$

Noting that x is small, we write

$$E = 1.44 - 0.059 \log \frac{5.0/160}{1.0/160} = 1.40 \text{ V}$$

Other values beyond the equivalence point are calculated in the same manner. The curve is plotted in Fig. 10.8, where its similarity to a strong-base titration curve may be seen. A large change in potential occurs in the vicinity of the equivalence point of the titration. The data are given in Table 10.1. □

10.6b. Titration of Tin(II) with Cerium(IV) Ion

Let us consider now an example where the reductant loses two electrons while the oxidant gains one.

EXAMPLE 10.8

Titration of Sn^{2+}
with Ce^{4+}

Calculate the potential at the equivalence point in the titration of tin(II) ion with cerium(IV) ion:

$$Sn^{2+} + 2Ce^{4+} \rightleftharpoons Sn^{4+} + 2Ce^{3+}$$

The potential is given by either of the following expressions:

$$E = 0.15 - \frac{0.059}{2} \log \frac{[Sn^{2+}]}{[Sn^{4+}]}$$

or

$$E = 1.44 - 0.059 \log \frac{[Ce^{3+}]}{[Ce^{4+}]}$$

Multiplying the first equation by 2 and adding it to the second gives

$$3E = 1.74 - 0.059 \log \frac{[Sn^{2+}][Ce^{3+}]}{[Sn^{4+}][Ce^{4+}]}$$

The logarithmic term is zero, since at the equivalence point[2]

$$[Ce^{4+}] = 2[Sn^{2+}] \quad \text{and} \quad [Ce^{3+}] = 2[Sn^{4+}]$$

Hence

$$E = \frac{1.74}{3} = 0.58 \text{ V}$$

For the case in which one redox system gains or loses one electron and the other loses or gains two, the potential of the equivalence point is

$$E = \frac{E_1^\circ + 2E_2^\circ}{3}$$

where E_1° is the standard potential of the first system and E_2° that of the second. It should be noted that the titration curve is not symmetrical about the equivalence point in this case as it is in the titration of iron(II) with cerium(IV) ion. This is always true when the two redox systems exchange a different number of electrons per molecule. □

A completely general expression for the potential at the equivalence point can be derived in the same manner that we have employed here (see Problem 8).

10.6c. Titration of Other Redox Couples

Figure 10.9 shows the variation of potential with fraction of reagent in the oxidized form for several redox couples. The curves for couples with electrode potentials greater than 1 V have been plotted on a scale to the right of those couples with potentials less than 1 V. This is done to emphasize the fact that a single titration curve is a combination of two of the branches plotted here.

In comparing Fig. 10.9 with the titration curves for acid-base reactions, one should keep in mind the fact that the acid-base properties of our solvent, water, are quite different from the redox properties. As noted earlier, we do not ordinarily encounter a leveling effect in redox titrations.

Several points about Fig. 10.9 should be noted.

---

[2] Remember that the concentrations are molarities. Obviously, the number of equivalents of cerium(IV) ion is the same as that of tin(II) ion. But there are twice as many moles of cerium(IV) as of tin(II) ion.

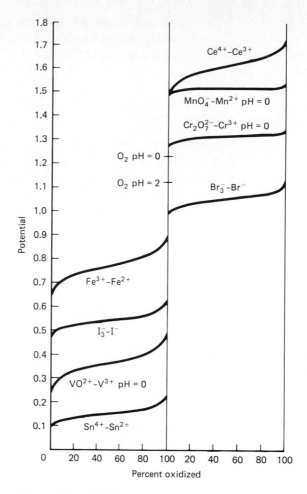

Figure 10.9 Variation of potential with percent oxidized. Curves for couples with potentials greater than 1V are plotted on scale to the right.

1. The change in potential at the equivalence point in the titration of, say, iron(II) ion depends upon the oxidant used. The oxidants are not leveled by water, as pointed out before.

2. The shape of a curve depends upon the value of n, the number of electrons gained or lost by the oxidant or reductant. Note that the iron(II)–iron(III) curve ($n = 1$) is steeper than the tin(II)–tin(IV) curve ($n = 2$). Also note the flatness of the permanganate and dichromate curves, where n is 5 and 6, respectively. The shape of a titration curve will obviously be determined by those of the two halves which make up the curve.

3. The curves are asymptotic to the vertical axis at zero and 100% oxidation. The curves are flattest near the midpoints (50% oxidation), which correspond to the standard potentials for couples such as Fe^{3+}–Fe^{2+}. The stabilization of the potential in this region is analogous to the buffering action of an acid-base pair about the pH region which corresponds to the pK_a. A redox couple is said to be *poised* in this region where the potential is stabilized.

4. In Fig. 10.9 the potential of the reaction

$$O_2 + 4H^+ + 4e \rightleftharpoons 2H_2O$$

is indicated at pH 0 and 2. Redox couples with standard potentials more positive than these values should not be stable in water at pH 0 and 2. Solutions of these couples are often stable, however, since the reactions with water to liberate oxygen are generally slow.

10.7

FEASIBILITY OF REDOX TITRATIONS

The magnitude of the equilibrium constant required for a feasible redox titration can be calculated as was done previously for acid-base (page 148), complexformation (page 208), and precipitation titrations (page 230). The following example illustrates this.

EXAMPLE 10.9

Calculation of the feasibility of a redox titration

For the redox reaction

$$Ox_1 + Red_2 \rightleftharpoons Red_1 + Ox_2$$

where

$$Ox_1 + e \rightleftharpoons Red_1 \quad E_1^\circ$$

$$Ox_2 + e \rightleftharpoons Red_2 \quad E_2^\circ$$

(a) Calculate the value of the equilibrium constant for the following conditions: 50 mL of 0.10 M Red_2 is titrated with 0.10 M Ox_1. When 49.95 mL of titrant is added, the reaction is complete. On the addition of two more drops (0.10 mL) of titrant, the value of $pRed_2$ changes by 2.00 units.

(b) What is the difference in the standard potentials of the two redox couples for this value of K?

(a) We start with $50 \times 0.10 = 5.0$ mmol of Red_2, and at 49.95 mL of titrant 0.0050 mmol remains unreacted. Hence

$$[Red_2] = \frac{0.0050 \text{ mmol}}{99.95 \text{ mL}} \cong 5.0 \times 10^5 \ M$$

$$pRed_2 = 4.30$$

For a change of 2.00 units, $pRed_2 = 6.30$ and $[Red_2] = 5 \times 10^{-7} \ M$ when the volume of titrant is 50.05 mL. The other concentrations are

$$[Ox_1] = \frac{0.05 \times 0.10}{100.05} \cong 5.0 \times 10^{-5}$$

$$[Red_1] = [Ox_2] \cong \frac{5.0 \text{ mmol}}{100.05 \text{ mL}} \cong 5.0 \times 10^{-2} \ M$$

Hence

$$K = \frac{(5.0 \times 10^{-2})(5.0 \times 10^{-2})}{(5.0 \times 10^{-5})(5.0 \times 10^{-7})}$$

$$K = 1.0 \times 10^8$$

(b) Since

$$E_1^\circ - E_2^\circ = \frac{0.059}{1} \log K$$

$$E_1^\circ - E_2^\circ = 0.059 \log 1 \times 10^8$$

$$E_1^\circ - E_2^\circ = 0.47 \text{ V}$$

Note that in this example the number of electrons, n, is 1 for both redox couples. See Problems 20 and 21 for calculations involving other values of n. □

10.8

REDOX INDICATORS

10.8a. Types of Redox Indicators

There are several types of indicators which may be used in redox titrations:

1. A colored substance may act as its own indicator. For example, potassium permanganate solutions are so deeply colored that a slight excess of this reagent in a titration can be easily detected.
2. A *specific* indicator is a substance which reacts in a specific manner with one of the reagents in a titration to produce a color. Examples are starch, which forms a deep blue color with iodine, and thiocyanate ion, which forms a red color with iron(III) ion.
3. External, or *spot test*, indicators were once employed when no internal indicator was available. The ferricyanide ion was used to detect iron(II) ion by formation of iron(II) ferricyanide (Turnbull's blue) on a spot plate outside the titration vessel.
4. The redox potential can be followed during a titration, and the equivalence point detected from the large change in potential in the titration curve. Such a procedure is called *potentiometric titration* (see Chapter 12), and the titration curve may be plotted manually or automatically recorded.
5. Finally, an indicator which itself undergoes oxidation-reduction may be employed. We shall refer to such a substance as a true redox indicator, and it is with such a reagent that the rest of our discussion is concerned.

For simplicity, let us designate the redox couple as follows:

$$In^+ + e \rightleftharpoons In$$

$$\text{Color A} \qquad\qquad \text{Color B}$$

where one electron is gained by the oxidant and no hydrogen ions are involved in the reaction. Let us also say that the colors of the oxidized and reduced forms are different, as indicated above. The equation for the potential of this system is

$$E = E_i^\circ - 0.059 \log \frac{[In]}{[In^+]}$$

where E_i° is the standard potential of the indicator couple.

Let us now assume that if the ratio $[In]/[In^+]$ is 10 to 1 or greater, only color B can be seen by the eye. (See page 145 for similar treatment of acid-base indicators.) Also, if the ratio is 1 : 10 or smaller, only color A is observed. That is,

Color B: $E = E_i^\circ - 0.059 \log 10/1 = E_i^\circ - 0.059$

Color A: $E = E_i^\circ - 0.059 \log 1/10 = E_i^\circ + 0.059$

Subtracting, $\Delta E = \pm 2 \times 0.059 = \pm 0.12$ V

Thus a change in potential of about 0.12 V is required to bring about a change in color of the indicator, if our assumptions are reasonable.

Table 10.2 lists some true redox indicators, with the observed colors and "transition potentials" of the redox couples. The treatment in the above paragraph supposes that the two colored forms of the indicator are equally intense to the eye. This is not always the case, and hence the transition potentials may not represent exactly 50% conversion of one indicator form to the other; i.e., they may not be the same as the formal potentials. With ferroin, for example, the formal potential in 1 M acid is about +1.06 V, but because the color changes from deep red to pale blue, the analyst will not perceive a visual end point until a potential of about +1.11 V is reached.

TABLE 10.2 Transition Potentials of Some Redox Indicators

INDICATOR	COLOR OF REDUCTANT	COLOR OF OXIDANT	TRANSITION POTENTIAL, V	CONDITIONS
Phenosafranine	Colorless	Red	+0.28	1 M acid
Indigo tetrasulfonate	Colorless	Blue	0.36	1 M acid
Methylene blue	Colorless	Blue	0.53	1 M acid
Diphenylamine	Colorless	Violet	0.76	1 M H_2SO_4
Diphenylbenzidine	Colorless	Violet	0.76	1 M H_2SO_4
Diphenylaminesulfonic acid	Colorless	Red violet	0.85	Dilute acid
5,6-Dimethylferroin			0.97	1 M H_2SO_4
Erioglaucin A	Yellow green	Bluish red	0.98	0.5 M H_2SO_4
5-Methylferroin			1.02	1 M H_2SO_4
Ferroin	Red	Faint blue	1.11	1 M H_2SO_4
Nitroferroin	Red	Faint blue	1.25	1 M H_2SO_4

10.8b. Selection of Indicator

Obviously, an indicator should change color at or near the equivalence potential. If the titration is feasible, there will be a large change in potential at the equivalence point, and this should be sufficient to bring about the change in color of the indicator. The following example illustrates more precisely the procedure that may be followed to select the proper indicator.

EXAMPLE 10.10 (a) In the titration of iron(II) iron with cerium(IV) sulfate in 1 M sulfuric acid, what indicator should be used? We have calculated the potential at the equivalence point to be 1.06 V (page 279). Referring to Table 10.2, it is seen that ferroin, with a transition potential of 1.11 V, is a suitable indicator. The standard potential of ferroin is 1.06 V, but the color change occurs at 1.11 V since it is necessary to have more of the indicator in the oxidized form (light blue) than the reduced form (dark red).

(b) Iron(II) ion is titrated with an oxidizing agent in a sulfuric-phosphoric acid medium. What should be the transition potential of an indicator which changes color when all but 0.1% of iron(II) ion is oxidized to iron(III)?

The formal potential of the Fe^{3+}–Fe^{2+} couple in 1 F H_2SO_4 and 0.5 F H_3PO_4 is 0.61 V. Hence

$$E = 0.61 - 0.059 \log \frac{[Fe^{2+}]}{[Fe^{3+}]}$$

$$E = 0.61 - 0.059 \log \frac{1}{1000}$$

$$E = 0.61 + 0.18 = 0.79 \text{ V}$$

The indicator diphenylaminesulfonic acid is frequently used when iron is titrated with potassium dichromate in sulfuric-phosphoric acid media. Note (Table 10.2) that its transition potential is 0.85 V, and hence it changes color when even less than 0.1% of Fe^{2+} remains unoxidized. $\qquad\square$

10.8c. *Structural Chemistry of Redox Indicators*

The redox indicators to which we have referred in this chapter are organic molecules that undergo structural changes upon being oxidized or reduced. There are fewer such indicators than there are acid-base indicators, and their chemistry has not been as widely studied. Nevertheless, the structural changes which account for the different colors are known for a number of substances. We shall consider only two examples here, sodium diphenylaminesulfonate and iron(II) *o*-phenanthroline (ferroin).

Diphenylamine was one of the first redox indicators to be widely used in titrimetric analysis. Since this compound is difficultly soluble in water, and since tungstate ion and mercury(II) chloride interfere with its action, the barium or sodium salt of diphenylaminesulfonic acid is more commonly used. The reduced form of this indicator is colorless, the oxidized form a deep violet. The mechanism of the color change has been shown to be as follows, using diphenylamine as an example.[3]

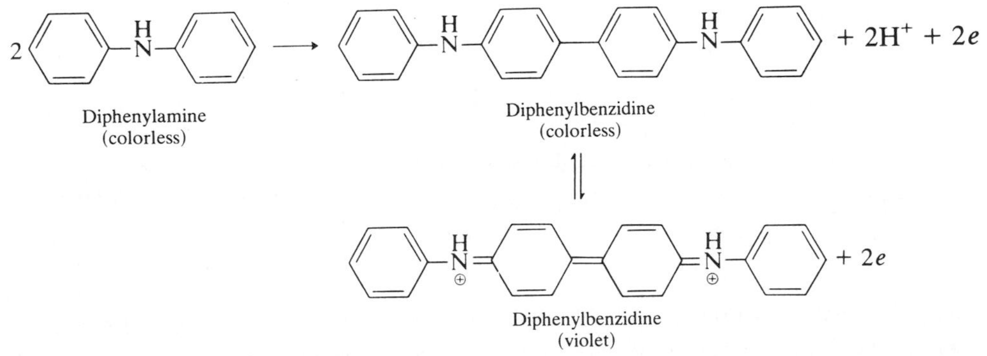

Diphenylamine
(colorless)

Diphenylbenzidine
(colorless)

$+ 2H^+ + 2e$

Diphenylbenzidine
(violet)

$+ 2e$

The presence of a long conjugated system, such as that in the diphenylbenzidine ion, leads to absorption of light in the visible region, and hence the ion is colored.

The indicator ferroin is the iron(II) complex of the organic compound 1,10-phenanthroline,

[3] A. M. Kolthoff and L. A. Sarver, *J. Am. Chem. Soc.*, **52**, 4179 (1930).

1,10-Phenanthroline   Iron(II) 1,10-Phenanthroline

Each of the two nitrogen atoms in 1,10-phenanthroline has an unshared pair of electrons that can be shared with iron(II) ion. Three such molecules of the organic compound attach themselves to the metallic ion to form a blood-red complex ion. The iron(II) ion can be oxidized to iron(III), and the latter ion also forms a complex with three molecules of 1,10-phenanthroline. The color of the iron(III) complex is light blue, and hence a sharp color change occurs when iron(II) is oxidized to iron(III) in the presence of 1,10-phenanthroline:

$$Ph_3Fe^{3+} + e \rightleftharpoons Ph_3Fe^{2+} \qquad E° = 1.06 \text{ V}$$

Light blue   Dark red

The indicator is prepared by mixing equivalent quantities of iron(II) sulfate and 1,10-phenanthroline. The complex salt is called *ferroin;* the complex salt of iron(III) ions is called *ferriin.* As previously mentioned, the color change occurs at about 1.11 V, since the color of ferroin is so much more intense than that of ferriin.

Substituted 1,10-phenanthrolines also form complexes with iron(II) and iron(III) ions and act as redox indicators. The redox potentials are different from that of the ferroin-ferriin system. A few examples are included in Table 10.2, where a partial list of redox indicators is given.

KEY TERMS

Ampere. The unit of current, abbreviated A. One ampere is a current flow of 1 C/s.

Anode. An electrode at which oxidation occurs.

Calomel electrode. A reference electrode based on the half-reaction $Hg_2Cl_2(s) + 2e \rightleftharpoons 2Hg + 2Cl^-$.

Cathode. An electrode at which reduction occurs.

Cell. A combination of two single electrodes, or half-cells.

Cell reaction. The redox reaction which occurs as a cell discharges.

Coulomb. The unit of electrical charge, abbreviated C. The charge on an electron is 1.60×10^{-19} C.

Current. The rate of flow of electrical charge through a circuit.

Electrolytic cell. A cell in which the cell reaction is forced to occur in the nonspontaneous direction.

Faraday. The electrical charge on 1 mol of electrons, or 96,500 C.

Formal potential. The potential of a redox couple when the analytical concentrations (formalities) of Ox and Red are the same and other constituents of the solution (buffers, acids, or whatever) are specified.

Galvanic cell. A cell in which the cell reaction occurs spontaneously, releasing energy which may be used for performing work.

Half-cell. A single electrode, such as a metal in contact with a solution of its ions.

Half-reaction. An oxidation or reduction reaction, the equation for which shows explicitly the electrons involved.

Inert electrode. A metal which serves as a conductor of electrons without entering into the reaction itself.

Liquid junction potential. A potential which develops at the interface between two ionic solutions of different composition.

Nernst equation. The equation which gives the relation between the potential of a single electrode or a cell and the activities of reactants. For the half-reaction

$$Ox + ne \rightleftharpoons Red$$

the equation is

$$E = E° - \frac{0.059}{n} \log \frac{a_{Red}}{a_{Ox}}$$

Poised solution. A solution which resists large changes in potential when an oxidizing or reducing agent is added.

Potentiometric titration. A titration in which the end point is determined by a measurement of potential.

Redox couple. The oxidized and reduced forms of a substance in equilibrium with one another, as $Fe^{3+} + e \rightleftharpoons Fe^{2+}$.

Redox reaction. A reaction in which electrons are lost by one reactant and gained by another.

Reference electrode. A single electrode whose potential is accurately known and can be used to measure the potential of another electrode.

Salt bridge. A solution of an electrolyte, such as KCl, used to make an electrical connection between the solutions in the anode and cathode compartments of a cell.

Single-electrode potential. The potential of a half-reaction referred to the standard H_2 electrode.

Specific indicator. A substance which reacts specifically with one of the reactants in a titration to produce a color.

Standard potential. The potential of a single electrode or a cell in which all reactants and products are at unit activity.

Transition potential. The potential at which a true redox indicator undergoes an observable color change.

True redox indicator. A substance which exhibits different colors in its oxidized and reduced forms.

Voltage. The "driving force" (also called emf or potential) of a galvanic cell for forcing electrons through an external circuit. One volt is the emf required to impart 1 J of energy to a charge of 1 C.

QUESTIONS

1. *Oxidants and reductants.* Consult the table of standard potentials (Table A-4, Appendix I). (a) List these reagents (all at unit activity) in order of *decreasing* strength as oxidants: Fe^{2+}, Ag^+, Cu^+, Ca^{2+}, and Co^{3+}. (b) List these reagents (all at unit activity) in order of *increasing* strength as reductants: Au, Cd, Cu^+, $S_2O_3^{2-}$, and Zn.

2. *Effect of pH.* Consider the two redox systems:

   (a) $ClO_4^- + 2H^+ + 2e \rightleftharpoons ClO_3^- + H_2O$
   $E° = +1.19$ V

   (b) $IO_3^- + 3H_2O + 6e \rightleftharpoons I^- + 6OH^-$
   $E° = +0.26$ V

   What is the effect on the potential (more positive, less positive, no effect) of increasing the pH?

3. *Leveling effect.* Explain why the leveling effect is not observed with redox reagents as it is with acids and bases.

4. *Effect of complexes.* The standard potential of the Fe^{3+}–Fe^{2+} couple is 0.77 V, whereas that of the ferroin-ferriin couple is +1.06 V. Which complex has the larger stability constant, ferroin or ferriin? Explain.

5. *Reference electrode.* Suppose that the 1 M silver–silver chloride electrode ($E° = +0.22$ V) be adopted as the primary reference electrode and assigned a value of 0.00 V. What would be the standard potentials of these redox couples referred to this electrode?

   (a) $2H^+ + 2e \rightleftharpoons H_2(g)$

   (b) $F_2 + 2e \rightleftharpoons 2F^-$

   (c) $Li^+ + e \rightleftharpoons Li$

Multiple-choice. In the following multiple-choice questions select the *one best* answer.

6. The emf of a cell made up of an electrode of unknown potential and an SCE is 0.65 V with the calomel electrode negative. The potential (V) of the unknown electrode referred to H_2 is (a) -0.90; (b) $+0.40$; (c) $+0.90$; (d) -0.40.

7. Given the redox couple

$$TiO^{2+} + 2H^+ + e \rightleftharpoons Ti^{3+} + H_2O$$
$$E^\circ = +0.10 \text{ V}$$

If the *pH* is decreased by 1 unit, the potential of this couple will (a) become more positive by 2×0.059 V; (b) become more negative by 2×0.059 V; (c) become more positive by $0.059/2$ V; (d) become more negative by 0.059 V.

8. In the titration of Fe^{2+} with Ce^{4+}, the equilibrium potentials developed by the Fe^{3+}–Fe^{2+} and the Ce^{4+}–Ce^{3+} couples will be equal (a) only at the EPt; (b) only halfway to the EPt; (c) throughout the titration; (d) never.

9. Which of these expressions is correct? (a) $2.3RT/F = 0.059$ at $298°C$; (b) $nE^\circ = +0.059 \log K$; (c) $E^\circ = -\Delta G^\circ/nF$; (d) all of the above.

10. Given:

$$A^{3+} + e \rightleftharpoons A^{2+} \qquad E^\circ = +1.42 \text{ V}$$
$$B^{4+} + 2e \rightleftharpoons B^{2+} \qquad E^\circ = +0.40 \text{ V}$$

In the titration of B^{2+} with A^{3+}, the potential at the EPt (V) is (a) 0.61; (b) 0.71; (c) 0.91; (d) 1.08.

PROBLEMS

1. *Cell potentials.* Calculate the potentials of the following cells. Write the cell reaction, give the direction of spontaneous reaction, and indicate the polarities of the electrodes.

(a) $Ni| Ni^{2+}(0.040\ M) || Co^{2+}(0.020\ M) |Co$

(b) $Ni| Ni^{2+}(0.0010\ M) || Co^{2+}(0.10\ M) |Co$

(c) $Hg| Hg_2Cl_2, KCl(0.020\ M) || H^+(0.10\ M), H_2(1\text{ atm}) |Pt$

(d) $Pt| Sn^{2+}(0.020M) + Sn^{4+}(0.010\ M) || Ce^{3+}(0.040\ M) + Ce^{4+}(0.060\ M) |Pt$
$Ce^{4+}(0.060\ M) |Pt$

2. *Cell potentials.* Calculate the potentials of the following cells. Write the cell reaction, give the direction of spontaneous reaction, and indicate the polarities of the electrodes.

(a) $Cr| Cr^{3+}(0.030\ M) || Zn^{2+}(0.0010\ M) |Zn$

(b) $Cr| Cr^{3+}(0.00010\ M) || Zn^{2+}(0.040\ M |Zn$

(c) $Hg| Hg_2Cl_2, KCl(1.0\ M) || Fe^{2+}(0.040\ M) + Fe^{3+}(0.10\ M) |Pt$

3. *Single-electrode potentials.* Calculate the following single-electrode potentials at the indicated concentrations.

(a) $2H^+(10^{-14}\ M) + 2e \rightleftharpoons H_2(1.0\text{ atm})$

(b) $2H_2O + 2e \rightleftharpoons H_2(1.0\text{ atm}) + 2OH^-(10^{-14}\ M), E^\circ = -0.83$ V

(c) $O_2(1.0\text{ atm}) + 4H^+(10^{-7}\ M) + 4e \rightleftharpoons 2H_2O$

(d) $H_3AsO_4(0.01\ M) + 2H^+(10^{-9}\ M) + 2e \rightleftharpoons HAsO_2(0.10\ M) + 2H_2O$

4. *Single-electrode potentials.* Suppose one could lower the Fe^{3+} concentration in a solution to only six ions per liter. (a) If the $[Fe^{2+}]$ is 0.010 *M*, what is the potential of the Fe^{3+}–Fe^{2+} couple? (b) Calculate the $[Fe^{3+}]$ needed to lower the potential of the couple to -2.40 V, $[Fe^{2+}]$ being 0.010 *M*.

5. *Equilibrium constant.* (a) Calculate the equilibrium constant of the reaction in the following cell:

$$Al| Al^{3+}(0.010\ M) || Sn^{2+}(0.010\ M) |Sn$$

(b) Suppose the cell is allowed to discharge until equilibrium is reached. Calculate the equilibrium concentrations of Al^{3+} and Sn^{2+}. (c) Calculate the single-electrode potentials of the aluminium and tin couples at equilibrium.

6. *Concentration cell.* Calculate the potential of the following galvanic cell:

$$\text{Hg} \mid \text{Hg}_2\text{Cl}_2, \text{KCl}(0.001 \; M) \parallel \text{KCl}(0.10 \; M),$$
$$\text{Hg}_2\text{Cl}_2 \mid \text{Hg}$$

7. *Free-energy change.* (a) Calculate the equilibrium constant of the reaction

$$\text{Hg}^2 + 2\text{Ag} \;\rightleftharpoons\; \text{Hg} + 2\text{Ag}^+$$

(b) In which direction is the reaction spontaneous if each reactant is at unit activity? (c) Calculate the value of $\Delta G°$. (d) Calculate the value of ΔG for these concentrations: $[\text{Hg}^{2+}] = 1 \times 10^{-6} \; M$; $[\text{Ag}^+] = 0.10 \; M$. In which direction is the reaction spontaneous at these concentrations?

8. *Equivalence point.* Show that the potential at the equivalence point in the titration of Red_1 with Ox_2 is

$$E = \frac{aE_1° + bE_2°}{a + b}$$

where

$$\text{Ox}_1 + ae \;\rightleftharpoons\; \text{Red}_1 \qquad E_1°$$

and

$$\text{Ox}_2 + be \;\rightleftharpoons\; \text{Red}_2 \qquad E_2°$$

9. *Equivalence point.* Show that the potential at the equivalence point in the titration of Fe^{2+} with MnO_4^- is

$$E = \frac{E_1° + 5E_2°}{6} - 0.08 \; \text{pH}$$

where $E_1°$ is the standard potential of the Fe^{3+}–Fe^{2+} couple and $E_2°$ is that of the MnO_4^-–Mn^{2+} couple.

10. *Mixture of solutions.* In the following examples the solutions are mixed, the reaction reaches equilibrium, and a platinum electrode is inserted in the solution to make a half-cell. This half-cell is connected to a standard H_2 electrode, and the potential of the cell is measured. Calculate the potential measured in each case.

(a) 50 mL of 0.040 M Sn^{2+} + 40 mL of 0.050 M Ce^{4+}.

(b) 50 mL of 0.050 M Fe^{2+} + 50 mL of 0.010 M MnO_4^-, $[\text{H}^+] = 1.0 \; M$

(c) 60 mL of 0.040 M Fe^{2+} + 20 mL of 0.015 M $\text{Cr}_2\text{O}_7^{2-}$, $[\text{H}^+] = 1.0 \; M$

(d) 30 mL of 0.010 M Sn^{2+} + 40 mL of 0.020 M Fe^{3+}

11. *Mixtures of solutions.* Repeat Problem 10 for the following:

(a) 60 mL of 0.050 M Fe^{2+} + 20 mL of 0.030 M KMnO_4, $[\text{H}^+] = 1.0 \; M$

(b) 50 mL of 0.050 M Sn^{2+} + 30 mL of 0.030 M Fe^{3+}

(c) 30 mL of 0.050 M Sn^{2+} + 40 mL of 0.10 M Fe^{3+}

(d) 100 mL of 0.020 M V^{3+} + 40 mL of 0.050 M Cr^{2+}

12. *Titration curve.* 50.0 mL of 0.0400 M Sn^{2+} is titrated with 0.0800 M Ce^{4+} in sulfuric acid solution. Calculate the potential of the solution (referred to H_2) after the addition of the following amounts (mL) of titrant: (a) 0.00; (b) 10.0; (c) 25.0; (d) 49.95; (e) 50.0; (f) 50.05; (g) 60.0.

13. *Fraction oxidized.* A solution of V^{2+} is titrated with an oxidizing agent. Calculate the potential of the V^{3+}–V^{2+} couple when the following percentages of V^{2+} have been oxidized: (a) 20%; (b) 30%; (c) 50%; (d) 80%; (e) 99%; (f) 99.9%.

14. *Fraction oxidized.* Repeat Problem 13 for the oxidation of Sn^{2+} to Sn^{4+}.

15. *Fraction oxidized.* A beaker containing 40.0 mL of solution is 0.0500 M each in Fe^{2+} and Fe^{3+} ions. When 10.0 mL of a reducing agent is added, the potential of the solution changes by -0.118 V. Calculate the normality of the reducing agent.

16. *Titration.* (a) Calculate the value of the equilibrium constant of the reaction

$$\text{Fe}^{2+} + \text{Ce}^{4+} \;\rightleftharpoons\; \text{Fe}^{3+} + \text{Ce}^{3+}$$

in sulfuric acid medium (Example 7, page 276). (b) Calculate the $[\text{Fe}^{3+}]/[\text{Fe}^{2+}]$ ratio at the EPt; (c) What percentage of the Fe^{2+} ion remains unoxidized at the EPt?

17. *Titration.* 5.00 mmol of Fe^{2+} is titrated with 0.020 M KMnO_4, the final volume being 100 mL and $[\text{H}^+] = 1.0 \; M$. (a) Calculate the equilibrium constant for the reaction between Fe^{2+} and MnO_4^- using single-electrode potentials; (b) Calculate the number of milligrams of Fe^{2+} remaining unoxidized when 0.10 mL of KMnO_4 is in excess.

18. *Indicator*. The standard potential of the indicator ferroin is 1.06 V. The indicator shows a color change when 85% of it is in the oxidized form. Calculate the transition potential of ferroin. A one-electron change is involved.

19. *Reference electrode*. The potential of a cell made up of an electrode of unknown potential and an SCE ($E = 0.25$ V) is 0.62 V. Calculate the potential of the unknown electrode if the polarity of the SCE is (a) positive; (b) negative.

20. *Feasibility of titration*. For the redox titration

$$A^{2+} + 2M^{3+} \rightleftharpoons A^{4+} + 2M^{2+}$$

(a) Calculate the value of the equilibrium constant for the following conditions: 50 mL of 0.050 M A^{2+} is titrated with 0.10 M M^{3+}. When 49.95 mL of titrant is added, the reaction is complete. On addition of two more drops (0.10 mL) of titrant the value of pA changes by 2.00 units. (b) What is the difference in the standard potentials of the two redox couples for this value of K?

21. *Feasibility of titration*. Repeat Problem 20 for the titration

$$A^{2+} + N^{4+} \rightleftharpoons A^{4+} + N^{2+}$$

Use 0.050 M A^{2+} and 0.050 M N^{4+}.

22. *Feasibility of titration*. (a) Calculate the equilibrium constant of the reaction

$$Fe^{2+} + B^{3+} \rightleftharpoons Fe^{3+} + B^{2+}$$

where the $E°$ of the B^{3+}–B^{2+} couple $+1.04$ V. Do you expect the titration to be feasible? (b) Calculate the number of milligrams of Fe^{2+} which remain unoxidized when 4.0 mmol of Fe^{2+} is titrated with 0.10 M B^{3+}, one drop (0.05 mL) of titrant in excess. The final volume is 100 mL.

23. *pH–cell potential*. Given the cell

Pt, H_2(1 atm)$|$ $H^+(x\ M)$ $\|$ KCl(satd.), Hg_2Cl_2 $|$Hg

(a) Derive an equation relating the potential of this cell to the pH of the solution in the left-hand electrode. (b) If the solution in the left-hand electrode is 0.010 M HCl, what is the potential of the cell? (c) If the solution in the left-hand electrode is 0.050 M NaOH, what is the potential?

24. *pH–cell potential*. Given the same cell as in Problem 23, calculate the cell potential of the solution in the left-hand electrode in (a) 0.050 M formic acid; (b) 0.10 M sodium formate; (c) 0.020 M methylamine.

25. *Buffer solution*. Given the same cell as in Problem 23 with the solution in the left-hand electrode consisting of 50 mL of a buffer which is 0.10 M in NH_3 and 0.20 M in NH_4Cl. (a) Calculate the potential of the cell. (b) If 25 mL of 0.20 M HCl is added to the buffer, calculate the potential of the cell after the reaction is complete.

26. *Dissociation constant*. The following cell has a potential of 0.56 V when the solution in the left-hand electrode is 0.20 M in salt BCl. B is a weak base.

Pt, H_2(1 atm)$|$ $H^+(x\ M)$ $\|$ HCl(1 M), AgCl$|$Ag

The Ag electrode is positive. Calculate the value of K_b of the base B.

27. *Equilibrium constants*. Using Table A-4 in Appendix I, select the two appropriate half-reactions and calculate the equilibrium constants of the reactions

(a) $2H_2O(l) \rightleftharpoons 2H_2(g) + O_2(g)$

(b) $PbCl_2(s) \rightleftharpoons Pb^{2+} + 2Cl^-$

(c) $3Fe^{2+} \rightleftharpoons 2Fe^{3+} + Fe$

(d) $3Mn^{2+} + 2MnO_4^- + 2H_2O \rightleftharpoons$
$\qquad\qquad 5MnO_2(s) + 4H^+$

28. *Solubility product constant*. Given the following cell:

Cu$|$ $Cu^{2+}(x\ M)$ $\|$ HCl(1 M), AgCl$|$Ag

Sodium hydroxide is added to the solution in the left-hand electrode, precipitating $Cu(OH)_2$. The final pH of the solution is 9.00. The voltage of the cell is now 0.14 V with the silver electrode positive. Calculate the K_{sp} of $Cu(OH)_2$.

29. *Solubility product constant*. Using the tables of standard potentials and solubility product constants in Appendix I, calculate the value of $E°$ for these half-reactions:

(a) $CuCl(s) + e \rightleftharpoons Cu + Cl^-$

(b) $Cu(OH)_2(s) + 2e \rightleftharpoons Cu + 2OH^-$

30. *Complex formation*. Given the cell

Zn$|$ $Zn^{2+}(0.010\ M)$ $\|$ $Cd^{2+}(0.020\ M)$ $|$Cd

(a) Calculate the potential and the direction

of the cell reaction and indicate the polarities of the electrodes. (b) Suppose the solution in the right-hand electrode contains 1.00 mmol of Cd^{2+}. Then 3.00 mmol of EDTA is added to complex the Cd^{2+}, and the pH is adjusted to 6.00. Repeat the calculations in part (a) under the new conditions.

31. *Complex formation.* (a) Calculate the potential of the cell

$$Zn|\,Zn^{2+}(0.10\,M)\,\|\,H^+(1.0\,M)\,|H_2(1.0\,atm), Pt$$

(b) Ammonia is added to the electrode containing Zn^{2+} until the final concentration of uncomplexed NH_3 is 0.10 M. Calculate the potential of the cell under these conditions.

32. *Complex formation.* A solution is prepared by mixing 30 mL of 0.10 M Hg^{2+} with 70 mL of 0.10 M EDTA. The pH of the solution is 4.00. Into the solution are inserted a mercury and a saturated calomel electrode ($E = 0.25$ V), forming the cell

$$Hg|\,Hg^{2+} + EDTA\,\|\,KCl(satd.), Hg_2Cl_2\,|Hg$$

Calculate the potential of the cell.

33. *Metabolic redox couples.* The following are two important redox couples in biochemistry:

$$Acetoacetate + 2H^+ + 2e \rightleftharpoons hydroxybutyrate$$
$$E^{\circ\prime} = -0.293\ (pH\ 7)$$

$$NAD^+ + H^+ + 2e \rightleftharpoons NADH$$
$$E^{\circ\prime} = -0.320\ (pH\ 7)$$

Suppose it is important in the metabolic af-

fairs of a living cell at pH 7 to maintain a $9:1$ molar ratio of acetoacetate of hydroxybutyrate. Also suppose the cell does this indirectly by regulating the molar ratio of NAD^+ to NADH. What should the latter ratio be in a healthy cell?

34. *Disproportionation.* (a) From the table of standard potentials, determine whether the following disproportionation occurs spontaneously or not when all reactants are at unit activity:

$$3Br_2(aq) + 3H_2O \rightleftharpoons BrO_3^- + 5Br^- + 6H^+$$

(b) Calculate the equilibrium constant of the reaction.

35. *Disproportionation.* In basic solution the reaction in Problem 34 is

$$3Br_2(aq) + 6OH^- \rightleftharpoons BrO_3^- + 5Br^- + 3H_2O$$

(a) Calculate the equilibrium constant for this reaction. (b) Calculate the pH at which the reaction is spontaneous. The activities of Br_2, BrO_3^-, and Br^- are each unity.

36. *Disproportionation.* (a) Refer to the table of standard potentials (Table A-4, Appendix I) and calculate whether or not copper(I) ion disproportionates (unit activities):

$$2Cu^+ \rightleftharpoons Cu^{2+} + Cu$$

(b) Calculate the equilibrium constant for this reaction. (c) How large a concentration of Cu^+ ions can exist in equilibrium with Cu^{2+} ions at 1 M concentration?

11

Applications of Oxidation-Reduction Titrations

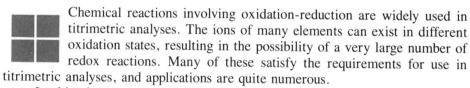

Chemical reactions involving oxidation-reduction are widely used in titrimetric analyses. The ions of many elements can exist in different oxidation states, resulting in the possibility of a very large number of redox reactions. Many of these satisfy the requirements for use in titrimetric analyses, and applications are quite numerous.

In this chapter we shall discuss some widely used redox reagents, their properties, methods used to standardize solutions, and applications to analyses.

11.1

REAGENTS USED FOR PRELIMINARY REDOX REACTIONS

In many analytical procedures the analyte is in more than one oxidation state and must be converted to a single oxidation state before titration. A common example occurs in the determination of iron in an ore. Once the ore is dissolved, iron is present in both the +2 and +3 oxidation states. It must be reduced completely to the +2 state before titration with a standard solution of an oxidizing agent. The redox reagent used in this preliminary step must be able to convert the analyte rapidly and completely into the desired oxidation state. Excess of the reagent is normally added, and one must be able to remove the excess conveniently so that it will not react with the titrant in the subsequent titration.

The following are a few of the common reagents used in preliminary steps.

11.1a. Oxidizing Agents

Sodium and Hydrogen Peroxide

Hydrogen peroxide is a good oxidizing agent with a large positive standard potential:

$$H_2O_2 + 2H^+ + 2e \rightleftharpoons 2H_2O \qquad E^\circ = +1.77 \text{ V}$$

In acidic solution it will oxidize Fe(II) to Fe(III). In alkaline solution it will oxidize Cr(III) to CrO_4^{2-} and Mn(II) to MnO_2. The excess reagent is easily removed by boiling the solution for a few minutes:

$$2H_2O_2 \rightleftharpoons 2H_2O + O_2(g)$$

Potassium and Ammonium Peroxodisulfate

The peroxodisulfate ion is a powerful oxidizing agent in acidic solutions:

$$S_2O_8^{2-} + 2e \rightleftharpoons 2SO_4^{2-} \qquad E^\circ = +2.01 \text{ V}$$

It will oxidize Cr(III) to $Cr_2O_7^{2-}$, Ce(III) to Ce(IV), and Mn(II) to MnO_4^-. The reaction is usually catalyzed by a trace of silver(I) ion. After the oxidation is complete the excess reagent can be removed by boiling the solution:

$$2S_2O_8^{2-} + 2H_2O \longrightarrow 4SO_4^{2-} + O_2(g) + 4H^+$$

Sodium Bismuthate

The formula of this compound is not known with certainty but is usually written $NaBiO_3$. It is a powerful oxidizing agent, oxidizing Mn(II) to MnO_4^-, Cr(III) to $Cr_2O_7^{2-}$, and Ce(III) to Ce(IV). The bismuth is reduced to Bi(III). The compound is sparingly soluble, and the solution of the substance to be oxidized is heated with excess solid. After the reaction is complete the excess bismuthate is removed by filtration.

11.1b. Reducing Agents

Sulfur Dioxide and Hydrogen Sulfide

Both of these gases are relatively mild reducing agents:

$$SO_4^{2-} + 4H^+ + 2e \rightleftharpoons H_2SO_3 + H_2O \qquad E^\circ = +0.17 \text{ V}$$
$$S + 2H^+ + 2e \rightleftharpoons H_2S \qquad E^\circ = +0.14 \text{ V}$$

They are readily soluble in water, and the excess reagent can be easily removed by boiling the solution. They will reduce Fe(III) to Fe(II), V(V) to V(IV), and Ce(IV) to Ce(III). Both gases are toxic and have an unpleasant odor. They are rarely employed in the elementary laboratory.

Tin(II) Chloride

This reagent is used almost exclusively for the reduction of Fe(III) to Fe(II) in samples which have been dissolved in hydrochloric acid. We shall discuss the reagent in detail later.

Metals and Amalgams

A number of metals can be used as reducing agents. Several metals, in particular silver, zinc, cadmium, aluminum, nickel, copper, and mercury, have been widely used in analytical procedures. Sometimes the metal can be used in the form of a rod or a coil of wire and is inserted directly in the solution of the analyte. When reduction is complete, the unused metal is removed from the solution and washed thoroughly. An alternative procedure which ensures more thorough contact between the solution and the metal is to prepare a *reductor,* a glass column containing granules of the metal used as the reductant. The solution to be reduced is poured through the column and caught in the titration flask.

Very active metals, such as zinc, cadmium, and aluminum, not only reduce the analyte but also dissolve in acidic solutions with the evolution of hydrogen. This side reaction is not desirable, since it can consume large amounts of the metal and introduce a considerable quantity of metallic ion into the sample solution. The reaction can be largely prevented by amalgamating the metal with mercury. Amalgamated zinc is used in the *Jones reductor.* Granulated zinc is treated with a dilute solution of mercury(II) chloride, and mercury is displaced, forming a coating of amalgam on the surface:

$$Zn + Hg^{2+} \rightleftharpoons Zn^{2+} + Hg$$

Hydrogen is not as easily displaced by zinc on the amalgamated surface because of the high overvoltage of hydrogen on mercury (Chapter 13). Amalgamated zinc can be used in very acidic solutions and makes ideal packing for a reductor.

Silver metal in the presence of hydrochloric acid is widely used as packing in a metal reductor. Silver is a poor reducing agent, but in the presence of hydrochloric acid its reducing ability is increased:

$$Ag(s) + Cl^- \rightleftharpoons AgCl(s) + e$$

Because it is not as strong a reductant as amalgamated zinc, silver is somewhat more selective than the latter metal.

The Jones and silver reductors are the two most widely used in analytical procedures. Table 11.1 shows some of the applications of the two reductors.

TABLE 11.1 Comparison of Reduction Products in the Jones and Silver Reductors

	REDUCTION PRODUCT	
Metal Ion	Silver Reductor[†]	Jones Reductor[‡]
Titanium(IV)	Not reduced	Titanium(III)
Vanadium(V)	Vanadium(IV)	Vanadium(II)
Chromium(III)	Not reduced	Chromium(II)
Molybdenum(VI)	Molybdenum(V)	Molybdenum(III)
Iron(III)	Iron(II)	Iron(II)
Copper(II)	Copper(I)	Copper(0)
Uranium(VI)	Uranium(IV)	Uranium(IV and III)

Source: I. M. Kolthoff and R. Belcher, *Volumetric Analysis,* Vol. 3, Interscience Publishers, Inc., 1957, New York, p. 12.
[†] $Ag(s) + Cl^- \rightleftharpoons AgCl(s) + e.$
[‡] $Zn(Hg) \rightleftharpoons Zn^{2+} + 2e + Hg.$

POTASSIUM PERMANGANATE

11.2a. Properties of Potassium Permanganate

Potassium permanganate has been widely used as an oxidizing agent for over 100 years. It is a reagent that is readily available, inexpensive, and requires no indicator unless very dilute solutions are used. One drop of 0.1 N permanganate imparts a perceptible pink color to the volume of solution normally used in a titration. This color is used to indicate excess of the reagent. Permanganate undergoes a variety of chemical reactions, since manganese can exist in oxidation states of $+2$, $+3$, $+4$, $+6$, and $+7$ (Example 2, Section 3.2.b).

The most common reaction encountered in the introductory laboratory is the one that takes place in very acidic solutions, 0.1 N or greater:

$$MnO_4^- + 8H^+ + 5e \rightleftharpoons Mn^{2+} + 4H_2O \qquad E° = +1.51 \text{ V}$$

Permanganate reacts rapidly with many reducing agents according to this reaction, but some substances require heating or the use of a catalyst to speed up the reaction. Were it not for the fact that many reactions of permanganate are slow, more difficulties would be encountered in the use of this reagent. For example, permanganate is a sufficiently strong oxidizing agent to oxidize Mn(II) to MnO_2 according to the equation

$$3Mn^{2+} + 2MnO_4^- + 2H_2O \longrightarrow 5MnO_2(s) + 4H^+$$

The slight excess of permanganate present at the end point of a titration is sufficient to bring about the precipitation of some MnO_2. However, since the reaction is slow, MnO_2 is not normally precipitated at the end point of permanganate titrations.

Special precautions must be taken in the preparation of permanganate solutions. Manganese dioxide catalyzes the decomposition of permanganate solutions. Traces of MnO_2 initially present in the permanganate, or formed by the reaction of permanganate with traces of reducing agents in the water, lead to decomposition. Directions usually call for dissolving the crystals, heating to destroy reducible substances, and filtering through asbestos or sintered glass (nonreducing filters) to remove MnO_2. The solution is then standardized, and if kept in the dark and not acidified, its concentration will not change appreciably over a period of several months.

Acidic solutions of permanganate are not stable because permanganic acid is decomposed according to the equation

$$4MnO_4^- + 4H^+ \longrightarrow 4MnO_2(s) + 3O_2(g) + 2H_2O$$

This is a slow reaction in dilute solutions at room temperature. However, one should never add excess permanganate to a reducing agent and then raise the temperature to hasten oxidation, because the foregoing reaction will then occur at an appreciable rate.

11.2b. Primary Standards for Permanganate

Arsenic(III) Oxide

The compound As_2O_3 is an excellent primary standard for permanganate solutions. It is stable, nonhygroscopic, and readily available in a high degree of purity. The oxide is dissolved in sodium hydroxide, and the solution is acidified with hydrochloric acid and titrated with permanganate:

$$5HAsO_2 + 2MnO_4^- + 6H^+ + 2H_2O \longrightarrow 2Mn^{2+} + 5H_3AsO_4$$

(The acid produced by dissolving As_2O_3 behaves as a monoprotic weak acid. We shall write its formula $HAsO_2$ rather than H_3AsO_3.) The reaction is slow at room temperature unless a catalyst is added. Potassium iodide, KI, potassium iodate, KIO_3, and iodine monochloride, ICl, have been used as catalysts.

Sodium Oxalate

This compound, $Na_2C_2O_4$, is also a good primary standard for permanganate in acid solution. It can be obtained in a high degree of purity, is stable on drying, and is nonhygroscopic. Its reaction with permanganate is somewhat complex, and even though many investigations have been made, the exact mechanism is not clear. The reaction is slow at room temperature, and hence the solution is normally heated to about 60°C. Even at an elevated temperature the reaction starts slowly, but the rate increases as manganese(II) ion is formed. Manganese(II) acts as a catalyst, and the reaction is termed *autocatalytic*, since the catalyst is produced in the reaction itself. The ion may exert its catalytic effect by rapidly reacting with permanganate to form manganese of intermediate oxidation states (+3 or +4), which in turn rapidly oxidize oxalate ion, returning to the divalent state.

The equation for the reaction between oxalate and permanganate is

$$5C_2O_4^{2-} + 2MnO_4^- + 16H^+ \longrightarrow 2Mn^{2+} + 10CO_2 + 8H_2O$$

For a number of years analysts employed the procedure recommended by McBride,[1] which called for the entire titration to be carried out slowly at elevated temperature with vigorous stirring. Later, Fowler and Bright[2] thoroughly investigated the reaction and recommended that almost all of the permanganate be added rapidly to the acidified solution at room temperature. After the reaction is complete, the solution is heated to 60°C and the titration completed at this temperature. This procedure eliminates any error caused by the formation of hydrogen peroxide.

Iron

Iron wire of a high degree of purity is available as a primary standard. It is dissolved in dilute hydrochloric acid, and any iron(III) produced during the dissolving process is reduced to iron(II). If the solution is then titrated with permanganate, an appreciable amount of chloride ion is oxidized in addition to the iron(II). The oxidation of chloride ion by permanganate is slow at room temperature. However, in the presence of iron, the oxidation occurs more rapidly. Although iron(II) is a stronger reducing agent than chloride ion, the latter ion is oxidized simultaneously with the iron. It is frequently said that iron "induces" the oxidation of chloride ion. No such difficulty is encountered in the oxidation of As_2O_3 or $Na_2C_2O_4$ in hydrochloric acid solution.

A solution of manganese(II) sulfate, sulfuric acid, and phosphoric acid, called "preventive," or Zimmermann-Reinhardt solution, can be added to the hy-

[1] R. S. McBride, *J. Am. Chem. Soc.,* **34,** 393 (1912).

[2] R. M. Fowler and H. A. Bright, *J. Res. Nat Bur. Standards,* **15,** 493 (1935).

drochloric acid solution of iron before titration with permanganate. Phosphoric acid lowers the concentration of iron(III) ion by formation of a complex, helping force the reaction to completion, and also removes the yellow color which iron(III) shows in chloride media. The phosphate complex is colorless, and the end point is made somewhat clearer.

11.2c. *Determinations with Permanganate*

Iron in Iron Ores

The determination of iron in iron ores is one of the most important applications of permanganate titrations. The best acid for dissolving iron ores is hydrochloric, and tin(II) chloride is frequently added to aid in the dissolving process.

Before titration with permanganate any iron(III) must be reduced to iron(II). This reduction can be done with the Jones reductor (Section 11.1.b) or with tin(II) chloride. The Jones reductor is preferred if the acid present is sulfuric, since no chloride ion is introduced.

If the solution contains hydrochloric acid, as is often the case, reduction with tin(II) chloride is convenient. The chloride is added to a hot solution of the sample, and the progress of the reduction is followed by noting the disappearance of the yellow color of the iron(III) ion:

$$Sn^{2+} + 2Fe^{3+} \longrightarrow Sn^{4+} + 2Fe^{2+}$$

A slight excess of tin(II) chloride is added to ensure completeness of reduction. This excess must be removed or it will react with the permanganate upon titration. For this purpose, the solution is cooled, and mercury(II) chloride is added rapidly to oxidize excess tin(II) ion:

$$2HgCl_2 + Sn^{2+} \longrightarrow Hg_2Cl_2(s) + Sn^{4+} + 2Cl^-$$

Iron(II) is not oxidized by the mercury(II) chloride. The precipitate of mercury(I) chloride, if small, does not interfere in the subsequent titration. However, if too large an excess of tin(II) chloride is added, mercury(I) chloride may be further reduced to free mercury:

$$Hg_2Cl_2(s) + Sn^{2+} \longrightarrow 2Hg(1) + 2Cl^- + Sn^{4+}$$

Mercury, which is produced in a finely divided state under these conditions, causes the precipitate to appear gray to black. If the precipitate is dark, the sample should be discarded, since mercury, in the finely divided state, will be oxidized during the titration. The tendency toward further reduction of Hg_2Cl_2 is diminished if the solution is cool and the $HgCl_2$ is added rapidly. Of course, if insufficient $SnCl_2$ is added, no precipitate of Hg_2Cl_2 will be obtained. In such a case, the sample must be discarded.

Tin(II) chloride is normally used for reduction of iron in samples which have been dissolved in hydrochloric acid. Zimmermann-Reinhardt preventive solution is then added if the titration is to be done with permanganate.

Determination of Other Reducing Agents

Many reducing agents in addition to iron(II) can be determined by titration with permanganate in acid solution. As mentioned in the discussion of standardization processes, arsenic can be titrated in hydrochloric acid. Antimony can also be determined, as well as nitrites (oxidized to nitrates) and hydrogen peroxide. The peroxide acts as a reducing agent in the reaction

$$2MnO_4^- + 5H_2O_2 + 6H^+ \longrightarrow 2Mn^{2+} + 5O_2(g) + 8H_2O$$

We have also seen that oxalates can be titrated with permanganate. The titrimetric determination of calcium in limestone is frequently used as a student exercise. Calcium is precipitated as the oxalate, CaC_2O_4. After filtration and washing, the precipitate is dissolved in sulfuric acid and the oxalate titrated with permanganate. The procedure is more rapid than the gravimetric one in which CaC_2O_4 is ignited to CaO and weighed. However, care must be taken in the precipitation to avoid contamination of the precipitate with other oxalates or oxalic acid. Such contamination, of course, leads to high results. Other cations which form insoluble oxalates can be determined in the same manner. These include manganese(II), zinc(II), cobalt(II), barium(II), strontium(II), and lead(II).

Table 11.2 summarizes some of the more common determinations that can be made by direct titration with permanganate in acid solution.

TABLE 11.2 Some Applications of Direct Titrations with Permanganate in Acid Solution

ANALYTE	HALF-REACTION OF SUBSTANCE OXIDIZED
Antimony(III)	$HSbO_2 + 2H_2O \rightleftharpoons H_3SbO_4 + 2H^+ + 2e$
Arsenic(III)	$HAsO_2 + 2H_2O \rightleftharpoons H_3AsO_4 + 2H^+ + 2e$
Bromine	$2Br^- \rightleftharpoons Br_2 + 2e$
Hydrogen peroxide	$H_2O_2 \rightleftharpoons O_2 + 2H^+ + 2e$
Iron(II)	$Fe^{2+} \rightleftharpoons Fe^{3+} + e$
Molybdenum(III)	$Mo^{3+} + 2H_2O \rightleftharpoons MoO_2^{2+} + 4H^+ + 3e$
Nitrite	$HNO_2 + H_2O \rightleftharpoons NO_3^- + 3H^+ + 2e$
Oxalate	$H_2C_2O_4 \rightleftharpoons 2CO_2 + 2H^+ + 2e$
Tin(II)	$Sn^{2+} \rightleftharpoons Sn^{4+} + 2e$
Titanium(III)	$Ti^{3+} + H_2O \rightleftharpoons TiO^{2+} + 2H^+ + 1e$
Tungsten(III)	$W^{3+} + 2H_2O \rightleftharpoons WO_2^{2+} + 4H^+ + 3e$
Uranium(IV)	$U^{4+} + 2H_2O \rightleftharpoons UO_2^{2+} + 4H^+ + 2e$
Vanadium(IV)	$VO^{2+} + 3H_2O \rightleftharpoons V(OH)_4^+ + 2H^+ + e$

Indirect Determination of Oxidizing Agents

A standard solution of $KMnO_4$ can also be employed indirectly in the determination of oxidizing agents, particularly the higher oxides of such metals as lead and manganese. Such oxides are difficult to dissolve in acids or bases without reduction of the metal to a lower oxidation state. It is impractical to titrate these substances directly, because the reaction of the solid with a reducing agent is slow. Hence the sample is treated with a known excess of some reducing agent and heated to complete the reaction. Then the excess reducing agent is titrated with standard permanganate. Various reducing agents may be employed, such as As_2O_3 and $Na_2C_2O_4$. The analysis of pyrolusite, an ore containing MnO_2, is a common student exercise. The reaction of MnO_2 with $HAsO_2$ is

$$MnO_2(s) + HAsO_2 + 2H^+ \longrightarrow Mn^{2+} + H_3AsO_4$$

COMPOUNDS OF CERIUM

11.3a. Properties

The element cerium (atomic number 58) can exist in solution in only two oxidation states, +4 and +3. In the quadrivalent state it is a powerful oxidizing agent, undergoing a single reaction,

$$Ce^{4+} + e \rightleftharpoons Ce^{3+}$$

The Ce(IV) ion is used in solutions of high acidity because hydrolysis leads to precipitation in solutions of low hydrogen ion concentration. The redox potential of the Ce(IV)/Ce(III) couple is dependent upon the nature and concentration of the acid present (page 271). The formal potentials in 1 M solutions of common acids are: $HClO_4$, +1.70 V; HNO_3, 1.61 V; H_2SO_4, +1.44 V; HCl, +1.28 V.

It is known that both cerium(IV) and cerium(III) ions form stable complexes with various anions. Some chemists name the acids and salts of cerium to indicate that the element is present as a complex anion rather than as a cation. For example, the salt $(NH_4)_2Ce(NO_3)_6$ is called *ammonium hexanitratocerate*. For simplicity, we shall call such a compound *cerium(IV) ammonium nitrate* and write the formula $Ce(NO_3)_4 \cdot 2NH_4NO_3$.

When Ce(IV) ion is used as a titrant, the compound ferroin (Section 10.9) is normally employed as an indicator. The ion can be employed in most titrations where permanganate is used, and it possesses properties which often make it a better choice as a titrant. Its principal advantages over permanganate are as follows:

1. There is only one oxidation state, Ce(III), to which the Ce(IV) ion is reduced.
2. It is a very strong oxidizing agent and, as pointed out above, one can vary the intensity of its oxidizing power by choice of the acid employed.
3. Sulfuric acid solutions of Ce(IV) ion are extremely stable.
4. The reagent can be used for iron titrations in hydrochloric acid solution without need of the Zimmermann-Reinhardt preventive solution, since chloride ion is not readily oxidized. However, Ce(IV) solutions in hydrochloric acid are unstable if the concentration of the acid is more than about 1 M.
5. A salt, cerium(IV) ammonium nitrate, sufficiently pure to be weighed directly for preparing standard solutions, is available.

11.3b. Standardization of Solutions

Solutions of Ce(IV) are usually prepared from cerium(IV) hydrogen sulfate, $Ce(HSO_4)_4$, cerium(IV) ammonium sulfate, $Ce(SO_4)_2 \cdot 2(NH_4)_2SO_4 \cdot 2H_2O$, or cerium(IV) hydroxide, $Ce(OH)_4$. The compounds are dissolved in 0.2 to 0.5 M strong acid to prevent hydrolysis and the formation of slightly soluble basic salts. The solution is then standardized against one of the primary standards listed below.

As mentioned above, the compound cerium(IV) ammonium nitrate is available as a primary standard, and standard solutions may be made by direct weighing, followed by dilution in a volumetric flask. This salt can also be obtained as ordinary reagent-grade material, in which case the solution must be standardized.

The same primary standards used for potassium permanganate, arsenic(III) oxide, sodium oxalate, and iron wire, can be used to standardize Ce(IV) solutions. A catalyst, such as OsO_4 or ICl, is often needed to speed up the reactions with arsenic(III) oxide and sodium oxalate.

11.3c. Determinations with Cerium(IV) Solutions

As previously mentioned, Ce(IV) solutions can be employed in most titrations where permanganate is used. The more important applications include the determination of iron, arsenic, antimony, oxalates, ferrocyanide, titanium, chromium, vanadium, molybdenum, uranium, and the oxides of lead and manganese.

11.4

POTASSIUM DICHROMATE

Potassium dichromate is a fairly strong oxidizing agent, the standard potential of the reaction

$$Cr_2O_7^{2-} + 14H^+ + 6e \rightleftharpoons 2Cr^{3+} + 7H_2O$$

being $+1.33$ V. It is not, however, as strong as potassium permanganate or the cerium(IV) ion. It has the advantages of being inexpensive, very stable in solution, and available in sufficiently pure form for preparing standard solutions by direct weighing. It is frequently used as a primary standard for sodium thiosulfate solutions (see the next section).

Dichromate solutions have not been so widely used as permanganate or cerium(IV) solutions in analytical procedures because they are not as strong oxidizing agents and because of the slowness of some of their reactions. The principal use has been in the titration of iron in hydrochloric acid solution, since no difficulty is encountered in the oxidation of chloride ion if the hydrochloric acid concentration is less than 2 M. The compound diphenylaminesulfonic acid is a suitable indicator when iron is titrated in a sulfuric-phosphoric acid medium (page 285). This indicator has a transition potential of $+0.85$ V and is oxidized to a deep purple color by excess dichromate. This color is sufficiently intense to be readily detected even in the presence of green chromium(III) ion produced by the reduction of dichromate during the titration. Sodium diphenylbenzidine sulfonate ($E° = +0.87$ V) is also a suitable indicator.

As mentioned above, the principal use of dichromate solutions is in the titration of iron in hydrochloric acid solutions. An indirect method for determining oxidizing agents involves treating the sample with a known excess of iron(II) and then titrating the excess with standard dichromate. Such oxidizing agents as nitrate, NO_3^-, chlorate, ClO_3^-, and hydrogen peroxide, H_2O_2, have been determined in this manner. Copper(I) has been determined by reaction with standard Fe(III) solution, followed by titration with dichromate of the resulting Fe(II).

The iodine (triiodide)-iodide redox system,[3]

$$I_3^- + 2e \rightleftharpoons 3I^-$$

has a standard potential of +0.54 V. Iodine, therefore, is a much weaker oxidizing agent than potassium permanganate, cerium(IV) compounds, and potassium dichromate. On the other hand, iodide ion is a reasonably strong reducing agent, stronger, for example, than Fe(II) ion. In analytical processes, iodine is used as an oxidizing agent (*iodimetry*), and iodide ion is used as a reducing agent (*iodometry*). Relatively few substances are sufficiently strong reducing agents to be titrated directly with iodine. Hence the number of iodimetric determinations is small. However, many oxidizing agents are sufficiently strong to react completely with iodide ion, and there are many applications of iodometric processes. An excess of iodide ion is added to the oxidizing agent being determined, liberating iodine, which is then titrated with sodium thiosulfate solution. The reaction between iodine and thiosulfate goes well to completion. This reaction is discussed below. It should be pointed out that some chemists prefer to avoid the term *iodimetry*, and instead, speak of direct and indirect *iodometric* processes.

11.5a. *Direct or Iodimetric Processes*

The more important substances that are sufficiently strong reducing agents to be titrated directly with iodine are thiosulfate, arsenic(III), antimony(III), sulfite, sulfite, tin(II), and ferrocyanide. The reducing power of several of these substances depends upon the hydrogen ion concentration, and only by proper adjustment of *p*H can the reaction with iodine be made quantitative.

Preparation of Iodine Solution

Iodine is only slightly soluble in water (0.00134 mol/liter at 25°C) but is quite soluble in solutions containing iodide ion. Iodine forms the triiodide complex with iodide,

$$I_2 + I^- \rightleftharpoons I_3^-$$

the equilibrium constant being about 710 at 25°C. An excess of potassium iodide is added to increase the solubility and to decrease the volatility of iodine. Usually about 3 to 4% by weight of KI is added to a 0.1 *N* solution, and the bottle containing the solution is well stoppered.

Standardization

Standard iodine solutions can be prepared by direct weighing of pure iodine and dilution in a volumetric flask. The iodine is purified by sublimation and is

---

[3] The principal species in a solution of iodine and potassium iodide is the triiodide ion, I_3^-, and many chemists refer to *triiodide* solutions rather than *iodine* solutions. For simplicity, we shall continue to use the term *iodine* solutions and write equations using I_2 rather than I_3^-.

added to a concentrated solution of KI, which is accurately weighed before and after the addition of iodine. Usually, however, the solution is standardized against a primary standard, As_2O_3 being most commonly used. The reducing power of $HAsO_2$ depends upon the pH, as shown by the following equation:

$$HAsO_2 + I_2 + 2H_2O \rightleftharpoons H_3AsO_4 + 2H^+ + 2I^-$$

The value of the equilibrium constant for this reaction is 0.17; hence the reaction does not go to completion at the equivalence point. However, if the hydrogen ion concentration is lowered, the reaction is forced to the right and can be made sufficiently complete to be suitable for a titration. Usually the solution is buffered at a pH slightly above 8, using sodium bicarbonate, and the titration gives excellent results.

Starch Indicator

The color of a $0.1 N$ solution of iodine is sufficiently intense that iodine can act as its own indicator. Iodine also imparts an intense purple or violet color to such solvents as carbon tetrachloride and chloroform, and sometimes this is utilized in detecting the end point of titrations. More commonly, however, a solution (colloidal dispersion) of starch is employed, since the deep blue color of the starch-iodine complex serves as a very sensitive test for iodine. The mechanism of the formation of the colored complex is not known, but it is thought that molecules of iodine are held on the surface of β-amylose, a constituent of starch. Starch solutions are easily decomposed by bacteria, and usually a substance, such as boric acid is added as a preservative.

Determinations with Iodine (Direct Titration)

Some of the determinations that can be done by direct titration with a standard iodine solution are listed in Table 11.3. The determination of antimony is similar to that of arsenic, except that tartrate ions, $C_4H_4O_6^{2-}$, are added to complex antimony and avoid precipitation of salts such as SbOCl when the solution is neutralized. The titration is carried out in a bicarbonate buffer of pH about 8. In the determination of tin and sulfites the solution being titrated must be protected from oxidation by air. The hydrogen sulfide titration is frequently used to determine sulfur in iron or steel.

TABLE 11.3 Determinations by Direct Iodine Titrations

ANALYTE	REACTION
Antimony(III)	$HSbOC_4H_6O_6 + I_2 + H_2O \rightleftharpoons HSbO_2C_4H_4O_6 + 2H^+ + 2I^-$
Arsenic(III)	$HAsO_2 + I_2 + 2H_2O \rightleftharpoons H_3AsO_4 + 2H^+ + 2I^-$
Ferrocyanide	$2Fe(CN)_6^{4-} + I_2 \rightleftharpoons 2Fe(CN)_6^{3-} + 2I^-$
Hydrogen cyanide	$HCN + I_2 \rightleftharpoons ICN + H^+ + I^-$
Hydrazine	$N_2H_4 + 2I_2 \rightleftharpoons N_2 + 4H^+ + 4I^-$
Sulfur (sulfide)	$H_2S + I_2 \rightleftharpoons 2H^+ + 2I^- + S$
Sulfur (sulfite)	$H_2SO_3 + I_2 + H_2O \rightleftharpoons H_2SO_4 + 2H^+ + 2I^-$
Thiosulfate	$2S_2O_3^{2-} + I_2 \rightleftharpoons S_4O_6^{2-} + 2I^-$
Tin(II)	$Sn^{2+} + I_2 \rightleftharpoons Sn^{4+} + 2I^-$

11.5b. Indirect or Iodometric Processes

Many strong oxidizing agents can be analyzed by adding potassium iodide in excess and titrating the liberated iodine. Since many oxidizing agents require an acidic solution for reaction with iodine, sodium thiosulfate is commonly used as the titrant. Titration with arsenic(III) (above) requires a slightly alkaline solution.

Sodium Thiosulfate

Sodium thiosulfate is commonly purchased as the pentahydrate, $Na_2S_2O_3 \cdot 5H_2O$, and solutions are standardized against a primary standard. Solutions are not stable over long periods of time, and frequently borax or sodium carbonate is added as a preservative.

Iodine oxidizes thiosulfate to the tetrathionate ion:

$$I_2 + 2S_2O_3^{2-} \longrightarrow 2I^- + S_4O_6^{2-}$$

The reaction is rapid, goes well to completion, and there are no side reactions. The equivalent weight of $Na_2S_2O_3 \cdot 5H_2O$ is the molecular weight, 248.17, since one electron per molecule is lost. If the pH of the solution is above 9, thiosulfate is oxidized partially to sulfate:

$$4I_2 + S_2O_3^{2-} + 5H_2O \longrightarrow 8I^- + 2SO_4^{2-} + 10H^+$$

In neutral, or slightly alkaline solution, oxidation to sulfate does not occur, especially if iodine is used as the titrant. Many strong oxidizing agents, such as permanganate, dichromate, and cerium(IV) salts, oxidize thiosulfate to sulfate, but the reaction is not quantitative.

Standardization of Thiosulfate Solutions. A number of substances can be used as primary standards for thiosulfate solutions. Pure iodine is the most obvious standard but is seldom used because of difficulty in handling and weighing. More often, use is made of a strong oxidizing agent which will liberate iodine from iodide, an iodometric process.

Potassium Dichromate. This compound can be obtained in a high degree of purity. It has a fairly high equivalent weight, it is nonhygroscopic, and the solid and its solutions are very stable. The reaction with iodide is carried out in about 0.2 to 0.4 M acid and is complete in 5 to 10 min:

$$Cr_2O_7^{2-} + 6I^- + 14H^+ \longrightarrow 2Cr^{3+} + 3I_2 + 7H_2O$$

The equivalent weight of potassium dichromate is one-sixth of the molecular weight, or 49. 03 g/eq. At acid concentrations greater than 0.4 M, air oxidation of potassium iodide becomes appreciable. For best results, a small portion of sodium bicarbonate or dry ice is added to the titration flask. The carbon dioxide produced displaces the air, after which the mixture is allowed to stand until the reaction is complete.

Potassium Iodate and Potassium Bromate. Both of these salts oxidize iodide quantitatively to iodine in acid solution:

$$IO_3^- + 5I + 6H^+ \longrightarrow 3I_2 + 3H_2O$$

$$BrO_3^- + 6I^- + 6H^+ \longrightarrow 3I_2 + Br^- + 3H_2O$$

The iodate reaction is quite rapid; it also requires only a slight excess of hydrogen ions for complete reaction. The bromate reaction is rather slow, but the speed can be increased by increasing the hydrogen ion concentration. Usually, a small amount of ammonium molybdate is added as a catalyst.

The principal disadvantage of these two salts as primary standards is that the equivalent weights are small. In each case the equivalent weight is one-sixth of the molecular weight,[4] that of KIO_3 being 35.67 and that of $KBrO_3$ being 27.84. In order to avoid a large error in weighing, directions usually call for weighing a large sample, diluting in a volumetric flask, and withdrawing aliquots. The salt potassium acid iodate, $KIO_3 \cdot HIO_3$, can also be used as a primary standard, but its equivalent weight is also small, one-twelfth the molecular weight, or 32.49.

Copper.   Pure copper can be used as a primary standard for sodium thiosulfate and is recommended when the thiosulfate is to be used for the determination of copper. The standard potential of the Cu(II)–Cu(I) couple,

$$Cu^{2+} + e \rightleftharpoons Cu^+$$

is +0.15 V, and thus iodine, $E^\circ = +0.53$ V, is a better oxidizing agent than Cu(II) ion. However, when iodide ions are added to a solution of Cu(II), a precipitate of CuI is formed,

$$2Cu^{2+} + 4I^- \longrightarrow 2Cu(s) + I_2$$

The reaction is forced to the right by the formation of the precipitate and also by the addition of excess iodide ion. The *p*H of the solution must be maintained by a buffer system, preferably between 3 and 4.

It has been found that iodine is held by adsorption on the surface of the copper(I) iodide precipitate and must be displaced to obtain correct results. Potassium thiocyanate is usually added just before the end point is reached to displace the adsorbed iodine.

Iodometric Determinations.   There are many applications of iodometric processes in analytical chemistry. Some of these are listed in Table 11.4. The iodometric determination of copper is widely used for both ores and alloys. The method gives excellent results and is more rapid than the electrolytic determination of copper. The classical method of Winkler[5] is a sensitive method for determining oxygen dissolved in water. To the water sample is added an excess of a manganese(II) salt, sodium iodide, and sodium hydroxide. White Mn(OH) is precipitated and is quickly oxidized to brown $Mn(OH)_3$. The solution is acidified,

---

[4] It should be noted that the iodate ion gains five electrons in the reaction with iodide ions, and therefore its equivalent weight *in this reaction* is one-fifth of the molecular weight. However, the reaction involved in the titration is that between iodine and thiosulfate. Since 1 mmol of iodate produces 3 mmol or 6 meq of iodine, the equivalent weight of iodate for the complete process is one-sixth of the molecular weight.

[5] L. W. Winkler, *Ber. Dtsbh. Chem. Ges.,* **21,** 2843 (1888); *Standard Methods for the Examination of Sewage,* 9th ed., American Public Health Association, New York, 1946, p. 124.

TABLE 11.4 Determinations by Indirect Iodine Titrations

ANALYTE	REACTION
Arsenic(V)	$H_3AsO_4 + 2H^+ + 2I^- \rightleftharpoons HAsO_2 + I_2 + 2H_2O$
Bromine	$Br_2 + 2I^- \rightleftharpoons 2Br^- + I_2$
Bromate	$BrO_3^- + 6H^+ + 6I^- \rightleftharpoons Br^- + 3I_2 + 3H_2O$
Chlorine	$Cl_2 + 2I^- \rightleftharpoons 2Cl^- + I_2$
Chlorate	$ClO_3^- + 6H^+ + 6I^- \rightleftharpoons Cl^- + 3I_2 + 3H_2O$
Copper(II)	$2Cu^{2+} + 4I^- \rightleftharpoons 2CuI(s) + I_2$
Dichromate	$Cr_2O_7^{2-} + 6I^- + 14H^+ \rightleftharpoons 2Cr^{3+} + 3I_2 + 7H_2O$
Hydrogen peroxide	$H_2O_2 + 2H^+ + 2I^- \rightleftharpoons I_2 + 2H_2O$
Iodate	$IO_3^- + 5I^- + 6H^+ \rightleftharpoons 3I_2 + 3H_2O$
Nitrite	$2HNO_2 + 2I^- + 2H^+ \rightleftharpoons 2NO + I_2 + 2H_2O$
Oxygen	$O_2 + 4Mn(OH)_2 + 2H_2O \rightleftharpoons 4Mn(OH)_3$
	$2Mn(OH)_3 + 2I^- + 6H^+ \rightleftharpoons 2Mn^{2+} + I_2 + 6H_2O$
Ozone	$O_3 + 2I^- + 2H^+ \rightleftharpoons O_2 + I_2 + H_2O$
Periodate	$IO_4^- + 7I^- + 8H^+ \rightleftharpoons 4I_2 + 4H_2O$
Permanganate	$2MnO_4^- + 10I^- + 16H^+ \rightleftharpoons 2Mn^{2+} + 5I_2 + 8H_2O$

and the $Mn(OH)_3$ oxidizes iodide to iodine, which is then titrated with a standard solution of sodium thiosulfate. The equations are given in Table 11.4.

11.6

PERIODIC ACID

The compound paraperiodic acid, H_5IO_6, is a powerful oxidizing agent which is extremely useful in performing selective oxidations of organic compounds with certain functional groups. The standard potential of the reaction

$$H_5IO_6 + H^+ + 2e \rightleftharpoons IO_3^- + 3H_2O$$

is about $+1.6$ to $+1.7$ V.

11.6a. *Preparation and Standardization of Solutions*

Three compounds are available commercially for the preparation of periodate solutions: H_5IO_6, paraperiodic acid; $NaIO_4$, sodium metaperiodate; and KIO_4, potassium metaperiodate. Of these compounds, $NaIO_4$ is generally preferred because of its relatively high solubility and ease of purification. Solutions as concentrated as $0.06\ M$ can be prepared. Periodic acid solutions slowly oxidize water to oxygen and ozone. Solutions containing an excess of sulfuric acid are the most stable.

Periodate solutions are standardized by an iodometric procedure:

$$IO_4^- + 2I^- + H_2O \longrightarrow IO_3^- + I_2 + 2OH^-$$

Excess potassium iodide is added to an aliquot of the periodate, and the liberated iodine is titrated with a standard solution of arsenic(III):

$$I_2 + AsO_2^- + 2H_2O \longrightarrow HAsO_4^{2-} + 2I^- + 3H^+$$

The solution is buffered at a pH of 8 to 9 with borax or sodium bicarbonate.

11.6b. *The Malaprade Reaction*

In 1928, Malaprade[6] reported that periodic acid could be used for the selective oxidation of organic compounds with hydroxyl groups on *adjacent carbon atoms*. The reaction with ethylene glycol is

$$\begin{array}{l} H_2C\!-\!OH \\ | \qquad\qquad + H_4IO_6^- \longrightarrow 2H_2C\!=\!O + IO_3^- + 3H_2O \\ H_2C\!-\!OH \end{array}$$

$$\text{Ethylene} \qquad\qquad\qquad\qquad\qquad\qquad \text{Formaldehyde}$$
$$\text{glycol}$$

The carbon-carbon bond in the glycol is broken, and two molecules of formaldehyde are produced. Excess periodate is used, and after the reaction is complete, the excess can be determined by the iodometric procedure used for standardization. The reaction is carried out at room temperature, usually requiring from 30 min to 1 h. At higher temperatures undesired side reactions may occur, and the selectivity of the periodate oxidation is not attained. Aqueous solutions which are neutral, slightly acid, or slightly basic can often be employed, although an organic solvent may be required if the compound is not soluble in water.

Organic compounds which contain carbonyl groups ($>C=O$) on adjacent carbon atoms are also oxidized by periodate. The reaction with glyoxal is

$$\begin{array}{l} H\!-\!C\!=\!O \\ | \qquad\qquad + H_4IO_6^- \longrightarrow 2H\!-\!C\!=\!O + IO_3^- + H_2O \\ H\!-\!C\!=\!O \qquad\qquad\qquad\qquad\quad | \\ \qquad\qquad\qquad\qquad\qquad\qquad\quad OH \end{array}$$

$$\text{Glyoxal} \qquad\qquad\qquad \text{Formic acid}$$

The carbon-carbon bond is broken, and two molecules of formic acid are produced. A compound which contains a hydroxyl and a carbonyl group on adjacent carbon atoms is oxidized to an acid and an aldehyde:

$$\begin{array}{l} \qquad H \\ \qquad | \\ CH_3\!-\!C\!-\!OH \\ \qquad | \qquad\qquad + H_4IO_6^- \longrightarrow CH_3\!-\!C\!=\!O + CH_3\!-\!C\!=\!O + IO_3^- + 2H_2O \\ CH_3\!-\!C\!=\!O \qquad\qquad\qquad\qquad\qquad\qquad\qquad\qquad\quad | \\ \qquad\qquad\qquad\qquad\qquad\qquad\qquad\qquad\qquad\qquad\qquad OH \end{array}$$

$$\text{Acetoin} \qquad\qquad\qquad\qquad \text{Acetaldehyde} \qquad \text{Acetic acid}$$

The compound glycerol is oxidized to 2 mol of formaldehyde and 1 mol of formic acid. The products can be rationalized by picturing the reaction as occurring in two steps. The first step produces two aldehydes:

$$\begin{array}{l} \quad H \\ \quad | \\ H\!-\!C\!-\!OH \qquad\qquad H \qquad\qquad\qquad\qquad\qquad\qquad\qquad H \\ \quad | \qquad\qquad\qquad | \qquad\qquad\qquad\qquad\qquad\qquad\qquad | \\ H\!-\!C\!-\!OH \longrightarrow H\!-\!C\!=\!O + H\!-\!C\!=\!O \longrightarrow H\!-\!C\!=\!O + H\!-\!C\!=\!O \\ \quad | \qquad\qquad\qquad\qquad\qquad\qquad\quad | \qquad\qquad\qquad\qquad\qquad\qquad\qquad | \\ H\!-\!C\!-\!OH \qquad\qquad\qquad\qquad\quad H\!-\!C\!-\!OH \qquad\qquad\qquad\qquad\quad OH \\ \quad | \qquad\qquad\qquad\qquad\qquad\qquad\quad | \\ \quad H \qquad\qquad\qquad\qquad\qquad\qquad\quad H \end{array}$$

$$\text{Glycerol} \qquad\quad \text{Formaldehyde} \quad \text{Glycolic} \qquad\quad \text{Formaldehyde} \qquad \text{Formic}$$
$$\qquad\qquad\qquad\qquad\qquad\qquad \text{aldehyde} \qquad\qquad\qquad\qquad\qquad\qquad \text{acid}$$

[6] L. Malaprade, *Compt. Rend.*, **186,** 382 (1928); *Bull. Soc. Chim. France*, (4)**43,** 683 (1928).

One of these aldehydes, glycolic aldehyde, contains a carbonyl and a hydroxyl group on adjacent carbon atoms and is therefore oxidized to an aldehyde and an acid.

11.7

POTASSIUM BROMATE

Potassium bromate, $KBrO_3$, is a strong oxidizing agent, the standard potential of the reaction

$$BrO_3^- + 6H^+ + 6e \longrightarrow Br^- + 3H_2O$$

being $+1.44$ V. The reagent can be employed in two ways, as a direct oxidant for certain reducing agents, and for the generation of known quantities of bromine.

11.7a. Direct Titrations

A number of reducing agents, such as arsenic(III), antimony(III), iron(II), and certain organic sulfides and disulfides can be titrated directly with a solution of potassium bromate. The reaction with arsenic(III) is

$$BrO_3^- + 3HAsO_2 \longrightarrow Br^- + 3HAsO_3$$

The solution is usually about 1 M in hydrochloric acid. The end point of the titration is marked by the appearance of bromine, according to the reaction

$$BrO_3^- + 5Br^- + 6H^+ \longrightarrow 3Br_2 + 3H_2O$$

The appearance of bromine is sometimes suitable for determining the end point of the titration. Several organic indicators which react with bromine to give a color change have been studied.[7] The color change is usually not reversible, and considerable care must be taken to obtain good results. There are three indicators which have been found to behave reversibly: α-naphthoflavone, quinoline yellow, and p-ethoxychrysoidine. These are commercially available.

11.7b. Bromination of Organic Compounds

A standard solution of potassium bromate can be employed for the generation of known quantities of bromine. The bromine can then be used to quantitatively brominate various organic compounds. Excess bromide (with respect to bromate) is present in such cases, so the quantity of bromine generated can be calculated from the quantity of $KBrO_3$ taken. Usually, bromine is generated in excess of the quantity required to brominate the organic compound in order to help force this reaction to completion.

The reaction of bromine with the organic compound is either one of substitution or addition. The reaction with 8-hydroxyquinoline is a substitution reaction:

---

[7] G. F. Smith and H. H Bliss, *J. Amer. Chem. Soc.,* **53,** 209 (1931).

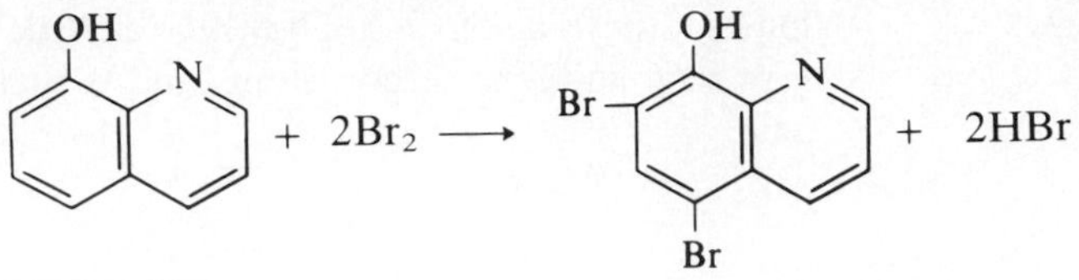

The reaction with ethylene is an addition reaction:

$$H_2C{=}CH_2 + Br_2 \longrightarrow H_2CBr{-}CBrH_2$$

In the analysis of an organic compound, a measured excess of a KBr–KBrO$_3$ mixture is added and the mixture acidified, liberating Br$_2$. After the bromination reaction is complete, the excess bromine is determined by the addition of potassium iodide, followed by titration of the liberated iodine with standard sodium thiosulfate:

$$Br_2 + 2I^- \longrightarrow I_2 + 2Br^-$$

$$I_2 + 2S_2O_3^{2-} \longrightarrow SI^- + S_4O_6^{2-}$$

A common application is the determination of metals with 8-hydroxyquinoline (page 81). A metal such as aluminum is precipitated with the organic reagent, and the precipitate is filtered, washed, and dissolved in hydrochloric acid. Potassium bromide and standard potassium bromate are then added. The reactions with aluminum (8-hydroxyquinoline abbreviated HQ) are as follows:

$$Al^{3+} + 3HQ \longrightarrow AlQ_3(s) + 3H^+ \qquad \text{(precipitation)}$$

$$AlQ_3(s) + 3H^+ \longrightarrow Al^{3+} + 3HQ \qquad \text{(redissolving)}$$

$$3HQ + 6Br_2 \longrightarrow 3HQBr_2 + 6HBr \qquad \text{(bromination)}$$

The number of equivalents of bromate is the same as that of aluminum. Here the equivalent weight of aluminum is one-twelfth its atomic weight, since $1Al^{3+} = 3HQ = 6Br_2 = 12$ electrons.

Addition reactions of bromine are used primarily in the determination of unsaturation in petroleum products and fats and oils. Many examples are found in the literature.

11.8

REDUCING AGENTS

Standard solutions of reducing agents are not so widely used as are those of oxidizing agents, because most reducing agents are slowly oxidized by oxygen of the air. Sodium thiosulfate is the only common reducing agent that can be kept for long periods of time without undergoing air oxidation. This reagent is used exclusively for iodine titrations, and its properties have already been discussed. The following are other reducing agents that are sometimes employed in the laboratory.

11.8a. Iron(II)

Solutions of iron(II) ions in 0.5 to 1 N sulfuric acid are only slowly oxidized by air and can be employed as standard solutions. The normality should be checked at least daily. Permanganate, cerium(IV), or dichromate solutions are suitable for titration of the iron(II) solution.

11.8b. Chromium (II)

The chromium(II) ion is a powerful reducing agent, the standard potential of the reaction

$$Cr^{3+} + e \rightleftharpoons Cr^{2+}$$

being -0.41 V. Solutions are oxidized rapidly by air, and extreme care must be employed in their use. Many substances have been determined by titration with either chromium(II) chloride or sulfate, including iron, copper, silver, gold, bismuth, uranium, and tungsten.

11.8c. Titanium(III)

Salts of titanium(III) are also strong reducing agents, the standard potential of the reaction

$$TiO^{2+} + 2H^+ + e \longrightarrow Ti^{3+} + H_2O$$

being $+0.04$ V. Solutions of these salts are readily oxidized by air but are easier to handle than solutions of chromium(II) salts. The principal use of titanium(III) solutions is in titrating solutions of iron(III). Other substances which can be determined include copper, tin, chromium, and vanadium.

11.8d. Oxalate and Arsenic(III)

The reactions of sodium oxalate and arsenic(III) acid have already been discussed. Standard solutions of oxalic acid are fairly stable; those of sodium oxalate are much less stable. Neutral or weakly acid solutions of $HAsO_2$ are fairly stable, but alkaline solutions are slowly oxidized by air.

KEY TERMS

Autocatalytic reaction. A reaction which is catalyzed by one of the products of the reaction.

Fowler-Bright procedure. A procedure for titrating oxalate with permanganate in which the permanganate is added rapidly to the acidified solution of oxalate at room temperature. The solution is then heated, and the titration completed at about 60°C.

Iodimetry. An analytical process in which a reducing agent is directly titrated with iodine (I_3^-), the iodine acting as the oxidizing agent.

Iodometry. An indirect process involving iodine. Excess iodide ion is added to an oxidizing

agent, liberating iodine which is then titrated with sodium thiosulfate.

Malaprade reaction. A reaction in which periodic acid oxidizes an organic compound which has hydroxyl groups on adjacent carbon atoms.

McBride procedure. A procedure for titrating oxalate with permanganate in which the entire titration is carried out slowly at elevated temperature with vigorous stirring.

Winkler method. An iodometric method for determining oxygen in water.

Zimmermann-Reinhardt preventive solution. A solution containing manganese(II) sulfate, sulfuric acid, and phosphoric acid used to prevent the oxidation of chloride ion by permanganate in the presence of iron.

QUESTIONS

1. *Oxidizing agents.* Explain how the excess is removed when each of these reagents is used in a preliminary redox step: (a) $NaBiO_3$; (b) $(NH_4)_2S_2O_8$; (c) H_2O_2; (d) SO_2.

2. *Potassium permanganate.* What special precautions must be taken in preparing a solution of $KMnO_4$ for use as a titrant? Why are these procedures needed?

3. *Equivalent weights.* Calculate the equivalent weights of these substances when determined by the reactions listed in Tables 11.2 and 11.3: (a) tin; (b) H_2S; (c) H_2O_2; (d) HNO_2; (e) $KBrO_3$.

4. *Iodine.* Explain the difference in *iodimetric* and *iodometric* processes. Why are there fewer iodimetric processes than there are iodometric ones?

5. *Zimmermann-Reinhardt reagent.* What is preventive solution and why is it needed? What is the function of the phosphoric acid in the solution?

Multiple-choice. In the following multiple-choice questions, select the *one best* answer.

6. Which of these reactions does not require a catalyst to obtain a convenient rate? (a) Titration of oxalate with cerium(IV); (b) titration of arsenic(III) with $KMnO_4$; (c) titration of arsenic(III) with cerium(IV); (d) titration of arsenic(III) with iodine at pH 8.5.

7. Which of these statements is *true*? (a) In the McBride procedure the titration is carried out rapidly at room temperature. (b) Acidic solutions of permanganate are not stable. (c) The oxidation of Cl^- by MnO_4^- is slow in the presence of Fe^{2+}. (d) Fe^{2+} is a weaker reducing agent than Cl^-.

8. Which of these statements is *true*? (a) In an iodometric process iodine acts as a reducing agent. (b) $K_2Cr_2O_7$ is a stronger oxidizing agent than $KMnO_4$ in 1 M acid. (c) The Ce^{4+} ion hydrolyzes readily in solutions of low pH. (d) I_2 is adsorbed on the surface of α-amylose to give a blue color.

9. When 1 mol of glycerol is oxidized by periodic acid, the organic products are (a) 2 mol of formaldehyde; (b) 2 mol of formic acid; (c) 1 mol of formaldehyde + 2 mol of formic acid; (d) 2 mol of formaldehyde + 1 mol of formic acid.

10. The reaction between arsenic(III) and I_2 is forced to the right by (a) raising the pH; (b) lowering the pH; (c) adding starch; (d) adding KI.

PROBLEMS

1. *Preliminary redox reactions.* Write a balanced equation for the following redox reactions which may be involved in a preliminary step of an analysis. Add H_2O, H^+, or OH^- as needed. (See Appendix II.)
   (a) Oxidation of Mn(II) to MnO_4^- by BiO_3^- in acid medium.
   (b) Oxidation of Cr(III) to CrO_4^{2-} by H_2O_2 in basic medium.
   (c) Oxidation of Cr(III) to $Cr_2O_7^{2-}$ by $S_2O_8^{2-}$ in acid medium.
   (d) Reduction of vanadium(V) to vanadium(IV) by SO_2 in acid medium.
   (e) Reduction of Cu(II) to Cu(I) in the silver reductor.

2. *Malaprade reaction.* Write the balanced half-reaction (Appendix II) for the following reactions:
   (a) Oxidation of acetoin, $C_4H_8O_2$, to 1 mol of acetic acid, $C_2H_4O_2$, and 1 mol of acetaldehyde, C_2H_4O, in acid medium.
   (b) Oxidation of glycerol, $C_3H_8O_3$, to 2 mol of formaldehyde, H_2CO, and 1 mol of formic acid, H_2CO_2, in acid medium. Calculate the equivalent weights of acetoin and glycerol in these reactions.

3. *Oxidation of ethylene glycol.* When ethylene glycol, $C_2H_6O_2$, is oxidized by acidic permanganate, the oxidation products are CO_2 and H_2O. When $Ce(ClO_4)_4$, is used, the

product is formic acid, H_2CO_2. When periodic acid is used, the glycol is oxidized to formaldehyde, H_2CO, and the periodic acid is reduced to IO_3^-. Write complete balanced equations for these three reactions and calculate the equivalent weight of ethylene glycol (MW 62.068) in each case.

4. *Standardization.* A solution of sodium thiosulfate is standardized against pure $K_2Cr_2O_7$. A sample of $K_2Cr_2O_7$ weighing 0.2153 g is dissolved, the solution acidified, and excess KI added. The liberated I_2 requires 44.86 mL of thiosulfate for titration. Calculate the molarity and normality of the dichromate solution.

5. *Analysis.* A 0.9882-g sample of iron ore is dissolved, the iron reduced to Fe^{2+}, and the solution titrated with 36.40 mL of 0.1065 N $KMnO_4$ in acid solution. (a) Calculate the percentage of Fe in the sample. (b) Calculate the percentage as Fe_2O_3.

6. *Potassium permanganate.* A solution of $KMnO_4$ is 0.0250 M. It is used for titration in a solution of pH 10 where the MnO_4^- is reduced to MnO_2. (a) Write the half-reaction that occurs. What is the normality of the solution? (b) If the solution is used in 1 M NaOH, the MnO_4^- is reduced to MnO_4^{2-}. Write the half-reaction. What is the normality of the solution?

7. *Preliminary reduction.* A sample of iron ore weighing 0.740 g and containing 24.0% Fe_2O_3 is dissolved, and the iron reduced to Fe^{2+} by the addition of 25.0 mL of 0.0500 M $SnCl_2$. The excess is oxidized to Sn^{4+} using 0.0500 M $HgCl_2$. How many milliliters of the latter solution are required?

8. *Titer.* A solution of $KMnO_4$ is 0.260 M. Calculate the titer of this solution for reduction to Mn^{2+} in terms of these substances (reactions in Table 11.2): (a) H_2O_2; (b) As_2O_3; (c) $Na_2C_2O_4$; (d) Fe_3O_4.

9. *Hydrogen peroxide.* What volume of 0.0500 M $KMnO_4$ is required to react in acid solution with 5.00 mL of H_2O_2 that has a density of 1.01 g/mL and contains 3.20% by weight H_2O_2? The permanganate is reduced to Mn^{2+}, and the H_2O_2 oxidized to O_2.

10. *Oxidation of sulfur dioxide.* The reaction between potassium permanganate and sulfur dioxide in acid solution is as follows:

$$MnO_4^- + SO_2 + H_2O \longrightarrow$$
$$Mn^{2+} + H^+ + SO_4^{2-}$$

(a) Balance the equation. (b) If 40.0 mL of 0.0400 M $KMnO_4$ is treated with SO_2, how many milliliters of 0.120 M NaOH are required to neutralize the acid formed? The excess SO_2 is removed by boiling.

11. *Oxalates.* A sample containing only $Na_2C_2O_4$ and KHC_2O_4 required three times the volume of 0.1000 N $KMnO_4$ for titration as of 0.1000 N base (same size sample in each case). Calculate the percentage of each salt in the mixture. The MnO_4^- is reduced to Mn^{2+}, and $C_2O_4^{2-}$ is oxidized to CO_2.

12. *Standardization of permanganate.* A sample of pure As_2O_3 weighing 0.2068 g is dissolved in NaOH, and the solution is acidified with HCl and titrated with 42.46 mL of a permanganate solution. Calculate the normality of the permanganate. MnO_4^- is reduced to Mn^{2+}, and As(III) is oxidized to As(V).

13. *Determination of MnO_2.* A 0.5000-g sample containing MnO_2 is treated with concentrated HCl, liberating Cl_2. The Cl_2 is passed into a solution of KI, and 30.24 mL of 0.1018 M $Na_2S_2O_3$ is required to titrate the liberated I_2. Calculate the percentage of MnO_2 in the sample. The MnO_2 is reduced to Mn^{2+}, and the $S_2O_3^{2-}$ is oxidized to $S_4O_6^{2-}$.

14. *Determination of chromium.* The chromium in a 0.2200-g sample of an ore is oxidized to CrO_4^{2-} with Na_2O_2. The solution is acidified and treated with 0.7642 g of pure $FeSO_4$. The excess Fe^{2+} requires 9.74 mL of 0.1000 N $K_2Cr_2O_7$ for titration in acid solution. Calculate the percentage of Cr_2O_3 in the sample. The Fe^{2+} is oxidized to Fe^{3+}, and the $Cr_2O_7^{2-}$ is reduced to Cr^{3+}.

15. *Calcium in limestone.* Calcium is determined in a limestone sample by precipitating CaC_2O_4, dissolving the precipitate in H_2SO_4, and titrating with standard $KMnO_4$. The CaC_2O_4 precipitate from a limestone sample weighing 0.4463 g requires 32.17 mL of 0.02272 M $KMnO_4$ for titration. Calculate the percentage of CaO in the sample. The $C_2O_4^{2-}$ is oxidized to CO_2, and the MnO_4^- is reduced to Mn^{2+}.

16. *Determination of copper.* A sample of copper oxide weighing 2.104 g and containing

12.04% CuO is dissolved in acid, the pH adjusted, and excess KI added, liberating I_2. How many milliliters of 0.1046 M $Na_2S_2O_3$ are required to titrate the I_2?

17. *Determination of Vitamin C.* Ascorbic acid (vitamin C, MW 176.126) is a reducing agent, reacting as follows:

$$C_6H_8O_6 \longrightarrow C_6H_6O_6 + 2H^+ + 2e$$

It can be determined by oxidation with a standard solution of I_2. A 200.0-mL sample of a citrus fruit drink is acidified, and 10.00 mL of 0.0500 M I_2 is added. After the reaction is complete the excess I_2 is titrated with 38.62 mL of 0.0120 M $Na_2S_2O_3$. Calculate the number of milligrams of ascorbic acid per milliliter of fruit drink.

18. *Bleaching powder.* Bleaching powder, $Ca(OCl)Cl$, reacts with iodide ion in acid medium, liberating I_2:

$$OCl^- + I^- + H^+ \longrightarrow I_2 + Cl^- + H_2O$$

(a) Balance the equation. (b) How many milliliters of 0.060 M $Na_2S_2O_3$ are required to titrate the iodine liberated from a 0.6620-g sample of bleaching powder which contains 10.72% Cl?

19. *Cerium(IV) solutions.* 32.0 g of the salt $Ce(SO_4)_2 \cdot 2H_2O$ is dissolved in 500 mL of solution. Calculate the number of milligrams of $Na_2C_2O_4$ which will react with 25.0 mL of the solution. The $Na_2C_2O_4$ is oxidized to CO_2.

20. *Cerium(IV) solutions.* 41.2 g of the salt $Ce(NO_3)_4 \cdot 2NH_4NO_3$ is dissolved in 750 mL of 0.2 M H_2SO_4. Calculate the number of grams of $K_2Cr_2O_7$ which should be dissolved in 750 mL of solution to make a solution of equal normality as an oxidizing agent. The $Cr_2O_7^{2-}$ is reduced to Cr^{3+}.

21. *Factor weight solution.* How many grams of $K_2Cr_2O_7$ should be dissolved in 400.0 mL of solution so that the number of milliliters used in a titration equals the %Fe_2O_3 in a 1.000-g sample. The iron is oxidized from Fe^{2+} to Fe^{3+}, and the $Cr_2O_7^{2-}$ is reduced to Cr^{3+}.

22. *Mixture of iron and vanadium.* A 50.00-mL sample of a solution containing iron(III) and vanadium(V) is passed through a silver reductor and then titrated with 30.58 mL of a 0.1016 N Ce(IV) solution. A second 50.00-mL sample is passed through a Jones reductor and then titrated with 42.85 mL of the same

Ce(IV) solution. Calculate the molarities of the iron(III) and the vanadium(V) in the original solution. See Table 11.1 for the reactions in the reductors. Assume that vanadium is reoxidized to vanadium(V) by Ce(IV).

23. *Mixture of chromium and iron.* A 25.00-mL aliquot of a solution containing iron(III) and chromium(III) is passed through a Jones reductor and then titrated with 35.64 mL of 0.02040 M $KMnO_4$ in 1 M H_2SO_4. A second 25.00-mL sample of the same solution requires 14.34 mL of the permanganate solution for titration after passing through a silver reductor. Calculate the molarities of the iron(III) and the chromium(III) in the original solution. See Table 11.1 for the reactions in the reductors. Chromium is reoxidized to Cr(III) by the permanganate.

24. *Malaprade reaction.* 50.00 mL of a solution containing 40.00 mg of ethylene glycol was treated with 50.00 mL of a 0.03000 M solution of paraperiodic acid, and the solution allowed to stand for 1 h. The solution was then buffered at pH 8 with $NaHCO_3$, and excess KI was added. How many milliliters of a 0.0640 N arsenic(III) solution were required to titrate the liberated I_2? The As(III) is oxidized to As(V).

25. *Determination of sodium formate.* Sodium formate, $NaHCO_2$, reacts with permanganate in neutral solution according to the equation (unbalanced)

$$CHO_2^- + MnO_4^- + H_2O \longrightarrow$$
$$MnO_2(s) + CO_2 + OH^-$$

(a) Balance the equation. (b) A 0.4680-g sample containing 30.10% $NaCHO_2$ is treated with 50.00 mL of 0.05000 M MnO_4^- (an excess) in neutral solution, and the reaction allowed to proceed to completion. The solution is filtered to remove the MnO_2, acidified with H_2SO_4, and titrated with 0.1020 M $H_2C_2O_4$. The MnO_4^- is reduced to Mn^{2+}, and the oxalate is oxidized to CO_2. How many milliliters of $H_2C_2O_4$ are required for the titration?

26. *Determination of aluminum.* The aluminum in a 0.2440-g sample of an ore is precipitated with 8-hydroxyquinoline and the precipitate filtered and then dissolved in acid. To this solution is added 30.00 mL of 0.0500 M $KBrO_3$

and 2 g of KBr. After reaction of the Br_2 with 8-hydroxyquinoline, 2 g of KI is added, and the liberated I_2 titrated with 25.60 mL of 0.1010 M $Na_2S_2O_3$. Calculate the percentage of Al_2O_3 in the sample.

27. *Determination of magnesium.* A 0.3000-g sample of an ore containing 16.8% MgO is dissolved, and the magnesium precipitated with 8-hydroxyquinoline. The precipitate is filtered and dissolved in acid, and 50.0 mL of 0.04000 M $KBrO_3$ and 3 g of KBr are added. After reaction between Br_2 and 8-hydroxy-quinoline is complete, 3 g of KI is added, and the liberated I_2 is titrated with 0.0500 M $Na_2S_2O_3$. How many milliliters of $Na_2S_2O_3$ are required for the titration?

28. *Potassium bromate.* How many milliliters of a 0.050 M $KBrO_3$ solution are required to furnish sufficient Br_2 to react with the precipitate formed between 40 mg of Th^{4+} and 8-hydroxyquinoline?

29. *Potassium iodate.* 30.00 mL of KIO_3 reacted with exactly 30.00 mL of 0.1000 M KI in acid medium according to the equation (unbalanced)

$$IO_3^- + I^- + H^+ \longrightarrow I_2 + H_2O$$

(a) Balance the equation and calculate the molarity of the KIO_3 solution. (b) 40.00 mL of this KIO_3 solution is treated with excess pure KI in acid solution, and the liberated I_2 is titrated with 0.1200 M $Na_2S_2O_3$. How many milliliters of titrant are required?

30. *Determination of potassium iodide.* A sample of impure KI weighing 0.640 g is dissolved in water, the solution is acidified, and 25.0 mL of 0.0420 M KIO_3 (an excess) is added. The iodate is reduced to I_2, and the iodide is oxidized to I_2. The I_2 is boiled off, the solution cooled, and an excess of pure KI is added to react with the unused KIO_3. The I_2 produced is titrated with 40.60 mL of 0.1042 M $Na_2S_2O_3$. Write the equations for the reactions which occurred and calculate the percentage of KI in the sample.

31. *Permanganate end point.* It was pointed out on page 295 that the sight excess of permanganate normally present at the end point of a titration is sufficient to cause the reaction

$$3Mn^{2+} + 2MnO_4^- + 2H_2O \longrightarrow$$
$$5MnO_2(s) + 4H^+$$

to occur. (a) Write the two half-reactions and calculate the equilibrium constant for the reaction. (b) Show that the reaction does tend to go to the right, assuming the following concentrations at the end point: $[Mn^{2+}] = 0.01$ M, $[MnO_4^-] = 1 \times 10^{-5}$ M, and $[H^+] = 1$ M.

32. *Arsenic(III)-I_2 reaction.* (a) Calculate the equilibrium constant for the reaction

$$HAsO_2 + I_3^- + 2H_2O \rightleftharpoons$$
$$3I^- + H_3AsO_4 + 2H^+$$

and compare it with the experimental value (page 302). (b) If 2.0 mmol of $HAsO_2$ is titrated with I_2 at pH 5.0, calculate (using the calculated K) the number of milligrams of the acid that remain unoxidized at the end point. Assume that at the end point: volume = 100 mL, $[I^-] = 0.10$ M, $[I_3^-] = 2 \times 10^{-5}$ M. (c) Why is the titration not carried out at pH 5.0?

12

Potentiometric Methods of Analysis

We saw in Chapter 10 how the voltages of galvanic cells respond to the activities of chemical species in solution according to the Nernst equation. In the present chapter, we shall see how to exploit this in a practical way to measure these activities. And sometimes, when activity coefficients can be controlled in both standard and unknown solutions, we can determine the corresponding analyte *concentrations*. In *direct potentiometry*, one sets up a galvanic cell whose voltage depends upon the activity of the analyte. In a *potentiometric titration*, progress toward the equivalence point is assessed by monitoring a cell voltage which depends upon the activity of one of the reactants or upon a ratio such as $a_{Fe^{2+}}/a_{Fe^{3+}}$ which changes during the titration. The cell will always have two electrodes, a *reference electrode* whose potential remains constant throughout the measurements and an *indicator electrode* whose potential responds to the activity changes in the test solution.

12.1

OVERVIEW OF POTENTIOMETRIC METHODS

12.1a. *Advantages of Potentiometric Methods*

The advantages of potentiometric methods include low cost. Voltmeters and electrodes are far cheaper than most modern scientific instruments. Models suitable for direct potentiometry in field work away from the laboratory are inexpensive, compact, rugged, and easy to use. Potentiometry is essentially nondestructive of the sample in that the insertion of electrodes does not alter the composition of the test solution (except for a slight leakage of electrolyte from the reference electrode). If the species to which the indicator electrode responds participates in a solution equilibrium, its activity is measured as it stands, without perturbing the equilibrium itself; thus direct potentiometry is often very useful for determining equilibrium constants. Stable potentials are often attained fairly rapidly, and voltages are easily recorded as functions of time. Thus potentiometry is sometimes useful for the continuous, unattended monitoring of such samples as public water supplies, industrial process streams, and effluent wastewater for pH and other ions such as fluoride, nitrate, sulfide, and cyanide. Finally, the wide range of analyte activities over which some of the available indicator electrodes exhibit stable, nearly Nernstian responses represents an important advantage (see Table 12.1).

TABLE 12.1   Some Commercially Available Ion-Selective Electrodes

Ion	Type	Concentration range, M	Some Interferences
Bromide	Solid state	5×10^{-6} to 1	S^-, I^-, CN^-
Cadmium	Solid state	10^{-7} to 1	Ag^+, Hg^{2+}, Cu^{2+}
Calcium	Liquid membrane	5×10^{-7} to 1	$Pb^{2+}, Na^+, Hg^{2+}, H^+,$ $Sr^{2+}, Fe^{3+}, Cu^{2+}, Ni^{2+},$ $Ba^{2+}, Zn^{2+}, Mg^{2+}, Li^+$
Chloride	Solid state	5×10^{-5} to 1	S^{2-}, Br^-, I^-, CN^-
Cupric	Solid state	10^{-8} to 1	Ag^+, Hg^{2+}
Cyanide	Solid state	8×10^{-6} to 10^{-2}	S^{2-}, I^-
Fluoride	Solid state	10^{-6} to 1	OH^-
Iodide	Solid state	5×10^{-8} to 1	$S^{2-}, CN^-, S_2O_3^{2-}$
Lead	Solid state	10^{-6} to 1	Ag^+, Hg^{2+}, Cu^{2+}
Nitrate	Liquid membrane	7×10^{-6} to 1	$ClO_4^-, I^-, CN^-, Br^-,$ NO_2^-, HS^-
Potassium	Liquid membrane	10^{-6} to 1	$Cs^+, NH_4^+, Tl^+, Ag^+, H^+$
Sodium	Glass	10^{-6} to 1	Ag^+, H^+
Tetraphenyl borate	Liquid membrane	10^{-6} to 1	$NO_3^-, Br^-, OAc^-, HCO_3^-,$ $F^-, Cl^-, OH^-, SO_4^{2-}$

12.1b. Accuracy of Potentiometric Methods

One sort of limitation on the accuracy of direct potentiometry is seen in the Nernst equation itself. Because the analyte activity appears in the log term, a tenfold change corresponds to a cell voltage change of only about 59 mV in the best case, where n (the number of electrons in the electrode process) is 1. (Recall from Chapter 10 that $2.3RT/F = 0.0592$ V at 25°C.) But it turns out that this is not really a problem. With a modern voltmeter and careful attention to shielding against stray electrical signals, voltages can be measured to about the nearest 0.01 mV. Suppose we have an electrode that responds to a_{ion}, the activity of some ion of interest, for the case where $n = 1$. Now, an error of 0.01 mV (0.00001 V) in the cell voltage corresponds to an error in log a_{ion} of $0.00001/0.0592 = 0.00017$. Comparing the antilog, $10^{0.00017} = 1.0004$, with the case for no error ($10^{0.00000} = 1.0000$), we obtain a relative error in a_{ion} of 0.04%, certainly not bad for a method that may work for analytes as dilute as 10^{-6} or 10^{-7} M.

EXAMPLE 12.1

Percent error in a potentiometric method

Suppose the activity of some ion, a_{ion}, really happens to be 1.0000×10^{-4}, but a researcher who doesn't know this wants to determine it potentiometrically. If the Nernst equation for the one-electron process at the indicator electrode is

$$E = E° - 0.0592 \log \left(\frac{1}{a_{ion}} \right)$$

what percent error in a_{ion} results from a 0.1-mV (0.0001-V) error in measuring the cell voltage? E_{corr} is the actual voltage, and E_{meas} is the experimental (erroneous) value.

$$E_{corr} = E° - 0.0592 \log \frac{1}{1.0000 \times 10^{-4}} = E° - 0.2368$$

$$E_{meas} = E_{corr} + 0.0001 = E° - .0592 \log \frac{1}{a_{ion_{meas}}}$$

$$\log \frac{1}{a_{ion_{meas}}} = \frac{E° - (E_{corr} + 0.0001)}{0.0592}$$

$$= \frac{E° - (E° - 0.2368 + 0.0001)}{0.0592}$$

$$= \frac{0.2367}{0.0592} = 3.9983$$

$$a_{ion_{meas}} = 10^{-3.9983} = 1.0038 \times 10^{-4}$$

$$\text{Percent error} = \frac{1.0038 - 1.0000}{1.0000} \times 100 = 0.4\% \qquad \square$$

The accuracy of the voltage measurement is usually the least of our worries in direct potentiometry. Liquid junction potentials (designated E_j) were introduced in Chapter 10. They were neglected for our purposes there, but if we wish to discuss the accuracy of potentiometry as an analytical tool, they must now be considered. At the least, there will always be a potential where the electrolyte solution of the reference electrode (or of a salt bridge therefrom) contacts the test solution. This represents the major limitation on the accuracy of determining analyte activities. The junction potential and the potential of the indicator electrode both contribute to the measured cell voltage, and there is in principle no way to sort them out experimentally. For simple, well-defined cases, approximate E_j values can be calculated from individual ion mobilities,[1] but this is normally impossible when dealing with samples from the real world. In extreme cases (say, that one of the ions is H^+, which has an exceptionally high mobility, and that we make an unfortunate choice of salt bridge), junction potentials can be terribly large: for the contact of 0.1 *M* HCl and 0.1 *M* KCl solutions, E_j is calculated to be about 27 mV. Fortunately, we can do much better than this: for a junction of 0.1 *M* HCl with a saturated KCl solution ($\sim$4 *M*), E_j is only about 4 mV, and for unexceptional cases, we probably confront E_j values ranging from a few tenths up to perhaps 2 or 3 mV.

[1] For example, see P. H. Rieger, *Electrochemistry*, Prentice-Hall, Inc., Englewood Cliffs, N.J., 1987, p. 161.

EXAMPLE 12.2

Effect of the junction potential

Someone is measuring pH values of unknown solutions using a hydrogen indicator electrode with $p_{H_2} = 1$ atm:

$$E_{H_2} = 0 - 0.0592 \log \left(\frac{1}{a_{H^+}} \right) = -0.0592 \, pH$$

Suppose the actual pH of one of the solutions happens to be exactly 5; i.e., $a_{H^+} = 1.0000 \times 10^{-5}$. The liquid junction potential between the test solution and the KCl of an SCE reference electrode is 1 mV (0.001 V), but the experimenter doesn't know what it is and ignores it. What is the percent error in a_{H^+} as determined using the electrode?

E_{H_2} is really $(-0.0592)(5.0000) = -0.2960$ V, but suppose it is thought, from the measurement, to be $-0.2960 - 0.001 = -0.2970$ V.

$$pH = -0.2970/-0.0592 = 5.0169$$

$$a_{H^+} = 10^{-5.0169} = 0.9618 \times 10^{-5}$$

$$\text{Percent error} = \frac{1.0000 - 0.9618}{1.0000} \times 100 = 3.8\% \qquad \square$$

Errors of several percent are to be expected in direct potentiometry, and except for very special circumstances where solution composition can be tightly controlled, there is no way to avoid them. Persons who are unaware of this may spend money on expensive voltmeters that provide meaningless decimal places at considerable cost.

12.1c. Activity and Concentration Considerations

For analytical chemists, there is another problem with direct potentiometry. In certain contexts (e.g., determining thermodynamic equilibrium constants), the *activity* of an electroactive species may be desired, but in analytical work we usually want to know the *concentration*. Unless the composition of the solution with respect to *all* ions is specified, converting an activity into a concentration is a risky game, and even then one may well be unable to find suitable activity coefficients for the stated conditions. Analytical chemists frequently take their samples where they find them, not where they have the easiest time. This problem can be countered by calibration if all the unknown samples have the same gross composition. Standards are prepared in which the analyte ion is varied but which are as similar as possible to the unknowns in every other regard. Voltage readings are then converted to concentrations using a graph of E vs. $\log C$ plotted from measurements on the standards. (Obviously this need not be done manually; a computer will find the slope and intercept of the graph and calculate the unknown C-value corresponding to any measured E. If we wish to see the graph to get a "feel" for how things are going, the computer can display it.) An example might be determining a minor metal ion in seawater; the standards would be synthetic seawater as close as possible to the real thing in regard to salt and other major solutes, spiked with known quantities of the metal ion. This type of problem—the effect of the overall composition of a sample, particularly with regard to major components, upon the analyte response—is sometimes called a *matrix effect*. We shall see examples of matrix effects with other techniques in later chapters; they are an important concern in many practical analytical situations. Sometimes, if variation in composition is too great within a set of samples, the analyst in effect creates the matrix by adding a large quantity of some solute to swamp out the differences.

Despite such problems, direct potentiometry is attractive because of the advantages enumerated earlier. Where matrix effects can be compensated for, and where the highest accuracy is not required, these methods find wide acceptance. We shall discuss below the types of electrodes employed. The very important determination of pH by direct potentiometry will then be examined, and finally we shall consider potentiometric end-point techniques in titrations.

12.2

INDICATOR ELECTRODES

12.2a. Metal–Metal Ion Electrodes

Electrodes of the First Kind

Some metals such as silver, mercury, copper, cadmium, zinc, and lead can act as indicator electrodes with respect to their own ions. For example, the single-electrode potential for a silver wire dipping into a solution of a silver salt varies with the activity of silver ion according to the Nernst equation:

$$Ag^+ + e \rightleftharpoons Ag \qquad E° = +0.80 \text{ V}$$

$$E = +0.80 - 0.0592 \log \left(\frac{1}{a_{Ag^+}} \right)$$

Such electrodes, where the analyte ion participates directly with its metal in a reversible half-reaction, are sometimes called *electrodes of the first kind*. These electrodes are far from universally useful for metals across the periodic table. In fact, more often than not, stable, reproducible potentials are not established, a few examples being iron, chromium, nickel, cobalt, tungsten, and aluminum. In some cases the problem probably relates to films of oxide or other products on the metal surface which hamper electron transfer; strains and irregular crystallinity that depend upon the metallurgical history of the electrode may be involved.

Electrodes of the Second Kind

The silver–silver chloride electrode described earlier (page 270) is an example of an *electrode of the second kind*. At constant a_{Cl^-} (e.g., saturated KCl), this serves as a *reference* electrode, as we saw, but it responds as an *indicator* electrode to *changes* in a_{Cl^-} and can be used to measure pCl values:

$$AgCl + e \rightleftharpoons Ag + Cl^- \qquad E° = +0.22 \text{ V}$$

$$E = +0.22 - 0.0592 \log a_{Cl^-} = 0.22 + 0.0592 \, p\text{Cl}$$

The electrode potential, of course, relates directly to a_{Ag^+}, but the latter reflects a_{Cl^-} via $K_{sp_{AgCl}}$, which is the equilibrium constant for the reaction $AgCl \rightleftharpoons Ag^+ + Cl^-$.

Electrodes of the Third Kind

A useful *electrode of the third kind* was described by Reilley and Schmid[2] and employed as indicator electrode in potentiometric EDTA titrations of 29 metal ions including alkaline and rare earths and transition and heavy metals.[3] (See Chapter 8 for a discussion of EDTA as a titrant for metal ions.) The electrode itself is a drop or small pool of mercury in a cup at the short end of a J-tube with a wire contact to the external circuit. A small quantity of mercuric-EDTA chelate, HgY^{2-}, is added to the solution of the analyte metal ion, M^{n+} (e.g., one drop of 10^{-3} M HgY^{2-} to perhaps 25 or 50 ml of buffered M^{n+} solution). The electrode can thus be represented as follows:

$$Hg \mid Hg^{2+} + HgY^{2-} + MY^{(4-n)-} + M^{n+}$$

The formation constant of HgY^{2-} is very large ($K_{abs} = 10^{21.8}$); thus there is little dissociation to Hg^{2+}, but yet enough to establish a potential according to

$$Hg^{2+} + 2e \rightleftharpoons Hg \qquad E° = +0.85 \text{ V}$$

$$E = +0.85 - \frac{0.0592}{2} \log \frac{1}{a_{Hg^{2+}}}$$

After rearranging the expressions for the two formation constants $K_{HgY^{2-}}$ and $K_{MY^{(4-n)-}}$, obvious substitution for $a_{Hg^{2+}}$ in the above equation leads to

$$E = +0.85 - \frac{0.0592}{2} \log \frac{K_{HgY^{2-}} a_{MY^{(4-n)-}}}{K_{MY^{(4-n)-}} a_{HgY^{2-}} a_{M^{n+}}}$$

(See Question 7.) Suppose now that we are titrating M^{n+} with EDTA. The K-values in the above equation are constants, and, in the important region near the equivalence point, $a_{MY^{(4-n)-}}$ will change only slightly; because the complex is so stable, $a_{HgY^{2-}}$ is virtually constant as well. Thus, lumping all of the constants and near-constants into one overall K, the equation boils down to

$$E = K - \frac{0.0592}{2} \log \frac{1}{a_{M^{n+}}} = K - \frac{0.0592}{2} pM$$

Essentially, then, we have a pM electrode. The original papers[2,3] may be consulted for information on titrating mixtures, interferences, etc.

12.2b. *Inert Electrodes*

An inert metal, usually platinum, serves well as the indicator electrode for certain redox couples such as $Fe^{3+} + e \rightleftharpoons Fe^{2+}$ (page 271). The role of the metal is simply to sense the tendency of the system to take up or release electrons; it does not itself participate appreciably in the redox reaction, and its potential is a Nernstian function of the activity ratio $a_{Fe^{2+}}/a_{Fe^{3+}}$. Of course, *inert* is a relative term, and platinum is not immune from attack by strong oxidants, especially in solutions where complexation can stabilize Pt(II) by forming species such as $PtBr_4^{2-}$ (compare $Pt^{2+} + 2e \rightleftharpoons Pt$, $E° = +1.19$ V, with $PtBr_4^{2-} + 2e \rightleftharpoons Pt + 4Br^-$, $E° = +0.58$ V). Platinum can also cause problems with very strong reductants: reduction of H^+ (or H_2O) is sometimes so slow that analytes can be

[2] C. N. Reilley and R. W. Schmid, *Anal. Chem.*, **30,** 947 (1958).
[3] C. N. Reilley, R. W. Schmid, and D. W. Lamson, *Anal. Chem.*, **30,** 953 (1958).

preferentially reduced in aqueous solution without interference from the solvent, but because $H^+ \; e = \frac{1}{2}H_2$ is catalyzed by platinum, this kinetic advantage may be lost.

12.2c. Membrane Electrodes

In terms of the mechanism by which the potential is established, membrane electrodes are different from the metal electrodes described above. Electrons as such are not directly involved; rather, the membranes function by some sort of differential penetration of *ions*. By far the most widely used of the membrane electrodes is the glass pH electrode, which we shall examine in some detail because of its general importance. Then we shall take a briefer look at some other types.

The Glass Electrode for pH Measurements

The measurement of pH is of fundamental importance in a wide variety of fields—biology and medicine, water treatment, agriculture, and food processing, to mention a few outside chemistry itself. It was observed shortly after the turn of the century that a potential develops across a thin glass membrane separating two solutions of different acidities. Studies on the effect of glass composition upon the potential, and electronic developments that allow voltages to be easily and accurately measured in cells of very high resistance, have led to the modern "pH meter." A typical experimental setup is shown in Fig. 12.1.

Composition of the Glass Electrode. The solution inside a glass electrode is sealed permanently in place and retains its constant hydrogen ion activity. The

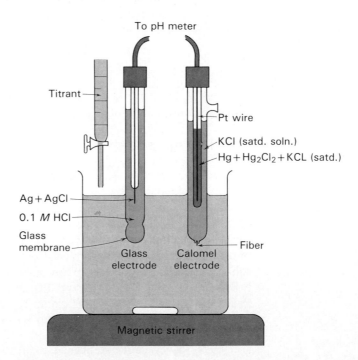

Figure 12.1 Apparatus for potentiometric acid-base titration with glass-calomel electrode pair.

reference electrode (usually silver–silver chloride, see page 270) in contact with the inner solution has a constant potential. The bottom portion of the glass electrode is immersed in the test solution along with a second, external reference electrode, frequently an SCE (page 268). Various physical arrangements provide electrolyte contact between the test solution and the filling solution of the SCE: a pinhole, a tiny fiber which soaks up solution, or a narrow annular space allows a very slow flow of electrolyte through the junction. The cell can be represented as follows:

$$\underbrace{\text{Hg} | \text{Hg}_2\text{Cl}_2, \text{KCl(satd.)}}_{\substack{\text{SCE reference} \\ \text{electrode}}} | \underbrace{\text{H}^+(a = x)}_{\substack{\text{Test} \\ \text{solution}}} | \underbrace{\text{glass} | \text{HCl}(0.1 \ M), \text{AgCl} | \text{Ag}}_{\text{"Glass electrode"}}$$

Note that we have simply placed a reference electrode in each of the solutions separated by the glass membrane. Thus any change in the cell voltage when we change the test solution must reflect a change in the potential developed across the glass membrane and any change in the liquid junction potential between the SCE and the test solution. We hope that the change in E_j is small, and, because saturated KCl is the tail that wags the dog at the junction, it probably is, since the mobilities of K^+ and Cl^- are approximately the same. It is found experimentally that the voltage of this cell follows the relation

$$E = k + 0.0592 \ pH$$

at 25°C over a pH range in the test solution from about 0 or 1 up to 10 or 13, depending upon the composition of the glass.

Advantages of the Glass Electrode. The glass electrode has important advantages over other pH electrodes. (For some other types, see Example 2 above and Problems 8 and 9.) Except for a slight leakage of KCl from the SCE, the test solution is uncontaminated and available for other measurements or for use as a reagent. Substances that are easily oxidized or reduced do not interfere, whereas they might react with H_2 (Example 2) or with the components of quinhydrone (Problem 9). Because potentials in general are independent of physical size, glass and calomel electrodes can be made small enough for insertion into very small volumes of solutions. There is no catalytic surface to lose its activity by contamination, as may the platinum in the hydrogen electrode. The pH values of poorly buffered solutions can be measured accurately. Finally, the electrode is well suited for continuous pH monitoring over extended time periods.

Limitations of the Glass Electrode. No device is perfect or universally useful. In solutions of very high pH (e.g., 0.1 M NaOH, $pH \sim$ 13, where $[H^+]$ is about $10^{-13} \ M$ and $[Na^+]$ about 0.1 M), the specificity for H^+ is lost; the dependence of voltage upon pH diminishes, and the potential becomes dependent upon a_{Na^+}. The upper pH limit depends upon the other cation (e.g., whether Na^+ or K^+) and upon the type of glass; with the common Corning 015 glass, it is about 10, but electrodes are available which are useful up to pH 13 or so based upon special glasses containing lithium oxide instead of sodium. In very acidic solutions (pH below about 0 or 1), errors are encountered that probably relate to changes in the activity of water in the gel layer of the glass (see below) and penetration of anions into the gel. From a practical standpoint, however, most solutions whose pH values are of interest lie within the useful range of the glass electrode.

Early in the century, the very high resistances (millions of ohms) across glass membranes created serious measurement problems, but modern electronic circuits and careful shielding of the electrode leads have effectively eliminated these. The pH meter is essentially an electronic voltmeter capable of measuring the voltage of a galvanic cell of very high internal resistance. The voltage varies linearly with pH, which is displayed directly in either analog (continuously variable position of a needle on a meter scale) or digital (like the answer on your electronic calculator) form. With the electrodes immersed in a buffer solution of known pH, the instrument is adjusted to display the correct value. Then, by turning a different knob, it is set to read correctly the pH of a second standard buffer. This essentially establishes the slope of the line relating voltage to pH. This slope will vary with temperature (recall the T in RT/F in the Nernst equation); it is also affected by the aging of the glass electrode, and it may vary from one electrode to another. Although people love short cuts, calibration of the pH meter with *two* buffers, not one, is essential for accurate work. The instrument is then ready for measuring unknown solutions.

Theory of the Glass Membrane Potential.   The earliest explanation for the origin of this potential is simple and easily understood but is now thought to be incorrect: if H^+ could migrate through the glass while other ions could not, then a potential just large enough to prevent further migration would develop; this would be analogous to an extreme case of a liquid junction potential (page 258) where only one ion could migrate at all.[4] However, in a sensitive tracer experiment, no movement of tritium ions (tritium is the radioactive hydrogen isotope ^3_1H) through a thin glass membrane could be detected when a small electric current was passed through the system.[5] This confirmed earlier studies which had failed to detect any net transport of H^+ across a glass membrane in an electrolysis circuit using classical techniques.[6] If H^+ does not migrate through glass in electrolysis experiments with appreciable currents, then there is little reason to assume such migration under the near-zero current conditions of potentiometry.

Current thought on the origin of the glass membrane potential is more complex but may be presented qualitatively as follows. Consider Corning 015 pH-sensitive glass, which chemically is a member of a class called *complex silicates*.[7] The material is a polymeric, irregular, three-dimensional network. Each tetrahedral silicon atom is surrounded by four oxygens. Some of the oxygens are bonded to two silicons, forming the bridges that hold the network together, while others, bonded to only one, provide anionic sites within the glass (cf. ion-exchange resins, Chapter 18). The cations Na^+ and Ca^{2+} occupy "holes" within the silicate

---

[4] For a student who had studied physical chemistry, we would say that an equilibrium was established where the chemical potential of H^+ (partial molar free energy, μ_{H^+}) on the less acidic side had been raised to that of H^+ on the other side of the membrane by the electrical potential resulting from the H^+ migration. That is to say, the work required to insert an additional H^+ would become the same for both solutions, despite their different a_{H^+} values, because an electrical work term would have to be included in order to calculate μ_{H^+}. A complete analysis leads to prediction of a Nernstian response.

[5] K. Schwabe and H. Dahms, *Z. Elektrochem.*, **65,** 518 (1961).

[6] G. Haugaard, *J. Phys. Chem.*, **45,** 148 (1941).

[7] The composition of glass is reported as though it were an oxide mixture; e.g., Corning 015 is said to be about 22% by weight Na_2O, 6% CaO, and 72% SiO_2. This does not imply that oxides are necessarily used in the manufacture; e.g., calcium may be added to the melt as $CaCO_3$, but loss of CO_2 at the temperature of molten glass makes it equivalent to CaO. Molten SiO_2 is an acidic oxide which dissolves and reacts with basic oxides like CaO and Na_2O to form the so-called complex silicates.

lattice where they are surrounded by oxygen atoms, but they are somewhat mobile, Na^+ more so than Ca^{2+} because of its lesser charge. There was, in fact, evidence of Na^+ movement through the glass in the electrolysis experiments mentioned above.[6]

Now the material emerging from the glass-making process is quite dry, but before use electrodes are "conditioned" by soaking, whereupon there is an uptake of water by the outside surface. (The inner surface has already been hydrated by the filling solution before purchase.) The depth of water penetration varies with the glass composition; here it may be on the order of 10^{-5} to 10^{-4} mm. Further, a cation exchange process occurs that replaces some of the Na^+ in the hydrated surface layer with H^+;

$$H^+(aq) + Na^+Gl^-(s) \rightleftharpoons Na^+(aq) + H^+Gl^-(s)$$

where Gl^- represents the silicate matrix of the glass with its anionic sites. The surface layer is sometimes described as a *silicic acid gel*.[8] We have, then, a thick interior region of dry glass in which Na^+ is the principal mobile cation, sandwiched between two very thin gel layers penetrated to some degree by H^+. This may be depicted as follows (the thicknesses of the layers are not to scale):

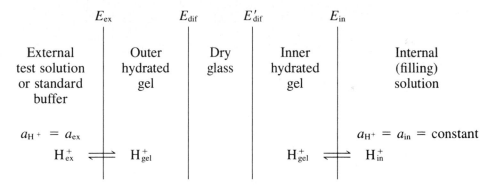

To see how the potential across the membrane changes when a_{ex} is changed, suppose we start from some arbitrary point and increase a_{ex} by dropping some acid into the test solution. Now the equilibrium $H^+_{ex} \rightleftharpoons H^+_{gel}$ in the above diagram shifts toward the right; i.e., additional H^+ moves into the gel, changing the potential designated E_{ex}. The outer gel layer becomes more positive than before, and the adjacent solution more negative. A new equilibrium has been established, the gel having become just enough more positive to counter the increased tendency of the solution to transfer H^+ into it. It is seen that this is analogous to a change in a liquid junction potential. Next, we note that any charge interacts electrostatically with other charges. The anions of the silicate lattice are immobile, but Na^+ can move, and in response to the increased a_{H^+} of the outer gel, Na^+ drifts through the dry glass toward the right in the diagram. This in turn displaces the equilibrium $H^+_{gel} \rightleftharpoons H^+_{in}$ toward the right, changing the potential designated E_{in}. Overall, the response to increased a_{H^+} of the test solution is a Na^+-mediated

---

[8] In general, a *gel* is a material which may appear to be solid when casually examined but which contains a large quantity of liquid. A good example is the white of a hard-boiled egg: heating converts the egg-white protein into an insoluble form with interlocked but still hydrophilic molecules that hold water (perhaps 90% of the egg white) in a fairly immobilized form; all the water in the original egg white is still there. Gelatin and agar are two other familiar gel formers; recall that a little gelatin powder and a lot of water go into gelatin desserts that are not "runny" at the table.

transfer of positive charge (H^+) from the external to the internal solution; the result is the same as in the H^+ migration theory proposed initially, but the mechanism is now consistent with what is known about ion mobilities in glass.

Presumably diffusion potentials designated E_{dif} in the diagram develop at the boundaries between gel layers and dry glass as a result of the differing mobilities of Na^+ and H^+, but they are opposite and approximately equal and tend to cancel. With such approximations, it can be shown that the potential across the membrane as it contributes to the voltage of the cell on page 321 is given by

$$E_{mem} = E_{in} - E_{ex} = 0.0592 \log \frac{a_{in}}{a_{ex}}$$

Finally, the constant value of a_{in} is combined with the constant potentials within the cell (E_{SCE}, etc.) to yield the equation for the cell voltage in terms of the pH of the test solution:

$$E_{cell} = k + 0.0592 \, pH$$

The high specificity for H^+ of the glass pH electrode is a notable attribute. Provided we remain within its useful range as noted on page 321, variations in the activities of other ions such as Na^+, K^+, and Ca^{2+} do not change the potential. Thus meaningful pH values can be easily obtained for an extraordinary variety of sample types. This behavior relates to the magnitudes of the equilibrium constants for ion-exchange processes like the one shown on page 323 for Na^+ and H^+. The equilibrium position in that reaction as written lies so far toward the right that increasing a_{Na^+} in the test solution causes virtually no Na^+ uptake by the gel layer unless, as we noted, a_{H^+} is very low.

Membrane Electrodes for Other Ions

Research on the effects of glass composition upon the interferences of other ions in pH measurements have brought us not only better pH electrodes, but also glass electrodes with poor pH responses which are useful indicator electrodes for other ions. The latter, along with some nonglass relatives, are often called **ion-selective electrodes** or ISEs. Note the word *selective*, not *specific*, in ISE. None of these electrodes, to our knowledge, is as uninfluenced by extraneous ions as is the pH electrode, but this does not mean that they are not very useful. Consider, for example, a sodium ISE based on the glass designated NAS 11–18 (the composition is about 11% Na_2O, 18% Al_2O_3, and 71% SiO_2 by weight): the electrode potential is unresponsive to H^+ above a pH of about 3; the response to Na^+ is near-Nernstian over a wide range (about 10^{-6} to $1 \, M$); K^+ *does* affect the potential, but near pH 7 it is about 0.003 times as effective as Na^+. Now suppose a physiologist needs blood sodium determinations for electrolyte balance studies in experimental animals or, let us say, a clinical researcher is looking at the effects of adrenal hormone therapy in patients with Addison's disease. The pH of the blood is normally well regulated; moreover, if it changes by more than a few tenths of a pH unit from the normal value of about 7.4, the animal or the patient will die. In other words, over the pH range encountered in biomedical research, the potential of a sodium ISE will be virtually unaffected by pH changes. Next, consider interference by K^+. The normal value of $[K^+]$ in blood serum ranges from about 3.5 to 5 mM. A little below about 2 mM, one expects congestive heart failure, and near 12 mM, cardiac arrest, with very serious problems at less extreme levels. Thus the researcher can use a sodium ISE to study the effects of various therapeutic interventions upon blood sodium levels without serious interference from K^+ provided $[K^+]$ remains within a range which is not physiologically threatening. Moreover, certain interfering ions may not be present at all in some kinds of samples. For example, Ag^+ interferes seriously with sodium determinations, but in blood samples, who cares? Unless the patient has been eating jewelry and tableware, Ag^+ will not be a problem.

Nonglass Membranes. The membrane in a membrane ISE need not be glass, and today there are more nonglass sensors than otherwise. Some of these are commercially available and may be found in catalogs; many more have been developed by researchers to solve special problems and, while serving well in narrow contexts, cannot be profitably marketed. Instead of describing numerous electrodes sketchily it may be more instructive to examine one in detail and to compare it with the glass pH electrode.

Ethidium salts (generally chloride or bromide) are drugs which have been used in veterinary medicine to treat trypanosome infections. The ethidium cation (hereafter Eth^+) has the following structure:

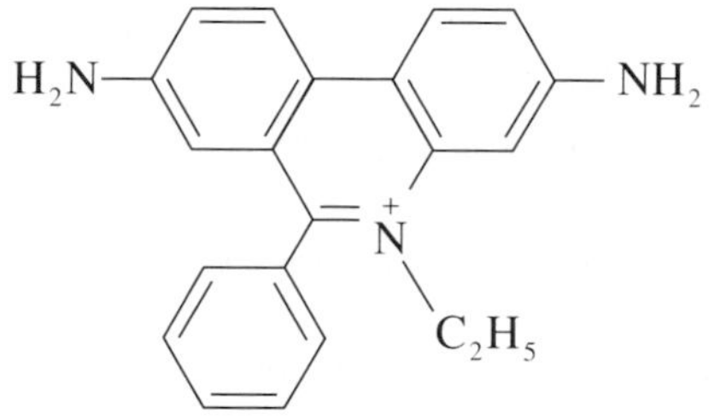

Although Eth^+ is not itself an approved drug for humans, it has nevertheless been important in research as a model for a class of compounds, including drugs and mutagens, called *intercalators*. Insertion of the flat, aromatic ring systems between the base pairs of helical DNA results in coiling changes and other structural responses in the biopolymer. Negative phosphate groups in the DNA surface provide additional binding sites for cationic species. Eth^+ has also been used as a fluorescent tag for locating DNA in biological separation procedures. Thus, for purposes that need not be detailed here but may be found in the biological literature, there have been numerous attempts to measure equilibrium constants for the binding of Eth^+ to DNA (and to smaller nucleotide fragments) in order to determine how many classes of binding sites can be sustained by the data and related to known features of DNA structure. The experimental approach is essentially a titration: known quantities of Eth^+ salt are added to a known quantity of DNA, and after each addition, the activity of *free* Eth^+ is determined, from which DNA-bound Eth^+ can be calculated. The mathematical analysis of the data to determine whether there is more than one type of interaction and to calculate equilibrium constants need not concern us here; our interest is the *analytical* problem of directly measuring a_{Eth^+}. For this purpose, an ethidium-ion-selective electrode was developed.[9]

One version of this electrode is shown in Fig. 12.2. Ethidium tetraphenylborate, $Eth^+(C_6H_5)_4B^-$, is a water-insoluble salt which dissolves in organic solvents, presumably forming pairs of associated ions. A solution of this salt, 0.5% by weight, is prepared in 3-nitro-o-xylene, and a 2- to 3-mm layer is drawn into

[9] B. B. Sauer, R. A. Flint, J. B. Justice, and C. G. Trowbridge, *Arch. Biochem. Biophys.*, **234**, 580 (1984).

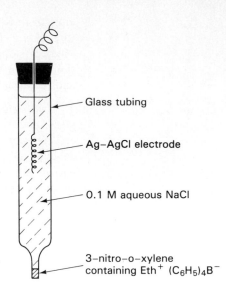

Figure 12.2 The ethidium ion-selective electrode.

- Glass tubing
- Ag–AgCl electrode
- 0.1 M aqueous NaCl
- 3–nitro–o–xylene containing Eth^+ $(C_6H_5)_4B^-$

the tip of a glass capillary. (Its retention is improved if the glass surface is made hydrophobic by silanization; see page 534.) As in a glass pH electrode, there is an internal reference electrode immersed in a filling solution. The nitroxylene is called a *liquid membrane* because, although its physical form is different, it plays the role of a membrane in separating the test solution from the filling solution. The cell is completed, of course, by inserting an external reference electrode in the test solution.

Recall the mechanism by which the glass membrane potential changed when we increased a_{ex} on page 323 and compare it with the result of increasing the activity of Eth^+ in the test solution in the present case, as shown below:

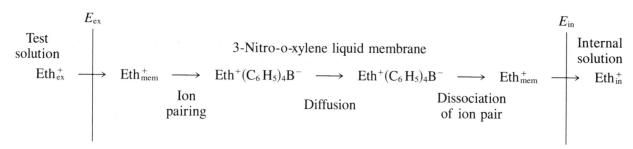

Just as more H^+ entered the outer gel and bound to anionic silicate sites, more Eth^+ will enter the organic liquid membrane. The mechanisms of charge transport, though, are different in the two cases. Fixed binding sites have been replaced by dissolved and mobile $(C_6H_5)_4B^-$, and because the ion pair Eth^+ $(C_6H_5)_4B^-$ can diffuse through the liquid membrane, transport of Eth^+ is not mediated by a second cation as Na^+ served earlier for H^+. But the result is the same. Increasing a_{Eth^+} in the test solution initiates a movement of positive charge toward the right in the above diagram. The potential across the membrane changes and appears as a change in cell voltage.

It is found experimentally that the relationship between E_{cell} and a_{Eth^+} is near-Nernstian down to $[Eth^+] \approx 10^{-6} M$. After a few days, the response deteriorates, but the electrodes are so inexpensive and easily made that this is a minor problem. A different version of the same electrode, prepared by incorporating the $(C_6H_5)_4B^-$ in an actual membrane of polyvinyl chloride rich in the plasticizer dioctylphthalate, is more durable. The electrode is subject to many interferences, but most of these are unimportant in the DNA-binding application because normally one would not introduce other organic cations that might be transportable across the membrane by $(C_6H_5)_4B^-$. The electrode does not work if the filling solution is KCl instead of NaCl, because K^+ binds to $(C_6H_5)_4B^-$ much more strongly than does Na^+, and K^+ must also be absent from the DNA solution.

Other Ion-selective Electrodes. The ethidium ISE is only one example of many such electrodes which have been developed for a large variety of analytes ranging from simple inorganic ions through biological metabolites, drugs, pesticides, and herbicides. Some of the electrodes reflect great ingenuity.[10] A few liquid membrane electrodes are available commercially for inorganic analytes, including K^+, Ca^{2+}, Cl^-, BF_4^-, NO_4^-, NO_3^-, and ClO_4^-. To take one case, the calcium ISE is based upon a layer of organic solvent containing an organic phosphate that binds Ca^{2+} more strongly than many other cations and provides a means for preferential transport of Ca^{2+} through the membrane; for durability, the liquid membrane is retained between porous glass plates or by an ion-permeable plastic membrane.

Solid-State Electrodes. In *solid-state electrodes,* a crystal or cast pellet functions as a membrane. For example, a fluoride ISE is based upon a crystal of quite water-insoluble lanthanum fluoride, LaF_3, doped with the europium salt EuF_2.[11] Ionic crystals tend to adsorb ions from solution upon their surfaces, and frequently an ion common to the crystal lattice, here F^-, is adsorbed preferentially (see the Paneth-Fajans-Hahn rule, page 72). Particles in crystals are not quite so immobile as we sometimes imagine, and a repulsive electrostatic driving force such as more F^- at one surface than at the other will initiate a motion of F^- through the crystal. Doping with Eu^{2+} leaves "holes" or vacancies in the lattice to accommodate fluoride ions as they hop along, leaving new holes behind them, and facilitates the movement. The result is that raising or lowering a_{F^-} in the test solution, in the face of a constant a_{F^-} in the filling solution, leads to a net migration of F^- through the crystal in one direction or the other and changes the potential across the crystal and thereby the voltage of the cell. Table 12.1 lists some ISEs that are currently available commercially. The discussion above and a glance at the column headed "Some Interferences" in the table will show that the prospective ISE user has available a marvelous tool but that it must not be applied in blind faith in any and all situations in the real world. Useful references are given below.[12]

---

[10] Good access to recent developments is provided by the fundamental reviews issues published in even-numbered years by the journal *Analytical Chemistry*. The most recent as this is written is R. L. Solsky, "Ion-Selective Electrodes," *Anal. Chem.,* **60,** 106R (1988).

[11] *Doping,* a common term in solid-state physics and chemistry, refers to the intentional incorporation of a small quantity of impurity, here Eu^{2+}, in an otherwise pure solid.

[12] R. A. Durst, Ed., *Ion-Selective Electrodes,* National Bureau of Standards Special Publication No. 314, U.S. Government Printing Office, Washington, 1969; J. Koryta and K. Stulik, *Ion-Selective Electrodes,* 2nd ed., Cambridge University Press, 1983; A. K. Covington, Ed., *Ion-Selective Electrode Methodology,* Vols. I and II, CRC Press, Inc., Boca Raton, FL, 1979.

DIRECT POTENTIOMETRY

12.3a. *Potentiometric Determination of pH*

We now examine briefly the meaning of the term pH as it is measured experimentally by direct potentiometry.[13] In 1909, before the concept of activity had been developed, the biochemist Sorensen defined pH in terms of the molar *concentration* of H^+:

$$pH = -\log [H^+]$$

This provided a convenient way to express $[H^+]$ values ranging over many orders of magnitude, and, from the Nernst equation, it was explicitly linear in the voltages of the cells used to measure H^+. In 1924, recognizing that electrode potentials reflect *activities* rather than concentrations, he redefined pH:

$$pH = -\log a_{H^+} = -\log [H^+] f_{H^+}$$

where f_{H^+} is an activity coefficient (page 96). This definition represents a more sophisticated view of electrolyte solutions, but at the same time it calls attention to a fundamental problem which in principle cannot be solved: in thermodynamic terms the activity of a single ionic species has no operational significance in regard to experiments that can be performed. The pH of a solution based on Sorensen's second definition is proportional to the work required to transfer H^+ reversibly from that solution into one where a_{H^+} is unity. But this is only a *thought* experiment. There is no way actually to transfer cations without transferring anions as well, and there is no thermodynamically valid way to break down the measured total work into the individual ion contributions.

The only way out is a pragmatic one which is theoretically nonrigorous: select on some basis (size, charge density, heat of hydration, phase of the moon, wisdom revealed in a dream, or whatever) two certain ions, M^+ and X^-, for which you believe the work terms ought to be about the same and assign to each of them one-half of the measured work. Then comparisons down the line (e.g., study of N^+X^- and M^+Y^-) will yield estimated single-ion activities for other species as well. There have been many arguments about the choice of M^+ and X^- and about the degree of uncertainty in the assumption that their activities in an MX solution are the same. Most people today seem reasonably content with the procedure which led to the practical or operational pH scale developed by the National Bureau of Standards.[14]

Let us return to a cell for measuring pH and, as on page 321, write (for 25°C)

---

[13] A thorough discussion of this topic which is still valid was given some years ago by the researcher who led the effort at the National Bureau of Standards to establish a useful pH scale and explain its significance: R. G. Bates, *Determination of pH: Theory and Practice,* 2nd ed., John Wiley & Sons, Inc., New York, 1973. An early paper entitled "Use and Abuse of pH Measurements" summarizes the problems and gives warnings (including some for biomedical workers) which should still be taken seriously: I. Feldman, *Anal. Chem.,* **28,** 1859 (1956).

[14] The National Bureau of Standards (NBS) was the name of this organization, within the U.S. Department of Commerce, at the time the work referred to above was done. Legislation changing the name to the National Institute of Standards and Technology (NIST) became effective on August 23, 1988.

$$E_u = k + 0.0592\,p\mathrm{H}_u \qquad \text{or} \qquad p\mathrm{H}_u = \frac{E_u - k}{0.0592} \tag{1}$$

$$E_s = k + 0.0592\,p\mathrm{H}_s \qquad \text{or} \qquad p\mathrm{H}_s = \frac{E_s - k}{0.0592} \tag{2}$$

E_s is the voltage observed when the test solution is a standard buffer whose $p\mathrm{H}$ is $p\mathrm{H}_s$; E_u and $p\mathrm{H}_u$ are the corresponding values for a test solution of unknown $p\mathrm{H}$. Subtracting Eq. (2) from Eq. (1) gives (see Question 5):

$$p\mathrm{H}_u = p\mathrm{H}_s + \frac{E_u - E_s}{0.0592} \tag{3}$$

Now there are two problems in using Eq. (3) to measure the $p\mathrm{H}$ of an unknown solution. First, when we subtracted Eq. (2) from Eq. (1), we assumed that k was the same for both cases and let it drop out. But one of the things lumped into k is the liquid junction potential at the interface between the test solution and the electrolyte in the reference electrode. How can we know that this is unchanged when we replace the standard buffer with the unknown solution? As noted earlier, if the reference electrode contains saturated KCl, we hope E_j is small and nearly constant. It is probable that variations in k are not much larger than about 1 mV under ordinary conditions, corresponding to an uncertainty of about 0.01 or 0.02 on a $p\mathrm{H}$ scale. By "ordinary" conditions in this context we mean:

1. The $p\mathrm{H}$ is not extreme, lying within the range of about 2 to 10.
2. The ionic strength (page 97) is not unusually high—less than, say, 2 or 3.
3. No ion of exceptional mobility (e.g., a very large organic ion or a very heavily hydrated ion such as Li^+) is present at an appreciable concentration.
4. Charged suspensions of macro-sized or colloidal particles (e.g., humus, clay, ion-exchange resin) are absent (but measurements on solutions of soluble macromolecules such as proteins or nucleic acids seem to give reasonable results).
5. $p\mathrm{H}$ values measured near charged surfaces such as those of cells or dental plaque are suspect on an absolute basis, although changes under controlled conditions may be meaningful.

The second problem with Eq. (3) is the assignment of a $p\mathrm{H}$ value ($p\mathrm{H}_s$) to the standard buffer as it relates to a single-ion activity, $a_{\mathrm{H}_s^+}$. Much of the argumentative dust has now settled, and most workers believe that the assumptions about activity coefficients decided upon by the National Bureau of Standards are such that $p\mathrm{H}_s = -\log a_{\mathrm{H}_s^+}$ as closely as we can come at this time. Examination of the internal consistency among six different $p\mathrm{H}$ standards suggests that the uncertainty in $p\mathrm{H}_s$ is about $\pm 0.006\,p\mathrm{H}$ units, which will frequently be a little less than the uncertainty from liquid junction potentials. The National Institute of Standards and Technology sells, as a part of its standard reference materials program, pure compounds for preparing $p\mathrm{H}$ standards within this limitation; some of these are shown in Table 12.2. Standards can, of course, be formulated in one's own laboratory from sufficiently pure materials, and chemical supply houses sell prepared buffer solutions of stated $p\mathrm{H}$.

Finally, what does all this mean to the many users of $p\mathrm{H}$ meters? The practical $p\mathrm{H}$ as measured is not exactly equal to $-\log a_{\mathrm{H}^+}$; a_{H^+} as calculated from a $p\mathrm{H}$ measurement usually differs from the true value by a few tenths of a percent up to several percent, depending upon the composition of the solution. No

TABLE 12.2 National Institute of Standards and Technology (NIST) Materials for Preparing Solutions of Known pH_s Values

Type	pH_s (25°C)
Potassium tetroxalate ($KHC_2O_4 \cdot H_2C_2O_4$)	1.681
Potassium hydrogen tartrate ($KHC_4H_4O_6$)	3.557
Potassium hydrogen phthalate ($KHC_8H_4O_4$)	4.006
Potassium dihydrogen phosphate (KH_2PO_4) + disodium hydrogen phosphate (Na_2HPO_4) (appropriate mixtures to be prepared)	6.863 to 7.415
Tris (hydroxymethyl) aminomethane ($C_4H_{11}NO_3$) + its hydrochloride salt	7.699
Sodium tetraborate decahydrate ($NA_2B_4O_7 \cdot 10\ H_2O$)	9.180
Sodium carbonate (Na_2CO_3) + sodium bicarbonate ($NaHCO_3$)	10.011

significance in terms of a_{H^+} can be attached to more than two decimal places in the pH value, no matter how good the voltmeter may be.

This is not to say that an additional decimal place can never be useful on an empirical basis. A clinician, for example, watching the blood pH of a patient with metabolic acidosis, might well decide to intervene if successive readings showed a distinct downward *trend* in the *third* decimal place, and if an upward trend accompanied the treatment, the physician might see no need to split hairs with physical chemists about the meaning of the individual values.

There are many situations where pH measurements of little fundamental significance in terms of a_{H^+} are very useful on a practical basis. Despite what we said above about humus and clay, measured pH values on soil suspensions may be correlated empirically with plant growth, or, to take another example, pH with the shelf life of canned foods. The people who use such information have no need to know the true hydrogen ion activities of their samples in order to use pH meters effectively, provided they know how pH numbers, measured under prescribed conditions, relate to other variables in their own game.

It should be clear by now that using measurements with a pH meter to calculate hydrogen ion *concentrations* is *very* risky indeed unless you are prepared to go deeply into the physical chemistry of electrolyte solutions. The problems described immediately above relate to errors in a_{H^+}, but since $a_{H^+} = [H^+]f_{H^+}$, conversion to a concentration basis requires information about f_{H^+} which is frequently unavailable. Theoreticians can sometimes estimate the error in assuming that the activity coefficient is unity for a real case; e.g., for an ionic strength of 0.16 (which is often of interest in biological chemistry), the error is thought to be about 25%.[15] In earlier chapters, the student *calculated* pH values of various solutions of acids, bases, and salts, using (perhaps) thermodynamic K_a or K_b values and assuming that all activity coefficients were unity. How close were these values to those which would have been obtained in the laboratory with a pH meter? The answer, of course, is that it depends upon the ions present in the solution and their concentrations (not to mention the reliability of the person who reported the K value). Typical discrepancies might range from a few hundredths up to a half a pH unit or more. When you need to prepare a solution of known pH, do a rough calculation as in Chapters 5 and 6 to get into the right ball park (you will not want to add a liter of reagent a drop at a time), but make the final adjustment with a pH meter; then anyone trying to repeat your work in another laboratory will be able to reproduce your pH value as closely as present circumstances allow.

[15] See I. Feldman, loc. cit.

12.3b. Ions Other Than H⁺

The above discussion of the accuracy and significance of potentiometric pH measurements (activity vs. concentration, single-ion activites, liquid junction potentials) is valid for measurements with other ion-selective electrodes. For analytical chemists, however, there is a difference in practice. People seldom attempt to convert measured pH values into concentrations, and frequently the focus is not even upon the relation to activity. Rather, pH values, simply as numbers which can be reproduced for corresponding cases all over the world, have become in practice variables dealt with directly in their own right. A biochemist reports the effect of pH on the catalytic activity of an enzyme; a drug is said to be stable in solution over a certain pH range; the pH of the medium affects the growth of microorganisms—in most fields, measured pH values are used, as is, for correlations with other properties of interest. But for other ions (F^- in drinking water, CN^- in an industrial discharge, or whatever), it is customary to report analytical results as *concentrations*. The empirical conversion of a measured cell voltage into a concentration using appropriate standard solutions has already been discussed (see page 317.) Clearly the major considerations are matching analyte activity coefficients in standards and unknown solutions and maintaining a constant junction potential throughout a series of measurements. In direct potentiometry, f_{ion} and E_j are the major targets of matrix effects, but frequently they can be controlled.

12.4

POTENTIOMETRIC TITRATIONS

We calculated titration curves for various sorts of reactions in earlier chapters. These were only approximations, because we neglected activity effects, although the general shapes were correct. From the preceding discussion in this chapter, it is clear that we can also obtain the curves experimentally whenever there is an indicator electrode responsive to one of the species in the titration reaction. A setup for potentiometric acid-base titrations is shown in Fig. 12.1. For redox reactions, the indicator electrode may be platinum, and ISEs may be used for various other cases. It is unnecessary to worry about the absolute validity of each measurement. So long as the titration curve is steep near the equivalence point (and we can decide at what titrant volume it is steepest), its absolute placement on the voltage axis as it may relate to activities or concentrations is of no concern. We can determine the end point from the *slope* of the curve. Thus we expect much better precision than in direct potentiometry. Small measurement errors may also be less serious because we will work with the curve that best represents a large number of individual data points.

12.4a. Manual Titrations

The principle here is simple: add successive increments of titrant solution; measure the cell voltage after each addition; prepare a table of titrant volumes and corresponding voltage values; plot a graph like the one in Fig. 12.3(a); decide where the graph is steepest and take that volume for the end point. For the best results, we desire for the last step a more precise and less subjective method than simply eyeballing the graph.

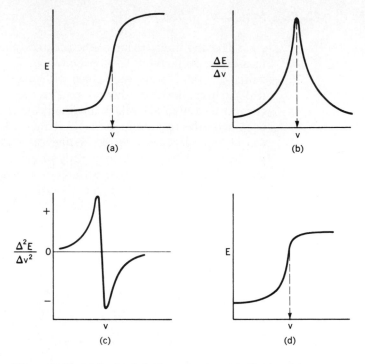

Figure 12.3 Methods of plotting potentiometric titration data.

Analysis of Slopes

Figure 12.3(b) is a plot of the *slope* of the titration curve against the volume of titrant, i.e., dE/dV vs. V. Of course this curve has a maximum at the volume where the curve in Fig. 12.3(a) is steepest. One instinctively feels that the end point determined from this graph will be more precise than one obtained by simply staring at the graph in Fig. 12.3(a) or even laying a ruler along it. A graph like that in Fig. 12.3(b) can be obtained experimentally: titrate quickly at first, obtaining just enough points so as not to run past the steep part; near the equivalence point, add a series of small, equal-volume increments, calculate $\Delta E/\Delta V$, the voltage *change* per volume increment, and plot it against the volume of titrant. Table 12.3 shows some actual data.

Next consider the second derivative, d^2E/dV^2, which goes through zero when dE/dV is a maximum, as shown in Figure 12.3(c). In practice, finite increments (Δ-values) will be calculated, as seen in the last column of the table. The end point can be calculated without actually drawing the graph. We assume that the center portion of the curve in Fig. 12.3(c), from a little above to a little below

TABLE 12.3 Potential Readings Near the Equivalence Point

Titrant, ml	E, mV	$\Delta E/\Delta V/0.1$ ml	$\Delta^2E/\Delta V^2$
24.70	210		
24.80	222	12	
24.90	240	18	+6
25.00	360	120	+102
25.10	600	240	+120
25.20	616	16	−224
25.30	625	9	−7

the zero crossing, is linear and calculate the end point by interpolation. For the data in the table, note that the sign change in the last column occurs between 25.00 and 25.10 ml. Between these points, a 0.10-ml increment in titrant caused a change of $120 - (-224) = 344$. Assuming linearity, we then calculate what volume increment would have changed $\Delta^2 E/\Delta V^2$ from $+120$ to just zero:

$$\frac{120}{120 + 224}(0.10) = 0.035 \text{ ml}$$

The end point, then, is $25.00 + 0.035 = 25.035$ ml.

For "unsymmetrical" reactions (different coefficients for analyte and titrant in balanced equations, e.g., $Sn^{2+} + 2Ce^{4+} \rightarrow Sn^{4+} + 2Ce^{3+}$), where curves like that in Fig. 12.3(d) are obtained, the true equivalence points do not correspond exactly to maxima in dE/dV, but the differences are insignificant for most purposes.[16]

Gran Plots

In the analysis of slopes above, the only data on the titration curve that were actually used to obtain the end point were near the end point itself. Measurements elsewhere were required only for orientation—to prevent running inadvertently through the end-point region. It would be convenient to use data away from this region to avoid the tedium of measuring a large number of voltages corresponding to very small volume changes. Further, a linear function relating the data to the end point would be better than any of those in Fig. 12.3. Gran developed a graphical end point method that meets these goals. His paper[17] discusses all types of titrations (acid-base, redox, precipitation, complex formation), but here we consider one case: titration of a weak acid with a strong base.

Let us titrate V_0 milliliters of a solution of the weak acid HB, ionization constant K_a, of initial concentration C_a, with a sodium hydroxide solution of concentration C_b. The volume of base solution added at any point is V_b, and the volume added at the equivalence point is V_e. Measured pH values will be good estimates of a_{H^+}, and K_a is given by

$$K_a = \frac{(a_{H^+})(a_{B^-})}{(a_{HB})} = \frac{a_{H^+}[B^-]f_{B^-}}{[HB]f_{HB}} \tag{1}$$

which rearranges to

$$a_{H^+} = \frac{K_a[HB]f_{HB}}{[B^-]f_{B^-}} \tag{2}$$

where f is an activity coefficient. We next obtain expressions for [HB] and [B$^-$] for any intermediate point in the titration based upon the reaction stoichiometry and taking account of the volume change, assuming that in the presence of excess HB, the reaction $HB + OH^- \rightarrow B^- + H_2O$ goes virtually to completion:

---

[16] Even for symmetrical reactions, there may be slight differences related to dilution by the titrant, but the error is negligible. For a detailed mathematical analysis, see L. Meites and J. A. Goldman, *Anal. Chim. Acta*, **29**, 472 (1963) and W. Lund, *Talanta*, **23**, 619 (1976).

[17] G. Gran, *Analyst*, **77**, 661 (1952). This paper antedates computers and hand calculators, and hence it contains some tips that would be ignored today (e.g., use of antilog graph paper.)

$$[HB] = C_a\left(\frac{V_0}{V_0 + V_b}\right) - C_b\left(\frac{V_b}{V_0 + V_b}\right) \qquad (3)$$

But

$$C_a V_0 = C_b V_e \qquad (4)$$

So Eq. (3) becomes

$$[HB] = \frac{C_b V_e}{V_0 + V_b} - C_b\left(\frac{V_b}{V_0 + V_b}\right) \qquad (5)$$

Factoring out C_b and combining terms with a common denominator then gives

$$[HB] = \frac{C_b}{V_0 + V_b}(V_e - V_b) \qquad (6)$$

Also,

$$[B^-] = \frac{C_b V_b}{V_0 + V_b} \qquad (7)$$

Substituting Eqs. (6) and (7) into Eq. (2),

$$a_{H^+} = \frac{K_a\left(\frac{C_b}{V_0 + V_b}\right)(V_e - V_b)f_{HB}}{\left(\frac{C_b V_b}{V_0 + V_b}\right)f_{B^-}} = K_a\left(\frac{V_e - V_b}{V_b}\right)\left(\frac{f_{HB}}{f_{B^-}}\right) \qquad (8)$$

Replacing a_{H^+} with 10^{-pH} and multiplying both sides of Eq. (8) by V_b gives

$$V_b \times 10^{-pH} = K_a V_e\left(\frac{f_{HB}}{f_{B^-}}\right) - V_b K_a\left(\frac{f_{HB}}{f_{B^-}}\right) \qquad (9)$$

Provided the activity coefficient ratio f_{HB}/f_{B^-} remains practically constant (we could ensure this by adding sufficient indifferent electrolyte, say, $NaNO_3$, to the solution), a graph of $(V_b \times 10^{-pH})$ vs. V_b will be linear, with a slope of $-K_a(f_{HB}/f_{B^-})$. Now notice what happens when $(V_b \times 10^{-pH})$ goes to zero: The right-hand side of Eq. (9) also becomes zero, which is to say that $K_a V_e = K_a V_b$ or, in other words, the value of V_b is V_e, the volume of titrant at the equivalence point.

Actually, the Gran plot may exhibit curvature in the early stages if K_a is large enough to invalidate Eq. (7) above, and it will curve near the other end because 10^{-pH} can only approach zero, not reach it. But there will be enough points on a virtually straight line to extrapolate to V_e on the V_b axis. (Even if f_{HB}/f_{B^-} is not really constant, it is usually sufficiently so over an adequate range to permit a linear extrapolation.) Figure 12.4 shows an acid-base titration curve and a Gran plot based upon the same data. For data past the equivalence point, Gran showed that a plot of $(V_0 + V_b) \times 10^{pH}$ vs. V_b is linear and also goes to zero when $V_b = V_e$.[17] Others have emphasized the value of Gran plots, particularly for titrations involving tricky mixtures.[18] Although an actual graph may be fun to look at, obviously there is no reason these days to use it for determining V_e, which can be obtained easily from the computed slope and intercept of a least-squares straight line (page 35) through the best of the data points.

[18] F. J. C. Rossotti and H. Rossotti, *J. Chem. Educ.* **42,** 375 (1965).

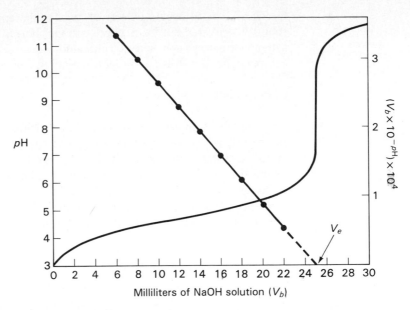

Figure 12.4 Potentiometric titration curve and Gran plot: 50.0 mL of 0.0500 M acetic acid titrated with 0.1000 M NaOH; note the two ordinates (right-hand ordinate is the axis for the Gran plot).

12.4b. *Automated Titrations*

A person who does a potentiometric titration only occasionally probably does it manually, as above. It would not pay to spend thousands of dollars to automate a rarely used procedure. On the other hand, for a laboratory doing hundreds or more a week, cost-effectiveness might demand automation; expensive instruments may cost less than people. There are various ways to eliminate in varying degree the human labor of potentiometric titrations.

Early Devices

The earliest methods involved automatic recording of the titration curve. A device called a *recording potentiometer,* or often just a *recorder,* plots with pen and ink on a moving chart the input voltage as a function of time. If titrant is delivered to the analyte solution at a constant rate and the cell voltage is fed to the recorder, the pen records a titration curve. While this is going on, the operator can be preparing the next sample or getting a drink or whatever. At the end, though, the old problem arises: where is the recorded curve steepest, and how precisely can that point be located? And the titration reaction must be rapid if the pen is to record equilibrium voltages continuously.

Computer-Controlled Devices

There are home-made devices that can do much better than this. Titrant delivery can be triggered by computerized voltage data. The computer watches the voltage as a function of time, and when it "decides" (within a preset limit) that stirring and reaction have achieved equilibrium in the cell, it tells a pump to deliver another portion of titrant. As it works its way along, it can also control the volumes of the titrant increments as the curve steepens and examine voltage differences so as to decide objectively where the end point is. The titrant in such setups is delivered not from a conventional buret, but perhaps from a device like a hypodermic syringe with a calibrated worm drive pushing the plunger. Any computer freak who is handy with plumbing can rig up such a device.

Commercially Available Devices

Finally, complete instruments are available commercially. A set of samples can be sucked up one at a time and successively titrated while the operator is out of town. Some instruments titrate to a predetermined end-point potential and record the result, while some determine at what titrant volume the curve is steepest by examining the slope as disclosed by successive voltages at small titrant increments. Others differentiate the voltage data electronically and stop the flow of titrant when the second-derivative signal changes sign.

KEY TERMS

Electrode of the first kind. An electrode in which the metal itself is a component of the redox couple, e.g., a copper electrode which participates in $Cu^{2+} + 2e = Cu$.

Electrode of the second kind. An electrode in which the potential associated with the electron-transfer reaction ($M^{n+} + ne = M$) is governed by the activity of another species which interacts with M^{n+}; e.g., see calomel electrode, page 268.

Glass electrode. A membrane electrode whose membrane is glass, originally developed for *p*H measurements. Such electrodes are now available which are selective for other ions, such as Na^+.

Indicator electrode. In potentiometry, the electrode in a galvanic cell whose potential varies with the analyte activity.

Ion-selective electrode (ISE). An electrode of the membrane or solid-state type whose potential responds selectively to the activity of a particular ion. None is totally *specific*; i.e., there are interferences from other ions, but if the response is selective enough and if other ions are present at low, or at least constant, activities, the potential is a measure of the analyte level.

Matrix effect. The effect of other components of the sample upon the behavior of the analyte in any analytical procedure (electroanalytical, spectroscopic, or whatever). The term relates to the definition of matrix in geology as a mass of material within which is embedded a small item of interest such as a gem. *Matrix* and *interfering substance* are not synonyms; e.g., a trace of some species may interfere if it mimics the analyte, while matrix effects are sometimes tolerable if reproduced in both unknowns and standards.

Membrane electrode. An electrode which incorporates a membrane across which a potential develops when the activities of a particular ion are different in the solutions on the two sides. The glass electrode for H^+ is an example. Sometimes a thin layer of a water-immiscible organic solvent serves as the membrane.

Potentiometric titration. A titration where the end point is found by measuring the emf of a galvanic cell.

Reference electrode. In potentiometry, the electrode in a galvanic cell whose potential is constant; thus the cell emf reflects only the response of the indicator electrode (q.v.) to the analyte.

QUESTIONS

1. *Definitions.* Define the following terms: indicator electrode; reference electrode; potentiometric titration; liquid junction potential; matrix effect.

2. *Activity and concentration.* We say that indicator electrodes really respond to activities rather than concentrations. Is it possible, then, to obtain analyte concentrations potentiometrically? If so, how?

3. *Matrix effect.* An analyst is using an ISE to determine [CN⁻] in water samples that vary greatly in salinity—all the way from fresh-water streams through estuaries to seawater. How can the effect of NaCl upon the activity coefficient of CN⁻ be overcome so that voltages can be converted into concentrations using a calibration curve based upon standard NaCN solutions?

4. *pH measurement.* Describe the basic problems in associating voltage measurements on a glass–SCE electrode pair with a_{H^+} and with [H⁺].

5. *Selection of electrodes.* Select an appropriate electrode pair for performing potentiometrically each of the following titrations: (a) acetic acid with NaOH; (b) Zn^{2+} with EDTA; (c) Sn^{2+} with Ce^{4+}.

6. *Interference from reference electrode.* A silver–silver chloride electrode is to be used as an indicator electrode for measuring a_{Cl^-} in water samples. An SCE is to be used as the reference electrode, but the analyst is afraid that leakage of Cl⁻ from this electrode will contaminate the test solutions. Suggest a solution to this problem.

7. *Algebra practice.* Refer to the derivation of the equation showing the dependence of E for the mercury electrode upon pM (page 319). Perform the missing steps suggested in the statement: "After rearranging the expressions for the two formation constants $K_{HgY^{2-}}$ and $K_{MY^{(4-n)-}}$, obvious substitutions for $a_{Hg^{2+}}$ in the above equation lead to _____ ."

8. *Algebra practice.* Write out all the intermediate steps that you personally would need in order to obtain Eq. (3) on page 329 by subtracting Eq. (2) from Eq. (1).

9. *Classification of electrode.* The hydrogen electrode is usually considered an electrode of the first kind for H⁺. Could this electrode be considered an electrode of the second kind for hydroxide ion? Explain.

10. *Classification of electrode.* Reilley and Schmid called the mercury electrode as described in this chapter a pM electrode for various metal ions. We adopted this in our discussion when we called it an electrode of the third kind (page 319). In another textbook, the same electrode is described as an important electrode of the second kind for measuring the EDTA anion Y^{4-}. Are both views correct? Explain. Can you present an equation giving the electrode potential as an explicit function of $a_{Y^{4-}}$?

PROBLEMS

1. *Potentiometric titration.* 100 mL of 0.0500 M HCl is titrated with 0.100 M NaOH. (a) Calculate the pH of the solution after addition of the following titrant volumes: 0.00; 10.0; 20.0; 30.0; 40.0; 49.0; 49.5; 49.6; 49.7; 49.8; 49.9; 50.0; 50.1; 50.2; 50.3; 50.4; 50.5; 52.0; 55.0 mL. (b) Calculate the voltage readings corresponding to the volumes in (a) for a hydrogen electrode–SCE pair in the solution during the titration. Assume p_{H_2} = 1.00 atm, E_{SCE} = +0.246 V. (c) Plot the titration curve of E vs. milliliters.

2. *Potentiometric titration.* Repeat Problem 1, substituting butyric acid ($CH_3CH_2CH_2COOH$, $K_a = 1.54 \times 10^{-5}$) for HCl.

3. *Derivative plots.* From the data in Problems 1 and 2, plot $\Delta E/\Delta V$ and $\Delta^2 E/\Delta V^2$, using 0.1-mL titrant increments, vs. volume of titrant.

4. *Potentiometric titration.* 0.1978 g of pure arsenious oxide, As_2O_3, is dissolved in NaOH, and the solution is acidified with HCl to form arsenious acid. The final volume is 100 mL, and the HCl is 1 M. The As(III) solution is titrated potentiometrically with a 0.1000 M ceric solution in 1 M HCl, using a Pt–SCE electrode pair. The formal potential for the arsenic system ($H_3AsO_4 + 2H^+ + 2e = HAsO_2 + 2H_2O$) in 1 M HCl is +0.557 V; for the ceric-cerous system, it is +1.23 V. Calculate the cell voltage after addition of the following volumes of titrant: 8.0; 15.0; 23.0; 30.0; 35.0; 39.0; 39.9; 40.0; 40.1; 41.0; 45.0 mL. Plot the titration curve.

5. *Potentiometric end-point calculation.* The following data were obtained near the end point in a potentiometric titration of a reducing solution with 0.1000 N oxidant, using a Pt–SCE electrode pair:

Milliliters of titrant	E, mV
38.70	541
38.80	547
38.90	555
39.00	566
39.10	583
39.20	844
39.30	1104
39.40	1121
39.50	1133

Calculate the end-point volume.

6. *Resistance in a glass electrode.* In a potentiometer circuit for measuring a cell voltage (page 263), a known external voltage is applied in opposition to that of the test cell and varied until an exact balance is attained where no net current can be detected with an ammeter. A certain glass electrode has a resistance through the glass of 4×10^6 ohms; all other resistance in the circuit is negligible compared with this. To obtain a precision of 0.02 pH unit, what is the smallest current that the ammeter would have to detect?

7. *Precision in potentiometry.* Suppose a cell with a cadmium indicator electrode and an SCE reference is used to monitor Cd^{2+} in water flowing from a smelter that processes metal ores. If the cell voltage can be measured to the nearest 0.1 mV, what is the smallest percentage change in $a_{Cd^{2+}}$ that can be detected, starting from a level of 2 μg/ml (2 ppm)?

8. *Measurement of pH.* Show that the potential of the following cell

$$M|M^{2+}(aq) + M(OH)_2(s)\|KCl(satd.),$$
$$Hg_2Cl_2|Hg$$

is related to the pH of the solution in the left-hand electrode by the equation

$$E = k + 0.059\ pH$$

where k is a constant. M^{2+} forms a slightly soluble hydroxide, $M(OH)_2$, and $a_{M^{2+}}$ depends upon a_{OH^-} via K_{sp}, the solubility product constant for $M(OH)_2$.

9. *The quinhydrone pH electrode.* The quinhydrone electrode was formerly used as an indicator electrode for hydrogen ions before the glass electrode became available. The test solution is saturated with quinhydrone, the term for a crystallizable 1 : 1 mixture of quinone (Q) and hydroquinone (H_2Q) that dissolves in water to release equimolar quantities of the two compounds. The electrode is platinum, and the redox couple in solution is

$E° = +0.70$ V

(Q)   (H_2Q)

Derive an equation relating the pH of the test solution to the voltage of the following cell:

$$Pt|H^+(x\ M) + Q + H_2Q\|SCE$$

10. *Quinhydrone electrode.* Answer the following questions about the cell described in Problem 9. (a) What is the polarity of the quinhydrone electrode at pH 4.0? (b) What is the polarity of the calomel electrode at pH 8.5? (c) At what pH is the voltage of the cell zero?

11. *Formation constant of a complex.* A solution containing 4.000 mmol of M^{2+} is titrated with X^- according to the reaction $M^{2+} + X^- = MX^+$. The voltage of the cell

$$Hg|Hg_2Cl_2, KCl(satd.)\|M^{2+}|M$$

is 0.328 V at the equivalence point, with the M electrode negative. Given that $E° = +0.392$ V for $M^{2+} + 2e = M$ and $E_{SCE} = +0.246$ V, calculate the formation or stability constant of the MX^+ complex. The volume at the equivalence point is 100.0 ml.

12. *Mercury electrode as a pM electrode.* Calculate the potential of the mercury indicator electrode at the equivalence point in the titration of 5.00 mmol of Zn^{2+} with EDTA at pH 6.00. The volume of the solution at the equivalence point is 100 ml, and the concentration of HgY^{2-} is $1.00 \times 10^{-5}\ M$.

13. *Effect of solubility on precipitation titrations.* A 50-ml portion of 0.10 M $AgNO_3$ is titrated with 0.10 M NaCl using a silver indicator electrode. (a) Calculate the potential of the indicator electrode after the addition of 49.9, 50.0, and 50.1 ml of titrant. (b) Repeat part (a) substituting NaBr for NaCl. (c) Repeat part (a) substituting NaI for NaCl. Calculate ΔE per 0.2 ml of titrant for each titrant.

13

Other Electroanalytical Methods

Electrochemistry is a very large and important field, including in its broadest aspects fuel cells, solar-based electrical energy, batteries, metal corrosion, commercial production of materials such as aluminum and chlorine, and electroorganic synthesis. Several electrochemical techniques, one of which we examined in Chapter 12, have found extensive application in analytical chemistry. We have selected for the present chapter a few other widely used electroanalytical methods which can be understood from the principles developed in Chapter 10.

13.1

ELECTROLYSIS

13.1a. Electrochemical Cells

Galvanic Cells

Consider the following galvanic cell:

$$Cu | Cu^{2+}(a = 1) \| Ag^+(a = 1) | Ag$$

$2Ag^+ + 2e$	$\rightleftharpoons$	$2Ag$	$E° = +0.80$ V
$Cu^{2+} + 2e$	$\rightleftharpoons$	Cu	$E° = +0.34$ V
$2Ag^+ + Cu$	$\rightleftharpoons$	$Cu^{2+} + 2Ag$	$E°_{cell} = +0.46$ V

The positive sign for E°_{cell} says, according to our convention in Chapter 10, that the silver electrode is the positive one, that the copper electrode is negative, and that the direction of spontaneous reaction, if the cell is allowed to discharge, is from left to right as written above.

Now let us connect this cell in series opposition to an external voltage source—a battery or electronic power supply—as shown in Fig. 13.1. The arrow through the source means that we can vary the external voltage applied to the galvanic cell. Circles A and V represent an ammeter and a voltmeter, respectively. If we adjust the applied voltage to exactly 0.46 V and close the switch, the ammeter will show no current because the external source and the galvanic cell, connected in opposition, just balance; neither can send electrons through the other. If we apply a voltage smaller than 0.46 V, a current will be observed: electrons will flow from the copper electrode through the external circuit and into the silver electrode; the cell reaction is proceeding spontaneously from left to right, with Ag^+ being reduced to Ag and Cu oxidized to Cu^{2+}. We are still in the galvanic mode.

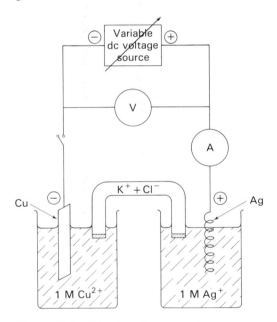

Figure 13.1 Galvanic cell connected to external voltage source (see discussion in text).

Electrolytic Cells

If, on the other hand, we apply against the galvanic cell a voltage greater than 0.46 V, again a current will flow, but it will be in the opposite direction. Electrons will flow from the negative side of the external source into the copper electrode, and they will flow away from the silver electrode into the external circuit. Cu^{2+} will be reduced to Cu, and Ag will be oxidized to Ag^+, which is to say that the cell reaction has been reversed; it is now proceeding from right to left as written above. This process is called *electrolysis,* and we say that the cell, which was galvanic earlier in our discussion, is now an electrolytic cell. In the spontaneous discharge of a galvanic cell, electrical energy is derived from the inherent tendency of a redox reaction to occur. In an electrolytic cell, an external energy source is used to force the chemical reaction to occur in the direction opposite to the spontaneous one.

Sign Conventions

Students sometimes experience confusion regarding signs. The cathode is *always* defined as the electrode at which *reduction* occurs, and the *anode* is the electrode where *oxidation* occurs. In the galvanic cell above, Ag^+ is reduced to Ag at the cathode, and, as we noted, the silver electrode is the positive one. In the electrolytic mode, the silver electrode is still positive, but now oxidation of Ag to Ag^+ occurs here; hence this electrode is the anode. In the galvanic case, the negative copper electrode is the anode (oxidation of Cu to Cu^{2+}), while in electrolysis, reduction occurs here and copper is the cathode. In summary:

	Cathode	Anode
Galvanic	+	−
Electrolytic	−	+

Current Flow

Note that a current flowing in the circuit in Fig. 13.1 involves electron flow in the external wiring and migration of ions in the cell solutions, including the salt bridge. This current is carried across the electrode-solution interface by electron-transfer reactions involving the redox couples Cu^{2+}–Cu and Ag^+–Ag. Such a current, directly associated with electrode reactions, is called a *faradaic current*. Faradaic currents are usually the major ones in large-scale electrolysis, but under special conditions a small *charging* or *capacitance current* may become significant; this is the electron flow associated with a changing electrical double layer at the electrode-solution interface (page 365).

The Electrolysis Process

In electrolysis with the cell in Fig. 13.1, we initially faced a preexisting galvanic cell whose electrodes were in equilibrium with their solutions; this cell possessed a definite voltage of its own. Sometimes in practice, however, we place into a solution a pair of inert electrodes (often platinum) which adopt the same potential. Suppose, for example, that two platinum electrodes are inserted into an aqueous solution of copper sulfate, as in Fig. 13.2. No galvanic cell really exists; if two identical electrodes are placed in the same solution, they have the same potential and there is no voltage between them.

Now let us turn up the source voltage until we begin to see something happen. As the ammeter shows that a current is flowing, we also observe that the negative electrode is becoming reddish-brown and we see bubbles of gas around the positive electrode. We are reducing Cu^{2+} ($Cu^{2+} + 2e \rightarrow Cu$) and oxidizing water ($2H_2O \rightarrow O_2 + 4H^+ + 4e$). The copper-plated platinum electrode is now a copper electrode; around the anode, where the solution was originally the same as that about the cathode, there are now molecules of O_2. In other words, when the electrolysis began, we created from its products a galvanic cell which we may write as follows:

$$Pt, O_2 | H^+ + H_2O + Cu^{2+} | Cu$$

Supposing for simplicity that the ions are at unit activity and that the partial pressure of O_2 above the solution is 1 atm, we can calculate the voltage of this cell from the standard potentials:

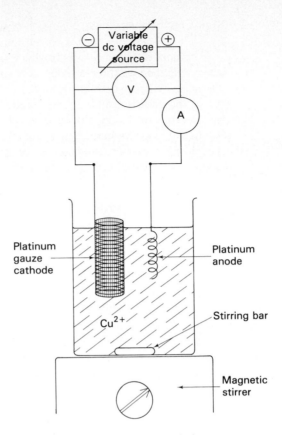

Figure 13.2 Apparatus for electrolysis.

$$2Cu^{2+} + 4e \rightleftharpoons 2Cu \qquad\qquad E° = +0.34\ V$$

$$\underline{O_2 + 4H^+ + 4e \rightleftharpoons 2H_2O \qquad\qquad E° = +1.23\ V}$$

$$2Cu^{2+} + 2H_2O \rightleftharpoons 2Cu + O_2 + 4H^+ \qquad E°_{cell} = -0.89\ V$$

The negative sign for $E°_{cell}$ indicates spontaneity from right to left as the cell reaction is written above. To make the electrolysis go (i.e., for the reaction to go from left to right) we must apply a larger voltage than the galvanic value of 0.89 V.

The galvanic voltage which must be overcome by the external power source in order to initiate electrolysis (0.46 V in the cell on page 339, 0.89 V in the case immediately above) is called the *decomposition potential* or sometimes the *back emf*, often abbreviated E_d. Values of E_d for particular cases are easily calculated, at least in principle, using the methods of Chapter 10, once the products of the electrolysis have been decided upon. The values will usually be estimates, because we will not in general know the *activities* of all of the species that will appear in the Nernst equation.

13.1b.   *Products of Electrolysis*

Sometimes several chemical substances are present in a solution, and we may wish to predict what the electrode reactions will be if the solution is electrolyzed. In principle, this is simple. For the cathode reaction, the rule is: Of all the substances which have access to the cathode, the one which is most easily reduced will be the one that *is* reduced. We evaluate the ease of reduction, of course, from the single-electrode potential. For reactions written $Ox + ne \rightleftharpoons$ Red, using the sign convention of this text, the most positive (least negative) potential indicates the greatest tendency to go from left to right.

EXAMPLE 13.1

Predicting products of electrolysis

A solution contains the following ions, each at a concentration of 1.0 M: Zn^{2+}, H^-, Cu^{2+}, Ag^+. Platinum electrodes are inserted, and the applied voltage is increased until electrolysis begins. What product is formed at the cathode?

The standard potentials are

$$Ag^+ + e \rightleftharpoons Ag \qquad E° = +0.80 \text{ V}$$
$$Cu^{2+} + 2e \rightleftharpoons Cu \qquad E° = +0.34 \text{ V}$$
$$H^+ + e \rightleftharpoons \tfrac{1}{2}H_2 \qquad E° = 0 \text{ V}$$
$$Zn^{2+} + 2e \rightleftharpoons Zn \qquad E° = -0.76 \text{ V}$$

Using the rule above, we select the reduction of Ag^+ as the most facile cathode reaction, and the product will be a plate of silver metal on the surface of the platinum cathode.

Of course, if the concentrations had not all been unity, then the actual potentials to be compared would have been calculated from the Nernst equation. For example, if $[Ag^+]$ had been 0.1 M, then

$$E = +0.80 - 0.0592 \log \frac{1}{0.1} = +0.74 \text{ V}$$

Since the concentrations appear in the log term, the $E°$ values are often adequate for predicting electrode reactions unless they are quite close together or the concentrations are quite disparate. □

The external voltage source is pumping electrons into the cathode, but if they are sufficiently consumed in the reaction $Ag^+ + e \rightarrow Ag$, then the cathode may not become negative enough to reduce, say, Cu^{2+}, which has a lower electron affinity than that of Ag^+.

At the anode, the substance which gives up electrons most readily is the one which will be oxidized. Again, we judge the ease of oxidation from the potential, but here we look for the least positive (most negative) value, that is, the value indicating the greatest tendency of $Ox + ne \rightleftharpoons$ Red to go from right to left.

EXAMPLE 13.2

Protecting products of electrolysis

A solution contains the following species, each at a concentration of 0.5 M: hydroquinone (H_2Q), Cl^-, Br^-, H^+. Platinum electrodes are inserted, and a current is passed. What product is formed at the anode?

The standard potentials are

$$PtBr_4^{2-} + 2e \rightleftharpoons Pt + 4Br^- \qquad E° = +0.58 \text{ V}$$
$$Q + 2H^- + 2e \rightleftharpoons H_2O \qquad E° = +0.70 \text{ V}$$
$$PtCl_4^{2-} + 2e \rightleftharpoons Pt + 4Cl^- \qquad E° = +0.73 \text{ V}$$
$$Br_2 + 2e \rightleftharpoons 2Br^- \qquad E° = +1.09 \text{ V}$$
$$O_2 + 4H^- + 4e \rightleftharpoons 2H_2O \qquad E° = +1.23 \text{ V}$$
$$Cl_2 + 2e \rightleftharpoons 2Cl^- \qquad E° = +1.35 \text{ V}$$

Note that we must not forget any possible reductant: We include the oxidation of water as a possibility, and the oxidation of the platinum anode itself must also be considered. Platinum, which is sometimes called an inert electrode, is really only relatively so, and platinum may be oxidized in electrolysis, depending upon the potential it is allowed to reach and the composition of the solution. The oxidation is facilitated in this example by the presence of Br^-, which converts Pt^{2+} into a stable complex ion.

Inspecting the standard potentials, we see that $+0.58$ is the least positive one. Strictly, of course, in this example the standard potentials cannot be used directly. For instance, because $[H^+] = 0.5\ M$, $+1.23$ V is not the correct value to use in predicting how easily water is oxidized. But it is easily verified that lowering $[H^+]$ from $1\ M$ to $0.5\ M$ will not shift this potential enough to make the oxidation of water a serious contender: a tenfold change in $[H^+]$ shifts the potential by only 0.0592 V. Thus, in this example, the oxidation of the platinum anode is the best possibility. Note that we cannot really calculate the potential for this, but that it does not matter. The Nernst equation for the process is

$$E = +0.58 - \frac{0.0592}{2} \log \frac{(0.5)^4}{[PtBr_4^{2-}]}$$

Before electrolysis, there is no $PtBr_4^{2-}$ in the solution according to the statement of the problem. But if we insert zero into the Nernst equation, we shall calculate that E is infinite, impossible physically. No one knows the real situation: perhaps there are a few of these ions in the solution, formed in some manner such as a tiny amount of air oxidation of Pt. In any event, we do not know the concentration and cannot calculate E. But we do know that if $[PtBr_4^{2-}]$ is very small, E will be even less positive than $+0.58$ V, and we still select this oxidation for the anode reaction. If we start the electrolysis, the platinum electrode will lose weight and $PtBr_4^{2-}$ will appear in the solution. Platinum is expensive; don't borrow platinum electrodes from your friends and treat them this way. ☐

The student should be cautioned not to inspect potentials blindly without a physical picture in mind of the electrochemical cell. For example, electrolysis experiments are often performed with a silver anode in a chloride solution, where the following reaction proceeds from right to left:

$$AgCl + e \rightleftharpoons Ag + Cl^- \qquad E° = +0.22\ V$$

AgCl is a solid which adheres to the surface of the silver anode, and it cannot reach the cathode. Thus the reduction of AgCl is not a possible cathode reaction even if $+0.22$ V is the largest positive number among those being contemplated. As a second example of a common mistake, consider

$$O_2 + 4H^+ + 4e \rightleftharpoons 2H_2O \qquad E° = +1.23\ V$$

If this reaction occurs at the anode, generating O_2, students sometimes seize upon the reduction of O_2 as the cathode reaction because $+1.23$ V is so large. But oxygen is not very soluble in water, and most of it bubbles out of the solution around the anode. In addition, special precautions are sometimes taken, such as using a blanket of nitrogen over the cathode compartment or a porous diaphragm between the electrodes to minimize transport of O_2 to the cathode by currents in the solution. In predicting electrode reactions, do not consider reactions which are impossible based on the physical arrangement of the cell.

Finally, there are sometimes problems in predicting electrolysis products which arise from kinetic effects. A detailed discussion is beyond the scope of this text, but it must be mentioned that occasionally an electrode reaction which is feasible thermodynamically (i.e., one which would be predicted to occur on the basis of the potential) does not occur appreciably during an electrolysis because of the very slow *rate*. The *tendency* for a reaction to occur and the *rate* at which it proceeds are two different things. Predicting electrolysis products from potentials will lead to an error once in a while, and we shall just have to live with this. Among examples which are often encountered, oxyanions such as ClO_4^-, NO_3^-, and SO_4^{2-} are well known for not reducing when the potentials indicate that they will.

It must be emphasized that in this section we have been talking about only the *initial* products of an electrolysis. As an electrode reaction proceeds in a prolonged experiment, concentrations change, potentials shift, and new electrode reactions may become possible. Aspects of this problem are considered later in the chapter.

13.1c. Voltage Requirements

We saw earlier that in order to initiate an electrolysis we had to exceed the galvanic back emf or decomposition voltage, E_d. This value, which is given by $E_d = E_{anode} - E_{cathode}$, is easily calculated, as we saw. Of course it changes as the electrolysis proceeds, but it can be calculated for any conditions by using the Nernst equation for E_{anode} and $E_{cathode}$. However, for an electrolysis actually to proceed, we must apply a larger voltage than E_d. The electrolytic cell offers a resistance, R, to the flow of current, and in order to pass a finite current, i, we know from Ohm's law that we must exceed E_d by a value of $i \times R$. For example, to pass a current of 0.1 A through a cell whose resistance is 5 ohms, we require an additional $5 \times 0.1 = 0.5$ V above E_d.

Concentration Overpotential

There is also another problem to consider. Suppose that our goal is to reduce Cu^{2+} to Cu at a cathode. Now, before the electrolysis is started, the concentration of Cu^{2+} is presumably uniform throughout the solution, as shown by curve A in Fig. 13.3. As the electrolysis proceeds, however, it is those ions near the cathode surface which are reduced to copper metal, thereby lowering the concentration of Cu^{2+} at the interface below that in the body of the solution, as suggested

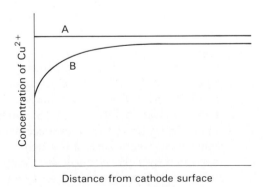

Figure 13.3 Concentration profiles. Curve A, before the start of electrolysis. Curve B, after electrolysis has proceeded.

in Fig. 13.3, curve B. Stirring the solution will help in this regard, but there exists at any such interface a stagnant film which is not well mixed with the bulk solution, even by vigorous stirring. Copper ions will tend to migrate into this depleted layer, of course, but some degree of depletion always persists if the electrolysis is being carried out at a finite rate. The depleted layer may vary from a few angstrom units to an appreciable fraction of a millimeter, depending upon the rate of electrolysis and the stirring conditions.

In general, we do not know the value of $[Cu^{2+}]$ at the electrode-solution interface, and in calculating E_d, we insert the bulk concentration into the Nernst equation for $E_{cathode}$. Thus, even if equilibrium between electrode and solution existed at the interface, our value for E_d would be too small. To take account of this in developing an equation for the required applied voltage, we shall add on a term designated $\omega_{cathode}$. It may be considered that this term arises from the slow diffusion of Cu^{2+} into the depleted layer; in this sense, $\omega_{cathode}$ represents a kinetic intrusion into the otherwise thermodynamic calculation of $E_{cathode}$. An ω-term originating in a nonuniform concentration profile as in Fig. 13.3, curve B, is called a *concentration overpotential.*

Activation Overpotential

A *chemical* step with slow kinetics in the overall electrode process may also cause an electrode potential to depart from the Nernstian value. For example, in the reduction of H^+ at a cathode, the initial electron-transfer step yields hydrogen *atoms:*

$$H^+ + e \longrightarrow H$$

To form the final product, hydrogen gas, the atoms must combine:

$$2H \longrightarrow H_2$$

Now the equilibrium for this combination lies *far* to the right. To dissociate H_2 into atoms requires temperatures in thousands of degrees. But the *rate* of combination may be slow, and to form the product H_2 at an appreciable rate, the cathode may have to be more negative than the Nernstian value suggests. This effect, which is often called *activation overpotential,* will be included in the ω-term in writing an equation for the required applied voltage.

Activation overpotential arising from a slow kinetic step in the overall electrode process may occur in many cases, but it is particularly pronounced where a gas is liberated. The two common examples in electroanalytical chemistry are reduction of H^+ to form H_2, used as an illustration above, and oxidation of water to form O_2, which is a frequent anode reaction. The magnitude of this overpotential depends upon the chemical nature of the electrode material and its physical state (e.g., surface area), the temperature, and the rate at which the electrolysis is performed, which is often stated as a current density (amperes per square centimeter of electrode surface). Some measured values of hydrogen activation overpotentials are given in Table 13.1. The differences between smooth and platinized platinum electrodes reflect the surface-area effect; platinized platinum is smooth platinum upon whose surface a rough platinum layer of large surface area has been deposited by rapid electrolysis. Comparing platinum and mercury is interesting; because platinum is a far better catalyst for the atomic combination reaction, the overpotential is much lower than with mercury. The high overpotential for hydrogen evolution on mercury is exploited in polarography, as we will see in a later section.

TABLE 13.1 Some Overpotentials for Hydrogen Evolution, V

CURRENT DENSITY, A/CM^2	SMOOTH PLATINUM	PLATINIZED PLATINUM	MERCURY
0.0001	—	0.0034	—
0.001	0.024	0.015	0.9
0.01	0.068	0.030	1.04
0.1	0.29	0.041	1.07
1.0	0.68	0.048	1.11

Source: International Critical Tables, Vol. VI, pp. 339–340. McGraw-Hill Book Co., Inc., New York, 1929.

To summarize overpotential for our purposes, then, let us say that in an electrolysis carried out at a finite rate, the electrodes are not in equilibrium with the bulk solution, and potentials calculated using bulk concentrations in the Nernst equation are incorrect. Where slow diffusion is the major effect, the electrode may be virtually in equilibrium with solution right at the interface, but in other cases chemical steps with slow kinetics may even preclude this. In any event, overpotential means that the cathode will be more negative than we would have predicted from the Nernst equation, and the anode more positive, if the electrolysis is actually proceeding at more than an infinitesimal rate.

Summarizing this section, we may write the following equation for the applied potential:

$$E_{applied} = E_{anode} - E_{cathode} + iR + \omega_{cathode} + \omega_{anode}$$

where $E_{anode} - E_{cathode} = E_d$ is the Nernstian decomposition potential.

13.1d. *Electrogravimetry*

Principles of Electrogravimetry

In its usual form, electrogravimetry involves plating a metal onto a weighed platinum cathode and then reweighing to determine the quantity of the metal. The experimental setup is similar to that shown in Fig. 13.2. The determination of copper serves as an example. The sample, perhaps a copper alloy, is dissolved in nitric acid. The platinum gauze cathode, having been cleaned in nitric acid, rinsed, dried in an oven, and weighed, is inserted into the solution, and an electrical connection is made using some sort of clamp. The applied voltage is increased until the ammeter shows a current and the cathode develops a coppery appearance; bubbles will be seen rising from the anode. In practice, the terms in the equation for $E_{applied}$ in the section above cannot be calculated accurately. We simply start with a rough idea of what $E_{applied}$ should be and add some extra voltage to make certain that the electrolysis goes. Of course, for a case as common as the deposition of copper, directions can be found in the literature. After a reasonable time, one can test for completeness of deposition by lowering the cathode and observing whether the fresh platinum surface acquires a copper plate. At the end, the cathode is removed from the solution with the voltage still applied (to prevent redissolving the copper plate by galvanic action) while being rinsed with distilled water. The cathode is then dipped into ethanol or acetone to facilitate drying, dried quickly in an oven to avoid surface oxidation of the copper, and finally cooled and weighed.

The nitric acid medium is desirable for the experiment described above. As the concentration of Cu^{2+} is lowered by the electroreduction, the cathode becomes increasingly negative until the reduction of nitrate begins:

$$NO_3^- + 10H^+ + 8e \longrightarrow NH_4^+ + 3H_2O$$

This stabilizes the cathode potential, which does not then become sufficiently negative to reduce certain other metals, such as nickel, which may be present in the sample. (It also prevents the reduction of H^+, which is undesirable in this case because concurrent hydrogen evolution tends to cause a spongy and nonadherent copper deposit.) Since the supply of nitric acid is large, this control of the cathode potential is maintained for a sufficient time.

Applications of Electrogravimetry

Simple electrogravimetric determinations as described above are widely used for metals. The technique works very well when a fairly noble metal such as silver or copper is to be determined in samples whose other constituents are less easily reduced than H^+. The intervention of hydrogen evolution prevents the cathode from becoming negative enough to reduce the other metals. Sometimes, as in the example with NO_3^- described above, other species (called *depolarizers* or *potential buffers*) are added to serve this same purpose. For example, the iron(III)–iron(II) system,

$$Fe^{3+} + e \rightleftharpoons Fe^{2-} \qquad E° = +0.77 \text{ V}$$

limits the cathode potential to a value no more negative than $+0.77$ V if the concentrations of the two ions are equal. If no substance is present that is more easily oxidized than iron(II) ion, then iron(III) ion is formed at the anode; the reduction of the latter at the cathode serves to limit the cathode potential while a fairly constant ratio of the two ions is maintained in the solution. When H^+ serves as a potential buffer, there is a wide latitude for cathode potential control by adjusting the pH of the solution. The instrumentation for this type of electrolysis is relatively inexpensive. The sensitivity of the method is determined largely by the weighing process.

Box 13.1   **Electrolysis on a Large Scale: Aluminum**

The major industrial application of electrolysis is the production of aluminum, which consumes about 5% of the electricity generated in the United States. Aluminum is the most abundant metal in the earth's surface, and every student knows such uses as aircraft construction and beer cans. The large, negative $E°$ value for $Al^{3+} + 3e = Al$ (-1.66 V) shows that it is a difficult metal to obtain from its compounds.

When Charles M. Hall (1863–1914) was an undergraduate at Oberlin College in Ohio, aluminum metal was a laboratory curiosity, prepared on a very small scale by reduction with alkali metals which were produced in small quantities by electrolysis. One of Hall's professors, who had seen a sample of the metal while studying in Europe, frequently told his classes that a fortune awaited anyone who devised a cost-effective process for winning aluminum from a plentiful ore on a large scale. Hall, at 21, began the work in a woodshed, using such apparatus as cast-iron cookware and glass fruit jars. He showed the professor a pellet of the metal early in 1886, and his process, patented in 1889, led to formation of the Aluminum Company of America (Alcoa). (A Frenchman, Paul-Louis Héroult, announced the same process in 1886; curiously, the years of birth and death were the same for both men.)

The modern solvent for Hall electrolysis is a less costly synthetic version of the expensive mineral cryolite, Na_3AlF_6, prepared from $NaAlO_2$ and HF, the fluorine source being fluorospar, CaF_2. The solute Al_2O_3 is obtained by processing the abundant mineral bauxite, a rock high in $Al(OH)_3$. Pure cryolite melts at 1012°C; the solute (2 to 6% by weight) and other additives such as CaF_2 and AlF_3 lower the melting point to 940 to 980°C. (Electrolysis of pure Al_2O_3 is not feasible because of the high melting point, 2050°C.) The cathode is a carbon liner at the base of the cell. Once electrolysis occurs, the molten aluminum (m.p. 660°C), which is denser than the cryolite melt and collects over the carbon, can be considered the cathode. Portions of the product are withdrawn from time to time. Carbon rods inserted into the melt serve as the anode. A dc voltage of 600 to 750 V is applied to 130 or more of the individual cells which are connected in series, providing for each cell an applied voltage of 4.5 to 5 V. Currents may be as high as 260 kA, the Joule heat maintaining the molten state.

Recently the industry has been producing over 5 million tons of alumium, about 99.7% pure, per year. The price was about 30¢ a pound in 1970, and 80¢ in 1980; perhaps Hall's work is better appreciated by comparing $100 a pound in 1885 with 50¢ in 1895. The electrolytic smelting uses 70 to 90% of the total energy of production, the rest going for mining and purifying bauxite, mill processing of the metal, etc. Recycling is much cheaper than production of new aluminum.

The mechanism of the electrolysis is still imperfectly understood because many chemical species are formed in the melt. Cryolite dissociates into sodium and AlF_6^{3-} ions, and the latter further dissociates to AlF_4^- and F^-. The Al_2O_3 forms oxyfluorides:

$$Al_2O_3 + 4AlF_6^{3-} \longrightarrow 3Al_2OF_6^{2-} + 6F^- \tag{1}$$

$$Al_2O_3 + AlF_6^{3-} \longrightarrow 3AlOF_2^- \tag{2}$$

After multiplying coefficients so as to balance electrons, the cathode reaction is thought to be

$$4AlF_6^{3-} + 12e \longrightarrow 4Al + 24F^- \tag{3}$$

while the oxidation at the anode is

$$6AlOF_2^- + 3C + 24F^- \longrightarrow 6AlF_6^{3-} + 3CO_2 + 12e \tag{4}$$

Adding Eqs. (3) and (4),

$$6AlOF_2^- + 3C \longrightarrow 4Al + 2AlF_6^{3-} + 3CO_2 \tag{5}$$

If $6AlOF_2^-$ in Eq. (5) is replaced by $2Al_2O_3 + 2AlF_6^{3-}$ [Eq. (2)], then the overall reaction amounts to

$$2Al_2O_3 + 3C \longrightarrow 4Al + 3CO_2$$

Note that the carbon anode is consumed in the process and must be replaced periodically.

There is a second commercial aspect of electrolysis with aluminum.

The reason that its corrosion is less serious than the rusting of iron despite its greater tendency to oxidize reflects the physical character of the oxides. The oxide film on aluminum is tightly held, dense, and impenetrable, hence protective. The thickness of the film can be deliberately increased by making the aluminum the anode of an electrolysis cell with a dilute H_2SO_4 electrolyte; if certain dyes are present, the oxide film can even be colored during the process. The product is called "anodized aluminum." After reading the information here, you will not be surprised that many aluminum plants are located near sources of hydroelectric power, in the TVA region, for example, and near some of the rivers of the west.

13.1e. *Controlled Potential Electrolysis*

Principles of Controlled Potential Electrolysis

In classical electrolysis as described in the last section, it is easy to measure the applied voltage using voltmeter V in Fig. 13.2, and it is easy to determine whether anything is happening by watching the ammeter. In many cases, however, this is not enough. In an electrolytic reduction, it is the potential of the cathode that really determines what will happen there. The student may refer again to the equation given earlier for $E_{applied}$ to see that the applied voltage does not tell us much about the cathode potential itself because we probably do not know other terms in the equation, such as the iR drop or the anode overpotential. Thus it is often desirable to measure the cathode potential against a reference electrode which is not in the electrolysis circuit and whose potential remains constant throughout the electrolysis. Then, by adjustments to the applied voltage, we may control the cathode potential, that is, prevent its becoming more negative than desired. (We happen to use electrolytic reductions as examples in this chapter, thereby focusing attention on the cathode, but it should be noted that the same considerations would apply in a case where oxidation at an anode is of primary interest.) An example will clarify the need for cathode potential control.

EXAMPLE 13.3

Electrolytic separation of silver and copper

A solution whose volume is 100 mL is about 0.1 M in both Ag^+ and Cu^{2+}. It is desired to determine the exact quantity of silver by electrodeposition on a weighed platinum cathode; the result will be high if copper also plates onto the cathode. Can this separation of silver from copper be done?

Suppose that the analytical balance is sensitive to 0.1 mg; then if 0.1 mg of Ag^+ is left in solution, we shall consider the electrodeposition of silver to be "complete." The molar concentration will then be

$$[Ag^+] = \frac{0.1 \text{ mg}}{100 \text{ mL} \times 108 \text{ mg/mmol}} = 9 \times 10^{-6} \, M$$

The potential which the cathode must attain in order to lower $[Ag^+]$ to this value is

$$E_{cathode} = +0.80 - 0.06 \log \frac{1}{9 \times 10^{-6}} = +0.50 \text{ V}$$

The concentration of Cu^{2+} is 0.1 M. The potential of an electrode which can be in equilibrium with this ion is given by

$$E_{cathode} = +0.34 - \frac{0.06}{2} \log \frac{1}{0.1} = +0.31 \text{ V}$$

That is, for reduction of Cu^{2+} to copper metal to occur, the cathode must be more negative than $+0.31$ V.

Since silver reduction is "complete" at $+0.50$ V, we see that it is possible to plate silver on the cathode without interference from Cu^{2+}, provided that we monitor the cathode potential and prevent it from becoming more negative than some value between $+0.50$ and $+0.31$ V, say, $+0.40$ V. □

Instrumentation

A manually operated experimental setup for limiting the cathode potential to the desired value is shown in Fig. 13.4. The apparatus is seen to include the features required for any electrolysis: variable voltage source, anode, and cathode. But there is, in addition, a third electrode in the solution, the reference electrode, frequently a saturated calomel electrode. The potential difference between the cathode and the reference electrode is measured with a voltmeter of high input resistance so as to draw little current.

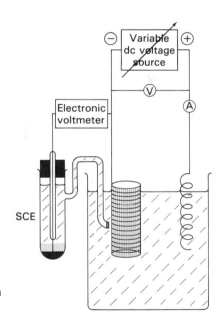

Figure 13.4 Apparatus for electrolysis with controlled cathode potential.

Returning to the example above, suppose that we decide to limit the cathode potential to $+0.40$ V. If the reference electrode is an SCE, its potential is $+0.25$ V. This means that the voltage between SCE and cathode is to be no smaller than $0.40 - 0.25 = 0.15$ V, with the cathode positive. The desired potential relationships are shown in Fig. 13.5. (Students are sometimes confused by the signs, but a little thought shows that there is no reason to be: In this example, the cathode is more positive than the SCE, but it is still negative with respect to the anode, about which we have said little but which is perhaps platinum at which O_2 is being evolved at a potential above $+1$ V, depending upon the acidity and overpotential effects.)

To perform the electrolysis, we turn up the applied voltage until an adequate current flows, but we watch the high-resistance voltmeter as the electrolysis proceeds. At first there is no problem: The high concentration of Ag^+ itself limits the cathode potential to a fairly positive value. But as $[Ag^+]$ decreases during the electrolysis, the SCE-cathode voltage will decrease. As it nears 0.15 V, we will cut back the applied voltage so that this value is not passed, and we will need to

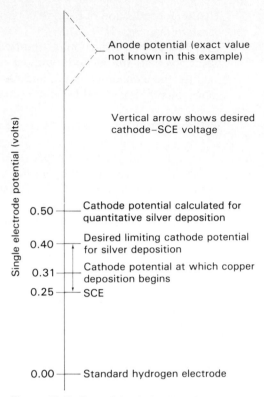

Figure 13.5 Potential relationships for controlled potential electrolysis.

make readjustments with increasing frequency as the electrolysis nears completion. Eventually, the current will decay to a very low value (because the potential control prevents the instrusion of other electrode reactions), and the run will be terminated.

Manual operation as described above is tedious; as the electrolysis proceeds, changes in concentrations, iR drop, and overpotentials require many adjustments to the applied voltage. As a result, instruments called *potentiostats* have been developed to perform the operation automatically. These commercially available units have the essential features of the manual setup in Fig. 13.4, but the human operator is replaced by a control link between the measuring device monitoring the cathode-SCE voltage and the variable voltage source. When the cathode-SCE voltage begins to move away from the predetermined value, an error signal is generated which, after amplification, changes the applied voltage via an electronic interaction.

Figure 13.6 shows how the applied voltage and the current changed during an electrolysis of copper from a tartrate solution, using a potentiostat to limit the cathode potential to -0.36 V vs. SCE.[1] Ideally, the current falls off exponentially according to

$$i_t = i_0 \times 10^{-kt}$$

[1] J.J. Lingane, *Electroanalytical Chemistry,* 2nd ed., Interscience Publishers, Inc., New York, 1958, p. 217.

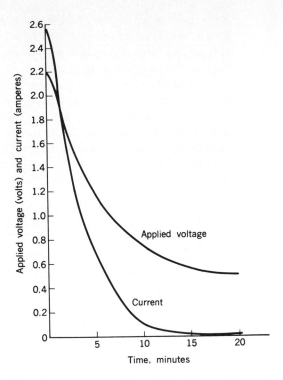

Figure 13.6. Change in applied voltage and current during controlled-potential electrolysis of copper from a tartrate solution. (After J. J. Lingane, *Electroanalytical Chemistry,* 2nd ed., Interscience Publishers, Inc., New York, 1958.)

where i_t is the current at time t, i_0 the initial current, and k a constant, but slight departures from this behavior are common in practice.

Applications

Some examples of successful separations of metals by controlled potential electrolysis are given in Table 13.2. The table is only illustrative, not exhaustive.

TABLE 13.2 Some Examples of Separations and Determinations by Controlled Potential Electrolysis

METAL DETERMINED	METALS FROM WHICH SEPARATED	ELECTROLYTE	$E_{cathode}$, V, vs. SCE
Ag	Cu and other, more active metals	Acetate buffer	+0.1
Cu	Ni, Sn, Pb, Sb, Bi	Tartrate with hydrazine and Cl^-	−0.3
Bi	Pb, Sn, Sb, and other, more active metals	Tartrate with hydrazine and Cl^-	−0.4
Sb	Sn	H_2SO_4 with hydrazine	−0.2
Cd	Zn	Acetate buffer	−0.8
Ni	Zn, Al, Fe	Ammoniacal tartrate	−1.1

Sometimes a separation which appears nonfeasible upon inspection of potentials for the reduction of uncomplexed metal ions can be accomplished by choosing a complexing electrolyte which selectively changes the ease of reduction of some of the metal ions in the sample solution.

A mercury cathode, because it is a liquid, is not a convenient electrode for electrogravimetry, but it offers advantages for certain separations. Because of the high overpotential for hydrogen evolution, some metals more active than hydrogen are easily reduced at this electrode, and the solubility of a number of free metals in mercury to form dilute amalgams further facilitates the reductions by lowering the activity of the metal below unity. An excellent discussion of the mercury cathode and its applications is available.[2]

Box 13.2 — Electrolysis as a Synthetic Tool

Electrolysis, especially with refined control of electrode potentials, was developed largely by analytical chemists, but applications extend into other fields. The ability to control the potential of an electrode at any desired value corresponds to the possession of a large number of oxidizing and reducing agents and sometimes offers clean pathways to desired reaction products. In some cases, a potentiostat is not required; something in the solution limits the electrode potential (page 348), or no competing reaction intrudes anyway. In others, full exploitation of the inherent selectivity requires very careful control.

The oldest example of a synthetically useful electrolysis is the Kolbe reaction, electrooxidation of an alkyl carboxylate to form a new carbon-carbon bond:*

$$2R—COO^- \longrightarrow R—R + 2CO_2 + 2e$$

The first step is removal of an electron at the anode to form a radical:

$$R—COO^- \longrightarrow R—COO\cdot + e$$

The electron transfer is followed by the chemical steps of decarboxylation and radical combination:

$$R—COO\cdot \longrightarrow R\cdot + CO_2$$
$$2\,R\cdot \longrightarrow R—R$$

Other follow-up reactions may occur, depending upon the compounds present, solvent, and other conditions. Some arise from another electron transfer that generates a cation ($R\cdot \to R^+ + e$) which may react with an available nucleophile.

Aromatic ketones can sometimes (depending upon substituents) be selectively reduced to either dimeric pinacols or secondary alcohols by proper choice of cathode potential and other conditions such as pH:

---

[2] J. A. Maxwell and R. P. Graham, *Chem. Rev.*, **46,** 471 (1950).

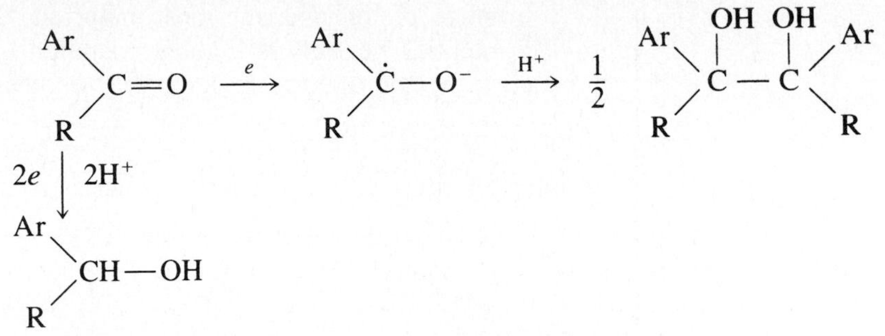

$$\underset{R}{\overset{Ar}{>}}C=O \xrightarrow{e} \underset{R}{\overset{Ar}{>}}\dot{C}-O^- \xrightarrow{H^+} \frac{1}{2} \underset{R}{\overset{Ar}{>}}\overset{OH}{\underset{}{C}}-\overset{OH}{\underset{R}{C}}\overset{Ar}{<}$$

$$2e \downarrow 2H^+$$

$$\underset{R}{\overset{Ar}{>}}CH-OH$$

Discussions of mechanisms for such reactions are available.**

Another example is the selective reduction of aromatic nitro compounds to either hydroxylamines or amines by appropriate choice of cathode potential:

$$AR-NO_2 \xrightarrow[-0.4 \text{ V vs. SCE}]{+4e} Ar-NOH$$

$$+6e \left| \begin{array}{c} -0.9V \text{ vs.} \\ \text{SCE} \end{array} \right.$$

$$Ar-NH_2$$

Other products are also obtainable under various experimental conditions.†

Intermediate free radicals are frequently encountered in organic electrochemistry. The following is another example of an electron-transfer reaction forming radicals which then combine to give a new chemical bond:‡

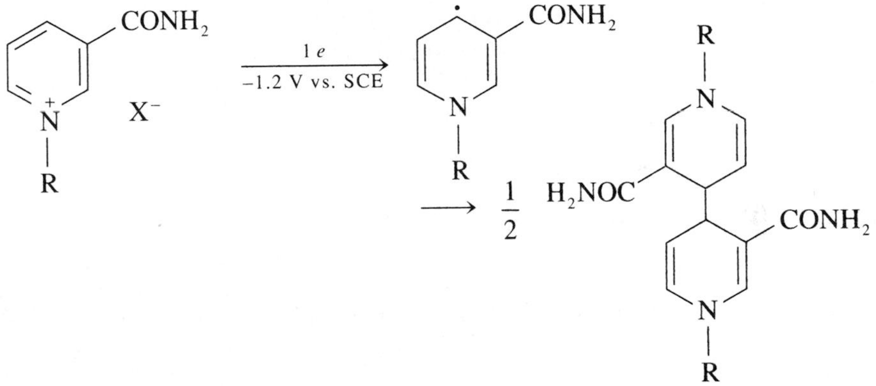

Ring closures can sometimes be effected in this way, e.g.:

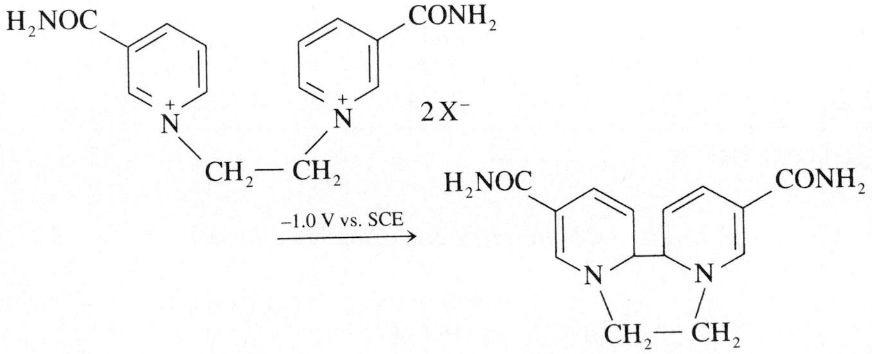

In terms of tonnage, the most important organic application of controlled potential electrolysis is the preparation of adiponitrile (used in the production of nylon) from acrylonitrile. The chemical problem of coupling two carbon atoms which possess positive character is circumvented by electrochemical generation of a free radical which attacks another molecule, forming an intermediate that further reduces to the product:

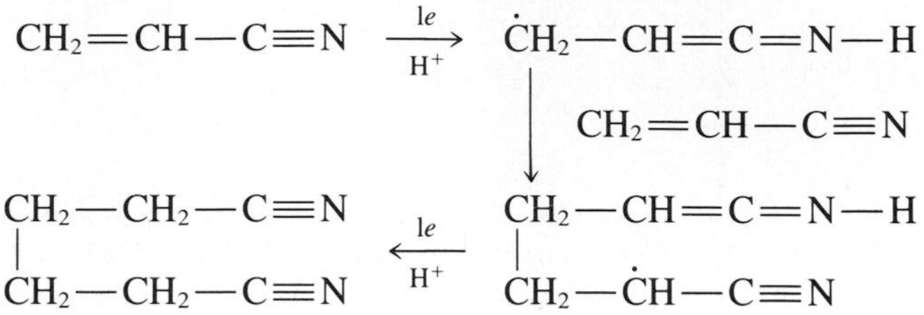

$$CH_2\!=\!CH\!-\!C\!\equiv\!N \xrightarrow[H^+]{1e} \dot{C}H_2\!-\!CH\!=\!C\!=\!N\!-\!H$$

$$\downarrow CH_2\!=\!CH\!-\!C\!\equiv\!N$$

$$\begin{array}{c} CH_2\!-\!CH_2\!-\!C\!\equiv\!N \\ | \\ CH_2\!-\!CH_2\!-\!C\!\equiv\!N \end{array} \xleftarrow[H^+]{1e} \begin{array}{c} CH_2\!=\!CH\!=\!C\!=\!N\!-\!H \\ | \\ CH_2\!-\!\dot{C}H\!-\!C\!\equiv\!N \end{array}$$

A plant for producing 15,000 tons of adiponitrile per year at about 90% yield opened in the TVA region (Decatur, Alabama) in 1965. This plant has been enlarged, and others are now in operation in various parts of the world.

Attractive aspects of synthetic steps based upon controlled potential electrolysis include frequent high yields and easier workup; side reactions are often minimized, and the products are not contaminated with excess chemical redox reagents. Application to organic chemistry goes back at least 47 years,* but the technique has still not been adopted in many laboratories. It has been employed by inorganic chemists to prepare intermediate oxidation states of various elements. Polarography (Section 13.3 below) is a convenient pilot technique which consumes little material for selecting the potential for electrolysis on a larger scale. An excellent general reference was revised not long ago.**

* H. Kolbe, *Liebigs Ann. Chem.*, **69**, 257 (1849).

** L. Mandell, R. M. Powers, and R. A. Day, Jr., *J. Amer. Chem. Soc.*, **80**, 5284 (1958).

† A. J. Fry, *Synthetic Organic Electrochemistry*, Harper & Row, Publishers, New York, 1972, Table 6.3, p. 233.

‡ A. L. Underwood and R. W. Burnett, "Electrochemistry of Biological Compounds," in A. J. Bard, Ed., *Electroanalytical Chemistry: A Series of Advances*, Vol. 6, Marcel Dekker, Inc., New York, 1973.

* J. J. Lingane, C. G. Swain, and M. Fields, *J. Amer. Chem. Soc.*, **65**, 1348 (1943).

** M. A. Baizer and H. Lund, Eds., *Organic Electrochemistry: An Introduction and a Guide*, 2nd ed., Marcel Dekker, Inc., New York, 1983.

13.2

COULOMETRY

13.2a. Coulometry and Faraday's Law

Coulometry refers to the measurement of coulombs, the coulomb being the unit for quantity of electricity. In analytical chemistry, the term implies measurement of the number of coulombs associated with an electron-transfer process that goes sufficiently to completion to be considered quantitative. The quantity of analyte is then calculated using Faraday's law. One coulomb (1 C) is the quantity of electricity associated with a current of one ampere flowing for one second:

$$C = A \times s$$

Fractional units such as millicoulombs are sometimes convenient. Faraday's law (1833) states that the number of coulombs associated with an electron-transfer process is directly proportional to the number of equivalents of chemical change at the electrode:

$$C = F \times \text{(no. of equivalents)} = F \times n \times \text{(no. of moles)}$$

where n is the number of electrons in Ox $+ ne \rightleftharpoons$ Red and F (which is called *the Faraday*) is the proportionality constant. The most recently recommended value for the Faraday (1986) is

$$F = 96,485.31 \text{ C}$$

Note that F could be defined as the quantity of electricity associated with an Avogadro number of electrons.

There are no exceptions to Faraday's law for faradaic processes—coulombs will always be proportional to equivalents of chemical change at the electrode. The only way, though, for this to be useful analytically is to ensure that the redox process of interest (or one tantamount to it) is the only one that occurs at the electrode during the measurement. When we oxidize analyte X at an anode and also oxidize some water, Faraday's law cannot sort out the coulombs associated individually with the two concurrent processes. We must use our talent as chemists to ensure that all of the electricity passed through the cell is, in fact, carried by the desired electrode reaction. Referring back to Example 3, instead of weighing the cathode to find the quantity of silver, we could have determined it by measuring the coulombs required for (virtual) completion of the reaction

$$\text{Ag}^+ + e \longrightarrow \text{Ag}$$

But, if reduction of Cu^{2+} or of any other species also occurred, then the coulometry could not have been related to the quantity of silver.

When all of the electricity passed through an electrolytic cell is associated with only the desired electrode reaction, then we say that the reaction proceeds at 100% *current efficiency*. If this condition is met, the quantity of analyte is given by

$$\text{Grams of analyte} = \frac{C \times \text{MW}}{n \times F}$$

(Of course, if g and MW are known, we can use this equation to determine n for an unknown electrochemical process.) There are two ways to achieve 100% current efficiency.

13.2b. Controlled-Potential Coulometry

An obvious way to prevent the intrusion of an undesired electrode reaction is to control the potential of the electrode, as described in the last section. This will require three electrodes in the solution (cathode, anode, and reference) and a potentiostat to control the potential of the electrode at which the desired reaction occurs. Sometimes the electrode where the reaction of interest occurs is called the *working electrode*. The second electrode in the electrolysis circuit is then called the *auxiliary* or *counter electrode;* some electrochemical process must occur here, of course, but we are not primarily concerned with it except to ensure that it does not introduce undesired substances into the test solution.

During a controlled-potential coulometric experiment, the current will decay somewhat as shown in Fig. 13.6. Since the current changes continuously, the total coulombs, C, associated with the desired reaction is given by

$$C = \int_0^t i \, dt$$

This integral represents the area under the current-time curve. There are many ways to obtain this area. In the past it was common to place a second electrolysis cell in series with the test cell so that the current through the two was the same. In this second cell, a well-characterized electrode reaction occurred at 100% current efficiency which generated a conveniently measured product—for example, a gas whose volume at known pressure could be determined or an ion which was easily titrated.[3] Nowadays, the integration is commonly performed electronically.

Controlled-potential coulometry has been applied to the determination of a number of metals, such as lead, copper, cadmium, silver, and uranium; certain organic compounds; and halide ions via the following reaction, proceeding from right to left at a silver anode:

$$AgX + e \rightleftharpoons Ag + X^-$$

The technique has also been used frequently to determine *n*-values in the investigation of new electrochemical reactions, particularly in organic chemistry.

13.2c. *Coulometric Titrations*

This technique has been used more widely than controlled-potential coulometry. In its basic form, the apparatus is relatively simple and inexpensive. Required is a constant-current source; essentially, this is a high dc voltage in series with such a large resistor that changess in the electrolytic cell resistance during operation are negligible and the current remains constant. Switching the current on and off is analogous to opening and closing a buret; we titrate an electroactive species by adding or removing electrons at a cathode or anode as the working electrode. The other requirement is a clock. The number of coulombs involved in the electrode reaction is then calculated from the value of the current and the time required for complete reaction.

Again, a major problem is to ensure that the desired electrode reaction proceeds at 100% current efficiency. From the discussion in previous sections, the student will appreciate the danger that the potential of the working electrode may shift sufficiently to permit the intrusion of an undesired electrode reaction which will contribute an unknown quantity to the total number of coulombs, thereby vitiating the analytical calculation for the desired constituent.

For this reason, *direct* coulometric titrations, where the analyte itself is to be the only reactant at the electrode, are seldom attempted. Much more common is an *indirect* titration, which involves the electrolytic generation of a reagent

---

[3] For a description of earlier coulometers. see J. J. Lingane, *Electroanalytical Chemistry,* 2nd ed., Interscience Publishers, Inc., New York, 1958, pp. 452–459.

which in turn reacts chemically with the analyte. The potential of the working electrode is kept fairly constant by maintaining a high concentration of the substance undergoing the electrode reaction to generate the titrant. For example, in the coulometric titration of iron(II) ion, electrolytically generated cerium(IV) ion can be used.[4] Cerium(III) ion, from which cerium(IV) ion is generated by anodic oxidation, is present in large concentration, and hence the anode potential is kept from becoming sufficiently positive for oxygen to be evolved. At the start of the electrolysis, iron(II) ion is directly oxidized at the anode; then the potential becomes more positive, reaching a value sufficiently positive to oxidize cerium(III) to cerium(IV) ion. The cerium(IV) ion, in turn, oxidizes any remaining iron(II) ion in the body of the solution. The quantity of electricity used is the same, of course, as if the iron(II) ion alone were directly oxidized at the anode. If the direct oxidation of iron(II) ion were attempted, however, oxygen would be liberated before the oxidation of iron(II) ion was complete, and the analysis would not be valid.

It is not necessary, of course, that the substance being titrated be itself electroactive, as it was in the example above. For instance, bromine generated by the anodic oxidation of bromide ion may be used to titrate phenol:

Since phenol does not react at either electrode, the process is totally indirect.

Figure 13.7 shows an example of a coulometric titration setup. The platinum generator electrode is placed directly in the test solution, but the auxiliary electrode is in a separate compartment, the bottom of which is a sintered-glass filter disk. This separation of the latter electrode tends to prevent the convective transport of any undesirable electrode product into the test solution. The clock and the current source are wired together so that one switch turns on both. The current source may provide for the selection of several values for the constant current, perhaps in the range of 1 to 200 mA. This is somewhat analogous to the availability of titrant solutions of vairous concentrations in classical titrimetry.

The titration is performed by operating the switch like a buret stopcock until the reaction of the desired constituent, directly at the electrode and/or indirectly with another substance formed at the electrode, is complete. The quantity is then calculated from the time required, the known constant current, and Faraday's law. Obviously this requires that we know when the reaction is complete; that is, we need to know, as in any titration, when to stop. In fact, any of the usual endpoint techniques can be employed in coulometric titrations—visual, potentiometric, photometric, amperometric, and so forth. For example, in the titration of phenol with electrolytically generated Br_2 mentioned above, a dye such as indigo carmine which is destroyed by excess Br_2 may be added for a visual end point. For a coulometric titration with a potentiometric end point, there will be four electrodes in the solution: an anode and a cathode for the electrolytic generation

---

[4] N. H. Furman, W. D. Cooke, and C. N. Reilley, *Anal. Chem.*, **23**, 945 (1951).

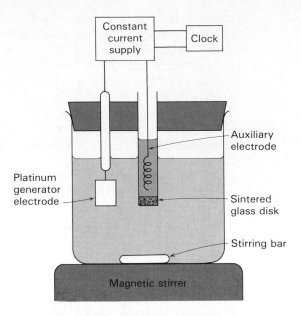

Figure 13.7 Cell for coulometric titration.

of titrant, connected to the constant current source, and an indicator-reference electrode pair, connected with an electronic voltmeter or recording potentiometer for the potentiometric readout. The titration cell can be positioned in the sample compartment of a spectrophotometer for a photometric end point.

Conditions have been worked out for the electrolytic generation of a large number of titrants for coulometric analysis. Some examples are given in Table 13.3; more complete compilations may be found elsewhere.[5] These include H^+

TABLE 13.3 Some Examples of Electrolytically Generated Titrants for Coulometric Titrations

TITRANT	SUBSTANCE TITRATED
Acids and bases	
H^+	Bases
OH^-	Acids
Oxidants	
Br_2	As(III), U(IV), olefins, phenols, aromatic amines
Ce(IV)	Fe(II), I^-, Ti(III), hydroquinone, metol (*p*-methylaminophenol sulfate)
Mn(III)	H_2O_2, Fe(II)
Ag(II)	Ce(III), As(III)
$Fe(CN)_6^{4-}$	Ti(I)
Reductants	
Fe(II)	MnO_4^-, Ce(IV), VO_3^-, Cr(VI)
Ti(III)	Fe(III), U(VI)
Precipitants	
Ag(I)	Cl^-, Br^-, I^-, mercaptans
Ce(III)	F^-

[5] For example, see D. D. DeFord and J. W. Miller, "Coulometric Analysis," Chap. 49 of I. M. Kolthoff and P. J. Elving eds., *Treatise on Analytical Chemistry*, Part I, Vol. 4. Wiley-Interscience, New York, 1963; see especially Table 49.1, p. 2516.

and OH^-, oxidizing and reducing agents, and reagents for precipitation and complexation.

It is advantageous that only the fundamental quantities current and time are required. Errors and inconveniences associated with the preparation and storage of standard solutions are eliminated. Since small currents and short times can be measured accurately with modern instruments, coulometric titrations are routinely applicable to smaller samples than can be titrated conventionally, perhaps by a factor of 10 to 100. Samples sizes from 100 mg down to a few hundredths of a microgram in volumes of 10 to 50 ml have been employed. Excellent results were reported with solution volumes as small as 10 μl.[6] The sensitivity and precision of coulometry can be appreciated from the following example: a quantity as small as 0.1 C can be measured with a precision of about 1 ppt; 0.1 C corresponds to $0.1/96,485 = 1.036 \times 10^{-6}$ eq or 1.036 μeq, which would represent 103.6 μg (about 0.1 mg) of an analyte whose equivalent weight is 100.

13.3

POLAROGRAPHY

13.3a. Overview

In *voltammetry,* current-voltage curves are recorded during electrolysis experiments. Polarography is a special case of voltammetry where the working electrode (usually the cathode) is a *dropping mercury electrode* or DME.[7] Figure 13.8 shows a typical experimental setup. The DME is a 5- to 10-cm length of very fine capillary tubing (internal diameter 0.04 to 0.08 mm) surmounted by a mercury column of adjustable height. As mercury flows through the capillary, drops grow at the tip until they become heavy enough to break away. The capillary length and the mercury height are adjusted so that a drop of roughly the following characteristics at maximal size falls once every 3 to 6 s: weight, 10 mg; diameter, 0.5 mm; surface area, 0.8 mm^2 (0.008 cm^2).

In the cell in Fig. 13.8, the SCE is the anode, and, as explained below, it also serves as a reference for monitoring the cathode potential. In the polarographic experiment, current through the cell is recorded as a function of the applied voltage, which is scanned from 0 to perhaps 2 V over a period of a few minutes. The process of interest in the example of the figure is the reduction of Cd^{2+} to the metal.

13.3b. Relation of Cathode Potential to Applied Voltage

Consider the magnitudes of the terms in the equation on page 347 in a polarography experiment:

[6] R. Schreiber and W. D. Cooke, *Anal. Chem.,* **27,** 1475 (1955).

[7] The invention of polarography by Jaroslav Heyrovsky at the Charles University, Prague, was reported in 1922. Heyrovsky received a Nobel Prize in 1959 for this and later work in the field, partly because of its wide applicability to research in diverse areas, e.g.., the redox properties of biological compounds. By now a "classical" technique, polarography was the forerunner of the many modern electroanalytical methods based upon microscale electrolysis (for some of these, see Box 13.3).

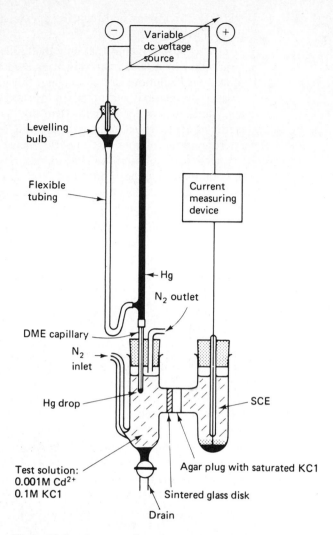

Figure 13.8. Apparatus for polarography with H-cell and DME in its simplest form. The apparatus is operated manually; the dc source may be a battery and voltage divider, and the current-measuring device a galvanometer adapted as a microammeter. In a recording instrument, the voltage source is programmed to vary $E_{applied}$ automatically with time, and the current-measuring device plots the polarogram with a pen-and-ink recorder.

$$E_{applied} = E_{anode} - E_{cathode} + iR + \omega_{cathode} + \omega_{anode}$$

First, the current i is very small, typically ranging from a fraction of a microampere up to a few microamperes. The resistance R is small because of the added strong electrolyte (KCl in Fig. 13.8). Much of the resistance reflects hampered ion migration through the sintered-glass disk and within the agar plug, but if the porosity is not too fine and the diameter at least a centimeter or so, R can be held to a few hundred ohms. Thus iR may typically be only a fraction of a millivolt (e.g., 1×10^{-6} A $\times$ 300 ohms = 0.0003 V), and it is usually permissible to neglect it except in special circumstances.

Second, E_{anode} is virtually constant. The SCE is a large electrode, and although there is an anode reaction during electrolysis ($2Hg + 2Cl^- \rightarrow Hg_2Cl_2 +$

2*e*), at very small current the consumption of Cl^- is negligible. (This is true even over the long run because of replenishment from excess solid KCl.) Furthermore, when the current density is very small (i.e., tiny current and large electrode area), the ω_{anode} term is negligible: there is no chemical or diffusion step in the anode process that is too slow to keep up with the very low rate at which electrons leave the SCE. Thus the potential of the SCE remains at the equilibrium value (+0.246 V) that would obtain in the potentiometric situation of no current drain.

In polarography, then, the equation above usually boils down to

$$E_{\text{applied}} = +0.246 - E_{\text{cathode}} + \omega_{\text{cathode}}$$

That is, a change in E_{applied} changes only the cathode potential, E_{DME}, which we may think of as comprising a Nernstian component, E_{cathode}, and an overpotential term, ω_{cathode}. Note that we do not need a third electrode as a reference, as we did in large-scale electrolysis, in order to monitor the cathode potential; the SCE is both anode and reference electrode, and E_{applied} is the same as E_{DME} vs. SCE. When iR is not negligible, e.g., in very precise measurements or in studies of organic compounds using solvents in which electrolytes are not very soluble, a three-electrode system may be used to obtain more accurate values of E_{DME}. (It is also possible to measure R and correct E_{applied} for the iR drop to obtain E_{DME}.)

Elsewhere in electrochemistry, as we saw in Chapter 10, single electrode potentials are almost always referenced to the standard hydrogen electrode, but in polarography, E_{DME} is not only *measured* against the SCE but also *reported* that way as a matter of custom.

13.3c. The Polarogram

The experimentally obtained current-voltage curve, or polarogram, for the system in Fig. 13.8 is shown in Fig. 13.9. Along LM, a small current, called the *residual current*, flows. Near M, the decomposition voltage of the cell is reached,

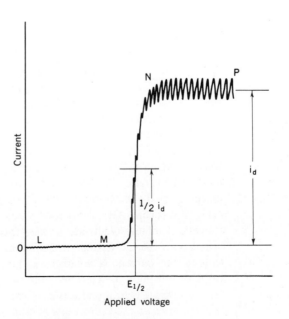

Figure 13.9 Typical polarogram.

and with further increase in applied voltage the current increases rapidly from M to N. The cell reaction is the following, proceeding from left to right:

$$Cd^{2+} + 2Hg + 2Cl^- \longrightarrow Cd + Hg_2Cl_2$$

The cadmium metal formed at the mercury cathode happens to be soluble in mercury. The resulting solution, called an *amalgam*, is often symbolized by Cd(Hg). For this reason, some writers formulate the cathode reaction as

$$Cd^{2+} + Hg + 2e \longrightarrow Cd(Hg)$$

The student should not confuse the Hg present in the SCE anode with the mercury flowing from the DME cathode.

Finally, as the voltage scan is continued, the current levels to a limiting value in the region NP. It is seen in the figure that the curve is not smooth. Rather, the current oscillates as the mercury drops grow and fall, reflecting the changes in surface area. Actually, at the instant a drop disengages, the electrode area decreases to the tiny cross-sectional area of the capillary bore, and the faradaic current falls nearly to zero, but the actual polarogram will not show this because of inertia in current-measuring devices. In working up polarographic data, a consistent method, such as the use of maximum current or average currents, must be adopted.

13.3d. *Important Features of the DME*

Advantages

The DME possesses several advantages over solid microelectrodes such as platinum.

First, the surface exposed to the solution is reproducible and continually renewed with fresh mercury. This leads to highly reproducible current-voltage curves, independent of the prior use of the electrode.

Second, the DME may be contrasted favorably with stationary electrodes in regard to the depleted layer adjacent to the electrode surface. With a stationary electrode, *peaks* rather than plateaus are found in the current-voltage curves. This results from the fact that the depletion extends progressively further from the electrode into the bulk solution as the electrolysis proceeds, and the electroactive species must diffuse further along shallower concentration gradients to reach the electrode as time goes by. Although there is nothing intrinsically bad about a peak as such, if, during the recording of a current-voltage curve, any jarring or vibration occurs, the diffusion setup will be disturbed and erratic currents observed. In principle, this problem can be countered by stirring the solution (which has the added advantage of yielding larger currents), but in practice it is difficult to reproduce the hydrodynamic variables in stirred solutions, and again, currents tend to be erratic. With the DME, the diffusion layer remains quite thin because of the periodic dropping of the mercury, and it is reestablished reproducibly at each new drop. The controlled, local stirring from the falling drops and the movement of the electrode surface toward the bulk solution as the drop grows, along with the increasing surface area, prevent the decay with time. The oscillation of the current is a nuisance, but it is so regular that in practice it is not difficult to obtain an average value in a reproducible manner.

Third, a very important advantage of the DME results from the high activation overpotential for hydrogen evolution on mercury. Once reduction of hydrogen ion begins, the supply from an aqueous solvent is so abundant that the current rapidly attains an enormous (by polarographic standards) value, whereupon it becomes impossible to observe the small currents associated with the processes we desire to study. The potential at which reduction of H^+ becomes appreciable depends, of course, upon the pH of the solution, but in any case it is much more negative on mercury than on platinum, gold or any other useful electrode material. This means that many reduction processes can be studied at the DME, even though thermodynamically hydrogen evolution should interfere.

Fourth, with many metals, amalgam formation lowers the activity of the metal, thereby facilitating the reduction of the metal ion. This, plus the overpotential effect mentioned above, means that many metals more active than hydrogen can still be dealt with in polarography. For example, polarograms have been obtained even for the alkali metals, although, to be sure, only in solutions of very high pH.

Limitations

One of the major disadvantages of mercury as an electrode, when compared with more noble metals such as platinum, is the oxidation of the mercury itself at a potential which is not very positive. Once mercury oxidation begins, an enormous current results from the ample supply, and the oxidation of other species can no longer be observed. This means that the DME has very limited use as an anode. The exact potential at which mercury oxidation becomes a problem depends upon the composition of the solution. For example, ions such as halides, which form slightly soluble mercury(I) salts, facilitate the oxidation. Roughly, the DME is seldom useful at potentials more positive than about $+0.4$ V vs. SCE, while a platinum anode can be used up to the potential where water begins to oxidize, perhaps $+1.4$ vs. SCE. Nevertheless, there is a wide potential range within which the DME is useful, and in this region it is unexcelled.

A second important disadvantage of the DME is the magnitude of the residual current. Part of this may sometimes be faradaic, resulting from electroactive trace impurities such as oxygen or easily reduced metal ions. However, cleaning up the solutions does not eliminate an appreciable residual current which is nonfaradaic in nature and which is much larger with the DME than with stationary microelectrodes of constant surface area.

Consider a mercury droplet in an electrolyte solution, say, KCl. At potentials more positive than about -0.5 V vs. SCE, the mercury surface is positively charged with respect to the adjacent solution, which contains an excess of Cl^- over Na^+ [Fig. 13.10(a)]. At more negative potentials, the surface double layer is reversed [Fig. 13.10(b)]. At about -0.5 V the electrode surface is uncharged with respect to the solution; in this condition, the mercury is said to be at its *electrocapillary maximum*.[8] We may think of the double layer as the analog of a

---

[8] The *maximum* in *electrocapillary maximum* reflects the fact that the surface tension at the mercury-solution interface passes through a maximum at this potential. On either side, excess positive or negative charge on the mercury droplet lowers the work of expanding the surface, i.e., lowers the surface tension. Actually, Heyrovsky was investigating this phenomenon when he happened to observe unexpected faradaic currents attributable to reducible impurities in nonelectroactive salts. Had modern reagent chemicals been available in 1920, it is likely that he would have done good, solid work along the intended lines, but the impurities (and his decision to investigate their effects) led to the Nobel Prize. The exact potential of the maximum depends upon the nature and concentration of the aqueous electrolyte; in 0.5 M Na^+ salts (NaX), it ranges from -0.40 V vs. SCE when X^- is F^- to -0.64 V for SCN^-, with other anions such as Cl^-, Br^-, and NO_3^- in between [A. L. Underwood and E. W. Anacker, *J. Colloid Interface Sci.*, **117**, 242 (1987), Table III].

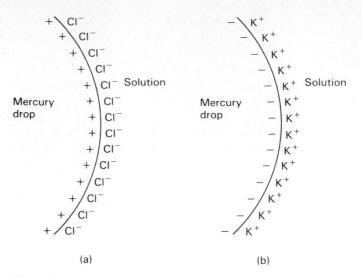

Figure 13.10 Schematic diagram of the electrical double layer at the surface of a mercury drop in a potassium chloride solution at two potentials on opposite sides fo the electrocapillary maximum.

charged capacitor in electronics. Once a fixed capacitor is charged in a dc circuit, no current need flow to sustain the charge, but the interface at a growing mercury drop is like a capacitor whose plates are continuously increasing in surface area. As the mercury drop in Fig. 13.10 (a) grows, the approach of additional Cl^- to the enlarging surface causes movement of electrons out of the DME into the external circuit. This *charging* or *capacitance current* is nonfaradaic: it does not depend upon an electro*chemical* process at the electrode surface—the electrons that leave the DME are not provided by the oxidation of a chemical species. It is conventional in polarography to call electron flow out of the DME a negative current.

When E_{DME} is more negative than the potential of the electrocapillary maximum [Fig. 13.10 (b)], the arrival of K^+ to maintain the double layer at the surface of the growing drop draws electrons into the DME from the external circuit, and the charging current becomes positive. Figure 13.11 shows the residual cur-

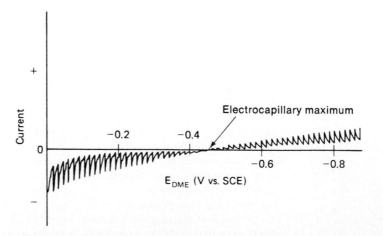

Figure 13.11. Portion of residual current from a polarogram (scale expanded as compared with that in Fig. 13.9).

rent from region LM in Fig. 13.9 on an expanded scale so that current profiles of the individual drops can be seen. The current is maximal when a new drop starts to form and decays toward zero with continuing growth. The rate of change in surface area with time, dA/dt, is greatest when the drop is smallest, corresponding to the highest current required to maintain the double-layer potential.

The residual current, typically 0.1 or 0.2 μA or so, presents the most serious limitation on the sensitivity of polarographic analysis. As the analyte concentration decreases, the limiting current (i_d in Fig. 13.9) decreases (Section 13.3e.) until eventually the question arises: how small a current can we measure? Electronics comes to mind—if it's too small, increase the amplification—but this helps only so far. The limiting current attributable to the analyte is the difference between the current on the plateau somewhere between N and P and our estimate of what the residual current would be at the same potential (perhaps extrapolated from LM). And we cannot amplify the current corresponding to reduction of the analyte without amplifying the residual current as well. No matter how good the electronics, the analyte *signal,* if small enough, will be buried in the *noise* of the residual background current.[9]

13.3e. *The Limiting or Diffusion Current*

Returning to the reduction of Cd^{2+} at the DME, there are three ways by which this ion could reach the electrode.

The *first* way in which a reducible ion may reach the electrode is through the process of convection. If the solution were vigorously stirred, Cd^{2+} would be swept toward the electrode by the bulk flow of solution. In polarography, though, the solution is kept as quiet as possible; stirring, as we mentioned, is difficult to control and reproduce, and it is likely as well to cause erratic currents by prematurely dislodging mercury drops.

Second, a cation like Cd^{2+} might feel the electrostatic pull of a negatively charged electrode. However, this is virtually eliminated by the large excess of supporting electrolyte (KCl in Fig. 13.9). The tendency of Cl^- to cluster about Cd^{2+} in the solution and the tendency of K^+ to collect at the negative electrode tend to shield Cd^{2+} from the electrostatic attraction of the electrode. This lowers the limiting current below what it would otherwise be, but it is advantageous in the long run.[10] The electrostatic migration of Cd^{2+} is subject to influence from other ions in the solution, and thus extraneous components of unknown samples might influence the current in an uncontrolled manner. By the deliberate addition of a large excess of KCl, such effects are swamped out, and minor differences from one sample to another may not affect the Cd^{2+} response. (As noted above, the KCl also serves to minimize the iR term.) It is sometimes recommended that

---

[9] Signal-to-noise ratio as a limitation on sensitivity is a concern in all measurements. An understandable analogy for the beginning student is desired music (signal) vs. static (noise) in a weak radio transmission. Turn up the volume and the music gets louder, but so do the roaring and crackling noises. This is not a perfect analogy because, strictly, noise usually means to most scientists *random* fluctuations, whereas the residual current is actually a definite, reproducible, nonrandom thing, but in the broader sense of a signal carrying desired information about the analyte as contrasted with an undesired intrusion, it serves our purpose.

[10] Note that elimination of electrostatic migration would *increase* the current for a reducible *anion* such as CrO_4^{2-} which would otherwise experience electrostatic *repulsion* from the cathode.

the supporting electrolyte be 50 to 100 times as concentrated as the analyte. This is another example of handling matrix effects (page 317). We have created our own matrix by adding a large quantity of material, which has the effect of making diverse samples more similar in gross composition. The KCl must, of course, be added to both calibration standards and unknowns.

The *third* way by which Cd^{2+} can reach the electrode is by diffusion. After the DME becomes sufficiently negative for reduction to begin (past M in Fig. 13.9), solution adjacent to the electrode becomes depleted in Cd^{2+} and a concentration gradient is established, leading to diffusion of Cd^{2+} from bulk solution to the electrode surface. The rate of diffusion is proportional to the concentration difference:

$$\text{Rate of diffusion} \propto [Cd^{2+}]_{bulk} - [Cd^{2+}]_{interface}$$

$$= k([Cd^{2+}]_{bulk} - [Cd^{2+}]_{interface})$$

On the limiting current plateau NP, the DME is sufficiently negative that virtually every Cd^{2+} ion reduces immediately upon reaching the surface. Thus $[Cd^{2+}]_{interface}$ falls to practically zero. Since the faradaic current depends upon the rate of supply of Cd^{2+} to the electrode, i.e., upon the rate of diffusion, we may write

$$i_d = k[Cd^{2+}]_{bulk}$$

where i_d is called the diffusion current. This equation shows, of course, the basis for quantitative analysis by polarography. A graph of i_d vs. $[Cd^{2+}]$, obtained with standard solutions, is a straight line. Once we know the slope, k, of the line, we can convert i_d values into concentrations for unknown samples.

13.3f. The Ilkovic Equation

The proportionality constant k in the above equation is really a collection of terms, some of which are real constants and some of which must be controlled to achieve linearity and reproducibility. An equation showing these terms was given in 1934 by Ilkovic:[11]

$$i_d = (607 \, nD^{1/2}m^{2/3}t^{1/6}) \, C$$

The terms are defined as follows:

■ i_d is the average[12] diffusion current in μA. The number 607 is the value at 25°C of a collection of terms including the Faraday and the density of mercury.

[11] Mathematical difficulties in treating radial diffusion toward the surface of an expanding sphere precluded a totally rigorous derivation. (Actually the mercury drop is not a perfect sphere; it is somewhat of a teardrop, pendant from its attachment to the mercury in the capillary.) The Ilkovic equation is considered valid within a few percent. It is accurate enough for many purposes, and in any case it highlights the important factors that must be controlled in quantitative work.

[12] Obviously the current depends upon the surface area of the mercury drop, hence the oscillations in Fig. 13.9 as the drops grow and fall. At break-off, the current actually goes nearly to zero; the break occurs slightly into the capillary bore, and the electrode area momentarily becomes only the cross-sectional area of the bore. Slower devices such as pen-and-ink recorders cannot track the current faithfully, and damped oscillations like those in the figure are seen in recorded polarograms. In ordinary polarography, average currents obtained from such recordings are satisfactory, and faster devices such as oscilloscopes are not required.

- n is the number of electrons in the process $Ox + ne \rightarrow Red$.
- D is the diffusion coefficient of the electroactive species in cm^2/s (this is a measure of how rapidly the species diffuses in a specified solution under the influence of a standard concentration gradient).
- m is the mass of mercury flowing through the capillary per unit time in mg/s.
- t is the drop time in s.
- C is the concentration of the electroactive species in mmol/liter or mM.

The terms m and t can be determined by measuring with a stopwatch the time required for a certain number of drops to fall and then weighing the mercury. This is done with the DME at a potential on the plateau NP (Fig. 13.9), with the capillary dipping into a solution of relevant composition. The value of $m^{2/3}t^{1/6}$ is constant in solutions of fixed composition over a modest potential range if the leveling bulb in Fig. 13.8 is always clamped at the same height. The diffusion coefficient D of the analyte is a constant if the gross composition of the solutions (supporting electrolyte, buffers, whatever) remains the same (cf. matrix effects, page 368). Thus, with proper care,

$$i_d = (607 n D^{1/2} m^{2/3} t^{1/6})\, C = k\, C$$

In practice, k is not calculated from the Ilkovic equation, primarily because of uncertainty in D values; rather, it is evaluated with standard solutions whose overall composition is as close as possible to that of the unknowns.

Most limiting currents in polarography are diffusion currents, but there are a number of exceptions. As an example, consider glucose. This sugar exists in aqueous solution as an equilibrium mixture of two forms, an open-chain aldehyde and a cyclic hemiacetal:

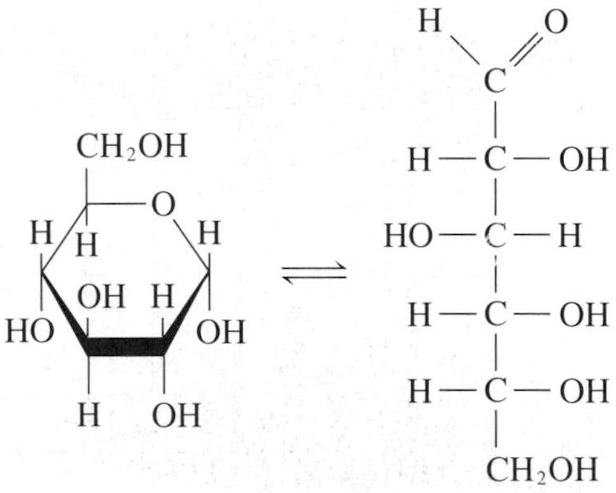

The ring form, which predominates in aqueous solution, is not reducible at the DME, while the free aldehyde group in the open-chain form is reducible to the alcohol. Studies of this system have shown that the limiting current is not diffusion-controlled but rather depends upon the rate of conversion of the nonreducible form into the aldehyde.[13] Such a current, sometimes very small, is said to be kinetically controlled, and polarography in such cases can be used to study the kinetics of conversion of nonreducible precursor to reducible form. For an easy experimental test, it can be shown that limiting currents which vary linearly with the square root of the height of the mercury column are diffusion-controlled.

[13] P. Delahay and J. E. Strassner, *J. Am. Chem. Soc.*, **74**, 893 (1952).

13.3g. Polarographic Maxima

Polarograms are frequently seen where the wave is not a smooth, S-shaped curve as in Fig. 13.9. Sometimes the current rises sharply to a peak and then rapidly decreases as shown in Fig. 13.12(a); in other cases, the current rises grad-

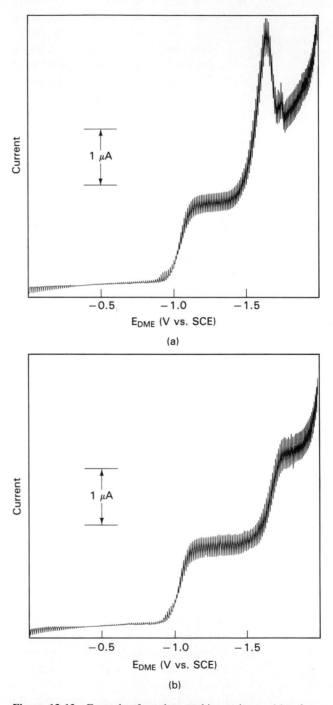

Figure 13.12 Example of a polarographic maximum: (a) reduction of a 1-alkyl-3-carbamidopyridinium salt with a pronounced maximum on the second wave; (b) same solution after addition of the maximum suppressor Triton X-100.

ually where it ought to be level and even gives the appearance of a second wave. These anomalous currents are called *maxima*, and they are undesirable in most polarographic work. There is still no completely satisfactory explanation for their origin. It is clear that an abnormally high current must result from an enhanced delivery of reducible material to the electrode, but the mechanism for this is not totally clear. Adsorption of the electroactive species as a function of potential used to be invoked, but most investigators believe that this is not an adequate explanation. There is evidence for an unusual streaming of solution past the surface of the mercury drop at certain potentials which could supply extra material to the electrode, but the cause of this is uncertain. It may be related to asymmetry in the electrical field around the drop arising from the fact that the drop is not a free sphere but rather a somewhat tear-shaped appendage to the capillary.

In any event, we know empirically how to eliminate maxima by adding so-called maximum suppressors to the solution. The commonest one used to be gelatin. More recently, the commercial surfactant Triton X-100 has become more widely used because it works as well as gelatin and its stock solutions are more stable. Typically, only very small concentrations of suppressors are required. For example, Triton X-100 is often used at a level of 0.002% in the final solution. The suppressors have some effect upon the diffusion current. In careful work, it is desirable to study the effect of suppressor concentration on the polarogram and then to employ the minimal concentration that is effective. Obviously, the suppressor concentration should be constant in all unknown and standard solutions which are to be compared.

13.3h. *The Half-wave Potential*

The potential at which the current is one-half the limiting current (i.e., where $i = i_d/2$) is called the *half-wave potential*, $E_{1/2}$ (see Fig. 13.9). At this potential, half of the Cd^{2+} ions that reach the cathode in a given time are reduced. $E_{1/2}$ is characteristic of the particular redox system and is independent of the concentration of the electroactive species in solution. This does not mean that it is very useful for qualitative identification except in very restricted circumstances, however, because too many systems exhibit $E_{1/2}$-values that are too close together.

For an electrode process which is fast enough for equilibrium to be established at the surface of the DME in the face of the changing potential, thermodynamic information is available; that is, $E_{1/2}$ is related to $E°$ for the redox couple.[14] Although it is understood in physical chemistry that such true reversibility cannot be achieved in a process occurring at a finite rate, there are many electrode pro-

---

[14] Even for a reversible electrode process, $E_{1/2}$ is not equal to $E°$ (or better $E°'$), although sometimes it may be close. When the reduction product is a metal like Cd which forms an amalgam, the difference between $E_{1/2}$ and $E°$ reflects the free energy of solution of the metal in mercury; i.e., converting the metal into a dilute solution in mercury facilitates the reduction. For two soluble species (e.g., $Fe^{3+} + e \rightleftharpoons Fe^{2+}$), $E_{1/2}$ may be close to a formal potential. When $E_{DME} = E_{1/2}$, then $[Fe^{3+}]_{surface} = [Fe^{2+}]_{surface}$, except for one thing: since the reduction product is not present in the bulk solution and forms only at the electrode, there is a concentration gradient promoting its diffusion away from the interface. The difference between $E_{1/2}$ and $E°'$, usually small, is related to the ratio of the diffusion coefficients of the two electroactive species. For a more sophisticated discussion, see: I. M. Kolthoff and J. J. Lingane, *Polarography*, 2nd ed., Interscience Publishers, New York, 1952, Vol. I, Ch. XI.

cesses that approach it so closely that we cannot tell the difference. For such a case, it can be shown that the following equation holds:

$$E_{\text{DME}} = E_{1/2} - \frac{0.059}{n} \log \frac{i}{i_d - i}$$

where the i values are currents at points on the polarographic wave corresponding to various values of E_{DME}, and n is the number of electrons gained by the reducible species. A plot of E_{DME} vs. $\log[i/(i_d - i)]$ is a straight line whose slope is $-0.059/n$; $E_{1/2}$ is the value of E_{DME} when the log term is zero. Figure 13.13 shows such plots for two processes of differing n-values.

It should be emphasized that *many* polarographic processes are not reversible, even in the sense explained above. Those that depart appreciably from reversibility can be recognized quickly by the fact that the wave is "drawn out," the current not rising steeply as it does in Fig. 13.9. Thermodynamic conclusions about ease of reduction of chemical species cannot be drawn from such data; there is a slow kinetic step in the electrode process which prevents equilibrium at any potential during the polarographic scan. In such cases, $E_{1/2}$-values reflect rate constants rather than positions of equilibrium.

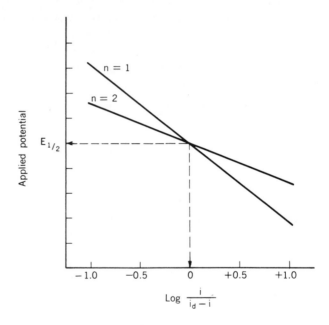

Figure 13.13   Plots of $i/(i_d - i)$ vs. E_{applied} for a polarographic wave.

13.3i.   Applications of Polarography

Because of the many substances that undergo redox reactions, a complete list of applications of polarography would be enormous; we can give only a brief sketch here. A polarographic scan is a quick way to determine the potential at which an ion or molecule will reduce, and it requires very little sample. Thus polarography is a useful pilot technique for establishing the conditions for a successful macroscale controlled potential electrolysis using a mercury cathode. Sometimes half-wave potentials for a series of organic compounds reflect electronic structure in a systematic manner to provide useful data on substituent effects that correlate nicely with other molecular properties.[15]

Applications to analysis include the determination of small quantities of metal ions—copper, lead, zinc, iron, cadmium, uranium, and many others. A number of inorganic anions yield useful waves; examples are iodate, bromate, selenite, and chromate.

Solutions containing O_2 exhibit two successive polarographic waves, the first representing a two-electron reduction to the level of H_2O_2, and the second the complete four-electron reduction to H_2O; $E_{1/2}$-values in acid solution are about $-.05$ V and -0.8 V vs. SCE. Diffusion current measurements on one or the other of these waves provide a useful method for the determination of dissolved oxygen in aqueous solutions; this has been used, for example, in photosynthetic studies involving algal suspensions and isolated chloroplasts. Because these two oxygen waves extend over most of the useful range for the DME, they interfere with the polarographic determination of most other substances, and the solutions must normally be free of oxygen. This is usually accomplished by bubbling pure nitrogen through the test solution; this is terminated before the polarogram is recorded to avoid stirring of the solution by the bubbles. A nitrogen inlet is seen in the apparatus in Fig. 13.8.

Many organic compounds reduce at the DME to yield analytically useful polarographic waves. Examples are carbonyl, nitro, azo, and olefinic compounds; many halogenated compounds reduce according to

$$RCH_2X + H^+ + 2e \longrightarrow RCH_3 + X^-$$

Polarography has been applied to the determination of many compounds of biological interest, including drugs, hormones, and vitamins. It has also been used, along with other techniques, to elucidate the redox properties of biological compounds such as flavin derivatives and pyridine coenzymes.

Table 13.4 lists some examples of polarographic analyses which have been reported, but it should be pointed out that this is by no means an exhaustive compilation.

In favorable cases, two or more components can be determined with a single polarogram; of course, the $E_{1/2}$-values must be sufficiently different for the individual waves to be distinguished. Figure 13.14 is a sketch from a polarogram obtained on a mixture of three reducible ions which illustrates this point. When the waves for two species in a sample are too close together, then a separation must precede the polarographic measurement step. Sometimes solution parameters such as the concentration of complexing agents or the pH can be manipulated in such a way as to improve the polarographic resolution of two components.

13.3j. Polarographic Techniques

Most work is performed with instruments which record the current-voltage curve automatically. The test solution is placed in the cell, nitrogen is bubbled through it for 10 to 20 min to remove oxygen, and the curve is then recorded in a few minutes. The method is commonly employed at concentrations in the range of

[15] See, for example, P. Zuman, *Substituent Effects in Organic Polarography,* Plenum Press, New York, 1967, and P. Zuman, *The Elucidation of Organic Electrode Processes,* Academic Press, Inc., New York, 1969.

TABLE 13.4 Some Examples of Polarographic Reductions

ION OR COMPOUND	SOLVENT[†] AND SUPPORTING ELECTROLYTE	$E_{1/2}$, V, VS. SCE	REDUCTION PRODUCT
Cd^{2+}	1 M HCl	−0.642	Cd(Hg)
Cu^{2+}	1 M HCl	−0.22	Cu(Hg)
Pb^{2+}	1 M HCl	−0.435	Pb(Hg)
Sn^{4+}	1 M HCl-4 M NH$_4$Cl	−0.25	Sn^{2+}
		−0.52	Sn(Hg)
CrO_4^{2-}	2 M K$_2$CO$_3$	−0.47	Cr^{3+}
Ni^{2-}	0.1 M KCl	−1.1	Ni(Hg)
Zn^{2+}	1 M KSCN	−1.05	Zn(Hg)
MoO_4^{2-}	1 M HCl	−0.63	$MoCl_6^{3-}$
Azobenzene	Aqueous ethanol, pH 2 to 3	−0.11	Hydrazobenzene
Benzaldehyde	50% ethanol, H$_3$PO$_4$, pH 1.3	−0.98	C_6H—CH—CH—$C_6H_5$5 $\qquad$ | $\quad$ | $\qquad$ OH $\quad$ OH
1,4-Naphthoquinone	Aqueous ethanol, phosphate buffer, pH 3	+0.12	Naphthohydroquinone
Nicotinamide-adenine dinucleotide	0.1 M tetra-n-butylammonium carbonate buffer, pH 9	−1.12	4,4′-Dimeric one-electron product from electro-generated free radical
Progesterone	50% ethanol buffered at pH 6	−1.36	—
Quinine	Acetate buffer, pH 3	−1.00	—
Riboflavin	H$_3$PO$_4$, pH 1.8	−0.16	—

[†] Solvent is water unless otherwise noted.

10^{-4} to 10^{-2} M. Errors of perhaps 2 to 5% are fairly common, although these may be reduced by a factor of 10 or so in favorable cases.

Since the concentration is proportional to the diffusion current, the main problem is to measure this current accurately. Correction must be made, of course, for the residual current. This current can be measured separately on a solution that contains the supporting electrolyte alone and subtracted from the cur-

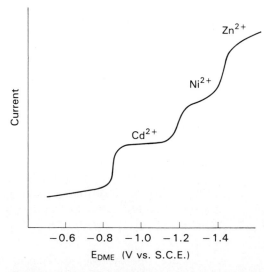

Figure 13.14 Sketch of the average current for a polarogram of three electroactive ions in an ammonia buffer.

rent measured on the diffusion plateau at the same voltage. Alternatively, the residual current along LM in Fig. 13.9 can be extrapolated to correct the current in region NP. Books on polarography should be consulted for details; that by Meites[16] is particularly valuable in helping inexperienced individuals attain a professional level. Standard solutions as well as unknowns are run, of course, in order to measure k in $i_d = kC$ as well as for reassurance that things are working.

A technique known as the *method of standard addition* is sometimes useful, especially in cases where variable compositions of unknown samples make it difficult to prepare standards with the same matrix as the unknowns. A polarogram is recorded for the unknown sample, and then a known volume of the unknown is spiked with a known quantity of a standard solution and a second polarogram is obtained. From the increase in i_d caused by the known addition of the desired constituent, the original unknown concentration can be calculated. If the spike is a small enough volume not to dilute the sample appreciably, then the added material is subject to the same matrix as existed in the original unknown solution. An example illustrates the calculation.

EXAMPLE 13.4

Calculation for the method of standard addition

A lead solution of unknown concentration yields a diffusion current of 1.00 μA. Then, to 10.00 ml of the unknown solution is added 0.500 ml of a standard solution of lead whose concentration is 0.0400 M. The diffusion current with the spiked solution is 1.50 μA. Calculate the lead concentration of the unknown solution.

Let C be the unknown concentration. Then the concentration of lead in the spiked solution is given by

$$\frac{10.00C + (0.500)(0.0400)}{10.500} = \frac{10.00C + 0.0200}{10.500}$$

Since i_d is proportional to concentration,

$$\frac{1.00}{C} = \frac{1.50}{(10.00C + 0.0200)/10.500}$$

$$C = 0.00348 \; M \qquad \square$$

| Box 13.3 | **Differential Pulse Polarography** |

In classical polarography, the voltage applied to the cell is a *linear ramp,* as shown in Fig. B13.3.1(a). The primitive electronics of an earlier age easily provided such a voltage scan and recorded the resulting current-voltage curve, as we have seen. With modern circuitry and signal processing, however, we can do much better (although classical polarography provides the basis for understanding recent developments). The practical lower concentration limit for ordinary polarography is usually about $10^{-5}\,M$ (10^{-6} in very favorable cases). Roughly, differential pulse polarography (DPP) extends this downward by a factor of about 1000 or so to about $10^{-8}\,M$. The method provides a great improvement in the signal-to-noise ratio (footnote 9) and is rapidly replacing classical polarography in many laboratories.

[16] L. Meites, *Polarographic Techniques,* 2nd ed., Wiley-Interscience, Inc., New York, 1965.

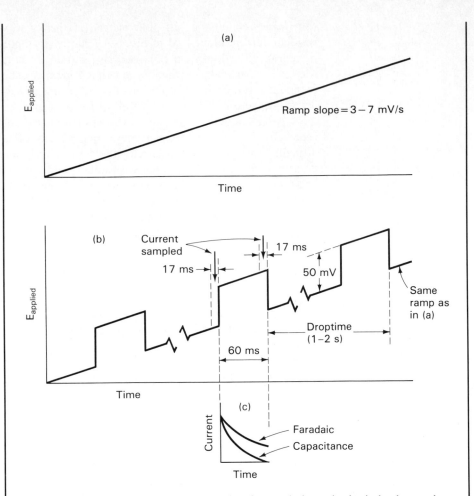

Figure B 13.3.1 (a) Applied voltage vs. time for a typical scan in classical polarography. (b) Applied voltage vs. time in DPP. (Note the breaks in the underlying ramp; otherwise the droptime and the pulse duration could not have been conveniently shown on the same time scale.) (c) Faradaic and capacitance currents vs. time for the duration of one pulse in a potential region where reduction can occur at the DME.

The voltage program for DPP, shown in Fig. B13.3.1(b), is seen to be a linear ramp as in Fig. B13.3.1(a) upon which are superimposed regular voltage pulses whose magnitude (modulation amplitude), frequency, and duration are selected by the operator. A representative example might be pulses of 50 mV lasting for 0.06 s (60 ms), applied every second or two. The pulses are synchronized with the DME so that each is applied late in the life of a drop; regularity is ensured by automatically dislodging each drop by a mechanical tap on the capillary just at the end of the pulse.

The current is sampled twice during the life of each drop, once just before application of the pulse and again near the end of the pulse. Essentially, the two currents are subtracted electronically and the amplified difference is plotted against the applied ramp voltage. The result is a peak at a voltage corresponding to $E_{1/2}$, with a height proportional to the concentration of the analyte.

It should be clear why a peak is obtained: the additional current caused by the voltage pulse is greatest in the region of greatest slope on a conventional dc polarogram like that in Fig. 13.9. The large increase in sensitivity, though, requires explanation. Two factors are involved. First, the current which is sampled just before application of the pulse [left-hand heavy vertical arrow in Fig. B13.3.1(b)] has only a very small capacitance component because it is measured late in the lifetime of the drop (page 367). Second, the desired faradaic signal is enhanced. To be sure, application of a pulse causes surges in both capacitance and faradaic currents; i.e., current must flow to reestablish correspondence between the double layer at the electrode surface and the new potential, and current must also flow (provided $E_{applied} > E_d$) as the reaction Ox + ne ⇌ Red seeks a new equilibrium corresponding to the new potential. These currents then decay for the duration of the pulse, but (importantly) not at the same rate. Their behavior during the lifetime of one pulse is shown in Fig. B13.3.1(c). The capacitance current decays exponentially, becoming nearly zero late in the pulse. The faradaic current, on the other hand, although it decays as the electron-transfer reaction "catches up" with the new potential, tends to approach the diffusion-limited value but is measured before it gets there. The result of all this is that we record the amplified difference between an enhanced faradaic current and a very low capacitance current.

Figures B13.3.2 and B13.3.3 show both ordinary dc polarograms and DPP curves for two examples. A glance shows that measuring DPP peak heights is far preferable to evaluating the limiting currents of ordinary polarograms. An additional feature of DPP relates to the resolution of mixtures. For successive waves such as those in Fig. 13.14, it becomes difficult to discern individual plateaus if the $E_{1/2}$ values are closer than about 200 mV, whereas DPP peaks separated by perhaps 40 mV can be distinguished.

Figure B13.3.4 shows a modern, multipurpose instrument which provides a selection of modern electroanalytical techniques including DPP. (The curve actually shown on the video display screen is a cyclic voltam-

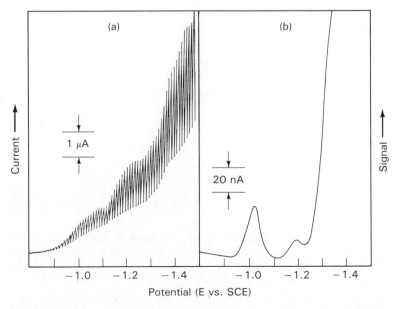

Figure B 13.3.2 (a) Ordinary dc polarogram on a buffered solution containing the antibiotic tetracycline hydrochloride at a concentration of 180 ppm. (b) Differential pulse polarogram for the same antibiotic under the same conditions but at a concentration of only 0.36 ppm. Reprinted with permission from J. B. Flato, *Analytical Chemistry*, **44**, 75A. Copyright 1972 American Chemical Society.

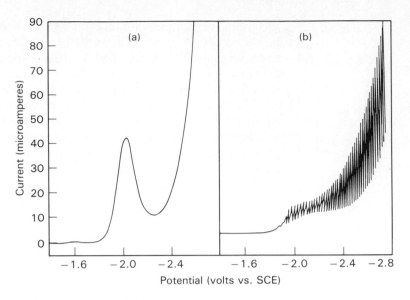

Figure B 13.3.3 Ordinary dc polarogram (b) and DPP curve (a) for 100 ppm acrylamide in aqueous methanol. Reprinted with permission from S. R. Betso and J. D. McLean, *Analytical Chemistry,* **48,** 766. Copyright 1976 American Chemical Society. Note the difficulty in measuring the analyte wave because electrolysis of the solvent itself is not far removed; it is easier to draw a baseline from which to measure a peak current in (a).

Figure B 13.3.4 A modern, multipurpose electroanalytical instrument (Model 270 Electrochemical Analysis System). With the permission of EG&G Princeton Applied Research Corporation.

mogram, where the voltage is scanned rapidly using a triangle-shaped voltage-time function; this has not proved very useful as an analytical tool, but it provides insight into the mechanisms of electrochemical processes.)

13.3k. Amperometric Titrations

Since the diffusion current is proportional to the concentration of the electroactive species, the polarographic technique can be used to follow the progress of a titration involving this substance. The voltage scan required for a complete polarogram is not needed here, of course. Once the region of the diffusion plateau is identified, the applied voltage is set at a value on this plateau and i_d-values are measured as the titration proceeds. The essential apparatus is some sort of polarographic cell, perhaps similar to the one shown in Fig. 13.8, with provision for the penetration of a buret tip into the solution. The titration curve is a graph of i_d vs. volume of titrant. The shape of the curve depends upon whether the substance titrated, the titrant, or both, can be reduced at the applied voltage which is selected.

Consider the titration of substance S with titrant T to form reaction product P:

$$S + T \longrightarrow P$$

Suppose that P is not electroactive; perhaps it is a precipitate, as in the titration of Pb^{2+} with CrO_4^{2-} to form insoluble lead chromate, $PbCrO_4$. Figure 13.15(a)

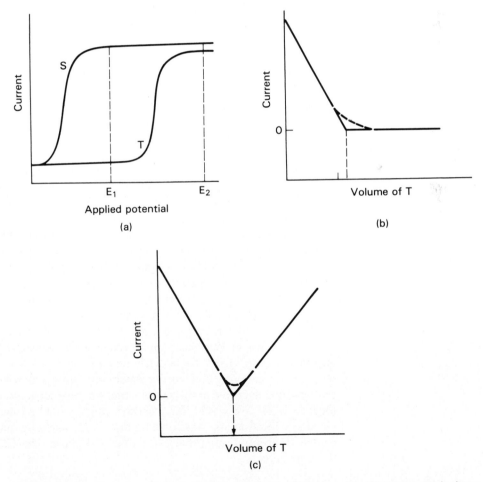

Figure 13.15 (a) Polarograms of substance titrated (S) and of titrant (T). (b) Amperometric titration of S with T at E_1. (c) Amperometric titration of S with T at E_2.

shows polarograms of S and T obtained separately. Now suppose that we select the applied voltage E_1, where S yields a diffusion current but T does not. As the titration proceeds, consuming S, the value of i_d drops, as seen in Fig. 13.15(b). Past the equivalence point, the current will level off to some small residual value. The break at the equivalence point may be sharp or more-or-less rounded, depending upon the extent to which the titration reaction goes to completion in a region where there is an excess of neither S nor T.

Suppose, on the other hand, that we perform the titration with an applied voltage E_2, where both S and T are reducible at the DME. The resulting titration curve is shown in Fig. 13.15(c). The first part of the curve is the same as in the previous case, but past the equivalence point the accumulation of excess titrant leads to increasing i_d-values.

Because errors in individual i_d-values tend to cancel when the "best" straight lines are drawn through a number of points, the error in locating the equivalence point is of the same order as in classical titrimetric analysis. Titrations can be performed rapidly because there is no need for the operator to "hit" the end point, which can be found by extrapolation from points on either side. The extrapolation technique also permits the use of reactions which do not go well to completion near the equivalence point. Substances titrated are usually in the concentration range 10^{-4} to 10^{-1} M.

It should be remembered that the titrant not only removes S through reaction but also, by its very addition, dilutes the solution. Measured i_d-values should be corrected for the dilution effect by multiplication by the factor $(v_0 + v)/v_0$, where v_0 is the initial volume of the solution and v is the volume of titrant added at the point where the correction is being applied. This correction can be minimized by using a very concentrated titrant solution delivered from a good microburet to keep the volume error small.

13.3l. *The Clark Oxygen Electrode*

We conclude our treatment of electroanalytical chemistry with a brief description of a very important application of amperometry—the determination of dissolved oxygen using a device often called the *Clark oxygen electrode*.[17] Dissolved-oxygen measurements are important in many areas including the following: ecological studies of natural waters; control of aerobic bacterial breakdown of sewage, where aeration is an operating expense and higher O_2 levels than necessary are wasteful; studies of photosynthesis; and control of oxidative corrosion in boilers. There have been thousands of publications in a wide variety of fields in which dissolved-oxygen determinations are mentioned. Despite its low solubility (the concentration of O_2 in fresh water in equilibrium with air at 20°C is only ~9 mg/liter, i.e., about 9 ppm), oxygen is a critical solute in many aqueous systems.

The classical Winkler method (page 304), developed in 1888, works adequately and is still used, particularly for calibration purposes. There are problems, however, in sampling and storing water containing a dissolved gas, more so if microflora remain active in the sample, and the titrimetric procedure is inconvenient for field work. We would like a sensor which could be lowered into a water sample at any location and which would rapidly furnish a readout related to the *in situ* oxygen concentration at that exact time and place. The Clark electrode provides this capability.

---

[17] L. C. Clark, Jr., "Electrochemical Device for Chemical Analysis," U.S. Patent 2,913,386, issued Nov. 17, 1959.

We saw earlier that O_2 yields two well-defined polarographic waves in ordinary voltage scans (page 373). The $E_{1/2}$-values are pH-dependent and hence vary somewhat with solution conditions; in 0.1 M KCl, the values are about -0.05 V and -0.9 V vs. SCE. Once we know approximately where the waves are on a voltage scale, we no longer need a voltage scan; we can fix the cathode potential at a value on a diffusion plateau and, after appropriate calibration, relate the current to the O_2 concentration. Figure 13.16 shows the essential features of a cell for this purpose. Although most people call the device a Clark electrode, it is seen actually to be a cell with two electrodes (cathode and anode). The cathode is usually gold, although platinum has also been used; the anode may be a silver-silver chloride electrode which, because only small currents flow, maintains a constant potential and thus provides a reference for the cathode. A power source, which for work in the field may be simply a battery pack, is adjusted to maintain the cathode at a potential well onto the plateau of the first O_2 wave, say, -0.6 V vs. Ag–AgCl. The current resulting from the reduction of O_2 is measured in the simplest instruments with a sensitive galvanometer acting as a microammeter.

Of course any chemical species which is reducible at the selected cathode potential represents a possible interference in the O_2 determination; i.e., the electrode as such is quite unselective for O_2 in the presence of other oxidants, some of which may appear in certain types of water samples. This problem is largely circumvented by the membrane which separates the electrode from the test solution. This membrane is a thin sheet of a hydrophobic material such as Teflon (polytetrafluoroethylene). We may think of the membrane as having tiny pores that are not wetted by water. The membrane is not only impermeable to water but also, then, to water-soluble ions and other polar substances. A gas such as oxygen, however, can gain access to the cathode by diffusion through the membrane.

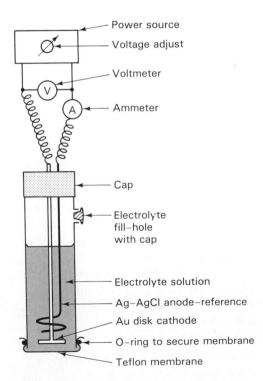

Figure 13.16 Sketch showing the essential features of the Clark oxygen electrode (see text for discussion of components).

The current depends upon the rate of diffusion of O_2, which in turn reflects the concentration of O_2 in the test solution.

There is great flexibility in the use of the Clark electrode. If circumstances permit, the amperometric signal can be amplified and recorded continuously for monitoring O_2 levels at a fixed location; on the other hand, the electrode can be carried onto a boat and lowered by cable to various depths in lakes or estuaries to obtain three-dimensional profiles for possible correlations with biotic observations.

KEY TERMS

Amalgam. A solution of a free metal (zero oxidation state) in mercury, which is a liquid at room temperature. The activity of the metal, expressed as its mole fraction in the solution, is less than that of the pure metal, i.e., amalgam formation, as contrasted with plating onto an electrode surface, facilitates the reaction $M^{n+} + ne = M$.

Amperometric titration. A titration where the end point is found by observing the effect of titrant addition upon a measured current.

Amperometry. Literally, measurement of current. In electroanalytical chemistry, the analyte concentration determines the value of the current. The determination of dissolved oxygen using the Clark electrode (page 380) is a well-known example of amperometry.

Anode. In any cell (galvanic or electrolytic), the electrode where oxidation occurs.

Cathode. In any cell (galvanic or electrolytic), the electrode where reduction occurs.

Controlled potential coulometry. Coulometry (q.v.) where undesired electrode reactions are prevented by controlling the potential of the working electrode (q.v.).

Controlled-potential electrolysis. Electrolysis in which the cathode potential is not allowed to become more negative than a predetermined value (if the reaction of interest is an oxidation, then the anode potential is prevented from becoming more *positive* than the preselected value). Three electrodes are required: anode, cathode, and a reference electrode against which the potential of the working electrode is monitored.

Coulomb. The quantity of electricity associated with the passage of a current of 1 A for 1 s.

Coulometer. A device for measuring the coulombs associated with an electrochemical process. Modern coulometers are electronic in nature.

Coulometric titration. Sometimes called "titration with electrons." A technique in which an analyte is determined by measuring the time required for an electrolytic process at constant current.

Coulometry. In electroanalytical chemistry, determination of an analyte based upon the quantity of electricity associated with an electrolytic oxidation or reduction of that analyte.

Counter electrode. The analyte or species of interest in electrolysis reacts at the *working electrode;* the other electrode in the electrolysis circuit, which is not of interest per se so long as the electrode reaction is harmless, is the *counter electrode*. May be an anode or a cathode depending upon the circumstances.

Current efficiency. The fraction of the total current carried by an electron-transfer reaction involving a particular species of interest. When no other reactions intrude, we say that the desired electrode process is proceeding at 100% current efficiency.

Decomposition potential. Symbol E_d. The voltage of a galvanic cell which must be overcome by the external power source in order for electrolysis to proceed. Sometimes called "back emf."

Diffusion current. Symbol i_d. A limiting current (q.v.) whose value is determined by the rate of diffusion of an electroactive species.

Direct coulometric titration. A coulometric titration (q.v.) where the analyte itself is the electroactive species at the working electrode.

Electrogravimetry. Analysis in which the analyte is determined by the change in weight of an electrode during electrolysis.

Electrolysis. Reversal of the cell reaction that occurs during spontaneous discharge of a galvanic cell by application of an external voltage. Electrical energy is converted into chemical energy in electrolysis.

Faradaic current. A current which is carried across an electrode-solution interface by means of an electron-transfer reaction involving the interaction of a chemical species with an electrode.

Faraday. The quantity of electricity (96,485 C) associated with one equivalent of chemical change in an electrochemical process. The value of the faraday is redetermined from time to time; actually 96,500 C is suitable for the calculations in this book. One could also say that the faraday is the quantity of electricity associated with Avogadro's number of electrons.

Faraday's law. The reaction of one equivalent of a chemical species in an electrochemical process is associated with 96,485 C (1 faraday) of electricity.

Half-wave potential. In polarography, the DME potential at which the current is one-half of its limiting value.

Ilkovic equation. Equation describing the polarographic diffusion current (see page 368).

Indirect coulometric titration. A coulometric titration (q.v.) where the analyte reacts with a chemical reagent which is generated at the working electrode rather than reacting at the electrode itself.

iR drop. According to Ohm's law ($E = iR$), there is a voltage drop E between any two points separated by a resistance R through which a current i is flowing. In electroanalytical chemistry, metal electrodes and wires usually have little resistance compared with that of a solution, and iR drop usually refers to the voltage drop across a solution between two electrodes.

Limiting current. The current (after subtracting the residual current) on the plateau of a polarographic wave, as in the region NP of the graph in Fig. 13.9.

Maxima. In polarography, unusually large currents in certain regions of the DME potential which interfere in the evaluation of diffusion currents. The problem can generally be controlled by the addition of maximum suppressors to the test solutions.

Ohm's law. The current flowing in a conductor is equal to the potential difference between any two points divided by the resistance between them, i.e., $i = E/R$ or $E = iR$. In the usual units, amperes = volts/ohms.

Overpotential. Additional voltage beyond the decomposition potential and iR drop that must be applied to overcome slow kinetic steps in electrolysis reactions. With *concentration overpotential,* the slow step is diffusion of electroactive species; with *activation overpotential,* the rate of a chemical reaction following electron transfer is involved.

Polarogram. A plot of current vs. DME potential (see polarography).

Polarography. A special form of voltammetry (q.v.) where the working electrode is a dropping mercury electrode (DME).

Potentiostat. An instrument for automating controlled potential electrolyses.

Residual current. In voltammetry, a nonfaradaic current, i.e., a current not carried by an electron-transfer reaction. In polarography, the major factor in the residual current is the capacitance or charging current associated with development of the double layer at the electrode-solution interface.

Standard addition, method of. Not restricted to polarography but useful in a number of areas for overcoming matrix effects. After a polarogram is recorded for an unknown solution, the solution is "spiked" with a known quantity of the analyte and another polarogram is recorded to determine the polarographic response of the spike in a solution of that particular composition. To maintain the matrix, the addition should not dilute the solution appreciably.

Voltammetry. Electrolysis in which the current is recorded as a function of the potential of the working electrode.

Working electrode. The electrode where the reaction of interest occurs. It will be the cathode if a reduction is being studied, and the anode if an oxidation is of primary interest.

QUESTIONS

1. *Products of electrolysis*. Predict the first products of electrolysis in the following cells. Each electrolyte is $0.10\ M$ in an aqueous solution which is $0.10\ M$ in H^+. Assume any gas would have a partial pressure of 1 atm.

Cathode	Electrolyte	Anode
Pt	NaCl	Pt
Pt	KI	Cu
Pt	HCl	Pt
Cu	$Cu(ClO_4)_2$	Cu

2. *Amalgam formation*. Explain why some metal ions (e.g., Cu^{2+}, Zn^{2+}, and Pb^{2+}) are more easily reduced at a mercury cathode than at a platinum cathode.

3. *Terms*. Define or explain each of the following terms: anode, cathode, working electrode, counter electrode, reference electrode, concentration overpotential, activation overpotential, decomposition potential, matrix effect.

4. *Combined reference-anode*. Explain why an SCE in polarography with aqueous solutions can simultaneously serve as both anode and reference electrode; i.e., why are three electrodes not required as they are in macroscale controlled-potential electrolysis? Under what circumstances might a three-electrode assembly be required if one planned to report accurate polarographic half-wave potentials?

5. *Establishing potential for an electrolytic preparation*. An organic chemist wants, as part of a synthetic sequence, to reduce a nitro compound to a hydroxylamine on a macroscale. What experiment, requiring only a very small quantity of material, would enable him to select a suitable limiting cathode potential for reducing several grams of the nitro compound at a mercury electrode?

6. *Controlled-potential electrolysis*. Explain why controlled-cathode-potential electrolysis is required to separate silver from copper, but electrolysis at constant *applied* voltage (involving a much simpler instrument) is adequate for separating silver from zinc in aqueous dilute acid solution.

7. *Amperometric titration*. Consider the amperometric titration of Y with X, where

$$Y + X = YX(s)$$

Sketch the titration curve obtained at a potential where (a) Y is reducible and X is not; (b) X is reducible and Y is not; (c) both Y and X are reducible.

Multiple-choice. In the following multiple-choice questions, select the *one best* answer.

8. If a current-time curve is recorded during a controlled-potential electrolysis, the number of coulombs is found from (a) the slope of the curve where $t = 0$ (the initial slope); (b) the time required to reach zero current; (c) the slope of a tangent to the curve when the electrolysis is halfway to completion; (d) the area under the curve.

9. Which factor represents the most severe limitation on the sensitivity of a conventional polarographic analysis? (a) the capacitance or charging current which flows as each mercury drop develops; (b) the sensitivity of galvanometers and other current-measuring devices; (c) the uncertainty in the potential of the DME resulting from the iR drop through the test solution; (d) liquid junction potentials of uncertain magnitude.

10. Concentration overpotential is associated with (a) a slow electron-transfer process; (b) concentration gradients in the solution near the electrode interface; (c) unusually low activity coefficients; (d) large iR drops through the solution.

11. Activation overpotential is associated with (a) inefficient stirring, leading to concentration gradients in the solution near the electrode surface; (b) slow mass transfer across liquid junctions; (c) a slow step in the overall electrode process, including chemical steps following electron transfer; (d) slow movement of electrons in aqueous solutions.

12. If an analyst ran a voltammogram on a dilute solution of cadmium chloride containing no other solutes, using a platinum microelectrode and stirring the solution, cadmium ion would be transported to the electrode by (a) convection; (b) diffusion; (c) electrical migration; (d) all the above processes.

13. The residual current in polarography (a) is always anodic; (b) is always cathodic; (c) may be anodic or cathodic depending upon the potential of the DME; (d) is eliminated by the supporting electrolyte.

14. The major purpose for employing the "method of standard addition" in analytical chemistry is to (a) increase sensitivity through an increase in the measured value; (b) decrease sensitivity when the analyte solution is too concentrated; (c) compensate for matrix effects; (d) compensate for operator errors.

15. In a controlled-potential electrolysis, the number of moles of electrons added to the reducible species at the cathode is given by (a) $n_e = CF$; (b) $n_e = C/F$; (c) $n_e = F/C$; (d) $n_e = iR$; where n_e = moles of electrons, C = number of coulombs, F = the Faraday, i = current, R = resistance.

16. Refer to Box 13.1. Why not prepare aluminum metal by electrolysis of a water-soluble aluminum salt in aqueous solution at room temperature instead of in a molten cryolite bath at over 900°C?

17. For students interested in economics only. Refer to Box 13.1. Why did the price of aluminum go from 30¢/lb in 1970 to 80¢/lb in 1980? (There was no shortage of bauxite or other starting materials.)

PROBLEMS

1. *Decomposition potential.* Calculate the equilibrium decomposition potentials for the following electrolytes, all with platinum electrodes; (a) 0.10 M HCl; (b) 0.010 M HCl; (c) 0.0010 M HCl.

2. *Decomposition potential.* Calculate the equilibrium decomposition potentials of the following cells:
   (a) $^-Pt|Ag^+$ (0.0010 M) + H^+ (0.010 M)|Pt$^+$
   (b) $^-Pt|KCl$ (0.10 M) + HCl (0.0010 M), $Hg_2CL_2|Hg^+$
   (c) $^-Pt|Cu^{2+}$ (0.010 M) + H^+ (0.010 M)|Cu$^+$
   (d) $^-Pt|Cu^{2+}$ (0.010 M) + H^+ (0.010 M)|Pt$^+$

3. *Decomposition potential.* Calculate the value of the equilibrium decomposition potential of the cell in part (d) of Problem 2 after the electrolysis has proceeded until [Cu^{2+}] has been reduced to 1.0×10^{-5} M. (Do not overlook the increase in [H^-] due to the anode reaction.)

4. *Applied voltage.* What voltage must be applied to a pair of smooth platinum electrodes immersed in a 0.050 M solution of Cu^{2+} in order for the initial current to be 0.80 A? The [H^+] is 0.10 M, the cell resistance is 1.50 ohms, and the overpotential terms at the anode and cathode are 0.58 and 0.21 V, respectively.

5. *Electrolysis.* A solution which is 0.20 M in Cu^{2+} and 0.20 M in H^+ is electrolyzed between platinum electrodes. (a) Assuming that ω_C for hydrogen evolution is negligible, what is the concentration of Cu^{2+} when H_2 just begins to be liberated at the cathode? (b) Repeat part (a), but assume that the activation overpotential for hydrogen evolution is 0.80 V.

6. *Faraday's law.* How many hours are required to add 0.0100 mol of electrons to a solution in a coulometric experiment using a constant-current power supply delivering 100 mA?

7. *Faraday's law.* A copper-containing alloy was dissolved in nitric acid, and the copper was plated onto a platinum cathode electrolytically. The weight of the cathode increased by 0.3029 g. If the reduction of Cu^{2+} proceeded at 100% current efficiency (page 357), how many coulombs of electricity were involved in the process?

8. *Electrogravimetry.* Brass is an alloy of copper and zinc with smaller amounts of lead, tin, and other elements. Generally, copper can be separated from the other components by electrolysis. (Tin, which might interfere, is separated by precipitation of the oxide before electrolysis.) A sample of brass is analyzed for copper by electrogravimetry. A portion of the alloy weighing 2.1049 g is dissolved in nitric acid, and the copper is plated onto a platinum

gauze cathode weighing 13.4963 g. After electrolysis is complete, the cathode is removed from the solution, washed, and dried. It now weighs 14.7402 g. Calculate the percentage of copper in the brass sample.

9. *Overpotential—pH*. A solution which is 0.10 M in Zn^{2+} has a hydrogen ion concentration of 0.100. It is electrolyzed using platinum electrodes. (a) Assuming that the overpotential term for H_2 evolution is zero, what is the first product at the cathode? (b) If the overpotential term for H_2 is 1.00 V, what is the first product at the cathode?

10. *Effect of pH*. (a) Theoretically, what must be the pH of a solution so that the concentration of Cd^{2+} can be reduced by electrolysis at a platinum cathode to 1.00×10^{-6} M before the evolution of H_2 commences? Assume that the overpotential for H_2 formation is zero? (b) Repeat part (a) for reduction of Al^{3+} instead of Cd^{2+}. (c) What do you think about the existence of an aqueous solution with the pH calculated in part (b)? What do you think about the significance of the term pH for such a solution?

11. *Mercury cathode and amalgam formation*. Repeat the calculation in part (a) in Problem 10, but use a mercury cathode in which Cd is soluble. The quantity of Hg and the initial $[Cd^{2+}]$ are such that when $[Cd^{2+}] = 1.00 \times 10^{-6}$ M, a_{Cd} in the amalgam is 0.00010.

12. *Concentration overpotential*. In a controlled-potential electrolysis of a silver solution, the cathode reaches a value of +0.21 V vs. SCE when $[Ag^+]$ in the body of the solution is 1.00×10^{-4} M ($E_{SCE} = +0.25$ V). (a) What is the value of the cathode overpotential term, ω_C? (b) Assuming that the electrode process itself is reversible (Nernstian behavior), what is the actual value of $[Ag^+]$ in the very thin layer of solution in direct contact with the electrode surface?

13. *Potential buffer*. In a 1 M HCl solution where $E^{\circ\prime}$ is +0.70 V for $Fe^{3+} + e = Fe^{2+}$, what should be the molar ratio of Fe^{3+} to Fe^{2+} (acting as a potential buffer) in order for this system to prevent a cathode potential from becoming more negative than +0.80 V?

14. *Indirect coulometric titration*. R. J. Meyers and E. H. Swift [*J. Amer. Chem. Soc.*, **70**, 1047 (1948)] found that arsenious acid could be titrated with electrolytically generated Br_2 using an amperometric end-point technique ($As^{3+} + Br_2 = As^{5+} + 2Br^-$). The electrodes in the generator circuit were platinum, and the arsenite solution was 0.1 M in H_2SO_4 and 0.2 M in NaBr. (a) What were the electrolysis products at the cathode and the anode? (b) Suppose that 2.476 min was required to titrate the arsenic in a certain solution with a constant current of 3.500 mA. How many micrograms of arsenic (As) were present?

15. *Indirect coulometric titration*. J. W. Sease, C. Nieman, and E. H. Swift [*Ind. Eng. Chem., Anal. Ed.*, **19**, 197 (1947)] showed that 2, 2'-thiodiethanol ($HOCH_2CH_2$—S—CH_2CH_2OH, $C_4H_{10}O_2S$, MW 122.2) could be oxidized quantitatively to the corresponding

$$\overset{\displaystyle O}{\underset{\displaystyle \|}{}}$$

sulfoxide ($HOCH_2CH_2$— $\overset{O}{\overset{\|}{S}}$ —CH_2CH_2OH) by electrolytically generated Br_2. The Br_2 was generated from Br^- at an anode with 100% current efficiency. Suppose in one case the titration with a constant current of 3.000 mA required 2.806 min. How many micrograms of thiodiethanol were present in the sample?

16. *Typical quantities in polarography*. Suppose that a Cu^{2+} solution is electrolyzed in the cell shown in Fig. 13.8 for 3.00 min with the DME at a potential on the limiting current plateau for the process $Cu^{2+} + 2e + H_g = Cu(Hg)$, the current being 0.60 μA. Calculate (a) the number of micrograms of copper reduced at the DME and (b) the number of micrograms of Hg_2Cl_2 formed at the SCE anode. (c) Assuming that the current remained constant at 0.60 μA, calculate the time required to lower $[Cu^{2+}]$ in 10 mL of solution from 4.00×10^{-4} M to 3.90×10^{-4} M. Comment on the statement that polarography is a nondestructive analytical technique in the sense that running a polarogram does not normally alter the concentration of the analyte.

17. *Half-wave potential*. A metal ion is reduced polarographically, the diffusion current being 10 μA. The following currents are obtained at the indicated potentials (all negative vs. SCE): 0.444 V, 1.0 μA; 0.465 V, 2.0 μA;

0.489 V, 4.0 μA; 0.511 V, 6.0 μA; 0.535 V, 8.0 μA; 0.556 V, 9.0 μA. Treat the data graphically (see page 372) to obtain (a) the half-wave potential and (b) the value of n.

18. *Method of standard addition*. Aromatic ketones can be reduced to alcohols under appropriate conditions at the DME in aqueous ethanol solution. An unknown solution of acetophenone (MW 120. 16) yielded a limiting current of 0.480 μA. To 6.00 mL of the unknown solution was added 0.100 mL of a standard solution of acetophenone, the concentration of which was 6.40×10^{-4} M. Another polarogram was recorded, and the limiting current was 0.528 μA. Calculate the molar concentration of acetophenone in the unknown solution.

19. *Diffusion coefficient*. In a polarographic reduction of Cd^{2+} to cadmium amalgam, Cd(Hg), the average value of the diffusion current for a 2.00×10^{-4} M solution (2.00×10^{-1} mM) was found to be 1.34 μA. The capillary characteristics were $m = 1.96$ mg/s and $t = 3.03$ s. Calculate D, the diffusion coefficient of Cd^{2+} in the particular solution involved.

20. *Reduction of oxygen*. The diffusion coefficient of O_2 in dilute aqueous solution is 2.6×10^{-5} cm^2/s. A 0.25 mM solution of O_2 in an appropriate supporting electrolyte gives a polarographic wave with a diffusion current of 5.8 μA. The DME constants are: $m = 1.85$ mg/s; $t = 4.09$ s. Calculate the value of n. To which product, hydrogen peroxide or water, is O_2 reduced under these conditions?

14

Spectrophotometry

14.1

INTRODUCTION

From the earliest times, color has been used to identify chemical substances, and modern spectrophotometry is an extension of this primitive technique. Replacing the human eye by other detectors removes restriction to the visible region of the spectrum, increases quantitative capability,[1] and provides electrical signals which can be processed automatically by modern data-handling systems.

In current usage, the term *spectrophotometry* suggests measuring the extent to which radiant energy is absorbed by a chemical system as a function of the wavelength (or the frequency) of the radiation ("spectral scanning"), as well as absorption measurements at a fixed, predetermined wavelength. In order to understand spectrophotometry, we must review the terminology employed in characterizing electromagnetic radiation, consider in an elementary fashion the interaction of radiation with chemical species, and see in a general way what the instruments do.

[1] The eye of a person with good vision using good sample illumination is actually *very* sensitive to small differences in radiant power transmitted through two colored solutions observed side by side, but it is virtually useless on a quantitative basis for measuring the transmission through either solution alone.

14.1a. The Electromagnetic Spectrum

Various optical phenomena (e.g., reflection, refraction, and diffraction) are best interpreted in terms of the idea that light is propagated in the form of transverse waves. By appropriate measurements, these waves may be characterized with regard to wavelength, velocity, and the other terms used to describe any wave motion. In Fig. 14.1, it is indicated that the *wavelength* refers to the distance between two adjacent crests (or troughs) of the wave. The reciprocal of the wavelength, which is the number of waves in a unit length, is referred to as the *wave number*. The wave front is moving with a certain *velocity*. The number of complete cycles or waves passing a fixed point in a unit time is termed the *frequency*. The relationship of these properties is as follows:

$$\frac{1}{\lambda} = \overline{\nu} = \frac{\nu}{c}$$

where λ is the wavelength, $\overline{\nu}$ is the wave number, ν is the frequency, and c is the velocity of light.

Figure 14.1 Transverse wave.

The velocity of light is about 3×10^{10} cm/s. Various units are employed for wavelength, depending upon the region of the spectrum: For ultraviolet and visible radiation, the angstrom unit and the nanometer are widely used, while the micrometer is the common unit for the infrared region. A micrometer, μm, is defined as 10^{-6} m, and a nanometer, nm, is 10^{-9} m, or 10^{-7} cm. One angstrom unit (Å) is 10^{-10} m or 10^{-8} cm. Thus there are 10 Å in 1 nm. Wave number is often used by chemists as a frequency unit because it has convenient numerical values ($\overline{\nu}$ and ν are related by a constant factor, c, the velocity of light); the common unit of wave number is the reciprocal centimeter, cm^{-1}.

Luminous bodies such as the sun or an electric bulb emit a broad spectrum comprising many wavelengths. Those wavelengths associated with *visible light* are capable of affecting the retina of the human eye and hence give rise to the subjective impressions of vision. But much of the radiation emitted by hot bodies lies outside the region where the eye is sensitive, and we speak of the *ultraviolet* and *infrared* regions of the spectrum which lie on either side of the visible. The entire electromagnetic spectrum is classified approximately as shown in Fig. 14.2.

Within the visible region of the spectrum, persons with normal color vision are able to correlate the wavelength of light striking the eye with the subjective sensation of color, and color is indeed sometimes used for convenience in designating certain portions of the spectrum, as shown in the rough classification in Table 14.1.

We "see" objects by means of either transmitted or reflected light. When "white light," containing a whole spectrum of wavelengths, passes through a medium such as a colored glass or a chemical solution which is transparent to certain wavelengths but absorbs others, the medium appears colored to the observer. Since only the transmitted waves reach the eye, their wavelengths dictate the color of the medium. This color is said to be *complementary* to the color that

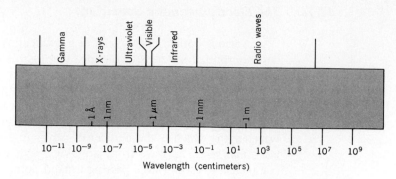

Figure 14.2 Approximate classification of the electromagnetic spectrum.

would be perceived if the absorbed light could be inspected, because the transmitted and absorbed light together make up the original white light. Similarly, opaque colored objects absorb some wavelengths and reflect others when illuminated with white light.

TABLE 14.1   Visible Spectrum and Complementary Colors

WAVELENGTH, nm	COLOR	COMPLEMENTARY COLOR
400–435	Violet	Yellow-green
435–480	Blue	Yellow
480–490	Green-blue	Orange
490–500	Blue-green	Red
500–560	Green	Purple
560–580	Yellow-green	Violet
580–595	Yellow	Blue
595–610	Orange	Green-blue
610–750	Red	Blue-green

14.1b.   The Interaction of Radiant Energy with Molecules

General Principles

We sometimes speak of "the dual nature of light." While the wave theory explains certain optical phenomena, there are other experimental results, such as the photoelectric effect, that are best interpreted in terms of the idea that a beam of light is a stream of particulate energy packets called photons. Each of these particles possesses a characteristic energy which is related to the frequency of the light by the equation

$$E = h\nu$$

where h is Planck's constant. Light of a certain frequency (or wavelength) is associated with photons, each of which possesses a definite quantity of energy. As explained below, it is the quantity of energy possessed by a photon which determines whether a certain molecular species will absorb or transmit light of the corresponding wavelength.

In addition to the ordinary energy of translational motion, which is not of concern here, a molecule possesses internal energy which may be subdivided into three classes. First, the molecule may be rotating about various axes and possess a certain quantity of *rotational energy*. Second, atoms or groups of atoms within the molecule may be vibrating, that is, moving periodically with respect to each other about their equilibrium positions, conferring *vibrational energy* upon the molecule. Finally, a molecule possesses *electronic energy,* by which we mean the potential energy associated with the distribution of negative electric charges (electrons) about the positively charged nuclei of the atoms.

$$E_{int} = E_{elec} + E_{vib} + E_{rot}$$

One of the basic ideas of quantum theory is that a molecule may not possess any arbitrary quantity of internal energy, but rather that it can exist only in certain "permitted" energy states. If a molecule is to absorb energy and be raised to a higher energy level, it must absorb a quantity appropriate for the transition. It cannot absorb an arbitrary quantity of energy determined by the experimenter and linger in an energy state intermediate between its permitted levels. This quantization of molecular energy, coupled with the concept that photons possess definite quantities of energy, sets the stage for selectivity in the absorption of radiant energy by molecules. When molecules are irradiated with many wavelengths, they will abstract from the incident beam those wavelengths corresponding to photons of energy appropriate for permitted molecular energy transitions, and other wavelengths will simply be transmitted.

The rotational energy levels of a molecule are quite closely spaced, as indicated schematically in Fig. 14.3. Thus pure rotational transitions require relatively little energy and are induced by radiation of very low frequency (long wavelength). It is in the far-infrared and "microwave" regions of the spectrum (wavelengths of perhaps 100 μm to 10 cm) that absorption of radiation is correlated with changes in rotational energy alone. Studies of absorption in this region have contributed fundamental information regarding molecular structure but have found relatively little application in analytical chemistry.[2]

Vibrational energy levels are farther apart (Fig. 14.3), and more energetic photons are required if absorption is to increase the vibrational energy of a molecule. Absorption due to vibrational transitions is seen in the infrared region of the spectrum, roughly from 2 to 100 μm. Pure vibrational changes are not observed, however, because rotational transitions are superimposed upon them. Thus a typical vibrational absorption spectrum is composed of complex bands rather than single lines. In practice, an infrared absorption spectrum consists not of discrete lines as might be supposed from Fig. 14.3, but rather of broad envelopes extending over wavelength spans, because of distortion of molecular energy levels by neighboring molecules and because of the inability of the instrument to resolve closely spaced lines.[3]

Absorption of visible light and ultraviolet radiation[4] increases the electronic energy of a molecule. That is, the energy contributed by the photons enables electrons to overcome some of the restraint of the nuclei and move out to new orbitals

---

[2] J. H. Goldstein, "Microwave Spectrophotometry," Chap. 62 of I. M. Kolthoff and P. J. Elving, eds., *Treatise on Analytical Chemistry*, Part I, Vol. 5, Wiley-Interscience, New York, 1964.

[3] Rotational "fine structure" can sometimes be observed in vibrational spectra, usually only partially resolved, particularly in gaseous samples where molecules perturb each other's energy levels only slightly. Examination of rotational lines in the vibrational spectrum of gaseous HCl is a common experiment in the physical chemistry laboratory.

[4] Many writers prefer that the word *light* be restricted to the visible region. It would then be improper to speak of "ultraviolet light" or "infrared light." For wavelengths to which the eye is unresponsive, *radiation* or *radiant energy* is preferred.

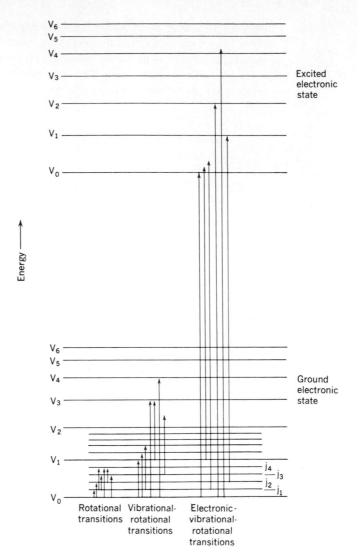

Figure 14.3 Schematic energy level diagram. Two electronic levels are shown: V_0, V_1, etc., are vibrational levels, and a few rotational levels represented by j-values are shown.

of higher energy. Vibrational and rotational effects are superimposed upon the electronic change, but the region where the absorption is found is determined by the electronic energy levels of the molecule. The vibrational and rotational changes introduce "fine structure" into the spectrum, so the absorption involves a band of wavelengths rather than a single line. The individual lines making up the band are usually not resolved under experimental conditions, and the observed visible or ultraviolet absorption spectrum generally consists of peaks exhibiting a smooth curvature.

Once the absorption has been recorded, the fate of the excited molecules is usually not of interest in ordinary spectrophotometry for analytical purposes, but we may briefly note, for the curious student, that the molecule tends not to remain in the excited state but rather to get rid of the excess energy. Commonly, the energy is degraded into heat by a stepwise process involving collisions with other molecules. (This heat is not noticeable in an ordinary spectrophotometric experiment.) Sometimes the energy is reemitted as radiation, usually of longer wavelength than was originally absorbed; this phenomenon is known as *fluorescence* (if there is a certain time delay in reemission, the term *phosphorescence* is used). Fluorescence can lead to errors in absorption measurements if the reemitted radiation reaches the detector of the instrument. Finally, in some cases the absorbed energy may cause the molecule to dissociate into free radicals or ions, which may then undergo chemical reactions.

Infrared (IR) Spectrophotometry

Infrared spectrophotometry is very important in modern chemistry, particularly (although not exclusively) in the organic area. It is a routine tool for detecting functional groups, identifying compounds, and analyzing mixtures. Instruments which record infrared spectra are commercially available and easy to use on a routine basis.

When we describe the structure of a molecule in terms of fixed bond lengths and bond angles, we are depicting a sort of average situation. Imagine a model of a complex molecule constructed of wooden balls connected by springs and suspended from a wire. Deliver a blow to the molecule, and it will become a quivering object with all its atoms in motion with respect to each other as perhaps dozens of springs stretch and compress and bend. This motion, which may at first appear hopelessly complicated, can be resolved into a series of individual vibratory modes whose natural frequencies depend upon the masses of the wooden balls and the characteristics of the springs. In a real molecule, analogous vibrations are occurring: pairs of atoms vibrating with respect to each other as the individual bonds lengthen and shorten, entire groups oscillating with respect to other atoms or groups, ring structures "breathing" (i.e., expanding and contracting), etc. Now, if there is an oscillating electric dipole associated with a particular vibratory mode, then there will occur an interaction with the electrical vector of electromagnetic radiation of this same frequency, leading to the absorption of energy which shows up as an increased amplitude of vibration.

TABLE 14.2 Some Infrared Group Frequencies

GROUP		FREQUENCY, cm^{-1}	WAVELENGTH, μm
OH	Alcohol	3580–3650	2.74–2.79
	H-bonded	3210–3550	2.82–3.12
	Acid	2500–2700	3.70–4.00
NH	Amine	3300–3700	2.70–3.03
CH	Alkane	2850–2960	3.37–3.50
	Alkene	3010–3095	3.23–3.32
	Alkyne	3300	3.03
	Aromatic	~3030	~3.30
C≡C	Alkyne	2140–2260	4.42–4.76
C=C	Alkene	1620–1680	5.95–6.16
	Aromatic	~1600	~6.25
C=O	Aldehyde	1720–1740	5.75–5.81
	Ketone	1675–1725	5.79–5.97
	Acid	1700–1725	5.79–5.87
	Ester	1720–1750	5.71–5.86
C≡N	Nitrile	2000–2300	4.35–5.00
NO₂	Nitro	1500–1650	6.06–6.67

Most groups, such as C—H, O—H, C=O, and C≡N, give rise to infrared absorption bands which vary only slightly from one molecule to another depending upon other substituents. Examples of these are shown in Table 14.2. In addition to these group frequencies, which can usually be definitely assigned, complex molecules may exhibit a myriad of absorption bands whose exact origins are difficult to ascertain but which are extremely useful for qualitative identification. Many of these occur in what is called the "fingerprint" region of the spectrum (ca. 6.5 to 14 μm). Figure 14.4 shows a sample infrared spectrum; note the richness of detail found even for a very simple molecule, and note how easy it is to spot the absorption band associated with the presence of a particular functional group.

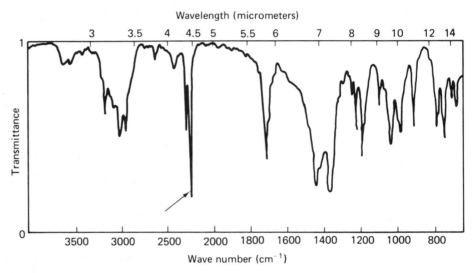

Figure 14.4 Infrared spectrum of acetonitrile (CH$_3$C≡N). Spectrum obtained on a thin film (0.005 mm) of the pure liquid. Arrow shows band due to the C≡N group.

Ultraviolet-Visible (UV-VIS) Spectrophotometry

The electronic spectra of compounds in the vapor phase sometimes show fine structure in which individual vibrational contributions are observed (Fig. 14.5), but in condensed phases, molecular energy levels are so perturbed by close neighbors that frequently only broad bands like that in Fig. 14.6 are seen. All molecules can absorb radiation in the UV-visible region because they contain electrons, both shared and unshared, which can be excited to higher energy levels. The wavelengths at which absorption occurs depend upon how firmly the electrons are bound in the molecule. The electrons in a single covalent bond are tightly bound, and radiation of high energy, or short wavelength, is required for their excitation. For example, alkanes, which contain only C—H and C—C single bonds, show no absorption above 160 nm. Methane shows a peak at 122 nm (Table 14.3) which is designated a σ-σ* transition. This means that an electron in a sigma-bonding orbital is excited to a sigma-antibonding orbital.

If a molecule contains an atom such as chlorine which has unshared electron pairs, a nonbonding electron can be excited to a higher energy level. Since nonbonding electrons are not so tightly bound as are sigma-bonding electrons, the absorption takes place at longer wavelengths. Note that such a transition occurs in

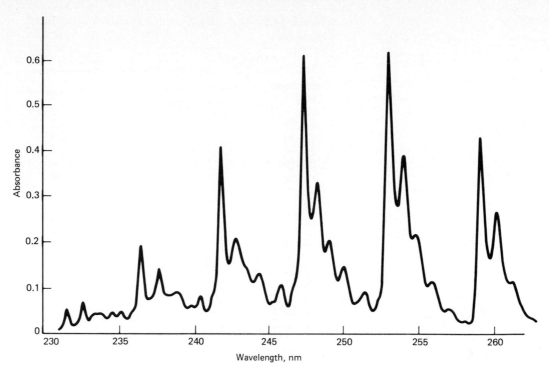

Figure 14.5 Portion of the ultraviolet absorption spectrum of benzene vapor. Benzene shows three π-π* transitions, of which this portion represents one; the others are near 184 and 204 nm. The individual peaks represent vibrational fine structure, much of which disappears as a result of molecular interactions when the benzene is dissolved in a solvent.

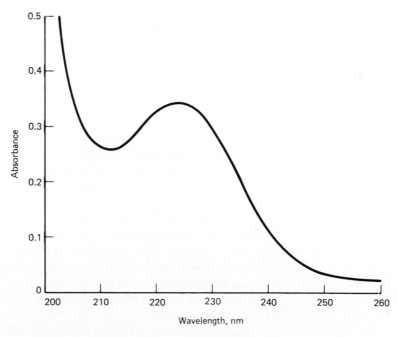

Figure 14.6 Ultraviolet absorption spectrum of hexadecyltrimethylammonium benzoate. Spectrum was obtained in a 1.00-cm cell with an 8.9×10^{-5} M solution of the salt in water; the band at 223.5 nm is due to the benzoate anion.

TABLE 14.3 Some Electronic Transitions in Organic Molecules

COMPOUND	WAVELENGTH, nm
Single bonds	
CH_4	122
$CH_3—CH_3$	135
CH_3Cl	173
CH_3Br	204
CH_3I	258
CH_3OH	184
CH_3OCH_3	184
Double bonds	
$CH_2=CH_2$	162
$—(CH=CH)_2—$	217
$—(CH=CH)_3—$	258
$—(CH=CH)_4—$	300
$—(CH=CH)_5—$	330
$(CH_3)_2C=O$	190, 280
$CH_3CH=CH—CHO$	217
$CH_2=CH—CH=CH—CHO$	263
Triple bonds	
$HC≡CH$	178
$HC≡N$	175

CH_3Cl at 173 nm (Table 14.3) and that the transitions in CH_3Br and CH_3I occur at even longer wavelengths. This is because the electrons are held less tightly by bromine and iodine. The transition described here is designated $n\text{-}\sigma^*$ to indicate that a nonbonding electron is raised to a σ-antibonding orbital.

Electrons in double or triple bonds are rather easily excited to higher pi-orbitals. A transition is designated $\pi\text{-}\pi^*$ when a pi-electron is raised from a pi-bonding orbital to a pi-antibonding orbital. The absorption of energy in such a transition is usually stronger than $\sigma\text{-}\sigma^*$ transitions. In conjugated molecules (i.e., those containing a series of alternating double bonds) the absorption is shifted to longer wavelengths, as can be seen in Table 14.3. Such molecules are described by writing resonance structures, saying that the electron is more "delocalized" than if it were confined to one bond between two atoms. The shift to longer wavelengths reflects the fact that the electron in a conjugated system is less tightly bound than one in a nonconjugated system.

It has been found that $\pi\text{-}\pi^*$ transitions in molecules which contain unsaturated groups are very similar irrespective of the atoms which make up the double bonds. Note (Table 14.3) that the absorptions in acetylene, $H—C≡C—H$, and hydrogen cyanide, $H—C≡N$, occur at about the same wavelengths. The same is true for the two conjugated systems $—C=C—C=C—$ and $—C=C—C=O$, both giving peaks around 217 nm. Many years ago the association between unsaturation and the absorption of light was recognized by organic chemists, and the term *chromophore* was introduced to describe the role of such groups as $C=C$, $C=O$, and $N=N$ in shifting the absorption of light toward the visible region.

Most applications of ultraviolet and visible spectrophotometry to organic compounds are based on $n\text{-}\pi^*$ or $\pi\text{-}\pi^*$ transitions and hence require the presence of chromophoric groups in the molecule. These transitions occur in the region of the spectrum (about 200 to 700 nm) which is convenient to use experimentally.

Commercial UV-VIS spectrophotometers usually operate from about 175 or 200 nm up to 1000 nm. The qualitative identification of organic compounds in this region is much more limited than in the infrared. This is because the absorption bands are broad and lacking in detail. However, certain functional groups, such as carbonyl, nitro, and conjugated systems, do show characteristic peaks, and useful information can often be obtained concerning the presence or absence of such groups in the molecule.

14.2

QUANTITATIVE ASPECTS OF ABSORPTION: THE BOUGUER-BEER LAW

Absorption spectra can be obtained using samples in various forms: gases, thin films of liquids, solutions in various solvents, and even solids. Most analytical work involves solutions, and we wish here to develop a quantitative description of the relationship between the concentration of a solution and its ability to absorb radiation. At the same time, we must realize that the extent to which absorption occurs will depend also upon the distance traversed by the radiation through the solution. As we have seen, absorption also depends upon the wavelength of the radiation and the nature of the molecular species in solution, but for the time being we may suppose that we can control these.

14.2a. *Mathematical Statement*

Bouguer's Law

The relationship between the absorption of radiation and the length of the path through the absorbing medium was first formulated by Bouguer (1729), although it is sometimes attributed to Lambert (1768). Let us subdivide a homogeneous absorbing medium such as a chemical solution into imaginary layers, each of the same thickness. If a beam of monochromatic radiation (i.e., radiation of a single wavelength) is directed through the medium, it is found that each layer absorbs an equal fraction of the radiation, or each layer diminishes the radiant power of the beam by an equal fraction. Suppose, for example, that the first layer absorbs half of the radiation incident upon it. Then the second layer will absorb half of the radiation incident upon *it,* and the radiant power emerging from this second layer will be one-fourth that of the original power; from the third layer, one-eighth, and so forth.

Bouguer's finding may be formulated mathematically as follows, where P_0 is the incident radiant power and P is the power emergent from a layer of medium b units thick:

$$-\frac{dP}{db} = k_1 P$$

The minus sign indicates that the power decreases with absorption. For the student who is unfamiliar with calculus, we may express this equation verbally as: The decrease in radiant power per unit thickness of adsorbing medium is proportional to the radiant power. For the student who has studied calculus, let us rearrange the preceding equation to

$$-\frac{dP}{P} = k_1 \, db$$

and integrate between limits P_0 and P and 0 and b:

$$-\int_{P_0}^{P} \frac{dP}{P} = k_1 \int_{0}^{b} db$$

$$-(\ln P - \ln P_0) = k_1 b$$

$$\ln P_0 - \ln P = k_1 b$$

$$\ln \frac{P_0}{P} = k_1 b$$

Usually the equation is written with base-10 logarithms, which simply changes the constant:

$$\log \frac{P_0}{P} = k_2 b$$

A verbal statement of this equation might be: The power of the transmitted radiation decreases in an exponential fashion as the thickness of the absorbing medium increases arithmetically. Some writers consider this integration step to be a "derivation" of Bouguer's law, but actually the two formulations are equivalent representations of what we are here taking as an experimental finding.

Bouguer's law appears to describe correctly, without exception, the absorption of monochromatic radiation by various thicknesses of a homogeneous medium. The student can convince himself that the law applies strictly only with monochromatic radiation by considering an extreme case. Pass two wavelengths through a medium, one of which is absorbed appreciably and the other not at all. According to Bouguer's law, if we allow the thickness of the medium to increase indefinitely, then the transmitted radiant power should approach zero. But it cannot fall to zero if an appreciable fraction is not absorbed at all.

It may be noted that Bouguer's law takes the same form as other familiar functions such as the rate expression for first-order kinetics or radioactive decay, and the compound interest law.

Beer's Law

The relationship between the concentration of an absorbing species and the extent of absorption was formulated by Beer in 1859. Beer's law is analogous to Bouguer's law in describing an exponential decrease in transmitted radiant power with an arithmetic increase in concentration. Thus

$$-\frac{dP}{dc} = k_3 P$$

which upon integration and conversion to ordinary logarithms becomes

$$\log \frac{P_0}{P} = k_4 c$$

Beer's law is strictly applicable only for monochromatic radiation and where the nature of the absorbing species is fixed over the concentration range in question. We shall comment further on this point in connection with so-called deviations from Beer's law.

Combined Bouguer-Beer Law

Bouguer's and Beer's laws are readily combined into a convenient expression. We note that, in studying the effect of changing concentration upon absorption, the path length through the solution would be held constant, but the measured results would depend upon the magnitude of the constant value. In other words, in Beer's law as written above, $k_4 = f(b)$. Similarly, in Bouguer's law, $k_2 = f(c)$. Substitution of these fundamental relationships into Bouguer's and Beer's laws gives

$$\log \frac{P_0}{P} = f(c)b \qquad \text{and} \qquad \log \frac{P_0}{P} = f(b)c$$

Bouguer   Beer

The two laws must apply simultaneously at any point, so

$$f(c)b = f(b)c$$

or, separating the variables,

$$\frac{f(c)}{c} = \frac{f(b)}{b}$$

Now, the only condition under which two functions of independent variables can be equal is that they both equal a constant:

$$\frac{f(c)}{c} = \frac{f(b)}{b} = K$$

or

$$f(c) = Kc \qquad \text{and} \qquad f(b) = Kb$$

Substitution into either the Bouguer or the Beer expression yields the same result:

$$\log \frac{P_0}{P} = f(c)b = Kbc$$

$$\log \frac{P_0}{P} = f(b)c = Kbc$$

14.2b Nomenclature and Units

Unfortunately, the development of the nomenclature regarding the Bouguer-Beer law has not been systematic, and a confusing array of terms appears in the literature. In analytical chemistry, the tendency in the United States has been to adopt recommendations of a Joint Committee on Nomenclature in Applied Spectroscopy, established by the Society for Applied Spectroscopy and the American Society for Testing Materials (ASTM).[5]

The symbols P_0 and P as used here are recommended for the incident and transmitted radiant powers, respectively.[6] The term $\log(P_0/P)$ is called the *ab-*

---

[5] For the report of this committee, see H. K. Hughes et al., *Anal. Chem.*, **24**, 1349 (1952).

[6] Many writers use I_0 and I for these terms, standing for *intensity* of the beam, but in ordinary spectrophotometers the quantity actually measured is the rate at which radiant energy is absorbed at the detector; this is best called "radiant power." Such units as watts or ergs per second might be employed, but in spectrophotometry we deal with a ratio (P/P_0) or the logarithm of a ratio $(\log P_0/P)$, and the units cancel.

sorbance and given the symbol *A*. Other terms which have been used synonymously with absorbance and which the student may encounter in the literature are *extinction, optical density,* and *absorbancy.*

The symbol *b* is accepted for the length of the path through the absorbing medium; it is ordinarily expressed in centimeters. Other writers have used the letter *l* for the same quantity, and, more rarely, the letter *d* or *t*.

Two different units for *c*, the concentration of absorbing solute, are often used, grams per liter and moles per liter. It is apparent that the value of the constant (designated *K* above) in the Bouguer-Beer law will depend upon which concentration system is used. When *c* is in grams per liter, the constant is called the *absorptivity*, symbol *a*. When *c* is in moles per liter, the constant is the *molar absorptivity*, symbol *ε*. Thus in the recommended system, the Bouguer-Beer law may take two forms:

$$A = abc_{\text{g/liter}} \qquad \text{or} \qquad A = \varepsilon bc_{\text{mol/liter}}$$

It is apparent that $\varepsilon = a \times \text{MW}$, where MW refers to the molecular weight of the absorbing substance in the solution. Other designations for *a* are *specific extinction, extinction coefficient, Bunsen coefficient,* and *specific absorption*. Similarly, some writers call *ε* the *molar extinction coefficient, molecular extinction,* and various other names.

The *transmittance*, $T = P/P_0$, is simply the fraction of the incident power which is transmitted by a sample. The *percent transmittance*, $\% T = P/P_0 \times 100$, is also encountered. If $A = \log (P_0/P)$ and $T = P/P_0$, then $A = \log (1/T)$. Since, from Beer's law, absorbance is directly proportional to concentration, it is clear that transmittance is not; $\log T$ must be plotted vs. *c* to obtain a linear graph. Figure 14.7 shows the situation. Analytical chemists prefer absorbance plots, but the student should be familiar with transmittance because it is encountered frequently. The detectors of most instruments generate a signal which is linear in transmittance, because they respond linearly to radiant power. Thus if an instrument is to be read in absorbance units, there must be a logarithmic scale on the readout device or the signal must be altered logarithmically by an electronic circuit or in some mechanical fashion.

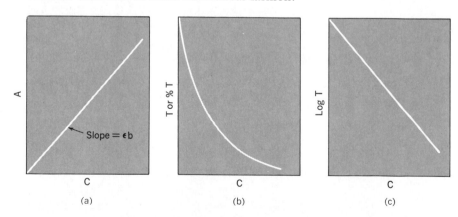

Figure 14.7 Appearance of Beer's law plots.

14.2c. Deviations from the Bouguer-Beer Law

According to the Bouguer-Beer law (or, as many writers say, simply Beer's law), a plot of absorbance vs. molar concentration will be a straight line of slope *εb*. Frequently, however, measurements on real chemical systems yield Beer's law plots which are not linear over the entire concentration range of interest. Such curvature suggests that ε is not a constant, independent of concentration, for such systems, but closer consideration leads to a somewhat more sophisticated view. The value of ε is expected to depend upon the nature of the absorbing species in solution and upon the wavelength of the radiation. Most deviations from Beer's law encountered in analytical practice are attributable to failure or inability to control these two aspects, and hence may be called *apparent* deviations because they reflect experimental difficulties more than any inadequacy of Beer's law itself.[7]

Chemically-Based Deviations

Consider, for example, absorbance measurements on a series of solutions of a weak acid, HB. The degree of dissociation of HB (fraction ionized) varies with the quantity of HB introduced into each solution if the final volumes are the same. Under this circumstance, it is possible to encounter either positive or negative Beer's law deviations, depending upon the ε-values of the two species, HB and B^-, at the wavelength employed. Since the fraction of the material present as B^- decreases with increasing analytical concentration of HB, a negative deviation from Beer's law will be seen if $\varepsilon_{B^-} > \varepsilon_{HB}$. On the other hand, if $\varepsilon_{HB} > \varepsilon_{B^-}$, a positive deviation will result. The system should follow Beer's law at a wavelength[8] where $\varepsilon_{HB} = \varepsilon_{B^-}$. These deviations from Beer's law may be circumvented not only by performing measurements at the isosbestic wavelength (which lowers the sensitivity because ε-values generally are not maximal here), but by adjusting all the solutions to a very low pH by addition of strong acid so as to repress the ionization of HB, or by addition of sufficient strong alkali to transform all the material into B^-.

As a second example, consider the spectrophotometric determination of chromium in an alloy based on absorption by Cr(VI). Two species of Cr(VI) are in equilibrium in aqueous solution:

$$2CrO_4^{2-} + 2H^+ \rightleftharpoons Cr_2O_7^{2-} + H_2O$$

Both chromate and dichromate are yellowish, but the colors are not identical; CrO_4^{2-} is a lemon yellow, while $Cr_2O_7^{2-}$ is more toward the orange, and the absorption spectra are different. Dilution favors formation of CrO_4^{2-} at the expense of $Cr_2O_7^{2-}$; i.e., the above equilibrium shifts with dilution, and the result may be a nonlinear plot of absorbance vs. total Cr concentration in the solution. Now note the participation of H^+ in the equilibrium: to obtain a linear plot, let the acid-

---

[7] There is another class of deviations which may be considered *real* rather than apparent, but they are not likely to be encountered in analytical chemistry. For example, it is shown in the theory of optics that ε for a substance in solution will change with changes in the refractive index of the solution. Since changes in refractive index attend concentration changes, Beer's law should not hold, even ideally. However, this effect is very small and is generally well within the experimental errors of spectrophotometry. Another real deviation from Beer's law sometimes occurs when relatively strong radiation passes through a medium containing only a few absorbing molecules. Under these conditions, all the molecules may be elevated to higher energy states by only a fraction of the available photons, and hence there will be no opportunity for further absorption regardless of how many more photons may be available. This situation, known as *saturation*, is ordinarily not encountered in analytical practice.

[8] A wavelength where two or more species in equilibrium with one another have the same ε-value is called an *isosbestic point*.

ity be high enough to force the reaction virtually completely toward $Cr_2O_7^{2-}$ or low enough for CrO_4^{2-} to predominate. Either one will result in a linear plot. [Usually the high acidity is preferred because the alloy is dissolved in acid to begin with and oxidation of chromium to Cr(VI) also requires acid.]

Many examples of this sort of Beer's law deviation are known. The general viewpoint here is that there is nothing wrong with Beer's law, that ε-values for individual species are constant over a wide concentration range, and that the deviations are predictable from a knowledge of the equilibria in which these species participate. Equilibria involving ions are often sensitive to added electrolytes, and failure to control the ionic strength may create problems in spectrophotometry. Temperature and various other factors may further complicate the situation.

Instrument-Based Deviations

Even with systems that are "well-behaved" chemically, deviations from Beer's law may occur because of characteristics of the instruments used in measuring absorbance values. In days past, such deviations sometimes resulted from fatigue effects in detectors, nonlinearity in amplifiers and readout devices, and instability in the sources of radiant energy. These problems have largely been solved in modern spectrophotometric instruments.

We pointed out earlier that the Bouguer-Beer law demands monochromatic radiation. Because ε-values depend upon wavelength, measured absorbance values reflect the wavelength distribution in the radiation, which, in a practical spectrophotometer, is never strictly monochromatic. Think again of an absorbing solution as a series of imaginary layers of equal thickness. Now if heterochromatic radiation passes through the first layer, the more strongly absorbed wavelengths are abstracted from the beam to a greater extent than the others. Thus the radiation impinging upon the second layer will be richer in the less strongly absorbed wavelengths, and the second layer will not absorb the same fraction of the radiation incident upon it as did the first layer. Since the Bouguer-Beer law states that each layer will absorb an equal fraction, deviation from the law will clearly result.[9]

14.3

INSTRUMENTATION FOR SPECTROPHOTOMETRY

A spectrophotometer is an instrument for measuring the transmittance or absorbance of a sample as a function of wavelength; measurements on a series of samples at a single wavelength may also be performed. Such instruments may be classified as manual or recording, or as single or double beam. In practice, single-beam instruments are usually operated manually and double-beam instruments generally feature automatic recording of absorption spectra, but it is possible to record a spectrum with a single-beam instrument. An alternative classification is based upon the spectral region, and we speak of infrared or ultraviolet spectrophotometers, etc. A complete understanding of spectrophotometers requires a detailed knowledge of optics and electronics which is far beyond the scope of this book. It is possible, though, for the student at this stage to understand what the

[9] A mathematical analysis of this type of deviation may be found in L. Meites and H. C. Thomas, *Advanced Analytical Chemistry*, McGraw-Hill Book Company, New York, 1958, p. 255.

instruments do. By combining this general, fundamental understanding with detailed instructions furnished in the form of manuals by the manufacturers, the chemist can obtain good data with modern spectrophotometers. Manually operated single-beam spectrophotometers will be discussed first, because this provides the background for appreciating the capabilities of the more complex instruments.

14.3a. Single-Beam Instruments

The essential components of a spectrophotometer are shown in the block diagram in Fig. 14.8. Both single- and double-beam spectrophotometers, and instruments which operate in various spectral regions, have these essential components, although the details are quite different in the several cases. In accord with the goal set forth above, we discuss these components briefly.

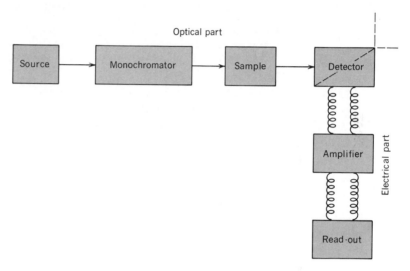

Figure 14.8 Block diagram showing components of a single-beam spectrophotometer. Arrows represent radiant energy, coiled lines electrical connections. The optical part and the electrical part of the instrument meet at the detector, a transducer which converts radiant energy into electrical energy.

Sources

The usual source of radiant energy for the visible region of the spectrum, as well as the near infrared and near ultraviolet, is an incandescent lamp with a tungsten filament. Under ordinary operating conditions, the output of this tungsten lamp is adequate from about 325 or 350 nm to about 3 μm. The energy emitted by the heated filament varies greatly with wavelength, as shown in Fig. 14.9. The energy distribution is a function of the temperature of the filament, which depends in turn upon the voltage supplied to the lamp; an increase in operating temperature increases the total energy output and shifts the peak of Fig. 14.9 to a shorter wavelength. (Practically, this cannot be exploited to provide more ultraviolet radiation because there is a tradeoff in the lifetime of the lamp filament at higher temperatures.) Thus the voltage to the lamp should be a stable one; a regulated power supply is incorporated into the instrument. The heat from a tungsten lamp may be troublesome; often the lamp housing is waterjacketed or cooled with a blower to prevent warming of the sample or of other instrument components.

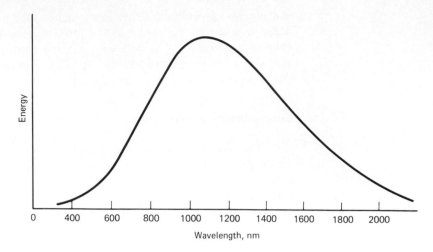

Figure 14.9 Output of a tungsten incandescent lamp as a function of wavelength. The sketch is approximately correct for the temperature of the lamp filament in a typical spectrophotometer (~2600 to 3000 K); note how little energy is available in the ultraviolet and infrared regions (except for the near infrared).

Below about 350 nm, the output of a tungsten lamp is inadequate for spectrophotometers, and a different source must be used. Most common is a hydrogen (or deuterium) discharge tube, which is used from about 175 to 375 or 400 nm. When a discharge between two electrodes excites emission by a sample of a gas such as hydrogen, a discontinuous line spectrum characteristic of the gas is obtained provided the pressure is relatively low. As the hydrogen pressure is increased, the lines broaden and eventually overlap, until at relatively high pressures, a continuous spectrum is emitted. The pressure required in a hydrogen discharge tube is lower than that with certain other gases; also, the tube runs cooler. The envelope is usually glass, but a quartz window is provided to pass the ultraviolet radiation. A high-voltage power supply is required for gaseous discharge tubes. In some spectrophotometers, provision is made for interchanging tungsten and hydrogen discharge sources in order to cover the visible and ultraviolet regions through which the instruments operate.

The source for infrared spectrophotometers, which commonly operate from about 2 to 15 μm, is usually the *Nernst glower*. This is a small rod of ceramic appearance fabricated from a special mixture of metal oxides, with platinum leads sealed into the ends. The rod is nonconducting at room temperature, but it is brought into a conducting state when heated, after which a flow of current maintains a glow which is rich in infrared radiation.

Monochromators

This is an optical device for isolating from a continuous source a beam of radiation of high spectral purity of any desired wavelength. The essential components of a monochromator are a slit system and a dispersive element. Radiation from the source is focused upon the entrance slit, then collimated by a lens or mirror so that a parallel beam falls upon the dispersing element, which is either a prism or a diffraction grating. By mechanically turning the prism or grating, various portions of the spectrum produced by the dispersive element are focused on the exit slit, whence, by a further optical path, they encounter the sample.

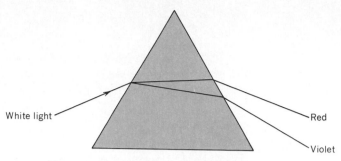

Figure 14.10   Dispersion of white light by a prism.

The student may recall from elementary physics the action of a prism in dispersing white light into a spectrum. When a beam of light passes through the interface between two different media, such as air and glass, bending takes place, which is called *refraction*. The extent of the bending depends upon the index of refraction of the glass. This index of refraction varies with the wavelength of the light; the blues are bent more than the reds, as shown in Fig. 14.10. As a result of the variation of refractive index with wavelength, the prism is able to disperse or spread out a beam of white light into a spectrum in which the various colors making up the white light may be recognized separately. Infrared and ultraviolet radiation are dispersed in the same manner, but here the words *light* and *color* are not used and the prism material is not glass. The material of choice represents a compromise between dispersive power and transparency in the desired wavelength region, along with several other factors. Spectrophotometers covering mainly the visible region of the spectrum have glass prisms, whereas quartz is the prism material for instruments covering the ultraviolet and near infrared as well as the visible; infrared spectrophotometers commonly have prisms of rock salt.

The spectral purity of the emergent radiation from the monochromator depends upon the dispersive power of the prism and the width of the exit slit. At first thought, one might suppose that monochromaticity could be approached as closely as desired by merely decreasing the slit width sufficiently, but this is not the case. Eventually, the slit becomes so narrow that diffractive effects at its edges create only a loss of radiant power with no increase in spectral purity (this is the so-called Rayleigh diffraction limit); actually, before this limit is approached in a typical spectrophotometer, the narrowed slit is passing insufficient energy to activate the detector.

With prism monochromators, a given slit width does not yield the same degree of monochromaticity throughout the spectrum. The wavelength dependence of the dispersion of a prism is such that the wavelengths in the spectrum are not spread out uniformly. The dispersion is greater for the shorter wavelengths, and hence wider slits may here achieve the same degree of spectral purity as would narrower slits at longer wavelengths.

Figure 14.11 is a schematic diagram of the optical system of a particular single-beam spectrophotometer with a quartz prism. The back of the prism is coated with a reflective metallized surface so that the radiation passes twice through the dispersive element. This not only enhances the dispersion but also is of great geometric convenience.

A diffraction grating (reflection) is made by ruling on a polished metal surface, such as aluminum, a large number of parallel lines. For the infrared region there are about 1500 to 2500 lines/in.; for the ultraviolet and visible regions,

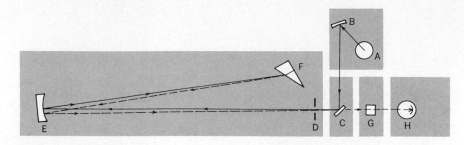

Figure 14.11 Schematic diagram of optical system of Beckman model DU spectrophotometer. *A*, Light source; *B*, *C*, mirrors; *D*, slit; *E*, collimating mirror; *F*, quartz prism with reflecting back surface; *G*, cell; *H*, phototube. (Courtesy of Beckman Instruments, Inc.)

about 15,000 to 30,000 lines/in. When light is reflected from this surface, that which strikes the rulings is dissipated by scattering; the unruled portions reflect regularly, acting as individual light sources. Overlapping of the waves from these sources establishes an interference pattern which results in the dispersion of the reflected light into its component wavelengths. The student is referred to elementary physics texts for a full explanation of this phenomenon.

The machines for ruling the lines on a grating must be constructed to very close tolerances, and original gratings are expensive. Much cheaper, and much more widely used, are *replica gratings,* large numbers of which can be prepared from a single master grating. The original is coated with a plastic material, which, after hardening, is stripped off to yield a replica. The plastic is made reflective by evaporating a film of metal, generally aluminum, onto the ruled face; the grating is mounted in the monochromator in such a way that rotation allows various portions of the spectrum to illuminate the exit slit.

Gratings differ from prisms in rendering a uniform dispersion throughout the entire spectrum; in other words, a single slit width yields the same degree of monochromaticity of the emergent radiation throughout the spectrum. Figure 14.12 shows the optical path through a widely used grating instrument.

As mentioned earlier, the radiation emergent from a monochromator is not monochromatic, although it is much more nearly so than is the original source. The wavelength distribution is somewhat as shown in Fig. 14.13. The terminology employed in describing the width of the band shown in the figure is not entirely standard, and advertisements for instruments often quote figures for

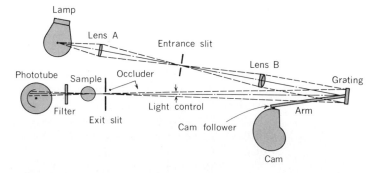

Figure 14.12 Schematic diagram of optical system of Spectronic 20. (Courtesy of Bausch and Lomb, Inc.)

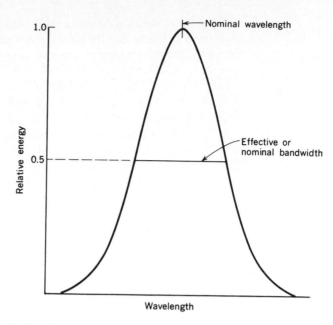

Figure 14.13 Wavelength distribution of the energy emergent from a monochromator.

"bandwidth" without specifying what is meant. A particular terminology which is widely understood is shown in the figure.

A problem in monochromators is so-called stray light, by which is meant radiation of unspecified wavelengths which is reflected about inside the monochromator and which may find its way to the exit slit. In good instruments, this is minimized by using dull black surfaces and by inserting baffles in appropriate positions. It is cut to an extremely low level in the finer instruments which employ double monochromators, generally combining both prism and grating. With ordinary instruments, spurious absorbance readings due to stray light may be obtained in spectral regions where very little energy of the desired wavelengths is available.

Until quite recently, instruments without true monochromators were widely used for absorbance measurements, mainly in the visible region, in laboratories where a low initial investment, simplicity, and speed were more important than the quality of the results. These instruments, designated *filter photometers,* utilized colored-glass filters to isolate fairly broad wavelength bands from the source. They served admirably for many routine analyses, but they have been largely displaced by inexpensive grating spectrophotometers.

Cells

Most spectrophotometry involves solutions, and thus most sample containers are cells for placing liquids in the beam of the spectrophotometer. The cell must transmit radiant energy in the spectral region of interest; thus glass cells serve in the visible region, quartz or special high-silica glasses in the ultraviolet, and rock salt in the infrared. It must be remembered that the cell, which in a sense is merely a container for the sample, is actually more than this; when in position, it becomes part of the optical path through the spectrophotometer, and its optical properties are important. In less expensive instruments, cylindrical test tubes are sometimes used as sample containers. It is important that such tubes be positioned reproducibly by marking one side of the tube and facing the mark in the same direction whenever the tube is placed in the instrument. The better cells have flat optical surfaces. The cells must be filled so that the light beam goes through the solution, with the meniscus entirely above the beam. Cells are generally held in position by kinematic design of the holder or by spring clips which ensure reproducible positioning in the cell compartment of the instrument.

Typical visible and ultraviolet cells have path lengths of 1 cm, but a wide variety is available, ranging from very short paths, fractions of a millimeter, up to 10 cm or even more. Special microcells may be obtained, by means of which minute volumes of solution yield an ordinary path length, and adjustable cells of variable path length are also available, particularly for infrared work. The variety of infrared cells currently on the market is beyond the scope of this discussion. Problems in the infrared are different from those in the ultraviolet and visible regions. Because solvents which are infrared-transparent are not available, the tendency is to run concentrated solutions at short path lengths (0.1 mm or even less) to minimize absorption by the solvent, and the cells are thus quite different from those employed at shorter wavelengths.

Detectors

In a detector for a spectrophotometer, we desire high sensitivity in the spectral region of interest, linear response to radiant power, a fast response time, amenability to amplification, and high stability or low "noise" level, although in practice it is necessary to compromise among these factors. Higher sensitivity, for example, can be bought only at the expense of increased noise. The types of detection that have been most widely used are based upon photochemical change (mainly photographic), the photoelectric effect, and the thermoelectric effect. Photography is no longer used in ordinary spectrophotometry; generally speaking, photoelectric detectors are employed in the visible and ultraviolet regions, and detectors based upon thermal effects are used in the infrared.

The commonest photoelectric detector is the *phototube*. This is an evacuated envelope, with a transparent window, containing a pair of electrodes across which a potential is maintained. The surface of the negative electrode is photosensitive; that is, electrons are ejected from this surface when it is irradiated with photons of sufficient energy. The electrons are accelerated across the potential difference to the positive electrode, and a current flows in the circuit. Whether or not electrons are emitted depends upon the nature of the cathode surface and the frequency of the radiation; the number of electrons emitted per unit time, and hence the current, depends upon the radiant power. A variety of phototubes are available which differ in the material of the cathode surface (also in the transparent window) and hence in their response to radiation of various frequencies. Some spectrophotometers provide for interchanging detectors so as to maintain a good response over a broad wavelength range.

Photomultiplier tubes are more sensitive than ordinary phototubes because of high amplification accomplished with the tube itself. Such a tube has a series of electrodes, each at a progressively more positive potential than the cathode. The geometry of the tube is such that the primary photoelectrons are focused into a beam and accelerated to an electrode which is, say, 50 to 90 V more positive than the cathode. The bombardment of this electrode (or *dynode*, as it is called) releases many more secondary electrons which are accelerated to a third, more positive, electrode, and so on, for perhaps ten stages. A regulated high-voltage power supply, furnishing about 500 to 900 V, is required to operate the tube. The output of the photomultiplier is still further amplified with an external electronic amplifier. The enhanced sensitivity of this detector permits narrower slit widths in the monochromator and hence better resolution of spectral fine structure.

The common infrared detector is the thermocouple. The student may recall the thermoelectric effect: If two dissimilar metals are joined at two points, a potential is developed if the two junctions are at different temperatures. Heating of one of the junctions by the infrared radiation is thus the basis of detection. This junction is specially designed to have a low heat capacity so that it will be warmed appreciably by radiant energy of the low power encountered in the instrument.

Amplification and Readout

It is beyond the scope of this text to discuss the detailed electronics of amplification and readout as they are accomplished in various spectrophotometers. To give an idea of what may be involved, we may briefly consider one possibility. Let us place a large load resistor in series with a phototube, as shown schematically in Fig. 14.14. Suppose the radiant power supplied to the cathode is such that

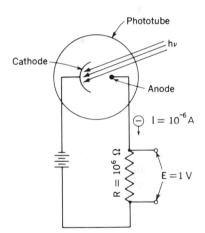

Figure 14.14 Simple phototube circuit (see text).

a current of 1 μA (10^{-6} A) flows in the circuit. If the resistor has a value of 1 MΩ (10^6 ohms) as shown in the figure, then according to Ohm's law the voltage across the resistor, $E = iR$, is $10^{-6} \times 10^6 = 1$ V. Although 1 V is a fair voltage, it cannot be measured by connecting an ordinary voltmeter across the resistor. As soon as such a connection is made, the meter becomes part of the circuit, establishing a parallel shunt around the resistor. Since the resistance of a typical voltmeter is very low as compared with 10^6 ohms, most of the current by passes the large resistance and flows through the meter, and the voltage across the resistor, although measured correctly as it is, is no longer 1 V, but perhaps only a few millivolts.

This problem is solved by using an amplifier of high input resistance so that the phototube circuit is not drained. Rather, the voltage across the load resistor is used to *control* a circuit which draws its power from an independent source and which has an output large enough to operate a meter or other readout device.

14.3b   *Operation of a Single-Beam Spectrophotometer*

Ordinary Operation

Typically, there is an opaque shutter, controlled by the operator, which may be placed in front of the phototube so that the tube is in darkness. With this shutter in position, a small current ("dark current") flows in the phototube circuit due to thermal emission of electrons by the cathode or perhaps a small leakage in the tube. By means of a knob on the instrument, the operator cancels out the dark current and sets the scale on the instrument to read infinite absorbance (zero transmittance). Next, with the wavelength set at the desired value and a cell containing a reference solution in the beam (the reference may be the pure solvent, a "blank" from an analytical procedure, etc.), the shutter is removed to expose the detector. Now, by adjusting the radiant power to the detector by means of the monochromator slit control, and/or by changing electronically the gain of the amplifier, the instrument scale is set to read zero absorbance (100% transmittance). With the scale thus established, the sample solution is placed in the beam and its absorbance or transmittance is read. (The scale is generally linear in transmittance, but most instruments have an absorbance scale alongside the transmittance scale and either one can be read.)

The scale, set up as described above, must be reestablished whenever the wavelength is changed in order to compensate for the variation in source output with wavelength and the wavelength dependence of the detector response, as well as any absorption by the reference solution or the cell. It is good practice to check the dark current and reference solution settings frequently because of possible drift in the circuit and in the output of the source. Usually two cells are used, one for the reference solution and one for the samples to be measured; it is obvious that these cells should be matched with regard to path length and optical qualities.

Differential Spectrophotometry

Ordinarily, the reference solution is the pure solvent or a "blank" solution of some kind which contains little or none of the analyte. Using this reference and the dark-current adjustment, the operator sets up a scale as described above and shown in the upper part of Fig. 14.15. Also shown along the scale are the absorbance and transmittance values of two solutions, one an unknown which is to be measured and the other a standard solution containing a known quantity of the substance being determined.

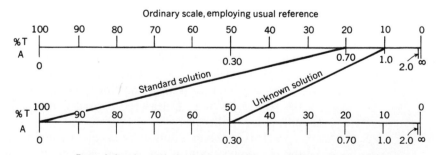

Figure 14.15   Scale expansion in differential spectrophotometry.

We must recognize that the instrument does not know anything about the sort of solution that is in the beam; it is capable only of producing a reading of zero absorbance (100% T) when a certain radiant power falls upon the detector and the amplification of the electronic circuit is appropriate. Thus the instrument can be set to read zero absorbance with a strongly absorbing solution, instead of the usual reference, in the beam by opening the slits of the monochromator and/or increasing the gain of the amplifier. Suppose, then, that we place the standard solution shown in Fig. 14.15 in the beam and set the absorbance reading at zero. As we have attempted to show in the figure by the lower scale and the lines tying it to the upper scale, we have now accomplished essentially a scale expansion. What was only a small portion of the upper scale becomes a much larger portion of the scale. Note that the difference in absorbance of the two solutions is 0.30 unit in each case, but 0.30 unit is a much larger portion of the lower scale than the upper. Hence we may expect that a constant instrumental error will result in a smaller relative concentration error. Actually, in some cases the error can be reduced to as little as a part per thousand, and the spectrophotometric measurement, normally not this good, can compete with ordinary titrimetric and gravimetric techniques, which are usually considered more precise. Several applications of differential spectrophotometry have been given by Bastian,[10] and a detailed mathematical analysis of the technique and its errors has been developed by Hiskey.[11]

To see more clearly how the error is reduced in differential spectrophotometry, let us consider an example. Suppose that we wish to determine copper by measuring the absorbance of blue copper(II) solutions. Let us simply assume, to establish a specific basis for our discussion, that a solution with a copper concentration of 2 mg/ml (solution A) can be measured against a pure water reference with an error of 1%. Now consider the error if a solution containing 20 mg of copper(II)/ml (solution B) is measured, using as a reference not water, but another cupric solution containing 18 mg of copper(II)/ml (solution C). The concentration difference between B and C is the same as the difference between solution A and water. Thus, if Beer's law is obeyed, the foregoing two measurements will give rise to the same absorbance value and the same error. In other words, if solution B were our unknown solution, we could determine how much it differed from solution C with a 1% error. But we know *accurately* the concentration of solution C, because it is a carefully prepared reference solution (by *accurately,* we mean that no spectrophotometric error is involved). Thus so far as errors in spectrophotometry are concerned, the concentration of solution B can be determined with an error of only 0.1%. (In measuring solution A vs. water, the error is $0.02/2 \times 100 = 1\%$; if B is measured vs. C, the error is still $0.02/2 \times 100 = 1\%$, but the error in the absolute concentration of B is only $0.02/20 \times 100 = 0.1\%$.)

The differential approach not only leads to lower errors, as explained above, but also permits the extension of spectrophotometry to the analysis of solutions which would be too highly absorbing for ordinary measurements.

It is possible to achieve even greater precision by setting both ends of the scale with standard solutions in the beam. In other words, the 100% T is set as

[10] R. Bastian et al., *Anal. Chem.*, **21**, 972 (1949); **22**, 160 (1950); **23**, 580 (1951).

[11] C. F. Hiskey et al., *Anal. Chem.*, **21**, 1440 (1949); **22**, 1464 (1950); **23**, 506 (1951); **23**, 1196 (1951); **24**, 342 (1952).

described above using a standard solution more dilute than the unknown, but the 0% T is set not with a shutter in the beam, but with a more concentrated standard solution. Reilley and Crawford, whose definitive paper[12] should be consulted for further details, refer to this as the "method of ultimate precision."

14.3c. Double-Beam Instruments

Single-beam instruments are generally operated manually as described in Section 14.3b. The first step toward automation was the development of UV-VIS and IR instruments that recorded absorption spectra. Once the operator set up the conditions, the spectral scan was automatically performed and the absorption spectrum displayed as a pen-and-ink plot on graph paper. With the more modern, computer-based instruments, electrical signals from the detector are processed, converted into the digital domain, and subjected to programs that permit storage, video display, calculations, comparisons with stored data, etc. The details of all this are clearly far outside the scope of an introductory textbook, but we can describe some general aspects.

As with the simplest instruments, we wish to compare the radiant power P reaching the detector through the sample solution with the power P_0 transmitted by a reference solution. One approach is depicted schematically in Fig. 14.16. The two beams to be compared come from one source and monochromator. This beam is directed alternately through the reference and the sample by a rotating mirror that allows passage first in one direction, then in another. The two beams thus pulsate out of phase as the mirror continuously rotates, and the detector sees first one beam and then the other, over and over again. The alternating electrical signals are processed, digitized, and compared, and the programmed calculations

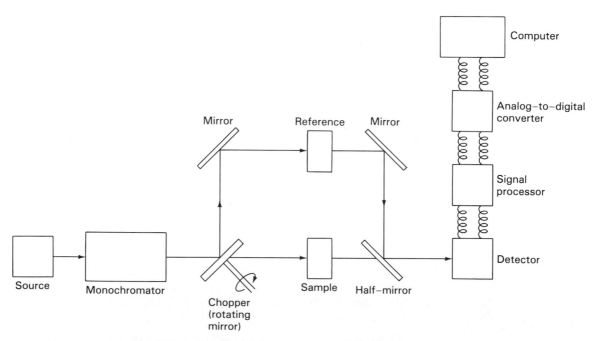

Figure 14.16 Schematic diagram of a possible double-beam spectrophotometer.

[12] C. N. Reilley and C. M. Crawford, *Anal. Chem.*, **27**, 716 (1955).

are performed. The monochromator is set for a motor-driven wavelength scan or for measurements at a selected wavelength by the operator. We may add that a person using one of these complex instruments with no more knowledge than provided by the above sketch should, when all else fails, read the manual.

14.3d. *Errors in Spectrophotometry*

Errors in spectrophotometric measurements may arise from a host of causes, many of which can be countered by care and common sense. Sample cells should be clean. Certain substances, e.g., proteins, sometimes adsorb very strongly to glass and other siliceous materials and are washed out of cells only with difficulty. Fingerprints may absorb ultraviolet radiation. The positioning of the cells in the beam must be reproducible. Gas bubbles must not be present in solutions. The wavelength calibration of the instrument should be checked occasionally, and drift or instability in electronic circuits must be corrected. It must not be assumed that Beer's law holds for an untested chemical system. Sample instability may lead to errors if the measurements are not carefully timed.

The concentration of the absorbing species is important in determining the error. It is intuitively reasonable that the analyte solution should not absorb too little or too much of the radiation. (Imagine your eye as a detector and pretend you are trying to decide something about a solution that looks almost like pure water or a solution that is so concentrated that it looks black; some intermediate concentration would certainly be better.) For the simplest case of a single-beam instrument, on the assumption that the source of error is a constant level of uncertainty in visually reading the position of a pointer on a meter scale which is linear in transmittance, an expression can be derived from Beer's law which shows where a minimal error in analyte concentration occurs.

Recall that

$$A = \log \frac{P_0}{P} = \frac{1}{2.3} \ln \frac{P_0}{P} = \varepsilon bc$$

Let the relative error in concentration be $dc/c = dA/A$. We want to obtain an expression for dc/c and then inquire where this expression has a minimum. Differentiating Beer's law, $A = (1/2.3) \ln (P_0/P)$, we obtain

$$dA = \frac{1}{2.3} d \ln \frac{P_0}{P} = \frac{(-P_0/P^2)\, dP}{2.3(P_0/P)}$$

Dividing numerator and denominator by P_0/P yields

$$dA = -\frac{(1/P)\, dP}{2.3} = -\frac{dP}{2.3P}$$

Dividing both sides by A, we obtain

$$\frac{dA}{A} = -\frac{dP}{2.3PA}$$

From Beer's law, $P = P_0 \times 10^{-A}$; substitution into the above equation gives

$$\frac{dA}{A} = \frac{dP}{2.3AP_0 \times 10^{-A}} = \frac{dc}{c}$$

It is convenient to normalize the equation by setting $P_0 = 1$, corresponding to the customary actual operation of setting the instrument to 100% T or zero absorbance with a reference solution in the beam. This gives

$$\frac{dc}{c} = -\frac{dP}{2.3A \times 10^{-A}}$$

The minimum in dc/c occurs when the term $A \times 10^{-A}$ is at a maximum. To find this maximum, we differentiate and set the derivative equal to zero:

$$\frac{d(A \times 10^{-A})}{dA} = 10^{-A} - 2.3A \times 10^{-A} = 0$$

or

$$10^{-A}(1 - 2.3A) = 0$$

If 10^{-A} is zero, A is infinite, and the error is infinite. Setting the other term equal to zero yields

$$1 - 2.3A = 0$$

$$2.3A = 1$$

$$A = \frac{1}{2.3} = 0.43$$

An absorbance value of 0.43 corresponds to 36.8% transmittance. (The student who is familiar with calculus may notice that the absorbance of 0.43 could, so far as we have actually shown above, represent either a maximum or a minimum error. Such a student will know that the first derivative may be tested for this; it turns out that we are dealing with a minimum error.)

The term dP in the equations above, following the usual practice in calculus, may be taken as an approximation of ΔP, the error in P. This is often called the *photometric error,* and for our purposes simply represents the uncertainty in reading the instrument scale. This uncertainty is considered constant in the present discussion, and it is probably roughly so with so many actual instruments. To find the relative error in concentration as a function of the photometric error at the optimal concentration, substitute 0.43 for A in the preceding equation for dc/c:

$$\frac{dc}{c} = -\frac{dP}{2.3 \times 0.43 \times 10^{-0.43}} = -2.72 \, dP$$

Thus, if a 1% error were made in reading the instrument, the relative error in c would be 2.72% at best; with absorbance values above and below 0.43, it would be even larger. Photometric errors may range from 0.1% to several percent, depending upon the instrument employed.

The relative error in concentration resulting from a 1% photometric error is plotted against percent transmittance in Fig. 14.17. The curve approaches infinity at both 0 and 100% T, passes through a minimum at 36.8% T, but is actually not very far from minimal over a fair range, say, 10 to 80% T (absorbance values of about 0.1 to 1.0).

In the error treatment above, it was supposed that the error in measuring transmittance was constant, independent of the value of the transmittance; the error was considered to arise entirely from uncertainty in reading the instrument scale. In some of the best modern instruments, on the other hand, the limiting

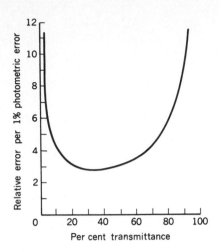

Figure 14.17 Error curve.

factor in the accuracy lies elsewhere, usually in the noise level of the detector circuit. In such cases, dP is not constant, and a different error function is obtained which is minimal not at 36.8% T, but at a lower T value. Actually, with a complex instrument it may be difficult to decide which of several factors limits the accuracy. Thus the way in which dP varies with P may not be clear, and it may not be legitimate to calculate a % T value corresponding to minimal error. With modern, high-quality UV-visible instruments, the smallest relative error probably occurs when the absorbance is between about 0.85 and a little over 1.

14.3e. *Plotting Spectrophotometric Data*

Absorption spectra are most frequently plotted as % T vs. wavelength (λ), A or ε vs. λ, and log A or log ε vs. λ. Comparison of these plots may be made clear by reference to Figs. 14.18, 14.19, and 14.20. Analytical chemists generally prefer absorbance to % T for the ordinate. Note that a minimum in % T corresponds to a maximum in A. The two curves are not mirror images, however, because A and % T are related logarithmically [$A = \log(1/T)$]. Sometimes ε-values are calculated from absorption data and plotted against λ.

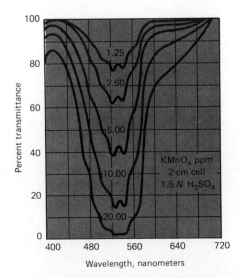

Figure 14.18 Transmittance-wavelength curves for solutions of potassium permanganate. (M. G. Mellon, ed., *Analytical Absorption Spectroscopy*, John Wiley & Sons, Inc., New York, 1950. Used by permission of the author and publisher.)

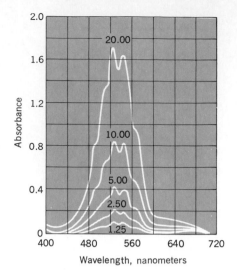

Figure 14.19 Absorbance-wavelength curves for solutions of potassium permanganate. (M. G. Mellon, ed., *Analytical Absorption Spectroscopy*, John Wiley & Sons, Inc., New York, 1950. Used by permission of the author and publisher.)

It may be seen from Fig. 14.19 that the shape of the absorption spectrum depends upon the concentration of the solution if the ordinate is linear in absorbance. That is, the curves in Fig. 14.19 are not superimposable by simple vertical displacement. This is clear from Beer's law, $A = \varepsilon bc$, which shows that changing the concentration changes the absorbance at each wavelength by a constant *multiple*. On the other hand, as seen in Fig. 14.20, the shape of the curve is independent of concentration if the ordinate is log A. That this should be the case is seen by taking logarithms of both sides of the Beer's law equation:

$$\log A = \log (\varepsilon bc) = \log \varepsilon + \log b + \log c$$

Now the concentration term is *added* rather than multiplied, and hence increasing the concentration adds a constant increment to log A at each wavelength across the spectrum. The curve for the higher concentration is thus displaced upward, but could be superimposed upon the lower one by a simple vertical movement. The same ε vs. λ plot should be obtained regardless of concentration provided the system follows Beer's law at all wavelengths. It is common practice, particularly among organic chemists, to plot log ε vs. λ.

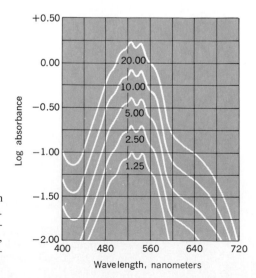

Figure 14.20 Log absorbance-wavelength curves for solutions of potassium permanganate. (M. G. Mellon, ed., *Analytical Absorption Spectroscopy*, John Wiley & Sons, Inc., New York, 1950. Used by permission of the author and publisher.)

APPLICATIONS OF SPECTROPHOTOMETRY

14.4a. *Identification of Chemical Substances*

The student is familiar with simple color tests which are used for identification purposes. The purple color of permanganate solutions, the blue of copper, the yellow of chromate, and many others might be mentioned. The absorption spectrum of a compound, determined with a spectrophotometer, may be considered as a more elegant, objective, and reliable indication of identity. The spectrum is another physical constant, so to speak, which, along with melting point, refractive index, and other properties, may be used for characterization. Like the others, absorption spectra are not infallible proof of identity, but simply represent another tool available for intelligent application.

It must be remembered that an absorption spectrum depends upon not only the chemical nature of the compound in question, but also other factors. Changing the solvent often results in shifts in absorption bands. The shape of a band and particularly the appearance of fine structure may well depend upon instrument characteristics such as the resolution of the monochromator, the amplifier gain, and the rate of scan as it relates to inertia in the recorder. Treating a recording spectrophotometer as a "black box" can lead to peculiar absorption spectra.

Spectra of many thousands of compounds and materials have been recorded, and locating the proper ones for comparison in connection with a particular problem may be difficult. Several catalogs and compilations are available. Increasingly, large laboratories are employing computerized data-handling techniques to store and retrieve spectra as well as other important information.

There is a large body of empirical data in the literature showing the effects of substituents upon the wavelengths of absorption bands in the spectra of parent molecules. Spectra-structure correlations in both UV-VIS and IR regions are frequently very useful in the identification of unknown compounds, especially if other information about the sample can place the search on the right track in the first place.

14.4b. *Multicomponent Analysis*

A spectrophotometer cannot *analyze* a sample. It becomes a useful tool only after the sample has been treated in such a way that the measurement is interpretable in unambiguous terms. In many cases, however, it is not necessary that each individual component of a complex sample be isolated from all others. In spectrophotometry, for example, it is sometimes possible to measure more than one constituent in a single solution. Let us suppose a solution to contain two absorbing constituents, X and Y. The complexity of the situation depends upon the absorption spectra of X and Y.

No Overlap

The spectra do not overlap, or at least it is possible to find a suitable wavelength where X absorbs and Y does not, and a similar wavelength for measuring Y. Figure 14.21 shows such a situation. The constituents X and Y are simply measured at wavelengths λ_1 and λ_2, respectively.

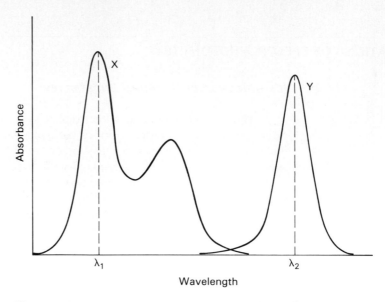

Figure 14.21 Absorption spectra of compounds X and Y. (There is no overlap at the two wavelengths which are to be employed.)

One-way Overlap

As shown in Fig. 14.22, Y does not interfere with the measurement of X at λ_1, but X does absorb appreciably along with Y at λ_2. The approach to this problem is simple in principle. The concentration of X is determined directly from the absorbance of the solution at λ_1. Then the absorbance contributed at λ_2 by this concentration of X is calculated from the previously known molar absorptivity of X at λ_2. This contribution is subtracted from the measured absorbance of the so-

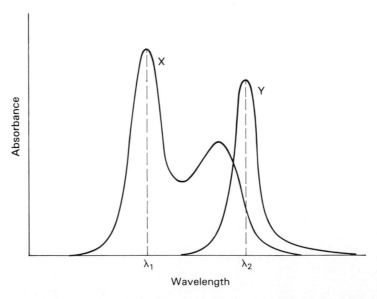

Figure 14.22 Absorption spectra of compounds X and Y. (One-way overlap: X can be measured with no interference from Y, but X interferes with the direct measurement of Y.)

lution at λ_2, yielding the absorbance due to Y, whose concentration is then calculated in the usual manner.

Two-way Overlap

When no wavelength can be found where either X or Y absorbs exclusively, as suggested in Fig. 14.23, it is necessary to solve two simultaneous equations in two unknowns. Let

$$A_1 = \text{measured absorbance at } \lambda_1$$

$$A_2 = \text{measured absorbance at } \lambda_2$$

$$\varepsilon_{X_1} = \text{molar absorptivity of X at } \lambda_1$$

$$\varepsilon_{X_2} = \text{molar absorptivity of Y at } \lambda_2$$

$$\varepsilon_{Y_1} = \text{molar absorptivity of Y at } \lambda_1$$

$$\varepsilon_{Y_2} = \text{molar absorptivity of Y at } \lambda_2$$

$$C_X = \text{molar concentration of X}$$

$$C_Y = \text{molar concentration of Y}$$

$$b = \text{path length}$$

Since the total absorbance is the sum of the contributions of the individual absorbing constituents of the solution,

$$A_1 = \varepsilon_{X_1} b C_X + \varepsilon_{Y_1} b C_Y$$

$$A_2 = \varepsilon_{X_2} b C_X + \varepsilon_{Y_2} b C_Y$$

C_X and C_Y are the only unknowns in these equations and hence their values can be

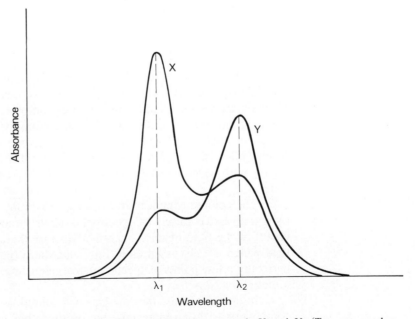

Figure 14.23 Absorption spectra of compounds X and Y. (Two-way overlap: There is no wavelength where either compound can be measured without interference by the other.)

readily determined. The ε-values must be known, of course, from measurements on pure solutions of X and Y at the two wavelengths.

Equations can be set up in principle for any number of components provided that absorbance values are measured at as many wavelengths. However, the importance of small errors in measurement is magnified as the number of components increases, and in practice this approach is generally limited to two- or possibly three-component systems. An exception to this is possible if a computer is available. Then, particularly if the spectrum is recorded, it becomes not too difficult to "overdetermine" the system (i.e., take absorbance values at many more wavelengths than there are components) and by a rapid series of successive approximations obtain reliable values for a large number of components.

14.4c. *Preparation of Samples for Spectrophotometric Analysis*

So far we have said little about chemistry in this chapter. But seldom will the analyst receive a sample which is ready to be measured without some sort of pretreatment. Often separations of interfering substances are necessary; some of the available techniques are considered in later chapters.

Many organic compounds absorb in the ultraviolet region of the spectrum, and pretreatment then involves only separation of interferences. Some elements in the periodic table absorb strongly in the visible or ultraviolet, at least in certain oxidation states, and the preliminary steps may involve redox reactions as well as separations. Manganese, for example, is often determined spectrophotometrically after oxidation to Mn(VII) by means of persulfate or periodate:

$$2Mn^{2+} + 5S_2O_8^{2-} + 8H_2O \longrightarrow 2MnO_4^- + 10SO_4^{2-} + 16H^+$$

The purple MnO_4^- solution is measured at about 525 nm. Chromium is determined similarly after oxidation to Cr(VI).

Development of absorption by means of inorganic reagents is occasionally possible. For example, iron may be determined by means of the red color obtained by treating iron(III) solutions with thiocyanate:

$$Fe^{3+} + SCN^- \longrightarrow (FeSCN)^{2+}$$

The system is complicated by the tendency to form higher complexes such as $[Fe(SCN)_2]^+$. Other examples of colored complexes formed with inorganic reagents are the blue tetraammine copper complex, $[Cu(NH_3)_4]^{2+}$, and the several complex heteropoly acids such as phosphomolybdic, $H_3P(Mo_3O_{10})_4 \cdot 29H_2O$, which are used to determine elements such as phosphorus and silicon. Iodide ion forms yellowish complex ions which exhibit absorption maxima in the ultraviolet region with several metals, including bismuth, antimony, and palladium (e.g., PdI_4^{2-}).

The colored complexes formed by metal ions with organic reagents offer the most impressive variety of spectrophotometric methods, and they are especially useful in the field of trace analysis. Most of these complexes are of the chelate type which are discussed more fully elsewhere (pages 81 and 466). We mention here only a few points of special interest regarding their adaptation to spectrophotometric analysis.

In some regards, the low aqueous solubility of many of the metal chelate compounds is disadvantageous, but on the other hand, extraction of metals into nonaqueous solvents by means of chelating agents may lead to very powerful analytical methods. In favorable cases, it may be possible to concentrate the metal, separate it from interferences, and develop the absorbing system in a single step.

For example, the chelates of 8-hydroxyquinoline (oxine, page 466) with such metals as aluminum, iron(III), cadmium, gallium, lead, and copper are soluble in chloroform, and extractions with this solvent generally precede the spectrophotometric determination. By controlling the pH of the aqueous phase and by adding complexing agents which mask certain metal ions, the extraction can be made quite selective. Reasonable absorbance values are generally obtained with chloroform solutions whose metal concentrations are on the order of a few micrograms per milliliter.[13] The solvent extraction process is described in Chapter 16.

The solvent used in spectrophotometric procedures poses a problem in some regions of the spectrum. The solvent must not only dissolve the sample but also must not absorb appreciably in the region in which the determination is made. Water is an excellent solvent in that it is transparent throughout the visible region and down to a wavelength of about 200 nm in the ultraviolet. However, since water is a poor solvent for many organic compounds, organic solvents are commonly employed for these substances. The transparency cutoff points in the ultraviolet region of a number of solvents are listed in Table 14.4. Aliphatic hydrocarbons, methanol, ethanol, and diethyl ether are transparent to ultraviolet radiation and are frequently employed as solvents for organic compounds.

TABLE 14.4 Solvents for Ultraviolet and Visible Regions

SOLVENT	APPROXIMATE TRANSPARENCY MINIMUM, nm	SOLVENT	APPROXIMATE TRANSPARENCY MINIMUM, nm
Water	190	Chloroform	250
Methanol	210	Carbon tetrachloride	265
Cyclohexane	210	Benzene	280
Hexane	210	Toluene	285
Diethyl ether	220	Pyridine	305
p-Dioxane	220	Acetone	330
Ethanol	220	Carbon disulfide	380

There is no single solvent which is transparent throughout the infrared region. Carbon tetrachloride is useful up to 7.6 μm, and carbon disulfide up to 15 μm. Not all substances are soluble in these solvents, however, and other liquids with more restricted ranges must be used. Water exhibits strong absorption in the infrared and is usually avoided, although much of the region can be covered if both H_2O and D_2O are available. The usual rock-salt cells are obviously unsuitable for aqueous solutions. Silver halide cell windows are sometimes used because of their resistance to water, although photochemical darkening is objectionable. Transparent forms of aluminum oxide such as sapphire can be used as windows for spectra of aqueous solutions up to about 6.5 μm.

Because of the problems encountered in working with solutions in the infrared, various other techniques have been employed. Liquids can be measured directly, using a very thin film placed between rock-salt plates. Another technique employs a liquid *mull*, made by dispersing the sample in a viscous hydrocarbon,

[13] See E. B. Sandell, *Colorimetric Determination of Traces of Metals*, 3rd ed., John Wiley & Sons, Inc., New York, 1959, for spectrophotometric methods for most metal ions; an expanded 4th edition of Part I has appeared (E. B. Sandell and H. Onishi), *Photometric Determination of Traces of Metals: General Aspects*, John Wiley & Sons, Inc., New York, 1978. For other elements, see D. F. Boltz and J. A. Howell, eds., *Colorimetric Determination of Nonmetals*, John Wiley & Sons, New York, 1978.

such as the mineral oil Nujol. The oil is transparent over much of the infrared region, although of course it interferes with sample bands representing vibrations such as C—H stretching modes which are exhibited by the oil itself. When sample C—H modes must be detected, substituted hydrocarbon oils or greases such as perfluorokerosene or fluorolubes are sometimes employed as mulling agents.

Still another technique employs solid potassium bromide as a material for dispersing samples. The salt is highly transparent to infrared radiation, and at high pressures (~100,000 psi), finely divided KBr is sufficiently plastic to form a clear disk in which absorption bands of dispersed sample components can be observed. The solid sample–KBr mixture is formed into the disk using a special die with polished faces in a hydraulic press. (For qualitative work, organic chemists often apply pressure to KBr by hand-turning two bolts with polished ends against each other in a threaded cylinder. The resulting disk is generally of poor quality, with cloudiness that scatters much radiation, especially at shorter wavelengths, but the infrared bands of incorporated sample components can be observed adequately for certain purposes.) The sample can be ground with KBr in a mortar, or sometimes a vibratory mixer as used by dentists for preparing amalgam fillings is employed. Excellent disks can be prepared from the solid obtained by freeze-drying aqueous sample solutions in which KBr has been dissolved. The particle size of the KBr obtained by freeze-drying is very small (probably on the order of 50-μm diameter), leading to very clear disks, and the dispersion of the sample within the KBr matrix is very uniform. The disadvantage is that freeze-drying is a time-consuming process. The disk, typically a centimeter or so in diameter and perhaps a millimeter thick, is mounted in a holder that positions it in the sample beam of the spectrophotometer, while a disk of pure KBr is placed in the reference beam.

14.4d. Photometric Titrations

Various properties of a solution may be measured in order to assess the progress of a titration toward the equivalence point. We have seen, for example, in Chapter 12 that the potential of an indicator electrode may be used for this purpose, and we have described another end-point technique, amperometric, in Chapter 13. The absorbance of a solution may likewise be measured during the course of a titration; we have available, then, still another end-point technique which may be useful in certain circumstances. Our discussion of photometric titrations will be brief; a more complete treatment may be found.

As a matter of fact, visual titrations are really photometric in nature.

The color change reflects a change in the absorption of light by the solution, accompanying changes in the concentrations of absorptive species. In a visual titration, one actually employs all the features of an automatic photometric titrator: Light passes through the solution to the eye, which is a photosensitive transducer responding with a signal to the brain. The brain is analogous to the circuitry of an instrument which amplifies the signal and otherwise renders it appropriate for transmission to an electromechanical shutoff system; traversing a motor neuron, the signal triggers a muscular response that closes the buret to terminate the titration. In visual titrations, the most complicated and expensive instruments of all—people—act as automatic photometric titrators.[14]

[14] A. L. Underwood, "Photometric Titrations," a chapter in *Advances in Analytical Chemistry and Instrumentation*, Vol. 3, C. N. Reilley, ed., Wiley-Interscience, New York, 1964, p. 31.

Photometric titrations often possess advantages of sensitivity of end-point detection and circumvention of interferences over visual titrations. Further, they are not restricted to the wavelength region where the human eye responds, and they are fairly easily automated. In comparison with potentiometric titrations, the photometric approach is often advantageous for borderline cases of titrations which are approaching nonfeasibility. While the potential of an indicator electrode responds to the logarithm of a concentration (or a concentration ratio), the absorbance of a solution is directly proportional to the concentration. If the equilibrium constant of a titration reaction is undesirably small, concentrations will not change as rapidly as one would like in the vicinity of the equivalence point, and the potential, because of the logarithmic compression, will change even less rapidly. The absorbance, on the other hand, will change just as rapidly as the concentration. This is not a unique advantage of photometric titrations, of course. The same consideration applies to amperometric and other end points obtained by linear extrapolation.

Just as with amperometric titrations (Chapter 13, page 380), it is necessary to correct measured absorbance values for dilution if the volume of titrant is appreciable compared with the initial volume of the solution.

Titrations without Indicators

Sometimes a substance directly involved in the titration reaction absorbs appreciably at an accessible wavelength, and the titration can be followed spectrophotometrically without adding an indicator. The shapes of the titration curves are predictable from the ε-values of the chemical species concerned. Some typical photometric titration curves of this type are shown in Fig. 14.24. If the titration reaction is appreciably incomplete in the vicinity of the equivalence point, the curve will become rounded, as shown by the dashed portion of curve A in Fig. 14.24. The end point is then located by the intersection of extrapolated straight lines drawn through points taken sufficiently before and after the rounded portion. Titration curves of this sort are easily calculated: one simply computes

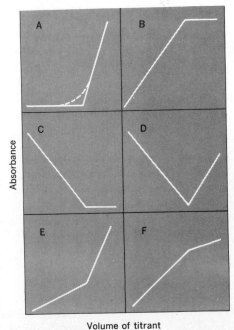

Figure 14.24 Typical photometric titration curves: (A) $\varepsilon_t > \varepsilon_s = \varepsilon_p$ (usually $\varepsilon_s = \varepsilon_p = 0$); (B) $\varepsilon_p > \varepsilon_s$, $\varepsilon_t = 0$; (C) $\varepsilon_s > \varepsilon_p$; $\varepsilon_t = 0$; (D) $\varepsilon_s > \varepsilon_p$, $\varepsilon_t > 0$; (E) $\varepsilon_p > \varepsilon_s$; $\varepsilon_t > \varepsilon_p$; (F) $\varepsilon_p > \varepsilon_s$, $\varepsilon_t < \varepsilon_p$. ε_s, ε_p, and ε_t are the molar absorptivities of the substance titrated, reaction product, and titrant, respectively.

the concentrations of absorbing species at any point using the equilibrium constant of the reaction and then calculates the contribution of each species to the absorbance of the solution from Beer's law using known ε-values and path length.

Titrations with Indicators and the Titration of Mixtures

In cases where none of the species involved in the titration reaction absorbs sufficiently, an indicator may be added to the solution. Figure 14.25 shows an example of an indicator titration, a chelometric titration of copper(II) ion with EDTA using the metallochromic indicator pyrocatechol violet. In this titration, a wavelength was selected where the free indicator absorbs more strongly than the copper-indicator complex. We see in the figure first the reaction of free copper(II) ion with EDTA, which does not affect the absorbance of the solution at this wavelength. Then, as the end point is approached, copper is pulled away from the indicator by the titrant, and the absorbance rises as free indicator accumulates, until finally all of the copper has been titrated and the absorbance becomes constant again.

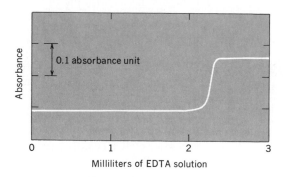

Figure 14.25 Titration of copper with EDTA using pyrocatechol violet indicator; 14.01 mg of copper in 150 mL titrated with 10^{-1} M EDTA at 440 nm. (Data of T. M. Robertson.)

Figure 14.26 shows a photometric titration of a mixture of bismuth and copper with EDTA. At the wavelength selected, the copper(II)–EDTA chelate absorbs strongly, while the other species (Bi^{3+}, bismuth–EDTA chelate, and EDTA) have ε-values of zero. The bismuth chelate is much more stable than the cop-

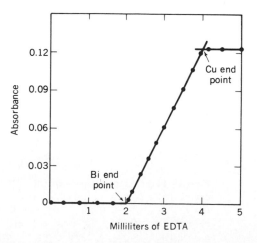

Figure 14.26 Titration of bismuth-copper mixture with EDTA; 41.8 mg of bismuth and 13.1 mg of copper in 100 mL (each 2×10^{-3} M) buffered at pH 2, titrated at 745 nm with 10^{-1} M EDTA.

per(II) complex. Thus, as EDTA is added to the Bi^{3+}–Cu^{2+} mixture, the bismuth chelate is formed first. When $[Bi^{3+}]$ has been reduced to a very low value, the copper(II)–EDTA chelate begins to form, and because this is the strongly absorbing species, the absorbance begins to rise. After the copper end point, the curve levels off as excess, nonabsorbing EDTA is added.

Instrumentation for Photometric Titrations

The simplest approach to a photometric titration is to titrate in a flask or beaker on the laboratory bench, taking samples out of the titration vessel for absorbance measurements as the titration proceeds. Of course, the samples must be returned each time, and this technique is inconvenient for more than an occasional single titration. On the other hand, any good spectrophotometer can be used without modification, and it is more obvious to a beginner exactly what is going on than might be the case if more elaborate instrumentation were employed.

Sometimes it is possible to fit a spectrophotometer with a modified cell compartment so that a titration vessel such as a beaker can be positioned in the light beam. It is convenient to stir by means of a magnetic stirrer underneath the compartment. The buret tip is introduced into the solution through a hole in the cover of the cell compartment; care must be taken that this arrangement be light-tight.

With a recording spectrophotometer, it is possible to record absorbance vs. time at a constant wavelength. If titrant is introduced into a titration vessel in the sample beam at a constant flow rate, if adequate stirring is provided, and also if the titration reaction is rapid, then the plot of absorbance vs. time readily becomes a photometric titration curve. With relatively simple on-the-spot modification, the output signal of a manual spectrophotometer can be recorded; thus an expensive double-beam recording instrument is not really required for this application.

Finally, photometric titrators which terminate titrant flow at the end point are on the market. The operator merely sets up and starts the titration and then later reads a buret. We described briefly in Chapter 12 an automatic potentiometric titrator based upon double electronic differentiation of the voltage from a pair of electrodes. The first derivatives of photometric titration curves like those in Fig. 14.24 exhibit the sigmoid shape associated with a typical potentiometric titration. An instrument was devised for automatic photometric titration that electronically differentiated a signal arising from a photodetector and then fed this into the circuit of the existing potentiometric titrator. Of course, it was necessary to add to the potentiometric titrator not only an additional differentiating circuit but also optical components, source, photodetector, etc.[15]

14.5

ATOMIC ABSORPTION SPECTROPHOTOMETRY

The earlier material in this chapter could be called molecular spectrophotometry; the infrared and UV-visible absorption bands that we considered involved polyatomic molecules. Individual atoms, however, also absorb radiation, leading to excited electronic energy states. The absorption spectra are simpler than molecular spectra because the electronic energy states do not have vibrational-rotational sublevels. Thus atomic absorption spectra are made up of lines which are much

[15] H. V. Malmstadt and C. B. Roberts, *Anal. Chem.*, **28**, 1408 (1956).

sharper than the bands observed in molecular spectroscopy. Atomic absorption has been known for many years. For example, dark lines at certain frequencies in the otherwise continuous spectrum of the sun were first noticed in 1802 by Wollaston; they were rediscovered and studied more thoroughly by Joseph von Fraunhofer, after whom they have been called *Fraunhofer lines*. The significance of the lines was not understood until 1859, when Kirchoff explained their origin after observing similar phenomena in the laboratory. The visible surface of the sun is much hotter than the surrounding gaseous envelope, and atoms in the atmosphere absorb specific frequencies from the emission continuum of the hotter surface. (The radiation is reemitted—otherwise the envelope would become steadily hotter—but the emission occurs in all directions; hence an observer on the earth is short-changed in terms of what passes directly to him.) Kirchoff and others, particularly Bunsen (of Bunsen burner fame), identified a number of elements in the sun's atmosphere by comparing frequencies of Fraunhofer lines with those of known elements in the laboratory.

14.5a.   *Principles*

For many years, mercury vapor detectors represented the major analytical application of atomic absorption. The vapor pressure of metallic mercury is large enough to represent a health hazard in an inadequately ventilated space. The detectors are basically primitive spectrophotometers in which the source is a low-pressure mercury vapor lamp. Mercury atoms excited in the electrical discharge of the lamp emit radiation when they revert to lower electronic levels; the radiation is not a continuum but rather comprises discrete frequencies representing electronic transitions in the mercury atom, with a particularly strong line at 253.7 nm corresponding to the energy difference between the ground state and the first excited electronic state for Hg. (A line representing a transition terminating in the ground state or, in absorption, excitation of a ground-state atom, is called a *resonance line*.) Mercury atoms in the laboratory air at room temperature are all in the ground electronic state, and of course 253.7 nm is exactly a wavelength which these atoms can absorb. Pure air is pumped through a flow cell positioned in the optical path of the instrument, and a galvanometer reading corresponding to P_0 is established, as in any spectrophotometer; then air from the test area is sampled to obtain P. The path length through the cell (cf. Bouguer's law) is selected to yield measurable absorbance values with mercury concentrations well below levels considered toxic to personnel. The use of a discontinuous line source in which much of the energy is concentrated in the 253.7 nm resonance line eliminates the need for a monochromator; the detector is not flooded with unabsorbable radiation, and a measurable ratio P/P_0 is obtained even at low sample concentrations. With a 15-cm path length, mercury vapor in air samples at a level of less than 1 ppb could be measured over 50 years ago.[16]

The extension of atomic absorption spectrophotometry to other elements was originally an outgrowth of flame emission spectroscopy (Chapter 15). Chemists have long used the emission of radiation by atoms excited in a flame as an analytical tool. In 1955, Walsh[17] pointed out that in a typical flame, most of the atoms are in the ground electronic state rather than in an excited one. For example, for the transition that yields the yellow sodium line at 589 nm, the ratio of

[16] T. T. Woodson, *Rev. Sci. Instrum.*, **10**, 308 (1939); V. F. Hanson, *Ind. Eng. Chem., Anal. Ed.*, **13**, 119 (1941).

[17] A. Walsh, *Spectrochim. Acta*, **7**, 108 (1955).

excited to ground-state atoms at 2700°C is about 6×10^{-4}. He further noted that the fraction of atoms excited varies exponentially with temperature, placing a premium upon flame regulation in emission studies, whereas, with so few atoms excited, the ground-state population is much less temperature-sensitive. Thus it was suggested that improved analytical methods would be possible based upon the absorption of radiation by ground-state atoms in the flame. Atomic absorption developed rapidly during the 1960s, commercial instruments became available, and the technique is now widely used for the determination of a number of elements, mostly metals, in a large variety of samples.

In principle, there are certainly no problems to be associated with measuring the absorbance of a ground-state atom population confined in a suitable space, but there are a number of difficulties in obtaining that population in a reproducible manner. Typically, an aqueous solution containing the metal to be determined— e.g., Pb^{2+} or Cu^{2+}—is introduced into the flame as an aerosol, i.e., a mist or fog of tiny droplets. As the droplets proceed through the flame, they desolvate, producing tiny specks of particulate matter. The solid then dissociates, at least partially, to yield the metal atoms. All these steps must occur within a distance of a few centimeters as the sample particles are carried at a high velocity by the flame gases. Under proper illumination, one can sometimes see unevaporated sample droplets issuing from the top of the flame, and the flame gases are diluted by inrushing air as a result of the low pressure created by high velocity. Furthermore, the optical system does not examine the whole flame but sees only a region lying a certain distance above the burner tip. There is at no point a stable, equilibrium atom population holding still for an absorbance measurement; kinetic parameters as well as sample concentration determine how many atoms have been introduced into the source beam at any instant. Efficiencies of far less than 1% are estimated for dispersal of a solution into an aerosol and conversion of the analyte into atoms actually in the optical path. This sounds pessimistic, but of course atomic absorption does work, else it would not be included in this chapter. The useful results that are obtained show what meticulous attention to reproducible conditions can sometimes do.

Because of the serious kinetic problems with flame atomization and because sensitivity is lowered very considerably by dilution of the analyte atom population with gases in the flame, special furnaces to replace the flame in atomic absorption spectrophotometry have been developed in recent years. The furnaces introduce problems of their own but offer advantages as well.

14.5b. Instrumentation

Figure 14.27 shows in schematic form the components of a basic atomic absorption spectrophotometer. It is seen that the same elements are present as are

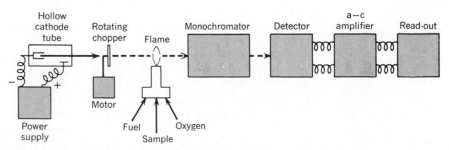

Figure 14.27 Components of an atomic absorption spectrophotometer. (The flame may be replaced by a furnace.)

found in Fig. 14.8—source, sample area, detector, etc.—but we call attention to differences in the nature of some of these elements.

Sources

First, the source is quite different. In molecular spectrophotometry, absorption bands are generally sufficiently broad to allow the use of a continuous source and a monochromator. As sketched in Fig. 14.28, the band of radiation emergent from the exit slit of a good monochromator is sufficiently narrow (bandwidth on the order of a nanometer or even somewhat less) to enable us to profile a typical ultraviolet absorption band whose width at half-height we have depicted as 10 nm. For quantitative analysis, if we set the monochromator scale to coincide with the peak of the absorption band we will obtain nearly maximal sensitivity and good adherence to Beer's law because most of the radiation is relatively very close to the wavelength of maximal sample absorption.

On the other hand, an atomic absorption line in a flame or furnace is very much narrower than the band which can be provided by any continuous source–monochromator combination (see the discussion of spectral purity in the *Monochromator* section, page 404). An atomic absorption (or emission) line is not a line in the mathematical sense of having no width, but it is still very narrow in the present context. There is a *natural width* on the order of 10^{-5} nm resulting from the probability distribution associated with each electronic energy level of an atom, but observed lines are considerably broader than this. *Doppler broadening* on the order of 10^{-3} nm at typical flame temperatures reflects the different ve-

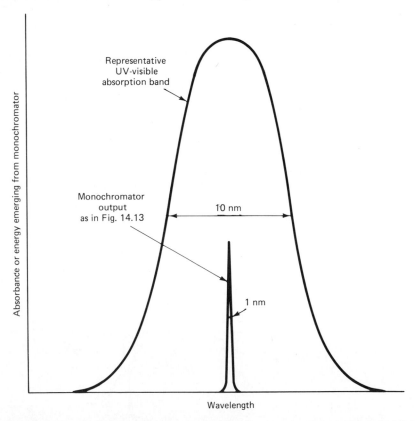

Figure 14.28 Sketch of a typical UV-visible absorption band on the same wavelength axis as the energy distribution emergent from a monochromator exit slit.

locity components of the atoms along the line of observation (this phenomenon is explained in elementary physics textbooks). *Pressure broadening,* which in flames is typically on the same order of magnitude as Doppler broadening, results from perturbations of atomic energy levels by neighboring atoms or molecules. Thus lines observed in atomic absorption spectrophotometry are typically between 10^{-3} and 10^{-2} nm in width. Hence most of the radiation from a monochromator exit slit lies outside the atomic absorption line and has no chance of being absorbed; the effect upon P/P_0 and thus the sensitivity is obvious.

Except for the special case of mercury vapor described earlier, the most widely used source in atomic absorption is the hollow cathode discharge tube as originally recommended by Walsh. This tube contains an anode and a hollow, cylindrical cathode in an inert gas atmosphere, often neon or argon, at low pressure. The tube is operated with a power supply furnishing voltages up to perhaps 300 V; currents through the tube are in milliampere range, seldom more than 20 or 30 mA and often less. Atoms of the gas are ionized in the electrical discharge and the energetic neon or argon ions are accelerated to the negative cathode, where collisions with the surface dislodge atoms of the cathode metal, a process sometimes called *sputtering*. In further collisions with energetic ions and atoms, the sputtered metal atoms are excited; then, in a cooler region of the discharge, they emit a line spectrum characteristic of the cathode metal which appears as a glow within the cavity of the hollow cathode. A resonance line is selected from this spectrum for the absorption measurement. The filler gas is at a low enough pressure (cf. pressure broadening) and the temperature is low enough (cf. Doppler broadening) that the lines in the emission spectrum of the lamp are narrower than the analyte absorption line in the flame or furnace, exactly what we wish.

A typical commercial hollow cathode source is limited to the determination of one metal—the metal of which the cathode, or at least its surface, is fabricated—although in a few cases alloys are employed which yield lines of several metals. Demountable tubes can be constructed in which various metal cathodes can be interchanged, although pumping out the air and reintroducing filler gas whenever the tube is opened is a nuisance. Some instruments provide a turret on which several different lamps are maintained under operating conditions; rotating the turret then enables the technician to switch from the determination of one metal to that of another by positioning the desired lamp in the optical path of the instrument.

As shown in Fig. 14.27, the source beam is usually "chopped" to pulsate at a certain frequency. This beam modulation, combined with the use of an ac amplifier tuned to the chopper frequency, frees the atomic absorption measurements from interference by light from unwanted sources such as the flame or furnace or even a little leakage of laboratory illumination.

Burners and Furnaces

The problem is to obtain from a sample solution a ground-state atom population whose absorbance is related in a simple manner to the analyte concentration in that solution. This is most frequently accomplished with a flame, and there are two types of burners which are employed. With the *premix nebulizer–burner* system, the flow of oxidant past the tip of a capillary tube draws sample from a container into a chamber and breaks it into droplets much as a perfume sprayer works. Fuel also feeds into the chamber, and the sample aerosol is swept past a series of baffles which send large drops to a drain and permit a fog or mist of tiny droplets to accompany the mixed fuel and oxidant gases into the burner. A "fishtail" burner design providing an optical path of 5 or even 10 cm through the flame is often employed to increase sensitivity (Bouguer's law).

The absorbance value obtained for a given analyte concentration in the original sample solution depends upon the nebulization efficiency in the premix chamber and then upon the temperature (and its variation with distance above the burner tip) and the gaseous flow rate through the flame, because these factors determine the ground-state atom population in the path of the source beam. With continuous flow of sample, we hope to obtain a constant, steady-state atom population of sufficient duration to permit measurement, and the conditions under which it is produced must be reproducible enough to permit a large number of samples as well as standard solutions to be compared. Obviously, gas pressures and flow rates of fuel and oxidant must be carefully regulated. The attainable temperature depends upon the gases employed, for which a few approximate values are: coal gas–air, 1800°C; natural gas–air, 1700°C; acetylene–air, 2200°C; acetylene–nitrous oxide, 3000°C.

The premix nebulizer–burner consumes much sample, since 80 or 90% is directed to the drain by the baffle system. If the rate of flame propagation becomes greater than the rate of gas flow into the burner, a dangerous explosion can occur in the premix chamber; prevention of this occurrence requires careful adherence to operating instructions and precludes the use of certain fuel-oxidant combinations such as hydrogen-oxygen. (H_2–O_2 is otherwise attractive because of reasonable cost, fairly high temperature—perhaps nearly 2700°C—and "cleanness" in terms of its own luminous emission.) The premix burner provides, among other advantages, a very uniform feed of small droplets to the flame and a long optical path.

The other type of burner sometimes employed in atomic absorption is the *total-consumption, integral aspirator-burner* originally developed for flame emission work; a diagram is found on page 454. Sample solution is drawn through a central capillary by the air or oxygen flow through the constricted tip of the surrounding inner annulus. At the burner tip, the liquid encounters strong shear forces which disperse it into droplets that are carried directly into the flame by the rushing gases. Sample consumption is much lower than with the premix burner, perhaps 0.5 to 2 ml/min vs. 10 to 30 mL/min. There is no danger of explosion from flashback of the flame, and mixtures such as H_2–O_2 can be safely employed. On the other hand, the sample aerosol is much less uniform, the path length is much shorter, and the burner emits a loud and thoroughly obnoxious noise. The premix burner has become much the more widely used except in special cases.

Atomization in an electrically heated graphite furnace is sometimes advantageous when the highest sensitivity is required. A much larger fraction of the analyte appears as atoms at one time than is the case with flames. The present generation of furnaces are cumbersome and beset by many problems, and they still do not provide what we would like: a stable ground-state atom population in thermal equilibrium with its surroundings, obtained by kinetic processes which have run their course prior to the absorbance measurement, and contained in a "cell" where it is easily and reproducibly introduced and from which it is easily flushed. Perhaps we want too much, although it is reasonable to hope for improvements. Typically, a few microliters of sample is placed on a tantalum ribbon or graphite rod or within the depression of a tiny graphite crucible. The solvent is evaporated outside the furnace under an infrared lamp, or internally by passing current through the sampling device; if desired, the temperature may be raised sufficiently to ash organic matter. Finally, the temperature is increased very rapidly to 2000 to 3000°C, producing, within a few seconds, a cloud of atomic vapor. Usually the vapor is introduced into a stream of inert gas such as argon; this gas prevents oxidation of the furnace materials themselves, prevents the formation of difficult-to-atomize oxides of certain metals in the samples, and, by moving the sample vapor rapidly through the furnace, partially prevents contamination of the graphite wall, which is somewhat penetrable to components of the vapor. The atom population shows up, then, as a transient pulse in a graphite tube positioned in the optical path of the spectrophotometer.

With flame atomization, manual operation of the spectrophotometer is possible. A P_0 value is obtained on a flame fed distilled water or an analytical blank of some sort; then the sample is introduced, and a meter reading corresponding to P is obtained. With furnaces, the sample pulse is quite short, and it is better to record the output of the detector circuit as a function of time. Sample introduction then yields a peak whose height or area relates to the analyte content of the sample.

Other Components

The other components of an atomic absorption spectrophotometer are conventional. The monochromator may be less costly than those of ordinary spectrophotometers of comparable quality because less is required of it. The only demand is that it pass the selected resonance line and no other lines in the source spectrum arising from the cathode metal or the inert gas. Usually a satisfactory source line can be found which is not close enough to other lines to require the best of monochromators for isolation. (Remember that we rely upon beam chopping and a tuned amplifier to eliminate the effects of radiative noise from flame or furnace.) The detector is typically a photomultiplier tube, since the lines with which we are dealing generally lie in the UV-visible region of the spectrum. The readout device may be a simple galvanometer, a digital voltmeter, or a pen-and-ink recording potentiometer; for laboratories with a large work load, the amplifier output may be digitized and processed by computer.

14.5c. *Applications of Atomic Absorption*

For detailed descriptions of atomic absorption methods, the interested reader should consult the literature.[18] The technique has been applied to the determination of about 60 elements, and it is a major tool in studies involving trace metals in the environment and in biological samples. It is also frequently useful in cases where the metal is present at a fair level in the sample but only a small sample is available for analysis; such is sometimes the case with metalloproteins, for example. The first report of an important biological role for nickel was based upon the determination by atomic absorption that the enzyme urease, at least from certain organisms, contains two nickel ions per protein molecule.[19] Frequently the first step in the analysis of biological samples is *ashing* to destroy organic matter. Wet ashing with nitric and perchloric acids is frequently preferable

[18] Up-to-date guides to the recent literature (including books, reviews, and original research papers) found in the biennial fundamental review issues of *Analytical Chemistry*, of which the most recent is J. A. Holcombe and D. A. Bass, *Anal. Chem.*, **60**, 226R (1988).

[19] N. E. Dixon, C. Gazzola, R. L. Blakeley, and B. Zerner, *J. Amer. Chem. Soc.*, **97**, 4131 (1975)

to dry ashing in regard to volatility losses of certain trace elements (dry ashing is simply heating the sample in a furnace to oxidize organic matter). Atomic absorption is then performed on the wet-ashed solution or on a solution prepared from the dry-ashed residue.

A major feature of atomic absorption is, of course, sensitivity. It is very unfortunate that many authors are careless in reporting this; in some papers it is even difficult to discern whether the author's statement of a lower concentration limit of so-many parts per million or billion refers to the level of the metal in the original sample or in the final solution after preliminary steps such as ashing! Table 14.5 gives a few examples of detection limits; in most cases, the values are the concentrations in the final solution presented to nebulization and measurement which yield absorbance values equal to two times the standard deviation of background absorbance values obtained with pure water. The values are presented only to give an idea of how sensitive atomic absorption is; actual detection limits will vary greatly with matrix (see below), nebulization conditions, temperature, path length, and other factors in a particular analysis with a certain instrument.

TABLE 14.5 Some Approximate Detection Limits in Atomic Absorption

ELEMENT	ACETYLENE-AIR FLAME[†]	GRAPHITE FURNACE[†]
Ca	0.002	0.0003
Cd	0.005	0.0001
Cr	0.005	0.0005
Cu	0.004	0.0001
Fe	0.004	0.003
K	0.004	0.001
Mg	0.003	0.00005
Ni	0.005	0.001
Zn	0.001	0.000006

Note: The values are rough averages of data from several sources and are given only to orient the reader regarding the magnitudes involved.

[†] Parts element per million parts solution.

In one regard, atomic absorption is remarkably free of interferences. The set of electronic energy levels for an atom is unique to that element. This means that no two elements exhibit spectral lines of exactly the same wavelength. There are often lines for one element which are very close to some of the lines of another, but it is usually possible to find a resonance line for a given element with which there is no direct spectral interference by other elements in the sample.

The major interferences in atomic absorption are matrix effects which influence the atomization process. Both the extent of dissociation into atoms at a given temperature and the rate of the process depend very much upon the overall composition of the sample. For instance, if a solution of calcium chloride is nebulized and desolvated, the tiny particles of solid $CaCl_2$ dissociate to yield Ca atoms much more readily than would particles of, say, calcium phosphate, $Ca_3(PO_4)_2$. We have noted earlier (page 317) that matrix effects are a frequent problem in analytical chemistry, and they are often of crucial importance in spectroscopy because the general, gross composition of the sample can exert a tremendous effect upon the extent and rate of the dissociation yielding the desired atomic vapor. It is particularly important that standard solutions be very similar to the unknown samples in general composition with regard to components that are present in large quantity. Where variation in overall composition is expected from one sample to another, it is generally desirable for the analyst to create his own matrix by adding enough of some material to swamp out sample variations.

KEY TERMS

Absorbance, A. $A = \log (P_0/P)$ where P_0 is incident radiant power and P is radiant power transmitted through a sample.

Absorptivity, a. The constant in the Bouguer-Beer expression when concentration is in grams per liter and path length is in centimeters. $A = abc$ or $a = A/bc$. The units of a are liters per gram per centimeter.

Angstrom unit, Å. After Anders Jonas Ångström, 1814–1874, Swedish physicist. A common wavelength unit, particularly in the UV-VIS region; also frequently used to specify atomic and molecular dimensions. 1 Å $= 10^{-10}$ m; thus 10 Å $= 1$ nm.

Ashing. Oxidative destruction of organic matter prior to determination of an inorganic analyte. Frequently employed to eliminate matrix effects (q.v.) with such samples as foods and biological materials. In *dry ashing,* organic matter is burned off in a furnace by gradually raising the temperature to (usually) 500–600°C; the initial charring is done slowly so that material will not be mechanically carried away by the boiling of moisture or by the flames of an actual fire. In *wet ashing,* the organic matter is destroyed in solution by hot, concentrated oxidizing acids such as H_2SO_4, HNO_3, and $HClO_4$, and the inorganic residue may remain in solution. With certain elements, e.g., As and Hg, there is danger of loss by volatilization, particularly in dry ashing, and sometimes it is difficult to dissolve certain dry ashed residues.

Atomic absorption spectrophotometry. Often abbreviated AA or AAS. A form of spectrophotometry in which the absorbing species are atoms.

Atomization. In atomic absorption spectrophotometry, the dissociation of the solid particles formed by desolvating nebulized (q.v.) sample solution. Note that this step, at typical flame or furnace temperatures, yields primarily *atoms,* not ions (appreciable ionization usually occurs only at higher temperatures than are commonly employed).

Beer's law. Absorbance, $\log (P_0/P)$, of monochromatic radiation is directly proportional to the concentration of an absorbing species in solution.

Bouguer-Beer law. Combination of Bouguer's and Beer's laws, often written $A = \varepsilon bc$, where A = absorbance, ε = molar absorptivity, and b = path length through a solution of solute molar concentration c.

Bouguer's law. Sometimes called Lambert's law. Divide a homogeneous absorbing medium into imaginary layers of equal thickness. Each layer absorbs the same fraction of monochromatic radiation which impinges upon it as does every other layer. Everything else the same, absorbance (q.v.) is directly proportional to path length through the medium.

Detector. In spectrophotometers, a device (transducer) that changes radiant energy into electrical energy, providing an electrical signal related to the radiant power absorbed by the sensitive surface.

Differential spectrophotometry. A technique in which the sample is compared with another absorbing solution instead of with a pure solvent or reagent blank. Can be more accurate than ordinary spectrophotometry.

Double-beam spectrophotometer. An instrument in which the beam is split to permit comparison of sample and solvent (or reagent blank) at the same time. Operation is usually highly automated.

Fraunhofer lines. Dark lines in the spectrum of the sun caused by atomic absorption in the sun's cooler envelope. Of historical interest in connection with atomic absorption spectrophotometry.

Group frequencies. Frequencies in absorption spectra, particularly in the infrared region, associated with specific functional groups, e.g., carbonyl. The functional group, in whatever molecule, is associated with absorption in a particular *region* of the spectrum, although the *exact* position of the band depends upon other features of the molecular structure.

Hollow cathode lamp. (Hollow cathode discharge tube) A source in atomic absorption spectrophotometers, chosen because emission lines of the cathode elements are narrower than the corresponding atomic absorption lines in flames and furnaces.

Hydrogen discharge tube. (Often deuterium) A source for ultraviolet spectrophotometry in which the emission lines of the filling gas (H_2 or D_2) are pressure-broadened sufficiently to provide a wavelength continuum through the ultraviolet region.

Infrared (IR) region. A portion of the electromagnetic spectrum from roughly 1 μm to 1 mm in wavelength. The region used routinely by chemists dealing with molecular structure extends from 1 or 2 μm to 15 μm; occasionally useful data are obtained at longer wavelengths, but seldom beyond 25 μm.

Matrix effects. The sample material in which the analyte is found is the *matrix*. Frequently, substances of the matrix influence the response of the analyte in an analytical measurement, and such interferences are called *matrix effects*. There are no hard and fast defined boundaries, but usually matrix effects refer to the influence of major constituents of the sample upon the measurement of a trace component. Matrix effects may be circumvented by separations preceding measurement, by the method of standard addition (page 375), or by creating a new "standard" matrix by addition of an overwhelming, "swamping" quantity of some substance.

Micrometer, μm. Formerly *micron, μ*. A common wavelength unit, particularly in the infrared region; also used to specify sizes of very finely divided materials such as aerosol droplets or chromatographic column packing materials. 1 μm = 10^{-6} m.

Molar absorptivity, ε. The constant in the Bouguer-Beer expression when concentration is molar and path length is in centimeters. $A = \varepsilon bc$ or $\varepsilon = A/(bc)$. The units of ε are liters per mole per centimeter.

Monochromator. In spectrophotometers, a device for isolating a narrow wavelength band from all the radiant energy which enters. Essential features are a dispersive element (prism or diffraction grating) and a slit system.

Nanometer, nm. Formerly millimicron, mμ. A common wavelength unit, particularly in the UV-visible region. 1 nm = 10^{-9} m.

Nebulization. In atomic absorption spectrophotometry, the dispersal of a solution into a fog or mist (aerosol) of tiny droplets which efficiently yield the analyte atom population when heated sufficiently.

Photometric titration. A titration where the end point is found by absorbance measurements.

Photomultiplier tube. A common photoelectric detector in UV-VIS and near-IR regions; more sensitive than ordinary phototubes, it is found in better instruments.

Phototube. A common photoelectric detector in the UV-VIS and near-IR regions, particularly in less expensive instruments.

Resonance line. A spectral line in atomic emission or absorption corresponding to a transition between two electronic energy levels one of which is the ground state.

Single-beam spectrophotometer. An instrument with one optical path. Sample and pure solvent (or reagent blank) are examined separately to establish P and P_0 for absorbance measurements. Usually operated manually.

Source. In spectrophotometry, the device providing the radiation for the instrument.

Transmittance, T. The fraction of incident radiant power transmitted by a sample. $T = P/P_0$. Often expressed as a percentage: $\% T = (P/P_0) \times 100$.

Tungsten lamp. An electric light bulb with an electrically heated filament, generally tungsten metal. Like other incandescent solids, the filament yields a wavelength continuum approximating "black-body radiation." Under usual operating conditions, the lamp is suitable as a source in the visible region of the spectrum and is useful for only short distances into the ultraviolet and infrared regions.

Ultraviolet (UV) region. A portion of the electromagnetic spectrum between the long-wavelength end of the X-ray region at roughly 40 nm (400 Å) and the violet boundary of the visible region at about 400 nm (4000 Å). Chemists routinely utilize absorption bands between about 200 and 400 nm.

QUESTIONS

1. A solution of potassium permanganate (KMnO₄) exhibits an absorption band in the vicinity of 525 nm, attributable to the MnO_4^- ion (Fig. 14.19). Suppose you did not remember from general chemistry that permanganate solutions are purple. Show how to figure out the color of a KMnO₄ solution using material presented in this chapter.

2. In chloroform solution, acetone, $(CH_3)_2C{=}O$, has an ultraviolet absorption band involving the carbonyl group, centered at 275 nm. What do we mean when we say that this band represents an $n{-}\pi*$ transition?

3. A student was given Bouguer's law, $\log(P_0/P) = k_2 b$, and Beer's law, $\log(P_0/P) = k_4 c$, and told to combine them into one Bouguer-Beer expression. He peeked ahead and knew that he was supposed to get something of the form $\log(P_0/P) = Kbc$. After trying several algebraic manipulations that led nowhere, he finally hit upon multiplying the two equations, obtaining $[\log(P_0/P)]^2 = k_2 k_4 bc$. Then he took square roots of both sides, obtaining $\log(P_0/P) = \sqrt{k_2 k_4 bc}$. This looked funny, and he was completely stumped. Explain in your own words why the square-root symbol on the right-hand side of his equation is not really so weird as he thought it was.

4. Define P, P_0, absorbance, transmittance, percent transmittance, absorptivity, molar absorptivity, absorption spectrum, and isosbestic point.

5. Define spectrophotometer, source, monochromator, and detector.

6. Complete the following table:

Region	Source	Prism material	Detector
		Instrument components	
Ultraviolet			
Visible		Glass	
Infrared			

7. The method of standard addition (page 375) can be used in atomic absorption spectrophotometry. Why is it important that the volume of standard solution be quite small relative to the volume of unknown solution to which it is added?

8. What do analytical chemists mean by *matrix effects*?

Multiple-choice: Select the *one best answer*.

9. The hydrogen or deuterium discharge tube can be used as a source of continuous ultraviolet radiation for spectrophotometers because of (a) the characteristics of chopper-modulated radiation; (b) pressure broadening of hydrogen or deuterium emission lines; (c) the great sensitivity of photomultiplier tubes; (d) the narrow band pass of modern grating monochromators.

10. Chopping the source beam in conjunction with the use of a tuned ac amplifier in an atomic absorption spectrophotometer accomplishes the following: (a) a recording potentiometer can be used instead of a voltmeter for the readout; (b) a less sensitive detector can be used instead of the usual photomultiplier tube; (c) a cooler flame can be used without decreasing the population of ground-state atoms; (d) radiation emitted by excited atoms in the flame will not interfere with the absorbance measurement.

11. Line spectra are emitted by (a) hot solids; (b) excited polyatomic molecules; (c) molecules in the ground electronic state; (d) excited atoms and monatomic ions.

12. Recording spectrophotometers sometimes operate with feedback loops that vary the power of the reference beam until it matches the power of the beam through the sample. This mode of operation (a) eliminates the need for a continuous source; (b) requires two monochromators; (c) eliminates the need for nonabsorbing solvents; (d) makes the detector a null device, with the result that nonlinear response to radiant power would not be deleterious.

13. Which of the following best explains why atomic absorption is sometimes more sensitive than flame emission spectroscopy? (a) At the temperature of a typical flame, the population of ground-state atoms is much greater than the population of excited atoms. (b) Detectors employed in absorption work are inherently more sensitive than those used to measure emission. (c) Hollow cathode

discharge tubes have a much greater radiant power output than do ordinary flames. (d) An absorption line in a flame is always much sharper than an emission line because of the Doppler effect.

14. The method of standard addition compensates for matrix effects provided (a) the addition does not dilute the sample appreciably and does not itself introduce appreciable quantities of interfering substances; (b) the addition dilutes the sample enough that the concentrations of interfering substances are lowered to negligible values; (c) the addition contains none of the substance being determined; (d) the addition contains a large enough quantity of some substance to swamp out sample variations.

15. Fe^{3+} and Cu^{2+} form complexes with EDTA. The ferric complex is colorless, as is Fe^{3+} itself, at the concentration involved here; the copper complex is a deep blue color (deeper than the color of Cu^{2+} itself), while EDTA is colorless. The photometric titration curve below was obtained when a solution containing both Fe^{3+} and Cu^{2+} was titrated with EDTA using a spectrophotometer set at a wavelength of 745 nm in the visible region of the spectrum.

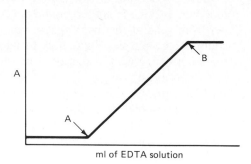

(a) A is the iron end point. (b) B is the iron end point. (c) The distance from A to B on the milliliter axis represents the quantity of iron. (d) The quantity of iron could not be calculated from a graph like this unless the quantity of copper in the solution was known ahead of time.

PROBLEMS

1. *Absorbance and transmittance*. The scale on the front panel of a certain manual spectrophotometer reads from zero to 100% transmittance. What are the corresponding absorbance values at these two ends of the scale?

2. *Absorbance and transmittance*. If absorbance were defined as $\ln (P_0/P)$ instead of $\log (P_0/P)$, (a) what would be the absorbance values in Problem 1; (b) what would be the absorbance of a solution with a percent transmittance of 36.8%? (c) If we actually decided to change to natural logarithms for all chemical usages and defined absorbance as $A = \ln (P_0/P) = \varepsilon bc$, would ε-values tabulated in the literature become larger or smaller or remain the same as the current ones?

3. *Absorbance and transmittance*. Convert the following values of % T to absorbance values: (a) 90; (b) 70; (c) 45; (d) 7.0; (e) 1.0.

4. *Absorbance and transmittance*. Convert the following absorbances into percent transmittance values: (a) 0.040; (b) 0.100;

(c) 0.500; (d) 1.000; (e) 1.400; (f) 3.000.

5. *Absorbance and transmittance*. The percent transmittance of a certain solution, measured at a certain wavelength in a 1.00-cm cell is 63.5%. Calculate the percent transmittance for this solution in cells with path lengths of (a) 1.00 mm; (b) 2.00 cm; (c) 10.00 cm.

6. *Absorptivity and molar absorptivity*. The organic compound 2,3-dimethoxybenzaldehyde ($C_9H_{10}O_3$, MW 166.2) exhibits an ultraviolet absorption band at 220 nm in ethanol solution. A solution containing 0.298 mg of the compound in 20 mL of ethanol gave an absorbance of 1.73 in a 1-cm cell. Calculate (a) the absorptivity and (b) the molar absorptivity of the compound at 220 nm.

7. *Absorptivity and molar absorptivity*. The compound in Problem 6 has another ultraviolet band at 259 nm, where the molar absorptivity is 8.64×10^3. What is the absorbance of the solution in Problem 5 at 259 nm?

8. The colorless, water-soluble organic reagent 1,10-phenanthroline (1,10-phen) forms a

deep-red chelate complex ion with Fe^{2+}, $[Fe(Phen)_3]^{2+}$. This is sometimes utilized in the spectrophotometric determination of iron. The wavelength of maximal absorption, λ_{max}, is 510 nm. The color is developed by treating iron-containing solutions with a reducing agent such as hydroxylamine to convert all the iron to Fe^{2+} and then adding an excess of Phen.

A standard iron solution was prepared, and a 5.00 mL aliquot of this solution, containing 20.00 μg of iron, was treated as described above and diluted to a final volume of 10.00 mL. The absorbance of this solution at 510 nm in a 1.00-cm cell was found to be 0.397. Calculate the molar absorptivity of $[Fe(Phen)_3]^{2+}$ at 510 nm.

9. *Beer's law.* Exactly 5 mL of an unknown iron-containing solution was treated in exactly the same manner as was the standard solution in Problem 8. The absorbance at 510 nm was 0.204 in a 1.00-cm cell. Calculate the iron concentration of the unknown solution in μg/mL.

10. *Dilution (blood volume) and Beer's law.* A person's blood volume can be estimated by injecting a known amount of a harmless dye into a vein and determining its concentration in the blood plasma spectrophotometrically a few minutes later, when it has been well mixed by the circulation. The assumptions are made that there are no unmixed, stagnant pools of blood in the body, that the dye remains totally in the vascular system, and that it is confined in the blood *plasma*; i.e., it does not penetrate the cellular elements of the blood. Having obtained the plasma volume, one calculates the total blood volume from independent information about the fractional volume of the plasma in whole blood.

A laboratory has a vial of sterile solution containing a dye called Evans blue. Exactly 1.00 mL of this solution is injected intravenously into a 75-kg human male subject. Ten minutes later, a blood sample is withdrawn from the man. The blood is centrifuged to separate plasma from cells, and it is found that the plasma represents 53% of the total volume. The absorbance of the plasma in a 1.00-cm cell at an appropriate wavelength is 0.323.

Another 1.00-mL sample of the original dye solution is diluted to exactly 1 liter in a volumetric flask. Then 10 mL of this solution is further diluted to 50 mL in a volumetric flask. The absorbance of the final solution is 0.200.

Calculate the man's blood volume in liters.

11. *Molecular weight by spectrophotometry.* Many amines form crystalline salts with picric acid. Solutions of the salts in ethanol all have about the same ultraviolet spectrum, with $\lambda_{max} = 380$ nm and $\varepsilon = 1.35 \times 10^4$, due to the picrate component. The molecular weight of picric acid (2,4,6-trinitrophenol, $C_6H_3N_3O_7$) is 229.1.

A sample of an unknown amine was converted into the picrate, and the salt was recrystallized and dried. Then 19.61 mg of the salt was dissolved in 100 mL of ethanol, and 10 mL of this solution was diluted to 100 mL. The absorbance of the resulting solution was 0.721 at 380 nm in a 1.00-cm cell. (a) Calculate the molecular weight of the original amine. (b) The organic chemist who had the amine suspected, on other grounds, that it was 4-nitroaniline ($C_6H_6N_2O_2$). Is the molecular weight determination consistent with this?

12. *Spectrophotometric analysis.* A 1.000-g sample of an unknown steel is dissolved in nitric acid, and the manganese is oxidized to permanganate with potassium persulfate. The solution is then diluted to 500 mL in a volumetric flask. Then 0.658 g of a steel sample from the National Bureau of Standards, certified to contain 0.31% Mn, is treated in exactly the same way. The absorbance of the unknown solution is 1.79 times as great as that of the standard. Calculate the percentage of Mn in the unknown steel.

13. *Spectrophotometric analysis.* A 0.570-g sample of an alloy steel is dissolved, the manganese is oxidized to permanganate, and the solution is diluted to 100 mL in a volumetric flask. The absorbance at 525 nm in a 1.00-cm cell is 0.523. The molar absorptivity of MnO_4^- at 525 nm is 2.24×10^3. Calculate the percentage of Mn in the steel.

14. *Relative error.* Suppose we are using a spectrophotometer for which the error treatment on page 413 is valid. Calculate the relative error in concentration per 1% photometric error

for the following values of the percent transmittance: (a) 1.0%; (b) 10%; (c) 40%; (d) 70%; (e) 98%.

15. *Differential spectrophotometry.* A solution of concentration c gives an absorbance reading of 0.250 when the instrument scale is set at $A = 0.000$ with pure solvent in the beam. (a) Calculate the absorbance readings for solutions of concentrations $2c$, $3c$, and $4c$. (b) Calculate the percent transmittances for the four solutions and examine the *differences* in % T values between the c and $2c$ solutions, the $3c$ and $2c$ solutions, and the $4c$ and $3c$ solutions. (c) Now suppose that the instrument is set to read $A = 0.000$ with the solution of concentration c in the beam. Repeat the calculations of parts (a) and (b).

16. *Multicomponent analysis.* The following problem has been simplified from a more sophisticated study by J. J. Lingane and J. W. Collat [*Anal. Chem.*, **22**, 166 (1950)]; we are neglecting some small corrections used by the authors to counter certain interferences. The table gives molar absorptivities for $Cr_2O_7^{2-}$ and MnO_4^- at two wavelengths:

	ε-values	
	440 nm	545 nm
$Cr_2O_7^{2-}$	369	11
MnO_4^-	95	2350

A steel sample weighing 1.000 g is dissolved in appropriate acids (H_2SO_4, H_3PO_4, and HNO_3) and treated with persulfate and periodate to oxidize Mn to MnO_4^- and Cr to $Cr_2O_7^{2-}$. The final solution from the preliminary treatment is diluted to 100 mL in a volumetric flask, and the absorbance values are determined in a 1.00-cm cell at 440 nm ($A = 0.108$) and 545 nm ($A = 1.296$). Calculate the percentages of Cr and Mn in the steel.

17. *Spectrophotometric determination of blood alcohol.* The compound nicotinamide-adenine dinucleotide, abbreviated NAD, is an important biological coenzyme which participates in many enzyme-catalyzed redox reactions in cellular metabolism. The basic redox couple may be written

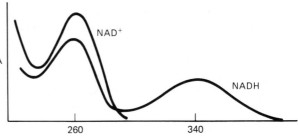

$$E_{pH=7}^{0'} = -0.32 \text{ V}$$

R is large (the molecular weight is over 600) but the redox activity of the molecule involves only the pyridine ring as depicted. $E^{0'}$ is a formal potential (page 271). We use NAD for the compound in general, and NAD^+ specifically for the oxidized form and NADH for the reduced form. NAD is a central compound in the redox affairs of living cells; also, it may be purchased for use in the laboratory, as may many of the enzymes which catalyze its various reactions.

Given below are a sketch of the ultraviolet spectra of NAD^+ and NADH and some actual molar absorptivity values:

	ε-values	
	260 nm	340 nm
NAD^+	18,000	0.000
NADH	14,400	6220

The redox reactions involving NAD are often *very* slow unless catalyzed, and the enzyme catalysts often provide a high degree of selectivity in allowing just one component of a complex mixture to interact at an appreciable rate with the NAD system. For example, with the enzyme alcohol dehydrogenase (alcohol:NAD^+-oxidoreductase) the following reaction is catalyzed:

$$CH_3CH_2OH + NAD^+$$
$$= NADH + CH_3CHO + H^+$$

A subject is brought to a laboratory for a blood alcohol test after an automobile accident. The lab takes a blood sample, separates the plasma from the cells by centrifugation, and pipets 0.200 mL of the plasma into a volumetric flask. Some pH 7 buffer solution is added, followed by a *large* excess of NAD^+ and some alcohol dehydrogenase enzyme. After a few minutes, the solution is diluted to exactly 25 mL and its absorbance at 340 nm is measured and found to be 0.849 in a 1.00-cm cell. Calculate the concentration of alcohol (MW 46.1) in milligrams per 100 mL of blood plasma.

18. *Absorbance and emf with NAD.* Use data on the NAD^+–NADH system given in Problem 17. A biochemist takes exactly 0.00100 mmol of NAD^+, some ethyl alcohol, some alcohol dehydrogenase enzyme, and some pH 7 buffer solution and dilutes to a final volume of 10 mL in a volumetric flask. After the solution has equilibrated, the absorbance at 340 nm is 0.480 in a 1.00-cm cell. NADH is the only absorbing species at this wavelength. (a) What is the single-electrode potential for the NAD^+–NADH couple in the solution? (b) What is the single-electrode potential for the system

$$CH_3CHO + 2H^+ + 2e = CH_3CH_2OH$$

in the same solution?

19. An enzyme called uricase or urate:O_2-oxidoreductase catalyzes the air oxidation of urate (salt of uric acid) to allantoin:

$$Urate + O_2 = allantoin + CO_2$$

The equilibrium lies far to the right, and the disappearance of urate is practically quantitative, but the rate is negligible in the absence of the enzyme. At pH 8.5, uric acid ($pK_A = 3.9$) is practically all in the urate form, with an ultraviolet band at 290 nm, $\varepsilon = 10,965$.

A certain unknown solution contains urate as well as a number of other substances, some of which absorb at 290 nm; urate is the only component, however, which is acted upon by the enzyme. To 5.00 mL of the sample solution are added a borate buffer of pH 8.5 and enough water to give a final volume of 10.00 mL. The absorbance of this solution is 0.881 in a 1.00-cm cell at 290 nm. To a

second 5.00-mL aliquot of the sample solution are added the same buffer, some enzyme solution, and water to make 10.00 mL. Air is bubbled through the solution for an adequate time to allow complete reaction, and the absorbance of this solution in the same cell is 0.827. Calculate the concentration in milligrams per 100 mL of urate ion (MW 157.1) in the unknown solution.

20. *Stability of a metal complex ion by spectrophotometry.* The metal ion M^{2+} forms a complex ion MX_3^- with the anion X^-. If a *very* large excess of X^- is present, we may assume that virtually all of the metal is in the MX_3^- form. There are no other complexes such as MX^+ to worry about for the purposes of this problem. The MX_3^- complex has an absorption band at 372 nm; no other species that may be present absorb at this wavelength. A volume of an M^{2+} solution furnishing 5.00×10^{-3} mmol of M^{2+} is pipetted into a 10-mL volumetric flask. Then 1.00 mL of a 2.000 M solution of NaX is added, followed by water to the 10-mL mark on the flask. The absorbance of the resulting solution at 372 nm in a 1.00-cm cell is 0.763. A second solution is prepared using the same quantity of M^{2+}, but only enough NaX solution is added to provide 2.00×10^{-2} mmol of X^-. After dilution to 10 mL and thorough mixing, the absorbance of this solution is 0.610 in the same cell. Calculate the stability or formation constant of MX_3^-, i.e., the equilibrium constant for the reaction $M^{2+} + 3X^- = MX_3^-$.

21. *Formula of a complex by method of continuous variations.* The formula of a complex formed by a reaction such as $M + nX = MX_n$ can be determined by the method of continuous variations. The absorbance values of a series of solutions of varying composition are measured at the wavelength at which the complex shows its maximal absorptivity. The *total* number of moles of M *and* X is kept constant, while the mole fractions of the reactants are varied.

The following data were obtained for the reaction of M and X. Solutions were prepared by mixing a 0.0100 M solution of M with a 0.0100 M solution of X, with volumes of each as shown in the table. Assume that absorbance values measure $[MX_n]$; i.e., no other species absorbs at the selected wavelength.

no.	mL of M	mL of X	absorbance
1	10.00	0.00	0.000
2	9.00	1.00	0.110
3	8.00	2.00	0.220
4	7.00	3.00	0.329
5	6.00	4.00	0.435
6	5.00	5.00	0.485
7	4.00	6.00	0.436
8	3.00	7.00	0.330
9	2.00	8.00	0.222
10	1.00	9.00	0.108
11	0.00	10.00	0.000

(a) Plot the absorbance vs. the milliliters of X and determine the formula of the complex, i.e., the value of n. (b) How do you interpret the curvature near the middle of the plot?

22. *Formula of a complex by the mole-ratio method.* The formula of the complex MX_n can also be determined by the mole-ratio method. Solutions are prepared in which varying quantities of X are added to a constant quantity of M, and the absorbance due to MX_n is measured.

To 4.00-mL portions of a 0.0100 M solution of M are added varying volumes of a 0.0100 M solution of X, and the solutions are diluted to 20 mL with an appropriate background solution. The data are shown below.

no.	mL of X	absorbance
1	0.00	0.000
2	2.00	0.125
3	3.00	0.185
4	4.00	0.248
5	5.00	0.310
6	7.00	0.435
7	8.00	0.498
8	9.00	0.559
9	11.00	0.685
10	12.00	0.738
11	13.00	0.771
12	14.00	0.775
13	15.00	0.775

(a) Plot the absorbance vs. milliliters of X and calculate the formula of the complex. (b) How do you interpret the slight curvature between about 11 and 14 mL? (c) *Mole-ratio method* is another term for what technique discussed in this chapter?

23. *Photometric titration.* Bi(III) forms a water-soluble chelate complex with EDTA which we abbreviate BiY^- (see page 201 for a complex of similar structure). The effective stability constant of the complex, K_{eff} (page 204), is about 3.2×10^{14} at pH 2 (based upon $K_{abs} \cong 8.7 \times 10^{27}$ and $\alpha_4 = 3.7 \times 10^{-14}$ at pH 2). The molar absorptivity of BiY^- is 9300 at its ultraviolet peak at 265 nm. No other species in the solution below absorb at this wavelength.

A 100-mL portion of 2.00×10^{-4} M Bi^{3+} containing a chloroacetate buffer of pH 2 was titrated with a 1.00×10^{-2} M EDTA solution in a special titration vessel which could be placed in the cell compartment of a spectrophotometer and provided a path length of 1.50 cm. Calculate the absorbance, corrected for dilution, after addition of the following volumes of EDTA solution: 0.20, 0.50, 0.80, 1.10, 1.50, 2.20, 2.50, and 3.00 mL. Plot the titration curve, absorbance vs. milliliters. Suppose an analyst decided that the absorbance values in the titration were going to be too large under the above conditions, yet did not wish to titrate smaller samples; also, he has no other titration cell that will fit in the spectrophotometer. What can he do to obtain absorbance readings which do not go beyond the useful part of the instrument scale?

24. pK_a *of an indicator by spectrophotometry.* To each of a series of 1-liter volumetric flasks are added exactly 1 mmol of a weak acid indicator, HIn, and 1 mmol of the salt NaIn. The solutions are diluted to the marks on the flasks with various aqueous buffer solutions. The pH values are measured with a pH meter, and the absorbance values are measured at 650 nm in a 1.00-cm cell. The data are given below.

pH	absorbance
1.00	0.000
2.00	0.000
3.00	0.000
7.00	0.557
10.00	0.796
11.00	0.796
12.00	0.796

Calculate the molar absorptivities of HIn and In$^-$ at 650 nm and calculate the pK_a of the indicator.

25. *Standard addition in atomic absorption spectrophotometry; determination of lead in plants.* As part of an investigation of lead in biological ecosystems, some weeds are gathered along a highway where they have been exposed to emissions from gasoline engines. A representative sample of the weeds weighing 6.250 g is ashed to destroy organic matter, and the inorganic residue is treated according to a standard recipe and diluted to 100 mL. A portion of this solution is aspirated into an acetylene-air flame, and an absorbance value of 0.125 is found using the 283.31-nm Pb line from a hollow cathode source. Next, 0.100 mL of a standard solution containing 55.0 μg of Pb per milliliter is added to 50 mL of the sample solution. When this "spiked" sample is aspirated into the flame, the absorbance is found to be 0.180. Calculate the Pb content of the plant material in micrograms per gram, i.e., parts per million (ppm).

15

Emission and Luminescence Spectroscopy

We saw in Chapter 14 that the absorption of radiation by atoms and molecules provides analytical methods for a wide variety of chemical species. In absorption spectroscopy or spectrophotometry, we were not concerned with the fate of the excited species; measuring the absorption process occupied our full attention. In this chapter, by contrast, we are very much concerned with the behavior of atoms and molecules following excitation. Just as there exists everywhere in nature the tendency to relieve a strain, excited species will normally revert to lower energy states. There are essentially three basic processes by which such deactivation may occur. First, by a series of steps involving collisions with other molecules, the energy may show up as heat. This will not be noticed in spectrophotometry under ordinary conditions; low radiant power input into dilute solutions will not warm the sample appreciably. Second, excited molecules may undergo chemical reactions. There is little of this to be seen in a spectrophotometer cell, but organic chemists use photoexcitation to induce unusual reactivity. A molecule in an excited electronic state is a different reactant from one in the ground state, and it will sometimes undergo reactions which would be impossible for ground-state molecules. Typically, the organic photochemist will use powerful ultraviolet sources to irradiate solutions that are much more concentrated than those encountered in analytical spectrophotometry.

The third process by which excited atoms and molecules revert to lower energy levels is the emission of radiation, and the measurement of this radiation is the subject of the present chapter. The introductory nature of this text, as well as space limitations, require that our treatment be brief. We are dealing with electronic transitions, and at room temperature, atoms and molecules in general exist in the ground electronic state. Thus before we can observe emission of radiation, we must provide excitation. There are essentially three modes of excitation.

First, some chemical reactions generate products in excited states, sometimes under what appear to be mild conditions. The most prominent of these are found in biological systems and are sometimes referred to as *bioluminescence*. A number of bacteria and fungi, for example, emit light; among higher organisms, the firefly is a well-known case. Compounds called *luciferins,* of different structures in different organisms, undergo enzyme-catalyzed oxidations to excited-state products which emit light. A few analytical chemists are investigating applications of *chemiluminescence*. Biochemists have used the firefly reaction for many years as an analytical tool: the reaction requires an input of energy provided by a compound called adenosine triphosphate, or ATP; an analytical method for determining ATP concentrations in biological samples is based upon measuring the intensity of light emission when ATP is added to buffered solutions prepared from dried firefly lanterns or freeze-dried extracts thereof. This can be done in a spectrophotometer with the source turned off, simply using the detector to measure the emission intensity from a cell containing the reactants.

Second, excitation may be accomplished by the absorption of radiation. We discuss this briefly under the heading of *fluorescence*. Third, we may employ thermal excitation, as in emission spectroscopy with such sources as flames, plasmas, arcs, and sparks. This is also described briefly.

Although we measure emission by excited species in all of these cases, regardless of the mode of excitation, the term *emission spectroscopy* by common usage generally refers only to methods where *thermal* excitation is employed. With the other two methods of excitation mentioned above, we speak of *luminescence,* sometimes using a prefix to specify the excitation mode, as in chemiluminescence. Fluorescence, then, is a form of photoluminescence, as is phosphorescence (see below).

15.1

MOLECULAR FLUORESCENCE[1]

15.1a. *Principles of Fluorescence*

The Absorption Process

The absorption process which leads to fluorescence usually involves a $\pi-\pi^*$ electronic transition (page 396) in an organic molecule. The process is shown in the simplified energy level diagram in Fig. 15.1. Rotational levels are omitted from the diagram; in condensed phases such as the solutions we ordinarily use, they are "smeared out" by neighboring molecules, and they would not be resolved by most instruments in any case. The molecule absorbs radiation, la-

[1] There is a technique called *atomic* fluorescence (AF) in which a ground-state atom population, obtained as in atomic absorption (Chapter 14), is irradiated and the emitted fluorescent radiation measured. Although continuing research suggests that AF, especially with laser excitation, may be the most sensitive version of atomic spectroscopy, it has not become a routine tool in analytical chemistry, nor are commercial instruments currently available. For an introduction, see J. D. Winefordner, *J. Chem. Educ.*, **55**, 72 (1978), and for recent literature, J. A. Holcombe and D. E. Bass, *Anal. Chem.*, **60**, 226R (1988).

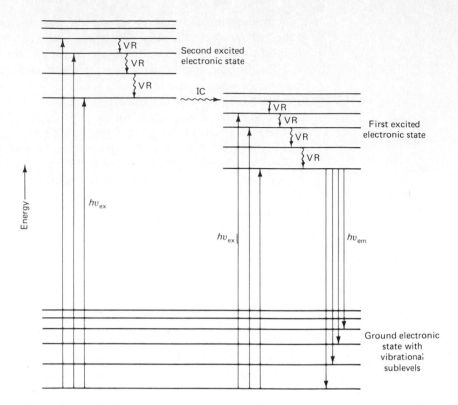

Figure 15.1 Energy diagram for fluorescence. Molecules are excited by absorption of photons of energies represented by the upward $h\nu_{ex}$ arrows; fluorescent emission is depicted by the downward $h\nu_{em}$ arrows. The wavy arrows represent nonradiative processes (VR = vibrational relaxation and IC = internal conversion, see text).

belled $h\nu_{ex}$; in this process, which probably takes no longer than 10^{-15} s, an electron is raised from the ground electronic state to an excited state. The Bouguer-Beer law (Chapter 14) describes the absorption situation. In Fig. 15.1, we show all the excitation transitions as originating from the ground vibrational level of the lowest electronic state. This is realistic; at room temperature, unperturbed molecules will all be in the ground electronic state, and here the lowest vibrational level will be, by far, the most heavily populated. Transitions can occur, though, into various vibrational levels of the excited electronic state, depending upon the exact energies of the absorbed photons. These transitions are depicted by the middle set of straight arrows in the figure. In a liquid phase, the excess vibrational energy is usually dissipated rapidly by collisions with solvent molecules, a process called *vibrational relaxation*. Nonradiative transitions of this sort are shown in the figure by the wavy arrows labelled VR. Thus fluorescent emission typically involves an energy transition between the lowest vibrational level of the excited electronic state and the ground electronic state. The transition can terminate in various vibrational levels of the latter, as shown by the right-hand set of arrows labeled $h\nu_{em}$, although subsequently most of the molecules will relax nonradiatively to the lowest one.

Excitation may also place molecules in a still higher electronic state, as shown on the left in the figure. Sometimes the lowest vibrational level of the higher excited electronic state and a higher vibrational level of the first excited electronic state are of comparable energy. Molecules in the higher electronic state, after relaxation to the lowest vibrational level, may then pass to an equally energetic vibrational level of the first excited electronic state, a process called *internal conversion*, shown by the wavy arrow labelled IC. They then relax to the lowest vibrational level of the first excited electronic state prior to fluorescent emission.[2] Thus, again, although the excitation was more energetic, the emission corresponds to the same transitions from the lowest vibrational level of the first excited electronic state to various levels of the electronic ground state. The net result of these processes is usually fluorescent emission of lower frequency or longer wavelength than that of the exciting radiation.

Relaxation Time: The Distinction between
Fluorescence and Phosphorescence

Usually fluorescent emission occurs very rapidly, perhaps from about 10^{-9} to 10^{-7} s after absorption of the exciting photon. With ordinary instruments, the observation of fluorescence ceases when the excitation is turned off. There are exceptions, however. In the ground state, most organic molecules (free radicals are an exception) have an even number of electrons and they are all spin-paired. It is possible, however, for one electron to flip its spin when the molecule is in an excited state. A molecule in this situation has a new set of excited energy levels, not indicated on the diagram in Fig. 15.1. The quantum mechanical description of this is far beyond the level of this book, but we may simply consider it a fact that a transition from an excited state with an unpaired spin to the ground state, where all electron spins must be paired, is an improbable one. Thus the excited state lifetime is much longer than in ordinary fluorescence, from, say, 10^{-4} s to perhaps 10 s or even longer, and emission may then persist for an appreciable time after excitation is discontinued. This phenomenon is called *phosphorescence*. Because of the time delay, there is more opportunity for nonradiative deexcitation by molecular collisions, and appreciable phosphorescence is seldom observed in solutions near room temperature; it is usually studied by dissolving organic molecules in solvents which set to rigid "glasses" at temperatures approaching $-200°C$. However, there have been interesting observations of room temperature phosphorescence, for example, by molecules incorporated in structured aggregates called *micelles* that are formed by surfactants in aqueous solution.[3]

Ideally, the relation between concentration c, of fluorescent molecules in solution and the emitted radiant power, P_{em}, would be linear:

$$P_{em} = kc$$

The "constant" k really represents a complicated mix of several factors. Because only absorbed radiation can possibly induce fluorescence, the incident radiant power is important, and there will be an ε-value and pathlength (page 400); also included will be a factor giving the fraction of excited molecules that deexcite by photon emission rather than by nonradiative processes. In a real instrument, the wavelength-dependent response of the detector to radiant power, as well as the fraction of the fluorescent emission that actually reaches the detector, will also be involved in the magnitude of the readout. With solutions which are sufficiently

---

[2] There are a few cases of emission from excited electronic states higher than the first (the organic compound azulene and some of its derivatives are notable examples), but this type of transition is rare because of the rapidity of the relaxation processes described above.

[3] L. J. Cline-Love, M. Skrilec, and J. G. Habarta, *Anal. Chem.*, **52**, 754 (1980).

dilute, a linear relation between electrical signal and concentration is actually observed in many cases. "Sufficiently dilute" will vary with the individual analyte, but usually this means something on the order of a few parts per million (μg/ml).

The Inner-Filter Effect

At higher concentrations, fluorescence becomes less than directly proportional to concentration and may even diminish as the concentration increases, as shown in Fig. 15.2. At high concentrations, the distribution of exciting radiation is not uniform throughout the solution. The first layers of solution may absorb enough to deprive deeper layers of full excitation; i.e., the exciting radiant power, P_0, will decrease appreciably across the width of the cell. This is sometimes called the *inner-filter effect;* it is ordinarily not serious if the solution absorbs no more than 5 or 10% of the incident radiation.

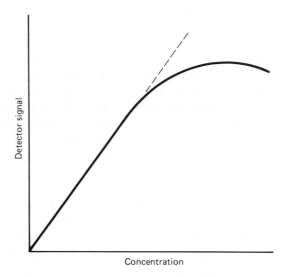

Figure 15.2 Fluorescent emission as a function of concentration. See text for discussion of the nonlinear region.

Quenching

Other processes that diminish fluorescent output may be lumped under the heading of *quenching*. There are a number of molecules that are very effective quenchers and hence interfere in fluorometric analyses. One such process may be written as follows:

$$\text{Excited analyte molecule} + \text{quencher} \longrightarrow \text{ground state analyte molecule} + \text{quencher} + \text{heat}$$

That is, the quencher induces a nonradiative deexcitation of the excited analyte molecule, and no photon is emitted. Oxygen, for example, is a serious quencher for some fluorescent aromatic hydrocarbons, and it is sometimes necessary to deoxygenate the solutions. In developing an analytical method based upon fluorescence, one must evaluate possible quenching activity by extraneous sample components.

Sensitivity

An outstanding attribute of fluorescence analysis is high sensitivity as compared with other common techniques such as spectrophotometry. It is inherently better to measure a little light vs. none than it is to measure a small diminution in a bright beam. The power of the fluorescent emission, P_{em}, can be measured independently of the incident power, P_0. Thus emission can be enhanced by increasing P_0. In spectrophotometry, increasing P_0 also increases P; thus the absorbance, $\log (P_0/P)$, is unchanged. Likewise, amplification of signals representing P_0 and P will not change the ratio P_0/P or its logarithm. This is quite different from a fluorescence instrument (see below), where the incident beam does not pass to the detector.

If we suppose that a very good spectrophotometer can detect a sample absorbance of 0.0001 (for most instruments this would be pushing to the limit), then for a compound with an ε-value of 10^5 (a very large one) in a 1-cm cell we would have a detection limit of

$$c = \frac{A}{\varepsilon b} = \frac{10^{-4}}{1 \times 10^5} = 10^{-9}\, M$$

We seldom do this well; $10^{-6}\, M$ would probably be a much more representative detection limit. Fluorescence detection limits, on the other hand, are frequently on the order of $10^{-9}\, M$ or less, and with special detection techniques, $10^{-12}\, M$ has been approached. As a rough guide, it would not be misleading to say that fluorescence is typically a thousand times more sensitive than spectrophotometry, although the actual values depend upon what compounds are involved and which instruments are available.

15.1c. *Instrumentation for Fluorescence Measurement*

The essential components of an instrument for measuring fluorescence are shown in Fig. 15.3. Note that the same components (source, monochromator, etc.) are found in spectrophotometers (Fig. 14.8), but notice that there are two monochromators and that sample emission is monitored by the detector at an angle of 90° to the exciting beam. (An actual instrument may have an external shape quite different from that of the schematic in Fig. 15.3 through the use of mirrors to send beams in directions that save space, but the right-angle configuration is retained at the sample cell.) Accessories such as lenses for transmitting exciting and emitted radiation efficiently through the system are not shown in the figure, and

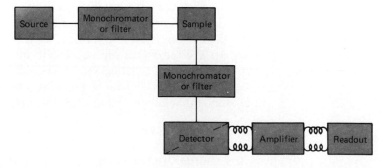

Figure 15.3 Block diagram showing components of a fluorometer or spectrofluorometer.

there may be other features such as scanning monochromator motor drives which are also omitted.

Sources

The best general-purpose source is the xenon arc lamp. The electrical discharge in the gas is initiated by high-voltage ionization of some of the Xe atoms, after which the current maintains itself at about 7.5 A and 20 V dc (150 W). The pressure is high enough to broaden the Xe emission lines to a continuum that is useful for excitation in the UV-visible region down to slightly above 200 nm. The lamp compartment is usually cooled by a fan, and the outflowing air contains ozone (O_3) from the ultraviolet irradiation of O_2, which must be vented properly because of its toxicity. An alternative source that is sometimes used is a high pressure mercury vapor lamp, which emits a continuum with superimposed Hg emission lines.

Wavelength Selection

Available instruments span a wide range of costs and sophistication. In some inexpensive models, designed for a few determinations where ultraviolet excitation leads to intense visible fluorescence, only simple filters are provided. Research instruments use monochromators, generally of the grating type, so that both excitation and emission can be well characterized. This provides maximal flexibility for analyses of samples containing mixtures of fluorescent species, because frequently a particular combination of excitation wavelength with wavelength-selective monitoring of emission provides an opportunity to discriminate among the several components.

Monochromatic Instruments

The better instruments provide for automatic scans of both excitation and emission wavelengths, with recording of the detector signal. The principle is as follows. Suppose we are developing a new method for determining a particular fluorescent compound. In a darkened room, we shine a hand-held ultraviolet lamp on a solution of the compound and observe the emission of a bluish glow. The monochromator between the sample and the detector in Fig. 15.3 is then set at a wavelength in the blue region of the spectrum, say, 450 nm. Now we use a motor drive on the excitation monochromator to record the detector signal as a function of exciting wavelength. Let us say that we obtain a graph showing maximal signal at 260 nm. Hence we set the excitation monochromator at 260 nm and scan the fluorescent emission as a function of wavelength. Perhaps we find several emission bands, with the largest at 428 nm. Now, if we are very conservative people, we may repeat the process: set the emission monochromator at 428 nm and redetermine the exciting wavelength that gives the maximal signal; then excite at that wavelength and reexamine the wavelength distribution on the emission side. We end up with optimal excitation and emission wavelengths, and at these settings we are ready to measure detector response as a function of concentration with a series of standard solutions and unknown samples.

Detection of Radiation

Research instruments with diode array detectors have been devised which rapidly yield a three-dimensional data set: the output is a fluorescence intensity surface upon a base of excitation and emission wavelengths which is particularly advantageous with mixtures of fluorescent compounds. A computer operates on the corresponding numerical array, called an excitation-emission matrix (EEM), to print out analytical results. With ordinary instruments, the detector is usually a photomultiplier tube (page 408). The readout of the amplified detector signal may involve a voltmeter, a pen-and-ink record of voltage vs. time (which in a spectral scan is voltage vs. wavelength), or a printout from an interfaced computer.

15.1d Applications of Fluorescence

There are only a few fluorescent inorganic ions, of which the best known example is the uranyl ion, UO_2^{2+}. Most fluorometric analyses involve organic molecules, some examples of which are listed in Table 15.1. There are some fluorescent metal chelate compounds which provide sensitive methods for several metal ions. Frequently the metal chelate is extracted from aqueous solution into an organic solvent prior to the measurement, a process which may simultaneously effect a separation from interfering ions and concentrate the fluorescent species (Chapter 16, page 466). There are, for instance, numerous fluorometric reagents for aluminum and beryllium. Heavier metals such as Fe^{3+}, Co^{2+}, Ni^{2+}, and Cu^{2+}, on the other hand, tend to kill the fluorescence which many chelating agents themselves exhibit; the presence of the metal in the complex somehow promotes nonradiative dissipation of absorbed energy.

Sometimes a nonfluorescent analyte is convertible into a strongly fluorescent molecule by a rapid and quantitative reaction which is easily incorporated into an overall analytical procedure. For example, the hormone epinephrine (adrenalin) is easily converted into adrenolutin:

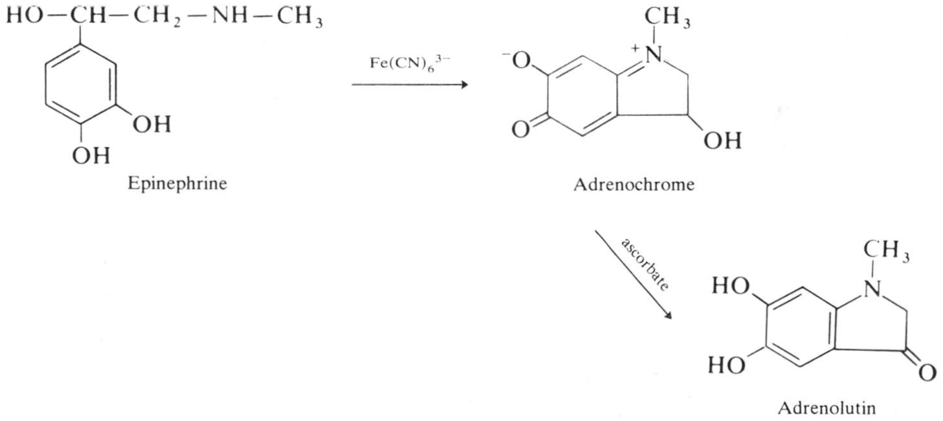

In alkaline solution, the phenolate anion of adrenolutin is strongly fluorescent (excitation, 360 nm; emission, 530 nm). Patients with certain tumors of the adrenal gland and also some hypertensives show elevated levels of epinephrine in the urine. The hormone (present at very low levels) can be concentrated from a large volume of urine by an ion-exchange procedure (Chapter 18) at a pH where the amino nitrogen is protonated to form a cation, $R—\overset{+}{N}H_2—CH_3$, eluted in a small volume by displacement with H^+, and treated as above to form the fluorophore.

TABLE 15.1 Some Compounds Exhibiting Analytically Useful Fluorescence

CLASS	COMMON EXAMPLE(S)	FORMULA
Polycyclic aromatic hydrocarbons	Pyrene	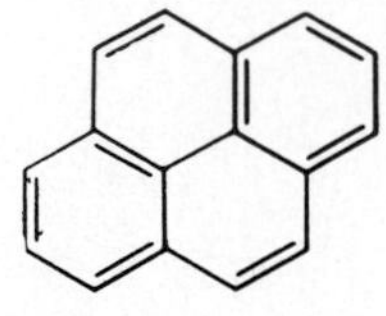
Aromatic amines	β-Naphthylamine	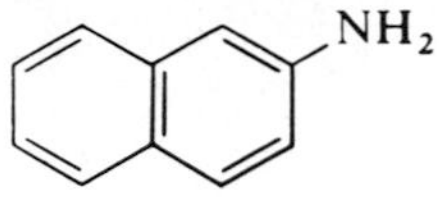
Phenols	Phenol	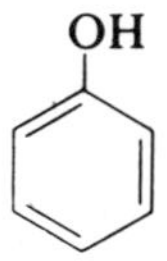
	β-Naphthol	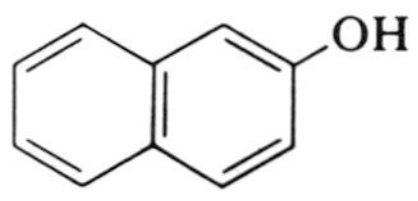
Metal chelates	Aluminum 8-hydroxyquinolate	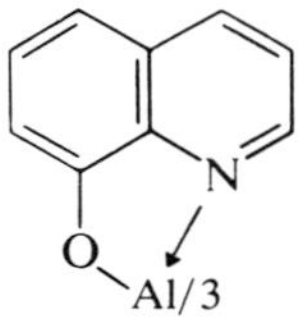
Vitamins	Pyridoxine (formula as hydrochloride salt)	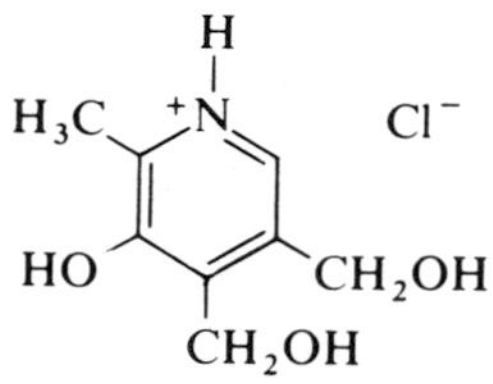
Drugs	Salicyclic acid	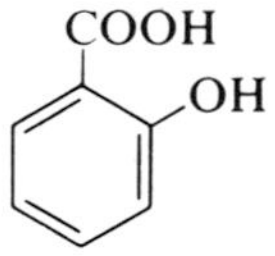
	Acetylsalicyclic acid (aspirin)	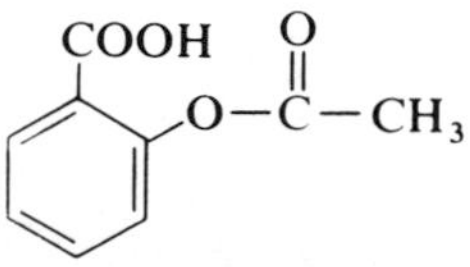
	Amphetamine	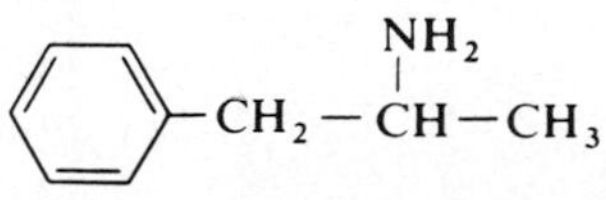

TABLE 15.1 *(cont.)*

CLASS	COMMON EXAMPLE(S)	FORMULA
	Barbiturates	
	Quinine	
Amino acids	Tyrosine	
	Tryptophan	
	(these are exceptions; most amino acids do not fluoresce)	

Several vitamins can be determined fluorometrically. Mild oxidation of thiamine (vitamin B_1) by $Fe(CN)_6^{3-}$, for example, yields a product called thiochrome which exhibits a blue fluorescence under appropriate conditions. If the fluorescent emission is measured on two portions of the sample, one treated with ferricyanide and the other untreated, one can subtract the contributions of fluorescent non-thiamine interferences to improve the selectivity. Riboflavin (vitamin B_2) and pyridoxine (B_6) are two other vitamins which can be determined by fluorescence.

Although most amino acids are nonfluorescent, they react readily with the reagent fluorescamine to form highly fluorescent compounds which have been used in biochemistry to detect fractional nanogram quantities. Fluorescence methods are very promising for determining several of the polycyclic aromatic hydrocarbons which have been classed as "priority pollutants" by the Environmental Protection Agency. It is mentioned in Chapter 18 that fluorescence provides very sensitive detection of certain sample components in liquid chromatography.

15.2

EMISSION SPECTROSCOPY WITH THERMAL EXCITATION

As we saw in the previous section, fluorescence is a special form of emission spectroscopy in which excited species are obtained by the absorption of electromagnetic radiation. The excitation is certainly not weak on a "per molecule" basis: In the UV-visible region (~200 to 700 nm), absorption represents an energy input on the order of 40 to 150 kcal/mol. To obtain the equivalent thermally would require a temperature of many thousands of degrees, which few molecules would survive. But selective excitation of particular molecular electronic transitions in dilute solutions by relatively low wattage power sources allows analytical exploitation of the transitions without the sample decomposition that would occur if all the molecules in the assemblage were subjected to the high temperatures of thermal excitation. Analytical chemists have long used thermal excitation, however, to induce the emission of radiation by *atoms*.[4] We saw in Chapter 14 that we could obtain an analyte atom population by the introduction of samples into flames or furnaces, but there we measured the absorption of radiation by ground-state atoms. The purpose of the thermal source was only to produce the atom population from the introduced sample. In emission spectroscopy, on the other hand, we measure the radiation emitted by that fraction of the atoms which are electronically excited at the temperature involved. A number of sources for sample atomization and excitation are employed, including flames, furnaces, electric arcs, sparks, plasmas, and laser beams. Our discussion will necessarily be very brief, with the emphasis on flames.

15.2a. *Flame Emission Spectroscopy (Flame Photometry)*

Many students have probably seen the emission of yellow light when a small amount of sodium chloride was introduced into a flame. Perhaps some have seen colored flames in fireplaces resulting from impregnation of the logs with various salts, or colored fireworks based upon the same idea. Bunsen and Kirchhoff discovered the elements cesium (1860) and rubidium (1861) by observing lines in flame emission spectra that were not ascribable to previously known elements. Quantitative analysis based upon measurement of the light owes its origin mainly to the Swedish agronomist Lundegardh, who developed improved burners and methods of sample introduction and called attention to the advantages of the

---

[4] Relatively few molecules are stable at the temperatures required for thermal excitation, and we confine our discussion to *atomic* emission. Under certain conditions, some metals can form oxides or other species which emit analytically useful molecular band spectra.

technique during the 1920s and 1930s. Flame emission spectroscopy, or flame photometry, as it is often called, developed into a routine analytical tool in Europe during the 1930s and in the United States shortly after World War II.

Of the common sources used in emission spectroscopy, the flame is the least energetic and excites the fewest elements, perhaps about 50 of the metallic ones.[5] A well-regulated flame is typically a more stable source than an arc or spark. Further, particularly with lower-temperature flames, the emission spectrum of an element is relatively simple; i.e., only a few of the lines that are seen with more energetic excitation are found in flame emission. This places a lighter burden upon the resolving power of the monochromator with regard to interferences: it is easier to find an emission line for a particular element which does not have lines of other elements as near neighbors. In fact, with a low-temperature flame source, the emission of an easily excited element such as sodium can be satisfactorily isolated using colored-glass filters.

Instrumentation for Flame Photometry

The essential components of a flame photometer are shown in the block diagram in Fig. 15.4. We looked briefly in Chapter 14 (page 429) at the nebulization of solutions by which we introduced into flames an aerosol or fog of tiny droplets containing the analyte. We do the same thing here, and the two types of aspirator-burner arrangements that we saw earlier are also employed in flame emission. After sample introduction, the process involves desolvation (loss of solvent by the droplets), followed by dissociation of the tiny solid particles to yield analyte atoms; a fraction of the atoms acquire sufficient energy by collisions with molecules of the hot flame gases to become electronically excited. Our goal then is to monitor emission of the radiant energy as the excited atoms revert to lower electronic energy levels.

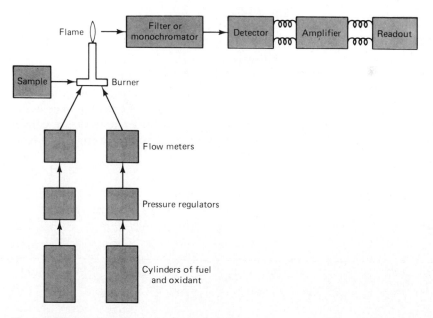

Figure 15.4 Block diagram of a flame photometer.

[5] There are, however, special flames (e.g., see cyanogen-oxygen, Table 15.2) whose temperatures are comparable to those of some of the other sources.

The temperature of the flame is obviously one of the most important variables in flame photometry. This is determined by the nature of the fuel and of the oxidant and their rates of flow, as well as the burner design and the rate of introduction of the sample solution. Table 15.2 shows approximate flame temperatures for some of the mixtures which have been employed. Of these, the commonest are natural gas or propane with air, widely used for determining easily excited elements such as sodium and potassium; hydrogen-oxygen for a hotter flame that is very "clean" with regard to background emission; and acetylene-air for a still higher temperature. It is seen in Fig. 15.4 that regulators for precise control of gas pressures and meters to monitor flow rates are used to establish reproducible flame conditions, although the less expensive instruments may dispense with the flow meters.

TABLE 15.2 Some Approximate Flame Temperatures

MIXTURE (FUEL-OXIDANT)	TEMPERATURE, °C
Natural gas–air	1700
Propane-air	1800
Hydrogen-air	2000
Hydrogen-oxygen	2650
Acetylene-air	2300
Acetylene-oxygen	3200
Acetylene–nitrous oxide	2700
Cyanogen (C_2N_2)–oxygen	4800

Both premix nebulizer-burners and total-consumption burners (Chapter 14, page 429) are employed in flame emission. The latter type must be used with certain flames (e.g., H_2–O_2) because of the explosion problem with a premix chamber. A sketch of a typical total-consumption, integral aspirator-burner is found in Fig. 15.5.

Commercial instruments over a wide price range represent compromises with regard to one or another of the components in order to provide adequate capability as well as marketability for applications of varying difficulty and sophistication. As noted above, inexpensive filters may take the place of a monochromator in an instrument that uses a low-temperature source for analyses of alkali metals because there are few emission lines in the output of the flame. A scanning monochromator and recorder readout are convenient for evaluating baseline effects arising from the flame background emission and for examining lines from

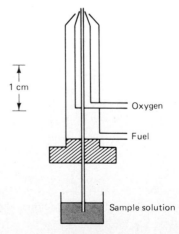

Figure 15.5 Schematic diagram of an integral aspirator-burner.

several elements. But a less expensive arrangement is possible in which a monochromator is set manually for a certain wavelength; a background reading is obtained by introducing distilled water or some sort of analytical blank into the flame, and then the sample is measured at the same wavelength. A simple electrical meter could then be used in place of a recording device. Because the emission lines lie in the UV-visible region, the detector is often a phototube or, for higher sensitivity, a photomultiplier tube.

Some instruments provide for a direct comparison of the radiant power emitted by two elements in the sample, thereby permitting the use of an *internal standard* to minimize the effects of variation in sample feed and fluctuations in the flame. For example, suppose sodium analyses were being performed. A constant quantity of a lithium salt might be added to all the standard and unknown sodium solutions. The choice of lithium would be based upon its similarity to sodium in its response to excitational variations and the unlikelihood of its occurrence in the particular set of unknowns at hand. The radiation emitted in one direction from the flame passes to a filter that transmits only the yellow sodium line to a detector. Radiation emitted in another direction goes to a filter that transmits only the lithium line, and thence to a second detector. A measuring circuit compares the two detector signals. The calibration curve for such a case would be a graph with the ratios of sodium-to-lithium emissions plotted against the sodium concentrations of the standard solutions.

Problems in Flame Photometry

The sequence of events required to produce an emission spectrum in the flame is almost forbidding enough to suggest that quantitative flame photometry should not work. Droplets of sample solution must evaporate; the resulting tiny solid particles must dissociate into atoms; the atoms must be excited. All these processes occur within a distance of a few centimeters as the sample droplets are being carried upward at a high velocity by the flame gases. Actually, it is more critical than this, because the optical system does not examine the whole flame but sees only the radiation from a region lying a certain distance above the burner tip. With the proper illumination, one may see unevaporated sample droplets issuing in profusion from the top of the flame. The flame gases are diluted by inrushing air as a result of the low pressure created by high velocity. There is at no point a stable, equilibrium atom population of the sort that might be obtained by holding the sample in a furnace for a sufficient time. Thus the observed emission intensity depends upon a series of kinetic effects—rate of evaporation, rate of dissociation, etc.

It turns out, however, that flame photometry does work provided conditions are very carefully controlled. Under the best circumstances, deviations of the order of perhaps 2% may be seen in replicate analyses. But there are many traps for the unwary, and errors of 50 or 100% are probably far more common than many people realize.

We shall describe briefly some of the commonest problems in quantitative flame photometry.

1. *Radiation from other elements*. Perhaps no two spectral lines have exactly the same wavelength, but some are *very* close together. Whether emission by one element will interfere in determining another will depend upon how close the lines are and the quality of the monochromator. With a good instrument, it is usually possible to measure an element without interference of this type. With

inexpensive filter instruments this would not be the case except for the fact that these instruments usually employ a low-temperature flame (often simply a Meker burner flame) with the result that there may not be many interfering lines in the emission spectrum. Band spectra emitted by excited molecules formed in the combustion process may also represent a problem in some cases.

2. *Cation enhancement*. In high-temperature flames, some of the metal atoms may ionize; for example,

$$Na \;\rightleftharpoons\; Na^+ + e$$

The ion has an emission spectrum of its own, with different frequencies from those of the atomic spectrum. Thus ionization decreases the radiant power of atomic emission. Sometimes a second metal, say potassium in this example, represses by its own ionization the ionization of the first (sodium). It is as though the partial pressure of free electrons had been increased in the equilibrium. The sodium atom emission is thereby enhanced. Sodium analyses on unknown samples containing varying quantities of potassium are thus subject to errors. One solution to this problem would be addition of a large quantity of a potassium salt to all the solutions—unknowns and standards—so as to swamp out variations from one sample to another.

3. *Anion interference*. Many examples of this are known. For instance, phosphate and sulfate ions lower the emission of calcium well below the level found for, say, calcium chloride solutions. Little is known of the detailed mechanism; presumably the solid residue resulting from solvent evaporation is less readily dissociated into atoms than is calcium chloride. Again, the swamping technique is often useful. For example, addition of 0.1 M EDTA to the solutions overcomes the effects of at least 0.01 M phosphate and sulfate on the emission from 6×10^{-4} M calcium.[6] Addition of a large excess of phosphate or sulfate would also overcome the effects of variations in these ions from sample to sample, but in this case EDTA is more satisfactory. For some reason which is not known, EDTA not only counters the effect of the other anions, but also enhances calcium emission, leading to improved sensitivity.

Because our goal is to provide an introduction to emission spectroscopy without unduly lengthening the book, there are many aspects of flame photometry which are not described adequately here. Anyone undertaking analyses based upon this technique will find additional reading essential.

Applications of Flame Photometry

The most important applications of flame photometry tend to involve analyses that are difficult or impossible to perform in other ways or where speed is somewhat more important than the highest accuracy. For many years the main forte of flame photometry has been analysis for alkali metals, mainly sodium and potassium, and to a lesser extent, the alkaline earths, primarily calcium. Analyses for these ions became important in connection with electrolyte balance studies in physiology and in clinical chemistry laboratories. The alkali metals form few compounds containing chromophores to provide a basis for UV-visible spectrophotometry, they are not electroactive at reasonable potentials for electroanalytical techniques such as polarography, they form few compounds insoluble

---

[6] A. C. West and W. D. Cooke, *Anal. Chem.*, **32**, 1471 (1960).

enough for gravimetry (at least in dilute solutions), and acid-base, redox, or chelometric titrations are in general not possible. Thus flame photometry has been exceedingly useful in studies involving these elements. The development of ion-selective electrodes for elements such as the alkali metals and calcium has led to the replacement of flame methods by potentiometry in some laboratories (for these electrodes, see Chapter 12). Flame photometry remains important, however, in biomedical research, clinical chemistry, agronomy, water analysis, nutrition studies, and other areas where elements which are difficult to determine by other means are important.

Recent years have seen the expanded use of lithium salts in psychiatric medicine because of their calming effect upon patients experiencing manic excitement. Because Li^+ is toxic at blood levels only somewhat higher than the therapeutically effective level, responsible management of the patients requires that the clinical laboratory perform lithium determinations, generally on serum samples.[7] Both atomic absorption (Chapter 14) and flame emission have been used for this purpose. Serum Li^+ levels during treatment generally range from 0.5 to 1.5 mmol/liter. A 50-fold dilution (say, 2 ml of serum diluted to 100 ml) will then give Li^+ concentrations of about 1×10^{-5} to 3×10^{-5} M, well above the detection limit for typical flame photometers in the clinical laboratory. Potassium is sometimes added to the samples as an internal standard (page 455); at the dilution employed, the potassium "spike" gives virtually a constant $[K^+]$ despite the possibility of small variations in serum K^+ levels from one patient to another. The dilution also minimizes possible matrix effects of proteins and other serum components.[8]

15.2b. *Emission Spectroscopy with Plasma and Arc Sources*

Inductively Coupled Plasma Sources
(Electrodeless Discharges)

These were first studied in the 1970s, and despite problems (including cost), they have proved very useful. Commercial instruments based upon the inductively coupled plasma (ICP) source have recently appeared, and the ICP promises to become the most widely used source for emission spectroscopy. Attractive features are very good stability, relatively low background emission, and very high energy. The sample is usually introduced into the plasma as an aerosol (i.e., tiny droplets suspended in a stream of argon) and upon reaching the observation site has experienced a temperature of 6000 to 8000°C for a couple of milliseconds. Desolvation is obviously extremely rapid, and dissociation of the solid matrix to yield an atomic analyte population is likely to be much more complete than with cooler sources. There is sometimes even ionization of analyte atoms, in which case an emission line in the spectrum of the excited *ion* is measured.

[7] Whole blood comprises the *formed elements* (red cells, white cells, and platelets) and the fluid phase called *plasma*. In normal males the cells represent about 47% of the blood volume when packed by centrifugation (the value is somewhat variable, and is a little lower in women). Normally the red cells occupy nearly all this volume. Many analyses can be performed on plasma if it has been treated to prevent clotting. Frequently, however, the clot is allowed to form; the liquid phase which separates from the clot, called *serum*, is then the analytical sample.

[8] A. L. Levy and E. M. Katz, Jr., *Clin Chem.*, **15**, 787 (1969).

The plasma torch is essentially a quartz tube through which argon flows, with an external induction coil wrapped around the tip. The water-cooled coil, powered by a radio-frequency generator, operates typically at about 27 MHz with an output of perhaps 2 kW. To start the torch, some "seed" electrons are formed by ionizing a little argon with a Tesla coil. These absorb energy from the electric field of the coil and, through collisions with argon atoms, induce additional ionization. Once enough electrons are present, interactions between charged particles and the magnetic field of the coil essentially induce a current spiralling peripherally about the plasma. Collisions of the particles represent resistance to current flow, which means heating. The transfer of power from the coil to the discharge is analogous to that from the primary to the secondary winding of a transformer. Radiation from the source normally passes to a monochromator, which selects the desired analyte emission line, and thence to a photomultiplier detector.

The Dc Arc

Although the ICP is rapidly gaining prominence, much spectroscopy has been done with the dc arc. An electrical arc is struck between two electrodes, using a voltage of perhaps 200 to 300 V. There may be events at the molecular level which are not thoroughly understood, but one supposes something like the following: there are always at least a few free electrons wandering about in a sample of any material, e.g., a gas (cf. thermal emission in phototubes, Chapter 14, page 410); in an electric field, these electrons may be accelerated to the degree that collisions with gaseous molecules induce additional ionization; as the process cascades and the gas becomes more conducting, power dissipation results in a high temperature. There may not be thermal equilibrium—the distribution of kinetic energies among various particles may vary with their masses and charges—but the ionized gas between the electrodes, sometimes called a *plasma,* is *hot*. Some readers may have seen electric arcs which melt steel in welding operations. Analyte atoms introduced into the plasma are excited, and their characteristic electronic emission spectra may be observed.

Collisions of energetic particles with the electrodes themselves eject electrode material into the arc, and, indeed, the atomic composition of alloys can sometimes be determined using the metal as an electrode. In general, though, an electrode with a low background emission of its own is desired. The electrodes are often graphite rods, to one of which the sample is applied in some manner. For instance, a depression may be machined into one end of a rod to hold a drop of sample solution. The rod is dried in an oven or under an infrared lamp. When an arc is formed between this rod and a second electrode, sufficient energy is provided to volatilize the sample residue, dissociate it into atoms, and excite these atoms, at least for many of the elements. Examination of the radiant energy emitted from the arc will disclose that it comprises discrete frequencies (or wavelengths) that serve to identify components of the sample. Under very carefully standardized conditions, quantitative analysis may be possible by measuring the radiant power at appropriate wavelengths, but there are many problems in quantitative emission spectroscopy with arc sources. Although very high temperatures are attained in the arc (say, 4000 to 8000°C), all the sample components may not be volatilized uniformly, and the time dependence of the emission spectrum may be tricky. Further, an arc tends to wander about as it plays upon an electrode surface, and much flickering is observed as well. Ac arcs, operated at 1000 V or more, are somewhat steadier than dc arcs. High-voltage sparks are also used to excite spectra in some cases.

Figure 15.6 shows schematically the nature of the major components of an instrument for recording an emission spectrum photographically. Radiation emitted from the source is focused on a slit and then directed to the dispersive element, a prism in the figure. The prism sorts out the individual wavelengths of the emission, and a device that is essentially a camera records the spectrum on a photographic plate. After the plate has been developed in the darkroom, a pattern of dark lines will be seen on the photographic negative, each representing an image of the slit photographed with radiation of a particular wavelength. By reference to wavelength tables or from experience, experts can easily decide what elements were present in the sample. If arc conditions, exposure, and photographic development are carefully controlled, the denseness of a dark line on the plate can be related to the percentage of a sample component. Photographic recording offers the advantage of simultaneously monitoring lines of all excited sample elements. For the determination of a few elements based upon intensities of selected lines, the photographic plate can be replaced by a series of carefully spaced slots behind each of which is a photoelectric detector. The outputs of the detectors over a definite number of seconds of exposure to the light from the arc can be integrated electronically and the results compared with data from standard reference materials of comparable composition.

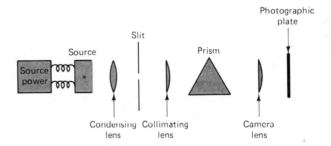

Figure 15.6 Schematic diagram of instrument for photographic recording of emission spectra.

Spectrochemical methods are among the most widely used of analytical methods in the real world outside the undergraduate laboratory. The treatment here has been brief; more background in physics and other areas than that possessed by most students using this textbook is required for a solid understanding. Yet even beginning students, if this is their last exposure to analytical chemistry, should be aware of these methods. Interested readers may find a much more thorough treatment in a recent text for advanced undergraduates and beginning graduate students.[9] One of the most difficult things to teach, though, is perspective: a wide variety of analytical tools is available, and the old ones are not necessarily poorer in certain situations; raise the money and buy an $80,000 instrument if that's the way to solve an important problem, but not if a nickel titration will do the job. Old methods are not abandoned as new ones emerge. We pick the best compromise among capability, required accuracy, cost, and the competence of laboratory personnel. Spectrometers will do things that cannot be done any other way, but burets still cost less than $50. Judgment in the selection of methods is probably the most important asset an analytical chemist can offer an employer.

[9] J. D. Ingle, Jr., and S. R. Crouch, *Spectrochemical Analysis*, Prentice-Hall, Inc., Englewood Cliffs, N. J., 1988.

KEY TERMS

Atomic fluorescence. Emission of radiant energy in the UV-visible region by atoms in excited electronic states produced by the absorption of radiant energy. The atom population is usually obtained in a flame. See footnote 1.

Burner (in flame photometry). See discussion of burners under *Atomic Absorption Spectrophotometry* in Chapter 14.

Chemiluminescence. Emission of radiant energy, often in the visible region, by molecules in excited electronic states produced in chemical reactions at ordinary temperatures. The emission is called *bioluminescence* if the reaction occurs in a biological system.

Dc arc. An electrical discharge between two electrodes in which very high temperatures are attained (the student may have seen arcs in watching arc welders at work). Used as a source in emission spectroscopy for applications where flame temperatures are inadequate for volatilization and excitation or for types of samples which are not easily introduced into flames.

Flame emission spectroscopy (flame photometry). Determinations, usually of metals, by measuring radiation produced by introduction of samples into flames. The source of the radiation is usually atoms resulting from sample dissociation and electronically excited by collisions with the flame gases.

Fluorescence. Emission of UV-visible radiation by atoms, or, more commonly, molecules which have been excited by the absorption of radiation and which ceases within ~0.1 μs of the termination of excitation (cf. *phosphorescence*, where emission persists for a longer period).

Inductively coupled plasma (ICP). A modern, very high temperature source for thermal excitation in emission spectroscopy. Energy is transferred by electromagnetic induction from a radio frequency coil to a partially ionized gas, usually argon. Analyte solutions are introduced into the plasma torch in the form of aerosols—fogs or mists of tiny droplets suspended in argon.

Inner-filter effect. In fluorescence, deprivation of some of the analyte molecules of full exciting radiant power when absorption appreciably attenuates the incident beam.

Instrumentation for fluorescence measurement. See Key Terms near the end of Chapter 14 for such terms as *monochromator* and *detector*.

Internal standard. In emission spectroscopy, an element added in known quantity to the sample; the ratio of analyte emission to emission by the standard is a better measure of analyte level than *its* emission alone if both elements are affected similarly by fluctuating conditions in flame, arc, etc.

Matrix effects. See Key Terms near the end of Chapter 14.

Phosphorescence. Photoluminescence (q.v.) which continues for a longer time period than does fluorescence after the exciting radiation is shut off. In fluorescence, emission continues for no more than about 10^{-7} s after excitation is terminated; in phosphorescence, emission may continue for several seconds, and cases of many minutes are known. The underlying cause of the longer excited-state lifetime in phosphorescence is nonradiative transfer of the excited molecule from a state where all electron spins are paired (*singlet state*) to one with an unpaired spin (*triplet state*). The transition from excited triplet to ground state is called *forbidden,* which really means improbable. Thus molecules linger in the excited state and slowly emit radiation after excitation has ceased. The transition is forbidden because it involves another electron "spin flip," since there is no triplet ground state.

Photoluminescence. Emission of radiation by atoms or molecules excited by photons; includes fluorescence and phosphorescence.

Plasma source. The newest source in emission spectroscopy. In physics *plasma* means a gas containing ions. Thus, literally, plasmas are common, because there are some ions even in flames and certainly in arcs. The term *plasma source,* however, usually refers to a very energetic source where ionization is a *major* factor. Commonest today is the inductively coupled argon plasma source (see above).

Quenching. An interference in fluorescence; by collisions, the quencher causes the excited molecule to lose its energy nonradiatively, and the emitted radiant power is less than in the absence of the quencher.

Resonance fluorescence. Fluorescence in which exciting and reemitted radiation are of the same frequency (or wavelength). Usually one of the two electronic states involved is the ground state.

Xenon arc lamp. A widely used source of excitation in fluorescence work. The lamp output is a relatively intense continuum over much of the UV-visible region. (There are some superimposed xenon emission lines which occur mainly in the near-infrared region and are not troublesome.)

QUESTIONS

1. A biochemist wants to use the firefly luminescence reaction (page 443) to determine the concentrations of ATP in samples that are important in some research activity. There are no instruments at hand by which the measurements can be made that are not already in full-time use, nor will the budget permit the purchase of a new commercial instrument, so it is decided to buy components and, with the help of the departmental shop, build one. The device must yield good ATP measurements but need possess no capability for doing anything else. Provide a rough sketch showing the essential components of an instrument that would do the job.

2. What is the basis for the distinction between molecular *fluorescence* and *phosphorescence*? Which phenomenon is currently more routinely useful in analytical chemistry? Why is phosphorescence usually very weak in solutions at room temperature?

3. Give reasons for the departure from linearity of the curve in Fig. 15.2.

4. In spectrophotometry, the incident radiant power, P_0, must obviously be great enough for the detector to "see" radiation adequate for generating a useful signal coming through the sample cell, but beyond that requirement, the level of P_0 does not affect the sensitivities of spectrophotometric methods. Increasing P_0, however, usually enhances fluorescent emission and thereby improves the sensitivity of fluorometric methods. Explain.

5. Why would an instrument for measuring fluorescence (spectrofluorometer) cost more than a spectrophotometer, assuming comparable features in terms of automated scans, etc., and components of equal quality in both instruments?

6. Describe the processes which must occur in a flame if we are to observe atomic emission by spraying a sample solution containing metal salts into that flame.

7. Could the method of standard addition (page 375) be used in flame photometry? What might one hope to accomplish by the use of this approach?

Multiple-choice: In the following multiple-choice questions, select the *one best* answer.

8. Line spectra are emitted by (a) incandescent solids; (b) excited molecules; (c) molecules in the ground electronic state; (d) excited atoms and monatomic ions.

9. Band spectra are emitted by (a) tungsten lamps and Nernst glowers; (b) excited molecules in the vapor phase; (c) excited atoms and monatomic ions; (d) incandescent solids.

10. A chemist is conducting research on the development of analytical methods based upon chemiluminescence. If a conventional spectrofluorometer is used for the measurements, and the instrument has separate switches for its various circuits, which switch will be left in the "off" position? (a) detector power supply; (b) amplifier; (c) source power supply; (d) power to the motor drive of the scanning monochromator on the emission side of the instrument.

11. Anions such as phosphate in the matrix sometimes interfere in atomic emission spectroscopy if metal standards are prepared using, say, chloride salts. If the interference relates to dissociation of desolvated "specks" of sample material, will the interference be more or less severe with an argon plasma source than with an oxyhydrogen flame? Explain.

16

Solvent Extraction

The partition of solutes between two immiscible liquids offers many attractive possibilities for analytical separations. Even where the primary goal is not analytical but rather preparative, solvent extraction may be an important step in the sequence that leads to a pure product in the organic, inorganic, or biochemical laboratory. Although complicated apparatus is sometimes employed, frequently only a separatory funnel is required. Often a solvent extraction separation can be accomplished in a few minutes. The technique is applicable over a wide concentration range and has been used extensively for the isolation of extremely minute quantities of carrier-free isotopes obtained by nuclear transmutation as well as industrial materials produced by the ton. Solvent extraction separations are usually "clean" in the sense that there is no analog of coprecipitation with such systems. Aside from its intrinsic interest, there is an important reason for discussing solvent extraction in this text: we shall use a particular approach to solvent extraction, the Craig pseudocountercurrent technique, as a model to aid our understanding of chromatographic processes in Chapter 17.

16.1

NERNST DISTRIBUTION LAW

16.1a. The Distribution Coefficient

When a solute distributes itself between two immiscible liquids, there is a definite relationship between the solute concentrations in the two phases at equilibrium. Nernst gave the first clear statement of the distribution law when he pointed out in 1891 that a solute will distribute itself between two immiscible liquids in such a way that the ratio of concentrations at equilibrium is constant at a particular temperature:

$$\frac{[A]_1}{[A]_2} = \text{constant}$$

$[A]_1$ represents the concentration of a solute A in the liquid phase 1.

Although this relationship holds fairly well in certain cases, in reality it is inexact. Strictly, in thermodynamic terms, it is the activity ratio rather than the concentration ratio that should be constant. The activity of a chemical species in one phase maintains a constant ratio to the activity of the same species in the other liquid phase:

$$\frac{a_{A_1}}{a_{A_2}} = K_{D_A}$$

Here a_{A_1} represents the activity of solute A in phase 1. The true constant K_{D_A} is called the *distribution coefficient* of species A. In approximate calculations, which are adequate for many purposes, concentrations rather than activities may be employed in problems involving K_D values, such as those at the end of this chapter.

Suppose that pure A is a solid and consider a solution of A in solvent 1 which is in equilibrium with excess solid; this is a saturated solution, and $[A]_1 = s_{A_1}$ where s_{A_1} is the molar solubility of A in solvent 1. At the same time, let this solution be in equilibrium with a solution of A in solvent 2. Now, if two phases are in equilibrium with each other, and one of these is in equilibrium with a third phase (in this case, solid A), then the other one is also in equilibrium with this third phase. In other words, $[A]_2 = s_{A_2}$. If $[A]_1 / [A]_2$ equals a constant, then this constant is also equal to s_{A_1} / s_{A_2}; i.e., K_{D_A} may be viewed as the ratio of the solubilities of A in the two solvents. (Remember, though, that the ratio $[A]_1 / [A]_2$ will remain constant if the two solutions have equilibrated with each other whether they are saturated with the solute or not.)

16.1b. The Distribution Ratio

Sometimes it is necessary or desirable to take into account chemical complications in extraction equilibria. For example, consider the distribution of benzoic acid between the two liquid phases benzene and water. In the aqueous phase, benzoic acid is partly ionized,

$$HBz + H_2O \rightleftharpoons H_3O^+ + Bz^-$$

In the benzene phase, benzoic acid is partially dimerized by hydrogen bonding in the carboxyl groups,

$$2HBz \rightleftharpoons (HBz)_2$$

Each particular species, HBz, Bz^-, $(HBz)_2$, will have its own particular K_D value. The system water, benzene, and benzoic acid may then be described by three distribution coefficients:

$$K_{D_{HBz}} = \frac{a_{HBz_{org}}}{a_{HBz_{aq}}}$$

$$K_{D_{Bz^-}} = \frac{a_{Bz^-_{org}}}{a_{Bz^-_{aq}}}$$

$$K_{D_{(HBz)_2}} = \frac{a_{(HBz)_{2org}}}{a_{(HBz)_{2aq}}}$$

Now it happens that the benzoate ion, in fact, remains almost totally in the aqueous phase, and the benzoic acid dimer exists only in the organic phase. Further, in a practical experiment, the chemist will usually want to know where the "benzoic acid" *is,* not whether part of it is ionized or dimerized. Also, one will be more interested in how much is there than in its thermodynamic activity and be better served, then, by an expression combining the concentrations of all the species in the two phases:

$$D = \frac{\text{total benzoic in organic phase}}{\text{total benzoic in aqueous phase}}$$

$$= \frac{[HBz]_{org} + 2[(HBz)_2]_{org}}{[HBz]_{aq} + [Bz^-]_{aq}}$$

The ratio D is called the *distribution ratio.*

It is clear that D will not remain constant over a range of experimental conditions. For example, raising the pH of the aqueous phase will lower D by converting benzoic acid into benzoate ion, which does not extract into benzene. The addition of any electrolyte may affect D by changing activity coefficients. However, the distribution ratio is useful when its value is known for a particular set of conditions.

16.2

EXTRACTIONS INVOLVING ADDITIONAL EQUILIBRIA

16.2a. *Weak Acids and Bases*

Consider a weak acid, HB. Assume for simplicity that the acid is monomeric in both solvent phases, and that the anion of the acid does not penetrate the organic phase. The pertinent equilibrium expressions then are

$$D = \frac{[HB]_{org}}{[HB]_{aq} + [B^-]_{aq}} \qquad (1)$$

$$K_{D_{HB}} = \frac{[HB]_{org}}{[HB]_{aq}} \qquad (2)$$

$$K_a = \frac{[H_3O^+]_{aq}[B^-]_{aq}}{[HB]_{aq}} \qquad (3)$$

Rearranging Eq. (3) gives

$$[B^-]_{aq} = K_a \frac{[HB]_{aq}}{[H_3O^+]_{aq}}$$

and substitution into Eq. (1) yields

$$D = \frac{[\text{HB}]_{\text{org}}}{[\text{HB}]_{\text{aq}} + (K_a[\text{HB}]_{\text{aq}}/[\text{H}_3\text{O}^+]_{\text{aq}})}$$

or, factoring out $[\text{HB}]_{\text{aq}}$ in the denominator,

$$D = \frac{[\text{HB}]_{\text{org}}}{[\text{HB}]_{\text{aq}}\{1 + (K_a/[\text{H}_3\text{O}^+]_{\text{aq}})\}}$$

Referring to Eq. (2), we see that

$$D = \frac{K_{D_{\text{HB}}}}{1 + (K_a/[\text{H}_3\text{O}^+]_{\text{aq}})} \tag{4}$$

Thus we have derived an expression showing explicitly the dependence of the distribution ratio upon the distribution coefficient of the weak acid, its ionization constant, and the pH of the aqueous phase. It might well be that we could capitalize upon inherent differences in the values of the appropriate constants to effect the separation of a mixture of acids by regulating the pH of the aqueous phase. It is apparent that the extraction process for a weak base B whose protonated form HB^+ is not extractable will be described by an analogous equation (see Question 2, page 486).

Equation (4), written as usually seen in analytical texts, is easily rearranged:

$$D = \frac{K_{D_{\text{HB}}}}{1 + (K_{a_{\text{HB}}}/[\text{H}_3\text{O}^+]_{\text{aq}})} = \frac{K_{D_{\text{HB}}}}{([\text{H}_3\text{O}^+] + K_{a_{\text{HB}}})/[\text{H}_3\text{O}^+]_{\text{aq}}}$$
$$= \frac{K_{D_{\text{HB}}}[\text{H}_3\text{O}^+]_{\text{aq}}}{K_{a_{\text{HB}}} + [\text{H}_3\text{O}^+]_{\text{aq}}} \tag{5}$$

Equation (5) is obviously no better than Eq. (4), but it is in a form more commonly seen in other contexts which the student may encounter. For example, the Michaelis-Menten equation in enzyme kinetics as given in biochemistry texts looks like Eq. (5); of course, each individual term has a different physical significance from that of the corresponding term here, but the mathematical formalism is the same.[1]

If D is plotted against $[\text{H}_3\text{O}^+]$, the graph is hyperbolic, as seen in Fig. 16.1. Remember that in our model the neutral species HB is extractable into the organic phase, while the anion B^- is not. The curve reflects the hyperbolic approach to saturation of B^- with protons to form the maximal quantity of the extractable species. D approaches $K_{D_{\text{HB}}}$ asymptotically as $[\text{H}_3\text{O}^+]$ increases, as seen in Eq. (5):

$$\lim_{[\text{H}_3\text{O}^+]_{\text{aq}} \to \infty} (K_{a_{\text{HB}}} + [\text{H}_3\text{O}^+]_{\text{aq}}) = [\text{H}_3\text{O}^+]_{\text{aq}}$$

and

$$\lim_{[\text{H}_3\text{O}^+]_{\text{aq}} \to \infty} D = \frac{K_{D_{\text{HB}}}[\text{H}_3\text{O}^+]_{\text{aq}}}{[\text{H}_3\text{O}^+]_{\text{aq}}} = K_{D_{\text{HB}}}$$

$K_{D_{\text{HB}}}$ and $K_{a_{\text{HB}}}$ can be evaluated from extraction data of the sort shown in Fig. 16.1. It is seen from Eq. (5), for instance, that when $D = K_{D_{\text{HB}}}/2$, $[\text{H}_3\text{O}^+]_{\text{aq}} = K_{a_{\text{HB}}}$:

$$\frac{K_{D_{\text{HB}}}}{2} = \frac{K_{D_{\text{HB}}}[\text{H}_3\text{O}^+]_{\text{aq}}}{K_{a_{\text{HB}}} + [\text{H}_3\text{O}^+]_{\text{aq}}}$$

[1] A. L. Underwood, *J. Chem. Educ.*, **61**, 143 (1984).

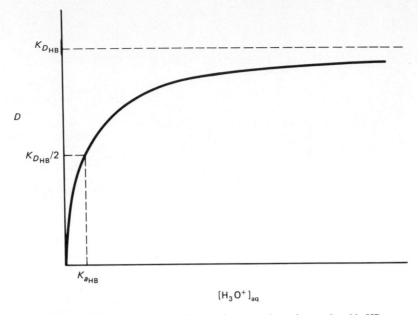

Figure 16.1 Distribution ratio, D, for extraction of a weak acid, HB, from an aqueous phase into an organic solvent as a function of $[H_3O^+]_{aq}$. The graph is a plot of Eq. (4) or Eq. (5) in text.

Dividing by $K_{D_{HB}}$, we have

$$\frac{1}{2} = \frac{[H_3O^+]_{aq}}{K_{a_{HB}} + [H_3O^+]_{aq}}$$

$$K_{a_{HB}} + [H_3O^+]_{aq} = 2[H_3O^+]_{aq}$$

$$K_{a_{HB}} = 2[H_3O^+]_{aq} - [H_3O^+]_{aq} = [H_3O^+]_{aq}$$

Because it is hard to decide, for real data, the limit toward which D is tending by inspecting a hyperbolic curve like that of Fig. 16.1, a linear plot is preferable. This can be obtained, among other ways, by taking reciprocals of both sides of Eq. (5):

$$\frac{1}{D} = \frac{K_{a_{HB}} + [H_3O^+]_{aq}}{K_{HB}[H_3O^+]_{aq}} = \left(\frac{K_{a_{HB}}}{K_{D_{HB}}}\right)\frac{1}{[H_3O^+]_{aq}} + \frac{1}{K_{D_{HB}}}$$

Now if we plot $1/D$ vs. $1/[H_3O^+]$, we obtain a straight line of slope $K_{a_{HB}}/K_{D_{HB}}$ with an intercept of $1/K_{D_{HB}}$ on the $1/D$ axis, as seen in Fig. 16.2. (This approach is helpful elsewhere; e.g., Fig. 16.2 is analogous to the Lineweaver-Burk plot in enzyme kinetics.)

16.2b. Extraction of a Metal as a Chelate Compound

Many important separations of metal ions have been developed around the formation of chelate compounds with a variety of organic reagents. As an example, consider the reagent 8-quinolinol (8-hydroxyquinoline), often referred to by the trivial name "oxine,"

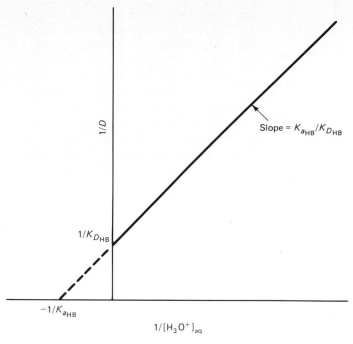

Figure 16.2 Double-reciprocal plot of Eq. (5) to linearize data shown in Fig. 16.1.

This reagent forms neutral, water-insoluble, chloroform- or carbon tetrachloride-soluble molecules with metal ions; the cupric oxinate chelate may be depicted as follows:

If we abbreviate oxine as HOx, we may write the chelation reaction as

$$Cu^{2+} + 2HOx \rightleftharpoons Cu(Ox)_2 + 2H^+$$

Another very important chelating agent for the solvent extraction of metal ions is diphenylthiocarbazone or "dithizone,"

Dithizone and its metal chelates are very insoluble in water but soluble in such solvents as chloroform and carbon tetrachloride. Solutions of the reagent itself are deep green, while the metal complexes are intense violet, red, orange, yellow, or other hues depending upon the metal ion. Metals forming dithizonates include Mn, Fe, Co, Ni, Cu, Zn, Pd, Ag, Cd, In, Sn, and Pb. The chelate concentrations in the extracts are normally determined spectrophotometrically.

We wish to develop an equation relating the distribution of a metal ion M^{n+} between an aqueous phase and an organic solvent containing a chelating agent HX to fundamental constants and experimental variables. We assume that $[M^{n+}]_{org}$ and $[MX_n]_{aq}$ are negligible compared with $[MX_n]_{org}$ and $[M^{n+}]_{aq}$, respectively. Further, MX_n is the only complex in the system; i.e., species such as MX_{n-1} are not present; this is a reasonable assumption if a large excess of HX as compared with M^{n+} is provided. In other words, if we simply keep track of metal ions, we have

$$D_M = \frac{[M^{n+}]_{org} + [MX_n]_{org} + [MX_{n-1}]_{org} + \cdots}{[M^{n+}]_{aq} + [MX_n]_{aq} + [MX_{n-1}]_{aq} + \cdots} = \frac{[MX_n]_{org}}{[M^{n+}]_{aq}}$$

A summary of the relevant equilibria is shown schematically below:

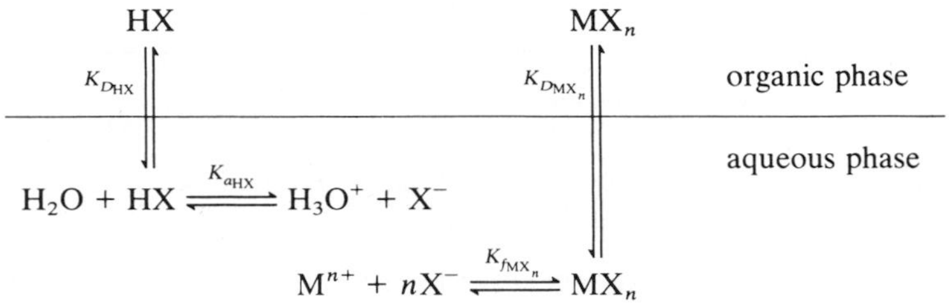

We formulate the overall reaction as

$$n\,H_2O + M_{aq}^{n+} + n\,HX_{org} \rightleftharpoons MX_{n_{org}} + n\,H_3O_{aq}^{+}$$

for which the equilibrium constant may be written

$$K_{ex} = \frac{[MX_n]_{org}[H_3O^+]_{aq}^n}{[M^{n+}]_{aq}[HX]_{org}^n} \tag{6}$$

We have the individual equilibria shown in the formulation above:

$$K_{a_{HX}} = \frac{[H_3O^+]_{aq}[X^-]_{aq}}{[HX]_{aq}} \tag{7}$$

$$K_{f_{MX_n}} = \frac{[MX_n]_{aq}}{[M^{n+}]_{aq}[X^-]_{aq}^n} \tag{8}$$

$$K_{D_{MX_n}} = \frac{[MX_n]_{org}}{[MX_n]_{aq}} \tag{9}$$

$$K_{D_{HX}} = \frac{[HX]_{org}}{[HX]_{aq}} \tag{10}$$

From Eq.(7),

$$[H_3O^+]_{aq} = \frac{K_{a_{HX}}[HX]_{aq}}{[X^-]_{aq}}$$

and

$$K_{ex} = \frac{[MX_n]_{org}K_{a_{HX}}^n[HX]_{aq}^n}{[M^{n+}]_{aq}[HX]_{org}^n[X^-]_{aq}^n}$$

Then from Eq. (10),

$$K_{ex} = \frac{[MX_n]_{org} K_{a_{HX}}^n}{K_{D_{HX}}^n [M^{n+}]_{aq} [X^-]_{aq}^n}$$

Now, from Eq. (8),

$$[X^-]_{aq}^n = \frac{[MX_n]_{aq}}{K_{f_{MX_n}} [M^{n+}]_{aq}}$$

Thus

$$K_{ex} = \frac{[MX_n]_{org} K_{a_{HX}}^n K_{f_{MX_n}} [M^{n+}]_{aq}}{K_{D_{HX}}^n [M^{n+}]_{aq} [MX_n]_{aq}} = \frac{[MX_n]_{org} K_{a_{HX}}^n K_{f_{MX_n}}}{K_{D_{HX}}^n [MX_n]_{aq}}$$

and from Eq. (9),

$$K_{ex} = \frac{K_{D_{MX_n}} K_{a_{HX}}^n K_{f_{MX_n}}}{K_{D_{HX}}^n} \tag{11}$$

From Eq. (6) and the definition of D_M,

$$K_{ex} = \frac{D_M [H_3O^+]_{aq}^n}{[HX]_{org}^n}$$

or

$$D_M = \frac{K_{ex} [HX]_{org}^n}{[H_3O^+]_{aq}^n} \tag{12}$$

where K_{ex} is the collection of constants shown in Eq. (11). We see that the distribution ratio for the metal, D_M, depends upon (a) constants which are fixed by the particular compounds in the selected system and (b) the variables $[H_3O^+]_{aq}$ and $[HX]_{org}$ which are subject to experimental manipulation.

The variation of D with pH is shown in Fig. 16.3 for a particular case. If we take logarithms of both sides of Eq. (12), we obtain

$$\log D_M = \log K_{ex} + n \log (HX)_{org} - n \log [H_3O^+]_{aq}$$

or

$$\log D_M = \log K_{ex} + n \log [HX]_{org} + n\,pH \tag{13}$$

Thus a plot of $\log D_M$ vs. pH should be a straight line of slope n and an intercept on the $\log D_M$ axis of ($\log K_{ex} + n \log [HX]_{org}$). Figure 16.4 shows such a plot for the data in Fig. 16.3. Also shown is a plot for the same system with a different value of $[HX]_{org}$; it is seen that varying the concentration of the chelating agent shifts the line along the pH axis.

Equation (12) shows that the extractability of a metal is a composite of factors which encompass both the tendency of the chelate compound to form and its relative solubilities in the two phases. There are many examples of data in the literature which follow the equation over a reasonable pH range. Eventually, however, other considerations assert themselves, and curves like those in Fig. 16.4 frequently become horizontal at high pH values. One may imagine, for example, a value of $[H_3O^+]_{aq}$ so low that, whatever the value of $K_{f_{MX_n}}$, chelate formation becomes virtually complete. The value of $[MX_n]_{org}$ is then limited only by $K_{D_{MX_n}}$. The intrusion of metal hydroxide precipitation at high pH values may be an additional factor.

The constants $K_{f_{MX_n}}$ and $K_{D_{MX_n}}$ vary from one metal ion to another, providing the basis for separations of metal ions by extraction. One of the attractive aspects of solvent extraction is that it works from the lowest levels (e.g., carrier-

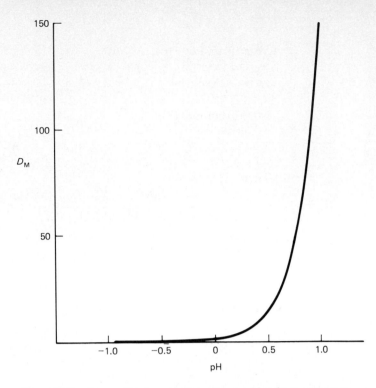

Figure 16.3 D_M vs pH_{aq} for extraction of a metal ion from aqueous solution into an organic solvent as a chelate compound. The curve is calculated, using Eq. (12), for the extraction of Cu^{2+} into CCl_4 with dithizone, where $K_{ex} = 1.5 \times 10^{10}$ and $[HX]_{org}$ is assumed to be constant at $1.0 \times 10^{-5}\ M$.

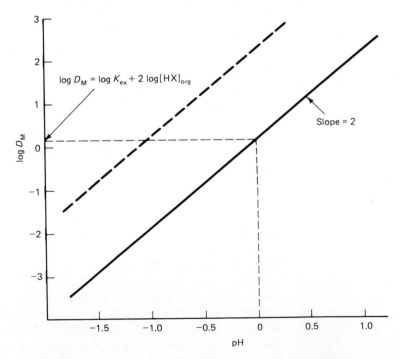

$$\log D_M = \log K_{ex} + 2 \log[HX]_{org}$$

Slope = 2

Figure 16.4 Plot of Eq. (13) for the extraction of Cu^{2+} from aqueous solution into CCl_4 as the dithizone chelate. Solid line: conditions are the same as for the curve in Fig. 16.3. Dashed line: calculated for $[HX]_{org} = 1.0 \times 10^{-4}\ M$.

free radioactive metal isotopes) up to macro quantities. Separations are "clean" in that other metals are not dragged into the organic solvent by the extracting species beyond the extent dictated by their own D_M values (cf. coprecipitation, page 75).

Because D_M is a concentration ratio, the extent of extraction depends upon the volume ratio of organic solvent to water. Let f be the fraction of the total metal ion in the organic solvent at equilibrium; then $(1 - f)$ is the fraction remaining in the aqueous phase. Also, let V_{org} and V_{aq} be the volumes of the two phases. Then

$$D_M = \frac{[MX_n]_{org}}{[M^{n+}]_{aq}} = \frac{f/V_{org}}{(1-f)/V_{aq}} = \frac{fV_{aq}}{(1-f)V_{org}} = \frac{fV_{aq}}{V_{org} - fV_{org}}$$

and

$$D_M V_{org} - D_M f V_{org} = f V_{aq}$$

$$f V_{aq} + D_M f V_{org} = D_M V_{org}$$

$$f(V_{aq} + D_M V_{org}) = D_M V_{org}$$

$$f = \frac{D_M V_{org}}{V_{aq} + D_M V_{org}} \tag{14}$$

or

$$\text{Percentage of metal extracted} = 100f = \frac{100 D_M V_{org}}{V_{aq} + D_M V_{org}}$$

$$= \frac{100 D_M}{D_M + (V_{aq}/V_{org})} \tag{15}$$

Figure 16.5 shows how the percentage extracted from water varies with pH for three metal ions using dithizone in CCl_4 as the extractant.

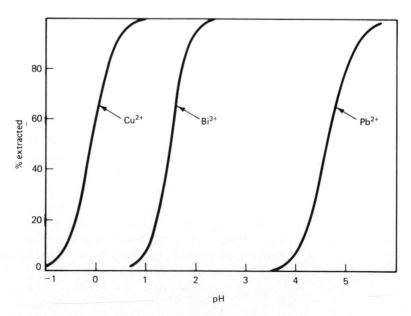

Figure 16.5 Extraction of metal-dithizonate chelates into CCl_4 as a function of pH. Curve for Cu^{2+} was calculated using Eqs. (12) and (15) for $K_{ex} = 1.5 \times 10^{10}$ and $[HX]_{org} = 1.0 \times 10^{-5}$ M, assuming $V_{org} = V_{aq}$; curves for Bi^{3+} and Pb^{2+} are sketches of experimental data under approximately the same conditions from the literature.

EXAMPLE 16.1

*Relationship of D_M
values to separation
efficiency*

What ratio of D_M-values for two metal ions would allow the extraction of 99% of metal A into $CHCl_3$ as the dithizone chelate while 99% of metal B remained in the aqueous phase if the volumes of aqueous phase and $CHCl_3$ are the same?

Using Eq. (15) for metal A, we have

$$99 = \frac{100 D_{M_A}}{D_{M_A} + 1} \quad \text{and} \quad D_{M_A} = 99$$

For metal B,

$$1 = \frac{100 D_{M_B}}{D_{M_B} + 1} \quad \text{and} \quad D_{M_B} = 0.01$$

$$\frac{D_{M_A}}{D_{M_B}} = 99/0.01 = 9900 \cong 10^4 \qquad \qquad \Box$$

Solvent extraction is like everything else in chemistry: It is not a panacea. With many metal-ion mixtures, separation will be incomplete in a one-stage extraction. There are cases where curves like those in Fig. 16.5 lie practically on top of each other. Where it works, though, extraction is a powerful technique, as seen in the figure for separating, say, bismuth and lead.

Values of K_{ex} for various combinations of metal ions, chelating agents, and solvents vary over many orders of magnitude but are frequently quite large. Also, extraction can be enhanced by low acidity and high concentration of chelating agent, as seen in Eq. (12). Thus it is often possible to concentrate the metal from a large volume of a very dilute aqueous solution into a small volume of organic solvent, raising the chelate concentration to a suitable level for measurement. Many metal chelates have absorption bands with large molar absorptivities (page 400) in the UV-visible region of the spectrum, permitting the measurement of microgram quantities of metals at levels of a few parts per million ($\mu g/mL$) or even less. Although UV-visible spectrophotometry is still most common, the extracts can also be subjected to atomic absorption and other spectroscopic measurements such as flame emission. An example is also seen in Chapter 17 (page 523) where a mixture of extracted metals is analyzed by gas-liquid chromatography, although most metal chelate complexes are insufficiently volatile to permit this approach.

In the discussion above, we emphasized the effect of pH upon the distribution ratio of a metal ion. This effect arises from the competition of H_3O^+ with M^{n+} for the chelating anion X^-. It is also possible to change a D_M value by the action of a species which competes with X^- for the metal ion. A complexing ligand such as CN^- or EDTA (page 201), which forms a stable, water-soluble complex ion with the metal, will lower D_M. Space does not allow a detailed treatment of such effects, but the student should be aware that differences among metals in the stabilities of their aqueous complex ions can sometimes be used to obtain better separations in solvent extraction of neutral chelates. It should also be remembered that as $[H_3O^+]$ decreases, $[OH^-]$ goes up; thus species such as $M(OH)^{(n-1)+}$ may intrude upon the distribution process, and since many metal ions form very insoluble hydroxides, precipitation of $M(OH)_n$ is also a possibility if the pH is too high.

16.3

EXTRACTIONS INVOLVING ION PAIRS AND SOLVATES

Generally, simple metal salts tend to be more soluble in a highly polar solvent like water than in organic solvents of much lower dielectric constant. Many ions

The separation of plutonium from uranium and the fission products by a precipitation process was described in Box 4.1, where it was pointed out that this method was later replaced by a solvent extraction procedure. In the Purex process, an aqueous solution of metal nitrates is extracted with a solution of tributyl phosphate (TBP, shown below) in

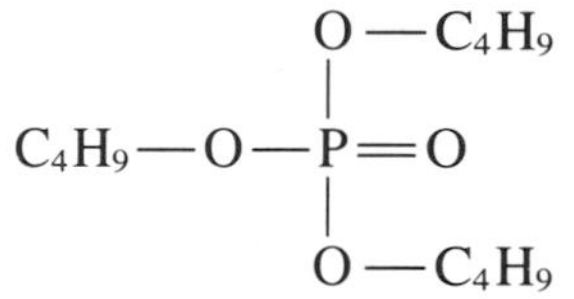

$$C_4H_9 - O - \overset{\displaystyle O - C_4H_9}{\underset{\displaystyle O - C_4H_9}{\overset{|}{\underset{|}{P}}}} = O$$

a hydrocarbon solvent such as kerosene. U(VI) and Pu(IV) are extracted, while fission products remain in the aqueous phase. Presumably the strong electron-donating tendency of the phosphoryl oxygen enables TBP to coordinate with the U(VI) of the uranyl (UO_2^{2+}) ion; the butyl groups then confer sufficient hydrophobicity to allow the complex to penetrate a hydrocarbon solvent, carrying NO_3^- along to maintain electrical neutrality (cf. ion pairs). Pu(IV) is similarly carried into the organic phase.

Pu(IV) is then reduced to Pu(III), which is easily extracted into an aqueous phase. U(VI), which is not so easily reduced as is Pu, remains in the organic solvent. It is later extracted back into dilute aqueous nitric acid. Thus plutonium and uranium are separated from each other and from the fission products. Since these solutions are highly radioactive, all operations must be carried out remotely behind thick concrete walls.

[†] S. Glasstone, *Sourcebook on Atomic Energy,* 3rd ed., D. Van Nostrand Co., Inc., Princeton, N.J., 1967, p. 619.

are solvated by water, and the energy of solvation contributes to the disruption of the crystal lattice of the solid salt. Furthermore, less work is required to separate ions of opposite charge in a high-dielectric solvent. Usually, then, the formation of an uncharged species is necessary if an ion is to be extracted from water into an organic solvent. We have seen an example of this in the extraction of metals converted into neutral chelates of 8-quinolinol. The metal ion is bound in the chelate by definite chemical bonds, often largely covalent in character.

Sometimes, on the other hand, an uncharged species extractable into an organic solvent is obtained through the association of ions of opposite charge. In point of fact, it must be admitted that it is difficult to draw a line between an ion pair and a neutral molecule. Probably, if the components stay together in water, it will be called a molecule; if the components are separated in water sufficiently that an entity cannot be detected, this entity will be called an ion pair if it does show up in a nonpolar solvent.

A common example of an extraction system involving ion-pair formation in the organic phase is found in the use of tetraphenylarsonium chloride to extract permanganate, perrhenate, and pertechnetate from water into chloroform. The species which passes into the organic phase is an ion pair, $[(C_6H_5)_4As^+, ReO_4^-]$. Similarly, the extraction of uranyl ion, UO_2^{2+}, from aqueous nitrate solutions into solvents such as ether (an important process in uranium chemistry) involves an association of the type $[UO_2^{2+}, 2NO_3^-]$. It is believed that the uranyl ion is solvated by ether as well as by water, a fact which doubtless facilitates penetration of the organic phase by an ion pair which then takes on more of the character of the solvent.

It has been known for many years that iron (III) ion can be extracted into ether from strong hydrochloric acid solution. This process is useful for the separation of bulk quantities of iron prior to the determination of other elements in ferrous alloys. Despite extensive study, the system water–ether–HCl–Fe^{3+} is still not completely understood. There is evidence that the extractable species is an ion pair of the type $[H_3O^+, Fe(H_2O)_2Cl_4^-]$; other equilibria may intrude, such as solvation of both proton and iron(III) ion by ether:

$$H_3O^+ + C_4H_{10}O \rightleftharpoons C_4H_{10}OH^+ + H_2O$$

$$Fe(H_2O)_2Cl_4^- + 2C_4H_{10}O \rightleftharpoons Fe(C_4H_{10}O)_2Cl_4^- + 2H_2O$$

Thus under certain conditions the species in the ether phase may be $[C_4H_{10}OH^+, Fe(C_4H_{10}O)_2Cl_4^-]$. The system is undoubtedly complicated, and mixtures of various solvated ion pairs probably participate under the usual conditions of the extraction.

16.4

MULTIPLE EXTRACTIONS

16.4a. Introduction to Multiple Extractions

In an ideal separation by solvent extraction, all of the desired substance would end up in one solvent and all the interfering substances in the other. Such all-or-none transfer from one solvent to another is rare, and we are much more likely to encounter mixtures of substances that differ only somewhat in their tendencies to pass from one solvent into another. Thus one transfer does not lead to a clean separation. In such cases, we must consider how best to combine a number of successive partial separations until we eventually achieve the desired degree of purity.

In considering how two phases may be brought together repetitively, four levels of complexity may be distinguished.[2] First would be the simple, one-shot contact as mentioned above. Second, one phase could be brought repeatedly into contact with fresh portions of a second phase. This would be applicable where one substance remained quantitatively in one phase, while another substance was distributed between the two phases. An example might be repeated extraction of an aqueous solution with successive portions of an organic solvent. The Soxhlet extractor would fall in this category, as would the technique of reprecipitation in gravimetric analysis.

Third, one phase may move while in contact with a second phase which remains stationary. The moving phase may move continuously, as in the various chromatographic techniques, or in a series of equilibrium steps, as in the Craig apparatus described below. Some techniques of this type have been designated *countercurrent,* but this is not really the case, since only one phase moves. The term *pseudocountercurrent* is sometimes applied to such processes.

[2] H. A. Laitinen and W. E. Harris, *Chemical Analysis,* 2nd ed., McGraw-Hill Book Company, New York, 1975, p. 407.

Fourth, we list true countercurrent methods, in which both phases move, continually in contact with each other, in opposite directions. Fractional distillation is an example of a true countercurrent process: Refluxing liquid runs continuously down the distilling column in contact with rising vapors. Countercurrent processes are extensively employed by chemical engineers in large-scale plant operations. Because of experimental difficulties, as well as problems in the theoretical treatment, however, they are rarely used in the research laboratory.

16.4b. *Multiple Extraction with Successive Portions*

Suppose that we have a weight W of solute A dissolved in water. For simplicity, let the extraction with an organic solvent be uncomplicated by other equilibria (i.e., A has the same chemical structure in both solvents). Let

$$K_{D_A} = \frac{[A]_{org}}{[A]_{aq}}$$

Let us extract the aqueous solution with an equal volume of organic solvent, perhaps ether. In the extraction, suppose that weight w of A moves from the aqueous solution into the organic solvent. Then

$$K_{D_A} = \frac{w}{W - w}$$

whence

$$WK_{D_A} - wK_{d_A} = w$$

$$w + wK_{D_A} = WK_{D_A}$$

$$w(1 + K_{D_A}) = WK_{D_A}$$

$$w = \frac{WK_{D_A}}{1 + K_{D_A}}$$

Now the fraction, f_{org}, extracted into the organic solvent is w/W. Dividing both sides of the equation above by W gives

$$f_{org} = \frac{w}{W} = \frac{K_{D_A}}{1 + K_{D_A}}$$

Since the fraction of A extracted into the organic solvent plus the fraction remaining in the aqueous solution must be equal to 1, the fraction of A remaining in the aqueous solution is

$$f_{aq} = 1 - \frac{K_{D_A}}{1 + K_{D_A}} = \frac{1 + K_{D_A} - K_{D_A}}{1 + K_{D_A}} = \frac{1}{1 + K_{D_A}}$$

Now let us separate the two solvent layers, set aside the organic solution, and extract the aqueous solution with a second, equal volume of fresh organic solvent. Of that fraction of A remaining in the aqueous solution after the first extraction, the fraction $K_{D_A}/(1 + K_{D_A})$ will be removed in the organic phase and the fraction $1/(1 + K_{D_A})$ will remain in the water. Thus, after two extractions,

$$f_{aq} = \left(\frac{1}{1 + K_{D_A}}\right)\left(\frac{1}{1 + K_{D_A}}\right) = \left(\frac{1}{1 + K_{D_A}}\right)^2$$

More generally, after n extractions of an aqueous solution with successive equal portions of organic solvent, the fraction of the original solute A unextracted (i.e., remaining in the water) is given by

$$f_{aq} = \left(\frac{1}{1 + K_{D_A}} \right)^n$$

The fraction extracted can always be obtained, of course, by subtracting the unextracted fraction from 1.

If the volume of organic solvent used each time is not the same as that of the water solution, it is easily shown that the expression for the unextracted fraction becomes

$$f_{aq} = \left(\frac{V_{aq}}{V_{aq} + K_{D_A} V_{org}} \right)^n$$

where V_{aq} and V_{org} are the volumes of aqueous and organic phases, respectively.

EXAMPLE 16.2

Calculation of percent extraction—Comparison of single and multiple extraction techniques

Suppose that $K_D(\text{org/aq})$ for a certain solute in a water-chloroform system is 10. Calculate the percentage of solute extracted from 50 ml of water by 100 ml of chloroform where (a) the chloroform is used all at once, and (b) the 100 ml of chloroform is divided into five 20-ml portions which are employed one after the other.

$$\text{(a) } f_{aq} = \frac{50}{50 + (10)(100)} = 0.04762$$

$$f_{org} = 1 - 0.04762 = 0.9524$$

$$\% \text{ extracted} = 95.24$$

$$\text{(b) } f_{aq} = \left(\frac{50}{50 + (10)(20)} \right)^5 = 0.00032$$

$$f_{org} = 1 - 0.00032 = 0.9997$$

$$\% \text{ extracted} = 99.97$$

Note that the extraction was more effective in removing the solute from the water when the chloroform was divided into several portions than it was when the same total volume of chloroform was employed in a one-shot process. ☐

16.4c. Craig Countercurrent Extraction

The Basic Concept

Consider an aqueous solution containing 1000 mg of some solute in a separatory funnel. Let it be a simple solute whose partition is uncomplicated by ionization, dimerization, etc. Add to the funnel an equal (for convenience) volume of an immiscible organic solvent. Also, for simplicity, suppose that the distribution coefficient of the solute is unity. After equilibration, there will be 500 mg of the solute in the aqueous phase and 500 mg in the organic phase.

Next, take a second funnel and transfer the lighter liquid (say, the organic phase) from the first funnel into it. Then add fresh aqueous solvent to this organic phase in the second funnel and add fresh organic phase to the first funnel. Now shake both funnels until equilibration is achieved. There will then be 250 mg of the solute in each layer in each funnel.

As the student probably suspects by now, we next secure a third funnel, and transfer the organic solution from the second funnel into the third and also introduce a fresh portion of the aqueous solvent. We replace the organic layer in the second funnel using the organic phase from the first one, and we add fresh organic phase to the latter. Then all three funnels are shaken to secure equilibrium.

After two or three transfers, it becomes easier to keep track of what is going on by introducing a simple schematic representation. Portions of aqueous and organic solvent are represented by boxes, and where an aqueous box adjoins an organic box we have equilibration of the phases. Within this box, we give the weight of solute present in that phase. Thus the first step is depicted as follows:

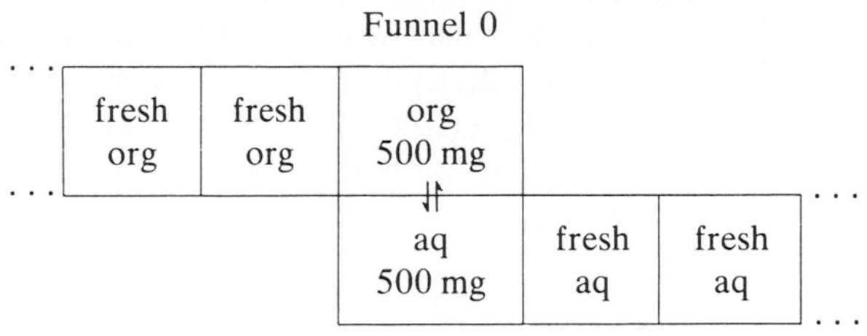

Funnel 0

...	fresh org	fresh org	org 500 mg		
...			⇅ aq 500 mg	fresh aq	fresh aq ...

In the second step noted above, we have performed one transfer ($n = 1$), and the resulting situation may be represented by

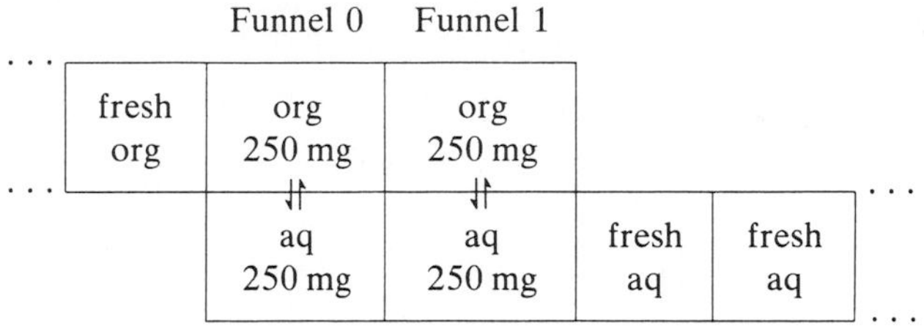

Funnel 0   Funnel 1

...	fresh org	org 250 mg	org 250 mg		
...		⇅ aq 250 mg	⇅ aq 250 mg	fresh aq	fresh aq ...

In effect, we are pushing the top row of boxes toward the right with each transfer.

For the third step, where n, the number of transfers, is 2, the resulting situation is

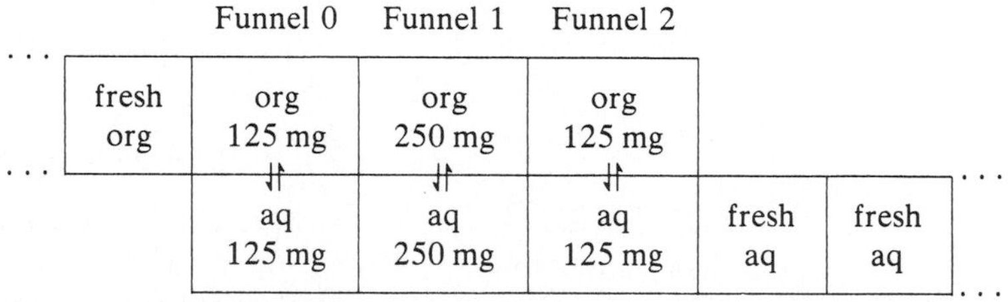

Funnel 0   Funnel 1   Funnel 2

...	fresh org	org 125 mg	org 250 mg	org 125 mg		
...		⇅ aq 125 mg	⇅ aq 250 mg	⇅ aq 125 mg	fresh aq	fresh aq ...

Note that the number of funnels is one more than the number of transfers, and hence the first funnel is labeled number 0. Let us write diagrams for two more steps, after equilibration takes place: fourth step, $n = 3$:

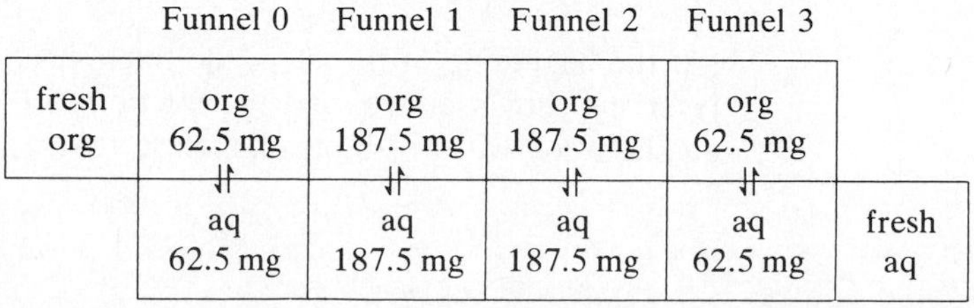

Funnel 0   Funnel 1   Funnel 2   Funnel 3

fresh org	org 62.5 mg	org 187.5 mg	org 187.5 mg	org 62.5 mg	
	⇅ aq 62.5 mg	⇅ aq 187.5 mg	⇅ aq 187.5 mg	⇅ aq 62.5 mg	fresh aq

fifth step, $n = 4$:

	0	1	2	3	4	
	31.25	125	187.5	125	31.25	
	⇅	⇅	⇅	⇅	⇅	
	31.25	125	187.5	125	31.25	

Perhaps we have carried this far enough to see what is happening. As the number of transfers is increased, the solute spreads out through more and more funnels, but it is "bunching up" toward the center (because $K_D = 1$) and the fraction of the solute in the extreme funnels is decreasing. It may also be surmised that for a different solute with a distribution coefficient, not 1, but favoring the aqueous phase, the peak concentration would not appear in the middle funnel but rather toward the left in our diagram. Likewise, a solute relatively more soluble in the organic layer would peak toward the right of the center.

Binomial Distribution in the Craig Extraction

Let us now formulate a more general mathematical treatment of the Craig countercurrent distribution. Actually, the mathematics is not difficult. In the first step of the treatment above, we distributed the solute between the two phases according to the distribution coefficient, K_D. Exactly as shown on page 475, the fractions of solute in the two phases are given by

$$f_{org} = \frac{K_D}{1 + K_D}$$

and

$$f_{aq} = \frac{1}{1 + K_D}$$

In the apparatus that Craig developed, and in the formulations encountered in the literature, the lighter phase (here, organic) is transferred from vessel 0 to vessel 1. Fresh organic phase is then introduced into vessel 0, and fresh aqueous phase into vessel 1. After equilibration of the two vessels,

$$f_{org_0} = \left(\frac{1}{1 + K_D}\right)\left(\frac{K_D}{1 + K_D}\right)$$

(The fraction $1/1 + K_D$ was present in the aqueous layer in the vessel after the first equilibration, as shown above, and the fraction $K_D/1 + K_D$ passed over into the fresh organic solvent upon equilibration. Thus the product of the two fractions gives the fraction of the original W now present in the organic layer of vessel 0.) Likewise,

$$f_{aq_0} = \left(\frac{1}{1 + K_D}\right)\left(\frac{1}{1 + K_D}\right)$$

(At the beginning of this step, the fraction $1/1 + K_D$ was present in the aqueous layer, as shown above, and the fraction $1/1 + K_d$ *of that* is what remains after equilibration with fresh organic phase.) Also,

$$f_{org_1} = \left(\frac{K_D}{1 + K_D}\right)\left(\frac{K_D}{1 + K_D}\right)$$

and

$$f_{aq_1} = \left(\frac{K_D}{1 + K_D}\right)\left(\frac{1}{1 + K_D}\right)$$

We have gone through these steps to make certain that the student understands what is happening. Actually, it is easier finally to consider the total solute in each vessel instead of focusing upon the two phases separately, although we may add what is in the two phases to get the total. Let us work this out for vessels 0, 1, and 2 where $n = 2$, including for practice all stages up to $n = 2$.

$n = 0$:

$$f_0 = \underbrace{\left(\frac{K_D}{1 + K_D}\right)}_{\substack{\text{Organic} \\ \text{phase}}} + \underbrace{\left(\frac{1}{1 + K_D}\right)}_{\substack{\text{Aqueous} \\ \text{phase}}} = \frac{1 + K_D}{1 + K_D} = 1$$

Since, where $n = 0$, all of the solute is in this vessel, the fraction contributed by the organic phase and the fraction in the aqueous phase must add up to 1.

$n = 1$:

$$f_0 = \left(\frac{1}{1 + K_D}\right)\left(\frac{K_D}{1 + K_D}\right) + \left(\frac{1}{1 + K_D}\right)\left(\frac{1}{1 + K_D}\right) = \frac{1}{1 + K_D}$$

$$f_1 = \left(\frac{K_D}{1 + K_D}\right)\left(\frac{K_D}{1 + K_D}\right) + \left(\frac{K_D}{1 + K_D}\right)\left(\frac{1}{1 + K_D}\right) = \frac{K_D}{1 + K_D}$$

Here, $f_0 + f_1 = 1$, and we may confirm our result:

$$\frac{1}{1 + K_D} + \frac{K_D}{1 + K_D} = \frac{1 + K_D}{1 + K_D} = 1$$

$n = 2$:

$$f_0 = \left(\frac{1}{1 + K_D}\right)\left(\frac{1}{1 + K_D}\right) = \left(\frac{1}{1 + K_D}\right)^2$$

$$f_1 = \left(\frac{1}{1 + K_D}\right)\left(\frac{K_D}{1 + K_D}\right) + \left(\frac{K_D}{1 + K_D}\right)\left(\frac{1}{1 + K_D}\right) = 2\left(\frac{1}{1 + K_D}\right)\left(\frac{K_D}{1 + K_D}\right)$$

$$f_2 = \left(\frac{K_D}{1 + K_D}\right)\left(\frac{K_D}{1 + K_D}\right) = \left(\frac{K_D}{1 + K_D}\right)^2$$

Now, examine the foregoing fractions:

$$f_0 = \left(\frac{1}{1 + K_D}\right)^2$$

$$f_1 = 2\left(\frac{1}{1 + K_D}\right)\left(\frac{K_D}{1 + K_D}\right)$$

$$f_2 = \left(\frac{K_D}{1 + K_D}\right)^2$$

The alert student will note that these three terms are the terms in the expansion of the binomial

$$\left(\frac{1}{1 + K_D} + \frac{K_D}{1 + K_D}\right)^2$$

In general, for any number of transfers, n, it may be shown that the fractions of the total solute to be found in the various vessels 0, 1, 2, . . . , n are given by the terms in the expansion of the binomial

$$\left(\frac{1}{1 + K_D} + \frac{K_D}{1 + K_D}\right)^n$$

In working with the formulas above, we assumed equal volumes of the two phases. In general, where the volumes are not necessarily equal (but the same in all vessels), it may be shown that the fractions of solute in the various vessels are given by the terms in the expansion of the binomial

$$\left(\frac{1}{1 + E} + \frac{E}{1 + E}\right)^n$$

where

$$E = K_D \times \frac{V_{upper}}{V_{lower}}$$

and V_{upper} and V_{lower} are the volumes of upper and lower phases, respectively.

If n is not small, expansion of the binomial becomes tedious. Fortunately, mathematical tables are available where it is worked out. Any single term in the binomial expansion can be obtained directly, using the formula

$$f_{n,r} = \left(\frac{n!}{r!(n - r)!}\right)\left(\frac{1}{1 + K_D}\right)^n K_D^r$$

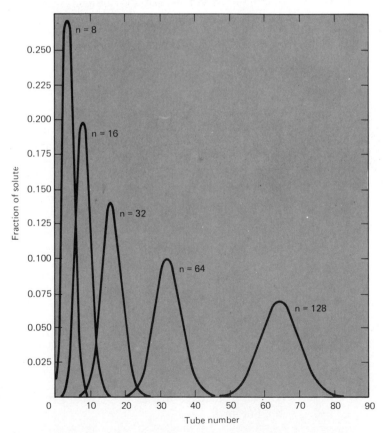

Figure 16.6 Theoretical distributions for solute with distribution coefficient $K_D = 1$ after various numbers of transfers, n, in a Craig experiment.

where $f_{n,r}$ is the fraction of solute in the r th tube after n transfers. Some writers use the form

$$f_{n,r} = \frac{n! K_D^r}{r!(n - r)!(1 + K_D)^n}$$

During a series of successive transfers, a solute moves through the vessels of a Craig apparatus as a sort of wave of diminishing amplitude. The solute spreads through more and more vessels as n increases, but at the same time the fraction of the vessels which contain the solute decreases. Figure 16.6 shows theoretical distributions of solute for various numbers of transfers, n, for the case where $K_D = 1$ and $V_{\text{org}} = V_{\text{aq}}$.

For a given number of transfers, substances with different distribution coefficients are distributed differently. Figure 16.7 shows theoretical distributions after 16 transfers for various values of the distribution coefficient. We see in the figure that solutes with different K_D-values are beginning to separate after the particular number of transfers indicated, but that separation is not complete in any case. This means that for these solutes more stages would be required in order to obtain pure components in quantitative yield.

It may be seen from the mathematical treatment given above why it is said that a Craig distribution is a *binomial* one. (Other types commonly encountered are *Gaussian*, as we saw in Chapter 2 for the distribution of random errors, and Poisson distributions.) However, as the number of transfers increases, the bino-

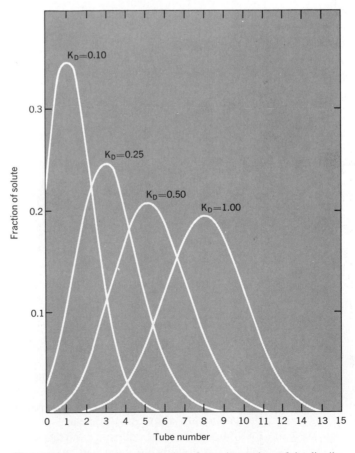

Figure 16.7 Theoretical distributions for various values of the distribution coefficient K_D in a 16-transfer Craig experiment.

mial distribution approximates more and more closely the Gaussian. (Poisson distributions also tend to Gaussian as n increases.) When n becomes large, say 50, although there is no definite demarcation, a Gaussian treatment is sufficiently accurate for describing a Craig distribution, and it is more convenient than the more cumbersome binomial theorem for such cases. The appropriate equations then are

$$r_{max} = \frac{nK_D}{K_D + 1}$$

$$f_{max} = \frac{1}{\sqrt{2\pi nK_D/(1 + K_D)^2}}$$

$$f_x = f_{max} \times e^{-[x^2/[2nK_D/(1+K_D)^2]]}$$

where r_{max} is the number of the tube containing the maximal quantity of the solute, f_{max} is the fraction of solute in this tube, f_x is the fraction of solute in a tube x tubes distant from r_{max}, and e is the base of natural logarithms. To use these equations for the case of unequal volumes of the two phases, simply replace K_D by E as defined earlier in the chapter.

Apparatus for Craig Extraction

For a large number of transfers, say 100 or 1000, manual operation with separatory funnels would be impossible in a practical sense. Craig, who expounded both the theory and practice of this type of extraction process, developed apparatus to take much of the labor out of the procedure. The typical Craig apparatus is based upon glass units shaped as shown in Fig. 16.8. Although most

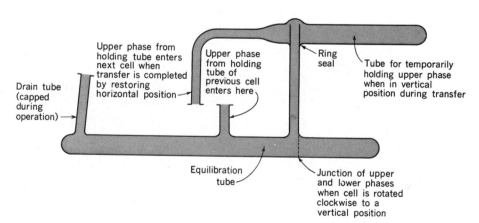

Figure 16.8 Typical glass cell for Craig apparatus.

people think in terms of vigorous shaking to equilibrate two liquid phases, Craig showed that a gentle sloshing of one liquid over another was actually more effective; further, troublesome emulsions were often avoided by the less vigorous technique. Thus it was unnecessary to adhere to the usual separatory funnel shape when designing a more automated apparatus. In the unit shown in Fig. 16.8, the two liquids are equilibrated by gently rocking the apparatus about 20 times, and the phases are allowed to separate. The apparatus is then rotated as indicated in the figure so that the cell is in a vertical position, whereupon the upper phase runs out of the equilibrium tube and into a temporary holding tube. Obviously, the volume of the lower phase must be such that the solvent interface occurs at the level of the runoff side arm. When the cell is returned to a horizontal position, the lighter phase runs out of the holding tube and into the equilibration tube of the next cell in line. (The liquid is prevented from returning to the equilibration tube whence it came by the design of the cell in the area labeled "Ring seal" in the figure.)

A battery of cells is clamped firmly to a metal frame so that all may be rocked and tilted together. The liquid-tight joints between adjacent cells are held together by spring clamps. In practice, the lower phase is introduced into all the cells at the beginning of the experiment. The solute mixture to be separated is placed in the first cell (number 0) in either phase. A solvent reservoir and metering device introduces the appropriate volume of the lighter phase into the first cell after each transfer. In the simpler instruments, the rocking and tilting are done manually by the operator. Larger outfits with as many as 1000 cells are operated by a motorized robot so that the entire distribution experiment requires no attention once it is set in motion. Two commercial Craig machines are shown in Figs. 16.9 and 16.10.

After the distribution has been completed, the bank of cells is tilted so that the liquids in the cells may be collected from the drain tubes shown in Fig. 16.8. If the experiment is a preparative one, the pooled solvents from tubes containing the desired solute may be evaporated to obtain the desired material. In an analytical run, the solutions from the individual cells may be analyzed by appropriate means, for example, by spectrophotometry, titrimetry, or measurement of refractive index.

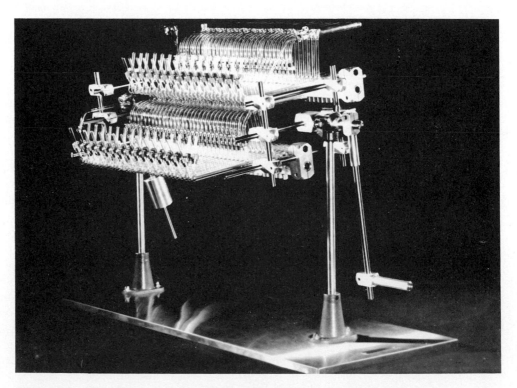

Figure 16.9 Manually operated 60-cell Craig countercurrent distribution apparatus. The cells are arranged in two banks of 30 each. (Courtesy of H. O. Post Scientific Instrument Co., Inc.)

Figure 16.10 1000-cell Craig countercurrent distribution instrument for automatic operation. The instrument can also be operated as two 500-cell units running concurrently. The driving mechanism is on the right-hand side under the table, with a driving shaft extending through the table top to the instrument; the unit is enclosed in a fume hood with glass doors and removable sides and top to afford access. (Courtesy of H. O. Post Scientific Instrument Co., Inc.)

Applications of Craig Extraction

Many of the applications of the Craig extraction process are found in the biochemical area. It has proved particularly useful for separating peptides in the molecular weight range of about 500 to 5000. Some antibiotics and certain hormones are polypeptides, and several examples of successful purification by the Craig method may be found. Crude tyrocidin from a bacterial source was separated after 673 transfers into three major components (A, B, and C) and several minor ones; 1600 transfers on component A led to a crystallizable material, and 2140 more transfers eliminated the last trace of a particularly troublesome impurity. Peptide hormones from the posterior lobe of the pituitary gland have been fractionated by Craig extraction; distribution between 2-butanol and 0.1 *M* aqueous acetic acid was employed in the final purification of synthetic oxytocin and in comparing this product with the naturally occurring hormone, and similar work was done on vasopressin. Craig distribution has been employed for isolating and characterizing adrenocorticotrophic hormone (ACTH) from the anterior pituitary. Although proteins present many difficulties, some have been successfully handled by the Craig technique, notably insulin, ribonuclease, lysozyme, and the serum albumins. The technique is used for purification of pharmaceutical preparations.

Although it retains its value for purifying macro quantities of materials, the Craig process is no longer used, to our knowledge, in analytical chemistry, having been replaced by chromatographic methods which are much more efficient. Do not ask, though, why it is discussed in this text before reading the next paragraph, and keep the Craig process in mind while studying Chapters 17 and 18.

Craig Extraction as a Model for Continuous-Separation Processes

In the Craig extraction experiment, the heavier liquid phase remains stationary, while the lighter phase is transported down the series of cells, carrying solutes with it to various extents in accord with their partition properties. In chromatography, which we shall consider in Chapter 17, there is likewise a stationary and a moving phase (not necessarily both liquid). But here the moving phase flows continuously, and equilibrium is not actually attained at any time during the experiment. This leads to great difficulty in formulating the separation process mathematically, and the difficulty appears in the form of complicated equations with many correction factors and adjustable parameters. It is easy for the student to lose sight of what is going on in such processes. In the Craig experiment, on the other hand, the intermittent flow of the moving phase permits equilibration in each step of the overall process, and as we have seen above, the theoretical treatment is not unusually difficult. The Craig extraction process then, while it is by no means real chromatography, is useful as a teaching tool in explaining chromatography to beginners. We shall try to show the similarities and differences in more detail in the next chapter.

KEY TERMS

Countercurrent extraction. A solvent extraction technique in which two immiscible liquids move in opposite directions in continuous contact with each other with a resultant separation of solutes. Rarely encountered in the laboratory (but sometimes processes which are not truly countercurrent are called this).

Craig extraction. Named for its developer, Lyman C. Craig. Sometimes called *countercurrent* distribution or extraction, but not a true countercurrent process because only one phase moves and hence sometimes termed *pseudocountercurrent*. One solvent, generally the lighter one, moves by steps through a series of equilibration vessels in which it encounters stationary portions of the other (heavier) phase. Solutes separate on the basis of differences in their partitioning between the two solvents. A separation technique in its own right, Craig extraction also serves in teaching as a model for chromatographic processes.

Distribution coefficient. Sometimes called *partition* coefficient. See under *distribution law*.

Distribution law. Sometimes called *Nernst distribution law*. A chemical species A will, at equilibrium, be distributed between two immiscible solvents in such a way that

$$\frac{a_{A_1}}{a_{A_2}} = \text{constant} = K_{D_A}$$

where a_{A_1} is the activity of A in solvent 1, etc., and K_{D_A} is the distribution coefficient for species A with those particular solvents. In actual practice, activities are often approximated by molar concentrations, and the law may be written

$$\frac{[A]_1}{[A]_2} = K_{D_A}$$

One solvent is usually aqueous and the other organic, and usually K_D is written with organic in the numerator; thus

$$\frac{[A]_{org}}{[A]_{aq}} = K_{D_A}$$

It is important to remember that K_{D_A} always pertains to the partitioning of the *single species* A and *not* to other species with which A may be in equilibrium.

Distribution ratio. Carefully distinguished from *distribution coefficient*. When a substance of interest in a solvent extraction situation participates in other equilibria in one or the other (or both) of the phases, a ratio D may be useful in which concentrations are summed over all relevant species in the two phases. See page 463 for a clarifying example involving benzoic acid.

Ion pair An overall neutral species formed by the electrostatic interaction of a cation and an anion in a solvent of low dielectric constant which does not encourage separation of ions as does water.

Metal chelate compound. An organic compound in which a metal ion becomes part of one or more heterocyclic rings by interacting with functional groups in the chelating agent which are favorably sited for ring formation with the metal. Chelation will generally yield a neutral molecule when the reaction involves both an acidic and a basic functional group, as with copper oxinate (page 467); oxine provides the acidic phenolic—OH group and a basic N-atom. Two basic functionalities, as in ethylenediamine, lead to a complex ion:

$$Cu^{2+} + 2H_2NCH_2CH_2NH_2 \longrightarrow$$

A neutral species is usually sought for solvent extraction. See also chelating agents as titrants (page 201) and precipitants (page 81).

QUESTIONS

1. Explain clearly in your own words the difference between a *distribution coefficient*, K_D, and a *distribution ratio*, D, in a solvent extraction study involving the base n-propylamine, $CH_3CH_2CH_2NH_2$, partitioning between an organic solvent and an aqueous phase the pH of which can be varied.

2. Look at Eq. (4) on page 465 for the effect of $[H^+]$ on the extraction of a weak acid from aqueous solution into an organic solvent. Derive an analogous equation showing how D for a weak *base* varies with $[H^+]$.

3. Consider a base B^- ($B^- + H_2O = HB + OH^-$) with dissociation constant K_b. Show that the fraction, f, of base molecules protonated is given by

$$f = \frac{K_b[H^+]}{K_w + K_b[H^+]}$$

Compare this with Eq. (5), page 465, and comment on the analogy in chemical terms. What would be the general shape of a graph of f vs. $[H^+]$? What, numerically, is the value of $\lim\limits_{[H^+] \to \infty} f$?

4. Myoglobin, abbreviated Mb, is an oxygen-binding protein molecule found in muscle tissue. The oxygenation of myoglobin may be represented as

$$Mb + O_2 = MbO_2$$

$$K = \frac{[MbO_2]}{[Mb][O_2]}$$

Derive an expression for f, the fraction of Mb molecules which are oxygenated, as a function of K and $[O_2]$. Compare this with Eq. (5), page 465 and with the equation of Question 3. Sketch a graph showing the general shape of the curve when f is plotted against $[O_2]$. What is the value of $\lim\limits_{[O_2] \to \infty} f$? What is the relationship between $[O_2]$ and K when $f = 0.5$?

5. For extraction of a metal, M^{n+}, from an aqueous phase into $CHCl_3$ as a chelate, MX_n, what would be the effect upon D_M of adding to the water a ligand Y^- which formed a water-soluble complex ion, $MY^{(n-1)+}$? That is, would D_M at a given pH increase, decrease, or be unaffected by the presence of Y^-?

6. The stability constants of complex ions can sometimes be determined by a solvent extraction method, as hinted in Question 5. A

chelating agent is employed, and the distribution ratio D_M is determined as a function of the concentration of the complexing agent in the aqueous phase. Consider a metal ion, M^{2+}, which forms a chelate MX_2 with HX and a complex ion MY^+ with Y^-. Derive an expression relating D_M to $[Y^-]_{aq}$ involving the stability (formation) constant K_f for the reaction

$$M^{2+}_{aq} + Y^-_{aq} = MY^+_{aq}$$

Assume that MX_2 is the only metal-containing species in the organic phase and that its concentration in the aqueous phase is negligible.

7. Repeat Question 6 except that Y^- is the anion of a weak acid, HY. The desired constant is that for the reaction

$$H_2O + M^{2+}_{aq} + HY_{aq} = MY^-_{aq} + H_3O^+_{aq}$$

8. Consider the partition of acetic acid, HOAc, between benzene and an aqueous solution, with all the significant equilibria shown, including dimer formation in the organic phase:

$$2\,HOAc \xrightleftharpoons{K_{dim}} (HOAc)_2$$

$$K_D \Big\| \qquad\qquad\qquad \text{benzene}$$

$$\overline{H_2O + HOAc \xrightleftharpoons{K_a} H_3O^+ + OAc^-} \quad \begin{array}{l}\text{aqueous}\\\text{solution}\end{array}$$

Show that this system is described by the following equation; i.e., derive the equation as though you were teaching a course and showing the class where it came from:

$$D = \frac{K_D(1 + 2\,K_{dim}[HOAc]_{org})}{1 + (K_a/[H_3O^+]_{aq})}$$

Notice the term $[HOAc]_{org}$. The value of D depends upon this concentration, and hence upon the total concentration of acetic acid in the system. Generally, when the degree of association of a solute in one phase differs from

that in the other, D will vary with the total concentration placed in the system. This contrasts with Eqs. (4) and (12) in the chapter where acids or metal ions were monomeric in both phases and D was independent of the total acid or metal concentration.

9. In Question 8, the dimerization of HOAc was written

$$2HOAc = (HOAc)_2$$

$$K_{dim} = \frac{[(HOAc)_2]_{org}}{[HOAc]^2_{org}}$$

Suppose that we had written the dimerization as

$$HOAc = \tfrac{1}{2}(HOAc)_2 \qquad K_{dim} = \frac{[(HOAc)_2]^{1/2}_{org}}{[HOAc]_{org}}$$

Show how to obtain the following equation for D:

$$D = \frac{K_D(1 + 2\,K^2_{dim}[HOAc]_{org})}{1 + (K_a/[H_3O^+]_{aq})}$$

Note that now K_{dim} is squared. Does this suggest that the value of D for a given set of conditions actually varies with the way we formulate the dimerization reaction on paper? Explain. (*Hint:* Compare K_{dim} in Question 8 with K_{dim} as written here.)

10. Explain in your own words why we say that the Craig solvent extraction process leads to a binomial distribution of a solute.

11. In Craig's apparatus for transferring a solvent through a row of vessels as described in this chapter, the *upper* (lighter) phase is moved when a transfer is made. Imagine you are an inventor and devise an apparatus where the *lower* (heavier) solvent is the mobile phase. Is your apparatus simpler to build or more complicated than Craig's? Does the issue of which phase moves have any bearing on the quality of a separation of a chemical mixture?

PROBLEMS

1. *Distribution coefficient.* (a) A chemist dissolves 155.13 mg of iodine (I_2) in exactly 1 liter of distilled water. The solution is equilibrated at 25°C with exactly 50 mL of carbon tetrachloride (CCl_4). Titration of the organic solvent requires 22.35 mL of 0.04474 *M*

$Na_2S_2O_3$ solution (titration reaction: $2\ S_2O_3^{2-} + I_2 \rightarrow S_4O_6^{2-} + 2\ I^-$). Calculate K_D, the distribution coefficient for I_2 between the two solutes (CCl_4/H_2O). (b) Sodium thiosulfate is a water-soluble salt. Explain how it is possible to titrate the I_2 in a CCl_4 solution with an immiscible aqueous solution of $Na_2S_2O_3$ (this is sometimes called a two-phase titration).

2. *Extraction.* (a) If 50.0 mL of 0.00100 M aqueous I_2 is extracted with 50.0 mL of CCl_4, calculate what percentage of the I_2 is removed from the aqueous phase. (Use K_D from Problem 1.) (b) Repeat part (a) with two successive extractions, each with 25 mL of CCl_4.

3. *Distribution ratio.* (a) When an aqueous solution of $FeCl_3$ in concentrated HCl is shaken with twice its volume of ether containing HCl, 99% of the iron is extracted. Calculate the distribution ratio (organic/aqueous) of Fe(III). (b) Iron(III) can be separated from other metals such as chromium by extraction into ether from a strong aqueous HCl solution. If 75 mL of aqueous HCl containing 0.390 g of Fe(III) is treated with 100 mL of ether in a one-shot process, how many milligrams of iron are left in the aqueous phase?

4. *Extraction of a weak base.* In its "free base" form, cocaine (MW 303.4) is a tertiary amine with a K_b of about 2.6×10^{-6} ($pK_b = 5.59$). In drug-testing laboratories, cocaine is sometimes extracted from alkaline urine samples into ether; the residue after evaporating the ether is then examined by a chromatographic technique, often thin-layer chromatography (page 555). (a) The pH of urine (unlike that of blood) is variable and depends strongly upon diet; it is usually on the acid side and may be as low as 4.5 or so. Why is the urine sample treated with alkali before the extraction? [Explain qualitatively, in words; see part (c) for calculation.] (b) According to one handbook, 1 g of cocaine will dissolve in 600 mL of water and in 3.5 mL of ether. Calculate K_d. (c) Calculate a distribution *ratio* (D) for cocaine at aqueous pH values of 5, 6, 7, 8, 9, 10, 11, and 12, and plot D vs. pH. Compare the actual data with your qualitative concept in part (a); do the data jibe with the "picture" based on your chemical instincts?

5. *Distribution coefficient.* Three extractions with 25-mL portions of an organic solvent removed 96% of a solute from 100 mL of an aqueous solution. Calculate the distribution coefficient (organic/aqueous) of the solute.

6. *Multiple extraction.* The distribution coefficient of a solute S between benzene and water is 12.0 (organic/aqueous). Calculate the percentage of S extracted into benzene from 50 mL of a 0.10 M aqueous solution of S if the extraction is performed with (a) one 50-mL portion of benzene; (b) two successive 25-mL portions of benzene; five successive 10-mL portions of benzene.

7. *Acid dissociation constant and distribution ratio.* An acid HB has a distribution coefficient of 9.8 between an organic solvent and water. At pH 5.00, exactly half of the material is extracted into the organic solvent from an equal volume of the aqueous phase. Calculate the dissociation constant of HB.

8. *Separation of acids.* Given two acids, HA and HB, with the following distribution coefficients and dissociation constants:

	K_D	K_a
HA	10	1.0×10^{-4}
HB	1000	1.0×10^{-9}

Calculate the distribution ratios (D-values) for the two cases at pH values of 4, 5, 6, 7, 8, 9, and 10. Assuming that the ratio of the two D's needs to be $10^6:1$ for a quantitative extraction of HB without extracting HA appreciably, what is roughly the lowest pH at which such a separation can be accomplished using one extraction with an equal volume of organic solvent?

9. *Extraction of a metal chelate.* A chelating agent, HT, dissolved in an organic solvent, extracts a metal, M^{2+}, from an aqueous solution according to the reaction

$$2H_2O + M_{aq}^{2+} + 2HT_{org} = MT_{2org} + 2H_3O_{aq}^+$$

The equilibrium constant for this reaction is 0.10. (a) Identify this equilibrium constant in terms of other constants (page 469). (b) Calculate the pH values at which 1, 25, 50, 75, and 99.9% of the metal is extracted from 10 mL of an aqueous phase with 10 mL of a 0.0010 M solution of HT in the organic solvent. (Assume that the metal concentration is so small that $[HT]_{org}$ remains constant.) (c) Plot the percentage of metal extracted vs. pH_{aq}.

10. *Extraction of a metal chelate.* Repeat Problem 9 for a metal N^{2+} where the equilibrium constant for the reaction as formulated above is 1.0×10^{-7}. Plot the data as above and choose as low a pH as possible for a quantitative separation of M^{2+} and N^{2+}.

11. *Extraction of a metal chelate.* A certain metal ion is extracted by a chelating agent as in Problems 9 and 10. The concentration of chelating agent in the organic phase is 0.010 M. The following data are obtained:

pH	1	2	3	4	5
D	10^{-8}	10^{-4}	1	10^4	10^8

Plot log D vs. pH (page 469) and evaluate n and K_{ex}.

12. *Craig extraction.* The distribution coefficient for the extraction of solute A from water into an organic solvent is 10 (organic/aqueous). 1.00 g of A is dissolved in water and placed in tube 0 of a Craig extraction apparatus and extracted with the organic solvent. The organic phase is transferred from tube 0 to tube 1, and so on. (a) Calculate the fraction of A remaining in tube 0 after five transfers. (b) Calculate the fraction of A in tubes 1, 2, 3, 4, and 5.

13. *Craig extraction.* Repeat Problem 14 except that the value of K_D is 100.

17

Gas-Liquid Chromatography

The resolution of mixtures into their components is important in all branches of chemistry and no less so in the many other fields where chemical techniques are employed in solving a wide variety of problems. Thus the impact of a powerful and versatile separation technique will be felt throughout much of modern science. In this connection, the significance of chromatography can scarcely be overstated. Utilizing chromatographic methods, separations in many cases are accomplished much more rapidly and effectively than before, and many separations which would never have been attempted by other techniques are routinely successful. Unparalleled breakthroughs in biochemistry—for example, in our understanding of the structure and function of enzymes and other proteins—have stemmed directly from the application of chromatography to biological research. Evaluating air and water pollution, determining pesticide residues on fruits and vegetables, identifying and classifying bacteria, monitoring respiratory gases during anesthesia, searching for organic compounds and living organisms on other planets, determining metabolic pathways and mechanisms of action of drugs—a list of all such studies based upon chromatography would be a lengthy one indeed.

Although forerunners can be found in the nineteenth century, it is generally considered that a paper published in 1906 by Michael Tswett, a lecturer in botany at the University of Warsaw, provided the first description in nearly modern terms of a chromatographic separation.[1] Tswett described the resolution of the chlorophylls and other pigments in a plant extract as follows:

> If a petroleum ether solution of [crude] chlorophyll is filtered through a column of adsorbent (I use mainly calcium carbonate which is packed firmly into a narrow glass tube), then the pigments are resolved from top to bottom into various colored zones according to an adsorption sequence, where the more strongly adsorbed pigments displace the more weakly adsorbed ones and force them further downward. This separation becomes practically complete if, following the pigment solution, a stream of pure solvent is passed through the column. Like a spectrum of light rays, the different components of the pigment mixture are systematically resolved on the calcium carbonate column and can be identified and also determined quantitatively. Such a preparation I term a chromatogram, and the corresponding method the chromatographic method.

Although the term *chromatography* is derived from Greek words meaning "color" and "write," the color of the compounds is obviously incidental to the separation process; Tswett himself anticipated applications to a wide variety of chemical systems. Had his work been immediately seized upon and extended, several sciences might have progressed more rapidly. As it was, chromatography remained dormant until about 1931, when separations of plant carotene pigments were reported by the prominent organic chemist Kuhn. This research attracted more attention, and adsorption chromatography became widely used in the field of natural product chemistry.

More recently, there have been four major developments: ion-exchange chromatography in the late 1930s, partition chromatography in 1941, gas chromatography in 1952, and gel-filtration chromatography in 1959. In addition to these major advances, which provided additional mechanisms to adsorption for distributing solutes between stationary and mobile phases, there have also appeared modifications in the geometry of the chromatographic system, as in paper and thin-layer chromatography.

Theoretical developments which permit a thorough understanding of the chromatographic process and hence clarify the factors which determine column performance appeared first in connection with gas chromatography. But certain of these insights have proved, with suitable adjustments, to be equally helpful in understanding chromatography where the moving phase is a liquid. Thus there began about 1968 a revolution in liquid chromatography which promised new speed and efficiency in the separation of nonvolatile compounds which do not lend themselves to the gas chromatographic approach.

Because of its great practical importance in many research areas, chromatography is a fast-moving field. Efforts continue along many lines, of which we may mention a few: better detectors, new column-packing materials, improved interfacing with other instruments (such as the mass spectrometer) which serve to identify the separated components, new data-processing techniques based upon computers, and new mathematical models which provide additional insight into the nature of the process. Our goal in this book is to present a basic introduction which will acquaint the student with the nature of the chromatographic process,

[1] M. Tswett, *Ber. Deut. Botan. Ges.*, **24**, 235 (1906).

explain in simple terms what the instruments fundamentally do, and show some of the applications which have made chromatography indispensable in so many fields.

17.1

DEFINITION AND CLASSIFICATION OF CHROMATOGRAPHY

Although the meaning of the term is largely understood by chemists, a good definition of chromatography is difficult to formulate. It is a collective term applied to methods which appear diverse in some regards but share certain common features. A definition should emphasize that components of the sample are distributed between two phases, but this alone is inadequate because we do not wish the term to embrace all separation processes. Keulemans' definition serves as well as any:

> Chromatography is a physical method of separation, in which the components to be separated are distributed between two phases, one of these phases constituting a stationary bed of large surface area, the other being a fluid that percolates through or along the stationary bed.[2]

The stationary phase may be either a solid or a liquid, and the moving phase may be either a liquid or a gas. Thus all the known types of chromatography fall into the four categories shown in Table 17.1: liquid-solid, gas-solid, liquid-liquid, and gas-liquid.

TABLE 17.1 Summary of Types of Chromatography

Stationary phase	SOLID		LIQUID	
Moving phase	Liquid	Gas	Liquid	Gas
Examples	Tswett's original chromatography, with petroleum ether solutions and CaCO₃ columns	Gas-solid chromatography, or GSC	Partition chromatography on silica gel columns	Gas-liquid chromatography, or GLC
	Ion-exchange chromatography		Paper chromatography	

In all the chromatographic techniques, the solutes to be separated migrate along a column (or, as in paper or thin-layer chromatography, the physical equivalent of a column), and of course the basis of the separation lies in different rates of migration for the different solutes. We may think of the rate of migration of a solute as the result of two factors, one tending to move the solute, and the other

[2] A. I. M. Keulemans, *Gas Chromatography*, 2nd ed., Reinhold Publishing Corp., New York, 1959, p. 2.

to retard it. In Tswett's original process, the tendency of solutes to adsorb on the solid phase retarded their movement, while their solubility in the moving liquid phase tended to move them along. A slight difference between two solutes in the firmness of their adsorption and in their interaction with the moving solvent becomes the basis of a separation when the solute molecules repeatedly distribute between the two phases over and over again throughout the length of the column.

In this chapter we consider gas-liquid chromatography, or GLC, by far the more important form of gas chromatography. In the next chapter we shall describe chromatographic techniques where the moving phase is a liquid.

17.2

BASIC APPARATUS FOR GLC

To orient readers who are totally unfamiliar with gas chromatography, we shall first describe the apparatus and technique of GLC briefly and in general terms. Then we shall consider the theory, next indicate more fully the functions of the components of the apparatus, and finally give some illustrative applications which show the power and versatility of the method.

17.2a. *The Carrier Gas and Sample Introduction*

Figure 17.1 is a schematic diagram of a common type of basic GLC instrument. Although gas chromatographs can become very complicated if additional features are included, the basic instrument is a fairly simple one. The moving phase in GLC is a gas, most commonly helium, hydrogen, or nitrogen. The choice of carrier gas depends primarily upon the characteristics of the detector, as we shall see later. The user buys a cylinder of the compressed gas and attaches a reducing value to it. Commercial gas chromatographs usually provide an additional regulating valve for good control of the pressure at the inlet of the column. With instruments of the type shown, employing the thermal conductivity detector, the carrier gas passes through one side of the detector and then enters the column. Near the column inlet is a device whereby samples may be introduced into the carrier-gas stream. The samples may be gases or volatile liquids. The injection port is heated so that liquid samples are quickly vaporized. Samples of a few microliters of liquid or a few milliliters of gas are commonly introduced through a rubber septum by means of a hypodermic syringe.

17.2b. *The Column*

The gas stream next encounters the column, which is mounted in a constant-temperature oven. This is the heart of the instrument, the place where the basic chromatographic process takes place. Columns vary widely in size and packing material. A common size is 6 ft long and $\frac{1}{4}$-inch internal diameter, made of copper or stainless steel tubing; to save space, it may be U-shaped or coiled into a spiral. The tubing is packed with a relatively inert, pulverized solid material of large surface area, frequently diatomaceous earth or firebrick. The solid, however, is actually only a mechanical support for a liquid; before it is packed into

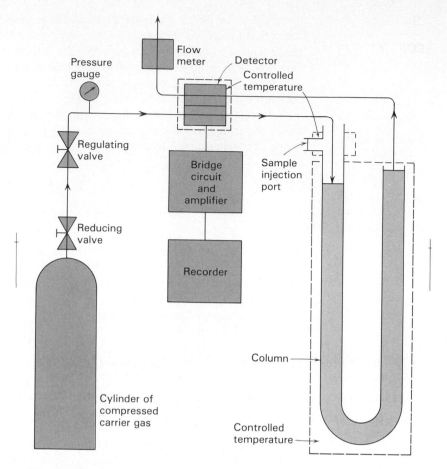

Figure 17.1 Schematic diagram of a gas chromatograph with a thermal conductivity detector. Large arrows indicate direction of gas flow.

the column, it is impregnated with the desired liquid which serves as the real stationary phase. This liquid must be stable and nonvolatile at the temperature of the column, and it must be appropriate for the particular separation.

17.2c. The Detector

After emerging from the column, the gas stream passes through the other side of the detector. Elution of a solute from the column thus sets up an imbalance between the two sides of the detector which is recorded electrically. The carrier-gas flow rate is important, and usually a flow meter of some sort is provided. There may be another regulating valve at the outlet end of the system, although normally the emerging gases are vented at atmospheric pressure. Because continual exposure of laboratory workers to the vapors of chromatographed compounds may be unwise even though the samples are usually small, attention should be paid to ventilation at the outlet of the instrument. Provision can be made to trap separated solutes as they emerge from the column if this is required for further investigation.

THEORY OF GLC

17.3a. *The Theoretical Plate Concept*

It is suggested that the reader review the Craig countercurrent solvent extraction distribution in Chapter 16; we shall approach GLC in essentially the same manner. Suppose that we have a series of small chambers as shown in Fig. 17.2, each containing a portion of a nonvolatile liquid, which serves as the stationary phase. Let us introduce into the first chamber a sample of the mobile phase, a gas such as nitrogen, containing vapor of an organic compound, say benzene. If the liquid is a suitable one for our purpose, some of the benzene will dissolve in it, and some will remain in the space above it. Now Henry's law, in its usual form, states that the partial pressure exerted by a solute in dilute solution is proportional to its mole fraction. Thus, for the equilibrium distribution of benzene between liquid and vapor phases in our chamber, we may write

$$p_{\text{benzene}} = kX_{\text{benzene}}$$

where p_{benzene} is the partial pressure of benzene in the vapor phase, X_{benzene} the mole fraction of benzene in the liquid, and k a constant. In gas chromatography, partial pressure and mole fraction are often replaced by concentration terms which yield a dimensionless *distribution coefficient*, K:

$$K = \frac{\text{concentration of benzene in liquid phase, wt/mL}}{\text{concentration of benzene in gas phase, wt/mL}} = \frac{C_L}{C_G}$$

It is customary for the liquid term to be the numerator. K is also called a *partition coefficient* by many writers.

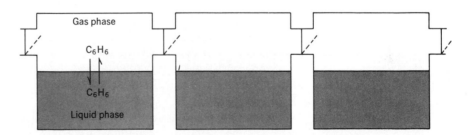

Figure 17.2 Imaginary chambers for a Craig model of a GLC experiment.

Now let us perform the same sort of operation that we employed in the Craig extraction. Transfer the gas (nitrogen plus benzene vapor) from the first chamber into the second one, where it encounters fresh liquid that contains no benzene. Introduce fresh nitrogen containing no benzene into the first chamber. Wait until equilibrium has been established in both chambers. Then transfer the gas from the second chamber into the third and from the first into the second, and introduce fresh nitrogen into the first. As the transfers proceed, the benzene will become distributed through the chambers in the same manner as a solute in the Craig solvent extraction system, illustrated in Fig. 16.6. The position of the benzene band on the horizontal axis will depend upon the number of transfers, the value of K, and the relative volumes of vapor and liquid in the individual chambers. (See the discussion accompanying Fig. 16.6.) Just as we saw in the Craig experiment, a second compound, say toluene, could be separated from benzene in this gas-liquid system if the K-values for the two compounds were different, provided sufficient stages were employed. (See Fig. 16.7.) After a large number of transfers, the binomial distribution will become virtually the symmetrical Gaussian one.

There is a major difference in practice between Craig solvent extraction and chromatography which should be pointed out, although it is not important with regard to the principle involved. With a Craig apparatus, it is customary to stop when the desired separation has been achieved and drain the solutions from the tubes containing the solutes. In modern chromatography, on the other hand, the flow of the mobile phase is continued until the solutes have migrated the entire length of the column, whence they emerge, one after the other, to enter the detector. The practice of terminating the Craig process as soon as possible simply reflects the more cumbersome nature of the apparatus and the time-consuming character of the process.

The equilibrium chambers in the apparatus described above are called *theoretical plates,* a term which originated in distillation theory and was later carried over into chromatography. Each cell in the Craig apparatus is a plate in this sense. Now a chromatographic column operates under conditions of continuous flow of the moving phase, and equilibrium is not attained at any point in the column. However, after traversing a certain length of column, a mixture will have been subjected to the same degree of fractionation as would have been achieved in one equilibrium step. That length of column which accomplishes this is called the *height equivalent of a theoretical plate* or HETP. The total length of a column divided by HETP is the number, n, of theoretical plates in the column, and it is customary to rate column performance in terms of number of plates. A good column will have more plates than a poor column of the same length. The great efficiency of GLC for performing difficult separations lies in the fact that large numbers of plates are fairly easily obtained with columns of reasonable length; columns with a couple of thousand plates may be only 5 or 6 ft long. (To give an idea of the power of the chromatographic technique, it may be noted that columns for fractional distillation are likewise rated in terms of theoretical plates; a fairly good 6-ft conventional distilling column may have something on the order of 20 or 30 plates.[3])

In gas chromatography, samples are injected as rapidly as possible so that a substance is placed on the column as a narrow "plug." However, just as in the Craig extraction technique, when the substance moves along, it spreads out through more and more plates but occupies a progressively smaller fraction of the total number of plates which it has encountered. In other words, as we increase the number of plates, the absolute width of the elution band increases, but there is

---

[3] The direct comparison of numbers of plates in distillation and chromatography is not quite fair unless it is qualified as follows. In a fractional distillation, all the plates in the column are utilized throughout the experiment. In chromatography, on the other hand, after the sample components have migrated along the column, that portion of the column already traversed might just as well not be there any longer. Thus chromatography basically requires more plates than distillation. But these plates are so readily obtained that even taking this factor into account, chromatographic columns are ordinarily far more efficient than distilling columns.

a decrease in width relative to the total base of the operation. In this perspective, elution bands look narrow as they emerge from a good column, broad from a poor column.

17.3b. Calculation of the Number of Theoretical Plates

It is often desirable to evaluate a chromatographic column by measuring n, the number of theoretical plates. With a Craig apparatus, the plates may simply be counted. But this cannot be done with a column, because the plates are now imaginary; a theoretical plate is a mental concept, part of a model developed to explain the chromatographic process in familiar terms. We can determine the apparent number of plates, however, because the model implies a relationship between the characteristics of a chromatographic elution band and the number of plates in the column. The mathematical derivation of an equation for calculating n from a chromatogram is beyond the scope of this book,[4] but the general idea is intuitively reasonable, and the result is a very simple formula. In the Craig extraction scheme, we were able to calculate how a solute distributed itself through a known number of tubes. We may readily imagine that if we had known instead the characteristics of the distribution, we might have turned our thinking around and calculated the number of tubes. Similarly, as implied in the preceding paragraph, the time required to elute a solute from a column and the width of the elution band should enable us to calculate n.

Figure 17.3 shows a Gaussian elution band and the parameters which are used to calculate n. The time from injection of the sample to the appearance of the peak of the elution band at the detector is called the *retention time, t_R*. The formula for calculating n from t_R and the bandwidth depends upon where the

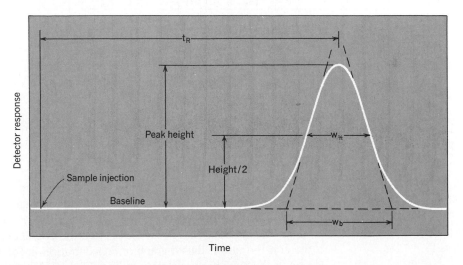

Figure 17.3 Chromatographic elution band showing measurement of t_R and w for estimating n, the number of theoretical plates.

[4] The interested reader may find a derivation in O. E. Schupp, III, *Gas Chromatography*, Vol. 13 of E. S. Perry and A. Weissberger, eds., *Technique of Organic Chemistry*, Wiley-Interscience, New York, 1968, pp. 39–46.

width is measured. The two commonest ways are given here. If the width is measured halfway between the baseline and the top of the band, we may designate it as $w_{1/2}$. Then the formula for obtaining n is

$$n = 5.54\left(\frac{t_R}{w_{1/2}}\right)^2$$

If the width is measured at the baseline using the construction shown in the figure, we designate it w_b; the formula then becomes

$$n = 16\left(\frac{t_R}{w_b}\right)^2$$

In the latter case, tangents to the band are drawn at the two inflection points; the width, w_b, is the distance between the intersections of these tangents with the baseline. The two formulas give comparable results, and the choice is a matter of personal preference. Although there may be some difficulty in locating accurately the inflection points of an elution band, the slopes of the tangents are not highly sensitive to this error.

If a mixture of several compounds with different K-values is injected into the chromatograph, then of course several elution bands will be recorded. These will show different retention times, but the bandwidths will also be different; thus comparable (although probably not identical) n-values will be calculated from the several bands. A component with a larger K-value will spend more time in the liquid phase and hence take longer to be eluted. Referring to a Craig type of model with discrete plates, we should say that more transfers would be required to elute such a compound. But the band is also broadened thereby, so that theoretically the n-value should be the same as that obtained for an earlier band.

For the purpose of determining n, the units for the abscissa in Fig. 17.3 make no difference. Thus, although t_R is defined as a time, we may actually measure distances on the recorder chart paper in centimeters or millimeters; of course, t_R and w must be measured in the same units.

The number of plates in a given column is found to vary with sample size in a fairly regular way. Overloading of any column leads to a deterioration of performance and poor separations. To obtain a value for n, sometimes extrapolation to zero sample size is performed on a graph of measured n-values vs. sample size.

17.3c. Nonideal Behavior: The van Deemter Equation

GLC as Linear Nonideal Chromatography

The plate model based upon a Craig type of solute distribution between gas and liquid phases, with equilibrium attained prior to each transfer from one plate to the next, represents what is sometimes called *linear ideal chromatography*. Linear in this connection means that the distribution coefficient, K, is independent of solute concentration; thus a graph of concentration in the liquid phase vs. concentration in the gas phase is a straight line (Fig. 17.4, curve 1a). Such a graph is often called an *isotherm*. In general, a linear isotherm leads to a symmetrical elution band as shown by curve 1b in Fig. 17.4. A departure from Henry's law behavior, shown by a nonlinear isotherm (Fig. 17.4, curves 2a and 3a), leads to a skewed elution band (Fig. 17.4, curves 2b and 3b). Referring to curves 2a and 2b, we may interpret the asymmetric elution band as follows: Where the solute concentration is high, a greater fraction of solute remains in the gas phase than is the

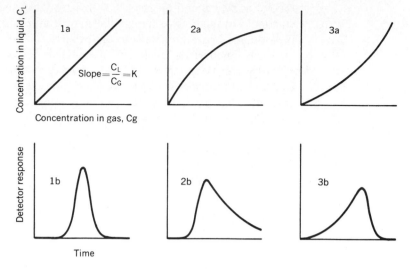

Figure 17.4 Linear and nonlinear isotherms (a) and the shapes of the corresponding elution bands (b).

case at lower concentrations. As a result, solute at the peak will move faster through the column, and hence the peak will tend to catch up to the leading edge of the elution band and leave a long tail at the trailing edge.

Nonlinear isotherms of the type shown in Fig. 17.4, curve 2a, are common in adsorption equilibria, and thus skewed elution bands with tailing as shown in curve 2b are often encountered in gas-solid chromatography and in liquid adsorption chromatography. At the low concentrations normally employed in GLC, deviations from Henry's law do not generally occur, and thus GLC commonly provides an example of linear chromatography.

Ideal behavior, on the other hand, is not attainable in any actual chromatographic process. First, ideality would require that the sample be placed initially in only the first plate. This can be done in the Craig apparatus, but in a column, where HETP may be only a fraction of a millimeter, it is impossible with a finite sample. Second, ideal chromatography would require a column whose packing was perfectly uniform in regard to particle size and shape, liquid loading, and geometric array; further, the velocity of the moving phase would have to be the same everywhere in the column. Third, in ideal chromatography, equilibrium would always exist at all points in the column between the stationary and moving phases with regard to the solute distribution; because the flow of the moving phase is continuous, this means that equilibration would have to be instantaneous. Finally, ideal chromatography would require that the solute move along the column as a result only of the motion of the moving phase; that is, the solute could not spread out in the column by its own tendency to diffuse.

Thus, under the usual conditions, GLC provides an example of *linear nonideal chromatography*. Departures from the requirements for the ideal process as listed in the paragraph above result in elution bands which are broader than the hypothetical bands for an ideal case. A detailed theoretical consideration of the kinetic factors which lead to this band broadening has given considerable insight into the nature of the chromatographic process in real columns. Not only has this been intellectually satisfying, but practical improvements have resulted in terms of better separations in less time, and these have spilled over from GLC to revitalize liquid chromatography as well. A mathematical analysis of chromatographic theory is far beyond the scope of this text, but the principal factors are understandable in qualitative terms. We may summarize these for what insights they provide and leave the details to the experts.

The pioneering treatment of nonideal behavior in GLC was presented in 1956 by van Deemter et al.[5] These workers considered three major factors that caused band broadening when a solute, initially a narrow plug, migrated along a column. An equation was derived containing three terms which represented the contributions of these three factors to HETP.

Eddy Diffusion

The first factor, which is usually called eddy diffusion, arises from the multiplicity of pathways for a gas flowing through a column which is packed with particles of various sizes and shapes arranged in an irregular manner. As it flows through the channels among the packing particles, the carrier gas divides into many streams which may merge and again split in a complex manner as the gas follows a myriad of routes through the column. Likewise, solute molecules, as they move along with the flowing carrier gas, will follow many paths, some shorter and some longer than the average distance. This means that the original solute plug will spread out, some molecules reaching the detector sooner, some later, with many at an average time. The magnitude of the contribution of eddy diffusion of band broadening ought to depend upon the size of the packing particles, their shape, and the uniformity of their distribution in the column.

There has been some dispute about the importance of eddy diffusion which appears now to be settled. Perhaps this factor was more severe with the column packing materials and techniques of the earlier years of GLC. Measurements with very carefully packed columns have suggested that it is less important than early workers supposed. Apparently, in a good column most of the pathways actually followed by the gas are about the same length. Also, it has been proposed that lateral diffusion of solute from one portion of the gas stream to an adjacent one might counter the effect of eddy diffusion upon the solute profile. Furthermore, it appears that some of the band broadening formerly attributed to eddy diffusion actually occurred outside the column, for example, in the glass wool that was sometimes used to plug the end of the column, in the tubing that led from the column to the detector, and in the detector itself. It now seems clear that eddy diffusion may be only a very small factor in a well-designed gas chromatograph with a good column, and indeed it may be practically negligible as compared with other band-broadening factors. Nevertheless, eddy diffusion is worth noting because it is negligible only if rendered so by proper attention to details in the design and fabrication of the chromatograph and the column.

Longitudinal Diffusion

The second factor contributing to band broadening is longitudinal diffusion of the solute in the gas phase. Solute molecules tend to diffuse along concentration gradients, and thus a solute band moving along a column will broaden as molecules spread into the regions of lower concentration ahead of the band and behind it. Solute molecules spend part of their time in the gas phase and part in the liquid phase, but diffusion is much faster in gases than in liquids (diffusion

[5] J. J. van Deemter, E. J. Zuiderwig, and A. Klinkenberg, *Chem. Eng. Sci.*, **5**, 271 (1956).

coefficients in the gas phase are typically on the order of 10^5 times as great as in liquids). Thus diffusion in the liquid phase is generally considered negligible in GLC. Diffusion takes time, and thus the contribution of diffusion to band broadening will increase with the length of time required to elute the band from the column. Hence the diffusion term in the van Deemter equation involves the velocity of the carrier gas, becoming smaller as the velocity increases.

Nonequilibrium in Mass Transfer

The last factor in nonideality arises from the fact that equilibrium cannot be attained for distribution of the solute between stationary and mobile phases in the face of continuous flow of the carrier gas. Using C_L and C_G for solute concentrations in the liquid and gas phases, respectively, we may write Henry's law as

$$C_L = KC_G$$

But Henry's law describes an equilibrium situation, and for cases where there is insufficient time or equilibration, we must write

$$C_L = KC_G \times f(t)$$

where $f(t)$ is some function of time which reflects the kinetics of the mass transfer process between the two phases. When t is large (i.e., when equilibrium is approached), then, of course, $f(t)$ must approach unity so as to yield Henry's law.

The band broadening that arises from the fact that equilibration is not instantaneous is depicted in Fig. 17.5, where both ideal and nonideal solute distributions are shown. The ideal case (upper portion of Fig. 17.5) is simply a Craig distribution, where equilibrium is attained in all of the plates before transfers are made. The lower part of Fig. 17.5 illustrates the case where the gas phase is moving continuously, equilibrium is never attained at any point, and hence in general $C_L \neq KC_G$. Where the front of the solute zone in the gas phase encounters fresh liquid, some of the solute dissolves in the liquid but, because we do not wait for equilibrium, $C_L < KC_G$. The result is that some of the solute which would have been retarded by the liquid phase if equilibrium had been attained actually continues to migrate along the column with the carrier gas. At the other extreme, the rear of the gas-phase solute zone, solute cannot leave the liquid rapidly enough to equilibrate with the fresh carrier gas, and $C_L > KC_G$. In other words, part of the solute is retarded to a greater degree by the stationary liquid than would have been the case if equilibration had been instantaneous. This nonideal behavior continues as the solute traverses the entire length of the column, with the obvious result that the elution band is broader than it would otherwise have been.

The extent of band broadening arising from the slow mass-transfer kinetics depends upon the time available for the process to move toward equilibrium. Thus the nonequilibrium term in the van Deemter equation ought to involve the carrier-gas velocity and, in contrast with the longitudinal diffusion term considered above, it should become larger as the velocity increases. Furthermore, we might expect that solute in a thin film of liquid would be closer to equilibrium with the moving gas stream than would be the case for solute which had penetrated deep pools of liquid. Thus the magnitude of the nonequilibrium term should depend upon the quantity of liquid with which the solid support is impregnated and the manner of its distribution within the nonuniform pores of the solid. In addition, the nonequilibrium term contains the diffusion coefficient of the solute in the liquid phase. The faster the solute molecules can move along concentration gradients in the liquid to and from the interface with the gas, the closer will be the ap-

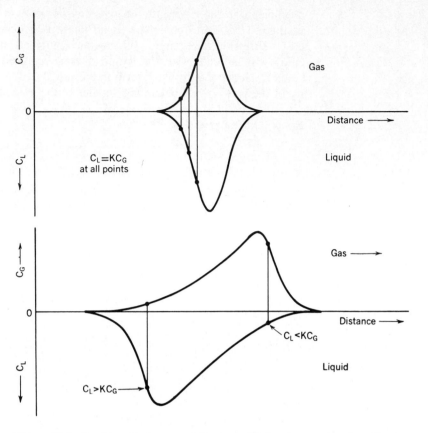

Figure 17.5 Band broadening caused by nonequilibrium in mass transfers: Concentration profiles in gas and liquid phases for a solute zone within a column. Upper: ideal distribution assuming instantaneous equilibration. Lower: actual distribution resulting from finite rate of mass transfer.

proach to equilibrium in the mass-transfer process, and the less the band spreading attributable to this factor.

The van Deemter Equation

Consideration of the three factors discussed above led to the following equation:

$$\text{HETP} = 2\lambda d_p + \frac{2\gamma D_G}{u} + \frac{8k d_f^2}{\pi^2 (1 + k)^2 D_L} u$$

where

λ = dimensionless parameter measuring the irregularity of the column packing
d_p = diameter of the packing particles
γ = correction factor accounting for the irregularity of diffusion pathways through the packing material
D_G = diffusion coefficient of the solute in the gas phase
u = linear velocity of the carrier gas
k = constant for a particular solute and a particular column
d_f = "effective film thickness," a measure of the liquid loading of the packing material
D_L = diffusion coefficient of the solute in the liquid phase

Theoretically minded people have scrutinized the three terms in the van Deemter equation very carefully and have proposed extensions of these as well as additional factors in order to gain a better understanding of the chromatographic process. Such work, which involves sophisticated mathematics, is far beyond the scope of an introductory textbook. The van Deemter equation as given here focuses attention upon the major factors which cause band broadening and serves quite well, for the interested reader, as a takeoff point for understanding the further refinements that have followed.

The van Deemter equation is often seen in the abbreviated form

$$\text{HETP} = A + \frac{B}{u} + Cu$$

where

$$A = \text{eddy diffusion term}$$

$$B/u = \text{longitudinal diffusion term}$$

$$Cu = \text{nonequilibrium in mass transfer term}$$

A graph of HETP vs. u is a branch of a hyperbola, as shown in Fig. 17.6. The dashed construction on the graph depicts the contributions of the A, B/u, and Cu terms to HETP at various carrier-gas velocities. The A-term remains constant, independent of velocity. At very low velocities, most of the band broadening is due to longitudinal diffusion, while at high velocities, the increasing departure from equilibrium in the mass-transfer process becomes dominant. There is an optimal velocity where the best balance of these factors is obtained; that is where HETP is a minimum or the number of plates in the column is a maximum. The minimum in the graph is a rather shallow one, and in practice it is not necessary to locate it exactly; it is obviously advantageous, however, to be in the right neighborhood.

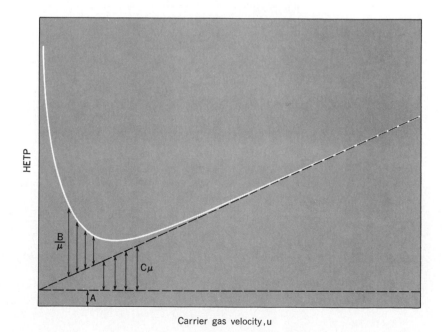

Carrier gas velocity, u

Figure 17.6 Depiction of the van Deemter equation $\text{HETP} = A + B/u + Cu$. Note that the contribution of the A-term to HETP is independent of velocity, that B/u increases as velocity decreases, and that Cu predominates at high velocities.

Although more recent work has led to a more sophisticated understanding of the processes that occur in the column, the van Deemter equation as presented here is fairly good, at least with regard to predicting the shape of a graph of HETP vs. u. Curves very similar to the one shown in Fig. 17.6 are in fact obtained in the laboratory. Such curves are not calculated because values are generally not at hand for the parameters such as λ and γ in the van Deemter equation.

Determination of the Van Deemter Coefficients. A, B, and C are easily obtained from experimental data by measuring HETP at three carrier-gas velocities and setting up three equations in three unknowns. The values will vary greatly from one sort of column to another; the following have been presented as typical for packed columns of the sort commonly used in analysis.[6]

$$A \cong 0 \text{ to } 1 \text{ mm}$$

$$B \cong 10 \text{ mm}^2/\text{s}$$

$$C \cong 0.001 \text{ to } 0.01 \text{ s}$$

$$\text{HETP}_{min} \cong 0.5 \text{ to } 2 \text{ mm}$$

$$u_{opt} \cong 1 \text{ to } 10 \text{ cm/s}$$

In summary, the plate theory of chromatography enables us to calculate a very useful measure of column performance, the number of theoretical plates or, if we wish, HETP. This is very simply done using one of the formulas on page 498. But the plate theory in itself does not suggest how the performance of a column may be improved. The so-called rate theory, as exemplified by the van Deemter equation, on the other hand, gives definite factors such as particle size, liquid loading, and carrier-gas velocity, over which we have some control, by which improved performance may be obtained.

17.3d. Resolution

In general, the positions of elution bands on the horizontal axis of the chromatogram and their widths will determine how complete a separation of the starting mixture has been accomplished. Although samples may have many components, we suppose that one pair of these will be the most difficult to separate and confine our discussion of resolving mixtures to two-component systems.

Expression of Resolution

The *resolution, R*, of two components is often defined as follows, using the terms shown in Fig. 17.7.

$$R = \frac{2(t_{R_2} - t_{R_1})}{w_{b_1} + w_{b_2}}$$

Alternatively, if the widths are measured halfway between the baseline and the tops of the bands, the equation becomes

$$R = \frac{2(t_{R_2} - t_{R_1})}{1.699(w_{1/2_1} + w_{1/2_2})}$$

[6] A. B. Littlewood, *Gas Chromatography: Principles, Techniques, and Applications*, 2nd ed., Academic Press, Inc., New York, 1970, p. 202.

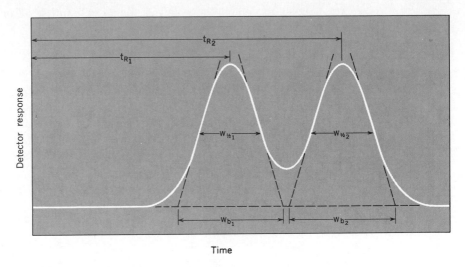

Figure 17.7 Measurements used to calculate the resolution, R, of two peaks.

Often, for two bands which are close together, the widths are about the same; in that case, the 2 can be removed from the numerator and one of the w-terms dropped from the denominator.

If $R = 1.5$, the two solutes are virtually completely separated; there will be only 0.3% overlap of the two elution bands. If $R = 1$, the separation is adequate for most analyses; the overlap is about 2%. As R decreases below 1, the overlap becomes progressively more severe until, at about $R = 0.75$ (50% overlap), the separation becomes unsatisfactory for most purposes. For quantitative analysis by GLC, the area under a solute peak is the best measure of the quantity of that solute in the sample. Complete resolution of a mixture is not required for this. If overlap is not too great and if the peaks are symmetrical, fairly good estimates of the areas can be made. This can be extended by computer methods for analyzing a complex shape into a family of discrete peaks. Of course, whatever technique is used, some point will be reached beyond which good results cannot be obtained. It then becomes necessary to consider the factors that determine resolution in order to improve the situation.

Resolution and Column Length

The number of theoretical plates in a column, everything else being the same, is proportional to its length, and hence one of the obvious ways to improve resolution is to employ a longer column. The separation of two peaks, $t_{R_2} - t_{R_1}$, is directly proportional to the distance that the two solutes migrate, whereas the width of an elution band increases directly with the square root of the distance. Thus as we lengthen a column, two bands will separate faster than they broaden, so to speak, and resolution will improve. There are limitations, however, on column length, of which we may mention two. If a column is too long, the pressure required to give a reasonable carrier-gas flow rate is excessive. Second, the longer the column, the longer will be the time required for elution. The efficiency of a busy laboratory handling many samples might be better served by improving the resolution in some manner that does not lead to great increases in t_R-values. Nevertheless, if preliminary experiments with an ordinary column, perhaps 4 to 6 ft long, showed a fair separation with a particular sort of sample, then it might be sensible to employ a somewhat longer column, say 8 or 10 ft, in order to achieve a really good separation.

Separation Factor

If a satisfactory separation is not obtained with a good column of reasonable length after careful attention to operating parameters such as temperature (see below) and carrier-gas flow rate, then the best approach is usually to try a different stationary liquid phase. In other words, if logical attempts to achieve resolution by narrowing the solute bands fail, then we must move the peaks farther apart by changing the K-values for the solutes.

The ratio of retention times, t_{R_2}/t_{R_1}, is called the *separation factor*, S (some writers use S.F. and some α for this ratio). Usually, the ratio of retention times is about the same as the ratio of the K-values for the two solutes. Thus

$$S = \frac{t_{R_2}}{t_{R_1}} \cong \frac{K_2}{K_1}$$

Note that S is not the same thing as the resolution, R. The ratio of retention times, measured at the peaks of the elution bands, does not in itself describe the effectiveness of the separation, because it tells nothing about the widths of the bands. However, it may be shown, as we might expect intuitively, that there is a relation between R and S if the number of the theoretical plates in the column is taken into account:

$$R = \frac{n^{1/2}(S - 1)/S}{4}$$

Taking $R = 1$, which as noted earlier represents a fairly good separation of two solutes, we may plot the number of plates required vs. the separation factor, S, which yields the curve shown in Fig. 17.8. The curve approaches the ordinate axis asymptotically, reflecting the fact that if $S = 1$, no separation is possible however long the column. As S increases above 1, the number of plates required decreases rapidly. In other words, if we can promote a rather small increase in S by changing the liquid phase, it may be far more effective in improving the resolution than even a large increase in the length of the column.

17.3e. *Factors in Retention*

Retention Volumes

For many purposes, the *retention volume* of a solute, V_R, is more convenient than the retention time. The retention volume is the product of the retention time and the carrier-gas flow rate; since there is an inverse relation between flow rate and retention time, V_R is independent of flow rate. The flow rate is ordinarily measured at a point beyond the column, as shown in Fig. 17.1. This measured value, F, should be corrected to account for the fact that the column is at a different temperature than the flow meter; a further correction is required if the flow meter introduces moisture into the gas or creates an appreciable pressure drop of its own. If F_c is the corrected flow rate, then

$$V_R = t_R \times F_c$$

Because the gas is compressible, its velocity is not uniform throughout the column—the gas moves faster near the outlet end—and V_R should be corrected to

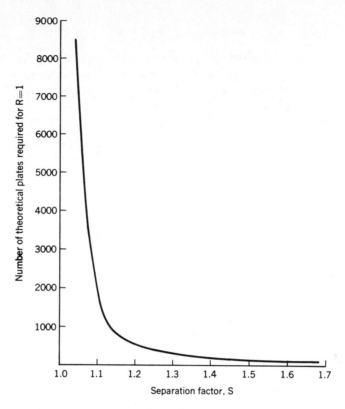

Figure 17.8 Number of theoretical plates required to achieve a resolution equal to one plotted as a function of the separation factor.

an average column pressure to yield the *corrected retention volume, V_R°*. The corrections are not discussed in detail in this text, but they may be found in books on gas chromatography.

Consider a compound which does not interact with the stationary phase (i.e., $K = 0$). This compound is not retarded by the liquid in the column, but its retention time is not zero; some time, which we may call t_G, will be required simply to wash the compound through the column. Using the same corrections as above, we may calculate a corrected retention volume, V_G°, for such a compound. This volume should amount to the same thing as the space within the column which is available to the gas, that is, the portion of the column which is not occupied by the packing material and its liquid load. Other terms, such as *interstitial volume* and *void volume,* have been used for the analogous space in other forms of chromatography such as ion exchange.

The fixed gases of the air are not appreciably soluble in most organic liquids that are used as stationary phases. Thus it is easy to obtain t_G by injecting a little air along with the sample. The thermal conductivity detector responds to air, and a small blip called the *air peak* appears on the recorder chart, from which V_G° can be calculated. (The other common detector, the flame ionization detector, does not respond to the gases of the air; another approach is then required.)

Distribution Coefficient and Liquid Load

Let V_L be the volume of the liquid phase in the column, and recall the distribution coefficient:

$$K = \frac{C_L, \text{ wt/ml}}{C_G, \text{ wt/ml}}$$

Now when the peak of an elution band for some solute appears at the detector, we suppose that half of that solute is still in the column and half has been eluted in a volume of $V_R^\circ - V_G^\circ$. Then we may cancel the weight terms in the preceding equation and write

$$K = \frac{1/V_L}{1/(V_R^\circ - V_G^\circ)} = \frac{V_R^\circ - V_G^\circ}{V_L}$$

whence

$$KV_L = V_R^\circ - V_G^\circ$$

and

$$V_R^\circ = V_G^\circ + KV_L$$

More rigorous derivations of this equation and a discussion of the assumptions involved may be found in monographs on gas chromatography.

V_G° is often quite small compared with V_R°. Thus it is frequently the case that the retention volume for a solute is almost directly proportional to the quantity of liquid phase in the column. Similarly, with a given liquid loading, the retention volume will vary directly with K. In other words, the experimenter has considerable control over retention in his choice of the liquid phase and the quantity used in preparing the column.

It may be noted that the time corresponding to a given retention volume is also under the control of the operator via the flow rate, but, as we have seen, the analysis time cannot be shortened at will in this manner without the penalty of poorer resolution. With certain types of mixtures, it may be advantageous to vary the flow rate during the chromatographic run, starting with a low value for optimal resolution of solutes whose K-values are small and close together, and continuously increasing the flow rate to accelerate the elution of laggard solutes which are easily resolved but are spending more time than necessary in the column. This technique is known as *programmed-flow gas chromatography*. It is mentioned here because it is instructive to consider it, but in practice it has been supplanted by temperature programming (see below).

Temperature

Virtually every aspect of GLC is sensitive to temperature to some extent. The volume of the liquid phase and hence the gas space in the column, the viscosity of the carrier gas and therefore the inlet pressure required for a given flow rate, diffusion coefficients of solutes—these are a few examples of factors which may be affected to some degree by the column temperature. But we are particularly interested here in the often very pronounced direct effect of temperature upon solute retention. The usual effect of a temperature increase is to lower the value of the distribution coefficient K; in other words, at the higher temperature, a solute is driven out of the liquid phase in accord with the general rule that an increase in temperature lowers the solubility of a gas in a liquid. Decreasing K in turn decreases the retention time and the retention volume. The magnitude of the effect depends upon the nature of the solute and of the liquid phase and the temperature region investigated, but roughly the change in retention volume is on the order of a few percent, say 3 to 12%, per degree. An analysis, then, is completed most rapidly at the highest column temperature compatible with the desired separation and sample stability.

On the other hand, the separation factor for a pair of solutes is generally larger the lower the temperature. As a crude rule of thumb, the higher the temperature, the more similar the behavior of two compounds in a GLC column. Thus the column temperature selected for an analysis ought to be low enough to achieve the necessary separation, but no lower than this so as not to waste time.

For a series of solutes which interact in the same manner with the liquid phase, for instance an homologous series, the distribution coefficients generally bear an inverse relation to the vapor pressures; the larger the vapor pressure, the smaller is K. In general, the lower the boiling point of a solute, the greater will be its vapor pressure at a given temperature. Thus the components of a mixture of such solutes will emerge from a column in order of increasing boiling point. (If some of the solutes interact in specific ways with the stationary liquid, e.g., by hydrogen bonding, then this simple rule may not hold.)

Figure 17.9(a) shows a chromatogram of a hydrocarbon mixture obtained in the ordinary manner with the column held at a certain temperature. The temperature selected was a compromise: It was too high to yield an optimal separation of the lower compounds in the hydrocarbon series and too low for the higher-molec-

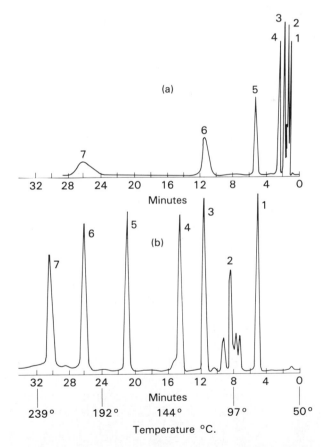

Figure 17.9 Gas chromatograms of a mixture of normal hydrocarbons. (a) Isothermal chromatogram of the following mixture at 168°C: (1) pentane, (2) hexane, (3) heptane, (4) 1-octene, (5) decane, (6) 1-dodecene, (7) 1-tetradecene. (b) Programmed-temperature chromatogram of the same mixture. [Reprinted from *Anal. Chem.*, **30,** 1157 (1956); copyright by the American Chemical Society. Reprinted by permission of the copyright owner.]

ular-weight compounds. A much nicer chromatogram of the same mixture is seen in Fig. 17.9(b): resolution is better, and in fact a number of impurities in the hydrocarbons that were mixed to prepare the sample may be seen which do not show up in curve (a). All the bands have about the same shape, which facilitates quantitative measurements; peak number 7 which was very low in curve (a) is now higher above the baseline, representing a better signal-to-noise ratio for improved quantitative accuracy. Curve (b) was obtained by the technique of *programmed temperature GLC*. Here the temperature of the column was raised during the chromatographic run, starting at a temperature that was suitable for the lower members of the series and finishing at a higher temperature where the elution of the higher boiling components was more satisfactory. The effect is rather similar to that of programming the flow rate as mentioned in the section above. Various temperature-time functions have been studied, but the commonest by far for ordinary work is a linear program: The temperature increases linearly with time at so many degrees per minute. Modern chromatographs often provide for this capability, and the operator can select on the panel an initial temperature, a rate of increase, and a final temperature.

17.4

EXPERIMENTAL ASPECTS OF GLC

Having seen what gas-liquid chromatography is, and after considering the theory, we may now discuss briefly some of the more practical aspects which will make the gas chromatograph less of a mysterious "black box."

17.4a. *Carrier Gas*

Various gases have been used in GLC, for example, hydrogen, helium, nitrogen, argon, carbon dioxide, and even water vapor. The lighter gases, hydrogen and helium, permit more longitudinal diffusion of solutes, which tends to lower column efficiency, particularly at lower flow rates. Thus nitrogen might be a better choice of carrier gas in order to accomplish a really difficult separation. In addition, it is cheaper than helium and safer in the laboratory than hydrogen. However, there is another consideration, the characteristics of the detector. It is obviously desirable that the response of the detector to the components of the sample differ greatly from its response to the ever-present carrier gas. In the case of the thermal conductivity cell, which is one of the most widely used detectors, this requirement is much better met by the lighter gases, hydrogen and helium, as we shall see below. Thus, with instruments employing this detector, helium is by far the commonest carrier gas in the United States, while in Europe, where helium is very expensive, hydrogen is more widely used. With the flame ionization detector, which has become a rather common one in recent years, nitrogen is probably the most widely used carrier gas.

17.4b. *Sampling System*

Liquid samples, typically ranging from a small fraction of 1 μl to perhaps 25 μl or more, are usually injected through a rubber septum by means of a hypodermic syringe. Special syringes delivering various volumes in the microliter range are on the market, sometimes equipped with mechanical devices that aid in reproducing sample size. Gaseous samples may also be injected, or they may be introduced by means of various gas-sampling devices designed for commercial chromatographs. The injection technique is important: The sample should be introduced as a sharp "plug" rather than being slowly bled into the carrier gas stream. Slow injection leads to much more band spreading than is necessary; actually, HETP calculated from the elution peak as described above is a function of the injection rate. Good injection technique requires practice.

It is important that the size of the sample not be too large for the apparatus. Overloading has an extremely deleterious effect upon column efficiency. The lower limit of sample size is determined by the detector: So far as the column is concerned, the smaller the sample the better. The sensitivity of the detector determines how small a sample can be handled.

The temperature at the injection port is very important. If a liquid sample evaporates slowly, the result is similar to that caused by injecting too slowly. The injection port is usually heated independently of the heating unit surrounding the column, and generally it should be held at a temperature above the boiling points of the sample components. On the other hand, the temperature should be below a level where the compounds would decompose.

17.4c. *Column*

Packed Columns

The stationary phase in GLC is a liquid, but it cannot be allowed simply to slosh around inside a tube. The liquid must be immobilized, preferably in the form of a thin layer of large surface area. This is most commonly accomplished by impregnating a ground-up solid material with the liquid phase before the column is packed. The solid should be chemically inert toward the substances which will be chromatographed, stable at the operating temperature, and of large surface area per unit weight. The pressure drop required for desirable gas flow rates should not be excessive. Mechanical strength is desirable so that the particles will not break and alter the particle size distribution with handling. Most of the solids employed as supports in GLC are highly porous, but the characteristics of the pores are very important. For example, the pores in silica gel tend to be narrow; they fill up with the liquid and provide an insufficient area of gas-liquid interface. The active adsorbents such as activated charcoal and silica gel are poor solid supports. Even when coated with the liquid film, these solids adsorb sample components, causing "tailing" of the elution bands as shown in curve 2b in Fig. 17.4. The commonest solid support materials are diatomaceous earth (a deposit formed on ocean bottoms from the siliceous residues of a certain type of algae) and firebrick. The materials are ground and carefully graded with respect to particle size and often subjected to various chemical pretreatments to improve their surface qualities. The preparation of the solid, its impregnation with the liquid phase, and the final packing into copper, stainless steel, or glass columns used to be an art that was cultivated by chromatographers. Today, it is much more common to buy ready-made columns from manufacturers who have made available a wide variety of very good columns.

Capillary Columns

There is another type of column for GLC called the open-tubular or capillary column. This is a long, thin tube of glass or other material such as stainless steel, perhaps 0.1 to 1 mm in diameter, sometimes several hundred feet long, coiled up to save space. The inner surface is coated with a very thin layer of the stationary liquid phase, just that quantity which will adhere as a film on the glass or metal; there is no column packing in the usual sense. The pathways through the column are practically the same length for all molecules of the sample, and hence eddy diffusion is virtually zero in open-tubular columns. The very thin liquid film, containing no deep, stagnant pools, promotes a rapid approach to equilibrium in the partition process. The columns are very long, and it has been argued that they are not much more efficient than packed columns of comparable length would be. On the other hand, packed columns of that length would require relatively enormous pressure drops to attain reasonable flow rates. Because of the very light liquid loading, open-tubular columns can handle only very small samples, and their widespread use awaited the development of very sensitive detectors. Very impressive separations are often obtained with these columns.

Selection of Liquid Phase

The stationary liquid phase must be selected with the particular separation problem in mind. The liquid should have a very low vapor pressure at the column temperature; a common rule of thumb suggests a boiling point at least 200°C above the temperature to which the liquid will be subjected. The two important reasons for desiring low volatility are, first, loss of liquid will eventually destroy the column, and second, the detector will respond to the vapor of the stationary phase with resulting drift of the recorder baseline and lowered sensitivity toward the components of the analytical sample.

Obviously, the liquid phase should be thermally stable at the column temperature, and, except in special cases, it should not react chemically with the sample components. The liquid must have an appreciable solvent power for the sample. Recalling the old rule that "like dissolves like," it may be stated rather crudely that generally there should be some chemical resemblance between the liquid substrate and the solutes to be separated. Thus the saturated hydrocarbon squalane ($C_{30}H_{62}$, MW 423, boiling point about 350°C) is a good liquid phase for the separation of low-molecular-weight alkanes on a column that will not be heated above about 150°C. For the separation of aromatic hydrocarbons, the aromatic liquid benzyldiphenyl, useful up to about 120°C, is sometimes recommended. A polyglycol column might be used to separate a mixture of alcohols. Of course, it is not required that the stationary liquid match the solutes functional group for functional group, and sometimes one liquid phase will serve for the separation of a variety of mixtures. This has led to the designation "general-purpose" for some columns, although this is misleading in that no liquid phase provides completely general effectiveness for separating all classes of solutes. Examples of general-purpose liquids are silicone oils and greases, useful for a wide variety of nonpolar solutes, and polyethylene glycols (Carbowaxes), widely used for mixtures of polar solutes. Lists of liquid phases, recommended temperature limits, and the types of compounds for whose separation they are useful may be found in monographs on GLC.

The quantity of liquid applied to the solid support is important. If too much liquid is present, solutes spend too much time diffusing through the liquid phase, and the separation efficiency is lowered. Too little liquid allows solutes to interact with the solid itself, in which case adsorption may cause "tailing" and consequent overlapping of elution bands. The liquid loading varies with the nature of the solid support, anticipated sample size, and other factors, but is generally in the range of 2 or 3 to perhaps 20% liquid by weight. Usually the solid is treated with a solution of the desired liquid in a volatile solvent, after which the solvent is removed by warming and, later, purging with carrier gas.

17.4d. *Detector Characteristics*

The separation process occurs in the column, and hence this component must be considered the heart of the instrument. On the other hand, the separation would be of little value without some way to detect and measure the separated solutes as they emerge from the column.

Integral Detectors

Two types of detectors are commonly distinguished, integral and differential. An integral detector provides at any instant a measure of the total quantity of eluted material which has passed through it up to that time. The first paper on GLC[7] described an example of an integral detector. A mixture of fatty acids was chromatographed, and the effluent gas from the column was bubbled through an aqueous solution containing a pH indicator. When an acid emerged from the column, the pH of the solution dropped, the indicator changed color, and a light beam of appropriate wavelength passing through the solution was attenuated by absorption. The resulting change in the electrical signal from a photodetector activated a relay which turned on a buret containing sodium hydroxide. The addition of the base restored the pH of the solution to its original value, and hence the indicator to its original color, whereupon the buret was automatically shut off. The volume of titrant was recorded as a function of time, and the resulting chromatogram, of the type shown in Fig. 17.10(a), consisted of a series of steps, each representing the titration of one of the acids in the original mixture. The quantity of each acid was easily found by measuring the height of the corresponding step. This detector, which seems crude and clumsy by hindsight, was useful at the time because it was easily assembled from available components, and it served its purpose very nicely at the birth of GLC. But integral detectors have been largely supplanted by differential detectors; the latter are found on the overwhelming majority of modern gas chromatographs, and it is these that we wish to emphasize in this chapter.

Differential Detectors

Differential detectors yield the familiar chromatograms consisting of peaks rather than steps, as shown in Fig. 17.10(b). Two major classes may be distinguished: first, those which measure the *concentration* of a solute by means of some physical property of the effluent gas stream, and second, those which respond to the solute directly and hence measure its *mass flow rate*. This distinction will be clarified as we examine one example of each type of differential detector. First, though, we list some general detector characteristics which are useful in evaluating any detector.

[7] A. T. James and A. J. P. Martin, *Biochem. J.*, **50**, 679 (1952).

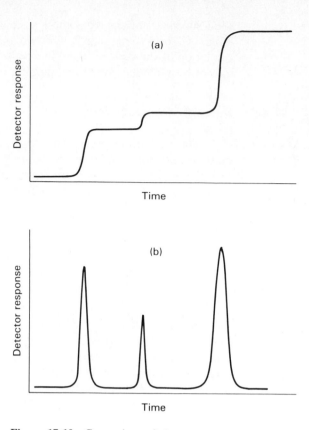

Figure 17.10 Comparison of chromatograms obtained with (a) integral and (b) differential detectors.

Sensitivity

As explained below, the sensitivity of the detector represents an important limitation upon the smallest quantity of a solute that can be determined by GLC, and the increased demand for trace analyses in many different fields has stimulated the development of more sensitive detectors. For example, our growing awareness of the impact of trace contaminants upon biological ecosystems has provided a market for very sensitive detectors in studies of water pollution and pesticide residues in food products. It is possible for detector sensitivity to have a direct economic, political, or legal implication. For example, certain federal agencies are empowered to establish permissible levels of various poisons in foods. However, an important exception was legislated: The so-called Delaney clause requires a level of zero for any compound which is known to be carcinogenic in humans or experimental animals. Now zero means zero to a politician, but to an analytical chemist it means a quantity smaller than can be detected with available methodology. It is perhaps not coincidental that the growing agitation by spokespersons for the food industry to repeal the Delaney clause has paralleled the increasing sensitivity of GLC and other analytical techniques. It may be argued that a level of zero means nothing outside the context of a particular analytical method employing a specified instrument.

Various measures of detector sensitivity are found in the literature, but basically, for our purpose, we may consider the sensitivity as the slope of a curve showing detector response as a function of the quantity measured, as shown in

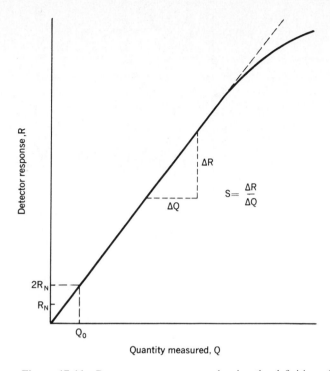

Figure 17.11 Detector response curve showing the definition of the sensitivity, S, and the relationship of peak-to-peak noise level, R_N, to limit of detection, Q_0.

Figure 17.11. A general expression for the sensitivity, then, is

$$S = \frac{\Delta R}{\Delta Q}$$

as shown in the figure.

Stability

The baseline of a chromatogram is subject to short-term fluctuations of a largely random nature which are called *noise*. A longer range upward or downward trend in the baseline is called *drift*. Noise and drift, illustrated in Fig. 17.12, may originate in various instrument components such as amplifiers or recorders and in fluctuations of the carrier-gas flow rate. Drift is seen in programmed temperature operation if the column reaches a temperature where the stationary liquid volatilizes. Much of the problem is eliminated by good circuit design, high-quality components, and proper operation of the chromatograph. There will always be, however, an inherent detector noise level which, along with the sensitivity, sets the lower limit for the quantity of a solute that can be detected.

The basic problem can be easily understood in qualitative terms. Suppose that a solute elution band is too small to be measured accurately. Perhaps the first thing to come to mind is this: Why not increase the amplification in order to enlarge the elution band on the recorder chart? The answer is, this can be done but there is a limitation upon the benefit to be derived from it because the noise is amplified too. The smallest elution band that can be distinguished from noise peaks corresponds to the limit of detection for a solute. This becomes essentially

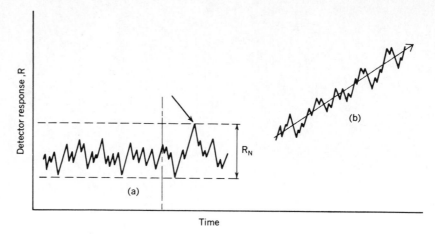

Figure 17.12 Expanded baselines. Left-hand portion illustrates noise and shows the measurement of the peak-to-peak noise level, R_N; the peak marked by the arrow might be ignored in estimating R_N (see text). Right-hand portion illustrates short-term noise superimposed upon upward drift.

a statistical problem: How much larger than the random baseline fluctuations must an elution band be in order to yield acceptable odds that we shall identify it correctly as an elution band and not confuse it with noise. This is a problem not unlike some of those encountered in Chapter 2; we need to estimate the level beyond which a recorder deflection is probably not noise but rather is due to a definite cause, the sample. Various analyses of this problem have been presented in the literature, and various recommendations may be found. Perhaps the commonest advice is to take as the limit of detection that quantity of a solute which gives an elution band whose height is twice the peak-to-peak noise level. The peak-to-peak noise level, R_N, is explained in Fig. 17.12.

The relationship among sensitivity, noise level, and limit of detection may be formulated as follows. Recalling the definition of sensitivity,

$$S = \frac{\Delta R}{\Delta Q}$$

If we associate the limit of detection, Q_0, with twice the peak-to-noise level, $2R_N$, then we may write

$$S = \frac{2R_N}{Q_0}$$

or

$$Q_0 = \frac{2R_N}{S}$$

In other words, low noise level and high sensitivity are favorable detector attributes with regard to the limit of detection.

Unusually large noise peaks, such as the peak marked by the arrow in Fig. 17.12, occur infrequently. If such peaks are excluded in estimating R_N, then the limit of detection will appear to be better. Along with this, of course, goes an increased risk of reporting an analytical result for a solute when, in fact, a noise peak was measured.

Linearity

The ideal detector response would be linear with respect to the quantity measured, Q. This is the case with commonly used detectors within certain concentration limits, but eventually, as shown in Fig. 17.11, the response generally falls off.

Versatility

It is obviously advantageous that a detector respond to a wide variety of chemical compounds. None of the components of a sample would then be overlooked, nor would it be necessary to change detectors in order to handle various types of samples.

Response Time

The detector should respond rapidly to the presence of the solute, or, as it is sometimes said, there should be a small "time constant." The total response time for a chromatograph is a function not only of the detector itself, but also of inertia in other components, for example, the recorder.

Chemical Activity

In many cases, this is not an important factor, but sometimes solutes which have been separated by GLC are subjected to further study. For example, a mass spectrum or an infrared spectrum may be desired in order to identify the solute with certainty. In that event, it is obviously important that the solute not be decomposed in the detection process.

There are additional desiderata which may be classed as nonfunctional but which may be important in certain circumstances, such as low cost, simplicity, safety, and ability to withstand abuse.

The geometry of the detector and the pathway to it are very important. Solutes which have been separated in the column must not remix in the tubing leading to the detector nor inside the detector itself. A small volume within the detector is also conducive to a fast response. Stagnant pockets of gas must be avoided, and the dead volume between the column and the detector should be as small as possible.

17.4e. *Types of Detectors*

Thermal Conductivity Detector

One of the most widely used detectors for general-purpose GLC is the thermal conductivity cell. This device contains either a heated metal filament (generally platinum, a platinum-rhodium alloy, or tungsten) or a thermistor. Thermistors are small beads prepared by fusing a mixture of metal oxides, generally of manganese, cobalt, nickel, and traces of other metals. There is usually a thin protective layer of glass on the surface, and fine platinum alloy wires provide electrical connections. The important property of thermistors in the present context is an unusually large temperature coefficient of electrical resistance.

The heated detector element, filament or thermistor, under steady-state conditions, adopts a certain temperature determined by the heat supplied to it and the rate at which it loses heat to the walls of the chamber which surrounds it. Although a small amount of heat is lost through radiation and by conduction through the metal electrical leads, the temperature of the element is determined primarily by the thermal conductivity of the gas in the space between the element and the walls. Detection is based upon the fact that different gases have different thermal conductivities. When the composition of the gas changes, the temperature of the element changes, and this is reflected by a change in the electrical resistance of the element.

As shown schematically in Fig. 17.1 the detector generally has two sides, each with its own element. The pure carrier gas traverses one side of the detector, which is ahead of the sample injection port, while the column effluent flows through the other side. This is seen in more detail in Fig. 17.13, where one type of detector employing thermistors is illustrated schematically.

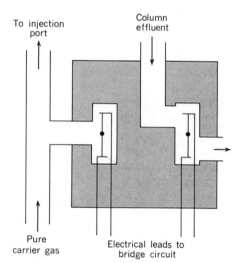

Figure 17.13 Schematic diagram of a thermal conductivity cell. The black dots are thermistor beads.

As we said, the detector elements are simply electrical resistances selected for their unusually large temperature coefficients of resistance. Thus the circuitry associated with the thermal conductivity detector is exactly what one would expect from elementary physics regarding resistance measurements. The two resistances in the two sides of the detector are two arms of a Wheatstone bridge circuit, as shown in Fig. 17.14. Before the injection of the sample into the chromatograph, pure carrier gas is flowing through both sides of the detector; the adjustable resistors are set so that the bridge is balanced, which establishes the baseline on the recorder chart. After injection, when a solute emerges from the column, the value of R_s in Fig. 17.14 changes, while the other resistances remain the same. The bridge goes out of balance, and a voltage appears across the leads labeled "To recorder" in the figure. After the solute has passed through the detector, the bridge returns to its original balance. Thus a record of the voltage across the bridge vs. time will exhibit a peak as shown in Fig. 17.3 for the elution of each separated component of the sample. Basically, the thermal conductivity detector responds to changes in the concentration of a solute in the carrier-gas stream, reflecting the way in which the thermal conductivity of the gas mixture depends upon the concentration.

Helium is an attractive carrier gas in conjunction with the thermal conductivity cell because its thermal conductivity, like that of hydrogen, is much greater

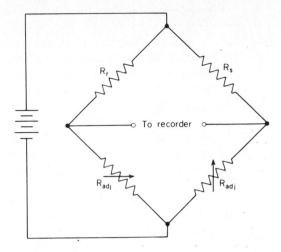

Figure 17.14 Wheatstone bridge circuit for thermal conductivity detector. R_r and R_s are the resistive elements in the reference and sample sides of the detector. R_{adj} is adjustable by the operator in order to balance the bridge.

than that of most organic compounds, while it does not represent an explosion hazard. Thus the appearance of an eluted solute at the detector causes a much greater change in the temperature of the resistive element than would be the case, say, with nitrogen as the carrier gas. This implies, of course, a greater sensitivity in detection, or a lower limit of detection. A few thermal conductivity values are given in Table 17.2.

The thermal conductivity detector is relatively simple and inexpensive, rugged, and reliable. Its sensitivity is adequate for many purposes. The sensitivity may be increased by operating the elements at a higher temperature by furnishing a larger bridge current, but this involves a trade-off with regard to the life expectancy of the elements. This detector, in general, is nondestructive; that is, solutes may be recovered unchanged and subjected to further investigation. As may be inferred from Table 17.2, the response is not the same for all compounds, and accurate quantitative work requires calibration with known quantities of the various solutes.

TABLE 17.2 Thermal Conductivities of Some Gases and Organic Vapors

Hydrogen	5.34
Helium	4.16
Methane	1.09
Nitrogen	0.75
Ethane	0.73
n-Butane	0.56
Ethanol	0.53
Benzene	0.44
Acetone	0.42
Ethyl acetate	0.41
Chloroform	0.25
Carbon tetrachloride	0.22

Note: Values are calories per second conducted through a 1-cm layer of gas 1 m^2 in area at 100°C, with a temperature gradient of 1°C/cm.

Flame Ionization Detector

This detector was developed in response to the need, in certain applications, for higher sensitivity and faster response time than are provided by the thermal conductivity cell. The sensitivity of a detector depends not only upon its type but also upon the specific design and the manner in which it is operated. Thus a definite numerical comparison is difficult, but very roughly we may state that the flame ionization detector is several hundred to a thousand times as sensitive as the thermal conductivity detector. This detector is in very wide use, although perhaps it still runs second to thermal conductivity in actual numbers. The circuitry associated with the flame ionization detector is more complicated than the simple bridge circuit just discussed, and gas chromatographs equipped with this detector are more expensive. Not only is the detector fast and sensitive, it is fairly stable, linear over a wide solute range, and responsive to almost all organic compounds. It is unresponsive to many inorganic compounds, including water.

The general principle of the flame ionization detector is as follows. The thermal energy in a hydrogen flame is sufficient to cause many molecules to ionize. The effluent gas from the column is mixed with hydrogen and burned at the tip of a metal jet in an excess of air. A potential is applied between the jet itself and a second electrode located above or around the flame. Ordinarily, the jet is the positive electrode. When ions are formed in the flame, the gas space between the two electrodes becomes more conducting, and an increased current flows in the circuit. This current passes through a resistor, the voltage across which is amplified to yield a signal which is fed to a recorder. Hydrogen may serve as the carrier gas, although it is more common to use nitrogen, in which case hydrogen is fed into the gas stream just ahead of the burner. Major aspects of the setup are shown in schematic form in Fig. 17.15.

With the flame ionization detector, the concentration of ions in the space between the electrodes and hence the magnitude of the current depends upon the

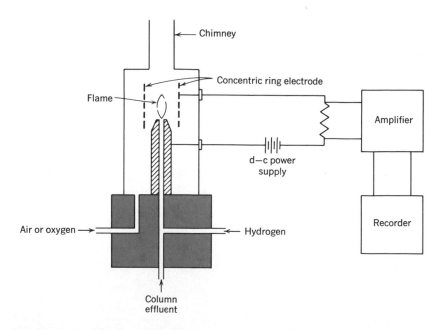

Figure 17.15 Schematic diagram of a flame ionization detector and the associated circuitry.

rate at which solute molecules are delivered to the flame. A given weight of solute reaching the flame in unit time will yield the same detector response regardless of the degree of dilution by the carrier gas. This is the basis for the statement that this detector responds not to solute concentration, but rather to the mass flow rate of solute. It should also be noted that the flame ionization detector is destructive of the sample components, in contrast with detectors such as the thermal conductivity cell which respond to some physical property of the gas related to solute concentration.

Although flame ionization is the most common, it may be mentioned that detectors are available in which solute molecules are ionized by radioactive sources. In one of these, the β-ray ionization detector, the source is a β-emitter such as tritium (^3H) or ^{99}Sr. The kinetic energy of the β-particle is much greater than that required to ionize a solute molecule; thus a series of ions is produced by one collision after another as each β-particle travels through the gas flowing through the detector. Isotopes which emit α-particles have also been employed as ionization sources. Attention must be given to safety with these detectors: Along with the dangerous voltages associated with ionization detectors, the radioactive source represents a potential health hazard in the event of leakage or inadequate shielding.

In recent years, the electron-capture detector has become important for certain purposes. The column effluent passes through a cell containing a source of β-particles (usually a metal foil containing ^{63}Ni). Collisions of carrier-gas molecules (often a mixture of N_2 and CH_4) with these particles produce ions and secondary electrons, which migrate to a positive electrode, yielding a certain current. When a solute which is capable of capturing electrons elutes from the column, there is a drop in this current which serves as the basis for detection. Very roughly, the electron-capture detector may be about 1000 times as sensitive as the flame ionization detector, but another important advantage in certain applications is selectivity. The detector is relatively insensitive to many hydrocarbons, alcohols, amines, and other compounds while responding 100,000 to 1 million times more strongly to certain other compounds such as heavily halogenated species. It has proved very useful for detecting certain pesticides (e.g., DDT, aldrin, and dieldrin) in samples where large excesses of other compounds tax the resolving ability of the column.

17.5

APPLICATIONS OF GLC

In discussing the theory of GLC, we emphasized that the column is a separation tool, but the gas chromatograph as a whole, because of the detection and recording of elution bands, is an analytical instrument which provides both qualitative and quantitative information about the components of a sample.

17.5a. Identification of Compounds

With a particular column, and with all the variables, such as temperature and flow rate, carefully controlled, the retention time or retention volume of a solute is a property of that solute, just as its boiling point or refractive index is a property. This implies that retention behavior could be used to identify a compound. It must be stated, however, that this is not the forte of the gas chromatograph. Instruments such as the mass spectrometer, the infrared spectrophoto-meter, and the nmr spectrometer provide far more information about the nature of an unknown compound. In fact, for an analyst starting from scratch, with no in-formation about the sample or whence it came, it would be virtually impossible to identify the components by their retention times alone. The thousands of known compounds simply provide too many possibilities from which to choose. In such a case, the best approach utilizes the capabilities of two instruments. For example, the chromatograph is used to separate the components of the sample mixture, and these are then introduced consecutively into the mass spectrometer. Various inter-facing devices for accomplishing this automatically have been described. The mass spectral data are handled by a computer that may display probable identities of the compounds. See Box 19.1 (page 573) for a brief introduction to the mass spectrometer and its application as a detector in so-called GC-MS. GC-IR is an-other "hyphenated technique" in which the detector provides information about the chemical nature of column eluents, in this case by monitoring infrared ab-sorption bands.

On the other hand, the analyst is not always faced with a totally unknown sample. The source of the sample and its history may permit reasonable guesses regarding some of the components. In such a case, a comparison of retention times with those of known compounds may confirm the identities of some of the components. Such an identification is quite likely to be correct if spiking the sam-ple with a known compound does not lead to an additional elution band with sev-eral different columns at several temperatures.

Sometimes, as in mass screenings for drug abuse, samples will be first subjected to ordinary, (relatively) inexpensive GLC; then those considered "possibles" on the basis of elution bands at suspicious retention times will be fur-ther examined by a technique that provides more definitive confirmation of chem-ical structure.

17.5b.   *Quantitative Analysis*

Quantitative analysis by GLC depends upon the relationship between the quantity of a solute and the size of the resulting elution band. In general, with differential detectors (which are almost always employed), the best measure of the quantity of a solute is the area under the elution band. Solutes with very low retention times yield sharp, narrow bands, in which case the height of the band may be an adequate measurement. Otherwise, some sort of integration is required to obtain the area. The detector sensitivity is different for various compounds; this may be inferred from Table 17.2 for the thermal conductivity cell, and the same is true for other types of detectors. Thus there is no way to relate the area of an elution band to quantity of solute other than by calibration with known sam-ples. Once this is done, we may write

$$\text{Quantity of a solute} = \text{calibration factor} \times \text{area under elution band}$$

The units in which the area is measured make no difference provided the calibra-tion factor is appropriate.

Measuring the areas under elution bands was tedious and imprecise in the early days of GLC. Nowadays, computer-based data systems have completely au-tomated this step. An analog-to-digital converter changes the electrical signal from the detector into digital "counts" which are processed by computer software to produce a printout of retention times, areas, and (based on stored calibration factors) concentrations.

17.5c. Versatility of GLC

There are far too many applications of GLC to permit, in the available space, anything more thorough than listing a few examples. Each year thousands of papers are published in which GLC is at least mentioned, and the technique has spread outside chemistry into many other fields.

Although news stories do not usually describe the methodology, GLC figures prominently in the laboratory work on topics of great current interest. For example, the Environmental Protection Agency (EPA) runs an extensive program of monitoring pesticide levels in soil, groundwater, and other sorts of samples. The general approach involves extracting the sample to concentrate the analytes in a suitable organic solvent and chromatographing the extract. Residues of hormones used to promote the growth of animals are measured in meat samples in a similar way, and extracts of urine specimens are likewise examined by GLC in drug-screening programs. And sometimes we need to be reminded that countless thousands of chromatograms are recorded every day in laboratories where GLC is a routine tool in more prosaic studies.

Limitations of GLC

The major limitation is volatility. The sample must have an appreciable vapor pressure at the temperature of the column, and this immediately eliminates many kinds of samples. An actual count is impossible, but it has been estimated that perhaps 20% of the known chemical compounds can be handled directly in a gas chromatograph. Biologists use GLC extensively, but unfortunately many of the most important biological compounds are insufficiently volatile, including amino acids, peptides, proteins, vitamins, coenzymes, carbohydrates, and nucleic acids. Sometimes, however, it is possible to convert nonvolatile sample components into volatile derivatives which can then be chromatographed. Obviously a difficult or time-consuming preliminary chemical step would nullify the speed and simplicity of the chromatographic analysis; hence there is a continuous search for reagents and reaction conditions which will derivatize all the components of the sample quickly, cleanly, and quantitatively. For example, many studies have been directed toward the preparation of volatile amino acid derivatives. These involve reactions both of the carboxyl group, such as the formation of methyl or other alkyl esters, and of the amino group, such as the formation of the trifluoroacetyl derivative. In recent years, volatile trimethylsilyl derivatives of acids, alcohols, amines, monosaccharides, and many other compounds have been studied extensively.

Most inorganic samples are not sufficiently volatile to permit the direct application of GLC, although some work has been done at very high temperatures using molten salts or eutectic mixtures as the stationary liquid phase. Halides of some elements such as tin, titanium, arsenic, and antimony are fairly volatile and have been separated by GLC. A number of metals such as beryllium, aluminum, copper, iron, chromium, and cobalt have been subjected to GLC in the form of fairly volatile chelate compounds with acetylacetone and its fluorinated derivatives.[8] For example, aluminum, iron, and copper have been determined in alloys by dissolution of the sample followed by extraction of the metals into a chloroform solution of trifluoroacetylacetone which is then chromatographed.[9] Relative errors on the order of 0.2 to 3% were reported.

17.5d. Pyrolysis Gas Chromatography

The technique called *pyrolysis gas chromatography* represents an exception to the requirement of sample volatility. Applications include characterization of tars, paints, rubbers, synthetic films and fibers, and other sorts of plastic materials. The sample is heated very rapidly to a high temperature in an inert (nonoxidizing) atmosphere. Heating rate and final temperature vary with the apparatus, sample type, and nature of the study. Several heating methods are employed, including zapping the surface of the sample with a laser pulse, simply heating in a furnace, and using the rapid temperature rise associated with the interaction of a high-frequency oscillator with a ferromagnetic metal. In the last case, an alternating magnetic flux is induced in a metal wire, often nickel, iron, cobalt, or an alloy of two or more of these; eddy currents in the surface of the wire and hysteresis losses cause a rapid temperature rise to several hundred degrees in only a few milliseconds. At a temperature characteristic of the wire material, called the *Curie point,* a transition from ferro- to paramagnetism occurs, energy absorption stops, and the temperature rise terminates. With Curie-point pyrolyzers, the sample is usually coated on the surface of the wire.

Probably the commonest pyrolysis method employs a metal filament, often platinum, heated by an electric current. The sample may be placed in a small depression in a segment of platinum ribbon. The most sophisticated units can provide a very rapid temperature rise in only a few milliseconds, followed by maintenance at a constant level, but times of a few seconds are more usual.

The gaseous products of the thermal decomposition of the sample are swept into the chromatographic column by diverting the carrier gas through appropriate valves, and a chromatogram is recorded in the usual manner. In some cases, pyrolysis products may be known compounds, but frequently a complex pattern is obtained with many unidentified elution bands. Despite the difficulty of a complete chemical interpretation, such a chromatogram may be reproducible and highly characteristic of the starting material, representing, so to say, a "fingerprint."

An interesting application of pyrolysis GC is in the identification of bacteria, a technique pioneered by Reiner.[10] In one example, the characterization of *Salmonella* organisms was described. The cultured cells were harvested, washed free of culture medium, and centrifuged. The wet, packed cells were freeze-dried, and a sample of about 80 μg of the dried bacteria was subjected to pyrolysis GC. Forty-seven species of *Salmonella* were correctly classified by examination of the chromatograms. Correlations were observed between GLC bands and groupings based upon traditional serological and biochemical classification tests. It is thought that some of the differences in the characteristic GLC band patterns

[8] R. W. Moshier and R. E. Sievers, *Gas Chromatography of Metal Chelates*, Pergamon Press, Inc., New York, 1965.

[9] R. W. Moshier and J. E. Schwarberg, *Talanta*, **13**, 445 (1966).

[10] See for example, E. Reiner, *Nature* (London), **206**, 1272 (1965); E. Reiner and W. H. Ewing, *ibid.*, **217**, 191 (1968); E. Reiner and G. P. Kubica, *Amer. Rev. Resp. Dis.*, **99**, 42 (1969); E. Reiner, *Anal. Chem.*, **44**, 1058 (1972).

arise from species differences in the hexose sugars which, in polymeric form, are part of the bacterial cell wall.

Figure 17.16 shows some representative pyrochromatograms from a study of several species of mycobacteria. One sees certain obvious differences at first glance, such as the absence of peak 29 in some organisms and variations in the vicinity of peak 24, but definitive identification is generally based upon careful examination of patterns involving not only the *presence* of peaks but *ratios* of one to another. Frequently the smaller peaks are the most definitive.

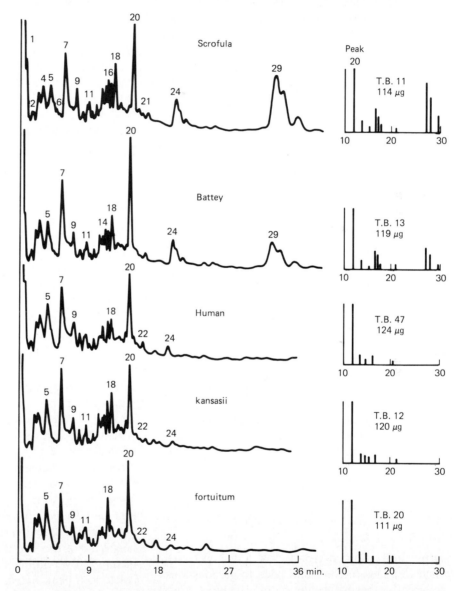

Figure 17.16 Representative pyrochromatograms of freeze-dried mycobacteria. The bar graphs at the right show some key peak centers on a chart-paper distance scale with peak heights normalized for sample weight. Pyrolysis: 10 s, terminating at 840°C. Column: 6 m × 0.75 mm i.d., stationary phase esterified Carbowax. Operation: N_2 carrier gas; flame ionization detector; temperature programmed to start at 0°C and increase at 12°C/min to 180°C, followed by an isothermal period. [Reproduced by permission from E. Reiner and G. P. Kubica, *Amer. Rev. Resp. Dis.*, **99**, 42 (1969).]

KEY TERMS

Carrier gas. The mobile phase in gas chromatography. The flow of carrier gas moves sample components along the column; the balance between this motion and retardation by the stationary phase (q.v.) differs for various sample components and provides the basis of the separation. The commonest carrier gases are He and N_2.

Chromatography. See definition, page 492.

Detector. A device which responds to changes in some property of the column effluent caused by the presence of sample components; the response takes the form of an electrical output different from that produced in the presence of pure mobile phase. An electrical response which bears information about the composition of the effluent is called a *signal*.

Differential detector. A detector which responds to either *concentration* or *rate of flow* of solute; its output at any instant contains no stored information about anything "seen" previously. The chromatogram is a series of bands or peaks, and the signal must be integrated over the transit time of an entire band to obtain a number proportional to *quantity* of solute.

Distribution coefficient. Also called *partition coefficient*. A dimensionless constant, K, obtained from Henry's law (q.v.) by replacing the partial pressure and mole fraction of a solute by two concentration terms in the same units:

$$K = \frac{\text{concentration of solute in solution}}{\text{concentration of solute in gas phase}} = \frac{C_L}{C_G}$$

The solution term is generally in the numerator. This is an equilibrium expression; it applies only when the vapor phase is in equilibrium with the solution.

Eddy diffusion. A band-broadening factor in chromatography arising from the varying distances solutes travel as they follow various pathways through a column.

Flame ionization detector. A GLC detector which responds to conduction due to gaseous ions produced by thermal excitation in a flame.

Height equivalent of a theoretical plate (HETP). That length of column which achieves the same degree of separation of two solutes as

would have occurred in one batch equilibration step. See *theoretical plate*.

Henry's law. The partial pressure of a solute in the vapor phase in equilibrium with a dilute solution of that solute in some solvent is proportional to its concentration in the solution. In physical chemistry, the concentration is often expressed as a mole fraction; thus

$$p_{i,A} = k_{i,A} X_i$$

where $p_{i,A}$ is the partial pressure of solute i above a solution of i in solvent A, and X_i is the mole fraction of the solute in the solution. The subscripts with the Henry's law constant, $k_{i,A}$, emphasize that the numerical value of k depends upon the solute i and the solvent A.

Integral detector. A detector whose output at any instant relates to the total *quantity* (*not* concentration or mass flow rate) of solute that traverses it. The chromatogram resembles a series of steps rather than peaks. Integral detectors are seldom used in modern practice. (See Fig. 17.10.)

Isotherm. A graph showing how the concentration of a solute in one phase varies with its concentration in a second phase when the two phases are in equilibrium. So called because all points on the curve are measured at the same temperature.

Limit of detection. That quantity of a solute that gives an elution band just large enough to be reliably distinguished from noise. To be sure, the detector really responds, not to *quantity* of solute, but to either its *concentration* in the carrier gas or its *rate of flow* through the detector, and hence, strictly, the quantity of solute relates to the *area* of an elution band rather than its height. However, especially in elementary treatments, we speak of the quantity of solute that gives a peak *height* above the noise level. This avoids much complexity, and we get away with it because, for good chromatograms with sharp elution bands, the band height is not too bad a measure of area and hence quantity. By relating the detectable *quantity* of a solute to a peak *height* relative to the noise level, we get at the lower limit for an analysis under our conditions, including both detector sensitivity and the degree of band spreading in the column. It is commonly recommended that limit of detection be associated with a solute peak height twice the peak-to-peak noise level of the detection system. See *noise*.

Longitudinal diffusion. A band-broadening factor in chromatography caused by the random motion of solute molecules; the gaseous solute migrates along concentration gradients into adjacent portions of the mobile phase. (The rate of diffusion is so much faster in the gas phase than in liquids that only gaseous diffusion is important in this aspect of band broadening.)

Noise. Electrical fluctuations which do not bear information regarding the composition of the chemical system of interest. (See Fig. 17.12 for "peak-to-peak" noise level.)

Nonequilibrium in mass transfer. A band-broadening factor in chromatography caused by the finite time required by a solute to equilibrate between two phases; with continuous flow in a column, equilibrium is never attained and a broader solute distribution than in a Craig-type experiment results. (See Fig. 17.5.)

Open-tubular column. Sometimes called *capillary column*. A long, narrow tube whose inner surface is coated with a layer of the stationary liquid phase; there is no column packing in the usual sense. Typical internal diameter (i.d.) values are in tenths of a millimeter, and lengths run to many meters, even hundreds.

Programmed-temperature GLC. Operation in which the column temperature is increased during the chromatographic run. The contrasting term is *isothermal GLC*, where the temperature remains constant.

Pyrolysis GLC. A technique in which nonvolatile materials are identified and characterized by patterns in the chromatograms yielded by gaseous products of thermal decomposition.

Resolution, *R*. In nonrigorous discourse, resolution of a mixture is often more or less synonymous with *separation*, but there is also a precise definition in regard to two solutes based upon their retention times and bandwidths:

$$R = \frac{2(t_{R_2} - t_{R_1})}{w_{b_1} + w_{b_2}} = \frac{2(t_{R_2} - t_{R_1})}{1.699(w_{1/2_1} + w_{1/2_2})}$$

where w_b and $w_{1/2}$ represent bandwidths at baseline and half-height, respectively. (See Fig. 17.7.)

Retention time, t_R. The time, usually in minutes or seconds, between injection of the sample and the appearance of the very peak of an elution band of a sample component at the detector. Separated components of a mixture obviously have different t_R values.

Retention volume, V_R. The product of a retention time and a mobile-phase flow rate; the volume of mobile phase which passes through the column between sample introduction and the appearance of the peak of a solute elution band at the detector.

Sensitivity (of a detector). The slope, dR/dQ, of a graph of electrical response of a detector vs. the quantity measured, such as solute concentration in the carrier gas at the peak of an elution band. Sometimes we relate sensitivity to the *quantity* of a solute rather than its concentration in the carrier gas in a manner that is not strictly correct (see discussion under *limit of detection*).

Solid support. The porous solid of large surface area which bears and immobilizes the stationary liquid phase in GLC.

Stationary phase. The immobilized liquid in GLC.

Theoretical plate. Sometimes shortened to *plate*. The term plate originated in fractional distillation theory at a time when it was incorrectly believed that equilibrium between liquid and vapor was attained in each of a series of structural regions of the column called plates. Each vessel of a Craig apparatus is a plate, as would be each chamber in Fig. 17.2. So far as true equilibrium is concerned, plates are entirely imaginary in any process with continuous flow. A chromatographic column will, however, achieve the same resolution of a solute mixture as would some number of equilibrium stages, and this number is the number, *n*, of theoretical plates in the column.

$$n = \frac{\text{length of column}}{\text{HETP}}$$

See *height equivalent of a theoretical plate*.

Thermal conductivity cell. A GLC detector whose electrical output varies with the ability of the gas flowing through it to conduct heat.

van Deemter equation. An equation with three terms showing the contributions of eddy diffusion, longitudinal diffusion, and nonequilibrium in mass transfer to the height equivalent of a theoretical plate (q.v.). In abbreviated form,

$$HETP = A + \frac{B}{u} + Cu$$

where u is carrier-gas velocity. The locus of the equation is hyperbolic, with a minimum representing an optimal gas velocity.

QUESTIONS

1. Consider the abbreviated form of the van Deemter equation:

$$HETP = A + \frac{B}{u} + Cu$$

   Sketch graphs of HETP vs. u that show the general shapes of the individual contributions of A, B/u, and Cu to HETP. That is, what would a graph of HETP vs. u look like if (a) $B/u = Cu = 0$; (b) $A = Cu = 0$; (c) $A = B/u = 0$?

2. If you walked into a laboratory and saw a Craig countercurrent solvent extraction apparatus, how would you determine the number of plates available for a separation?

3. Explain what is meant by *height equivalent of a theoretical plate* in connection with continuous-flow separation processes such as GLC.

4. Look at the isothermal chromatogram in Fig. 17.9(a) and note the three elution bands to the left of the others, with retention times in the vicinities of 5, 11, and 26 min. (a) Using on these three bands a ruler that can measure to the nearest 0.5 mm or so, obtain three rough estimates of the number of theoretical plates in the column under the experimental conditions employed. (b) In the paper from which the figure was reproduced, the column is described as 4 ft by 5 mm i.d. with a stationary phase of 25% silicone oil on diatomaceous earth support. Estimate values of HETP in millimeters for this column. (1 in. = 2.54 cm.)

5. A manufacturer redesigned a thermal conductivity detector, taking advantage of new developments in solid-state devices. With the new detector, the peak-to-peak noise level was found to be exactly one-half of what it was in the older model, and the sensitivity had increased by a factor of 1.50. What was the numerical effect of these improvements upon the limit of detection for a certain organic compound?

6. Operator A injects samples into a gas chromatograph much more slowly than does operator B. How could you tell these operators apart by simply inspecting chromatograms that they turned out?

Multiple-choice: In the following multiple-choice questions, select the *one best* answer.

7. In chromatography, a substance for which the distribution coefficient, K, is zero may be used to estimate (a) the volume within the column occupied by the packing material; (b) the total volume of the column; (c) the volume within the pores of the packing material; (d) the volume within the column available to the mobile phase.

8. The purpose of the solid support material in a GLC column is to (a) immobilize the stationary liquid phase; (b) adsorb sample components that are insufficiently soluble in the stationary liquid phase; (c) provide a "backup" stationary phase in the event that the liquid is lost by evaporation; (d) remove impurities from the carrier gas.

9. Helium, rather than nitrogen, is sometimes used as the carrier gas in GLC because (a) being lighter than nitrogen, helium elutes the sample components more rapidly; (b) helium is less expensive than nitrogen; (c) nitrogen has stable isotopes which separate and cause anomalous column behavior; (d) of its much higher thermal conductivity.

10. An important feature of open-tubular GLC columns is: (a) they can accept much larger samples than can packed columns because of their great length; (b) solute partitioning between stationary and mobile phases is very rapid and the C term in the van Deemter

equation is relatively small; (c) they permit the use of a wider variety of carrier gases; (d) they can be operated at temperatures closer to the boiling point of the stationary liquid phase than can packed columns and thus handle less volatile samples.

11. Raising the column temperature in GLC decreases solute retention times primarily because (a) solute diffusion coefficients in the liquid phase decrease with increasing temperature; (b) van der Waals interactions between solutes and stationary phase are stronger at higher temperatures; (c) gases are generally less soluble in liquids at higher temperatures; (d) detector sensitivity is a function of temperature, especially with a thermal conductivity cell.

12. Which of the following would have practically no effect upon the retention volume of a solute in GLC? (a) changing the carrier-gas flow rate; (b) increasing the stationary liquid loading of the column packing from 5 to 10% by weight; (c) increasing the column temperature; (d) changing the chemical nature of the stationary liquid.

13. The separation factor, S, in chromatography depends upon (a) the length of the column; (b) the square root of the length of the column; (c) the nature of the stationary liquid phase; (d) the number of theoretical plates in the column.

14. In GLC, interaction of solutes with the solid support will often cause (a) unusually narrow elution bands; (b) asymmetric elution bands with "tailing"; (c) excessive eddy diffusion; (d) decreased detector sensitivity.

15. Increasing the quantity of stationary liquid phase applied to the column packing will, with everything else the same, (a) increase t_R for a solute; (b) decrease t_R for a solute; (c) not influence t_R for a solute; (d) decrease the nonequilibrium term in the van Deemter equation.

18

Liquid Chromatography

In this chapter, we shall consider several forms of chromatography where the mobile phase is a liquid. Perhaps as many as 80% of all chemical compounds are insufficiently volatile for gas chromatography, including major classes which are of central importance in biology and medicine. Thus, although gas chromatography (GC) is usually the method of choice where volatile compounds are involved, liquid chromatography (LC) is potentially more important. Historically, LC predated GC by many years, but the theoretical ideas and the modern instrumentation that led to highly efficient performance were developed first in GC. In the late 1960s, the effort began to upgrade LC utilizing concepts which had been successful in GC. This process continues, and its accelerating pace makes LC today one of the dynamic research areas in analytical chemistry.

As we saw in the last chapter, the retention behavior of a solute in GC depends upon its interaction with only one phase—the stationary liquid—plus the operating parameters of temperature and flow rate. Because, at least to a first approximation, the behavior of solute molecules in the vapor phase is independent of the presence of other gases, the distribution coefficient, K, does not depend upon the nature of the carrier gas; the mobile phase (nitrogen, helium, hydrogen, etc.) is selected on other grounds, usually safety, cost, and detector characteristics. In LC, by contrast, solutes may interact strongly with the liquid mobile phase, and furthermore, the interaction of the latter with the stationary phase may have a great effect upon solute retention. Thus manipulating the composition of the mobile phase in LC provides a mode of control over retention which is lacking in GC.

In GC, only one phase distribution process is important, which we characterized in terms of the solubility of a gas in a liquid utilizing Henry's law. [Gas-solid chromatography (GSC) based upon adsorption from the vapor phase onto a solid surface, is relatively little used.] In LC, on the other hand, there are more modes of phase distribution available for separations involving various classes of solutes. We shall first describe some of the more important of these. Next we shall describe some of the older versions of LC which, despite recent developments, still have their place. Finally, we shall describe modern practice as it has evolved so far. Concepts presented in the previous chapter will be useful here.

18.1

PHASE DISTRIBUTION PROCESSES

18.1a. Adsorption

Imagine a solid material with a clean, dry surface. If this surface is exposed to a fluid—gas, liquid, or solution—there is a tendency for molecules of gas, solvent, or solute to interact with the surface. If the solid material is very finely divided or is highly porous—in other words, if there is a large surface area—then the extent of adsorption may be appreciable. For example, if a good adsorbent such as a specially prepared charcoal in introduced into a vessel containing a gas, the decrease in pressure as the surface attracts gas molecules is easily measurable. Similarly, the adsorption of acetic acid from aqueous solution onto charcoal is easily observed by titrating the equilibrated solution with sodium hydroxide, and the removal of dark-colored impurities from organic preparations by charcoal treatment during recrystallization is a common operation.

An atom, ion, or molecule in the surface layer of a solid, unlike its counterparts in the interior, does not have neighboring particles on all sides. Thus residual attractive forces are exerted upon components of the fluid which bathes the surface, and the free energy of the system may be minimized if such components concentrate at the interface. In certain systems under special conditions, the adsorbed layer may be only one molecule thick (monolayer adsorption), but it is more common for adsorbed molecules to hold others in turn so that a multi-molecular layer is built up. The force responsible for adsorption depends upon the chemical nature of the surface and the structure of the adsorbed species. An obvious electrostatic effect is involved in the adsorption of ions onto the surface of an ionic solid. In other cases, we may encounter an interaction of a polar group in an organic molecule, say, carbonyl or hydroxyl, with a polar adsorbent. Sometimes the formation of hydrogen bonds may be involved. A polar surface may induce a complementary charge separation in a polarizable molecule possessing, for example, an aromatic ring system. In other cases, a nonpolar surface may adsorb hydrophobic molecules from a polar solvent.

Adsorption Equilibrium

The isotherm which describes an adsorption equilibrium is usually nonlinear. Many systems follow the Freundlich equation, at least if the concentration is not too high. This equation, dating from the late 1800s, is an empirical one which was not derived from any particular model but simply happens to fit experimental data in a number of cases. It may be given in the form

$$C_S = kC_L^{1/n}$$

where C_S is the concentration of an adsorbed solute on a solid phase in equilibrium with a solution of solute concentration C_L. Typical units for C_S are millimoles of solute per gram of adsorbent, and for C_L, molarity; k and n are constants. It is seen that if $n = 1$, the Freundlich equation reduces to the form of other equilibrium expressions such as Henry's law or the Nernst distribution law for a solute in solvent extraction. In general, however, $n > 1$, and hence a graph of C_S vs. C_L (called an adsorption isotherm) resembles curve 2a in Fig. 17.4. To evaluate k and n, one may take logarithms of both sides of the Freundlich equation, obtaining

$$\log C_S = \log k + (1/n) \log C_L$$

A graph of $\log C_S$ vs. $\log C_L$ is a straight line of slope $1/n$ with an intercept of $\log k$ on the C_S axis.

k and n are constants only for a given system and, of course, only for a stated temperature. They vary with the nature of the adsorbent and its surface character and with the solvent and the solute. k is more sensitive than is n to the nature of the solute, and separations based upon adsorption depend largely upon differences in the k-values for various solutes. As we saw in the last chapter, nonlinear isotherms are associated with skewed chromatographic elution bands. Frequently, the bands obtained with adsorption columns resemble curve 2b in Fig. 17.4, although at low concentrations the isotherms may be nearly linear, leading to fairly symmetrical bands with little tailing.

Adsorbent Characteristics

For many years, variability in the surface properties of commercial adsorbents was a problem for chromatographers. In some cases, simply washing an adsorbent such as alumina with acid or alkali considerably modifies its behavior, and the temperature at which it is dried may also be very important. Different batches of adsorbents, even from the same producer, may exhibit troublesome variability. The marketing of specially prepared adsorbents which are tested and labelled "for chromatography" has been helpful to workers who make their own columns. Finally, recent years have seen the appearance of commercial columns, ready for use, with specified performance characteristics.

For many years, the commonest adsorbents were solids that can be roughly characterized as *polar*. These included such inorganic materials as calcium and magnesium carbonates, silica gel, and aluminum oxide, as well as organic substances like sucrose, starch, and cellulose. Such adsorbents display a high affinity for polar solutes, particularly if the polarity of the solvent is low. On the basis of experience with such systems, several general rules emerged: (1) everything else being the same, the more polar a compound, the more strongly it will be adsorbed; (2) other factors being equal, high molecular weight favors adsorption; (3) the more polar the solvent, the greater is its tendency to occupy surface sites in competition with the solute, and hence that solute will be adsorbed less. The use of polar stationary phases with mobile phases which were no more polar than necessary in order to elute polar solutes in a reasonable time was considered for many years to be the normal form of adsorption chromatography.

A different approach utilizing nonpolar adsorbents and more polar mobile phases has in recent years become dominant, accounting for perhaps 80% or more of the LC separations based upon adsorption. Although it seems now to be "normal," when it first appeared, this type of chromatography was called *reversed phase*, and the designation, sometimes shortened to RPLC, has stuck. Space does not permit a description of all the available materials, and in any case, rapid current developments would very quickly render such a treatment obsolete. We describe here two examples of nonpolar adsorbents.

Polymer-Based Adsorbents

First is a copolymer of styrene and divinylbenzene. The polymerization of styrene yields a linear polystyrene:

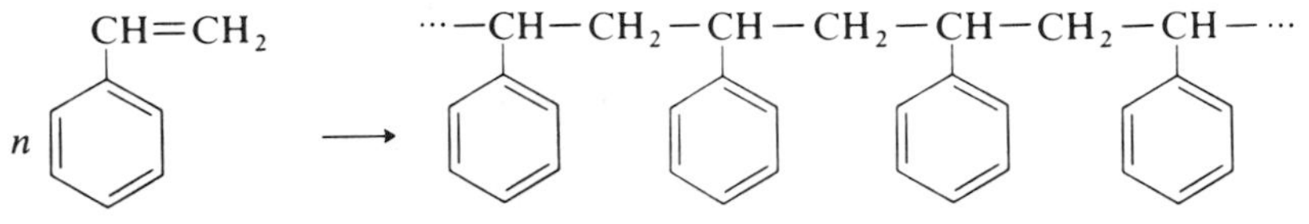

$$CH\!=\!CH_2 \qquad \cdots\!-\!CH\!-\!CH_2\!-\!CH\!-\!CH_2\!-\!CH\!-\!CH_2\!-\!CH\!-\cdots$$

Addition of the bifunctional monomer divinylbenzene to the polymerization mix, on the other hand, links together the polystyrene chains and yields a material with a three-dimensional network structure:

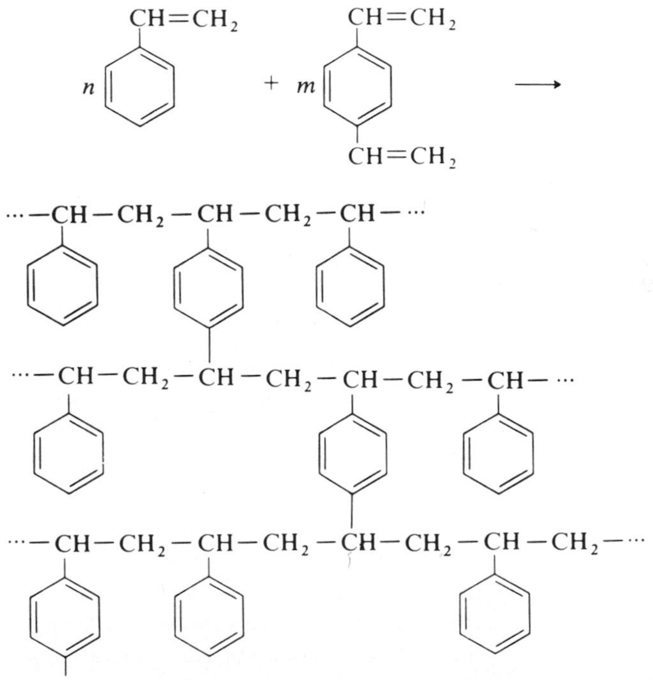

$$CH\!=\!CH_2 \qquad\qquad CH\!=\!CH_2$$

$$n \qquad + \quad m \qquad\qquad CH\!=\!CH_2 \qquad\longrightarrow$$

$$\cdots\!-\!CH\!-\!CH_2\!-\!CH\!-\!CH_2\!-\!CH\!-\cdots$$

$$\cdots\!-\!CH\!-\!CH_2\!-\!CH\!-\!CH_2\!-\!CH\!-\!CH_2\!-\!CH\!-\cdots$$

$$\cdots\!-\!CH\!-\!CH_2\!-\!CH\!-\!CH_2\!-\!CH\!-\!CH_2\!-\!CH\!-\!CH_2\!-\cdots$$

The degree of cross-linking, and hence the porosity, of the polymer depends upon the divinylbenzene content of the starting material, generally between 2 and 12%; the polymerization technology leads to a product in the form of beads of controlled diameter. A typical resin (Amberlite XAD-2, Rohm and Haas Chemical Co.) has a surface area of 300 m²/g and an average pore diameter of about 90 Å.

Bonded-Phase Adsorbents

A second approach involves chemically bonding a layer of nonpolar adsorbent to the surface of a solid such as silica, providing a hydrophobic surface on a solid possessing mechanical strength and other properties desirable for column operation. Shown below is the preparation of a bonded material by reaction of a monochlorosilane with a silanol group in the silica surface:

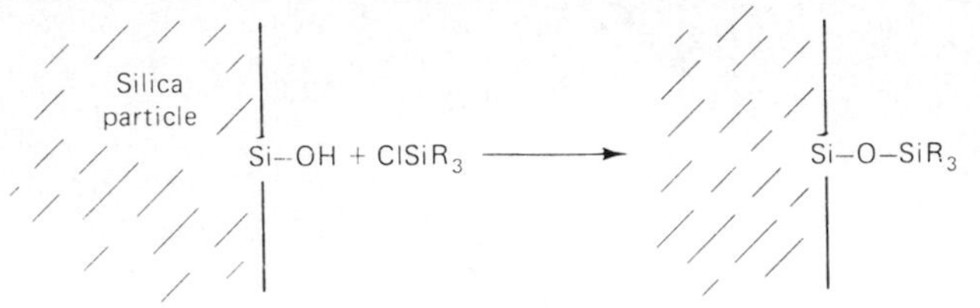

Two of the R groups are generally —CH₃, and the third may be varied to attain the desired degree of hydrophobicity; for example —CH_3, —$(CH_2)_7CH_3$, and —$(CH_2)_{17}CH_3$ are commercially available. With long chains, e.g., C_{18}, steric effects may prevent some of the surface sites from reacting, but most of these undesirably polar regions can be capped with methyl groups to produce a fairly uniform hydrophobic surface. The image of hydrocarbon chains anchored to a macrosurface, something like trees in a forest, has led to the term "brush" materials for these preparations.

Some writers consider the bonded surface not as an adsorbent, but rather as an immobilized liquid phase where solute distribution resembles the solvent extraction situation (see liquid-liquid partition below). A monolayer of hydrocarbon chains covalently attached to a siliceous surface is scarcely a liquid in the usual sense, nor is it quite a solid. More chain motion can doubtless occur than in a solid hydrocarbon polymer, but such mobility must be far less than that of molecules in a typical liquid phase. The student who remains in science will become accustomed to uncertainties whose origins lie in the human compulsion to think in terms of neat categories which may not exist in the real world.

It is commonly stated that van der Waals forces are responsible for the adsorption of nonpolar molecules on hydrocarbon surfaces. These weak interactions are real for closely juxtaposed molecules, but in reversed phase, with a polar solvent, the phenomenon may also be viewed as an example of the so-called hydrophobic effect, which may be briefly explained as follows. A hydrocarbon residue on a molecule in aqueous solution occupies space; i.e., it requires a cavity in the solvent, and in contrast with an ion or a polar group, it experiences no favorable electrostatic interaction with water molecules to compensate for the creation of the cavity. There is evidence for the existence of a highly organized cage of hydrogen-bonded water molecules surrounding a nonpolar intruder, and transfer of the latter from the solvent to the hydrophobic surface allows the water to destructure into a less ordered state; the attendant entropy increase provides the driving force for adsorption.

In reversed phase, adsorption is stronger the less polar the solute and the more polar the solvent, in contrast with the "normal" situation described earlier. The very high polarity of water is often moderated by the addition of other solvents such as methanol and acetonitrile to the mobile phase in order to control solute retention by the column.

18.1b. Ion Exchange

Ion-Exchange Resins

A wide variety of materials, organic and inorganic, exhibit ion-exchange behavior, but in the research laboratory, where uniformity is important, the preferred ion-exchangers are usually synthetic materials known as ion exchange resins. The resins are prepared by introducing ionizable groups into an organic polymer matrix, of which the commonest is the cross-linked polystyrene described above as an adsorbent. Products are available with varying degrees of cross-linking; a typical general-purpose resin is "8% cross-linked," meaning that the divinylbenzene content is 8%. The resins are produced in the form of spherical beads, usually with diameters of 0.1 to 0.5 mm, although other sizes are available. Figure 18.1 shows a photomicrograph of a commercial product.

Figure 18.1 Photomicrograph of Dowex 50W-X8 strong-acid cation-exchange resin, 20 to 50 mesh. Most of the particle diameters are in the range of about 0.3 to 0.8 mm. (Photo kindly provided by Jeffrey P. Jackson, Dept. of Biology, Emory University.)

To prepare a typical cation-exchange resin, the polymer is sulfonated to introduce $-SO_3H$ groups onto the aromatic rings:

There is probably an average of about one sulfonic acid group per aromatic ring, mostly in the para position. Sulfonation dramatically alters the character of the polymer, upon which the polar substituents confer a high affinity for water. Upon suspension in water, the resin particles swell as a result of the water uptake, with the degree of swelling limited by the extent of cross-linking. (In a sense, a linear polymer, i.e., a non-cross-linked material, would swell indefinitely, yielding a molecular dispersion called a *polyelectrolyte solution*.)

The arysulfonic acids are strong acids. Thus these groups are ionized when water penetrates the resin beads:

$$R-SO_3H \longrightarrow R-SO_3^- H^+$$

But, in contrast with ordinary electrolytes, the anion is permanently attached to the polymer matrix; it cannot migrate through the aqueous phase within the resin pores, nor can it escape to the external solution. The fixation of the anion in turn restricts the mobility of the cation, H^+. Electrical neutrality is maintained within the resin, and H^+ will not leave the resin phase unless it is replaced by some other cation, which *is* the ion-exchange process. The exchange is stoichiometric; i.e., one H^+ is replaced by one Na^+, two H^+ by one Ca^{2+}, etc. As discussed below, ion exchange is an equilibrium process and seldom does it go to completion, but regardless of the extent to which it proceeds, the stoichiometry is exact in that one positive charge leaves the resin for each one that enters. The exchangeable ion, i.e., the ion which is not fixed to the polymer matrix, is called the *counterion*.

The sulfonated resin described above is called a *strong acid cation exchanger*. Other types can be prepared. For example, the functional group may be the weakly acidic —COOH; here, the resin does not exhibit ion-exchange properties unless the *p*H is high enough to convert the neutral free acid into the carboxylate anion, —COO⁻.

The introduction of basic groups into the polymer yields anion-exchange materials. One of the common strong base anion exchangers may be represented as

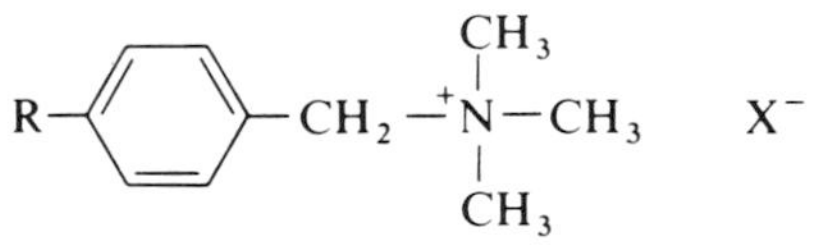

where X^- is an exchangeable counterion such as OH^-, Cl^-, or NO_3^-.

Resins with functional groups which are chelating ligands can also be prepared. One which is commercially available contains iminodiacetic acid groups:

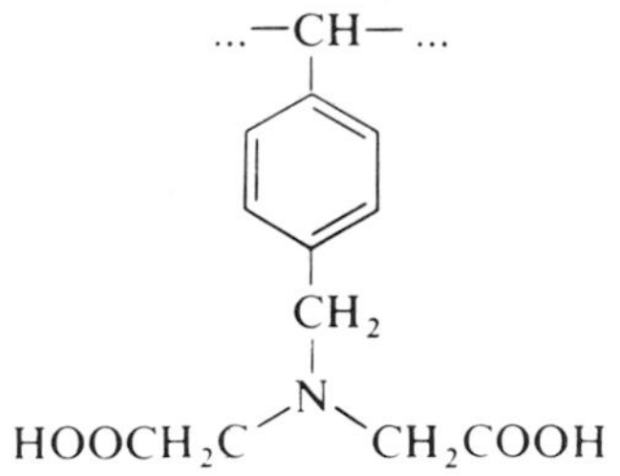

The binding of a metal ion like Cu^{2+} by a resin may be depicted as follows:

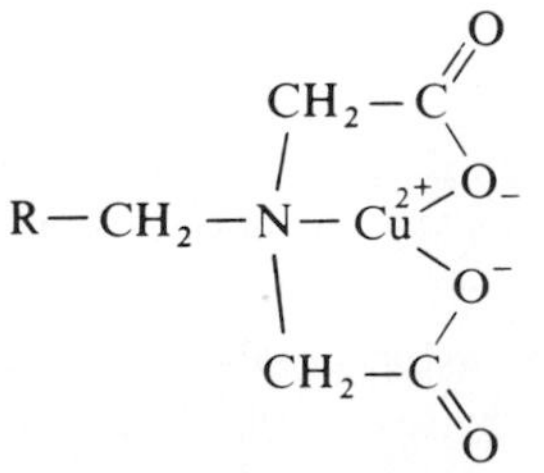

The resin has a very much greater affinity for chelate-forming di- and trivalent cations than for ions like Na^+ or K^+ and has proved useful for collecting transition metals from solutions containing much larger concentrations of alkali metal ions. The metal ion can be displaced from the resin by disrupting the chelate complex with hydrogen ion, as seen in Fig. 18.2, where the distribution ratio (see below) is plotted as a function of pH for several cations. The magnitudes of the D-values, requiring a logarithmic scale, are noteworthy.

The total exchange capacities of typical ion-exchange resins are in the range of about 3 to 6 meq/g of dry resin.

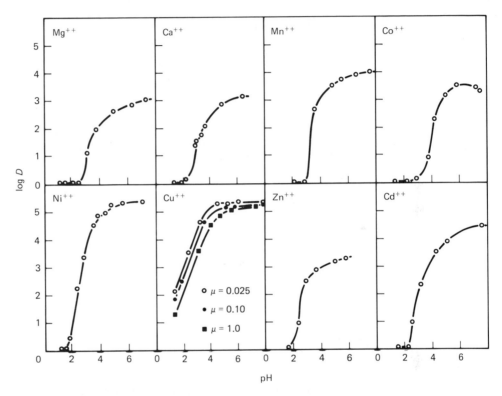

Figure 18.2 Distribution of several cations as a function of pH for the iminodiacetic acid chelating resin Dowex A-1. Ionic strength = 0.10 for all cases except as noted for Cu^{2+}. [Reprinted with permission from D. E. Leyden and A. L. Underwood, *J. Phys. Chem.*, **68**, 2093 (1964); copyright 1964 by the American Chemical Society.]

In addition to totally synthetic resins, available materials include a number of modified natural polymers. For example, biochemists make extensive use of modified celluloses and other polysaccharides. One of the most common is diethylaminoethylcellulose (DEAE-cellulose), in which some of the hydroxyl groups of the natural polymer have been derivatized as follows:

$$R—OH + Cl—CH_2—CH_2—N(C_2H_5)_2 \longrightarrow$$

$$R—O—CH_2—CH_2—N(C_2H_5)_2$$

Protonation of the amino nitrogen in aqueous solutions of pH lower than 9 or so gives a cationic site which binds anions, including protein molecules, which bear net negative charges above their isoelectric pH values.

Ion-Exchange Equilibrium

Suppose that a cation-exchange resin containing the exchangeable counter-ion B^+ is placed in contact with a solution containing A^+. The exchange reaction

$$A^+ \quad + \quad RB \quad \rightleftharpoons \quad B^+ \quad + \quad RA$$

Solution	Resin phase	Solution	Resin phase

occurs, equilibrium is attained with some of each ion in the resin phase and some in solution, and in principle we may write the following equilibrium constant:

$$K = \frac{a_{A_r^+} \times a_{B_s^+}}{a_{B_r^+} \times a_{A_s^+}}$$

where $a_{A_r^+}$ represents the activity of ion A^+ in the resin phase, and $a_{A_s^+}$ its activity in the external solution, and so forth. This expression is frequently written in the form

$$K = \frac{X_{A_r^+} \times m_{B_s^+}}{X_{B_r^+} \times m_{A_s^+}} \times \frac{\gamma_{A_r^+} \times \gamma_{B_s^+}}{\gamma_{B_r^+} \times \gamma_{A_s^+}}$$

or

$$K = Q \times \frac{\gamma_{A_r^+} \times \gamma_{B_s^+}}{\gamma_{B_r^+} \times \gamma_{A_s^+}}$$

where $X_{A_r^+}$ is a mole fraction expressed as the number of moles of A^+ in the resin phase per mole of fixed anionic binding sites, $m_{A_s^+}$ is the molal concentration of A^+ in the solution, and $\gamma_{A_r^+}$ and $\gamma_{A_s^+}$ are the activity coefficients of A^+ in resin phase and solution, respectively. The Q term is called the *concentration quotient* or *practical selectivity coefficient*. Some writers use the term *selectivity coefficient*, symbolized by Q_γ, to represent the product of Q and the activity coefficient ratio of the ions in solution:

$$Q_\gamma = Q \times \frac{\gamma_{B_s^+}}{\gamma_{A_s^+}} = K \times \frac{\gamma_{B_r^+}}{\gamma_{A_r^+}}$$

At low ionic strengths, $Q_\gamma \cong Q$, since the γ_s-values approach unity. It is clear why Q_γ-values are called selectivity coefficients. If Q_γ is large, the resin is showing a preference for A^+; if Q_γ is small, the resin selectivity favors the binding of ion B^+.

It is not easy to obtain good thermodynamic equilibrium constants for the ion-exchange process. Activity coefficients of individual ions in solution can be estimated by one means or another, although not without significant uncertainty, but we know very little about the microenvironment of an ion in the resin phase or how these surroundings affect its activity. A detailed discussion of this problem is found in monographs on ion exchange which require more depth in physical chemistry than this textbook presupposes.

It should be kept in mind that we may not speak of the tendency of a resin to pick up a certain ion without noting that there is already another ion in the resin; i.e., we should consider not the tendency of the resin to pick up ion A^+ in an absolute sense, but rather the tendency to pick up A^+ at the expense of B^+. The tendency to pick up A^+ will be different if the resin phase contains some other ion C^+ instead of B^+ as the counterion. However, we can prepare a resin containing a certain counterion and then compare a series of other ions using this counterion as a reference. For the ions in this series we may simply write distribution ratios:

$$D = \frac{\text{concentration of an ion in the resin}}{\text{concentration of the same ion in solution}}$$

The conventional units of D are

$$\frac{\text{amount/kilogram of dry resin}}{\text{amount/liter of solution}}$$

The "amount" term may be in milligrams, moles, or whatever, since its units cancel in the D ratio.

D-values are generally determined in batch experiments by shaking weighed portions of resin with solutions containing the ions of interest until equilibrium is attained. Because the isotherm is nonlinear, reported D-values are usually limiting slopes at very low concentrations, as shown in Fig. 18.3. Radioactive isotopes are conveniently employed in such experiments; the D-value is obtained by simply counting the solution before and after equilibration with the resin.

A distribution ratio with different units is sometimes used, with the symbol D_v:

$$D_v = \frac{\text{amount/liter of wet resin bed}}{\text{amount/liter of solution}}$$

The conversion factor for D to D_v is the bed density, ρ,

$$D_v = D \times \rho$$

where ρ is in kilograms of dry resin per liter of wet resin bed. The significant as-

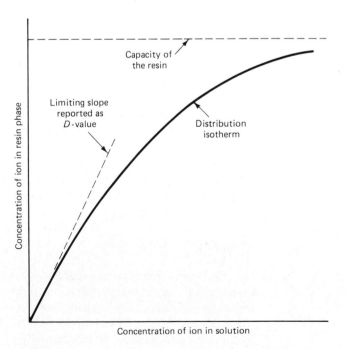

Figure 18.3 Typical distribution isotherm for an ion-exchange resin.

pect of ion exchangers is, of course, their selectivity; that is, D-values are different for various ions, and hence separations may be accomplished by ion exchange.

Neutral molecules can find their way into the pores of an ion-exchange resin, but they are not subject to forces so strong as those acting on ions, and in general they can be washed out by water or some other solvent. Solutes in the resin which are not so strongly held as ions are said to be *sorbed;* the pickup of such solutes by the resin is called *sorption.* Sorption can sometimes be used to effect separations, but in this discussion we are concerned only with legitimate ion exchange. Of course, the sign of the charge on an ion is important in selectivity, but this is so obvious as to be trivial. A cation cannot participate in exchange on an anion-exchange resin; it might find its way into the resin pores by some sort of general electrolyte sorption, but it would not be strongly held and could be washed out with water.

In a series of ions which have the proper sign to act as true counterions, the magnitude of the charge is important. Normally, the resin prefers the ion of higher charge. Thus the extent of exchange with, say, H_3O^+, would decrease in the order

$$Th^{4+} > Al^{3+} > Ca^{2+} > Na^+$$

provided proper allowance was made for other factors such as concentration. There are exceptions to this, but it is a good rule of thumb under ordinary conditions.

With a series of ions of the same charge, the resin still shows selectivity. For example, with alkali metals, the following order is generally found with cation-exchange resins:

$$Cs^+ > Rb^+ > K^+ > Na^+ > Li^+$$

The important factor here is probably the radius of the ion; the smaller an ion of given charge, the more strongly it will be held by the resin. At first glance, the above order may not appear consistent with this statement, which would imply that Cs^+ is a smaller ion than Li^+. The usual values of ionic radii, however, are obtained by X-ray diffraction studies of solid crystals, and these "crystallographic" or "naked" radii are not the right ones to use here. The ions in solution are hydrated, and it is the radius of the hydrated ion that determines the ion-exchange behavior. Such hydrated radii are much more difficult to measure, but estimates are available. While the naked radius of Li^+ is 0.68 Å, the hydrated radius is about 10 Å; the naked hydrated radii for Cs^+ are 1.65 Å and 5.05 Å, respectively.

Exceptions to the above rule regarding radius are seen with large organic ions, say, naphthoate or tetrabutylammonium, where hydrophobic interactions (page 534) with the organic matrix enhance the binding of the ion in the resin phase.

Applications of Ion Exchange.

Obviously, ions can be separated by ion exchange if their D-values are different, and there are more applications than one could recite, including many of a nonanalytical nature. The student should remember that the principle of a separation holds regardless of the reason for doing it. Preparing a pure compound in the organic, physical, or biochemical laboratory is a separation problem, as is the isolation of a pure penicillin from fermentation broth. We provide here some examples that illustrate the utilization of ion exchange for various purposes.

Sometimes differences in the solution chemistry of several elements can be combined with small differences in D-values to effect better separations. For example, the retention behavior of metal ions on a cation exchanger can be manipulated by adding complexing ligands such as citrate or tartrate to the mobile phase. A notable case is the preparation of very pure lanthanide metals (rare earth elements) by applying a mixture of the cations to the top of a cation-exchange column and eluting with citrate buffers in the mobile phase; scaled up to the kilogram level, the separation provided inorganic chemists in the 1940s with the first very pure samples of several of these elements which had been virtually impossible to purify by older techniques.

Another example involves interaction of certain metal ions with *anion-exchange* resins in the presence of HCl by the formation of anionic complexes, as shown in Fig. 18.4. Consider the curve for Fe(III): at low HCl concentrations, the metal is cationic, unretained by the resin, but as the acid level is raised, D increases with formation of the anionic complexes $FeCl_4^-$, $FeCl_5^{2-}$, and $FeCl_6^{3-}$; finally, chloride ions furnished by the HCl begin to compete with the metal species for resin sites, and the D-value for the metal is lowered. Possible separations can be predicted by examining the figure. For instance, a small volume of a solution containing Fe(III) and Co(II) is pipetted onto the top of the column. After this has penetrated into the resin bed, an elution with 4 M HCl is begun. Co(II) does not form anionic chloro complexes at this HCl concentration, and the cationic form Co^{2+} is rapidly washed through the column while Fe(III) is held quite strongly by the resin. Finally, the Fe(III) is eluted by lowering the HCl concentration to, say, 0.2 M, where the D-value is very low.

One of the most dramatic successes of the ion-exchange technique is seen in the modern amino acid analyzer. Proteins, which play a central role in biochemistry, are condensation polymers of amino acids. The first step in the structure elucidation of a protein is to determine the amino acid composition. The protein

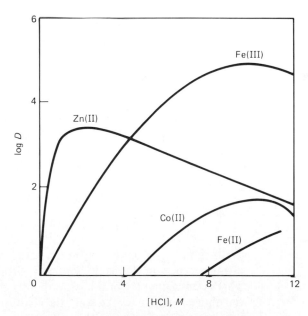

Figure 18.4 Distribution ratios for several metal ions with an anion-exchange resin in hydrochloric acid solution. (Reprinted with permission from H. F. Walton, *Principles and Methods of Chemical Analysis,* 2nd ed., Prentice-Hall, Inc., Englewood Cliffs, N.J., 1964.)

is hydrolyzed, breaking amide linkages to release about 20 different amino acids, which we may depict in their ionic forms as follows:

$$
\begin{array}{ccc}
\text{COOH} & \text{COO}^- & \text{COO}^- \\
| & | & | \\
\overset{+}{\text{H}_3\text{N}}\!-\!\text{C}\!-\!\text{H} & \overset{+}{\text{H}_3\text{N}}\!-\!\text{C}\!-\!\text{H} & \text{H}_2\text{N}\!-\!\text{C}\!-\!\text{H} \\
| & | & | \\
\text{R} & \text{R} & \text{R}
\end{array}
$$

The determination of all the amino acids in a protein hydrolysate was virtually impossible before the advent of chromatography, utilizing such classical techniques as precipitation of metal salts from aqueous ethanol or fractional distillation of methyl esters; in too many cases the R-groups are too similar. Through the dissociable α-amino and carboxyl groups and, in some cases, additional ionizations of R-group functionality, the net charge on an amino acid molecule can be manipulated via the pH; hence ion exchange is applicable to the separation. A solution containing the amino acid mixture is applied to the top of a cation-exchange column, and the flow of an eluting solution containing a replacement counterion is started, initially at a low pH. The cationic amino acids migrate downward at different rates reflecting their individual abilities to compete with the eluting ion for the anionic sites of the resin. A single set of conditions is not optimal for separating all the sample components; when conditions are right for efficiently separating the first few ions to emerge from the column, others require an excessive elution time. Thus a *gradient elution* is performed, where the concentration of the eluting ion is increased during the run; also, for this particular separation, the pH and the temperature are modified as the elution proceeds in order to optimize the separation. The solution flows from the column through a mixer where ninhydrin (a reagent that reacts with amino acids to produce a purple product) is added, then through a heater for complete reaction, and finally through a photometer where the absorption of light by the colored product is converted into an electrical signal which is recorded. The result is a chromatogram like that in Fig. 18.5, where the quantity of each amino acid is obtained from the area under the corresponding peak after suitable calibration. The electrical signal can be integrated electronically, converted into a digital form, and transmitted to a computer for processing, thereby totally automating the analysis.

Deionization of water for the laboratory is an everyday application of ion exchange. Tap water is passed through a bed containing a strong-acid cation exchanger in the hydrogen form (H^+ counterions in the resin phase) and a strong-base anion exchanger in the hydroxyl form (OH^- counterions). Using Na^+ and Cl^- as examples of ionic impurities in the water, we have the processes

$$
\begin{array}{l}
\text{RH} + \text{Na}^+ \rightleftharpoons \text{RNa} + \text{H}^+ \\
\hspace{4cm} \searrow \\
\hspace{5.5cm} \text{H}_2\text{O} \\
\hspace{4cm} \nearrow \\
\text{ROH} + \text{Cl}^- \rightleftharpoons \text{RCl} + \text{OH}^-
\end{array}
$$

With the two resins packed in the same column, H^+ released by the cation exchanger reacts immediately with OH^- from the anion exchanger to form water. This combination to form a very slightly dissociated final product drives each of the above reactions well toward the right, providing a much more nearly complete uptake of Na^+ and Cl^- by the resins than would passage of the water through two resin beds separately. Plastic cartridges containing the resins, ready for attachment to the water tap, are commercially available. The quality of the effluent water is very high in regard to its ion content—generally higher than that of ordinary distilled water—but other methods are obviously required for removing nonionic

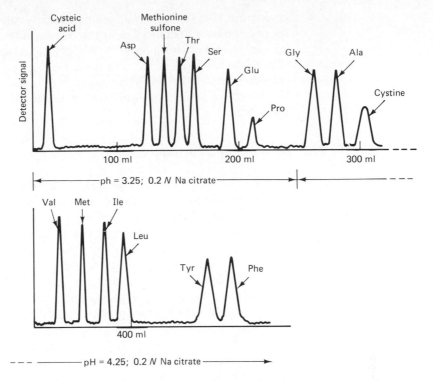

Figure 18.5 Analysis of an amino acid mixture by ion-exchange chromatography with photometric detection based upon postcolumn reaction with ninhydrin. [Adapted with permission from D. H. Spackman, W. H. Stein, and S. Moore, *Anal. Chem.*, **30**, 1190 (1958); copyright 1958 by the American Chemical Society.]

impurities. Emergency kits on life rafts may contain small resin cartridges for obtaining potable water from the sea. In situations where it is cost-effective, the resins can be regenerated by reversing the two reactions above, using strong acid and strong base respectively; this requires that the cation and anion exchangers be separated and then recombined after the regeneration, and in many laboratories the resins are considered expendable after they have become exhausted.

Many situations arise, in analytical chemistry and elsewhere, when it is desired to replace one ion by another. Perhaps phosphate interferes in the determination of a cation, say, Fe^{3+} or Ca^{2+}. Then, passing a solution of the sample through an anion-exchange column replaces the undesired anion by an innocuous one, perhaps Cl^- or NO_3^-. The cation is completely recovered, of course, by simply washing the column.

Suppose that a chemist desires a sample of hexyltrimethylammonium benzoate. He follows a standard organic method for alkylating an amine, using the alkyl halide:

$$CH_3(CH_2)_5Br + (CH_3)_3N \longrightarrow CH_3(CH_2)_5 - \overset{+}{N}(CH_3)_3Br^-$$

By passing an aqueous solution of the bromide salt through an anion-exchange column which has been previously conditioned by washing with a solution of sodium benzoate, Br^- can be exchanged for the desired counterion, and the quaternary ammonium benzoate salt can be recovered from the effluent solution.

We can ensure a nearly complete exchange in processes like those above by using sufficient resin in a long enough column. Even if the ion we wish to remove from solution is one for which the resin has a relatively high affinity, simply stirring the resin with our solution in a batch operation will usually yield at equilibrium only partial exchange; resins typically display only moderate preferences for one ion over another. But in the column mode, the most depleted resin near the top encounters fresh solution, while depleted solution contacts fresher resin as it moves downward; if the column is long enough, never will depleted solution encounter depleted resin. Thus one pass of solution through a column is equivalent to a large number of equilibrium batch treatments with successive portions of fresh resin.

18.1c. Liquid-Liquid Partition

If a solution of sodium silicate is acidified under the proper conditions, the precipitated silicic acid takes the form of a gel, a hydrophilic network structure which contains a large quantity of water. If this water is then driven out by heating the gel to an appropriate temperature, the silica which remains is a hard solid with a highly porous structure of very large surface area known as silica gel. Silica gel has a high affinity for water and is widely used as a desiccant. In 1941, Martin and Synge employed silica gel as a solid support to immobilize water as the stationary phase in a chromatographic column.[1] With a mixture of n-butanol and chloroform as the moving phase, acetylated amino acids were separated on this column. The basis for the separation was considered to be partition of the solutes between the stationary water phase and the mobile organic phase, the same process utilized in Craig's countercurrent solvent extraction technique adapted to the chromatographic mode. This particular form of chromatography was at first not widely adopted because shortly after it was proposed, variants such as paper chromatography appeared which were more convenient in view of the technological limitations on chromatography in the 1940s. Liquid-liquid partition has returned, however, in newer forms which will be mentioned later when we describe modern HPLC. As noted above, chemically bonded stationary phases are viewed by many writers as liquid films on solid supports, or at least analogs thereof; reversed-phase operations with such column packings are then treated in terms of liquid-liquid partition. The bonded phases eliminate problems related to stripping of stationary liquid from the column by the mobile phase. In principle, this is circumvented by presaturating the mobile phase with the stationary liquid, but difficulty ensues if it is desired to alter the composition of the mobile phase during the experiment (gradient elution).

18.1d. Size Exclusion

Chromatography based upon size exclusion apparently originated at the Biochemical Institute in Uppsala, Sweden, in 1959[2]. As applied to the separation of water-soluble macromolecules of biological importance, the technique is usually called *gel-filtration chromatography* (GFC). Five years later, a similar approach was developed for synthetic polymers soluble in organic solvents;[3] workers in this area often refer to the process as *gel permeation chromatography* (GPC).

---

[1] A. J. P. Martin and R. L. M. Synge, *Biochem. J.,* **35**, 91 (1941). The Nobel Prize in chemistry was awarded to Martin and Synge in 1952 for this and related work (e.g., paper chromatography), in large part because of the revolutionary impact of chromatography in biology and medicine. Coincidentally, Martin first described GLC (Chapter 17) in 1952.

[2] J. Porath and P. Flodin, *Nature,* **183**, 1657 (1959).

[3] J. C. Moore, *J. Polymer Sci., Part A,* **2**, 835 (1964).

Materials for Size Exclusion Processes

Two important types of column packing materials for GFC in aqueous media are cross-linked dextrans and polyacrylamides, sold under the registered names Sephadex G (Pharmacia, Inc.) and Bio-Gel P (Bio-Rad Laboratories), respectively.

Dextrans are polysaccharides produced by certain microorganisms. They are glucose polymers with molecular weights of roughly 10 to 300 million in the native state. The glucose residues are joined by α-1,6 linkages. Chain branching occurs to a certain extent, involving 1,2 or 1,3 or 1,4 glucosidic linkages; in the material of commerce, produced by a strain of the bacterium *Leuconostoc mesenteroides,* the branches are joined to the main chains by 1,3 links. The Sephadex producer purifies the crude dextran and partially hydrolyzes it to a lower-molecular-weight material which is fractionated to yield a suitable product with a narrow molecular weight distribution. A portion of such a polymer is shown below:

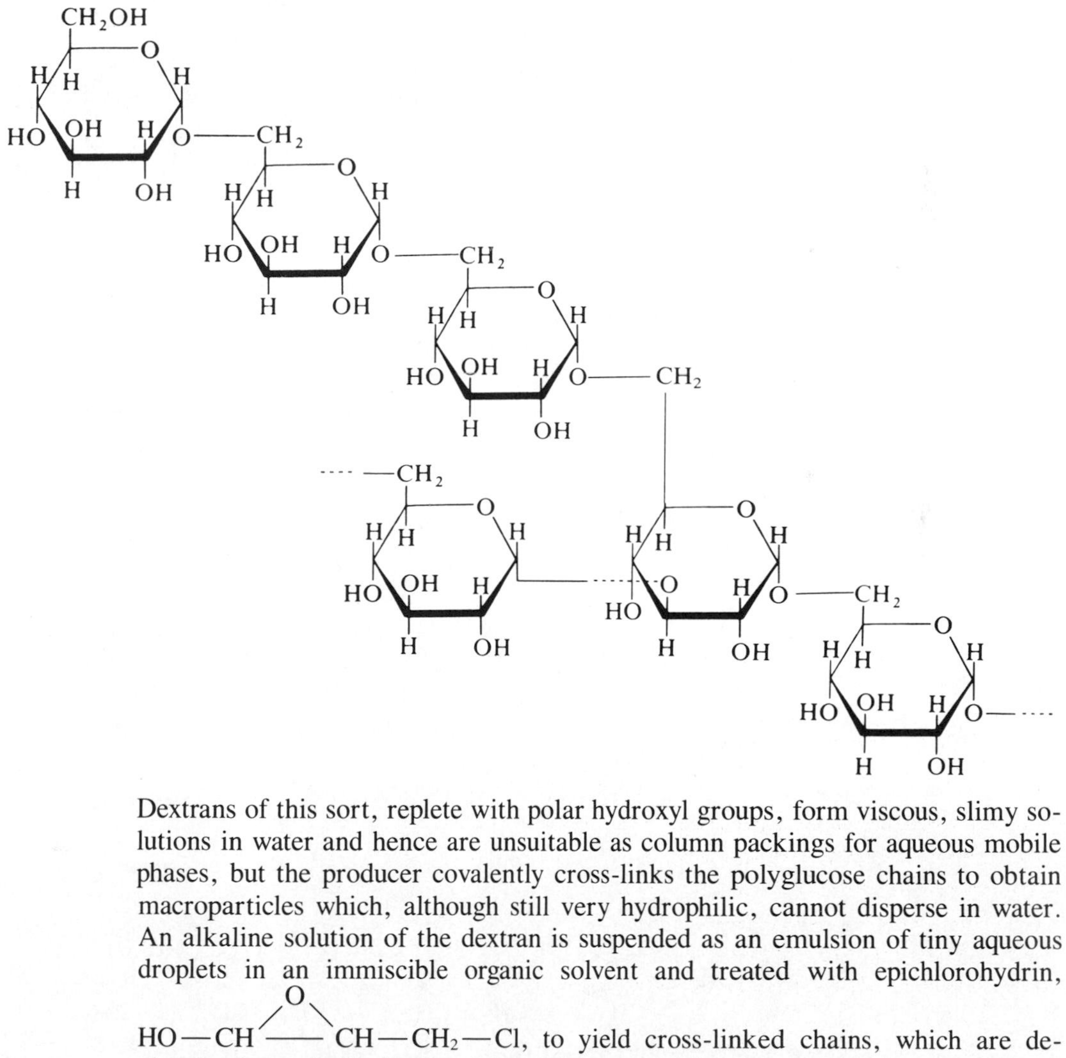

Dextrans of this sort, replete with polar hydroxyl groups, form viscous, slimy solutions in water and hence are unsuitable as column packings for aqueous mobile phases, but the producer covalently cross-links the polyglucose chains to obtain macroparticles which, although still very hydrophilic, cannot disperse in water. An alkaline solution of the dextran is suspended as an emulsion of tiny aqueous droplets in an immiscible organic solvent and treated with epichlorohydrin,

$$HO—CH \overset{O}{\diagup \diagdown} CH—CH_2—Cl,$$ to yield cross-linked chains, which are depicted schematically in Fig. 18.6. The material is washed, dried, and sold as a free-flowing powder in which the individual particles retain the spherical shape of the aqueous droplets before the cross-linking reaction.

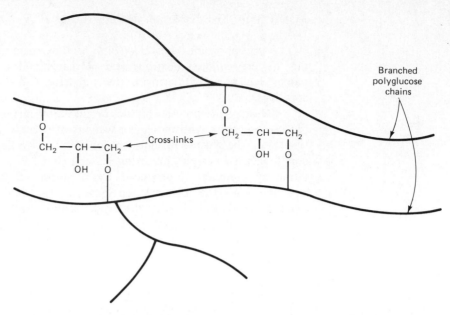

Figure 18.6 Schematic depiction of branched dextran chains with covalent cross-links.

The Sephadex materials are very hydrophilic. When the beads are suspended in water, there is a large uptake, which is called the *water regain;* this is accompanied by swelling of the particles to form the Sephadex gel. The water regain varies with the degree of cross-linking, ranging from 1 g of water per gram of dry Sephadex in a heavily cross-linked material to 20 g or more in less cross-linked preparations. The suspension of swollen beads is poured into a column and allowed to settle; the final volume of the bed per gram of dry Sephadex ranges from perhaps 2 to 40 ml, depending upon the degree of cross-linking.

The polyacrylamide gels are prepared by copolymerizing acrylamide with the cross-linking agent N,N'-methylenebisacrylamide:

Like the Sephadexes, these Bio-Gel P materials are produced in the form of spherical beads which imbibe large quantities of water because of their porous nature and the presence of the polar amide groups.

The materials described above are not the only ones available for gel filtration. Others, for example, are based upon agarose, a polysaccharide found in the seaweed product agar, and the inorganic materials silica gel and porous glass. Such materials as the styrene-divinylbenzene copolymers described earlier (page 533) provide hydrophobic size exclusion capability; these polymers swell in solvents like toluene and methylene chloride, forming useful gel phases for chromatography of organic-soluble macromolecules. The technique provides information on molecular weight distributions in synthetic polymers such as polyesters, polyolefins, and epoxy resins.

Mechanism of the Size Exclusion Process

The process with which we are concerned is the distribution of solutes between the aqueous phase within the gel particles and external water. At first glance, transferring a solute from one water phase into another is no process at all. But selectivity with regard to different solutes is achieved on the basis of the pore size. A solute molecule which is small enough may freely enter the gel phase; i.e., the water within the particles is available to that molecule (or at least most of it is; some internal water hydrates polar groups in the gel matrix and may not interact freely with solutes). In a column operation, this will have a retarding effect; molecules which are able to penetrate the gel will spend part of their time sheltered from the moving phase. At the other extreme, if a molecule is large enough, it cannot penetrate the pores at all; such a molecule will spend all of its time in the mobile phase and will move most rapidly through the column. In between, there will be molecules of intermediate size which can penetrate in some degree and whose progress down the column will be somewhat retarded. Thus we have the basis for separating molecules of different sizes. Figure 18.7 shows schematically three stages in the chromatographic separation of two extreme sizes of molecules. It is seen that penetration of the gel particles by the small molecules retards their progress down the column, resulting in a separation from the large molecules, which keep up with the flow of the mobile phase.

Let V_R be the retention volume for a solute in a chromatographic experiment with a Sephadex column. Let V_0 be the interstitial volume or void volume, that is, the volume within the column which is available for the moving phase (this is analogous to V_G in Chapter 17), and let V_L be the volume of the water within the gel particles which is available for accepting solutes. Then we can write an equation of the same sort that we had in GLC:

$$V_R = V_0 + KV_L$$

where K has the form of a distribution coefficient. If a solute is completely excluded from the interior of the gel particle, then $K = 0$ and $V_R = V_0$; colored marker substances which approximate this behavior are available from the Sephadex supplier. For a solute which can freely enter the gel particle, there should be no preference for water inside or outside the gel, and hence $K = 1$ and $V_R = V_0 + V_L$. For molecules of intermediate size which can penetrate the gel to some extent but not freely, K values should fall between 0 and 1. If sieving based upon molecular size were the only phenomenon occurring, then K values greater than 1 would never be encountered. In fact, however, such values are sometimes obtained, suggesting that solutes may interact with the dextran matrix itself; effects

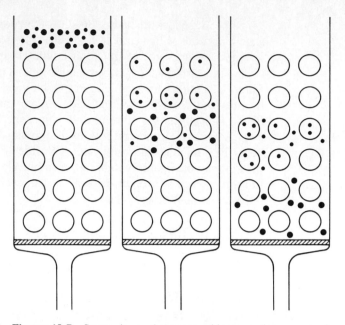

Figure 18.7 Stages in a chromatographic separation on a gel-filtration column. Large open circles represent the porous stationary phase; larger black circles represent a molecule with a K-value of zero; smaller black circles represent molecules which can penetrate the gel particles and are retarded in their progress down the column.

such as adsorption, hydrogen bonding, and ion exchange would explain such behavior. Figure 18.8 shows the manner in which solute retention volume typically varies with molecular weight.

Molecules of differing size in the sloping central region of the curve in Fig. 18.8 can be separated chromatographically. Varying the degree of cross-linking

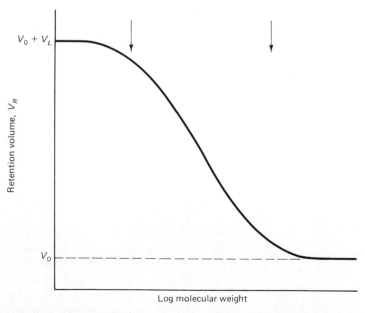

Figure 18.8 Relation between retention volume and molecular weight for solutes on a gel-filtration column. The steep region between the arrows is called the *fractionation range*.

shifts the curve horizontally; thus materials are available which fractionate molecules in various molecular weight ranges. For example, Sephadex G-25 is useful for separating molecules ranging from about 1000 to 5000, while the more lightly cross-linked Sephadex G-100 is recommended for 4000 to 150,000. Using a column which has been calibrated by measuring the retention volumes of several known substances, an analyst can estimate the molecular weight of an unknown molecule from its behavior on the column; biochemists often obtain approximate molecular weights of protein molecules in this manner. In order for such a measurement to work well, all the molecules involved should be of approximately the same overall shape, e.g., spherical or ellipsoidal.

Box 18.1. Preparative Chromatography: Purification of an Enzyme

Naturally, analytical applications of chromatography are emphasized in a quantitative analysis text, but chromatography, often on a larger scale, is a powerful separation tool for nonanalytical purposes. The purification of the enzyme triose phosphate isomerase provides a biochemical example of preparative-scale gel filtration and ion-exchange chromatography.[†]

An extract was prepared from 450 g of ground rabbit muscle by homogenization of the tissue with an aqueous buffer in a food blender, followed by centrifuging and filtering off particulate matter. This extract contained 18.6 g of total protein, including the desired enzyme. Crude separations by fractional precipitation with acetone and by differential thermal denaturation yielded 2.25 g of total protein which contained most of the enzyme. This product, in 96 mL of solution, was applied to a Sephadex G-100 column 6 cm in diameter and 185 cm (a little over 6 ft) long. Elution with an aqueous buffer yielded the chromatogram shown in Fig. 1. Eluent fractions were tested for both total protein and for the enzyme, with the result seen in the figure. (Only a portion of the chromatogram is shown; presumably additional extraneous proteins eluted in other fractions.) Notice the

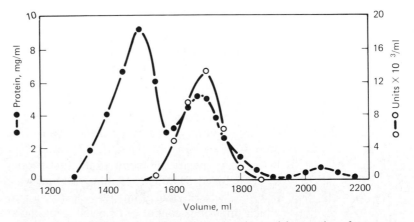

Figure 1 Portion of a chromatogram showing the partial separation of an enzyme from other proteins on a Sephadex G-100 gel-filtration column. Solid circles are total protein concentrations in the column effluent; open circles are measures of catalytic activity associated with the desired enzyme. [Reprinted with permission from I. L. Norton, P. Pfuderer, C. D. Stringer, and F. C. Hartman, *Biochemistry*, **9**, 4952 (1970); copyright 1970 by the American Chemical Society.]

incomplete separation of the enzyme band from nonenzyme protein, to which we say, "Welcome to the real world—it happens." The enzyme-containing fraction, between 1550 and 1800 mL in the figure, had 608 mg of total protein. This solution was treated to induce crystallization, yielding 169 mg of crystalline protein.

A solution of this material was applied to a 2 × 57 cm (~2 ft) cation-exchange column. (The enzyme molecules were positively charged at the *p*H employed.) Figure 2 shows a portion of the chromatogram resulting from a gradient elution with increasing sodium ion concentration to displace the enzyme molecules.

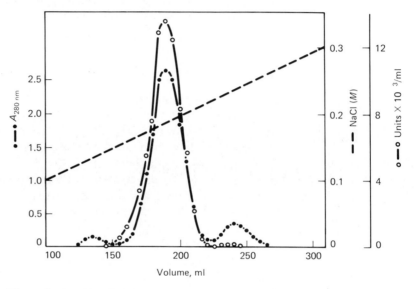

Figure 2 Portion of a chromatogram showing the gradient elution of an enzyme from an ion-exchange column. Solid circles are total protein concentrations in the column effluent; open circles are measures of catalytic activity associated with the desired enzyme; the dashed straight line depicts the gradient in concentration of the displacing cation. [Reprinted with permission from I. L. Norton, P. Pfuderer, C. D. Stringer, and F. C. Hartman, *Biochemistry,* **9**, 4952 (1970); copyright 1970 by the American Chemical Society.]

Finally, solutions from those receiving vessels containing most of the catalytic activity were pooled and the enzyme was again crystallized, yielding 85 mg of a product that was free of impurities by the most sensitive available tests for protein homogeneity.

In connection with this example, the student may reflect upon what we mean by "pure" in chemistry. In the 1930s, most biochemists considered a protein in the form of beautiful crystals as pure. For this reason, Nobel Prizes were awarded to some of the first people to show that proteins could be crystallized. But the crystalline material which yielded the chromatogram in Fig. 2 was obviously impure. Clearly, a pure substance is simply one in which impurities cannot be detected using the analytical methodology of the day. An investigator in any discipline is, in a sense, peforming analytical chemistry when deciding whether a substance is pure.

†I. L. Norton et al., *Biochemistry,* **9**, 4952 (1970).

TRADITIONAL LIQUID CHROMATOGRAPHY TECHNIQUES

18.2a. Column Chromatography

We saw in Chapter 17 that the gas chromatograph is a complete analytical instrument in the sense that components of the sample are not only separated but are measured as well. Liquid chromatography developed earlier, in a period when speed and automation were not major concerns in the laboratory and before the technology was available for a "systems" approach to analytical instrumentation. The great power of liquid chromatography as a separation tool was widely recognized by the mid-1930s, but the idea of building an efficient, self-contained analytical instrument around it goes back about 30 years. The more recent developments will be better appreciated if we examine briefly the older approaches, which, as a matter of fact are still widely used when appropriate in terms of scale, cost, and convenience.

Various sizes of columns may be employed, the major consideration being adequate capacity to accept the sample without overloading the stationary phase. It is a common rule of thumb that the length of the column should be at least ten times its diameter. For a typical case, let us say that we have a column 20 cm long and 1 or 2 cm in diameter, something like the one shown in Fig. 18.9. The

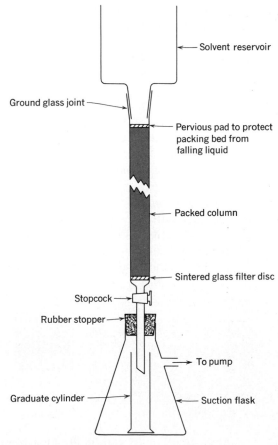

Figure 18.9 Possible setup for conventional liquid chromatography.

packing material, an adsorbent like alumina or perhaps an ion-exchange resin, is added in the form of a suspension in a portion of the moving phase and allowed to settle into a wet bed with a little liquid remaining above the surface. The stop-cock is opened, and the liquid level is allowed to fall just to the top of the bed; then a small portion of sample solution (a few tenths of a milliliter up to perhaps a couple of milliliters) is carefully pipetted onto the top of the bed. The liquid reservoir is positioned, and the flow of the mobile phase is started. The desired flow rate is obtained by gravity alone, by inserting the outlet end of the column into an evacuated vessel as shown in the figure, or by pumping liquid in at the top of the column. A typical flow rate might be a few tenths of a milliliter per minute, possibly faster if the separation is not a difficult one.

The effluent solution is collected in a series of fractions of convenient volume. The solution may drip into a graduated cylinder which the operator dumps into a beaker or test tube each time a certain volume, say 5 or 10 ml, has accumulated. It is not uncommon for the chromatographic elution process to require several hours, even all day or overnight. In such cases, a mechanical device called a *fraction collector* is convenient. The operator sets up a series of tubes on a turntable which positions a new tube under the column when the desired volume has been collected in the previous one; activation can be based upon time, drop counting, or the deflection of a light beam by the rising meniscus in the tube.

It sometimes happens that no single moving phase is well suited for the elution of all the components of a sample. In adsorption, for example, a fairly non-polar solvent may be ideal for eluting some of the less polar solutes, whereas the more polar solutes may then show an inordinately long retention. In such cases, the technique of gradient elution may be useful. The composition of the mobile phase is changed continuously by allowing a more polar solvent to flow into the reservoir containing the less polar one, whence the mixture flows to the column. The reservoir is stirred. Figure 18.10 shows such an arrangement schematically. Now the laggard solutes will move along faster as the eluting power of the solvent mixture increases. Bands with serious "tailing" may be sharpened, since the tail sees a stronger solvent than the front of the band. The result of gradient elution is similar to that of temperature programming in GLC, which was mentioned in the last chapter.

The individual fractions of the effluent solution are examined by whatever means is appropriate—spectrophotometry, polarography, radioactive assay, titrimetry, etc.—to locate the desired components of the sample and to determine their quantities.

As seen above, the conventional column operation is slow and tedious, and it is poorly suited to automation and modern methods of data processing. It is not totally without merit, however; for example, fairly large columns can be used

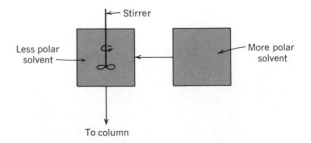

Figure 18.10 Arrangement for gradient elution in liquid chromatography involving adsorption. Arrows show solvent flow.

which have sufficient capacity for preparative-scale work. But for handling large numbers of analytical samples, recent developments enable us to do better. Before describing these developments, we shall complete our discussion of conventional liquid chromatography by considering briefly two widely used forms which do not employ a column in the usual sense.

18.2b. Paper Chromatography

In 1944, again from Martin's laboratory, the separation of a mixture of amino acids was reported using paper chromatography.[4] In this technique, a small volume of sample solution is applied near one end of a strip of filter paper, and the spot is allowed to dry (blowing on it with a hair dryer is convenient). The end of the strip is then placed in a trough containing a suitable solvent within a closed chamber. In ascending paper chromatography, seen in Fig. 18.11, the paper is suspended from the top of the chamber so that it dips into the solvent at the bottom, and the solvent creeps up the paper by capillarity. In the descending form, the paper is anchored in a solvent trough at the top of the chamber, and the solvent migrates downward by capillarity assisted by gravity. After the solvent front has moved almost the length of the paper, the strip is removed, dried, and examined. In a successful case, solutes from the original mixture will have migrated along the paper at different rates, forming a series of separated spots. If the compounds are colored, of course, the spots can be seen. If not, they must be found in some other way. Some compounds fluoresce, in which case glowing spots may be seen when the paper is held under an ultraviolet lamp. For amino acids, the paper is usually sprayed with a solution of ninhydrin, a reagent which reacts with

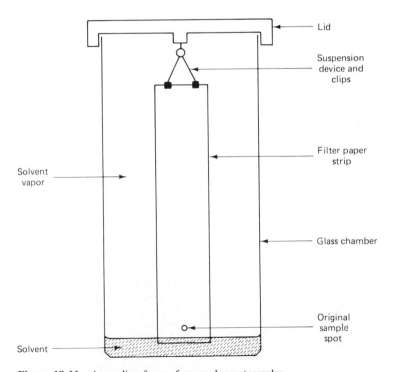

Figure 18.11 Ascending form of paper chromatography.

[4] R. Consden, A. H. Gordon, and A. J. P. Martin, *Biochem. J.*, **38**, 224 (1944).

the amino group to yield a purple compound. For quantitative analysis, the spots may be cut out with scissors, the solutes leached from the paper by appropriate solvents, and the solutions examined by a suitable technique, often spectrophotometry.

For identification purposes, spots are often characterized by their R_f-values. An R_f-value is the ratio of the distance moved by a solute to the distance the solvent front moved during the same time. Identical R_f-values for a known and an unknown compound using several different solvent systems provides good evidence that the two are identical, especially if they are run side by side along the same strip of paper.

It sometimes happens that all the components of a sample cannot be separated using any one solvent system; some components separate better in one system, some in another. Two-dimensional paper chromatography may then be employed. The sample is spotted near one corner of a square filter paper sheet. After migration of the solutes parallel to one edge of the paper using one solvent system, the paper is turned 90°, and a second solvent system carries the solutes into the unused portion of the paper. The pattern of ninhydrin-stained spots that results from applying this technique to the amino acids in a protein hydrolysate is often called a "fingerprint" of the protein. (See Fig. 18.12)

Paper chromatography was first considered simply a form of liquid-liquid partition. The hydrophilic cellulose fibers of the paper can bind water; after exposure to a humid atmosphere, filter paper that appears dry may actually contain a large percentage of water, say 20% or more by weight. Thus the paper was considered to be the analog of a column containing a stationary aqueous phase. Sol-

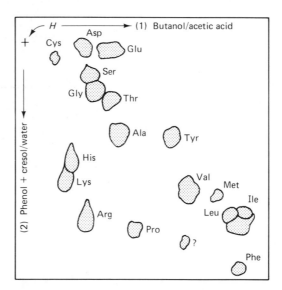

Figure 18.12 Two-dimensional paper chromatogram of a protein hydrolysate. The sample was placed on the paper at point *H* and run toward the right with an acetic acid-butanol solvent; after drying, the solvent was changed to a phenol-cresol-water mixture for migration downward in the figure; shaded areas are purple ninhydrin spots, identified by the three-letter amino acid abbreviations. (From F. Haurowitz, *Biochemistry: An Introductory Textbook,* John Wiley & Sons, Inc., New York, 1955. Reprinted by permission of John Wiley & Sons, Inc.)

utes were then partitioned between this water and the moving immiscible organic solvent. It was soon realized, however, that this model was too simple. Separations were obtained where the moving phase was miscible with water or in some cases was itself an aqueous solution. Thus, although liquid-liquid partition may indeed play a role in some cases, the mechanism of paper chromatography is often more complicated than that. Interactions between solutes and the cellulose support are probably involved, for example, adsorption and hydrogen bonding. Carboxyl and other ionizable groups are introduced into cellulose during the pulping and bleaching operations of paper manufacture, and hence the paper may also act as an ion exchanger.

18.2c. Thin-Layer Chromatography

Thin-layer chromatography (TLC), like paper chromatography, is inexpensive and simple to perform. It has an advantage of speed over paper chromatography: The process may require only a half hour or so, whereas a typical separation on paper requires several hours. TLC is very popular and is used routinely in many laboratories.

The separation medium is a layer perhaps 0.1 to 0.3 mm thick of an adsorbent solid on a glass, plastic, or aluminum plate. A typical plate is 8×2 in. Typical solids are alumina, silica gel, and cellulose. Workers used to prepare their own plates by coating the glass with an aqueous suspension of the solid, which usually contained a binder such as plaster of Paris, and then drying the plates in the oven. Precoated glass plates and sheets of plastic and aluminum foil which can be cut to size with scissors are commercially available, and the majority of workers use these today.

The sample, generally a mixture of organic compounds, is applied near one end of the plate as a small volume of solution, usually a few microliters containing microgram quantities of the compounds. A hypodermic syringe or a small glass pipet may be used. The sample spot is dried, and then the end of the plate is dipped into a suitable moving phase. The solvent moves up the thin layer of solid on the plate, and as it moves, sample solutes are carried along at rates which depend upon their solubilities in the moving phase and their interactions with the solid. After the solvent front has migrated perhaps 10 cm, the plate is dried and examined for solute spots as in paper chromatography. A two-dimensional run using two different moving phases is often performed; here a square plate is used rather than a narrow one. The separation may be followed by a quantitative determination; where a spot is located, the adsorbent can be scraped from the plate with a spatula, the solute eluted from the solid material with a suitable solvent, and the concentration of the solution determined by a technique such as spectrophotometry.

Along with paper chromatography, TLC is sometimes called *planar chromatography*. There is no easy way to elute sample components from the plate (or paper) for transit through a detector, but instrumentation has been developed for scanning plates for such sample properties as ultraviolet absorption and fluorescence. Thus it is not fair to say that TLC is inherently only a qualitative technique, but there are background problems in quantitation related to such factors as scattering of radiation by the solid coating. Recent developments in TLC include the introduction of fine-particle layers with a narrow size distribution which are very efficient in regard to HETP. More sensitive detection allows the application of smaller samples (say, 0.1 or 0.2 μL) to the TLC plates. Effective separations may be obtained after solvent migration of only 3 to 6 cm, and 18 to

36 samples may be run simultaneously in parallel lanes on the plate. The current status of TLC has been described recently in more detail.[5]

18.3

HIGH-PERFORMANCE LIQUID CHROMATOGRAPHY (HPLC)

The development of gas chromatography as an analytical technique ran parallel with an explosion in technology which brought a revolution in scientific instrumentation. By the late 1960s, analysts were becoming accustomed to excellent separations of complex mixtures in minutes or even seconds, with electronic integration to obtain areas under elution bands as well as computer printouts of complete analyses. Nanogram, and sometimes even picogram, quantities could be detected in favorable cases (1 nanogram = 10^{-9} g; 1 picogram = 10^{-12} g). It was natural for people dealing with nonvolatile samples, for which GC was inappropriate, to desire a similar capability in liquid chromatography. Modern HPLC has emerged from the confluence of need, the human desire to minimize work, technological capability, and the theory to guide development along rational lines. (In some writings, the "HP" in HPLC has represented "high pressure," and some of those with budgetary problems may think it stands for "high-priced," but to most it means "high performance.")

It was clear before the modern instrumentation era that LC possessed great separating power even for closely related components; successful ion-exchange separations of rare earths and amino acids had demonstrated this. But bringing to LC the efficiency and sensitivity which had become commonplace in GC required new approaches. LC had to be speeded up, automated, and adapted to much smaller samples. Several hours' elution time, individual manual assays of collected fractions, one or two samples containing milligrams or even grams of material through a column per day, liters of solvent—all this had to be changed. Although certain developments in LC may seem esoteric to nonspecialists, the student, drawing upon what was learned about GC in the last chapter, can easily understand qualitatively the most important aspects of the modern approach.

18.3a. Column Technology

The first thing that came to mind in connection with speed in LC was to increase the mobile phase velocity. This meant better pumps as well as plumbing connections that did not leak under higher pressures. By and large today's pumps and fittings are adequate for most work. But there was more to consider than this. With conventional stationary phases, it had been known for years that high mobile phase velocity was incompatible with good solute resolution. Efficiency in regard to both resolution and speed required different column packings.

It is instructive in this regard to recall the three terms in the van Deemter equation (page 502):

$$\text{HETP} = A + \frac{B}{u} + Cu$$

Although this equation was developed for GC, it is reasonable that band spreading

---

[5] C. F. Poole and S. K. Poole, *Anal. Chem.*, **61**, 1257A (1989).

in LC arises from the same kinds of factors, although their relative importance might differ from liquid to gaseous mobile phases. The eddy diffusion term, A, is probably quite small with any good column; thus, while we continue to be careful about it, we would not expect a major breakthrough in efficiency from further studies of eddy diffusion.

The longitudinal diffusion term, B/u, is important in GC, but because diffusion is very much slower in liquids than in gases, band broadening arising from solute diffusion in the mobile phase is significant in LC only at exceedingly slow flow rates. In seeking high performance by combining speed with good separations, we desire *faster* flow rates, not slower ones; thus for all practical purposes, we may neglect longitudinal diffusion in LC.

The nonequilibrium in mass transfer term, Cu, holds the key to high performance. High flow rates are inconsistent with high resolution unless we can achieve faster transfer of solutes between stationary and mobile phases; if we wish to increase mobile phase velocity, u, without increasing HETP and losing plates, then we must lower C. One way to obtain faster mass-transfer kinetics is to raise the temperature. This increases D_L in the longer version of the van Deemter equation (page 502) and hence lowers the Cu term. If the sample is stable at, say, 50, 60, or 80°C, we could operate the column at such a temperature to accelerate the approach to equilibrium. The higher temperature would also lower the viscosity of the mobile liquid, with the result that the desired flow rate would require a somewhat smaller pressure drop. The boiling points of our common solvents represent a limitation on temperature increases, however, and, especially with biological samples, solute instability may preclude the use of elevated temperatures.

The most successful approach has involved the development of stationary phases in which the space available within the particles for solute diffusion is severely limited as compared with conventional column packings. The older materials—resin beads, silica gel, dextrans, and the like—are highly porous particles with diameters typically in the range of 40 or 50 μm to perhaps 150 or 200 μm. (If 0.1 or 0.2 mm seems small to the student, it is large by present standards in LC columns and enormous on a molecular scale.) Deep pores in such materials fill with stagnant portions of mobile phase, and it is possible for solute molecules, as they wander about, to penetrate so deeply into the packing particles that their return by diffusion into the moving liquid is slow. This can be tolerated in conventional LC with slow flow rates, but high velocities accentuate the problem, leading to excessive solute band broadening of the sort depicted in Fig. 17.5 (page 502).

The most obvious route to faster mass transfer is to grind the conventional packing materials to a much smaller particle size, perhaps 2 to 5 μm in diameter. This clearly decreases the volume of mobile phase trapped within the individual particle and shortens the distance from the deepest recess to the external moving liquid. The first efforts in this direction, however, were unpromising. Packing irregularities were more pronounced with the small, irregularly shaped particles, and the available pumps could not provide adequate inlet pressures for the desired flow rates. (The student can easily imagine the analogy that it is easier to blow through a tube packed with marbles than one packed with sand or talcum powder.)

The first successful modern packings have been called *superficially porous supports, porous layer beads,* or *pellicular supports.* These provide a combination of large overall particle size for easy column packing and a small pressure drop with a very small pore depth for rapid mass transfer of solutes. DuPont's materials under the tradename ZIPAX provide examples. Each bead is a nonporous, im-

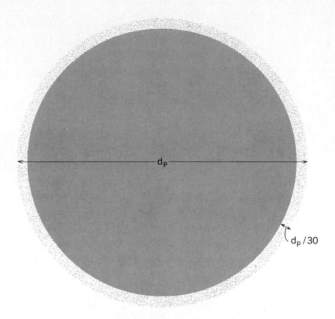

Figure 18.13 Schematic section through the center of a porous layer bead. d_p is typically about 30 μm.

pervious sphere of siliceous material, 30 μm in diameter, with a thin, porous surface coating as depicted in Fig. 18.13. The surface layer is made up of spherical silica microbeads whose diameters (e.g., 200 nm) control the porosity. A sintering process (i.e., incipient melting) results in adhesion of the particles without actual flow that would destroy the pores. This basic approach can be adapted to various phase distribution processes: The porous silica surface is itself a polar adsorbent; the pellicle can retain enough liquid for liquid-liquid partition with an immiscible mobile phase; control of the pore size allows production of size-exclusion materials; the porous surface can be coated with a thin layer of ion-exchange resin; and a hydrocarbon layer, e.g., C_{18}, can be chemically bonded to the silica to provide a reversed phase packing (page 534).

Although the superficially porous materials are still available, recent years have seen increasing utilization of a new generation of packing materials of very small particle size which are totally porous. As noted above, simply grinding conventional packings into very small particles is not totally satisfactory, and the newer materials are produced by a different technology. (Better pumps have also facilitated the recent developments.) Du Pont's ZORBAX materials provide an example. Very small silica particles with sizes in the colloidal range (page 72) are formed into spherical algomerates which are initially held together by an organic polymer coating. By a heating process, the urea-formaldehyde polymer is burned off and the tiny silica particles are brought to incipient melting sufficient to bind them together. The result is a spherical silica particle which is porous throughout, as seen in Fig. 18.14. The overall diameter is generally in the neighborhood of 5 or 6 μm. Various pore sizes are obtained by starting with colloidal silica particles of different diameters, say, from 80 to 2000 Å. Because of their porosity, the particles provide a very large surface area, typically about 350 m²/g. For perspective, we may note that the modest quantity of 12 or 13 g will possess roughly the surface area of an American football field.

The spherical form and narrow particle size range lead to very uniform chromatographic columns whose performances are predictable. The basic type of

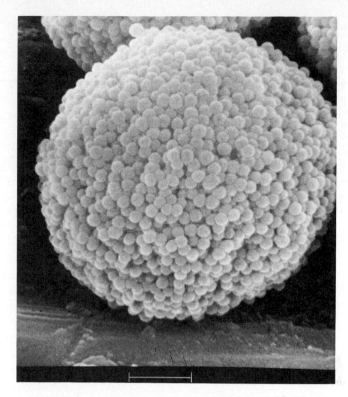

Figure 18.14 Scanning electron micrograph of a ZORBAX particle. The length of the line between the bars below the picture is 1.0 μm. (Photo courtesy of Du Pont Bioresearch Systems Division.)

porous silica microsphere can be adapted to various forms of chromatography. Control of pore size provides size-exclusion materials that fractionate in various ranges of molecular weight; the silica can be silanized to yield a surface which is wetted by organic solvents for chromatography of water-insoluble polymers. Chemically bonded hydrocarbon chains provide a reversed phase packing. Bonded phases which contain cationic or anionic sites are available for ion-exchange chromatography.

18.3b. *Mobile Phases*

As noted earlier, mobile phase composition in LC provides a dimension for experimental manipulation that is lacking in gas chromatography. In GC, the separation factor, *S*, for a pair of sample components (page 506) depends upon only the nature of the stationary phase; in LC, it depends upon the mobile phase as well. When the solvent composition is unchanged during a chromatographic run, we call the process an *isocratic elution*. The contrasting term is *gradient elution* (page 552); here the solvent composition is changed during the elution process. The enormous number of possible mobile phases (solvents *and* solutions) coupled with gradient possibilities yields abundant opportunity for optimizing separations of complex mixtures in regard to both resolution and time. Normally an isocratic elution will be preferred over a gradient, everything else being the same, because a less expensive instrument with a simpler solvent delivery system can be employed. It belabors the obvious to say that the mobile phase must provide the proper kind of interaction for the particular distribution process involved; for instance, ions of the sample will not migrate down an ion-exchange column unless the eluent contains displacing ions.

Frequently a mixed solvent is a better mobile phase than a pure liquid for separating complex mixtures, and optimization of solvent composition by trial and error could be a formidable and tedious task. Fortunately there are rational approaches which minimize the effort. A pioneering study published in 1980[6] showed how to establish in a few experiments a very nearly optimal mixture of methanol, acetonitrile, tetrahydrofuran, and water for the efficient separation of a nine-component mixture of substituted naphthalenes on a reversed phase column. Many additional examples have appeared illustrating both reversed and normal phase separations. The four solvents are selected initially by a combination of common sense, experience with similar separations, and tabulated parameters which define solvent-solute and solvent-stationary phase interactions in terms of basic properties such as polarity and ability to form hydrogen bonds. After a few preliminary experiments with extreme mixtures that provide widely differing S-values, the program allows a rapid approach in only a few successive approximation steps to a solvent blend which is nearly optimal in regard to both resolution and short elution time. Although a four-solvent mixture at first seems needlessly complex, it actually simplifies the search for an optimal mobile phase composition, and computer-interactive instrumentation permits automation of the preliminary experiments leading to an efficient chromatographic analysis. The use of an appropriate mixed solvent often eliminates the need for a gradient elution, simplifying the apparatus required for routine analyses.

This is not to say that complex solvent mixtures are always required. For example, size-exclusion chromatography with biopolymers, e.g., proteins, often requires only a simple aqueous mobile phase, perhaps buffered if the sample components are pH-sensitive, or containing electrolyte if the sample solubility is sensitive to ionic strength.

18.3c. Detectors

The detector, as in GC, is an integral part of a modern analytical LC instrument. Several types are in use, with the choice generally based upon sensitivity requirements, the type of compounds in the sample, and other factors such as cost. The most general detector is based upon the refractive index of the column eluate, since almost any solute will yield a solution of refractive index different from that of the pure solvent. The detector senses the difference and generates a proportional electrical signal which is amplified and recorded to yield the chromatogram. The major limitation is sensitivity; the detection limit will vary with circumstances, but typically is in the neighborhood of a microgram (10^{-6} g) of solute. An additional problem is the high temperature dependence of the refractive index, which makes necessary very careful temperature control of both detector and entering solution.

Spectrophotometric Detectors

Spectrophotometric detection, usually in the ultraviolet region, is widely used. Ideally, a real spectrophotometer with a complete choice of wavelengths would provide maximal flexibility for detecting various kinds of solutes with the

[6] J. L. Glajch, J. J. Kirkland, K. M. Squire, and J. M. Minor, *J. Chromatog.*, **199**, 57 (1980).

greatest sensitivity; the usual sort of sample cell would be replaced by a flow cell to pass the column effluent solution through the sample beam of the instrument. A modern ultraviolet spectrophotometer is a very expensive detector, however, and compromises are frequently required. Compared with many thousands of dollars for even a second-rate ultraviolet spectrophotometer, $2500 or so will buy a detector to monitor the column effluent solution at either 254 or 280 nm (1989 prices). The lower cost reflects the use of a mercury vapor lamp line spectrum rather than a continuous source and monochromator; simple filters isolate the desired line in the mercury spectrum. The limited wavelength choice seems restrictive, but these simple detectors serve well in many cases. For instance, proteins all absorb at 280 nm due to aromatic amino acid side chains, and most aromatic compounds including many of biological interest (e.g., purines, pyrimidines, nucleosides, nucleotides, and nucleic acids) can be detected at 254 nm. The sensitivity varies with the match between a solute absorption band and the available detector wavelength as well as the intensity of the band and the path length through the detector cell, but as a rough guide the ultraviolet detector will be able to "see" nanogram quantities; let us say it is on the order of 1000 times more sensitive than the refractive index detector. In addition, it is relatively insensitive to temperature.

Fluorometric Detectors

Detectors based upon fluorescence are becoming more common. The most versatile type provides for continuously variable excitation across a broad wavelength range by utilizing a continuous source and monochromator; usually simple filters are employed to transmit the fluorescent emission to the photodetector while blocking the exciting radiation. Less expensive versions employ filters on both excitation and emission sides and utilize sources with more limited excitation wavelengths. A large number of compounds can be detected by fluorescence, including many of environmental concern, such as polycyclic aromatic hydrocarbons, and of biological interest, such as vitamins, drugs, and neurotransmitters. Sometimes the mobile phase passes through a postcolumn reactor where nonfluorescent sample components are converted into fluorescent derivatives. A well-known example is the detection of amino acids at the subnanogram level after reaction with the reagent fluorescamine.

Electrochemical Detectors

Finally, we mention electrochemical detectors. Typically, the effluent solution from the column enters a cell where it flows over the surface of an electrode held at a potential where sample components undergo electron-transfer reactions. This type of detection has been applied, for example, to neurotransmitters and their metabolites in extracellular fluid from brain tissue of experimental animals; compounds such as dopamine, norepinephrine, serotonin, and homovanillic acid yield oxidation currents at a glassy carbon electrode held at +0.6 V vs. a silver-silver chloride reference electrode. The reference electrode, of course, must be in electrolytic contact with the analyte solution, generally through a salt bridge of some sort.

Detector Design

Detector design is obviously very important. The dead volume, including the connection to the column, should be very small, and solution must flow smoothly through the device without turbulent mixing. Typically, the detector volume is only a few microliters.

18.3d. Instruments

Many companies sell complete instruments for HPLC; some workers assemble their own setups, particularly when they have special needs, from a mixture of commercial components with others made in their own shops. Prices vary greatly with such factors as pressure provided by the pump, whether only isocratic capability is provided or solvent gradients can be programmed, type and quality of detector, and computer interfacing capability. Probably the range of $20,000 to $50,000 includes most commercial chromatographs in 1990. A pump that provides a pressure of 1000 lb/in.^{2} (about 68 atm) will serve many purposes, although for broad flexibility in regard to flow rates through columns of varying sizes with all sorts of packings one might prefer a pump capable of providing 7000 or 8000 lb/in.^{2} There are many technical aspects of the instruments that cannot be covered in detail in our brief discussion. For instance, if the solvent contains dissolved gases, as from the air, bubble formation may be troublesome at the detector, and the instrument may include an outgassing capability. Sometimes a "precolumn," a small and inexpensive device with a suitable stationary phase, is placed ahead of an expensive analytical column to pick up particulate matter and solvent impurities that might foul the costly column during a series of analyses. A basic liquid chromatograph is shown as a block diagram in Fig. 18.15.

Columns may be purchased ready for use or packed locally using bulk material. Although modern packings are often expensive on a "per gram" or "per column" basis (perhaps several hundred dollars for a single column), the fact that the column can be used hundreds of times if properly cared for makes the cost *per analysis* quite small for a laboratory with many samples to run. High detector sensitivity, permitting the analysis of small samples, and packings providing a large number of theoretical plates per centimeter allow the use of quite small

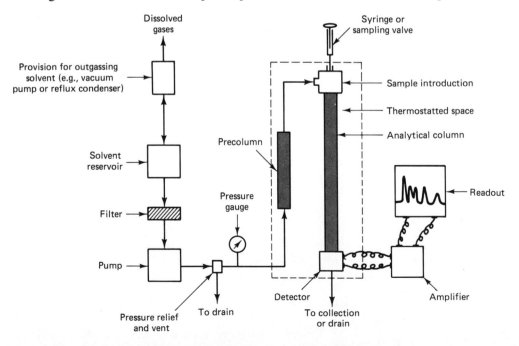

Figure 18.15 Block diagram showing the components of an HPLC instrument.

columns; thus only small quantities of mobile phase are consumed, and solvent costs are much lower than in classical LC. The automation possible with a modern instrument may also lower personnel costs.

Samples are introduced into the flowing mobile phase just ahead of the column as in GC. This may be done by direct microsyringe injection through a self-sealing elastomeric septum or by insertion into the loop of a special sampling valve which is later swept out by a mobile phase; the latter is generally considered to be more accurate and reproducible.

Figures 18.16 and 18.17 show actual chromatograms. Figure 18.17 is particularly striking as an example of optimization; the resolution is quite good, and the student should note the time scale. If columns are treated carefully and experimental conditions are scrupulously reproduced, retention times (or volumes) can serve to identify sample components if enough is known of the nature of the mixture. The liquid chromatograph as such, though, does not directly provide structural information; this is better obtained by other means such as infrared spectrophotometry or mass spectrometry. There is much current interest in developing suitable interfaces so that eluted sample components can be examined by such techniques. Quantitative analysis is basically done as in GC, i.e., by comparing the area under the elution band with areas obtained with known quantities of standard materials. By careful attention to details, it is probably possible to hold errors as low as 0.5%, although errors of a few percent are certainly more common.

The literature of LC is large and growing rapidly. A book by two of the leaders in the field is recommended for students desiring a more thorough and authoritative, but still introductory, treatment.[7] There are a host of books, some practical and some more theoretical, on all aspects of chromatography.[8]

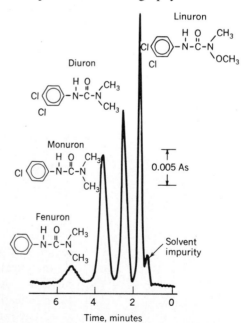

Figure 18.16 Separation of substituted urea herbicides. Column: 50 cm, 2.1 mm i.d., 1% β,β'-oxydipropionitrile on 37 to 44 μm controlled surface porosity support. Mobile phase: di-*n*-butyl ether. Flow rate: 1.14 mL/min. Sample: 1 μL of a solution of 67 μg/ml each in carrier. (Courtesy of the publisher of the *Journal of Chromatography Science*.)

[7] L. R. Snyder and J. J. Kirkland, *Introduction to Modern Liquid Chromatography*, 2nd ed., John Wiley & Sons, Inc., New York, 1979.

[8] GC, HPLC, and modern TLC are well treated at about the first-year graduate level in C. F. Poole and S. A. Schuette, *Contemporary Practice of Chromatography*, Elsevier, Amsterdam, 1984.

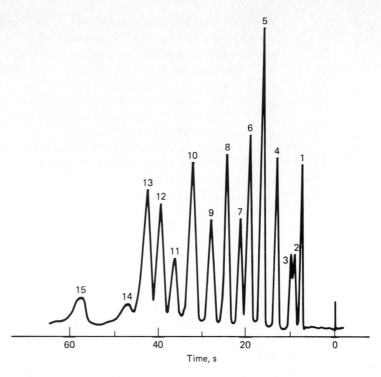

Figure 18.17 Separation of aromatic hydrocarbons. Column: 6.5 cm, 0.4 cm i.d.; spherical porous silica packing of 4.4 μm diameter. Mobile phase: *n*-pentane. Sample: 3 to 50 μg of each compound in chloroform solution. Temperature: 25°C. Pressure drop through column: 72 atm. HETP = 20 to 25 μm (number of theoretical plates: 2600 to 3250). Detector: ultraviolet absorption, 254 nm. Flow rate: 0.93 cm/s. Peaks: 1, methylene chloride; 2, chloroform (solent); 3, benzene; 4, naphthalene; 5, biphenyl; 6, anthracene; 7, pyrene; 8, fluoranthene; 9, *o*-terphenyl; 10, 1,2-dibenzanthracene; 11, 3,4-benzpyrene; 12, perylene; 13, 1,12-benzperylene; 14, coronene; 15, 1,2,5,6-dibenzanthracene. [Reprinted by permission from 1. Halasz, R. Endele, and J. Asshauer, *J. Chromatog.*, **112**, 37 (1975); copyright 1975 by the Elsevier Science Publishers B.V.]

| Box 18.2 | **Affinity Chromatography** |

Affinity chromatography is an important tool in biomedical research. It is more often a preparative than an analytical technique, but separations are separations whatever their purpose, and a brief sketch is not out of place in an analytical book. Affinity chromatography has been applied to a variety of materials, up to and including whole cells, but we shall consider here only the purification of proteins.

The interactions of a protein molecule of interest with other molecules (including other proteins) are frequently noncovalent. Complexes are formed which involve electrostatic attractions of charged groups, hydrogen bonding, hydrophobic bonding (page 534), and van der Waals forces. Frequently these interactions are specific, or at least highly selective; the term "lock-and-key" has been used in this context. The shape of the protein molecule, the way in which the polypeptide chain is coiled and folded, is a critical factor in creating a binding site by favorable positioning of functional groups which act in concert for complex formation. Furthermore, complexes which are rather stable under one set of conditions can often be dissociated by rather modest changes in such variables as pH and ionic strength. A second molecule which bonds to the protein of interest is called a *ligand;* which of two binding partners is to be called the ligand is arbitrary, depending simply upon the interests of the researcher. Examples of biospecific interactions are formation of antigen-antibody complexes, combination of enzymes with certain inhibitors, and binding of messenger molecules to cell membranes.

The purification of bovine (cattle) prothrombin provides an interesting example of the power of affinity chromatography.[†] With classical fractionation techniques, very low yields (~15%) are obtained by a very tedious procedure; an improvement in time, labor, and yield was desired. Omitting details that are not essential to our understanding, we summarize the successful procedure. Some bovine prothrombin obtained by the traditional method was injected into a goat. Now proteins are species-specific, and bovine prothrombin is not identical with goat prothrombin (although they are close enough to be functionally the same in blood clotting). The bovine prothrombin is perceived as a foreign invader [antigen or (Ag)] by the goat's immune system, and part of the immune response is generation of an antibody (Ab). The antibody is a protein molecule which circulates in the goat's blood and forms a complex very specifically with bovine prothrombin (Ag-Ab complex). The goat donated a blood sample, and the immunoglobulin fraction of the blood protein was obtained by standard precipitation techniques. This material contained the anti-prothrombin molecules but also, of course, the other antibodies generated during a lifetime of exposure to all sorts of antigens.

More of the prothrombin obtained the hard way was covalently bonded to a polysaccharide gel-filtration material (agarose) by an organic reaction gentle enough not to alter the "active site" where the antibody binds. The goat antibody mixture was dissolved and run onto the column under conditions where the Ag-Ab complex was stable. Eluent flow was started, and the immobilized prothrombin plucked from the mixture the one antibody to which it could bind, while the other Ab molecules were simply washed through the column and discarded. Then the composition of the eluent was changed to dissociate the Ag-Ab complex, and the released pure antibody was washed through the column and collected.

You guessed it! This antibody was covalently attached to a gel matrix, providing a column that removed prothrombin from a mixture of beef blood proteins, after which the pure clotting factor was eluted by changing the eluent composition. The initial supply of prothrombin was small, and only 180 mg of pure antibody was obtained, but by repeated use of the affinity column, as much pure prothrombin as desired was obtained from the beef blood. The yield was 75%, and the product was homogeneous by standard tests for protein purity. The process is shown schematically in Fig. 1.

Protein molecules cannot be roughly treated without losing the conformations required for biospecificity, and the —OH groups of the polysaccharide gel matrices are not reactive enough to permit protein attachment under mild conditions. Thus they are converted into other functionalities which

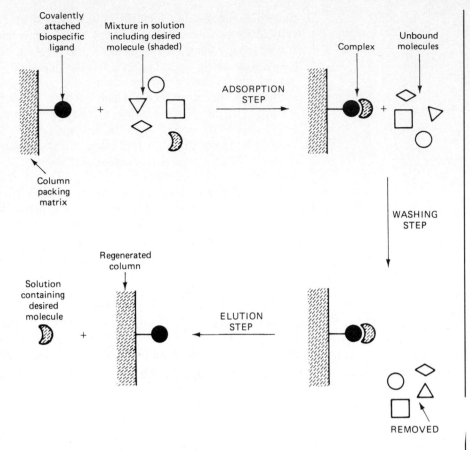

Figure 1 Depiction of the idea of an affinity column separation.

are more reactive, a process called *activation* of the gel. One way of doing this is shown below:

The *coupling* of a protein is then accomplished by a very facile reaction involving a free amino group:

If we are lucky, the covalently attached protein will retain enough structure in the critical region where bioreactivity resides to yield a useful column.

Actually, immobilized proteins are sometimes more stable than is the case in free solution.

Further information may be found in books;[‡] new applications are seen frequently in the biochemical research literature.

[†] R. Wallin and H. Prydz, *FEBS Lett.*, **51**, 191 (1975).

[‡] For example, W. H. Scouten, *Affinity Chromatography*, John Wiley & Sons, Inc., New York, 1981.

KEY TERMS

Adsorption. As used in this chapter, the tendency of molecules or ions in solution to collect on the surface of a solid. A contrasting term is *absorption,* which implies penetration of solute into the solid, but the distinction is often lost because it is difficult to determine whether penetration occurs. Because the term adsorption itself does not imply a mechanism, many writers use it for the uptake of solutes by any second phase in contact with a solution, and it is not incorrect usage to extend the term to ion exchange, bioaffinity interactions, etc.

Affinity chromatography. Sometimes *bioaffinity, biospecific,* or one of several other terms. A separation process utilizing very highly selective or even specific interactions among biological macromolecules. The interactions generally relate more to conformational aspects of structure than to charge, size, solubility, or other properties utilized in more conventional techniques. The ligand which complexes the desired molecule is immobilized on a gel matrix.

Bonded phase. Stationary phase materials with interactive layers which form the basis for the chromatographic process chemically bonded to a support material, as with hydrocarbon layers bonded to glass particles.

Capacity of a resin. The total number of fixed binding sites, usually expressed as milliequivalents per gram of dry resin.

Chelating resins. Resins whose functional groups are chelating ligands which form multiple bonds with complex-forming metal ions; the resins have *much* higher affinities for, say, transition metals than for alkalies such as Na^+, and the binding is highly pH-dependent.

Counterion. The mobile ion in an ion-exchange resin or analogous system which is exchangeable for other ions of the same sign.

Distribution ratio. In ion exchange, a measure of the uptake of an ion by the resin at the expense of a reference counterion with which the resin is initially loaded. See Fig. 18.3.

Eluate. The liquid emerging from an LC column.

Eluent. The mobile phase that causes elution.

Elution. Moving solutes down the column and out the bottom.

Gel. Generally, a three-dimensional molecular network holding a large quantity of solvent in a seemingly semisolid state (e.g., gelatin desserts, agar culture media). In LC, frequently applied to water-swollen size-exclusion materials such as Sephadexes.

Gradient elution. An LC technique in which the composition of the mobile phase is varied continuously during the elution. Sometimes the composition is changed in discrete steps, and it becomes questionable whether to use the term gradient for such a case.

High performance LC (HPLC). Modern LC, similar in efficiency to GLC, in which both high resolution and speed of analysis are obtained by means of technologically innovative stationary phases which permit high mobile phase velocities. Sensitive detectors often allow analysis of small samples on columns which, although much smaller than traditional ones, provide several thousand theoretical plates in only a few centimeters. Data handling is often computerized.

Hydrophobic effect. Tendency of water to expel a nonpolar solute, leading to adsorption of the latter on a nonpolar surface.

Ion-exchange resin. An organic polymer containing covalently attached charged groups which can interact electrostatically with mobile ions of the opposite sign.

Isocratic elution. A chromatographic run in which the mobile phase composition remains constant (cf. gradient elution.)

Normal phase chromatography. A form based upon adsorption on a *polar surface* from a relatively *nonpolar solvent*.

Paper chromatography. Separation of solutes by differential migration through a planar paper medium which is the analog of a column.

Partition chromatography. Same as liquid-liquid partition chromatography. A form of LC in which the differential migration of solutes is based upon differences in relative solubilities in two liquids, one stationary and the other mobile.

Reversed-phase chromatography. A form employing relatively *nonpolar stationary phases* and more *polar solvents*, as contrasted with normal phase.

Selectivity coefficient. In ion exchange, a measure of how effectively an ion competes with a second ion for resin binding. See page 538 for basis of definition.

Size-exclusion chromatography. An LC separation technique based upon differential penetration of a retarding stationary phase by solute molecules of varying size.

Strong-acid cation exchanger. A resin with fixed anionic sites, usually $-SO_3^-$, which are not protonated even by contact with solutions of low pH; analogously, a *strong base anion exchanger* contains groups, generally quaternary ammonium, which are cationic even at high pH.

Thin-layer chromatography. Similar to paper chromatography except that the stationary phase is an adsorbent or other interactive material coated on a base of glass or plastic.

Weak-acid cation exchanger. A resin with weakly dissociated functional groups, often $-COOH$, which are anionic ($-COO^-$) only if the solution with which they are in contact has a high enough pH: analogously, a weak base anion exchanger has fixed groups, e.g., $-NH_2$, which provide cationic sites only if the pH is low enough to form $-NH_3^+$.

QUESTIONS

1. Aspirin (2-acetoxybenzoic acid) is adsorbed near the top of a column packed with alumina, Al_2O_3. Briefly explain which of the following five solvents will elute the aspirin most rapidly, everything else being the same: hexane, benzene, chloroform, diethyl ether, ethanol.

2. A six-component mixture of steroids is obtained by extracting a biological sample with CCl_4 and subjected in an endocrinology laboratory to two-dimensional thin-layer chromatography. A portion of the extract is applied near one corner of a square thin-layer plate, and solvent A moves the components in one direction [(a) below]. Then the plate is dried, turned at right angles, and a different solvent B is allowed to flow perpendicular to the direction of the first solvent [(b) below]. The black dot represents the original sample starting spot, and the arrows depict the directions of solvent flow and solute migration.

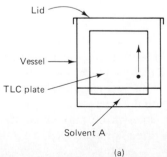

(a)

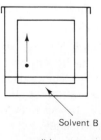

(b)

Finally, the plate is sprayed with a reagent that yields a red color when it reacts with steroids, and the researcher sees six reddish spots on the plate. If R_f-values of these six steroids have previously been measured using pure, individual steroid samples, then the analyst can tell which spot on the plate represents which steroid. The R_f-values are given in the table below, and the final chromatogram is also given. Decide which spot goes with each of the numbered steroids.

| | R_f-value | |
Steroid	Solvent A	Solvent B
1	0.0	0.5
2	0.8	0.0
3	0.5	0.5
4	0.8	0.8
5	0.2	0.4
6	0.8	0.5

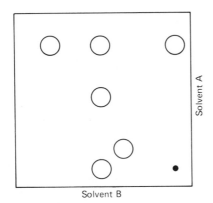

Solvent A

Solvent B

3. In what sort of situation might a gradient elution be useful in LC? What is accomplished with a solvent gradient that would not be accomplished without it? What common experimental approach in GLC leads to an analogous result? Name two organic solvents which, if appropriately blended, might yield a useful gradient for separating a mixture of organic compounds of differing polarities on an adsorption column.

4. Select from the list of solutions the best one for each of the following problems.

Problems

Separate 0.2 mg of Cu^{2+} from 1 g of Na^+.

Separate 1 mg of ribonuclease (MW 13,700) from a few milligrams of another protein (MW 150,000).

Remove Na^+, K^+, Ca^{2+}, Mg^{2+}, Cl^-, SO_4^{2-}, and NO_3^- from tap water.

Separate an antibody from immunologically inactive proteins of about the same molecular weight.

Decide rather quickly which adsorbent and which solvent system to use for a large-scale chromatographic separation of an organic reaction product from materials formed in side reactions.

Separate ten organic compounds which exhibit a wide range of polarities.

Estimate the molecular weight of a protein.

Replace phosphate with Cl^- prior to measuring Fe^{3+} by a method in which phosphate interferes.

Solutions

Affinity chromatography	Thin-layer chromatography
Adsorption chromatography with gradient elution	Paper chromatography
Gel filtration with the proper Sephadex	Determine retention volume with an appropriate Sephadex column calibrated with known proteins.
Ion exchange with a chelating resin	Ion exchange with a strong-acid cation-exchange resin
Ion exchange with a mixed bed of cation- and anion-exchange resins	Ion exchange with a strong-base anion exchanger containing chloride counter-ions

Multiple-choice: In the following multiple-choice questions, select the *one best* answer.

5. A neutral molecule such as ethanol or sugar which has found its way into the pores of a typical anion-exchange resin can be eliminated (a) only by replacement with a cation; (b) only by replacement with an anion; (c) only if replaced by another organic molecule on a one-for-one exchange basis; (d) by flushing out with water.

6. In chromatography, a substance for which the distribution coefficient, K, is zero may be used to estimate (a) the volume within the

column available to the moving phase; (b) the volume within the column occupied by the packing; (c) the volume within the pores of the packing material; (d) the total volume of the column.

7. Which of the following statements is false in normal phase adsorption? (a) The more polar a compound, the more strongly it will be adsorbed from a solution. (b) A high molecular weight favors adsorption, other factors being equal. (c) The more polar the solvent, the stronger the adsorption of the solute. (d) The adsorption isotherm is usually nonlinear.

8. The best measure of the quantity of a solute in LC is (a) the height of the elution band; (b) the area of the elution band; (c) the baseline width of the elution band; (d) the retention volume.

9. Which of the following would be the fastest way to decide which adsorbent and what solvent system to use for a large-scale chromatographic separation of an organic reaction product from materials formed in side reactions? (a) Paper chromatography; (b) affinity chromatography; (c) TLC; (d) adsorption chromatography with gradient elution.

10. To deionize tap water by ion exchange for laboratory use, the best approach employs (a) a column containing a strong-acid cation exchanger in the hydrogen form; (b) a column containing a strong-base anion exchanger in the hydroxyl form; (c) a mixed-bed column containing a strong-acid cation exchanger in the sodium form and a strong-base anion exchanger in the chloride form; (d) a mixed-bed column containing a strong-acid cation exchanger in the hydrogen form and a strong-base anion exchanger in the hydroxyl form.

PROBLEMS

1. *Resin capacity.* Typical ion-exchange resins hold water avidly. Resins that are air-dried by equilibrating thin layers of material with the laboratory air contain varying quantities of water depending upon the humidity. We would like to compare results from one laboratory (or one climate) with results from another source, but uncertain moisture content would then be a problem. If resins are *thoroughly* dehydrated, there is always a suspicion that the required heating may have modified the organic polymer. The following procedure is usually used: Air-dry a resin batch in your laboratory and bottle it. Determine the water content on a sample by a procedure that will drive off most of the moisture without altering the resin very much, but discard the dried product just in case. Do ion-exchange studies with the rest of the batch and report results on a dry-weight basis.

A sample weighing 1.6987 g of air-dried anion-exchange resin in the bromide form was heated in vacuo over P_2O_5 at 70°C for 17 h. The weight loss was 0.2331 g. (a) Calculate the percentage of moisture in the resin batch. (b) A portion of the air-dried batch weighing 0.8576 g was washed repeatedly with portions of nitric acid to remove the Br^- by replacement with NO_3^-. The collected washings were titrated with 0.2000 M $AgNO_3$ solution, requiring 11.84 mL. Calculate the capacity of the anion-exchange resin in milliequivalents per gram of dry resin.

2. *Distribution ratio.* A 20.00-mL solution containing a radioactive monovalent cation was shaken for 48 h with .500 mg (dry weight) of a cation-exchange resin in the sodium form. Before contact with the resin, the solution had radioactivity expressed as 80,000 cpm/mL. After equilibration with the resin, the activity was 65,200 cpm/mL. Calculate the distribution ratio for the cation with dry resin in the sodium form. The half-life of the radioactive isotope is *very much* longer than 48 h.

3. *Distribution ratio.* An air-dried anion-exchange resin in the bromide form contains 12.00% water. The weight 1.1351 g of this material is shaken for 72 h with 10.00 mL of a 1.000×10^{-3} M solution of sodium salicylate (sodium o-hydroxybenzoate). After equilibration, the pBr of the solution is measured potentiometrically, using a bromide ion-selective electrode, and found to be 3.039. Calcu-

late the distribution ratio for salicylate ion (Sal⁻) with dry resin in the bromide form.

4. *Solution standardization by ion exchange.* A chemist has a pure solution of calcium chloride ($CaCl_2$). He wants to use this as a source of known quantities of Ca^{2+} for preparing standards for flame emission analysis. But $CaCl_2$ is hygroscopic, and when he weighed it he did not know how much water he was weighing, so he needs to standardize the solution. He pipets exactly 10 mL of this solution onto a column of adequate capacity packed with a strong-acid cation-exchange resin in the hydrogen form and then passes good deionized water through the column to flush out unbound ions. The effluent, including all washings, is titrated with 0.1209 M NaOH, using phenolphthalein indicator; this requires 23.98 mL. Calculate the molar concentration of $CaCl_2$ in the original solution.

5. *Analysis for total cations by ion exchange.* A sample of water from the Caribbean Sea was brought into an oceanographic laboratory. A 10.00-mL portion was placed on a column packed with an ample quantity of Dowex 50, a strong-acid cation exchanger, in the hydrogen form. Some deionized water was then passed through the column to flush out any unbound ions. The column effluent, including all washings, was titrated with 0.1282 M NaOH; 48.56 mL was required. Calculate the total milliequivalents of cations (Na^+, Mg^{2+}, etc.) per liter of seawater.

6. *Elution volumes in gel filtration.* A gel-filtration column is prepared using a Sephadex gel packing which excludes molecules whose molecular weights are larger than about 500,000. The volume within the gel particles which is available for accepting solutes is 210 mL. The interstitial or void volume (V_0) is 65 mL. (a) What is the expected retention volume, V_R, for a protein of molecular weight 750,000? (b) What is the expected retention volume for a small protein such as ribonuclease (MW 13,700) which can freely penetrate all the pores of the gel phase? (c) Comment on the behavior of a compound of molecular weight 15,000 which elutes with a retention volume of 442 mL.

19

Perspectives

In the first analytical course, students are guided through a series of topics and laboratory exercises selected for beginners, and there is obviously much of modern analytical chemistry that is not presented. The purpose of this closing chapter is to compensate in some measure—to broaden your perception of the field. The boxes, which provide brief "snapshots" of topics that cannot be covered in depth in an elementary textbook, will illustrate some of the points to be made.

The most significant recent trend in science has been the erosion of traditional boundaries, and, along with other fields, analytical chemistry has prospered from the breakdown of the barriers separating disciplines. In this century, there have been three major contributions to the present vitality of analytical chemistry. First came the flow of theory from physical chemistry into analytical chemistry, transforming a highly empirical field into a much more rational one which attracted larger numbers of bright people. Since perhaps 1940 or 1950, most good analytical chemistry has been in great measure applied physical chemistry. This process has also facilitated the application of developments in basic physics to analytical problems. Once a barrier comes down and a few people become familiar with a new idea or technique, they develop it for their own ends and the new methodology spreads throughout their field. There is no better example of this than mass spectrometry (Box 19.1). Chemists took this technique from physics and carried it far beyond anything physicists had envisioned.

Box
19.1.

Physics into Chemistry: The Mass Spectrometer, or Weighing a Molecule and Smashing It to Learn What It Is

The first mass spectrometers were developed by the Cambridge physicists J. J. Thompson and F. W. Aston in the early 1900s for separating isotopes of the elements and determining their masses. After refinement, the instrument attracted the attention of a few chemists during the 1950s; perhaps 10 years later it had become a useful (if not quite routine) tool in chemistry, and today it is a widely used analytical instrument. There is no dominant form; a variety of components and conformations meet various needs.

In its simplest version, still frequently employed, a volatile compound is vaporized into a chamber where the molecules are bombarded by electrons emitted by a hot filament (Fig. 1). The acceleration of these electrons in an electric field determines their energies. Certain collisions between a sample molecule and an electron may simply knock another electron out of the molecule to form a positive ion of (virtually) the same mass, the so-called molecular or "parent" ion.

$$\text{Molecule} + e \longrightarrow (\text{parent ion})^{+} + 2\,e$$

Other encounters, depending upon the energy of the electrons, lead to fragmentation of the molecule into smaller groups, some of which may engage in further reactions. Positive ions are directed through electric and/or magnetic fields in the mass analyzer, and the conditions under which the detector perceives the arrival of a particular ion define a trajectory which in turn depends upon the mass to charge ratio (m/e) of that ion.

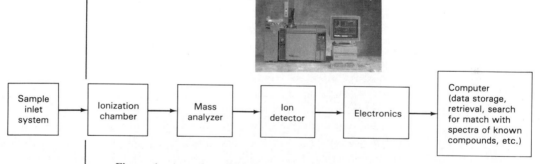

Figure 1 A modern GC-MS instrument and block diagram. (©1990, Hewlett-Packard Co. Reproduced with permission. Auxiliary components such as vacuum systems are not shown.)

The first chemists who so bombarded organic molecules were considered eccentric; their colleagues predicted an uninterpretable complexity of fragment formation and much brown goo on the inner walls of the instrument. Not so. It is apparent that fragmentation patterns (and sometimes even reaction products of the "hot" intermediate species) make sense, and that the mass spectra (ion beam intensity at the detector vs. m/e) can be used to deduce molecular structure. The parent peak gives the molecular weight.

Interfaces have been developed to link the mass spectrometer to GLC columns as a detector, providing one of the first so-called hyphenated techniques, GC-MS; this detector not only perceives the arrival of a sample component eluted from the column but provides structural information as well. The pressure in the ionization chamber must be low enough (perhaps on the order of 10^{-6} torr) to prevent reactions of the desired ions with other species. The interface for GC-MS takes advantage of the rapid diffusion of the light carrier gas helium by pumping most of it off through a leak system that permits most of the slower sample molecules to enter the ionization chamber. Interfaces are also being developed for HPLC-MS.

Although the conventional electron source is still widely used, ions can now be generated by other modes of bombardment which may be more useful for studying compounds of higher molecular weight (e.g., fast atom bombardment or FAB). Probes are available by which even solids can be introduced into the instrument.

The modern mass spectrometer is not cheap. The least expensive GC-MS instrument is probably $50,000 or so (most of which represents the MS detector); $150,000 in 1989 dollars will buy average capability. The most expensive MS setup probably costs about $1.2 million. This is a tandem instrument, sometimes referred to as MS-MS, in which ions selected by the first unit can be allowed to react with other molecules; the mass spectra of the reaction products are then obtained in the second unit, yielding information about ion-molecule reactions, currently of great interest in chemistry. If the laboratory work load does not justify purchase of MS equipment, a mass spectrum can be obtained commercially for about $150.

A second barrier was breached by analytical chemists who taught themselves electronics or learned it during military service in World War II. This meant that instrumentation for new, faster, or more sensitive measurements on chemical systems could be assembled before instrument manufacturers were assured of profitable markets; the success of the measurements in turn created such markets.

The third major development has made computers commonplace in the analytical laboratory for both data processing and automated control of instruments.

Such interactions among several disciplines have resulted in a virtual explosion in regard to the kinds of samples that can be analyzed and the sensitivity of analyte detection, as well as a great decrease in the time and human labor required for each determination. Enormously enhanced analytical capability has coincided with society's need for increasingly difficult studies in areas where methodology was formerly unavailable. In simpler times, analytical chemists were concerned largely with determinations of the major components of a few ores and minerals and with quality control of simple chemical products, steels and other alloys, and the like. No more. Sophisticated studies of air and water quality, contamination of the food chain by the complex wastes of a modern society, effects of trace elements on the properties of solid-state devices—these and countless more require analytical methods that have been developed only recently. Table 19.1 gives some examples of analytical problems addressed in 1989 in two analytical research journals; the entries are only illustrative—hundreds of others could have been selected.

Another way to see the importance of analytical chemistry is to examine the origins of analytical research publications. Table 19.2 gives some examples. It should be noted that the sampled journal does not publish routine applications or trivial modifications of standard methods; all of the papers are expected to report research and to display originality or imagination. We have purposely omitted organizations whose main goal is to develop analytical methods in order to empha-

TABLE 19.1 Some Recently Reported Determinations

SAMPLE	ANALYTE(S)	METHOD	DETECTION LIMIT	SIGNIFICANCE
Blood	selenium	HPLC with fluorescence detection	0.15 ng	Se is an essential trace element in the body, but higher levels are toxic
High-purity O_2	Ar, N_2, Kr, CH_4, Xe	GC with photoionization detection	0.01–0.4 ppm	Very pure O_2 is used in semiconductor manufacturing
Sediment	Tributyltin (TBT)	Mass spectrometry	0.2 μg Sn per gram of sample	TBT is toxic to shellfish
Aqueous solutions	Amphetamine as amphetaminium cation	Potentiometry with ion-selective electrode	3 ppm	Amphetamine is a potent stimulant in the central nervous system
Natural waters and biological tissues	Lead	Atomic absorption spectrophotometry	1 pg/ml in waters	Pb is a toxic environmental contaminant
Urine	Polyamines such as putrescine, cadaverine, and spermidine	HPLC with chemiluminescent detection based on enzymic oxidation releasing H_2O_2	5 pmol (putrescine)	Polyamine levels are elevated in certain cancer patients
Muscle, liver, and kidney of slaughtered animals	Nitroxynil	HPLC with mass spectrometry detection	2 ng/g tissue	Nitroxynil is used in veterinary medicine to control liver flukes; permissible levels in meat products must not be exceeded
Blood serum	Interleukin-2 (IL-2)	Immunoassay	100 pg/ml	Il-2 is a growth factor for certain cells in the immune system; it is undergoing evaluation for treatment of certain AIDS and cancer victims
Vegetables and herbage	Arsenic and selenium	X-ray fluorescence spectroscopy	0.1 μg	Ash from power plants used as land fill contains As and Se which may appear in the food chain
Urine	Methadone	Voltammetry	0.3 μg/ml	Methadone is a narcotic analgesic used in treating heroin and morphine addicts

TABLE 19.2 Origins of Some of the Research Papers Published in the Journal *Analytical Chemistry* in August, September, and October 1989

Department of Physiology and Biophysics, New York University Medical Center

Health and Environmental Chemistry, Los Alamos National Laboratory

Department of Earth Sciences, University of California, Riverside, and Earth Sciences Department, Lawrence Livermore National Laboratory

Division of Geological and Planetary Sciences, California Institute of Technology

Department of Laboratory Medicine, Faculty of Medicine, University of Tokyo, and the Medical Equipment Technical and Engineering Division, JEOL, Ltd., Tokyo

The Research Institute, King Fahd University of Petroleum and Minerals, Dhahran, Saudi Arabia

Tokyo College of Pharmacy and Gasukuro Kogyo, Inc., Saitama, Japan

Laboratory for Technical Development, National Heart, Lung, and Blood Institute, National Institutes of Health

Institüt für Klinische Chemie und Laboratoriumsmedizin, Katharinenhospital, Stuttgart, FRG

U.S. Environmental Protection Agency, National Enforcement Investigations Center, Denver

Department of Biochemistry, Purdue University

Farmaceutisch Instituut, Vrije Universiteit Brussel, Brussels

Chemistry and Physics Division, National Institute for Environmental Studies, Ibaraki, Japan

Federal Institute for Geosciences and National Resources, Hanover, FRG

Department of Agricultural Chemistry and the Environmental Health Sciences Center, Oregon State University

Academy of Mining and Metallurgy, Institute of Material Science, Cracow, Poland

Department of Polymer Chemistry, Tokyo Institute of Technology

Physical and Analytical Chemistry Research, The Upjohn Co.

Institute of Statistics and Decision Sciences, Department of Mathematics, and Department of Chemistry, Duke University

Department of Chemistry and Biochemistry and Cooperative Institute for Research in Environmental Sciences, University of Colorado

National Institute of Standards and Technology, Gaithersburg

Chemical and Laser Science Division and Material Science and Technology Division, Los Alamos National Laboratory

Division of Contaminants Chemistry, Food and Drug Administration

Department of Chemical Engineering, Yale University

Center for Drug Design and Delivery, College of Pharmacy, University of Florida

---

size the importance of the methods in other fields such as medicine, environmental science, geology, and agriculture.

The breakdown of barriers obviously does not mean that all scientists now have exactly the same interests. There is indeed an identifiable group of people who call themselves analytical chemists. It does mean that you might not be able to identify them simply by looking at their laboratories. Even publication practices may blur distinctions: analytical chemists, when appropriate, perform studies that may appear in the journals of other fields. Analytical chemistry is difficult to define, then, by looking only at what researchers are doing. Any reasonable definition must be based on the vague grounds of viewpoints, goals, and what people find interesting. For example, an implantable sensor to monitor continuously the glucose concentration in the blood plasma of a diabetic patient (and also control the release of insulin by a miniature pump) might represent a vast improvement over the wide swings in sugar level associated with insulin injections by timetable. The development of such a system is a project an analytical chemist might enjoy; the end use, on the other hand, would be rewarding to a physician who would rather work with patients than in methods development. The two might collaborate, then, and the results might appear in the medical literature.

In the pecking order of science, we generally think of a flow of information in the sequence physics → chemistry → biology. Yet one of the most sensitive analytical techniques came from biology into analytical chemistry. When barriers come down, anything can happen. See Box 19.2.

| Box 19.2. | **Biology into Chemistry: Radioimmunoassay** |

Medical researchers needed an analytical method for determining insulin concentrations in blood plasma from normal people and patients with various forms of diabetes. Normal concentrations of this small protein hormone (MW 5700) are in the neighborhood of 50 ng/mL (1 ng $= 10^{-9}$ g), and there was no way to detect this molecule in a matrix containing, by comparison, massive levels of other constituents including the plasma proteins. The solution[†] led to the Nobel Prize for Rosalyn Yalow in 1977. Interestingly, Yalow began her career with a Ph.D. in physics; much of her work cut across disciplinary lines [applications of radioisotopes in medicine, endocrinology, and finally radioimmunoassay (RIA)]. The technique of radioimmunoassay is now routinely used for determining a large number of hormones, drugs, vitamins, and other compounds at exceedingly low concentrations. The outstanding attributes of the method are *sensitivity* (nanogram and even 10^{-12} or picogram levels), due to the inherent sensitivity of radioactivity measurements, and *specificity* (or at least very high selectivity), based upon the high degree of selectivity of immunological reactions.

Proteins are generally species-specific, and the injection of a protein from one kind of animal into another species elicits an immune response. The recipient's immune system perceives the foreign protein or antigen (Ag) as "nonself," and one part of its overall immune response is the production of antibodies. An antibody (Ab) is a protein molecule which circulates in the blood and which possesses the property of forming, very specifically, a complex with the antigen that triggered its formation.

Small molecules, although they may possess biological activity as drugs, poisons, etc., do not elicit immune responses. However, if a small molecule is covalently linked to a protein (bovine serum albumin is a favorite), then antibodies that react very selectively with the small molecule may be produced. The small molecule is called a *hapten*.

Immunologists provide protocols that optimize antibody yields by attention to dosage and timing of antigen injection, use of adjuvants to stimulate the immune response, and bleeding schedules. When necessary, the desired antibody can sometimes be separated by affinity chromatography (see Box 18.2) from all the others the animal carries from a lifetime of exposure to a host of antigens.

Also required for the assay is a sample of the antigen or hapten analyte that has been labelled with a radioactive isotope. The nature of the label depends upon the analyte, availability of a suitable radioisotope in regard to half-life and cost, and the complexity of the chemistry required to insert the label. With insulin, it is relatively easy to introduce an iodine isotope (e.g., ^{125}I, $t_{1/2} = 60$ days) by iodination of aromatic amino acid side chains under conditions mild enough not to destroy the ability of the hormone to combine with its antibody. For hapten analytes (e.g., drugs, steroids, vitamins), the label is often tritium (^{3}H, $t_{1/2} = 12$ years) or ^{14}C ($t_{1/2} = 5700$ years). For many common assays, kits are commercially available containing antibody preparation, radiolabelled analyte, and analyte standards for calibration. Because few in the testing game want to obtain their own antibodies from in-house animal colonies or synthesize labelled organic compounds, commercial materials are attractive, although analytical chemists are instinctively uneasy about the possibility of blind use in the absence of very rigorous quality assurance programs.

The principle of RIA is simple: the analyte in either a standard solution for calibration or in the unknown sample is allowed to compete with a measured quantity of radiolabelled analyte for complexing with a limited quantity of antibody. and the more analyte in the sample, the less radioactivity in the Ag-Ab (or hapten-Ab) complex. The working curve obtained with standards is generally nonlinear but nonetheless useful, as shown in Fig. 1.

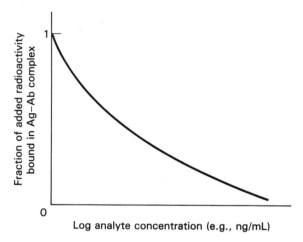

Figure 1 Calibration curve for a typical radioimmunoassay.

There is one other aspect of the story. A radioactive isotope decays independently of its *chemical* state. The radioactivity measurement cannot distinguish complexed from uncomplexed radioactive material, and this means that the Ag-Ab complex must be separated from unbound Ag before the counting step. There are various ways to accomplish this, including several forms of chromatography and electrophoresis (migration of charged macromolecules in an electric field). The Ag-Ab complex is usually stable enough not to dissociate appreciably as it moves away from excess Ag.

[†] R. S. Yalow and S. A. Berson, *Nature,* **184,** Suppl. No. 21, 1648 (1959).

A major problem for people who use analytical results for their own ends, people who do not intrinsically enjoy developing analytical methods, is that there are too few analytical chemists. Well over half of the research papers in analytical journals are authored by researchers whose main interests lie elsewhere. (This clearly shows why analytical chemistry should be a required chemistry course.) People in other fields are perfectly capable of evaluating the analytical methods they use; there is no priesthood with a lock on the expertise. But if the work of analysis is a necessary evil and the real goal is not obtaining the results but rather using them in some other field, then human impatience may intrude and the analyses may be of lower quality than need be. What should a scientist worry about when analytical results are obtained, either directly or from someone else?

The response to this question is obviously: What is the quality of the data? Are they good enough to use as a basis for action? Shall we fine a company for its toxic emissions on the basis of these air and water analyses? Shall we purchase tons of this material because the laboratory results suggest that it is suitable for the intended use? Are we confident enough of these results from our clinical laboratory to risk a treatment that sometimes has dangerous side effects? Shall we ban this food shipment from the marketplace because of what a laboratory says about the level of a pesticide or a hormone? Should this pilot be removed from the flight deck because a testing lab has found a drug metabolite in a urine sample? Are we sure enough about the drug test to rescind an Olympic win? Clearly, many people have a stake in the quality of analytical data.

Possible consequences of these decisions are not analytical chemistry questions. They lie in such realms as economics, medicine, and ethics. Where does the burden of proof lie with regard to a toxic substance? Do we expose people to it if it is not known to be harmful, or do we ban it if it is not known to be safe? This is not an analytical chemistry question, nor is the analysis of risk-benefit tradeoffs in public policy our concern. But analytical chemistry finds itself at the center of such controversies in any case, the more so as improved methodology lowers limits of detection. The question the analyst must be able to answer—the question that goes to the heart of professional competence and integrity—is: What is the quality of the data? Any laboratory that performs analyses providing a basis for action by others must have a solid quality assurance program.

Such a program cannot be described in detail here. Some aspects (e.g, statistical control and the use of control charts) were discussed in Chapter 2. Fortunately there is an excellent book that may be consulted for both the general philosophy and practical aspects of quality assurance.[1] We mention briefly some of the important considerations.

First come end use requirements. How good do the data need to be? There is no point in spending money determining an element to the nearest 0.1% if $\pm 10\%$ is good enough. What is high quality in one context might be unacceptable in other circumstances. Once we decide how good the data need to be, then potential problems can be addressed. There are two major aspects, which relate to *precision* and *accuracy* as defined (and carefully distinguished) in Chapter 2.

1. *Statistical control.* This means that the analytical results should be statistically stable, which is to say that the mean and standard deviation of replicate results on the same sample (perhaps an SRM, Box 19.3) should not vary greatly from day to day. Central tendency and scatter of replicate measurements should not depend upon the phase of the moon or whether the analyses were done on your grandmother's birthday. This does not speak to accuracy, but it is important because it is silly to ask about the accuracy of a result that cannot be reproduced anyway.
2. *Bias.* The analytical results should not be systematically high or low (see determinate errors, Chapter 2) unless a known, reliable, correction factor can be applied. The use of reference materials where known analyte levels are found in appropriate matrices is obviously important here.

A quality assurance program to convince end users of a laboratory's ability to generate accurate results over and over again has many aspects and entails costs. There will be a person with high professional standards who understands

[1] J. K. Taylor, *Quality Assurance of Chemical Measurements,* Lewis Publishers, Inc., Chelsea, Mich., 1987.

the chemistry of the methods and the pitfalls of the measurement process. Instruments will be calibrated, maintained, and repaired when necessary. Operators will be well-trained, motivated, and tactfully monitored. There will be an iron-clad system to prevent misidentification of samples, and records will be scrupu-

---

Box 19.3.   Standard Reference Materials

For over 80 years, NIST, the National Institute of Standards and Technology [until 1988 called the National Bureau of Standards (NBS)], has sold standard reference materials (SRMs). Some of these are compounds of high purity for use as primary standards, buffers for calibrating pH meters, etc., but many are samples containing constituents (sometimes major and in other cases minor or trace) whose levels are certified on the basis of analyses by carefully evaluated methods, often two or more. In some cases, the values are obtained (as the NIST catalog puts it) from "a network of cooperating laboratories, technically competent and thoroughly knowledgeable with the material being tested." The catalog of, say, 50 years ago contained a small offering of ores, minerals, glasses, ferrous and nonferrous alloys, and the like, with mostly major constituents certified. The 1988–89 catalog lists over 900 SRMs, states that about 100 new ones are in preparation, and mentions a backlog of many more that customers would like. The table below lists a few of the available SRMs.

Some Examples of Standard Reference Materials Available from NIST

Type of sample	Some of the analytes certified	Levels[†]
Stainless steel	C	0.310 wt%
	Mn	0.330 wt%
	Ni	58.1 wt%
Nickel-base alloy	Cr	19.3 wt%
	Mn	0.019 wt%
High-purity gold wire	Cu	0.1 ppm by wt
Nonfat milk power	Cu	0.70 μg/g
	I	3.38 μg/g
	Se	0.11 μg/g
Rice flour	As	0.41 μg/g
	Zn	19.4 μg/g
Tomato leaves	As	0.27 μg/g
	Cr	4.5 μg/g
	U	0.061 μg/g
	Zn	62 μg/g
Simulated rainwater	NO_3^-	0.501 mg/L
	SO_4^{2-}	0.205
Residual fuel oil	S	0.719 wt%
Coal (bituminous)	S	1.85 wt%
Methanol	2,4,6-Trichlorophenol	20.4 μg/mL
Freeze-dried urine	Pb	1.109 μg/mL

[†] These "nominal values" are guides for selecting samples. The certified values, not far from these, as supplied with the samples are to be used.

Glancing at the table, note the concern for matrix effects. For instance, the last entry is a lead standard, but it is not simply a dilute solution of a Pb^{2+} salt. The constituents normally found in urine at macro levels (constituents that might affect the volatilization of lead in atomic absorption spectrophotometry, for example, or perhaps the diffusion coefficient of Pb^{2+} in an electroanalytical procedure) are present. If a laboratory runs this SRM and gets the right answer (or one close enough), people will be reassured regarding the analytical results in a study of children's exposure to lead in the environment. It will be especially reassuring if the experiment is blind, i.e., if the laboratory personnel do not know they are being tested, if they think the SRM is just another sample, and if they still get it right. Whether the children are at risk is up to a toxicologist, but what can he say about it without a quality assurance program in the analytical laboratory?

lously kept. Samples with known analyte levels will be submitted frequently, indistinguishable from unknowns by laboratory personnel.

Sometimes quality assurance programs are mandated by outside agencies such as the Public Health Service, the Food and Drug Administration, and the Environmental Protection Agency, which have oversight authority in certain situations. Laboratories that perform drug testing are under increasing quality control pressure because of the possibly serious consequences of mistakes. Reports of simple positive or negative findings used to suffice, but the courts are beginning to require quantitative data. Imagine a laboratory director with no quality assurance program squirming in the witness box under the cross-examination of a good defense lawyer.

We recommend that readers look at a paper entitled "Analytical Chemistry at the Games of the XXIIIrd Olympiad in Los Angeles, 1984."[2] This paper beautifully illustrates many of the points made in this chapter. In 15 days, nearly 10,000 analyses for more than 200 drugs and drug metabolites were performed on 1510 urine specimens furnished by Olympic competitors. Techniques included GLC, HPLC, GC-MS, and RIA. The paper emphasizes the quality assurance program, including standard samples, careful calibration, and record keeping, and, at least by implication, highlights the ideals to which analysts should aspire when the stakes are high.

Probably few users of this textbook will become professional analytical chemists, but none will be unaffected by the analytical results required by a complex technological society.

[2] D. H. Catlin et al., *Clin. Chem.*, **33**, 319 (1987).

20

General Laboratory Directions

In determining the constitution of an unknown sample, a sound method must first be selected. It then remains a matter of technique alone whether the ultimate measurement is performed upon *all* or part of the desired constituent. The heart of *quantitative technique* is simply to carry a sample through a number of manipulations without accidental losses and without introducing foreign material. Since every conceivable fortuity cannot be anticipated in writing a text, the student must develop independent judgment in connection with his or her laboratory work. Common sense plus awareness of the danger spots are the main requirements of the beginning student in this regard.

20.1

NEATNESS AND CLEANLINESS

Good analysts are usually scrupulously neat. The student with an orderly laboratory bench is not likely to mix up samples, add the wrong reagents, or spill solutions and break glassware. Neatness in the laboratory must extend, of course, from the student's own bench to the shelves where materials are available for the whole class. Neatness also includes stewardship over the more permanent laboratory fixtures, such as ovens, hot plates, hoods, sinks, and the benches themselves.

No analysis should ever be performed using anything but clean glassware. Glassware that looks clean may or may not be clean as an analyst understands the term. Surfaces on which no visible dirt appears are often still contaminated by a thin, invisible film of greasy material. When water is delivered from a vessel so contaminated, it does not drain uniformly from the glass surface, but leaves behind isolated drops that are troublesome or sometimes impossible to recover. Glassware into which a brush can be inserted, such as beakers or Erlenmeyer flasks, are best cleaned with soap or synthetic detergent. Pipets, burets, and volumetric flasks may require hot detergent solutions for thorough cleaning. If the glass surface still does not drain uniformly, it may be necessary to use *cleaning solution*,[1] whose strong oxidizing properties ensure clean glass surfaces in most cases. After cleaning, apparatus should be rinsed several times with tap water, then with small portions of distilled water, and finally allowed to drain. Cleaning solution is usually avoided in biological work because many microorganisms are sensitive to traces of chromium which remain on the glass even after thorough rinsing.

20.2

REAGENTS

The reagents found in the analytical laboratory can be roughly classified as follows, although some of the designations have not been precisely defined:

1. *Technical* (or *commercial*)-*grade* chemicals are used mainly in industrial processes on a large scale and are seldom employed in the analytical laboratory except for such purposes as the preparation of cleaning solution.
2. *U.S.P.* reagents meet purity standards that can be found in the *United States Pharmacopoeia*. These standards were established primarily for the guidance of pharmacists and the medical profession, and in many cases, impurities are freely tolerated that are not incompatible with the use to which these persons will put the compounds. Thus U.S.P. reagents are not usually suitable for analytical chemistry.
3. *C.P.* reagents are often much more pure than U.S.P reagents. On the other hand, the designation C.P. (standing for "chemically pure") has no definite meaning; standards of purity for reagents of this class have not been established. Thus C.P. reagents may often be used for analytical purposes, but there are many situations where they are not sufficiently pure. In many analyses it is necessary to test these reagents for certain impurities before they may be used.
4. *Reagent-grade* chemicals conform to the specifications established by the Committee on Analytical Reagents of the American Chemical Society.[2] The

---

[1] Many different formulas are available for the preparation of cleaning solution, and the choice seems largely personal. A satisfactory solution can be made as follows: In a 600-ml beaker place 20 to 25 g of technical-grade sodium dichromate and 15 ml of water. Then add slowly and carefully about 450 ml of technical-grade sulfuric acid with occasional stirring. Cool the solution and store it in a glass-stoppered bottle. The appearance of the green chromium(III) ion indicates that the solution is exhausted. Cleaning solution is very corrosive; take special precautions not to get it on your skin or clothing. If any is spilled on the skin, rinse the affected parts quickly and thoroughly with tap water. Cleaning solution is most effective when heated to about 60 to 70°C. Consult your instructor for directions before using this solution. The solution should not be discarded after use. It is returned to the storage bottle until the color turns green as mentioned above.

[2] *Reagent Chemicals, American Chemical Society Specifications*, 5th ed., Washington, D.C., 1974.

label on such a reagent bears a statement such as "Meets A.C.S. specifications," and generally furnishes information regarding the actual percentages of various impurities or at least the maximal limits of impurities. Many reagent-grade chemicals are used as primary standards in the introductory laboratory. A label from a bottle of potassium acid phthalate (potassium biphthalate) is shown in Table 20.1. Note that the purity is 99.95 to 100.05%, well within the usual requirements for a primary standard.

TABLE 20.1 Label of a Reagent-Grade Chemical

Potassium Biphthalate, Crystal, Primary Standard	
Meets A.C.S. Specifications	
Assay	99.95–100.05%
Insoluble matter	0.005%
pH of a 0.05 M solution at 25°C	4.00
Chlorine compounds (as Cl)	0.003%
Sulfur compounds (as S)	0.002%
Heavy metals (as Pb)	0.0005%
Iron (Fe)	0.0005%
Sodium (Na)	0.0005%

Substances of even higher degrees of purity are available for research and work in such areas as solid-state physics. These ultrapure materials may be obtained from the National Institute of Standards and Technology (NIST) and various commercial supply houses.

Certain acids and bases are provided in the laboratory as concentrated solutions. Table 20.2 gives the composition of most of the common reagents. Dilute solutions are prepared as needed by adding the concentrated solution to water. The degree of dilution is often indicated by the ratio of the volume of concentrated solution to that of water. For example, 1 : 4 nitric acid means that 1 volume of nitric acid is added to 4 volumes of water.

The chances of contamination increase enormously when a reagent bottle is placed in the laboratory for the use of a large number of people. Thus it is most important that students carefully adhere to certain rules governing the use of the reagent shelf. In addition to the following instructions, any further suggestions of the instructor should be heeded. (1) The reagent shelf should be clean and orderly. (2) Any spilled chemicals must be cleaned up immediately. (3) The stoppers of reagent bottles should not be placed on the shelf or laboratory bench. Stoppers may be placed on clean towels or watch glasses, although it is best to

TABLE 20.2 Composition of Concentrated Common Acids and Bases

REAGENT	DENSITY g/mL	PERCENT BY WEIGHT	APPROXIMATE MOLARITY
Acetic acid, $HC_2H_3O_2$	1.057	99.5	17
Ammonium hydroxide, NH_4OH	0.90	28 (as NH_3)	15
Hydrochloric acid, HCl	1.18	36	12
Nitric acid, HNO_3	1.42	72	16
Perchloric acid, $HClO_4$	1.68	71	12
Phosphoric acid, H_3PO_4	1.69	85	15
Sodium hydroxide, NaOH	—	50	16
Sulfuric acid, H_2SO_4	1.83	85	15

hold them between two fingers while reagents are being withdrawn. (4) The mouths of reagent bottles should be kept clean. (5) Pipets, droppers, or other instruments should never be inserted into reagent bottles. Rather, a slight excess of reagent should be poured into a clean beaker from which the pipetting is done and the excess discarded, not returned to the bottle. (6) Fingers, spatulas, or other implements should not be inserted into bottles of solid reagents.

20.3

APPARATUS

In addition to the usual equipment found in any chemistry laboratory, there are certain items of special interest to the analytical chemist. Some of the more important ones are described in this section, and advice is given regarding their use.

20.3a. Wash Bottle

Each student should have a wash bottle of reasonable capacity, capable of delivering a stream of distilled water from a tip connected to the main part of the bottle. A convenient type, shown in Fig. 20.1(a), is made of polyethylene, and the body is squeezed to force water from the tip. The wash bottle is used whenever a fine, directed stream of distilled water is needed, as when rinsing down the sides of a glass vessel to ensure that no droplets of sample solution are lost.

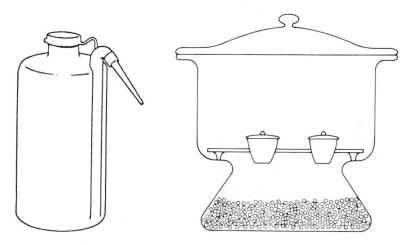

Figure 20.1   (a) Wash bottle. (b) Dessicator.

20.3b. Stirring Rods

As the name implies, stirring rods are used for stirring solutions or suspensions, generally in beakers. In addition, they are used in transferring solutions from one vessel into another. When an aqueous solution is poured from the lip of a vessel such as a beaker, there is a tendency for some of the liquid to run down the outside surface of the glass. This is prevented by pouring the solution down a stirring rod, the rod being held in contact with the lip of the vessel and directing the flow of liquid into the receptacle (see Fig. 20.8). Stirring rods also serve as handles for "rubber policemen" (sections of rubber tubing sealed together at one end, with the other end slipped over a stirring rod, used to salvage small quantities of precipitates from the walls of beakers).

20.3c. Desiccator

A desiccator is a vessel, usually of glass but occasionally of metal, which is used to equilibrate objects with a controlled atmosphere. Since the desiccator usually stands in the open, the temperature of this atmosphere generally approaches room temperature. It is normally the humidity of this atmosphere which is of interest. Objects such as weighing bottles or crucibles, and chemical substances, tend to pick up moisture from the air. The desiccator provides an opportunity for such materials to come to equilibrium with an atmosphere of low and controlled moisture content so that errors due to the weighing of water along with the objects can be avoided. A common type of desiccator is shown in Fig. 20.1(b).

The nature of the drying agent placed in the bottom of the desiccator determines the equilibrium partial pressure of water vapor in the space above. Substances commonly used in desiccators are anhydrous calcium chloride, calcium sulfate (Drierite), and magnesium perchlorate (Anhydrone or Dehydrite). Calcium chloride, while rather poor in regard to the equilibrium water vapor pressure, is inexpensive and is adequate for most work in the beginning laboratory.

After reagents or objects such as crucibles have been dried in the oven, or perhaps at even higher temperatures, they are usually cooled to room temperature in the desiccator prior to weighing. When a hot object cools in the desiccator, a partial vacuum is created, and care must be taken in opening the vessel lest a sudden rush of air blow material out of a crucible or disturb the desiccant itself. For this reason, and also because glass is a very poor conductor of heat, it is usually best to allow a very hot object to cool well toward room temperature before placing it in the desiccator. After placing a hot object in the desiccator, it is well to cover the vessel in such a way as to leave a small opening at one side. This allows air displaced by the warm object to reenter as the object cools, and hence minimizes the tendency to form a vacuum. The desiccator is completely closed during the final stages of cooling.

The desiccator cover should slide smoothly on its ground-glass surface. This surface should be lightly greased with a light lubricant such as Vaseline (never stopcock grease!) Needless to say, the desiccator should be scrupulously clean and should never contain exhausted desiccant. After filling the desiccant chamber, beware of dust from the desiccant in the upper part of the desiccator.

20.3d. Pipets

Some common types of pipets are shown in Fig. 20.2. The *transfer pipet* is used to transfer an accurately known volume of solution from one container to another. The pipet should be cleaned if distilled water does not drain uniformly but leaves droplets of water adhering to the inner surface. Cleaning can be done with a warm solution of detergent or with cleaning solution (consult the instructor).

The pipet is filled by gentle suction to about 2 cm above the etch line [Fig. 20.3(a)], using an aspirator bulb. Alternatively, a water aspirator can be used to apply suction. A long rubber tube is attached from the top of the pipet to the trap shown in Fig. 20.10. The tip of the pipet should be kept well below the surface of the liquid during the filling operation. The forefinger is then quickly placed

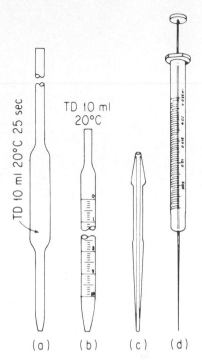

Figure 20.2 Pipets: (a) transfer pipet,
(b) measuring pipet, (c) lambda pipet, and
(d) microliter syringe.

over the top of the pipet [Fig. 20.3(b)], and the solution is allowed to drain out until the bottom of the meniscus coincides with the etched line. Any hanging droplets of solution are removed by touching the tip of the pipet to the side of the beaker, and the stem is wiped with a piece of tissue paper to remove drops of solution from the outside surface. The contents of the pipet are then allowed to run into the desired container, with care being taken to avoid spattering. With the pipet in a vertical position, the solution is allowed to drain down the inner wall for about 30 s after emptying, and then the tip of the pipet is touched to the inner side of the receiving vessel at the liquid surface. A small volume of solution will remain in the tip of the pipet, but the pipet has been calibrated to take this into account; thus this small final quantity of solution is *not* to be blown out or otherwise disturbed. Pipets with damaged tips are not to be trusted.

Measuring pipets are graduated much like burets and are used for measuring volumes of solutions more accurately than could be done with graduated cylinders. However, measuring pipets are not ordinarily used where high accuracy is required.

Two types of micropipets are also shown in Fig. 20.2. So-called *lambda* pipets are available in capacities of 0.001 to 2 mL, where 0.001 mL = 1 lambda. They are filled and emptied using a syringe. Those calibrated to *contain* a certain volume are rinsed with a suitable solvent. Those calibrated to *deliver* are not rinsed, but the last drop is forced out of the pipet with the syringe. Microliter syringes are widely used for delivering small volumes in such operations as gas chromatography. They can be bought equipped with stainless steel tips for use in injecting a sample into a closed system. The syringe shown in Fig. 20.2 has a capacity of 0.025 mL (25 μL, or 25 lambdas), and the smallest divisions correspond to 0.0005 mL. "Pushbutton" pipets, which make the transfer of liquids rapid and easy, are now available. Such a pipet consists of a syringe with a piston which can be operated by pressing a button at the top. Liquid is drawn into a disposable plastic tip and is then delivered by reversing the direction of the piston. Tips which deliver volumes of 0.001 to 1 mL (1 to 1000 μL) are available.

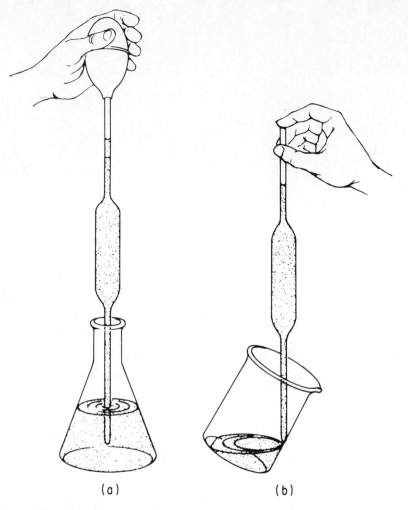

Figure 20.3 (a) Filling pipet—liquid drawn above graduation mark and (b) use of forefinger to adjust liquid level in pipet.

The National Institute of Standards and Technology specifies 20°C as the standard temperature for calibration of volumetric glassware. The use of such glassware at other temperatures leads to errors. However, the errors are normally small, and pipets can be used at "room temperature" without special precautions except in work of highest accuracy.

20.3e. Burets

A common form of buret is shown in Fig. 20.4(a). The buret is used to deliver accurately known but variable volumes, mostly in titrations. The stopcock plug is made of either glass or Teflon. The Teflon stopcock requires no lubrication, but the glass plug should be lightly greased with stopcock grease (not one containing silicones). If too heavy a coating is applied, the stopcock may leak; also, some of the grease may plug the buret tip. To lubricate a stopcock, remove the plug and wipe old grease away from both plug and barrel with a cloth or paper tissue. Make sure the small openings are not plugged with grease (pipe cleaners are helpful in this event). Then spread a thin, uniform layer of stopcock grease over the plug, keeping the application especially thin in the region near the hole in

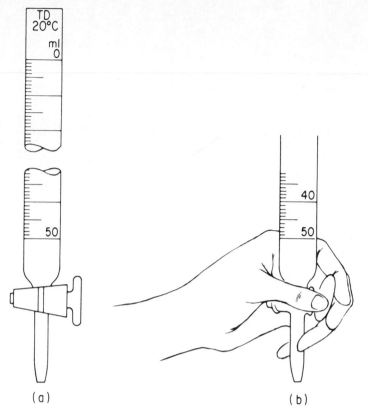

Figure 20.4 (a) Buret and (b) method of grasping stopcock.

the plug. Finally, insert the plug in the barrel and rotate it rapidly in place, applying a slight inward pressure. The lubricant should appear uniform and transparent, and no particles of grease should appear in the bore.

Burets must be cleaned carefully to ensure a uniform drainage of solutions down the inner surfaces. A hot, dilute detergent solution may be used for this purpose, especially if used in conjunction with a long-handled buret brush. Cleaning solution may also be used, applied hot for a few minutes or overnight at room temperature. The instructor should be consulted for directions on the proper use of cleaning solution. When not in use, the buret should be filled with distilled water and capped (paper cups or small beakers are convenient) to prevent the entry of dust.

It is poor practice to leave solutions standing in burets for long periods. After each laboratory period, solutions in burets should be discarded, and the burets rinsed with distilled water and stored as suggested above. It is especially important that alkaline solutions not stand in burets for more than short periods of time. Such solutions, which attack glass, cause stopcocks to "freeze," and the burets may be ruined.

The beginner must be very cautious in reading burets. In order to become familiar with the graduations and adept at estimating between them, much practice is needed early in the laboratory work. An ordinary 50-mL buret is graduated in 0.1-mL intervals and should be read to the nearest hundredth of a milliliter. An aqueous solution in a buret (or any tube) forms a concave surface referred to as a *meniscus*. In the case of solutions that are not deeply colored, the position of the bottom of the meniscus is ordinarily read (the top is taken if the solution is so intensely colored that the bottom cannot be seen, e.g., with permanganate solu-

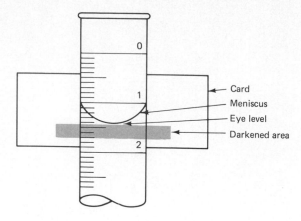

Figure 20.5 Card for help in reading a buret. The eye should be level with the bottom of the meniscus. Here the reading is 1.42 mL.

tions). It is most helpful to cast a shadow on the bottom of the meniscus by means of a darkened area on a paper or card held just behind the buret with the dark area slightly below the meniscus. (See Fig. 20.5.) Great care must be taken to avoid parallax errors in reading burets: The eye must be on the same level with the meniscus. If the meniscus is near a graduation that extends well around the buret, the right eye level can be found by seeking a position so that the graduation mark seen at the back of the buret merges with the same line at the front. A loop of paper encircling the buret just below the meniscus serves the same purpose.

Before a titration is started, it must be ascertained that there are no air bubbles in the tip of the buret. Such bubbles register in the graduated portion of the buret as liquid is delivered if they escape from the tip during a titration, and hence cause errors. When a solution is delivered rapidly from a buret, the liquid running down the inner wall is somewhat detained. After the stopcock has been closed, it is important to wait a few seconds for this "drainage" before taking a reading.

In performing titrations, the student should develop a technique that permits both speed and accuracy. The solution being titrated, generally in an Erlenmeyer flask, should be gently swirled as the titrant is delivered. One way to accomplish this, while retaining control of the stopcock and permitting ease of reading the buret, is to face the buret, with the stopcock on the right, and operate the stopcock with the left hand from behind the buret while swirling the solution with the right hand [Fig. 20.4(b)]. The thumb and forefinger are wrapped around the handle to turn the stopcock, and inward pressure is applied to keep the stopcock seated in the barrel. The last two fingers push against the tip of the buret to absorb the inward pressure.

20.3f. *Volumetric Flasks*

A typical volumetric flask is shown in Fig. 20.6. The flask contains the stated volume when filled so that the bottom of the meniscus coincides with the etched line. If the solution is poured from the flask, the volume delivered is somewhat less than the stated volume, and volumetric flasks are never used for measuring out solutions into other containers. They are used whenever it is desired to make a solution up to an accurately known volume.

When solutions are made up in volumetric flasks, it is important that they

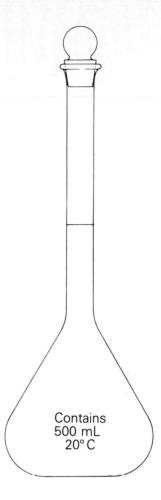

Figure 20.6 Volumetric flask.

be well mixed. This is accomplished by repeatedly inverting and shaking the flask. Some analysts make a practice of mixing the solution thoroughly before the final volume has been adjusted, and mixing again after the flask has been filled to the mark: It is easier to agitate the solution vigorously when the narrow upper portion of the flask has not been filled.

Solutions should not be heated in volumetric flasks, even those made of Pyrex glass. There is a possibility that the flask may not return to its exact original volume upon cooling.

Most volumetric flasks have ground-glass or polyethylene stoppers, screw caps, or plastic snap caps. Alkaline solutions cause ground-glass stoppers to "freeze" and thus should never be stored in flasks equipped with such stoppers.

When a solid is dissolved in a volumetric flask, the final volume adjustment should not be made until all the solid has dissolved. In certain cases marked volume changes accompany the solution of solids, and these should be allowed to take place before the volume adjustment is made.

20.3g. *Funnels and Filter Paper*

In gravimetric procedures the desired constituent is often separated in the form of a precipitate. This precipitate must be collected, washed free of undesirable contaminants from the mother liquor, dried, and weighed, either as such or after conversion into another form. Filtration is the common way of collecting precipitates, and washing is often accomplished during the same operation. Filtration is carried out with either funnels and filter paper or filtering crucibles. The important factors in choosing between the two are the temperature to which the precipitate must be heated to convert it into the desired weighing form and the ease with which the precipitate may be reduced.

The cellulose fibers of filter paper have a pronounced tendency to retain moisture, and a filter paper containing a precipitate cannot be dried and weighed as such with adequate accuracy. It is necessary to burn off the paper at a high temperature. During the burning, reducing conditions due to carbon and carbon monoxide prevail in the vicinity of the precipitate. Thus precipitates that cannot be heated to high temperatures or that are sensitive to reduction are normally not filtered using filter paper; filtering crucibles of the types described in a later section are employed. Some of the techniques given here, however, will apply to all types of filtration.

Various types of filter paper are available. For quantitative work, only paper of the so-called ashless quality should be used. This paper has been treated with hydrochloric and hydrofluoric acids during its manufacture. Thus it is low in inorganic material and leaves only a very small weight of ash when it is burned. (A typical figure for the ash from one circular paper 11 cm in diameter is 0.13 mg.) The weight of ash is normally ignored; for very accurate work, a correction can be applied, since the weight of ash is fairly constant for the papers in a given batch.

Within the ashless group, there are further varieties of paper that differ in porosity. The nature of the precipitate to be collected dictates the choice of paper. "Fast" papers are used for gelatinous, flocculent precipitates such as hydrous iron(III) oxide and for coarsely crystalline precipitates such as magnesium ammonium phosphate. Many precipitates that consist of small crystals (e.g., barium sulfate) will pass through the "fast" papers. "Medium" papers require a longer time for filtration but retain smaller particles and are the most widely used. For very fine precipitates such as silica, "slow" paper is employed. Filtration at best is rather time-consuming, and the analyst should use the fastest paper consistent with retention of the precipitate.

Filter paper is normally folded so as to provide a space between the paper and the funnel, except at the top of the paper, which should fit snugly to the glass. The procedure is shown in Fig. 20.7. The second fold is made so that the ends fail to match by about $\frac{1}{8}$ in. Then the paper is opened into a cone. The corner of the outside fold on the thicker side is torn off in order to fit the paper to the funnel more easily and to break up a possible air passage down the fold next to the funnel. With the paper cone held in place in the funnel, distilled water is poured in. A clean finger, applied cautiously to prevent tearing the fragile wet paper is used to smooth the paper and obtain a tight seal of paper to glass at the top. Air does not enter the liquid channel with a properly fitted paper, and thus the drainage from the stem of the funnel establishes a gentle suction which facilitates filtration. A malfunctioning filter can seriously delay an analysis; it is preferable by far to reject such a filter and prepare a new one.

Filter paper circles are available in various diameters. The size to be used depends upon the quantity of precipitate, not the volume of solution to be filtered. Larger paper than necessary should be avoided: The paper and the funnel should match in size. It is especially important that the paper not extend above the edge of the glass funnel, but come within 1 or 2 cm of this edge. The precipitate should occupy about one-third of the paper cone and never more than one-half.

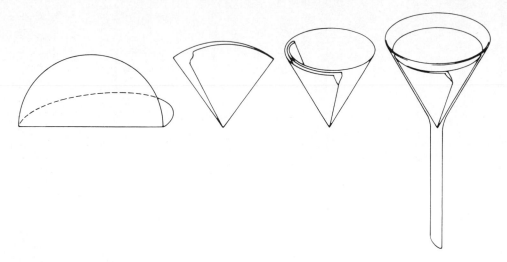

Figure 20.7 Folding filter paper.

Technique of Washing and Filtering a Precipitate

Usually a precipitate is washed, either with water or a specified wash solution, before it is dried and weighed. The washing is generally carried out in conjunction with the filtration step (Fig. 20.8), wherein the precipitate is separated from its mother liquor in compact form. Once the precipitate is in the filter, it can be washed by passing wash solution through the filter. However, this technique is often rather inefficient; the wash solution does not penetrate uniformly into the compact mass of precipitate. It is usually preferable to wash the precipitate by decantation, at least in cases where the precipitate settles rapidly from suspension. The supernatant mother liquor is carefully poured off through the filter while as much of the precipitate as possible is retained in the beaker. The precipitate is then stirred with wash solution in the beaker, and the washings are decanted through the filter. The washing is repeated as often as desired,[3] until, in the final instance, the precipitate is not allowed to settle but is poured into the filter along with the wash solution. Residues of precipitate remaining in the beaker are usually transferred to the filter by a directed jet from a wash bottle, as shown in Fig. 20.9. If the precipitate tends to adhere to the glass, the last traces may be scavenged by means of a rubber policeman. The precipitate is then wiped from the policeman with a small bit of filter paper, which is added to the paper in the funnel for ignition.

The stem of the funnel should extend well into the vessel receiving the filtrate, and the tip of the stem should touch the inner surface of the vessel to prevent spattering of the filtrate. All transfers into the funnel should be made with the aid of a stirring rod, and care must be taken that no drops of solution are lost. The filtrate should be examined for turbidity; sometimes small amounts of precipitate run through the filter early in the filtration, but can be caught by refiltering through the same filter after its pores have been somewhat clogged by collected precipitate.

[3] It should be noted that several washings with small volumes of solution are more effective than one washing with the same total volume of wash solution.

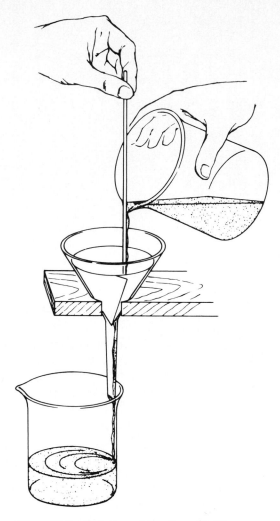

Figure 20.8 Technique of filtration with filter paper.

Technique of Ignition of Precipitate with Filter Paper

After the filter paper has drained as much as possible in the funnel, the top of the paper is folded over to encase the precipitate completely. Great care must be used to avoid tearing the fragile, wet paper when the bundle of paper is transferred from the funnel to the prepared crucible (see the discussion of crucibles below). It is better to handle the paper where it is three layers thick rather than by the other side. The next steps in the ignition of the material in the crucible are generally as follows:

1. *Drying the Paper and Precipitate.* This may be done in an oven at temperatures of 100 to 125°C if the schedule permits setting aside the experiment at this stage. If the ignition is to be followed through immediately, the drying may be accomplished with a burner. The covered crucible is positioned in a slanted position in a clay or silica triangle, and a small flame is placed beneath the crucible at about the middle of the underside. Too strong a heating must be avoided; the flame should not touch the crucible, and the drying should be leisurely.

Figure 20.9 Use of wash bottle in transferring precipitate.

2. *Charring the Paper*. (See Fig. 20.10.) After the precipitate and paper are entirely dry, the crucible cover is set ajar to permit access of air, and the heating is increased to char the paper. The size of the flame is increased slightly and moved back under the base of the crucible. The paper should smoulder, but must not burn off with a flame. If the paper bursts into flame, the crucible must be covered immediately to extinguish it. Small particles of precipitate may be swept from the crucible by the violent activity of escaping gases; also, under these conditions, carbon of the paper may reduce certain precipitates which can be safely handled in filter paper under less vigorous conditions. Care must be taken that reducing gases from the flame are not deflected into the crucible by the underside of the cover. During the charring, tarry organic material distills from the paper, collecting on the crucible cover. This is burned off later at a higher temperature.

3. *Burning Off the Carbon from the Paper*. After the paper has completely charred and the danger of its catching fire is past, the size of the flame is increased until the bottom of the crucible becomes red. This should be done gradually. The carbon residue and the organic tars are burned away during this stage of the ignition. The heating is continued until this burning is complete, as evidenced by the disappearance of dark-colored material. It is well to turn the crucible from time to time so that all portions are heated thoroughly. Sometimes it is necessary to direct special attention to the underside of the cover to remove the tarry material collected there.

4. *Final Stage of Ignition*. To conclude the ignition, the crucible is placed up-

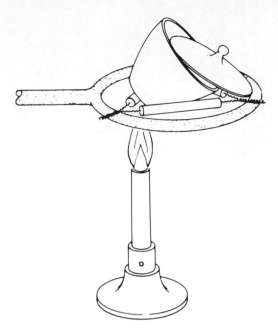

Figure 20.10 Ignition of precipitate.

right, the cover is removed to admit air freely, and the crucible is heated at the temperature recommended for the particular precipitate. A Tirrill burner will heat a covered porcelain crucible to about 700°C, and a Meker burner will give a temperature roughly 100°C higher. With platinum crucibles, temperatures about 400°C higher can be obtained. The ignition is continued until the crucible has reached constant weight, that is, until the difference between two weighings with a heating period in between is less than about 0.5 mg.

20.3h. Filtering Crucibles

Certain precipitates cannot be handled with filter paper, either because they are too easily reduced or because they cannot be heated to a temperature adequate to burn off the paper. Such precipitates are filtered by means of *filtering crucibles*, several types of which are pictured in Fig. 20.11. The *Gooch*, which is seldom used today, is a porcelain crucible with a number of small holes in the bottom. A mat of asbestos is formed on the inside of the bottom when the crucible is prepared for use, and this mat is the filtering medium. It is prepared by forming a suspension of specially cut and treated asbestos fibers in water and pouring the suspension through the crucible under suction. Gooch crucibles can be ignited at high temperatures and are adequate for many different precipiates.

Sintered-Glass Crucibles

For precipitates that need not be heated above 500°C or so, *sintered-* or *fritted-glass crucibles* may be used. These crucibles are made of glass and have a bottom of sintered ground glass fused to the body of the crucible. They are mounted in a suitable holder (as shown in Fig. 20.11), and suction is applied. Sintered-glass crucibles are available in varying porosities for handling various types of precipitates. In the Corning Glass Co. Pyrex line, the three porosities

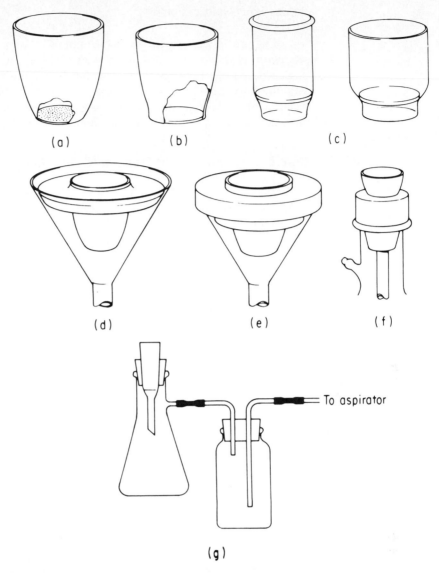

Figure 20.11 Filtering crucibles: (a) Gooch, (b) porous porcelain, (c) fritted glass, high and low forms, (d) Sargent holder, (e) Bailey holder, (f) Walter holder, and (g) filtration with suction, Walter holder.

designated coarse, medium, and fine (the latters C, M, and F appear near the upper edge of the glass crucible) serve most analytical uses. Although temperatures up to 500°C are said to be safe, sintered-glass crucibles must be heated very gradually if such a temperature is to be approached without damage. It must be kept in mind that strong alkalies attack the crucibles, especially the sintered filtering disks. The crucibles are cleaned with solvents appropriate to the particular contamination at hand. For three reasons sintered-glass crucibles are never heated directly over a flame. First, they may be broken; second, carbon, which is very difficult to remove, may be deposited in the fritted disk; and third, reducing gases from the flame may penetrate to the precipitate via the porous bottom. If a burner must be used to attain the desired temperature, the sintered-glass crucible is placed within an ordinary porcelain crucible.

Porous Porcelain Crucibles

For precipitates which must be ignited at very high temperatures, *porous porcelain crucibles* may be employed. These are porcelain crucibles with unglazed, porous bottoms through which precipitates can be filtered off under suction. As with sintered-glass crucibles, porous porcelain crucibles must not be heated with an open flame. They are ignited within an ordinary porcelain crucible unless a furnace is used. The porous bottom is readily attacked by strong alkali.

20.3i. *Crucibles for Ignition of Precipitates in Filter Paper*

Two kinds of crucibles are widely used for the ignition of precipitates in filter paper: platinum and porcelain. Platinum is normally preferred, but because of its cost is rarely used in the beginning laboratory. Platinum is an inert metal and is resistant to certain reagents which attack porcelain. It also reaches a higher temperature over a flame than does porcelain. Porcelain crucibles can be heated to high temperatures (about 1200°C in an oven), and their weight changes very slightly with strong and prolonged heating. They are attacked by alkaline substances and by hydrofluoric acid, but are resistant to most other reagents.

The idea of *constant weight* needs to be explained more thoroughly. The weight of a precipitate in a crucible is normally obtained by difference. For that reason the empty crucible is first ignited exactly as the crucible plus precipitate will be ignited later. It is then cooled, weighed, reheated, and weighed again. If the two weighings agree (within, say, 0.2 to 0.5 mg), the crucible has been ignited to "constant weight." If not, the procedure is repeated until such agreement is obtained. Then the precipitate is added, and the crucible ignited at the same temperature until it is again at constant weight. Then the difference in the two weights gives the weight of the precipitate.

20.3j. *Special Instruments*

Not too many years ago, the analytical chemist used only the apparatus described above, plus the analytical balance, for almost all determinations. A striking change has taken place, and nowadays pH meters, colorimeters, spectrophotometers, polarographs, and still more elaborate and complex instruments are found in most analytical laboratories. Directions for the use and care of instruments are necessarily specialized and are best obtained from the manufacturer's bulletins and from personal instruction by experienced people. We include here only a few general remarks to fill out our discussion of analytical apparatus.

The rule of greatest importance is that no instrument should ever be touched by a person unfamiliar with the directions for its proper use and the precautions against damaging it. Some instruments contain fragile components which may be injured by improper handling, and sometimes a carefully worked out calibration may be ruined by manipulation of the wrong knobs.

The other rule that must always be remembered is that an instrument should never be used by a person who has not thought through its advantages and limitations for the job at hand, who does not have a proper estimate of the reliability of the data obtained, and who cannot interpret correctly the significance of the instrumental measurement and apply it with intelligence. Meaningless measurements are made every day by imposters masquerading as chemists. Anyone can learn to turn knobs and read galvanometer, but the assurance that a measurement has been made on the best possible system must come from a well-trained chemist.

CALIBRATION OF VOLUMETRIC GLASSWARE

20.4a. *Characteristics of Volumetric Glassware*

Table 20.3 shows some of the tolerance values established for volumetric glassware by the National Institute of Standards and Technology.[4] It may be noted that glassware meeting these specifications is adequate for all but the most exacting work of the analytical laboratory. Such glassware, stated by the manufacturer to conform to NIST standards, may be purchased.[5] For an extra fee, it is also possible to obtain glassware that has actually been tested by the NIST. Because of the expense, however, the beginning laboratory will rarely be equipped with glassware guaranteed to meet NIST tolerance specifications. Less expensive glassware stated by the manufacturer to meet tolerances about twice those of the NIST is available and may be furnished to the student.[6]

Since most analytical work involves dilute, aqueous solutions, water is generally used as the reference material in the calibration of volumetrc glassware. The general principle in calibration is to determine the weight of water contained in or delivered by a particular piece of glassware. Then, with the density of water known, the correct volume is found.

The units of volume commonly employed in analytical chemistry are the *liter* and the *milliliter*. The liter was formerly defined as the volume occupied by 1 kg of water at the temperature of its maximum density (about 4°C) under a pressure of 1 atm. In 1964 the Twelfth General Conference on Weights and Measurements, meeting in Paris, France, abolished this definition and instead made the liter a special name for the cubic decimeter. This new definition eliminates the previous discrepancy of 28 parts per million between the milliliter and cubic centimeter (1 mL was 1.000028 cc), and these two units are now equivalent.

TABLE 20.3 Tolerances for Volumetric Glassware, mL

CAPACITY, LESS THAN AND INCLUDING	VOLUMETRIC FLASKS	TRANSFER PIPETS	BURETS
2		0.006	
5		0.01	0.01
10		0.02	0.02
25	0.03	0.03	0.03
50	0.05	0.05	0.05
100	0.08	0.08	0.10
200	0.10	0.10	
500	0.15		
1000	0.30		

[4] Nat. Bur. Standards Circ. 434 (1941).

[5] For example, in the Kimball Glass Co., the Kimax Class A Inc line and the Corning Glass Works Pyrex line of volumetric glassware.

[6] Kimball Glass Co. Kimax volumetric glassware.

The National Institute of Standards and Technology has specified 20°C as the temperature at which glassware is calibrated. Since the laboratory temperature will usually not be exactly 20°C, glassware must, strictly speaking, be corrected when used at other temperatures because of errors due to expansion (or contraction) of both the glass vessel itself and the solution contained therein. The coefficient of expansion of glass is sufficiently small that the correction required for this factor is negligible for most work (it amounts to the order of 1 part per 10,000 for a change of 5°C). The change in the volume of the solution itself, on the other hand, is more important, but it can still be ignored in many cases if the working temperature is not far removed from 20°C (the volume change is on the order of 1 part per 1000 over a 5°C range).

As noted above, calibration data are secured by converting weight of water into volume via the density. Tables showing the density of water at various temperatures are available in handbooks. However, the data in such tables are usually given on the basis of weights *in vacuo,* while the actual weighings are made in air. Since the water being weighed generally displaces more air than do the weights, it is necessary to correct the weighings for the buoyancy effect of the air if such handbook tables are to be used. (See Chapter 21 for a fuller discussion of this buoyancy effect). On the other hand, we may change *in vacuo* densities into densities that would be obtained in air with steel weights and use these directly with our weighings obtained in air. The values in Table 20.4, which are the reciprocals of such adjusted densities, may be used directly by the student without taking the buoyancy effect into consideration.

TABLE 20.4 Volume of 1 g of Water Weighed in Air with Steel Weights at Various Temperatures

°C	mL	°C	mL
10	1.0013	21	1.0030
11	1.0014	22	1.0033
12	1.0015	23	1.0035
13	1.0016	24	1.0037
14	1.0018	25	1.0040
15	1.0019	26	1.0043
16	1.0021	27	1.0045
17	1.0022	28	1.0048
18	1.0024	29	1.0051
19	1.0026	30	1.0054
20	1.0028		

20.4b.   *Methods of Calibration*

There are general approaches to the calibration of volumetric glassware that are in wide use and with which the student should be familiar.

1. The first method, which we may designate as a *direct, absolute calibration,* is based on the principles outlined above. The volume of water delivered by a buret or pipet, or contained in a volumetric flask, is obtained directly from the weight of the water and its density.
2. Volumetric glassware is sometimes calibrated by comparison with another vessel previously calibrated directly. We may refer to this as an *indirect, absolute calibration,* or *calibration by comparison.* This method is especially conve-

nient if many pieces of glassware are to be calibrated, and it is sufficiently accurate for all ordinary usages provided the reference vessel itself has been accurately calibrated. Calibrating bulbs are not available in many student laboratories, and specific directions for their use are not included in this chapter. They are not difficult to use if proper care is taken. The student may obtain directions from the instructor if he or she is to use such equipment.

3. Sometimes it is necessary to know only the relationship between two items of glassware without knowing the absolute volume of either one. This situation arises, for example, in taking an aliquot portion of a solution. Suppose that it is desired to titrate one-fifth of an unknown sample. The unknown might be dissolved, appropriately treated preparatory to the titration, and diluted to volume in a 250-mL volumetric flask. A 50-mL pipet would then be used to withdraw an aliquot for titration. For the calculations in this analysis, it would not be necessary to know the exact volume of the flask or the pipet, but it would be required that the pipet hold exactly one-fifth as much solution as the flask. The method used for a *relative calibration* of this sort simply involves discharging the pipet five times into the flask and marking the level of the meniscus on the flask.

20.5.

RECORDING LABORATORY DATA

There are three main requirements for the recording of data obtained in the analytical laboratory. These may be briefly expressed as follows: The record (1) should be complete, (2) should be intelligible to any reasonably competent chemist, and (3) should be easy to find on short notice. The requirements may be met by adherence to the following rules.

1. The student should have a bound notebook for recording laboratory data, calculations, results, and all other matters pertinent to the analysis of a sample. The pages of the notebook should be numbered, and a table of contents should be developed so that any given experiment can be quickly found.

2. All data obtained in the laboratory should be recorded directly in the notebook at the time the work is performed. Especially forbidden is the recording of data on loose paper with the idea of copying it into the notebook later. While neatness may be sacrificed somewhat by taking the notebook directly into the laboratory, the prevention of loss of data and errors in transcribing them more than counterbalances this.

3. Entries should be recorded in ink. If a mistake is made and a recorded value is invalidated, it is not to be erased but is crossed out so as to be still legible. A notation as to why it has been rejected is made in the notebook.

4. The data in the notebook should be organized and recorded in a systematic way. This benefits the student because it is then relatively easy to locate errors in the analytical calculations; the student may thus be saved repeating an entire determination in order to obtain satisfactory results. To facilitate an orderly presentation of the data, the student should plan how best to record it before the experiment is actually begun. It is especially helpful to arrange beforehand a table in which the experimental data, the calculations, and the final results can be entered systematically.

5. The rules regarding significant figures (see Chapter 2) should be followed in recording data in the notebook.

TABLE 20.5 Sample Notebook Page for Soda Ash Analysis

Soda Ash Analysis

p. 25
10/18/91
Unknown no. 186

Method: Sample dissolved in water and titrated with standard
HCl solution, using modified methyl orange indicator

Reaction: $Na_2CO_3 + 2HCl \longrightarrow CO_2 + H_2O + 2NaCl$

	I	II	III
Wt of sample	0.3276 g	0.3342 g	0.3128 g
Final buret reading, HCl	35.86 mL	36.32 mL	34.56 mL
Initial buret reading, HCl	0.18 mL	0.10 mL	0.38 mL
Volume HCl	35.68 mL	36.22 mL	34.18 mL
Initial buret reading, NaOH	0.37 mL	0.57 mL	0.92 mL
Final buret reading, NaOH	0.06 mL	0.37 mL	0.57 mL
Volume NaOH	0.31 mL	0.20 mL	0.35 mL

Volume relation of NaOH to HCl:
1.00 mL NaOH = 0.95 mL HCl

ML HCl equiv. to NaOH.	0.29 mL	0.19 mL	0.33 mL
Total HCl req. for sample	35.39 mL	36.03 mL	33.85 mL

Normality of HCl: 0.1076

$$\%Na_2CO_3 = \frac{mL\ HCl \times N \times EW}{mg\ sample} \times 100$$

	61.60%	61.49%	61.70%

Average % Na_2CO_3 61.60%

Average deviation, ppt   1.2 ppt

Table 20.5 shows an example of a satisfactory laboratory record for a typical volumetric analysis. The opposite page in the notebook may be used for arithmetic calculations and other material which does not fit well into the tabulation. The student should, of course, follow any style suggested by the individual instructor. Some instructors grade unknowns directly from the notebook, while others prefer that students summarize their results on 3-by 5-inch cards somewhat as shown in Table 20.6. Still others may ask students to report their results on a computer terminal.

TABLE 20.6 Sample Report Card for Soda Ash Unknown

10/18/91 Name ———————————————————— Soda Ash Unknown No. 186

Normality of HCl: 0.1076

	Wt sample	mL HCl	% Na_2CO_3	Deviation
1	0.3276 g	35.39 mL	61.60	0.00
2	0.3342 g	36.03 mL	61.49	0.11
3	0.3128 g	33.85 mL	61.70	0.10

Average % Na_2CO_3: 61.60

Average deviation: 0.07 in 61.60 or 1.2 ppt

21

The Analytical Balance

The analytical balance used in the introductory laboratory is a precision instrument capable of detecting the weight of an object of 100 g to within ±0.0001 g (±0.1 mg). This is an uncertainty of only 1 part per million. Until the 1950s most of these balances were *two-pan* balances, also referred to as *equal-arm* balances. Then the *single-pan* or *unequal-arm* (sometimes called *constant-load*) balance essentially replaced the two-pan balance. Today the *electronic* balance (also called the *electromagnetic force* balance) is rapidly replacing the mechanical, or single-pan, balance. It uses an electric current to generate a magnetic force which balances the load placed on the balance pan. The current required to generate the force is directly proportional to the mass of the object on the pan.

We shall first describe the determination of mass and weight and then consider the three types of balances mentioned above.

21.1

MASS AND WEIGHT

The equal-arm balance is a lever of the first class; that is, the fulcrum (B in Fig. 21.1) lies between the points of application of forces (A and C in Fig. 21.1). Since the arms are equal in length, $l_1 = l_2$. Pans are suspended from A and C, and the object to be weighed (mass M_1) is placed on the left-hand pan while

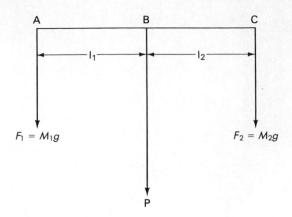

Figure 21.1 Principle of weighing on a two-pan balance.

known weights (mass M_2) are placed on the right-hand pan. Both M_1 and M_2 are attracted by the earth (gravity), the forces being, according to Newton's second law,

$$F_1 = M_1 g$$

$$F_2 = M_2 g$$

where g is the acceleration of gravity. The operator adjusts the value of M_2 until the pointer, P, returns to its original position. Then, according to the principle of moments,

$$F_1 l_1 = F_2 l_2$$

and since $l_1 = l_2$, $F_1 = F_2$, and hence

$$M_1 g = M_2 g \quad \text{or} \quad M_1 = M_2$$

Since M_2 is known, M_1 is determined.

It should be noted that the *weight* of an object is the force exerted on the object by gravitational attraction (F_1 above). The *mass* is the quantity of matter of which the object is composed. The weight of an object is different at different locations on the earth's surface, whereas the mass is *invariant*. It is mass that the analyst determines, but since the term g cancels in the case of the equal-arm balance, the ratio of weights is the same as the ratio of masses. Hence it is customary to use the term *weight* instead of mass, and we commonly speak of the process of determining the mass of an object as *weighing*.

21.2

TWO-PAN BALANCES

The beam of a two-pan balance contains three prism-form *knife-edges*, located at A, B, and C in Fig. 21.1. The Scottish chemist Joseph Black (1728–1799) is credited with first introducing the use of knife-edges, which are made of agate, a hard, brittle material. It was only after knife-edges were used that analytical weighing measurements could be attained with the two-pan balance.

Weighing on a two-pan balance was always tedious and time-consuming, and scientists constantly sought ways to make the operation easier. In addition, there are two sources of error which are inherent in such a balance. First, in order to obtain accurate results, it is necessary for the two balance arms to be equal in length. In practice, it is very difficult to make beams in which the balance arms are of identical length. If one arm is only 1/100,000 longer than the other, an error of 1/100,000 of the weight is introduced. With a load of 100 g, this amounts to an error of 1 mg. Second, for constant scale deflection per unit of weight added (sensitivity), the three knife-edges must lie exactly on a straight line. An increasing load tends to bend the beam slightly, bringing the terminal knife-edges below the plane of the central knife-edge. This results in a smaller scale deflection per unit weight at higher balance loads.

21.3

SINGLE-PAN BALANCES

The two sources of error inherent in the two-pan system are eliminated in the single-pan, or unequal-arm, balance. A schematic diagram of a typical mechanical balance of this type is shown in Fig. 21.2. There are two knife-edges rather than three, and the balance arms are unequal in length. The balance pan and a full complement of weights are suspended from the short arm, and the longer arm has a constant counterweight (plus a damping device) built into the balance beam. Hence the empty balance is fully loaded. When the object to be weighed is placed on the pan, weights are removed from the shorter arm by turning knobs on the outside of the balance case. A set of dials shows the sum of the weights removed. Thus weighing is done by *substitution,* leaving the same load on the beam at all times. Constant loading of the beam provides constant balance sensitivity, a characteristic not found in two-pan balances.

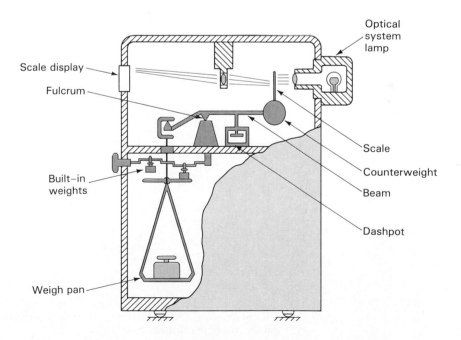

Figure 21.2 A modern single-pan mechanical analytical balance. (Reprinted from R. M. Schoonover, *Anal. Chem.,* 54, 937A (1982). Reprinted by permission of the author and the American Chemical Society.)

Once the sum of weights removed is within 0.1 g (but on the low side) of the weight of the object, the beam is fully released and allowed to come to rest. A *reticle*, consisting of a scale inscribed on a glass plate, is attached to the longer arm of the balance. The deflection of the beam from its original point of rest is greatly magnified by projecting optically the image of this scale onto a glass plate on the front of the balance. The scale displacement in milligrams can be read directly, and some device, such as a vernier, is employed to read to the nearest 0.1 mg.

A number of commercial balances contain so-called *tare* devices, which enable the operator to set the balance at zero with an empty weighing bottle, or beaker, on the pan. The operator can then weigh the sample by pouring it directly into the container and avoid the operation of subtracting the weight of the container from that of the container plus sample.

21.4

ELECTRONIC BALANCES[1]

The *electronic,* or *electromagnetic force,* balance is based on the principle that when a current is passed through a wire placed between the poles of a permanent magnet, a force is generated which moves the wire outside the magnetic air gap. The application of this principle in an analytical balance is best understood by examining an *electromagnetic servo system,* shown schematically in Fig. 21.3. In this system the force associated with the object being weighed is coupled mechanically to a servo motor that generates the opposing magnetic force. The system contains a null, or position, indicator which checks the position of the wire in the magnetic field. This may be an optical device, consisting of a vane attached to the

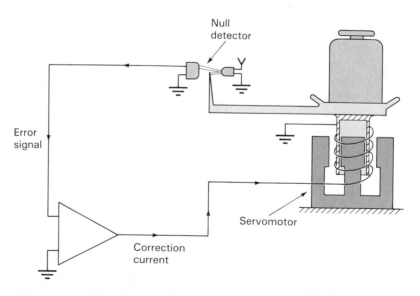

Figure 21.3 Simplified electromagnetic servo system (Reprinted from R. M. Schoonover, *Anal. Chem.,* 54, 937A (1982). Reprinted by permission of the author and the American Chemical Society.)

[1] R. M. Schoonover, *Anal. Chem.,* **54**, 973A (1982).

beam, a small lamp, and a photo detector. When the two forces are in equilibrium, the "error" indicator is at the reference position, and the average current in the servo motor coil is proportional to the resultant force holding the mechanism at the reference position. When the beam is displaced from its balanced position, the amount of radiation reaching the detector changes and causes a very rapid change in current through the coil. An error signal is sent to the circuit that generates a correction current. This current flows through the coil attached to the base of the balance pan, creating a magnetic field and restoring the indicator to its reference position. The correction current needed to restore the system is proportional to the mass of the object on the balance pan. Calibration is performed by placing a known weight on the balance pan and adjusting the circuitry to indicate the mass of the calibrating weight.

Figure 21.4 shows a schematic diagram of a top-loading electronic balance. No balance beam is needed, but the weight pan is attached to a solid parallelogram load constraint to prevent torsional forces, caused by off-center loading, from perturbing the alignment of the balance mechanism. Figure 21.5 shows the force balance cell in a classical balance case, where the weighing pan is placed below the cell rather than above it, as in Fig. 21.4. This configuration gives better axial alignment, or minimum off-center loading, and a reduction in servo motor force with a corresponding drop in capacity.[1]

There are many models of electronic balances available today with numerous optional features. A typical balance will have a way of indicating the zero setting with no load on the pan, and a digital display to give the weight of the object. Some characteristics listed by the Mettler Instrument Corporation for its Model AE 100 are as follows: weighing range, 0 to 109 g; readability, 0.1 mg; reproducibility (s), 0.1 mg; linearity, ± 0.2 mg; stabilization time, ~ 5 s; built-in calibration weight, 100 g. The semimicro Model AE240 has a readability of 0.01 mg and a reproducibility of 0.02 mg.

Electronic balances do have limitations and potential errors. These will be discussed in the next section.

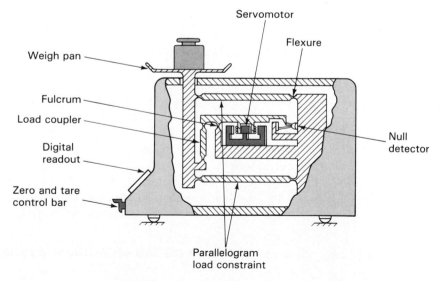

Figure 21.4 Top-loading balance. (Reprinted from R. M. Schoonover, *Anal. Chem.*, 54, 937A (1982). Reprinted by permission of the author and the American Chemical Society.)

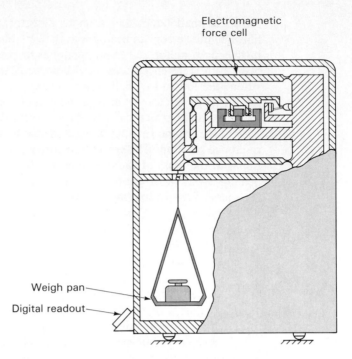

Electromagnetic
force cell

Weigh pan

Digital readout

Figure 21.5 The force balance in a classical enclosure. (Reprinted from R. M. Schoonover, *Anal. Chem.*, 54, 937A (1982). Reprinted by permission of the author and the American Chemical Society.)

21.5

ERRORS IN WEIGHING

21.5a. General Sources of Error

Students should obviously avoid careless errors in weighing, such as spilling the sample and misreading the value of a weight, the dial, or the position of a vernier. In addition the following possible errors should be avoided.

1. Samples which may take up water vapor or carbon dioxide from the air during the weighing process should be kept in closed containers or weighing bottles. Ignited precipitates are usually weighed in closed crucibles.
2. A glass vessel should not be wiped with a dry cloth before the vessel is weighed. The object may acquire a charge of static electricity and cause an error in weight.
3. The object weighed should be at the same temperature as the balance. Crucibles that have been heated and samples that have been dried should always be cooled to room temperature before weighing.

21.5b. Sources of Error in the Use of Electronic Balances

Electronic balances deserve special comments. Schoonover[1] pointed out three possible causes of inaccuracy or imprecision in using such balances: (1) interference by ferromagnetic or magnetized samples; (2) interference by electromagnetic radiation from nearby equipment; and (3) dust which may lodge between the coil and permanent magnet of the servomotor. Johnson and Wells[2] discussed these effects in some detail. They also found that small, but significant, errors may occur if samples are placed off-center on the pan of a top-loading electronic balance.

Finally, it should be noted that physical "weights" are not added or removed when weighing on an electronic balance. This balance measures the force (weight) on the pan directly and may be more sensitive to buoyancy effects than a single-pan balance. The effect of buoyancy is discussed in the next section.

21.5c. *Buoyancy Errors*

In the normal weighing process, both the object and the weights are buoyed up by the weight of air displaced. This is in accordance with the principle of Archimedes. If the object and weights displaced the same amount of air, no error would be introduced by this effect. This is not usually the case, however, since the density of the weights is normally different from that of the object. In quantitative analysis the density of the weights is usually larger, and hence the object displaces a greater volume of air than do the weights. The weight of the object is therefore less in air (the apparent weight) than it would be in a vacuum (the true weight). The true weight, W_v, is given by

$$W_v = W_a + (V_o - V_w)d_a$$

where W_a is the weight in air, V_o and V_w are the volumes of the object and weights, respectively, and d_a is the density of air (about 0.0012 g/ml under usual conditions).

Since $V_o = W_v/D_o$ or approximately W_a/D_o, and $V_w = W_a/D_w$, where D_o and D_w are the densities of the object and weights, respectively,

$$W_v = W_a + \left(\frac{W_a}{D_o} - \frac{W_a}{D_w}\right)0.0012$$

or

$$W_v = W_a\left\{1 + \left(\frac{1}{D_o} - \frac{1}{D_w}\right)0.0012\right\}$$

In the usual weighings made in the analytical laboratory, the errors caused by buoyancy are quite small. Most analytical results are expressed in terms of the ratio of two weights. If the densities of the sample and final precipitate are nearly equal, no appreciable error is introduced by using weights in air. In such operations as the calibration of volumetric apparatus by weighing large volumes of water, the error is appreciable, since the density of water is much less than that of the weights.

The following example illustrates the magnitude of the error due to buoyancy in weighing a liquid sample.

EXAMPLE 21.1

Buoyancy effect

A sample of benzene, density 0.88 g/mL, is pipetted into an empty glass vessel which weighs 12.2480 g. The bottle plus benzene is found to weigh 14.46204 g. A single-pan balance with stainless steel weights, density 7.8 g/mL, is used. Calculate the true weight (W_v) of the benzene.

---

[2] B. B. Johnson and J. D. Wells, *J. Chem. Ed.*, **63**, 86 (1986).

The weight in air, W_a, is $14.4624 - 12.2480 = 2.2144$ g. Substituting in the expression above,

$$W_v = 2.2144[1 + \left(\frac{1}{0.88} - \frac{1}{7.8}\right)0.0012]$$

$$W_v = 2.2144[1 + (1.1364 - 0.1282)0.0012]$$

$$W_v = 2.2171 \text{ g} \qquad \qquad \square$$

Direct-reading electronic balances are calibrated by the manufacturer to indicate what is known as "apparent mass vs. 8.0 g/mL." The calibration procedure for such a balance is essentially the setting of its sensitivity so that the apparent mass in air of a standard "weight" of density 8.0 g/mL is indicated correctly.[2] Since the force of gravity at the factory may not be the same as that in the user's laboratory, the user should calibrate a balance with a standard mass in his or her own laboratory. This can be done on a balance such as the Mettler AE 100 using the built-in calibration weight which is automatically lowered by a lever. If the operator so desires, he can use his own external standard mass for calibration.

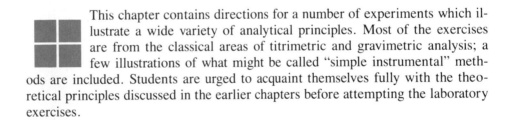

Laboratory Procedures

This chapter contains directions for a number of experiments which illustrate a wide variety of analytical principles. Most of the exercises are from the classical areas of titrimetric and gravimetric analysis; a few illustrations of what might be called "simple instrumental" methods are included. Students are urged to acquaint themselves fully with the theoretical principles discussed in the earlier chapters before attempting the laboratory exercises.

22.1

ACID-BASE TITRATIONS

Directions are given for the preparation and standardization of hydrochloric acid and sodium hydroxide solutions, and for these solutions to be used in several analyses. The theoretical principles of acid-base titrations are discussed in Chapters 3 and 6.

Experiment 1. ***Preparation of 0.1 N Solutions of Hydrochloric Acid and Sodium Hydroxide***

Procedure *(a) Hydrochloric Acid.* Measure into a clean, glass-stoppered bottle approximately 1 liter of distilled water. With a graduated cylinder or measuring pipet, add to the water about 8.5 mL of concentrated hydrochloric acid. Stopper the bottle, mix the solution well by inversion and shaking, and label the bottle.

(b) *Sodium Hydroxide*. Carbonate-free sodium hydroxide can be prepared most readily from a concentrated solution of the base, because sodium carbonate is insoluble in such a solution. A 1 : 1 solution of sodium hydroxide in water is available commercially, or may have been prepared by the instructor (Note 1). Carefully add 6 to 7 mL of this solution to approximately 1 liter of distilled water (Note 2) in a clean bottle, using a graduated pipet and rubber bulb. Close the bottle with a rubber stopper (Note 3), shake the solution well, and label the bottle.

Notes

1. If such a solution is not available, dissolve about 50 g of sodium hydroxide in 50 mL of water in a small, rubber-stoppered Erlenmeyer flask. Be careful in handling this solution, as considerable heat is generated. Allow the solution to stand until the sodium carbonate precipitate has settled. If necessary, the solution can be filtered through a Gooch crucible. Alternatively, carbonate-free base can be prepared by dissolving 4.0 to 4.5 g of sodium hydroxide in about 400 mL of distilled water and adding 10 mL of 0.25 M barium chloride solution. The solution is well mixed and then allowed to stand overnight so that barium carbonate will settle out. The solution is then decanted from the solid into a clean bottle and diluted 1 liter.
2. Some directions call for boiling the water about 5 min to remove carbon dioxide. If this is done (consult the instructor), be sure to protect water from the atmosphere as it cools.
3. Glass-stoppered bottles should not be used, since alkaline solutions cause the stoppers to stick so tightly that they are difficult or impossible to remove. Polyethylene bottles, if available, are excellent for storing dilute base solutions. It may be desirable to protect the solution from atmospheric carbon dioxide (consult the instructor). This can be done by fitting a two-holed siphon stopper with a siphon and soda-lime tube, as shown in Fig. 22.1.

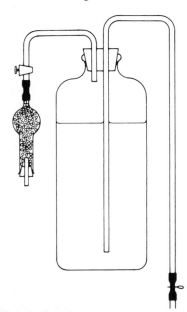

Figure 22.1 Bottle for storing carbonate-free base.

Experiment 2. *Determination of the Relative Concentrations of the Hydrochloric Acid and Sodium Hydroxide Solutions*

In this experiment the ratio of the concentrations of the acid and base solutions is determined. Following standardization of either solution, the normality of the other can be calculated from this ratio.

Procedure   Rinse two clean burets and fill one with the hydrochloric acid and the other with the sodium hydroxide solution prepared in Experiment 1. Remove any air bubbles from the tips, lower the liquid level to the graduated portions, and record the initial reading of each buret.

Now run about 35 to 40 mL of the hydrochloric acid solution into a clean 250-mL Erlenmeyer flask and record the buret reading (Note 1). Add two drops of phenolphthalein indicator (Note 2) and about 50 mL of water from a graduated cylinder, rinsing down the walls of the flask. Now run into the flask the sodium hydroxide solution from the other buret, swirling the flask gently and steadily to mix the solutions. As an aid in preventing overrunning of the end point, notice the transient, local, pink coloration as it becomes more persistent with the progress of the titration. Finally, when the color first pervades the entire solution even after thorough mixing, stop the titration and record the buret reading. The color should persist for at least 15 s or so, but may gradually fade because of the absorption of atmospheric carbon dioxide. It is well to rinse down the inside of the flask and also the buret tip with distilled water just before the termination of the titration so that stray droplets will not escape the reaction. If the end point is accidentally overrun, the titration can still be salvaged: Run enough hydrochloric acid solution into the flask to turn the phenolphthalein indicator colorless, record the buret reading again, and then approach the end point once more with the sodium hydroxide solution.

Repeat the titration at least two more times (Note 3). Finally calculate the volume of acid equivalent to 1 mL of base:

$$1.000 \text{ mL of base} = \frac{\text{volume of acid}}{\text{volume of base}}$$

Use buret corrections if necessary (consult instructor).

Notes

1. This volume is recommended in order to minimize errors in reading a buret. The acid is usually titrated with base instead of base with acid to minimize absorption of carbon dioxide during the titration.
2. Methyl red, bromthymol blue, and other indicators can also be employed. Solutions are prepared as follows: *Phenolphthalein*: Dissolve 2 g of phenolphthalein per liter of 95% ethanol. *Methyl red*: Dissolve 1 g of the sodium salt of methyl red per liter of water. *Bromthymol blue*: Dissolve 0.1 g bromthymol blue in 8 mL of 0.02 M NaOH and dilute to 100 mL with water.
3. Do not allow the sodium hydroxide to remain in the buret any longer than necessary. As soon as the titrations are finished, drain the base from the buret and rinse thoroughly, first with dilute hydrochloric acid and then with water.

Experiment 3.   *Standardization of Sodium Hydroxide Solution with Potassium Acid Phthalate*

A number of good primary standards are available for standardizing base solutions. Directions are given here for the use of potassium acid phthalate, but these can be readily modified to suit another standard.

Procedure   Place about 4 to 5 g of pure potassium acid phthalate in a clean weighing bottle and dry the sample in an oven at 110°C for at least 1 h. Cool the bottle and its contents in a desiccator (Note 1). Weigh accurately into each of three clean, numbered Erlenmeyer flasks about 0.7 to 0.9 g of the potassium acid phthalate (Note 2). Record the weights in your notebook.

To each flask add 50 mL of distilled water (Note 3) from a graduated cylinder and shake the flask gently until the sample is dissolved. Add 2 drops of phenolphthalein to each flask. Rinse and fill a buret with the sodium hydroxide solution. Titrate the solution in the first flask with sodium hydroxide to the first permanent pink color. Your hydrochloric acid solution, in a second buret, may be used for back-titration if required. Repeat the titration with the other two samples, recording all data in your notebook.

Calculate the normality of the sodium hydroxide solution obtained in each of the three determinations. Average these values and compute the average deviation in the usual manner. If the average deviation exceeds about 2 ppt, consult the instructor. Finally, calculate the normality of the hydrochloric acid solution from the normality of the sodium hydroxide and the volume ratio of the acid and base obtained in Experiment 2.

Notes

1. Potassium acid phthalate is relatively nonhygroscopic, and the drying process may be omitted (consult the instructor).
2. Since 4 meq of potassium acid phthalate weights $4 \times 204.2 = 816.8$ mg, the quantity recommended should require 35 to 45 mL of a 0.1 N base solution for titration.
3. Some directions call for boiling the water for about 5 min to remove carbon dioxide before use. If this is done (consult the instructor), the water should be cooled to room temperature before the titration, and it should be protected from the atmosphere while cooling.

Experiment 4.   *Standardization of the Hydrochloric Acid Solution*

The hydrochloric acid solution can be standardized against a primary standard if so desired. Sodium carbonate is a good standard and is particularly recommended if the acid solution is to be used to titrate carbonate samples. The organic base tris(hydroxymethyl)aminomethane, $(CH_2OH)_3CNH_2$, also called TRIS or THAM, is also an excellent primary standard for acid solutions (Section 6.7.b). Directions are given for the use of each of these compounds.

Procedure   *(a) Sodium Carbonate.* Accurately weigh three samples (about 0.20 to 0.25 g each) of pure sodium carbonate (Note 1), which has been previously dried, into three Erlenmeyer flasks. Dissolve each sample with about 50 mL of distilled water and add two drops of methyl red[1] or methyl orange (see Note 2 and consult the instructor).

(a-1) Methyl Red. Titrate each sample with the hydrochloric acid solution. Methyl red is red in acid and yellow in basic solution. As soon as the solution becomes distinctly red, add an additional 1 mL of hydrochloric acid and remove the carbon dioxide by boiling the solution gently for about 5 min (Note 3). Cool the solution to room temperature and complete the titration. Back-titration can be done with the sodium hydroxide solution previously prepared. If the color change is not sharp, repeat the heating to remove carbon dioxide.

[1] The color changes shown by these indicators can be modified by addition of a suitable dye. Methyl red is frequently mixed with methylene blue, the color change then being red-violet (acid) to green (base) with an intermediate shade of gray. The dye xylene cyanole FF is added to methyl orange to give "modified methyl orange." The color change is then pink (acid) to green (base) with a gray intermediate. See I. M. Kolthoff and C. B. Rosenblum, *Acid-Base Indicators*, Macmillan Publishing Co., New York, 1937.

(a-2) Methyl Orange. Prepare a solution of pH 4 by dissolving 1 g of potassium acid phthalate in 100 mL of water. Add two drops of methyl orange to this solution and retain it for comparison purposes. Now titrate each sample with hydrochloric acid until the color matches that of the comparison solution.

Calculate the normality of the hydrochloric acid solution obtained in each of the three titrations. Average these values and compute the average deviation in the usual manner. If this figure exceeds 2 to 3 ppt, consult the instructor. The normality of the sodium hydroxide solution can be calculated from the normality of the acid and the relative concentrations of acid and base obtained in Experiment 2.

(b) TRIS. TRIS is a monoprotic base, $pK_b = 5.93$, which can be considered a derivative of ammonia. It reacts with a single hydrogen ion,

$$RNH_2 + H_3O^+ \longrightarrow RNH_3^+ + H_2O$$

and the pH at the EPt is about 4.8. Bromocresol green (Note 4) is a suitable indicator for the titration. The EW of TRIS is 121.14 g/eq.

Accurately weigh three samples (about 0.40 to 0.50 g each) of TRIS (Note 5), which has been previously dried, into three Erlenmeyer flasks. Dissolve the first sample in about 50 mL of distilled water and add enough bromocresol green to give the solution a distinct blue color. Immediately titrate with the hydrochloric acid solution until the color changes from blue to green. If the end point is overrun, the color may become yellow. If necessary, back-titration can be done with the sodium sodium hydroxide solution previously prepared.

Repeat the titration on the other two samples. Calculate the normality of the hydrochloric acid solution obtained in each of the three titrations. Average these values and compute the average deviation in the usual manner. The normality of the sodium hydroxide solution can be calculated from the normality of the acid and the relative concentrations of acid and base obtained in Experiment 2.

Notes

1. Analytical-grade sodium carbonate (assay value 99.95%) can be used after drying for about $\frac{1}{2}$h at 270 to 300°C.
2. Various indicators and mixed indicators have been suggested for this titration. The pH at the equivalence point of the reaction

$$CO_3^{2-} + 2H^+ \rightleftharpoons H_2CO_3$$

is about 4, and methyl orange changes color near this pH. The titration curve is not very steep, however, and hence it is often suggested that excess acid be added and carbon dioxide removed by boiling or vigorous shaking. The subsequent titration of excess acid with base involves only strong electrolytes, and a sharp end point is obtained if the carbon dioxide is completely removed. An indicator blank must then be determined, since methyl orange changes color at a pH appreciably different from 7. If methyl red (pH 5.4) is employed, no indicator blank is necessary. Directions are given here for the titration to the methyl orange end point without removal of carbon dioxide and for the titration to the methyl orange end point without removal of carbon dioxide. The latter procedure is more rapid and is recommended where a high degree of accuracy is not required.
3. If insufficient acid is present to completely convert bicarbonate into carbonic acid, the indicator will turn back to its basic color as the carbon dioxide is expelled and the pH rises. The titration is then continued with acid. If excess acid is added, the indicator will retain the acid color and the titration is continued by addition of base.

4. Dissolve 0.1 g of bromocresol green in 1.45 mL of 0.100 M NaOH and dilute to 100 mL with distilled water.
5. TRIS is available commercially at a purity of about 99.95%. It should be dried for about 1 h at 100 to 105°C. Solutions of TRIS will absorb CO_2 from the air, and hence a sample should be titrated immediately after it is dissolved.

Experiment 5. *Determination of the Purity of Potassium Acid Phthalate*

Directions are given for the titration of a solid acid sample. Commercial unknowns usually contain potassium acid phthalate or sulfamic acid, either of which can be titrated with phenolphthalein indicator.

Procedure Dry the unknown sample in a weighing bottle (unless otherwise directed) at 110°C for at least 1 h and cool in a desiccator. Weigh three samples of appropriate size (Note 1) into clean Erlenmeyer flasks and dissolve in 75 to 100 mL of distilled water (Note 2). Add two drops of phenolphthalein indicator to each flask.

Titrate the contents of the first flask with the standard sodium hydroxide solution to the first permanent pink color. Record all buret readings as usual and then titrate the other two samples in the same manner.

Calculate the percentage of potassium acid phthalate in each sample and obtain the average purity and average deviation (Note 3). Report the percentage purity as suggested on page 602.

Notes

1. The instructor will specify the size sample required to use 30 to 45 mL of a 0.1 N base for titration. If this weight is greater than 1 g, as it may be with potassium acid phthalate samples of low percentage purity, the samples need be weighed only to the nearest milligram. Why?
2. Some directions call for boiling the water for about 5 min to remove carbon dioxide before use. If this is done (consult the instructor), the water should be cooled to room temperature before the titration, and it should be protected from the atmosphere while cooling.
3. The precision that can be obtained depends upon the homogeneity of the sample as well as the technique of the student. Consult the instructor for the expected precision.

Experiment 6. *Determination of Acetic Acid Content of Vinegar*

The principal acid in vinegar is acetic acid, and federal standards require at least 4 g of acetic acid per 100 mL of vinegar. The total quantity of acid can be readily determined by titration with standard base using phenolphthalein indicator. Although other acids are present, the result is calculated as acetic acid.

Procedure Pipet 25 mL of vinegar into a 250-mL volumetric flask, dilute to the mark, and mix thoroughly (Note). Pipet a 50-mL aliquot of this solution into an Erlenmeyer flask and add 50 mL of water and two drops of phenolphthalein indicator. Titrate with standard base to the first permanent pink color. Repeat the titration on two additional aliquots.

Assuming all the acid to be acetic, calculate the number of grams of acid per 100 mL of vinegar solution. Assuming that the density of vinegar is 1.000, what is the percentage of acetic acid by weight in vinegar? Average your results in the usual manner.

Note

The quantity should require a reasonable volume of 0.1 N base for titration. If a more concentrated base solution is employed, a 100-mL volumetric flask can be used. The dilution with water prevents the color of the vinegar solution from interfering in detection of the end point.

Experiment 7. ## Determination of the Alkalinity of Soda Ash

Crude sodium carbonate, called soda ash, is commonly used as a commercial neutralizing agent. The titration with standard acid to the methyl orange end point gives the total alkalinity, which is mainly due to sodium carbonate. Small amounts of sodium hydroxide and sodium bicarbonate may also be present. The results are usually expressed as percentage of sodium carbonate or sodium oxide.

Since the samples are frequently nonhomogeneous, the method of aliquot portions is employed. Either methyl red or methyl orange can be employed as the indicator.

Procedure   Weigh accurately into a clean 250-mL beaker a sample of the dried unknown of appropriate size (Note). Dissolve the sample in about 125 mL of distilled water. Place a clean funnel in a 250-mL volumetric flask and transfer the solution from the beaker to the flask. Rinse the beaker, add the rinsings to the flask, and finally dilute to the mark. Mix the contents of the flask thoroughly by inversion and shaking.

Pipet a 50-mL aliquot into an Erlenmeyer flask and add two drops of methyl red or methyl orange. Titrate with standard acid according to procedure (a-1) or (a-2), page 614. Repeat the titration with two other 50-mL aliquots. At the end of the titrations, be sure to empty and thoroughly rinse the volumetric flask. An alkaline solution should not be left in a volumetric flask for a long period of time.

Report the percentage of sodium carbonate or sodium oxide (consult the instructor) in the sample. A precision of 3 to 5 ppt is not unusual for the titration.

Note

The instructor will specify the size of sample required to use 30 to 45 mL of 0.1 N acid for titration.

Experiment 8. ## Titrations in Nonaqueous Media

The principles of titrations in media other than water are discussed in Chapter 6. Such titrations are rather widely used today, and it is appropriate to include one or two in a beginning course in volumetric analysis. The procedures given below employ a solution of perchloric acid in glacial acetic acid. This solution can be standardized against pure potassium acid phthalate, which acts as a base in acetic acid solvent. The solution is then used to titrate an amine, which is too weak a base to be titrated in water. Methyl violet is used as the indicator, or the end point can be detected potentiometrically using a glass and calomel electrode pair. These directions are based upon the recommendations of Fritz.[2]

Procedure   **Preparation of Solutions**

(a) 0.1 N $HClO_4$. Add 4.3 mL of 72% perchloric acid to 150 mL of glacial acetic acid, mix well, add 10 mL of acetic anhydride, and allow the solution to stand for 30 min (Note 1). Dilute to 500 mL with glacial acetic acid and allow the solution to cool to room temperature.

[2] J. S. Fritz, *Acid-Base Titrations in Nonaqueous Solutions*, G. F. Smith Chemical Co., Columbus, Ohio, 1952.

(b) *0.1 N sodium acetate*. Dissolve 4.1 g of anhydrous sodium acetate in glacial acetic acid and dilute to 500 mL with the acid.

(c) *Methyl violet indicator*. Dissolve 0.2 g of methyl violet in 100 mL of chlorobenzene.

Relative Concentrations. Fill two burets with the perchloric acid and sodium acetate solutions (Note 2). Withdraw about 35 to 40 mL of perchloric acid, add two drops of methyl violet indicator, and titrate with the sodium acetate solution. Take the first permanent violet tinge as the end point. Repeat the titration with two additional samples and calculate the relative concentrations of the two solutions.

Standardization. Weigh samples of about 0.5 to 0.6 g of pure potassium acid phthalate into three Erlenmeyer flasks and add about 60 mL of glacial acetic acid to each flask. Heat the first flask cautiously until the sample is in solution. Then cool and add two drops of methyl violet. Titrate with perchloric acid solution to the disappearance of the violet tinge. Use the sodium acetate solution for back-titration if needed.

Repeat the titration with the other two samples and then calculate the normality of the perchloric acid solution. From the relative concentration obtained above, calculate the normality of the sodium acetate solution (Note 3).

Analysis. Amino acids or amines can be titrated with perchloric acid in acetic acid. If the sample is an amino acid, it is recommended that excess perchloric acid be added and the excess titrated with sodium acetate. Amines can be titrated directly with perchloric acid.

Weigh three samples of the substance to be determined, taking about 3 meq for each sample (consult the instructor). If the sample is an amino acid, dissolve in exactly 50 mL (pipet) of standard perchloric acid. Add two drops of methyl violet indicator and back-titrate with sodium acetate, taking the first permanent violet tinge as the end point. If the sample is an amine, dissolve it in about 50 mL of acetic acid and titrate with standard perchloric acid to the first appearance of the violet color.

Calculate the number of milliequivalents of amino acid or amine in the sample and report as desired by the instructor (Note 4).

Notes

1. The acetic anhydride is added to react with water in the perchloric acid.
2. Avoid contact of acetic acid solutions with the skin. If these solutions are spilled on the hands, rinse the hands immediately with tap water.
3. Room temperature should be noted when these titrations are carried out. If there is a large change in temperature between the time of standardization and analysis, a correction should be applied to the volume of acetic acid solution.
4. The instructor may prefer to give pure samples of the amine or amino acid as unknowns. The student then reports the equivalent weight of the unknown.

22.2

PRECIPITATION AND COMPLEX-FORMATION TITRATIONS

The theoretical principles of complex formation and precipitation titrations are discussed in Chapters 8 and 9. In this section, laboratory directions are given for precipitation titrations involving the reaction of silver cation with chloride and thiocyanate anions. The Mohr, Volhard, and Fajans methods are illustrated. Titrations involving complex formation include the use of EDTA in determining the hardness of water and analyzing for calcium in calcium carbonate and in a drug.

Experiment 9. *Preparation of 0.1 M Solutions of Silver Nitrate and Potassium Thiocyanate*

Silver nitrate can be weighed as a primary standard. However, a solution of approximately the desired molarity is usually prepared and then standardized using the indicator which will be employed in the analysis.

Procedure   *(a) Silver Nitrate.* Weigh on a trip balance about 8.5 g of silver nitrate (Note 1), dissolve the salt in distilled water in a beaker, and transfer the solution to a clean bottle (Note 2). Dilute the solution to 500 mL with distilled water, mix thoroughly, and label the bottle. Protect the solution from the sunlight as much as possible.

(b) Potassium Thiocyanate. If the Volhard method is to be used, prepare a 0.1 *M* thiocyanate solution by weighing about 4.9 g of potassium thiocyanate (Note 3) and dissolving the salt in water. Dilute the solution to about 500 mL and store in a clean bottle.

   Notes

1. The molecular weight is 169.87. To prepare 500 mL of a 0.1 *M* solution, 8.5 g are needed, which is ample for standardization and determination of an unknown. If 1 liter of solution is desired, take 17 g.
2. A brown-glass bottle is recommended to protect the solution from sunlight.
3. The molecular weight is 97.18. If 1 liter of the solution is desired, take 9.7 g of the salt.

Experiment 10. *Standardization of Silver Nitrate and Potassium Thiocyanate Solutions*

Sodium or potassium chloride can be used as a primary standard for a silver nitrate solution. Either potasssium chromate (Mohr method) or dichlorofluorescein (Fajans, or adsorption indicator, method) can be used as the indicator. Directions are given for both indicators below.

   After standardization of the silver nitrate, this solution can be used to standardize the potassium thiocyanate solution. The two solutions are titrated directly, using iron(III) ion as the indicator.

Procedure   *(a) Mohr Method.* Weigh accurately three samples of pure, dry sodium chloride of about 0.20 to 0.25 g each (Notes 1 and 2) into 250-mL Erlenmeyer (Note 3) flasks. Dissolve each sample in 50 mL of distilled water and add 2 mL of 0.1 *M* potassium chromate solution (Note 4). Titrate the first sample with silver nitrate, swirling the solution constantly, until the reddish color of silver chromate begins to spread more widely through the solution, showing that the end point is almost reached. The formation of clumps of silver chloride is also an indicator that the end point is near. Continue the addition of silver nitrate dropwise until there is a permanent color change from the yellow of the chromate ion to the reddish color of silver chromate precipitate. Run an indicator blank (Note 5) if desired (consult the instructor).

Titrate the other two samples in the same manner. Calculate the molarity of the silver nitrate solution and its chloride titer.

(b) Fajans Method. Weigh accurately three samples of pure, dry sodium chloride of about 0.20 to 0.25 g each (Notes 1 and 2) into 250-mL Erlenmeyer flasks. Dissolve each sample in 50 mL of water and add ten drops of dichlorofluorescein indicator (Note 6) and 0.1 g dextrin (Note 7). Titrate the first sample with silver nitrate to the point where the color of the dispersed silver chloride changes from yellowish white to a definite pink (Note 8). The color change is reversible, and back-titration can be done with a standard sodium chloride solution. Titrate the other two samples in the same manner. Calculate the molarity of the silver nitrate solution and its chloride titer.

Notes

1. The molecular weight is 58.44. If potassium chloride (MW 74.55) is used, take 0.25 to 0.30 g. The method of aliquot portions may be used if desired. Weigh accurately 1.1 to 1.2 g sodium chloride (1.5 to 1.6 g potassium chloride), dissolve the salt in a 250-mL volumetric flask, and withdraw 50-mL aliquots for titration.
2. Dry the salt at 120°C for at least 2 h. For complete removal of the last traces of water, these salts should be heated to about 500 to 600°C in an electric furnace. This is not necessary except for very precise work.
3. Porcelain casseroles are sometimes recommended. The reddish color of silver chromate is more readily distinguished against a white background.
4. Dissolve 19.4 g in 1 liter of water.
5. To determine the indicator blank, add 2 mL of the indicator to about 100 mL of water to which is added a few tenths of a gram of chloride-free calcium carbonate. This gives a turbidity similar to that in the actual titration. Swirl the solution and add silver nitrate dropwise until the color matches that of the solution that was titrated. The blank should not be larger than about 0.05 mL.
6. This solution is prepared by dissolving 0.1 g of sodium dichlorofluoresceinate in 100 mL of water (or 0.1 g of dichlorofluorescein in 100 mL of 70% alcohol). Fluorescein, or its sodium salt, can be used in place of dichlorofluorescein.
7. The function of dextrin is to prevent coagulation of colloidal silver chloride.
8. The end point is easier to detect in diffuse light. Avoid direct sunlight.

(c) Titration of Silver Nitrate with Potassium Thiocyanate. Pipet 25 mL of standard silver nitrate solution into a 250-mL Erlenmeyer flask. Add 5 mL of 1:1 nitric acid (Note 1) and 1 mL of iron(III) alum solution as the indicator (Note 2). Titrate with thiocyanate, swirling the solution constantly, until the reddish brown color begins to spread throughout the solution. Then add the thiocyanate dropwise, shaking the solution thoroughly between addition of drops. The end point is marked by the permanent appearance of the reddish color of the iron-thiocyanate complex.

Titrate two additional portions of the silver nitrate solution with thiocyanate. Calculate the molarity of the thiocyanate solution.

Notes

1. If the nitric acid has a yellow tinge indicating the presence of oxides of nitrogen, boil the solution until the oxides are expelled.
2. This is a saturated solution of iron(III) ammonium sulfate in 1 *M* nitric acid.

Experiment 11. **Determination of Chloride**

Chloride can be determined by titration with standard silver nitrate. A direct titration using either chromate ion or dichlorofluorescein as the indicator can be employed. Alternatively, excess standard silver solution can be added to the unknown and the excess titrated with standard potassium thiocyanate. Directions are given below for all three procedures. It is assumed that the sample is water-soluble.

Procedure *(a) Mohr Method.* Accurately weigh three samples of the dried material (Note 1) into three 250-mL Erlenmeyer flasks (Note 2). Dissolve each sample in about 50 mL of distilled water (Note 3) and add 2 mL of 0.1 M potassium chromate solution. Titrate the first sample with standard silver nitrate as directed in Experiment 10(a)

Titrate the other two samples in the same manner. Calculate the percentage of chloride in the sample.

(b) Fajans Method. Accurately weigh three samples of the dried material of appropriate size (Note 1) into three 250-mL Erlenmeyer flasks. Dissolve each sample in about 50 mL of distilled water (Note 3) and add ten drops of dichlorofluorescein indicator and 0.1 g of dextrin. Titrate the first sample with standard silver nitrate as directed in Experiment 10(b).

Titrate the other two samples in the same manner. Calculate the percentage of chloride in the sample.

Notes

1. Consult the instructor regarding the size of the sample. A large sample may be dissolved in a 250-mL volumetric flask and aliquot portions titrated, if desired.
2. Porcelain casseroles are sometimes recommended.
3. The solution should be nearly neutral. Dissolve a small portion of the material in about 10 mL of water and place a drop of the solution on a piece of litmus paper. If the solution is basic, add one drop of phenolphthalein to each solution and then add dilute nitric acid (about 1 mL to 150 mL of water) dropwise until the pink color of the indicator is just discharged. If the solution is acidic, add one drop of phenolphthalein to each solution and then add 0.1 N sodium hydroxide dropwise until the solution is barely pink. Then add one or two drops of dilute nitric acid until the pink color of the indicator is just discharged.

(c) Volhard Method. Accurately weigh three samples of appropriate size (Note 1) into three 250-mL Erlenmeyer flasks. Dissolve each sample in about 50 mL of distilled water and then add 5 mL of 1:1 nitric acid (Note 2). To the first sample add standard silver nitrate solution from a buret until an excess of about 5 mL is present (Note 3). Add 1 to 2 mL of nitrobenzene (Note 4), stopper the flask with a rubber stopper, and shake the flask vigorously until the silver chloride is well coagulated (about 30 s). Now add 1 mL of iron(III) alum indicator (Note 5) and titrate the excess silver nitrate with standard potassium thiocyanate solution. The end point is marked by the permanent appearance of the reddish color of the iron-thiocyanate complex.

Titrate the other two samples in the same manner. Calculate the percentage of chloride in the sample.

Notes

1. Consult the instructor. If desired, a larger sample may be taken, dissolved in a 250-mL volumetric flask, and aliquot portions titrated.
2. See Note 1 of Experiment 10(c).
3. The silver chloride coagulates near the equivalence point. Shake the solution well, allow it to stand for a few moments for the precipitate to settle, and then add a few drops of silver nitrate to the supernatant liquid. If no precipitate forms, silver nitrate is in excess.

4. Instead of adding nitrobenzene, one may filter off the silver chloride with a Gooch or sintered-glass crucible. The precipitate is then washed with 1% nitric acid solution, and the washings are added to the filtrate. The indicator is then added to the filtrate and the latter titrated with potassium thiocyanate.
5. See Note 2 of Experiment 10(c).

Experiment 12. *Determination of Silver in an Alloy*

Silver can be determined by direct titration with potassium thiocyanate using iron(III) ion as the indicator.

Procedure   Accurately weigh three samples of a silver alloy of appropriate size (consult the instructor) and place each in a 250-mL Erlenmeyer flask. Dissolve the first sample in 15 mL of 1:1 nitric acid. Boil the solution until all the oxides of nitrogen are removed and then dilute the solution to about 50 mL. Add 1 mL of iron(III) alum indicator (Note) and titrate with standard thiocyanate solution to the appearance of the reddish color of the iron-thiocyanate complex.

Dissolve and titrate the other two samples in the same manner. Calculate the percentage of silver in the alloy.

Note

See Note 2 of Experiment 10(c).

Experiment 13. *Preparation and Standardization of Sodium EDTA Solution*

Titrations involving the use of the chelating agent EDTA are described in Chapter 8. Directions are given below for the preparation of a 0.01 M solution of sodium EDTA and the standardization against calcium chloride.

Procedure   Weigh about 4 g of disodium dihydrogen EDTA dihydrate and about 0.1 g of $MgCl_2 \cdot 6H_2O$ into a clean 400-mL beaker. Dissolve the solids in water, transfer the solution to a clean 1-liter bottle, and dilute to about 1 liter (Note 1). Mix the solution thoroughly and label the bottle.

Prepare a standard calcium chloride solution as follows. Weigh accurately about 0.4 g of primary standard calcium carbonate that has been previously dried at 100°C. Transfer the solid to a 500-mL volumetric flask, using about 100 mL of water. Add 1:1 hydrochloric acid dropwise until effervescence ceases and the solution is clear. Dilute with water to the mark and mix the solution thoroughly.

Pipet a 50-mL portion of the calcium chloride solution into a 250-mL Erlenmeyer flask and add 5 mL of an ammonia–ammonium chloride buffer solution (Note 2). Then add five drops of Eriochrome Black T indicator (Note 3). Titrate carefully with the EDTA solution to the point where the color changes from wine red to pure blue. No tinge of red should remain in the solution.

Repeat the titration with two other aliquots of the calcium solution. Calculate the molarity of the EDTA solution and the calcium carbonate titer.

Notes

1. If the solution is turbid, add a few drops of 0.1 M sodium hydroxide solution until the solution is clear.
2. Prepare this solution by dissolving about 6.75 g of ammonium chloride in 57 mL of concentrated ammonia and diluting to 100 mL. The pH of the buffer is slightly above 10.
3. Prepare by dissolving about 0.5 g of reagent-grade Eriochrome Black T in 100 mL of alcohol. If the solution is to be kept, date the bottle. It is recom-

mended that solutions older than 6 weeks to 2 months not be used. Alternatively, the indicator may be used as a solid, which has a much longer shelf life. It is prepared by grinding 100 mg of the indicator into a mixture of 10 g of NaCl and 10 g of hydroxylamine hydrochloride. A small amount of the solid mixture is added to each titration flask with a spatula.

Alternatively, Calmagite may be used as the indicator. A solution is prepared by dissolving 0.05 g of the indicator in 100 mL of water. Add four drops of the indicator to each flask. The color change is from red to blue, as with Eriochrome Black T.

Experiment 14. *Determination of Calcium by EDTA Titration*

The standard EDTA solution can be used to determine the purity of calcium carbonate samples. In the procedure below the method of aliquot portions is employed.

Procedure The unknown sample should contain about 350 to 400 mg of $CaCO_3$ in order that 35 to 40 mL of titrant be used. Consult the instructor as to the size of sample to be taken. Weigh the sample accurately into a clean 400-mL beaker and carefully add 5 to 10 mL of 1:1 hydrochloric acid until the sample dissolves. Then add 50 mL of distilled water and boil the solution gently for a few minutes to remove CO_2. Cool the solution, add two drops of methyl red indicator (Note 1), and neutralize the solution with 6 M NaOH. Transfer the solution to a 500-mL volumetric flask, dilute with water to the mark, and mix thoroughly.

Pipet a 50-mL portion of the solution into a 250-mL Erlenmeyer flask and add 5 ml of an ammonia–ammonium chloride buffer solution (Note 2). Then add five drops of Eriochrome Black T indicator (Note 3). Titrate carefully with standard EDTA solution to a point where the color changes from wine red to blue. No tinge of red should remain in the solution.

Repeat the titration on two other aliquots of the unknown solution. Calculate the percentage of CaO in the sample.

Notes

1. See Note 2 of Experiment 2.
2. See Note 2 of Experiment 13.
3. See Note 3 of Experiment 13.

Experiment 15. *Determination of the Total Hardness of Water*

The standard EDTA solution can be used to determine the total hardness of water.

Procedure Obtain the water to be analyzed from the instructor and pipet a portion into each of three 250-mL Erlenmeyer flasks (Note 1). To the first sample add 1.0 mL of the buffer solution (Note 2) and five drops of indicator solution (Note 3). Titrate with the standard EDTA solution to a color change of wine red to pure blue.

Repeat the procedure on the other two portions of water. Calculate the total hardness of the water as parts per million (ppm) of calcium carbonate. This is done as follows:

$$\text{Volume EDTA (mL)} \times CaCO_3 \text{ titer (mg/mL)} = \text{mg } CaCO_3$$

$$\frac{1000 \text{ mL/liter} \times \text{mg } CaCO_3}{\text{mL sample}} = \text{mg } CaCO_3/\text{liter, or ppm}$$

Notes

1. The volume of water titrated should be chosen so that 40 to 50 mL of the EDTA solution will be used for titration.
2. See Note 2 of Experiment 13.
3. See Note 3 of Experiment 13.

Experiment 16. *Determination of Calcium in Calcium Gluconate*[3]

Of all the minerals in the body calcium is the most abundant. Ninety-eight percent of the body's calcium is in the bones, and 1% is in the teeth. The other 1% present in all the other tissues is essential for certain metabolic reactions, such as the contraction of muscles.

Blood serum contains 4.5 to 5.5 mmol/liter of calcium. A significant increase or decrease produces dire pathological symptoms. Hypocalcemia results from a deficiency in parathyroid hormone, resistance to the hormone, or a deficiency in vitamin D. Severe tetany occurs when the calcium concentration falls to 3.5 mmol/liter.

Standard treatment of a hypocalcemic patient consists of intravenous injection of calcium gluconate, along with parathyroid extract and vitamin D. In less acute cases 1 to 5 g of calcium gluconate may be given orally three times daily.

Calcium gluconate is a pleasant-tasting, white compound available as tablets or in solution. The anion is from a weak organic acid and is not as likely to evoke acidosis as is chloride ion. Calcium gluconate thus is accepted as the prototype for calcium preparations in medicine.

Procedure Calcium in calcium gluconate will be determined by titration with a standard solution of EDTA. It is recommended that the concentration of EDTA be about 0.025 M. Follow the procedure for preparing the solution given in Experiment 13, except use about 10 g of $Na_2H_2Y \cdot 2H_2O$. If some solid fails to dissolve, add NaOH pellets one at a time and shake the solution until no undissolved solid remains.

Prepare the standard calcium chloride solution as in Experiment 13, except use about 1.0 g of primary standard grade calcium carbonate. Titrate three 50-mL aliquots of this solution with the EDTA solution as directed in Experiment 13.

Secure your calcium gluconate tablets from the instructor (Note 1). Crush them with the blade of a stainless steel spatula and mix the powder. Dry the powder at 100°C for 1 h. Weigh accurately a 0.4-g sample into each of three clean 250-mL Erlenmeyer flasks and dissolve each sample in 50 mL of distilled water. Some heating may be required. To the first sample add 5 mL of the ammonia–ammonium chloride buffer (Note 2 of Experiment 13) and five drops of Eriochrome Black T indicator (Note 2). Titrate with the standard EDTA solution to a color change of wine red to pure blue.

Repeat the procedure on the other two samples. Calculate the percentage of calcium in the drug.

Notes

1. The tablets are available without a prescription from a drug store.
2. See Note 3 of Experiment 13.

---

[3] This experiment was designed by Professor Hubert L. Youmans of Western Carolina University, who has kindly given us permission to use it here.

GRAVIMETRIC METHODS OF ANALYSIS

Gravimetric methods of analysis and the principles of separation by precipitation are discussed in Chapters 4 and 9. In this section, directions are given for the precipitation of silver chloride, barium sulfate, and hydrous iron(III) hydroxide. These three compounds illustrate the common types of precipitates discussed in Chapter 4: curdy, crystalline, and gelatinous. Directions are also given for the homogeneous precipitation of calcium oxalate and for the precipitation of nickel using the organic precipitant dimethylglyoxime.

Experiment 17. *Determination of Chlorine in a Soluble Chloride*

The usual samples given to students are readily soluble in water, and no interfering ions are present. Hence this is a very simple determination that requires a relatively short time, and it is quite suitable for the introduction to gravimetric techniques. See Chapter 4 for a discussion of the properties of silver chloride and the possible errors encountered in its use in gravimetric analysis.

Procedure The sample should be dried for 1 to 2 h in an oven at 100 to 120°C. Clean three (Note 1) porous-bottom filtering crucibles (either sintered glass or porous porcelain) of "medium" porosity. Much of the dirt may be removed with detergents, although acids or other "strong" reagents may be necessary at times, depending upon the nature of the contamination. It is often well to draw a little concentrated nitric acid slowly through the porous filtering disc to eliminate certain material with which the disk may be impregnated. It is not wise to use cleaning solution on a porous-bottom crucible. The cleaned and thoroughly rinsed crucibles should be dried to constant weight at the same temperature at which the silver chloride is later to be dried. An oven is recommended, and temperatures of 100 to 150°C are suitable for ordinary student work.

Weigh out accurately three portions of the dried sample of about 0.5 to 0.7 g each (Note 2). Dissolve each portion in a 250-mL beaker, using 100 to 150 mL of water to which about 1 mL of concentrated nitric acid has been added. Prepare an approximately 0.1 M solution of silver nitrate (15 to 20 mg of $AgNO_3$/mL). Now heat the first of the chloride solutions to boiling, and with constant stirring add silver nitrate slowly to precipitate silver chloride. Obviously, in the case of an unknown chloride solution, the quantity of silver nitrate to be added cannot be specified. The student must determine when precipitation is complete. This is done by adding the silver nitrate in small portions, stirring vigorously, allowing the precipitate to settle somewhat, and noting whether a new cloud of precipitate appears upon further addition. After precipitation is complete, add about 10% more silver nitrate solution.

After the precipitate has coagulated well, remove the beaker from the heat, cover it with a watch glass, and set it aside to cool in the laboratory bench (for protection from light) for at least 1 h. Precipitate the other two samples in the same way. After the first one has been done, the student will know the approximate amount of silver nitrate solution to add to the others and hence can proceed more rapidly.

After the solution has cooled, filter it through a weighed crucible with suction, retaining the bulk of the precipitate in the beaker. It is wise to test the filtrate once again for completeness of precipitation, using a few drops of silver nitrate solution. Wash the precipitate in the beaker by decanting with three 25-mL portions of 0.01 M nitric acid (about two drops of concentrated nitric acid in 100 mL of water). Pour the washings through the filter. Finally, stir up the precipitate in a small portion of wash solution and transfer it into the crucible. Carefully rinse out any precipitate remaining in the beaker into the crucible, using a rubber policeman if necessary to remove precipitate adhering to the walls of the beaker. Now wash the precipitate in the crucible three or four times more with small portions of wash solution, allowing it to drain each time. Collect the last portion of wash solution separately and test it for the absence of silver ion with a drop of hydrochloric acid. Finally, drain the crucible completely with strong suction and place it in a covered beaker for drying in the oven. Raise the watch glass covering the beaker with small glass hooks so as to permit the circulation of air over the precipitate.

After the samples have dried for about 2 h, cool them in a desiccator and weigh. Return them to the oven for about 30 min, cool them in a desiccator, and reweigh. Drying may be considered complete if no more than 0.4 mg is lost during the second drying period.

Calculate the percentage of chlorine in the unknown sample. Report the result in the manner prescribed by the instructor.

Notes

1. Consult the instructor concerning the number of replicates to run. Ordinarily, it is considered adequate to perform the analysis in triplicate.
2. The sample size should be such that a convenient quantity of precipitate is obtained. Consult the instructor. Portions of 0.5 to 0.7 g are appropriate for most student unknowns purchased currently.

Experiment 18. *Determination of Sulfur in a Soluble Sulfate*

See Chapter 4 for a discussion of the properties of barium sulfate and the possible errors encountered in its use in gravimetric analysis.

Procedure The sample should be dried in an oven at 100 to 120°C (Note 1). Weigh out accurately three portions of about 0.5 to 0.8 g each (Note 2) and dissolve each sample in about 200 mL of distilled water and 1 mL of concentrated hydrochloric acid. Prepare a 5% solution of barium chloride by dissolving 5 g of $BaCl_2 \cdot 2H_2O$ in 100 mL of water. Calculate the volume of this solution required to precipitate the sulfate in each sample, including a 10% excess (Note 2). Measure this volume of solution into a clean beaker using a graduated cylinder.

Now heat both the sample solution and the solution of barium chloride nearly to boiling. Pour the hot barium chloride solution quickly but carefully into the hot sample solution and stir vigorously. Allow the precipitate to settle and test the supernatant liquid for completeness of precipitation by adding a few more drops of the barium chloride solution.

After precipitation is complete, cover the beaker with a watch glass and allow the precipitate to digest for 1 to 2 h, keeping the solution hot (80 to 90°C) on a steam bath, on a hot plate, or using a low flame.

Either a fine porous porcelain filtering crucible or a funnel with filter paper can be used to collect the precipitate. The ignition temperature is too high to permit the use of sintered-glass crucibles. If filter paper is used, the slow type such as Whatman No. 42 should be selected unless the precipitate has been unusually well digested, in which case No. 40 may suffice.

Clean, rinse, and heat to constant weight three crucibles and lids of porous porcelain (or, if filter paper is to be used, ordinary porcelain). Use the highest temperature of the Tirrill burner for an ordinary crucible. Remember that a porous porcelain crucible must be heated within another crucible to prevent damage and the admission of gases from the flame through the porous bottom. Thus the full temperature of a Meker burner is required in this case.

The directions for washing and filtering the precipitate apply whichever type of filter is used. The solution must be hot at the time of filtration. Decant the clear supernatant solution through the filter. Discard the clear filtrate so that if precipitate later runs through the filter only a small volume need be refiltered. Then rinse the precipitate into the funnel or filtering crucible with hot water. Remove any precipitate from the walls of the beaker with a rubber policeman and rinse such particles into the filter with hot water. If the filtrate is cloudy, it must be refiltered, in which case the second passage generally clears it up. Continue to rinse the precipitate in the filter with hot water until a drop of silver nitrate solution added to a test portion of the washings collected in a test tube shows that chloride is absent.

After washing is complete, if filter paper was used, transfer the paper and precipitate carefully to the previously prepared crucible. Dry the precipitate slowly over a low flame (an oven may be used if desired). Then, increasing the heat, char the paper carefully and finally burn it off completely. Ignite the precipitate for about 20 min at the highest temperature of the Tirrill burner. The crucible should be uncovered and in a slanted position for free access of air to prevent reduction of barium sulfate to barium sulfide. (If such reduction is suspected, moisten the cooled precipitate with a little concentrated sulfuric acid and carefully raise the temperature, finishing the ignition once again at the highest temperature of the burner.) Cool and weigh the crucible and its contents; then reignite for a second 20-min period. Repeat this procedure until constant weight is attained.

In the event that a filtering crucible was used instead of filter paper, ignition is the same except that there is no paper to be burned off; the crucible is heated inside an ordinary crucible, and a Meker burner is used. Be sure the precipitate is dried at a low temperature, because steam formed upon strong ignition of wet material may carry precipitate out of the crucible.

Sulfur is usually reported as sulfur trioxide. Calculate the percentage of sulfur trioxide in the unknown sample and report the results in the style required by the instructor.

Notes

1. Consult the instructor. Certain samples of hydrated salts should not be dried in the oven. The common samples purchased from commercial suppliers of student unknowns are usually to be dried.
2. Consult the instructor regarding the weight of the sample.

Experiment 19. ## *Determination of Iron in an Oxide Ore*

See Chapter 4 for a discussion of the properties of iron(III) hydroxide and the possible errors encountered in its use in gravimetric analysis.

Procedure Accurately weigh out three samples of about 0.5 g each (Note 1) of the dried iron oxide unknown. Treat each sample in a clean 250-mL beaker with about 20 mL of 1 : 1 hydrochloric acid. Cover the beaker with a watch glass and heat the solution nearly to boiling. Continue this careful heating until solution of the sample is complete, or, if there is a persistent insoluble residue, until this residue (generally silica) shows no red-brown color.

Rinse down the watch glass and sides of the beaker with water from the wash bottle. Add 10 mL of saturated bromine water to the solution and boil gently until the bromine vapors are completely expelled. [If desired, the oxidation of iron(II) ion may be accomplished with nitric acid rather than bromine: Add 1 to 2 mL of concentrated nitric acid dropwise to the solution and boil for 1 or 2 min to expel the nitrogen oxides.]

If there is an undissolved siliceous residue in the sample, the solution should be filtered at this point. A fast paper is adequate. Collect the filtrate in a clean 400-mL beaker. Wash the filter paper with dilute hydrochloric acid until no yellow remains, collecting the washings in the same beaker.

Now dilute the solution to about 200 mL, bring it nearly to a boil, and slowly add 1:2 ammonium hydroxide solution until a definite odor of ammonia persists in the vapors (Note 2). Hold the solution at the near-boiling temperature for another minute or so and then allow the precipitate to settle. Test for completeness of precipitation with a drop or two of ammonium hydroxide.

The precipitate is now to be separated from the mother liquor, washed, redissolved, and reprecipitated. In the first filtration, use decantation to the greatest extent possible, retaining the bulk of the precipitate in the original beaker for redissolving. Much time is saved by this technique. Carefully decant the supernatant solution through a fast filter paper (Whatman No. 41, for example), retaining the precipitate in the beaker as much as possible. Then wash this precipitate with about 25 mL of 0.1 M ammonium nitrate wash solution (8 or 10 g of NH_4NO_3/liter). Allow the precipitate to settle and decant the washings through the filter. Repeat this washing a second time. Discard the filtrate and washings and place the beaker containing the precipitate under the funnel. Now pour through the funnel about 50 mL of 1:10 hydrochloric acid solution. This should completely dissolve any precipitate in the funnel and the bulk of the precipitate in the beaker. Make certain that the acid washes the filter paper thoroughly. This can be accomplished by adding the acid from a pipet, directing a slow stream near the upper edge of the paper all the way around its circumference. No trace of yellow should remain on the paper.

The iron is now reprecipitated with ammonium hydroxide exactly as before. Allow the precipitate to settle, test for completeness of precipitation, and decant the supernatant liquid through the same filter paper as used above after the first precipitation. Wash the precipitate by decantation three times with hot 0.1 M ammonium nitrate wash solution, finally bringing the precipitate onto the filter paper. Use a rubber policeman and rinsing to effect a quantitative transfer. (Alternatively, the inside of the beaker may be wiped with a small piece of ashless filter paper which is then added to the precipitate in the funnel.) The precipitate is now washed in the funnel with the hot ammonium nitrate solution until the filtrate shows only a faint test for chloride with silver nitrate.

After the precipitate has drained as much as possible, fold down the edges of the filter paper, and transfer the paper with its contents to a previously prepared porcelain crucible. Dry and char the paper in the usual manner and finally ignite the precipitate at bright-red heat (a Meker burner is recommended, although a Tirrill burner may be adequate). There is some tendency for reduction of iron(III) oxide to the magnetic oxide, Fe_3O_4; hence the ignition should be performed with plentiful access of air into the crucible. After ignition for about 45 min, cool the crucible in the desiccator and weigh. Reignite for 20-min periods until a constant weight is attained.

Calculate the percentage of iron in your unknown sample, Report the results as requested by the instructor.

Notes

1. Consult the instructor regarding recommended weight of sample.
2. Unless it was perfectly clear, the ammonium hydroxide should have been previously filtered to remove silica, which is often suspended in the solution as a result of attack on the glass container.

Experiment 20. *Gravimetric Determination of Calcium*

In this experiment calcium will be precipitated as the oxalate, $CaC_2O_4 \cdot H_2O$, and ignited to CaO. Since it is possible to weigh the oxalate as such, alternative directions are given for this procedure. The method of homogeneous precipitation (Section 4.3.d) will be employed.

Procedure   Weigh accurately three samples of the unknown of about 0.5 to 0.7 g each and transfer them to 400-mL beakers (Note 1). Add about 75 mL of water to each beaker and then 10 mL of 1:3 hydrochloric acid to dissolve the sample. Then boil the solution gently for about 2 min to expel any CO_2 and dilute to about 150 mL with distilled water. Add 30 mL of a solution containing about 1 g of ammonium oxalate in 30 mL of water containing 2 mL of concentrated hydrochloric acid. Also add three drops of methyl red indicator (Note 2, Experiment 2).

Now heat each solution to a gentle boil and add, with stirring, about 15 g of solid urea. Hold the solution at boiling temperature (the evolution of CO_2 helps prevent bumping) until the indicator acquires its alkaline color. This generally requires 30 to 60 min. Test for completeness of precipitation by adding a few drops of ammonium oxalate solution.

Wash each precipitate with cold 0.1% ammonium oxalate solution, decanting through an ashless filter paper of medium porosity (for example, Whatman No. 40) (Note 2). Finally, collect the precipitate on the paper, rinsing the beaker with the two 10-mL portions of the ammonium oxalate solution and pouring the washings over the precipitate.

Transfer each precipitate and paper to a porcelain crucible (weighed with lid) placed upright in a triangle (Fig. 20.10). Heat the crucible slowly with a Tirrill burner as described in Section 20.2.h until all the paper has burned. Then place each crucible in a muffle furnace at 1000°C for 30 min. Remove the crucible, allow it to cool below redness, place it in a desiccator, and allow the desiccator to cool. Weigh each crucible as quickly as possible as the calcium oxide gains weight readily when exposed to air. Repeat the procedure until the crucible is at constant weight. Report the percentage of CaO in the sample.

Notes

1. Consult the instructor as to size of sample to take. Solutions containing $CaCO_3$ dissolved in HCl may be used as unknowns. In this case the unknown samples will be pipetted into 400-ml beakers.
2. If it is desired to weigh the calcium oxalate monohydrate, the precipitate should be collected in a sintered-glass crucible of medium porosity and filtered with suction [Fig. 20.11(g)]. Each crucible should be previously dried for 1 h at 105°C in an oven, cooled in a desiccator, and then weighed to constant weight. The precipitate should be dried with aspirator suction for about 2 min, and then in an oven at 105°C for 1 h. It is then cooled in a desiccator for 30 min and

brought to constant weight. The oxalate is hygroscopic, and the weighings should be done rapidly. Report the molarity of Ca^{2+} if the unknown is a solution, the weight percent CaO if a solid is used.

Experiment 21. *Determination of Nickel in Steel*

One of the most common applications of the organic precipitants is the precipitation of nickel by dimethylglyoxime (Chapter 4). In the determination of nickel in steel, an acid solution containing iron in the +3 oxidation state is treated with tartaric acid and an alcoholic solution of dimethylglyoxime. The solution is made slightly basic with ammonia, precipitating the nickel quantitatively. Iron is not precipitated because it forms a soluble complex with tartrate ion.

Procedure Weigh accurately three samples of about 1 g each (Note 1) of the steel into 400-mL beakers. Dissolve each sample by warming it with about 60 mL of 1:1 hydrochloric acid. Add cautiously 10 mL of 1:1 nitric acid and boil the solution gently to expel oxides of nitrogen. Dilute the solution to 200 mL, heat it nearly to boiling, and add 25 mL of a 20% solution of the tartaric acid (Note 2).

Now neutralize the solution with concentrated ammonia until a definite odor of ammonia persists in the vapors. Then add 1 mL excess ammonia. If any insoluble material is evident, remove it by filtration (filter paper), and wash it with a hot solution containing a little ammonia and ammonium chloride. Combine the washings with the remainder of the solution.

Make the solution slightly acidic with hydrochloric acid and heat it to about 70°C. Add 20 mL of a 1% solution of dimethylglyoxime in alcohol (Note 3). Make the solution slightly alkaline with ammonia (note odor) and add 1 mL of excess. Allow the precipitate to digest for 30 min at 60°C and then cool the solution to room temperature.

Filter the solution through a weighed fritted-glass or Gooch crucible and wash the precipitate with cold water until the washings are free of chloride ion. Dry the precipitate by heating the crucible in an oven at 110 to 120°C until a constant weight is obtained. Calculate the percentage of nickel in the steel.

Notes

1. Consult the instructor. The sample should contain about 25 to 35 mg of nickel.
2. The solution is prepared by dissolving 25 g of tartaric acid in 100 mL of water. It should be filtered before use if it is not clear.
3. Dissolve 10 g of dimethylglyoxime in 1 liter of 95% ethanol. For best results, about 10 mL of this solution should be used for each 1% of nickel present. If too much solution is added, dimethylglyoxime itself may precipitate.

22.4

OXIDATION-REDUCTION TITRATIONS

The principles of oxidation-reduction titrations are discussed in Chapter 10, and applications to analyses are treated in Chapter 11. In this section directions are given for the preparation and standardization of solutions of potassium permanganate, cerium(IV) sulfate, potassium dichromate, iodine, and sodium thiosulfate. A number of experiments illustrating the applications of these reagents to analysis are included.

Experiment 22. *Preparation of a 0.1 N Potassium Permanganate Solution*

The permanganate titrations described here will be carried out in acid solution. Hence the equivalent weight of potassium permanganate is one-fifth the molecular weight, or 31.61. Directions are given for the preparation of 1 liter of a 0.1 N solution, thereby requiring 31.61×0.1 or about 3.2 g of the salt.

Procedure   Since potassium permanganate solutions are susceptible to decomposition (page 295), special precautions are recommended for preparing the solution if it is to be used over a period of several weeks. If the solution is to be prepared, standardized, and used the same day, the special precautions are not necessary (Note).

Weigh approximately 3.2 g of a good grade of potassium permanganate and place it in a clean 250-mL beaker. Dissolve the salt by adding 50 mL of water and stirring. Decant the solution into a large beaker and add 50 mL of additional water to dissolve any crystals remaining in the first beaker. Repeat this procedure until all the crystals are dissolved. Dilute the solution to about 1 liter, transfer to a glass-stoppered bottle, and label properly.

If the instructor recommends removal of manganese dioxide, proceed as follows: Before transferring the solution to the bottle, heat it just to boiling and keep it slightly below the boiling point for 1 h. Then allow the solution to cool, and filter it through a sintered-glass crucible using suction. Transfer the solution to a glass-stoppered bottle and label properly.

Note

Consult the instructor. The instructor may have prepared in advance a large quantity of stock solution which has stood for a week or so. If the manganese dioxide has settled, the clear solution can be withdrawn through an all-glass siphon, avoiding filtration.

Experiment 23. *Standardization of Potassium Permanganate Solution*

Procedure   *(a) Sodium Oxalate. Fowler-Bright Method.* Weigh accurately three samples of about 0.25 to 0.30 g each (Note 1) of the dried salt into clean 500-mL Erlenmeyer flasks. Add 250 mL of dilute sulfuric acid (12.5 mL of concentrated acid diluted to 250 mL) which has previously been boiled 10 to 15 min and then cooled to 24 to 30°C. Swirl the flask until the solid dissolves and then titrate with permanganate. Steadily add the permanganate directly to the oxalate solution (not down the walls of the flask) and stir slowly. Add sufficient permanganate (35 to 40 mL) to come within a few milliliters of the equivalence point, adding it at a rate of about 25 to 35 mL/min. Let the solution stand until the pink color disappears (Note 2) and then heat the solution to 55 to 60°C. Complete the titration at this temperature, adding permanganate slowly until one drop imparts to the solution a faint pink color that persists at least 30 s. Add the last milliliter slowly, allowing each drop to be decolorized before adding another. The solution should be as warm as 55°C at the end of the titration.

To about 300 mL of previously boiled dilute sulfuric acid, add permanganate solution dropwise until the color matches that of the titrated solution. Subtract this volume (usually about 0.03 to 0.05 mL) from the volume used in the titration.

After titration of three samples, calculate the normality obtained in each titration and average the results. With care, the average deviation should be as small as 2 ppt (consult the instructor).

Notes

1. The equivalent weight of sodium oxalate is 67.00. This weight is sufficient to react with about 35 to 45 mL of a 0.1 N solution.
2. This may take 30 to 45 s, since the reaction is not instantaneous. If the color does not disappear, indicating excess permanganate, discard the solution and add less permanganate to the next sample.

McBride Method. In place of the more lengthy Fowler-Bright procedure, the McBride procedure may be preferred for beginning students.

Weigh accurately three samples of about 0.25 to 0.30 g each of dried sodium oxalate into clean 250-mL Erlenmeyer flasks. Dissolve each sample in about 75 mL of 1.5 N sulfuric acid (20 mL of concentrated sulfuric acid added to 400 mL of water). Then heat the first solution almost to boiling (80 to 90°C) and titrate slowly with the permanganate with constant swirling. The end point is marked by the appearance of a faint pink color that persists at least 30 s. The temperature should not drop below 60°C during the titration. Titrate the other two solutions in the same manner.

To about 100 mL of the 1.5 N sulfuric acid, add permanganate solution dropwise until the color matches that of the titrated solution. Subtract this volume from the volume used in the titration.

Calculate the normality of the permanganate solution. The average deviation should be as small as about 2 ppt (consult instructor).

(b) Arsenic(III) Oxide. Weigh accurately three portions of about 0.2 g each of pure arsenic(III) oxide, previously dried in the oven, into each of three 250-mL Erlenmeyer flasks. Add to each flask 10 mL of a cool sodium hydroxide solution made by dissolving 20 g of sodium hydroxide in 80 mL of water. Allow the flask to stand for 8 to 10 min, stirring the solution occasionally until the sample has completely dissolved. Then add 100 mL of water, 10 mL of concentrated hydrochloric acid, and one drop of 0.0025 M potassium iodide solution as a catalyst (Note). Titrate with the permanganate solution until a single drop imparts to the solution a faint pink color which persists for at least 30 s after the liquid is swirled. Add the last milliliter slowly, allowing each drop to be decolorized before adding another. Finally, run a blank to determine the volume of permanganate required to color the solution and to react with any reducible material in the reagents. This should amount to no more than one drop of permanganate.

Titrate the other two samples and calculate the normality of the permanganate solution. The average deviation should be about 1 to 2 ppt.

Note

Potassium iodate or iodine monochloride can also be used as catalysts if desired. The potassium iodide solution is prepared by dissolving about 0.4 g in 1 liter of water.

(c) Iron Wire. Pure iron wire, free of rust, can be used as a primary standard for permanganate. Weigh accurately three samples of wire of about 0.2 g each into 150-mL beakers. Add 20 mL of 1:1 hydrochloric acid to the first sample and warm the solution on a steam bath or over a low flame until all the iron dissolves.

Since the sample is in hydrochloric acid solution, it is convenient to employ tin(II) chloride as the reducing agent and then add preventive solution before titration with permanganate. Directions for this precedure are given in Experiment 26.

Repeat the procedure on the other two samples and calculate the normality of the permanganate solution. The deviation should be about 2 ppt or less.

Experiment 24. *Determination of an Oxalate*

Procedure   Weigh accurately three portions of the dried material of appropriate size (Note). If the Fowler-Bright procedure is followed, dissolve each sample in 250 mL of dilute sulfuric acid in a 500-mL Erlenmeyer flask. The acid (12.5 mL of concentrated acid diluted to 250 mL with water) should have been previously boiled and then cooled to 24 to 30°C. Titrate as directed on page 631.

If the McBride procedure is used, dissolve each sample in 75 mL of 1.5 N sulfuric acid (20 mL of concentrated acid to 400 mL of water) in a 250-mL Erlenmeyer flask. Heat the solution almost to boiling and titrate as directed on page 632.

Calculate and report the percentage of sodium oxalate in the sample in the usual manner.

Note

Consult the instructor. The equivalent weight of sodium oxalate is 67.00. For material containing about 25% sodium oxalate, a 1-g sample is a convenient amount.

Experiment 25. *Determination of Hydrogen Peroxide*

Procedure   Pipet 25 mL of the peroxide solution (Note 1) into a 250-mL volumetric flask, dilute the solution to the mark, and mix thoroughly. Transfer a 25-mL aliquot of this solution to a 250-mL Erlenmeyer flask to which has been added 5 mL of concentrated sulfuric acid and 75 mL of water. Titrate with standard permanganate solution to the first appearance of a permanent pink tinge.

Repeat the titration on two additional aliquots of the solution. Calculate the weight of hydrogen peroxide in the original 25-mL sample and, assuming that the weight was exactly 25 g, calculate the percentage by weight of hydrogen peroxide (Note 2).

Notes

1. If the density of the solution is about 1, this volume gives $25 \times 0.03 = 0.750$ g, or 750 mg H_2O_2. Since the equivalent weight of H_2O_2 is 17.01, this is about 44 meq. Thus one-tenth of this solution will use about 44 mL of a 0.1 N permanganate solution for titration.
2. Commercial hydrogen peroxide often contains small amounts of organic compounds such as acetanilide, which are added to stabilize the peroxide. These compounds may react with permanganate to some extent, causing incorrect results.

Experiment 26. *Determination of Iron in an Ore*

The determination of iron in iron ores by titration with permanganate is discussed in Section 11.2.c. Directions are given here for dissolving the sample in hydrochloric acid and reduction of iron(III) ion, prior to titration, by tin(II) chloride. In this case Zimmermann-Reinhardt preventive solution is added to prevent oxidation of chloride ion by permanganate.

Procedure   *Dissolving the Sample.* Weigh three samples of iron ore (Notes 1 and 2) of appropriate size (consult the instructor) into three 150-mL beakers. Add 10 mL of concentrated hydrochloric acid and 10 mL of water to each beaker. Cover the beakers with watch glasses and keep the solution just below the boiling point on a steam bath, hot plate, or wire gauze until the ore dissolves (hood) (Note 3). This may require 30 to 60 min. At this point the only solid present should be a white residue of silica. If an appreciable amount of colored solid remains in the beaker, the sample must be fused to bring the remainder of the iron into solution (Note 4).

Notes

1. Some commercial samples are made from iron oxide and are easily soluble in acid. Such samples will not require lengthy heating to effect solution and will not leave a residue of silica. The instructor will alter the directions if such samples are used.
2. If the ore is not finely ground, it should be ground in an agate mortar before drying. If it is suspected that the ore contains organic matter, the sample, after weighing, should be ignited in an uncovered porcelain crucible for 5 min. This oxidizes organic matter.
3. If tin(II) chloride is to be used to reduce the iron, add successive portions of tin(II) chloride until the solution changes from yellow to colorless. This will aid in dissolving the sample. Avoid an excess of tin(II) chloride.
4. If fusion is required (consult the instructor), dilute the solution with an equal volume of water and then filter. Wash the residue with 1% hydrochloric acid and then wash with water to remove the acid. Transfer the paper to a porcelain crucible and burn off the carbon. If the residue is white, it may be disregarded, as the color was probably caused by organic matter. If the color remains, add about 5 g of potassium pyrosulfate and heat carefully until the salt just fuses. Maintain this temperature for 15 to 20 min or until all the iron reacts. Then cool and dissolve the residue in 25 mL of 1 : 1 hydrochloric acid and add this solution to the main filtrate.

Reduction with Tin(II) Chloride. Adjust the volume of the solution to 15 to 20 mL by evaporation or dilution. The solution should be yellow in color because of the presence of iron(III) ion (Note 1). Keep the solution hot and reduce the iron in the first sample (Note 2) by adding tin(II) chloride (Note 3), drop by drop, until the color of the solution changes from yellow to colorless (or very light green). Add one or two drops of excess tin(II) chloride. Cool the solution under the tap and rapidly pour in 20 mL of saturated mercury(II) chloride solution (Note 4). Allow the solution to stand for 3 min and then rinse the solution into a 500-mL Erlenmeyer flask. Dilute to a volume of about 300 mL and add 25 mL of Zimmermann-Reinhardt solution (Note 5). Titrate slowly with permanganate, swirling the flask constantly. The end point is marked by the first appearance of a faint pink tinge which persists when the solution is swirled (Note 6).

Reduce and titrate the second and third samples in the same manner (Note 7). Calculate and report the percentage of iron in the sample in the usual manner.

Notes

1. If tin(II) chloride has been used during dissolving, and the solution is colorless or almost so at this point, add a small crystal of potassium permanganate and heat a little longer until the yellow color is distinct. The reduction can then be followed more readily.
2. Reduce only one sample and finish the titration before reducing the second. Why?
3. This is prepared by dissolving 113 g of $SnCl_2 \cdot 2H_2O$ (free of iron) in 250 mL of concentrated hydrochloric acid, adding a few pieces of mossy tin, and diluting to 1 liter with water.
4. If the mercury(II) chloride were added slowly, part of it would be temporarily in contact with an excess of tin(II) chloride, which might reduce the substance to metallic mercury. Also, if the solution were hot, there would be a danger of forming mercury. The precipitate here should be white and silky and not large

in quantity. If the precipitate is gray or black, indicating the presence of mercury, the sample should be discarded. If no precipitate is obtained, indicating insufficient tin(II) chloride, the sample should be discarded.

5. This is prepared as follows: Dissolve 70 g of $MnSO_4$ in 500 mL of water and add slowly, with stirring, 110 mL of concentrated sulfuric acid and 200 mL of 85% phosphoric acid. Dilute to 1 liter.

6. The color may fade slowly because of oxidation of mercury(I) chloride or chloride ion by permanganate.

7. If desired (consult the instructor), a blank can be determined by carrying a mixture of 10 mL of concentrated hydrochloric acid and 10 mL of water through the entire procedure. The blank normally will be about 0.03 to 0.05 mL of 0.1 N potassium permanganate.

Experiment 27.   *Determination of Oxygen in Pyrolusite*

Procedure   Weigh accurately three samples of about 0.5 g of the finely ground and dried ore into 250-mL Erlenmeyer flasks. Calculate approximately the number of milliequivalents of oxygen in the sample (Note 1) and then the weight of sodium oxalate required to react with the sample. Add to each flask about 0.25 g of pure sodium oxalate in excess of the calculated amount, weighing this accurately on the balance. Add to each flask about 100 mL of sulfuric acid (10 mL of concentrated acid to 100 mL of water), cover the flask with a watch glass, and heat each flask on a steam bath or gently over a low flame. Shake the flask occasionally. Keep the temperature just below the boiling point and avoid allowing the solution to evaporate very much. After the samples have completely digested, as indicated by the disappearance of all the black or brown particles (Note 2) and the cessation of evolution of carbon dioxide, rinse the watch glass and rinse down the walls of the flask with distilled water. Titrate the hot solution with standard permanganate to the first appearance of a faint pink tinge.

After digestion and titration of all three samples, calculate the percentage of oxygen in the sample. The average deviation may be as high as 4 ppt (consult the instructor). Remember that the milliequivalents of oxygen are given by the relationship

$$\text{Meq oxygen} + \text{meq } KMnO_4 = \text{meq } Na_2C_2O_4$$

Notes

1. Consult the instructor for approximate pecentage of oxygen in the sample. For example, if the sample is about 10% oxygen, a 500-mg sample would contain 50 mg of oxygen, or about 6 meq. Six milliequivalents of sodium oxalate (EW 67) weigh about 0.4 g. Hence one should take about 0.65 g of sodium oxalate.

2. There may be a residue of white or light-brownish silica. This should be disregarded.

Experiment 28.   *Preparation of a 0.1 N Cerium(IV) Solution*

Procedure   *(a) Cerium(IV) Sulfate.* Weigh on a trip balance about 33 g of cerium(IV) sulfate or 63 g of $Ce(SO_4)_2 \cdot 2(NH_4)_2SO_4 \cdot 2H_2O$ (Note 1). Add this solid with stirring to a sulfuric acid solution made by adding 28 mL of concentrated sulfuric acid to 500 mL of water. Stir until the solid dissolves and dilute the solution to 1 liter in a clean bottle (Note 2). Label the bottle appropriately.

Notes

1. If cerium(IV) oxide is used, weigh 21 g of the solid into a 1500-mL beaker. Add to the solid with stirring a hot sulfuric acid solution made by adding 78 mL of

concentrated sulfuric acid to 300 mL of water. After the oxide has dissolved, dilute the solution to 1 liter and transfer to a clean bottle.

2. If the solution is at all turbid, it should be filtered before use. The precipitation of cerium(IV) phosphate is quite slow in acid solution, and if the solution is to be used soon after standardization, the filtration may be omitted. However, if the solution is to be kept for some time, it is preferable to allow it to stand for 1 or 2 weeks and then filter before standardization.

(b) Cerium(IV) ammonium nitrate. Weigh accurately 54.83 g of primary standard grade cerium(IV) ammonium nitrate into a 1-liter beaker (Note 1). Then pour 56 mL of concentrated sulfuric acid over the salt and stir for about 2 min. Add carefully 100 mL of water and stir well (Note 2). Continue the addition of water slowly with stirring until the volume is about 600 mL. Then transfer the cool solution to a 1-liter volumetric flask, dilute to the mark, and mix the solution thoroughly.

Notes

1. The equivalent weight of the salt is 584.23. It is sufficiently accurate to weigh to only the second decimal place. Why? If the ordinary grade of salt is used, weigh about 56 g on a trip scale and dissolve as indicated. Dilute to approximately 1 liter and then standardize the solution against a primary standard.
2. Ordinarily, of course, it is not advisable to add water to concentrated sulfuric acid. In the present case, however, this is recommended (see H. Diehl and G. F. Smith, *Quantitative Analysis,* John Wiley & Sons, Inc., New York, 1952, p. 274). Perform the operation very cautiously to prevent spattering of the acid solution.

Experiment 29. *Standardization of the Cerium(IV) Solution*

Procedure *(a) Arsenic(III) Oxide.* Weigh accurately three portions of about 0.2 g each of pure arsenic(III) oxide, previously dried in an oven, into each of three 500-mL Erlenmeyer flasks. Add to each flask 10 mL of a cool solution of sodium hydroxide made by dissolving 20 g of sodium hydroxide in 80 mL of water. Allow the flask to stand 8 to 10 min, stirring occasionally until the sample has completely dissolved. Then add 100 mL of water and make the solution acidic by adding 25 mL of 1:10 sulfuric acid. Add two drops of 0.01 M osmium tetroxide solution (Note 1) as a catalyst and two drops of ferroin [iron(II) 1,10-phenanthroline sulfate] as the indicator (Note 2).

Titrate the pale pink solution with cerium(IV) solution. The pink color becomes more pronounced as the titration proceeds. The end point is approached without warning, at which point the solution changes from pink to a very faint blue (or colorless). The color change is very sharp.

Titrate the other samples in the same manner and calculate the normality of the cerium(IV) solution.

Notes

1. This solution is made by dissolving 0.125 g of OsO_4 in 50 mL of 0.1 N sulfuric acid.
2. This solution is prepared by dissolving 1.5 g of 1,10-phenanthroline monohydrate, $C_{12}H_8N_2 \cdot H_2O$, in 100 mL of 0.025 M iron(II) sulfate (freshly prepared). The indicator can be purchased from the G. F. Smith Chemical Co., Columbus, Ohio.

(b) Iron Wire. Weigh accurately three samples of pure iron wire, free of rust, of about 0.2 g each. Place each sample in a 150-mL beaker and add 20 mL of 1:1

hydrochloric acid to each beaker. Warm the solution on a steam bath or over a low flame until all the iron dissolves.

Adjust the volume of the solution to 15 to 20 mL by evaporation or dilution. The solution may not be intensely yellow if very little of the iron has been oxidized to the +3 state during the dissolving process. Add tin(II) chloride (Note 1) drop by drop to the first solution (hot) until the solution is colorless or a very light green, and then add one or two drops excess. The first drop or so may be sufficient to decolorize the solution. Now cool the solution under the tap and rapidly pour in 20 mL of saturated mercury(II) chloride solution (Note 2). Allow the solution to stand for 3 min and then rinse it into a 500-mL Erlenmeyer flask. Add 10 mL of concentrated hydrochloric acid, dilute the solution to about 300 mL, and add two drops of ferroin indicator (Note 3). Titrate with the cerium(IV) solution until the color changes from pink (or reddish orange) to pale yellow.

Reduce and titrate the other two samples in the same manner. Calculate the normality of the cerium(IV) solution.

Notes

1. See Note 3, page 634.
2. See Note 4, page 634.
3. See Note 2 of part (a) above.

Experiment 30. *Determination of Iron in an Ore*

Procedure Weigh and dissolve the samples of iron ore as directed in Experiment 26. Then reduce the first sample with tin(II) chloride as directed in that experiment. Cool the solution under the tap after reduction is complete and rapidly add 20 mL of saturated mercury(II) chloride solution. Allow the solution to stand for 3 min and then rinse into a 500-mL Erlenmeyer flask. Dilute to a volume of about 300 mL with 1 M HCl and add two drops of ferroin indicator. Titrate with the standard cerium(IV) solution until the color changes from pink (or reddish orange) to pale yellow.

Reduce and titrate the second and third samples in the same manner. Calculate and report the percentage of iron in the sample.

Experiment 31. *Preparation and Standardization of a 0.1 N Potassium Dichromate Solution*

Procedure Weigh accurately a sample of about 4.9 g of pure potassium dichromate (Note 1) that has been previously dried in the oven. Dissolve the sample in water in a 400-mL beaker and transfer the solution quantitatively to a 1-liter volumetric flask. Dilute to the mark and mix the solution thoroughly. Calculate the normality of the solution.

If the solution is to be standardized against iron (consult the instructor), proceed as follows.

Weigh and dissolve samples of iron wire as directed in Experiment 23. Reduce the iron(III) ion with tin(II) chloride as directed on page 634, and rinse the solution into a 500-mL Erlenmeyer flask. Add 250 mL of water, 5 mL of concentrated sulfuric acid, 5 mL of 85% phosphoric acid, and eight drops of sodium diphenylaminesulfonate indicator (Note 2). Titrate slowly with the dichromate solution, swirling the flask constantly. The solution becomes green, since the reaction produces chromium(III) ions. The oxidized form of the indicator is purple or violet. As the end point is approached, indicated by a transient purple color, add the dichromate dropwise until the color changes permanently to purple.

Calculate the normality of the dichromate solution. Compare this with the normality calculated from the weight of dichromate dissolved in the solution.

Notes

1. The equivalent weight is 294.18/6, or 49.03.
2. The indicator solution is made by dissolving 0.3 g of the sodium salt in 100 mL of water. If the barium salt is used, dissolve 0.3 g in 100 mL of hot water, add an excess of 0.1 *M* sodium sulfate solution, and let the solution stand until barium sulfate settles. Either filter or decant the solution from the precipitate.

Experiment 32.   *Determination of Iron in an Ore Using Dichromate*

Procedure   Weigh and dissolve the first sample and reduce it with tin(II) chloride as directed under Experiment 26. Then cool the solution and rapidly add 20 mL of saturated mercury(II) chloride solution. Allow the solution to stand for 3 min and then rinse into a 500-mL Erlenmeyer flask. Add 250 mL of water, 5 mL of concentrated sulfuric acid, 5 mL of 85% phosphoric acid, and eight drops of sodium diphenylaminesulfonate indicator (see Note 2 of Experiment 31). Titrate with dichromate solution as directed in Experiment 31.

Reduce and titrate the other two samples in the same manner. Calculate and report the percentage of ion in the ore.

Experiment 33.   *Preparation and Standardization of a 0.1 N Iodine Solution*

Procedure   Weigh on a trip balance about 12.7 g of reagent-grade iodine (Notes 1 and 2) and place this in a 250-mL beaker. Add to the beaker 40 g of potassium iodide, free of iodate (Note 3), and 25 mL of water. Stir to dissolve all the iodine and transfer the solution to a glass-stoppered bottle. Dilute to about 1 liter (Note 4).

Standardize the iodine solution as follows: Weigh accurately a sample of reagent-grade arsenic(III) oxide (Note 5) of about 1.25 g into a 250-mL beaker. Add a solution made by dissolving 3 g of sodium hydroxide in 10 mL of water. Allow the beaker to stand, swirling it occasionally, until the solid has completely dissolved. Then add 50 mL of water, two drops of phenolphthalein indicator, and 1 : 1 hydrochloric acid until the pink color of the indicator just disappears. Then add 1 mL of hydrochloric acid in excess. Transfer the solution to a 250-mL volumetric flask, dilute to the mark, and mix thoroughly.

Transfer with a pipet 25 mL of the arsenite solution to a 250-mL Erlenmeyer flask and dilute with 50 mL of water. Then carefully add small portions of sodium bicarbonate to neutralize the acid. When vigorous effervescence ceases, add 3 g of additional sodium bicarbonate to buffer the solution. Add 5 mL of starch indicator (Note 6) and titrate with iodine until the first appearance of the deep blue color which persists for at least 1 min. If the end point is overrun, some of the standard arsenic(III) solution can be used for back-titration.

Titrate two other aliquots in the same manner. Calculate the normality of the arsenic(III) solution from the weight of arsenic(III) oxide taken. Then calculate the normality of the iodine solution from the volumes of the two solutions used.

Notes

1. If the solution is to be standardized and only one unknown analyzed, 500 mL of solution will suffice. The quantities of iodine and potassium iodide can be halved (consult the instructor).

2. If desired, the iodine can be weighed accurately and the solution made up to a definite volume. To prepare 500 mL of solution, first place about 20 g of potassium iodide in a large weighing bottle and dissolve this in 10 mL of water. After the solution has come to room temperature, weigh the bottle accurately (to the nearest milligram is sufficient). Take the bottle to a trip balance and there add about 6.4 g of iodine and stopper the bottle tightly. Do not open the bottle near the analytical balance, as iodine fumes are corrosive. Reweigh the bottle and record the weight of iodine taken. When the iodine has dissolved, transfer the solution to a 500-mL volumetric flask, dilute to the mark, and mix thoroughly. Store the solution in a glass-stoppered bottle.

3. Test the salt for iodate as follows: Dissolve about 1 g in 20 mL of water and add 1 mL of 6 N sulfuric acid and 2 mL of starch solution. No blue color should develop in 30 s.

4. A bottle of brown glass is preferable for storing this solution. In any event, the solution should be kept from the light as much as possible.

5. Pure arsenic(III) oxide is not hygroscopic and may not need drying unless it has been exposed to air for some time. If drying is necessary, it is usually preferable to place the material in a desiccator over sulfuric acid for about 12 h. The oxide (octahedral variety) tends to sublime when heated as high as 125 to 150°C.

6. This solution is prepared as follows: Make a paste of 2 g of soluble starch and 25 mL of water and pour this gradually (with stirring) into 500 mL of boiling water. Continue the boiling for 1 or 2 min, add 1 g of boric acid as a preservative, and allow the solution to cool. Store the solution in a glass-stoppered bottle. Alternatively, a solid complex of starch and urea can be prepared by melting urea in a small beaker and adding soluble starch. The ratio of weights of urea to starch should be 4 : 1. The solution is mixed and allowed to cool and solidify, and then the solid is ground in a mortar. A small scoopful is used in place of starch solution. The complex dissolves readily in water and is quite stable.

Experiment 34. *Determination of the Purity of Arsenic(III) Oxide*

Procedure Weigh three samples of the unknown of appropriate size (Note) into 500-mL Erlenmeyer flasks. Add to each flask 50 mL of water and 1 g of sodium hydroxide and warm until the sample dissolves. Cool the solution, add two drops of phenolphthalein indicator, and add 1 : 1 hydrochloric acid until the pink color of the indicator just disappears. Then add small portions of sodium bicarbonate to neutralize the acid. When vigorous effervescence ceases, add 3 g of additional bicarbonate to buffer the solution. Dilute to about 150 mL with water and add 5 mL of starch solution (Note 6, Experiment 33). Titrate the first sample with iodine solution to the first appearance of the deep blue color which lasts for at least 1 min.

Titrate the other two samples in the same manner. Calculate and report the percentage of arsenic(III) oxide in the sample.

Note

Consult the instructor as to size of sample and method of drying. Commercial unknown samples are usually about 3 to 15% arsenic(III) oxide.

Experiment 35. *Determination of Vitamin C*

Vitamin C (ascorbic acid) is a reducing agent and can be determined by titration with standard iodine solution:

$$\underset{\text{Ascorbic acid}}{CH_2OH-CHOH-\overset{\displaystyle\lceil\qquad O\qquad\rceil}{CH-COH}=COH-C=O} + I_2$$

$$\longrightarrow \underset{\text{Dehydroascorbic acid}}{CH_2OH-CHOH-\overset{\displaystyle\lceil\qquad O\qquad\rceil}{CH-\underset{\displaystyle O}{\overset{\displaystyle\|}{C}}-\underset{\displaystyle O}{\overset{\displaystyle\|}{C}}-C=O}} + 2H^+ + 2I^-$$

Since the molecule loses two electrons in this reaction, its equivalent weight is one-half the molecular weight, or 88.07 g/eq.

Procedure Weigh accurately three 100-mg vitamin C tablets (Note 1) and place them in a 250-mL Erlenmeyer flask. Dissolve the tablets in about 50 mL of water, swirling the flask to hasten dissolution (Note 2). Add 5 mL of starch indicator (Note 6 of Experiment 32) and titrate immediately (Note 3) with standard I_2 solution to the first appearance of the deep blue color which persists for at least 1 min.

Repeat the above procedure twice, remembering to carry out the titration as soon as the sample is dissolved. Calculate the number of milligrams of vitamin C per tablet.

Notes

1. 300 mg of ascorbic acid should require about 34 mL of 0.1 N I_2 for titration.
2. A stirring rod can be used to break up the tablets. A small amount of binder in the tablets will not dissolve and will remain as a suspension.
3. A solution of vitamin C is readily oxidized by oxygen in the air, so the titration should be carried out as soon as the sample is dissolved. During the titration the flask can be covered by a piece of cardboard having a small hole for the tip of the buret.

Experiment 36. ## Preparation and Standardization of a 0.1 N Sodium Thiosulfate Solution

Procedure Dissolve about 25 g of sodium thiosulfate pentahydrate crystals in 1 liter of water that has been recently boiled and cooled. Add about 0.2 g of sodium carbonate as a preservative and store in a clean bottle.

Standardization *(a) Potassium Dichromate.* Weigh three portions of pure, dry potassium dichromate (Note 1) of about 0.2 g each into 500-mL Erlenmeyer flasks. Dissolve each sample in about 100 mL of water and add 4 mL of concentrated sulfuric acid. To the first sample carefully add 2 g of sodium carbonate (Note 2) with gentle swirling to liberate carbon dioxide. Then add 5 g of potassium iodide dissolved in about 5 mL of water (Note 3), swirl, cover the flask with a watch glass, and allow the solution to stand for 3 min (Note 4). Dilute the solution to about 200 mL and titrate with thiosulfate solution until the yellowish color of iodine has nearly disappeared (Note 5). Then add 5 mL of starch solution and continue the titration until 1 drop of titrant removes the blue color of the starch-iodine complex. The final solution will be clear emerald green, the color imparted by chromium(III) ion.

Treat the other samples in the same manner and calculate the normality of the thiosulfate solution. The average deviation should be about 1 to 3 ppt.

Notes

1. The material can be obtained from the National Institute of Standards and Technology, but the best grade available from commercial supply houses is sufficiently pure for most purposes. Dry at 150°C if necessary.

2. The purpose of this is to remove air from the flask and lessen the danger of air oxidation of iodide. Do not add too much carbonate, as this will use up too much acid. The final concentration of acid is about 0.4 M if these quantities are used.
3. Do not allow this solution to stand, as the iodide may be oxidized by air. The iodide should be free of iodate (see Note 3, Experiment 33), or a blank must be determined.
4. The reaction is somewhat slow but should be complete within this time.
5. Starch should not be added to a solution that contains a large quantity of iodine. The starch may be coagulated, and the complex with iodine may not easily break up. A recurring end point will then be obtained.

(b) *Copper*. Secure some clean copper wire or foil, weigh three pieces of about 0.20 to 0.25 g each (Note 1), and place them in 250-mL Erlenmeyer flasks. Add to each flask 5 mL of 1 : 1 nitric acid and dissolve the copper by warming the solution on a steam bath or over a low flame. Add 25 mL of water and boil the solution for about 1 min. Then add 5 mL of urea solution (1 g in 20 mL of water) and continue boiling for another minute (Note 2). Cool the solution under the tap and neutralize the acid with 1 : 3 ammonia, adding the ammonia carefully until a pale blue precipitate of copper(II) hydroxide is obtained (Note 3). Add 5 mL of glacial acetic acid and cool the solution if it is warm. To the first sample, add 3 g of potassium iodide, cover the flask with a watch glass, and allow to stand for 2 min. Then titrate with thiosulfate solution until the brownish color of iodine is almost gone (Note 4). Add 5 mL of starch solution and 2 g of potassium thiocyanate (Note 5). Swirl the flask for about 15 s and complete the titration, adding thiosulfate dropwise. At the end point the blue color of the solution disappears and the precipitate appears white, or slightly gray, when allowed to settle (Note 6).

Treat the second and third samples in the same manner and titrate with thiosulfate. Calculate the normality and copper titer of the thiosulfate.

Notes

1. The equivalent weight of copper is the atomic weight, 63.546.
2. Boiling removes oxides of nitrogen which result from the following reactions:

$$3Cu + 8HNO_3 \longrightarrow 3Cu(NO_3)_2 + 2NO + 4H_2O$$

$$2NO + O_2 \longrightarrow 2NO_2$$

Nitrous acid is formed by the reaction

$$NO_2 + NO + H_2O \longrightarrow 2HNO_2$$

and is decomposed by urea according to the equation

$$2HNO_2 + (NH_2)_2CO \longrightarrow CO_2 + 2N_2 + 3H_2O$$

3. If excess ammonia is added, copper(II) hydroxide redissolves, forming the deep blue copper(II) ammonia complex, $Cu(NH_3)_4^{2+}$. If the excess of ammonia is large, a large quantity of ammonium acetate will be produced later; this may keep the reaction between copper(II) and iodide ions from being complete. The excess ammonia can be removed by boiling, and the precipitate will re-form.
4. After the first titration, calculate the approximate volume required for the other samples. Then titrate to within 0.5 mL of this volume before adding starch and thiocyanate.
5. Thiocyanate displaces adsorbed iodine from the precipitate.
6. The precipitate is seldom completely white. Do not continue addition of thiosulfate until the precipitate is white, since this will require considerable excess titrant.

Experiment 37. Determination of Copper in an Ore

Procedure Accurately weigh three samples of appropriate size (consult the instructor) of the finely ground, dried ore into 250-mL beakers. Add 5 mL of concentrated hydrochloric acid and (slowly) 10 mL of concentrated nitric acid (Note 1). Cover the beaker with a watch glass and heat over a low flame until only a white residue of silica remains. Remove the watch glass and add 10 mL of 1:1 sulfuric acid. Then evaporate the solution (hood) until white fumes of sulfur trioxide appear (Note 2). Cool the solution and carefully add 25 mL of water.

Next add 5 mL of saturated bromine water and boil the solution gently for several minutes to expel excess bromine (Note 3). Then add 1:3 ammonium hydroxide carefully until a slight precipitate of iron(II) hydroxide is formed, or, if no iron is present, the deep blue color of the copper-ammonia complex is just formed. Add 5 mL of glacial acetic acid and 2 g of ammonium acid fluoride (Notes 4 and 5) and stir until the iron(III) hydroxide redissolves.

Dissolve about 3 g of potassium iodide in 10 mL of water and add to the copper solution. Titrate the liberated iodine at once with thiosulfate until the brownish color of the iodine is almost gone (Note 6). Then add 5 mL of starch solution and 2 g of potassium thiocyanate (Note 7). Swirl the flask gently for about 15 s and complete the titration, adding thiosulfate dropwise. At the end point the blue color of the solution disappears, and the precipitate appears white, or slightly gray, when allowed to settle (Note 8).

Treat the other two samples in the same manner and titrate with thiosulfate. Calculate the percentage of copper in the ore.

Notes

1. Some commercial unknowns are prepared from copper oxide and are easily dissolved by warming for 10 to 15 min with 15 mL of 1:2 sulfuric acid. No evaporation is then required. Continue the procedure with the addition of bromine water.
2. This removes nitric acid, which would react with iodide ion.
3. Bromine is added to ensure that arsenic and antimony are in the +5 oxidation state. These elements may be reduced back to the +3 state by sulfur in a sulfide ore. Unless the excess bromine is removed, it will, of course, oxidize iodide ion. A test to see if bromine has been completely expelled is to hold in the vapors a piece of filter paper moistened with a starch solution which contains some potassium iodide. If the paper darkens, bromine is still present. If arsenic and antimony are absent, the bromine treatment can be omitted. Consult the instructor.
4. The container will be etched by hydrofluoric acid. It is preferable to transfer the solution at this point to a flask or beaker which is already etched or chipped. This container can be used for each sample after addition of fluoride and then discarded.
5. Samples of copper oxide (Note 1) do not normally contain iron, and it may not be necessary to add fluoride. Add glacial acetic acid and continue the procedure. Consult the instructor.
6, 7, 8. See Notes 4, 5, and 6 of Experiment 36(b).

Experiment 38. Iodometric Determination of Hydrogen Peroxide

Procedure Pipet 25 mL of the peroxide solution into a 250-mL volumetric flask, dilute to the mark, and mix thoroughly. Transfer a 25-mL aliquot of this solution to a 250-mL Erlenmeyer flask and add 8 mL of 1:6 sulfuric acid (about 6 N), 3 g of potassium iodide in 10 mL of water, and 3 drops of 3% ammonium molybdate solution. Titrate with thiosulfate until the brown color of iodine has almost disappeared. Then add 5 mL of starch solution and finish the titration to the disappearance of the deep blue color.

Titrate two other portions of the peroxide solution in the same manner. Calculate the weight of hydrogen peroxide in the original sample. Assuming that the weight of the sample was exactly 25 g, calculate the percentage of hydrogen peroxide by weight.

Experiment 39. ## Determination of Bleaching Power by Iodometry

Commercial bleaching products contain oxidizing agents such as hypochlorites or peroxides. The oxidizing power can be determined by the same procedure used in Experiment 38.

Procedure Place an accurately measured sample of the bleach in a 250-mL Erlenmeyer flask. Liquid bleaches, such as Clorox and Purex, contain sodium hypochlorite, and 2.00 mL is a convenient sample if 0.1 N thiosulfate is used as the titrant. Solid products, such as Clorox II and Snowy Bleach, contain peroxides. A sample of 0.7 to 0.8 g is usually a convenient size for titration (Note 1). Add to the Erlenmeyer flask 75 mL of distilled water, 3 g of KI, 8 mL of 1:6 sulfuric acid, and 3 drops of 3% ammonium molybdate solution (Note 2). Titrate the liberated iodine with 0.1 N thiosulfate until the brown color of iodine has almost disappeared. Then add 5 mL of starch solution and finish the titration to the disappearance of the deep blue color.

Titrate at least two portions of each bleaching product being compared. The oxidizing ability of a bleach is usually reported as percent chlorine. That is, the calculation assumes that chlorine is the oxidizing agent, although in fact it may not be. Report the percent by weight of chlorine in each product, assuming that the liquid bleaches have a density of 1.000 g/mL. The equivalent weight of chlorine is the atomic weight.

Notes

1. The solids may not be homogeneous. If duplicate samples do give widely different results, a large sample can be taken, dissolved in a 500-mL volumetric flask, and 50-mL aliquots titrated.
2. Some bleaches react slowly with iodide ion. Molybdate ions catalyze the reaction. Alternatively, the solutions can be heated to about 60°C to speed up the reaction.

22.5

OPTICAL METHODS OF ANALYSIS

The principles of spectrophotometry are discussed in some detail in Chapter 14. The experiments given in this section illustrate some of these principles as they are applied to chemical analysis and to the determination of physical constants.

Experiment 40. ## Determination of Manganese in Steel

Manganese in steel can be determined spectrophotometrically after oxidation to the purple permanganate ion.[4] The steel is dissolved in nitric acid and the oxidation is effected with potassium periodate:

[4] H. H. Willard and L. H. Greathouse, *J. Am. Chem. Soc.*, **39**, 2366 (1917).

$$2Mn^{2+} + 5IO_4^- + 3H_2O \longrightarrow 2MnO_4^- + 5IO_3^- + 6H^+$$

Since the yellow iron(III) is present, phosphoric acid is added to form the colorless iron-phosphate complex.

If the steel contains cobalt or nickel, the color of these ions interferes in the manganese determination. This interference can be canceled by the addition of about the same amounts of these elements to the standards as are in the unknown. Alternatively, a sample can be carried through the entire procedure except for the periodate oxidation and then used as a blank in setting the spectrophotometer to read zero absorbance. In the directions below, a steel of known manganese content which contains about the same quantities of nickel and cobalt as the unknown is used as the standard.

The experiment first calls for determining a spectral-transmittance curve to find the wavelength at which to perform the analysis. A check of Beer's law is then made by measuring the absorbances of permanganate solutions at several concentrations. The unknown is determined by comparison with the Beer's law plot.

Procedure

(a) Spectral-Transmittance Curve. Weigh accurately a sample of steel of known manganese content (Note 1). Dissolve the sample in 50 mL of 1 : 3 nitric acid, using heat and finally boiling gently for 1 to 2 min to remove oxides of nitrogen. Then remove the flask from the burner, add about 1 g of ammonium persulfate, and boil the solution gently for 10 to 15 min (Note 2). If a precipitate of an oxide of manganese forms or if the permanganate color develops, add a few drops of sodium sulfite or sulfurous acid (Note 3) and boil the solution a few minutes more to expel sulfur dioxide.

Dilute the solution with water to about 100 mL and add about 10 mL of 85% phosphoric acid and 0.5 g of potassium periodate. Boil the solution gently for about 3 min to effect oxidation to permanganate. Then cool the solution and dilute to 250 mL in a volumetric flask.

Measure the absorbance of the prepared solution or a suitable dilution thereof (Note 4) using the directions given for the spectrophotometer employed. Cover the range from about 440 to 700 nm,[5] taking readings at 20-nm intervals. In the region of maximum absorbance, take readings at 5-nm intervals. Plot the absorbance vs. wavelength and connect the points to form a smooth curve. Select the proper wavelength to use for the determination of manganese (Note 5).

(b) Beer's Law Check. Secure four clean, dry 100-mL beakers or Erlenmeyer flasks and number them 1 to 4. In beaker 1 place some of the solution prepared in part (a) and do not dilute it. In beaker 2 place 30 mL of the standard solution (use a 10-mL pipet) and then add 10 mL of water. In beaker 3 place 10 mL of the standard solution and 10 mL of water, and in beaker 4 place 10 mL of the standard and 30 mL of water. Mix the solutions which were diluted and then measure the absorbance of each against distilled water at the wavelength of maximum absorbance. Plot the absorbance vs. concentration for the four solutions. Draw the best straight line between the points. Is Beer's law obeyed?

(c) Analysis of Sample. Weigh accurately a sample of steel whose manganese content is to be determined, dissolve it, and oxidize the manganese to permanganate as directed in part (a). The solution is finally diluted to 250 mL in a volumetric flask. Measure the absorbance of some of the solution against distilled wa-

---

[5] The abbreviation nm is for a *nanometer*, 10^{-9} m or 10^{-7} cm. It is synonymous with millimicron, mμ, a term still widely used by chemists.

ter at the same wavelength used in part (b) to check Beer's law. Using the Beer's law plot, read off the concentration of permanganate and calculate the percentage of manganese in the sample.

Notes

1. Consult the instructor for the rough manganese content of the sample. Calculate the size sample needed to give a final solution whose absorbance falls in the range 0.7 to 0.8 if the molar absorptivity of permanganate is 2360 liters/mol-cm at 525 nm.
2. The ammonium persulfate is added to oxidize carbon or carbon compounds. The excess persulfate is destroyed by boiling the solution.
3. This reduces manganese to the bivalent state. The solution should be clear.
4. Any spectrophotometer or filter photometer can be used. Follow precisely the operating directions given by the instructor or in the operation manual.
5. This wavelength is 525 nm, but the wavelength calibration of the spectrophotometer is often not reliable and the value found with the instrument should be used.

Experiment 41. *Determination of Iron with 1,10-Phenanthroline*

The reaction between iron(II) ion and 1,10-phenanthroline to form a red complex serves as a good, sensitive method for determining iron. The molar absorptivity of the complex, $[(C_{12}H_8N_2)_3Fe]^{2+}$, is 11,100 at 508 nm. The intensity of the color is independent of pH in the range 2 to 9. The complex is very stable, and the color intensity does not change appreciably over very long periods of time. Beer's law is obeyed.

The iron must be in the $+2$ oxidation state, and hence a reducing agent is added before the color is developed. Hydroxylamine, as its hydrochloride, can be used, the reaction being

$$2Fe^{3+} + 2NH_2OH + 2OH^- \longrightarrow 2Fe^{2+} + N_2 + 4H_2O$$

The pH is adjusted to a volume between 6 and 9 by addition of ammonia or sodium acetate. An excellent discussion of interferences and of applications of this method is given by Sandell.[6]

Procedure[7] **Prepare the Following Solutions:**

(a) Dissolve 0.1 g of 1,10-phenanthroline monohydrate in 100 mL of distilled water, warming to effect solution if necessary.
(b) Dissolve 10 g of hydroxylamine hydrochloride in 100 mL of distilled water.
(c) Dissolve 10 g of sodium acetate in 100 mL of distilled water.
(d) Weigh accurately about 0.07 g of pure iron(II) ammonium sulfate, dissolve in water, and transfer the solution to a 1-liter volumetric flask. Add 2.5 mL of concentrated sulfuric acid and dilute the solution to the mark. Calculate the concentration of the solution in milligrams of iron per milliliter.

Into five 100-mL volumetric flasks, pipet 1-, 5-, 10-, 25-, and 50-mL portions of the standard iron solution. Put 50 mL of distilled water in another flask to

[6] E. B. Sandell, *Colorimetric Determination of Traces of Metals,* 3rd ed., Interscience Publishers, Inc., New York, 1959.

[7] H. Diehl and G. F. Smith, *Quantitative Analysis,* John Wiley & Sons, Inc., New York, 1952.

serve as the blank, and a measured volume of unknown in another (Note). To each flask add 1 mL of the hydroxylamine solution, 10 mL of the 1,10-phenanthroline solution, and 8 mL of the sodium acetate solution. Then dilute all the solutions to the 100-mL marks and allow them to stand for 10 min.

Using the blank as the reference and any one of the iron solutions prepared above, measure the absorbance at different wavelengths in the interval 400 to 600 nm. Take readings about 20 nm apart, except in the region of maximum absorbance where intervals of 5 nm are used. Plot the absorbance vs. wavelength and connect the points to form a smooth curve. Select the proper wavelength to use for the determination of iron with 1,10-phenanthroline.

Using the selected wavelength, measure the absorbance of each of the standard solutions and the unknown. Plot the absorbance vs. the concentration of the standards. Note whether Beer's law is obeyed. From the absorbance of the unknown solution, calculate the concentration of iron (mg/liter) in the original solution.

Note

Prepared solutions may be used as unknowns. Consult the instructor concerning size of sample to be used. If a natural water is used, be sure that it is colorless and free of turbidity.

Experiment 42. *Spectrophotometric Determination of the pK_a of an Acid-Base Indicator*[8]

In this experiment, spectrophotometry is employed to measure the pK_a of bromthymol blue, an acid-base indicator. The indicator (HIn) is a monoprotic acid, and we can represent its dissociation as follows:

$$HIn \rightleftharpoons H^+ + In^-$$

As is shown in Chapter 5, the equilibrium expression for such a dissociation can be written

$$pH = pK_a - \log \frac{[HIn]}{[In^-]}$$

This can be rearranged to give

$$\log \frac{[In^-]}{[HIn]} = pH - pK_a$$

This is in the slope-intercept form of an equation for a straight line,

$$y = mx + b$$

where y is $\log [In^-]/[HIn]$, $m = 1$, and $b = -pK_a$. Hence, if the log term is plotted vs. pH, the slope is 1, the intercept is $-pK_a$, and the line should cross the pH axis at $pH = pK_a$ (Fig. 22.2). At the latter point $[In^-] = [HIn]$, and hence the log of the ratio of these terms is zero, making $pH = pK_a$.

The ratio $[In^-]/[HIn]$ can be determined spectrophotometrically. First, a solution of bromthymol blue is prepared in acidic solution (low pH) where es-

---

[8] C. N. Reilley and D. T. Sawyer, *Experiments for Instrumental Methods,* McGraw-Hill Book Company, New York, 1961.

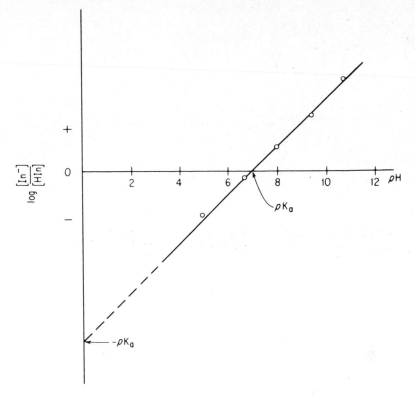

Figure 22.2 Plot to determine pK_a.

sentially all the indicator is in the HIn form. The absorption spectrum is then determined. Second, a solution is prepared in basic solution (high pH) where essentially all the indicator is in the In^- form. The absorption spectrum of In^- is then determined. From the two absorption spectra the wavelengths of maximum absorbance of HIn and In^- are selected for further measurements.

Buffered solutions with pH values on either side of the pK_a of bromthymol blue are then prepared, and the absorbances measured at the selected wavelengths. The solutions contain the same total concentration of indicator, [HIn] + [In^-], but the ratios vary with pH. Figure 22.3 shows a typical plot of absorbance vs. pH at the wavelength of maximum absorbance for the In^- species. The terms used in this figure are as follows:

$$A_a = \text{absorbance of HIn}$$

$$A_b = \text{absorbance of In}^-$$

$$A = \text{absorbance of mixture}$$

From the graph it is evident that

$$\frac{[\text{In}^-]}{[\text{HIn}]} = \frac{A - A_a}{A_b - A}$$

If the wavelength used is the one at which HIn shows maximum absorbance, the curve will be similar to that shown in Fig. 22.3, except that it will start at a high absorbance and curve down to a low absorbance value at high pH.

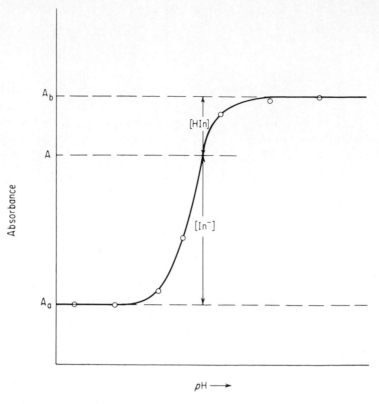

Figure 22.3 Plot of absorbance vs. pH.

Procedure   **Prepare the Following Solutions:**

(a) Dissolve 0.1 g of bromthymol blue in 100 mL of 20% ethanol.

(b) Dissolve 2.4 g of NaH_2PO_4 in water and dilute to 100 mL. The solution is 0.2 M.

(c) Dissolve 3.4 g of K_2HPO_4 in water and dilute to 100 mL. The solution is 0.2 M.

(d) Prepare a small amount of 3 M NaOH. Three grams of NaOH dissolved in 25 mL of solution will provide more than enough of this reagent.

Prepare a series of buffered solutions as follows: Secure seven clean 50-mL volumetric flasks (Note 1). Pipet 2.0 mL of the bromthymol blue solution into each flask. Then add the following volumes of phosphate solutions to these flasks. (These volumes can be measured with a graduated cylinder.)

Flask	NaH₂PO₄, mL	K₂HPO₄, ml
1	5	0
2	5	1
3	10	5
4	5	10
5	1	5
6	1	10
7	0	5

To flask 7 add 2 drops of the 3 M NaOH solution. Now dilute each solution to the mark and mix thoroughly. Measure and record the pH value of each solution using a pH meter.

To determine the absorption spectrum of bromthymol blue at low pH (the spectrum of HIn), measure the absorbance of solution 1 from 400 to 640 nm using water as a reference (Note 2). Read the absorbance at 20-nm intervals except in the vicinity of the maximum in the spectrum, where readings should be taken every 5 nm. Plot the absorbance vs. wavelength.

To determine the absorption spectrum at high pH (the spectrum of In$^-$), measure the absorbance of solution 7 in the same manner as directed above. Plot the results on the same piece of graph paper.

Using the spectra obtained above, select two wavelengths at which further absorbance measurements will be made. Choose wavelengths at which HIn and In$^-$ exhibit maximal differences in absorbance.

Now measure the absorbances of each of the seven solutions at the two wavelengths selected (Note 3). Prepare a graph of absorbance vs. pH for each of the two wavelengths. (See Fig. 22.3.) Determine the [In$^-$]/[HIn] ratios in solutions 2 to 6 as explained above and shown in Fig. 22.3. Plot log [In$^-$]/[HIn] vs. pH (Fig. 22.2) and obtain a value of pK_a from the graph (Note 4).

Report the value of the pK_a of bromthymol blue in the manner desired by the instructor. Turn in all your graphs.

Notes

1. If it is more convenient to use 100-mL or 25-mL volumetric flasks, the amounts of reagents should be scaled up or down proportionately.
2. The pH of this solution is sufficiently low that most of the indicator is in the HIn form. If the absorbance reading at any wavelength is very high (above 0.8 to 1.0), the solutions should be diluted or less bromthymol blue used in each flask.
3. The instructor may wish measurements made at one wavelength only and should be consulted before making the measurements.
4. Note that the value of pK_a can also be obtained from the graph in Fig. 22.3 by reading the pH at which $A_b - A = A_a - A$ (i.e., the absorbance halfway between A_a and A_b). Consult the instructor; he or she may wish you to determine the pK_a by several methods and report the average.

Experiment 43.   *Ultraviolet Analysis of Benzene in Cyclohexane*[9]

Spectrophotometry in the near-ultraviolet region (200 to 390 nm) can be applied to the qualitative and quantitative determination of many organic and some inorganic compounds (Chapter 14). Such compounds as aldehydes, ketones, aliphatic nitro compounds, and nitrate esters absorb in this region, although the intensities are so low that the spectra are useful only under special circumstances. Molecules containing conjugated double bonds, such as benzene and toluene, have rather high molar absorptivities in the near-ultraviolet and can be determined spectrophotometrically. Other molecules which show strong absorption in this region include azo, diazo, and nitroso compounds, quinones, and nitrile esters.

This experiment involves the spectrophotometric determination of benzene in cyclohexane. The differential method (page 410) is employed. A recording spectrophotometer is desirable for making the measurements.

[9] This experiment was designed by Professor Hubert L. Youmans of Western Carolina University, who has kindly given us permission to use it here.

Procedure   *Spectrum of Benzene Vapor.* Record a spectrum of benzene vapor against air in the reference cell from 220 to 290 nm. To obtain the sample of benzene vapor, carefully put one drop of benzene in a cuvette. Do not get any liquid benzene on the side of the cuvette. Cap the cuvette, wait about 2 min for liquid-vapor equilibrium to be established, and then record the spectrum.

Preparation of Standards. Secure a stock solution of 10.0 v/v% benzene in cyclohexane from the instructor. With this solution prepare six standards in the 0.2 to 1.0% range. Use a 10-mL buret with a Teflon stopcock to measure the stock solution into 25-mL volumetric flasks. Dilute to volume with spectrographic-grade cyclohexane. Use only solvents furnished by the instructor (Note 1).

Standard Curve and Analysis. Record a spectrum of the 1% standard against the 0.2% standard as the reference. Choose a wavelength where the absorbance falls between 0.7 and 1.0. Then measure the absorbance of all the standard solutions at this wavelength. Take a 25-mL volumetric flask to your instructor for an unknown sample. Dilute it to volume with cyclohexane and measure its absorbance at the wavelength used for the standards (Note 2). Prepare a calibration curve and use it to determine the v/v% benzene in the unknown after it has been diluted to volume.

Explain the difference between the spectra of benzene vapor and benzene in cyclohexane.

Notes

1. These standards can also be used in Experiments 44 and 51.
2. The unknowns will contain 0.2 to 1.0% benzene when diluted to volume. The instructor may give more than one sample for analysis.

Experiment 44.   *Infrared Analysis of Benzene in Cyclohexane* [10]

The use of infrared spectrophotometry in analytical chemistry is discussed in Chapter 14. This experiment involves the determination of benzene in cyclohexane using an infrared spectrophotometer.

Direct application of Beer's law is very difficult in the infrared region because of instrument limitations. The source and detector have rather limited output. The instrument thus uses a rather large slit width and a wide band of wavelengths, whereas most infrared spectral peaks are rather narrow. For this reason most infrared spectrophotometers cannot measure the actual height of the peak chosen for analysis but measure an average or integrated height across the peak in the range of wavelengths passed by the instrument. Another instrumental factor that produces deviation from Beer's law is the large amount of stray radiation.

A calibration curve must be prepared and checked frequently for direct quantitative work in the infrared. A baseline technique is generally used. An internal standard method is also frequently employed.

Procedure   Secure a stock solution of 10.0 v/v% benzene in cyclohexane from the instructor. With this solution prepare six standards in the 0.2 to 1.0% range (Note 1). Use a 10-mL buret with a Teflon stopcock to measure the stock solution into 25-mL volumetric flasks. Dilute to volume with spectrographic-grade cyclohexane.

Record spectra of 0.2% and 1.0% benzene in cyclohexane over the full range of the instrument. Use the attenuator in the reference beam. Start each

[10] This experiment was designed by Professor Hubert L. Youmans of Western Carolina University, who has kindly given us permission to use it here.

spectrum with the pen set at 90% T and scan at fast speed. Compare the spectra and determine a band suitable for quantitative analysis. There is really only one.

Record spectra of the standards and unknown (Note 2). You need record only a small portion of the spectra near the transmittance minimum of the analytical band. Set 100% T in a flat portion of the spectrum and record through the transmittance minimum. Start at the same wavelength each time. You should be able to record all spectra on one sheet of recorder paper by shifting the drum. Use slow scan speed.

Plot % T vs. concentration on semilog paper or log % T vs. concentration on regular graph paper. Connect the points with an appropriate curve or straight line and determine the concentration of your unknown from this graph.

Notes

1. The standards prepared in Experiment 43 can be used here.
2. Take a 25-mL volumetric flask to your instructor for an unknown sample and dilute it to volume with cyclohexane.

Experiment 45. *Determination of Copper by Atomic Absorption Spectroscopy[11]*

The principles of atomic absorption spectroscopy are discussed in Chapter 14. Your instructor will furnish directions on the operation of the atomic absorption spectrophotometer you will use. The experiment involves the determination of copper in Monel, an alloy from which nickel coins are made. It contains 20 to 30% copper.

Procedure   Copper shot is used as the primary standard. Weigh on the analytical balance about 0.3 g of copper shot and dissolve it in 20 mL of 1 : 1 water–nitric acid. After the alloy has dissolved, transfer the solution to a 1-liter volumetric flask and dilute to volume with distilled water.

Pipet 0, 2, 4, 5, 7, 9 and 10 mL of the standard solution into a series of 100-mL volumetric flasks. To each flask add 10 mL of 0.2% Sterox solution. Sterox is a nonionic detergent. Dilute the solutions to volume with deionized water and mix thoroughly. Express the concentrations as ppm Cu.

Weigh duplicate samples of the Monel unknown on the analytical balance and dissolve each in 20 mL of 1 : 1 water–nitric acid. Dilute each sample to 1 liter in a volumetric flask. Pipet 10 mL of each solution and 10 mL of 0.2% Sterox into 100-mL volumetric flasks and dilute to volume with deionized water.

The atomic absorption spectrophotometer is operated in the absorption mode. Set 100% T with the blank and measure % T of the standard and unknown. Plot % T vs. concentration on semilog paper. Report the percentage of Cu in the Monel alloy.

Experiment 46. *Flame Photometric Determination of Calcium in Water[12]*

The principles of flame photometry are discussed in Chapter 15. Your instructor will furnish directions on the operation of the flame photometer you will use. The experiment involves the determination of calcium in water. The method of stan-

[11] This experiment was designed by Professor Hubert L. Youmans of Western Carolina University, who has kindly given us permission to use it here.

[12] This experiment was designed by Professor Hubert L. Youmans of Western Carolina University, who has kindly given us permission to use it here.

dard addition is used. In this method known amounts of a standard solution of the analyte are added to aliquots of the unknown. Each aliquot of the unknown is of the same size, but the amount of standard is varied. The emission intensity is plotted against the concentration of standard (right side of ordinate). If the points fall on a straight line (of course there may be some "scatter"), the line may be extrapolated to the abscissa on the left side of the ordinate. The intercept gives the concentration of the unknown. If the right side of the line is curved, the standard addition method should not be used.

Procedure   All solutions should be made with deionized water. Distilled water contains a measurable amount of metal ions. Glassware should be rinsed with deionized water.

Prepare a standard solution containing 10.0 ppm of Ca^{2+} by dissolving 0.250 g of dried $CaCO_3$ in 5 mL of 6 M HNO_3. Dilute this solution to 1 liter in a volumetric flask. Then pipet 10 mL of this solution into a 100-mL volumetric flask and dilute to the mark.

Number five 100-mL volumetric flasks B, 0, 5, 10, and 15. To the flasks pipet in order 0, 0, 5, 10, and 15 mL of the 10.0 ppm Ca standard. To each flask pipet 10 mL of 0.2% Sterox solution. To all flasks except B pipet 10 mL of the water sample. Dilute each flask to volume with deionized water and then mix.

Measure the emission intensities and make a standard addition method graph. Set 0 emission with solution B. Determine if the instrument operates linearly over the range of the analysis. Determine the ppm (w/v basis) of Ca in the unknown. Report the concentration to three significant figures.

22.6

POTENTIOMETRIC TITRATIONS

It was pointed out in Chapter 12 that a potentiometric titration involves measurement of the difference in potential between an indicator electrode and a reference electrode during a titration. The difference in potential can be measured with a potentiometer or a pH meter. Generally, precise measurements of potential differences are made with a potentiometer. However, for the precision required in titrations a pH meter gives satisfactory results and is more convenient to use. A brief discussion of the pH meter is given below, followed by directions for performing several experiments.

pH Meters   An ordinary potentiometer cannot be employed with a glass electrode because of the high resistance, 1 to 100 megohms, of this electrode. The so-called pH meter is a voltage measuring device designed to be used with cells of high resistance. There are two common types available commercially, the potentiometric and the direct-reading. The former is basically a potentiometer, but since the off-balance currents are so small because of the high resistance of the cell, the current is amplified electronically so that it will affect a galvanometer or microammeter. The direct-reading instruments are electronic voltmeters of very high input resistance; the circuit is so arranged as to give a meter reading proportional to pH. The voltage of the glass-reference electrode pair is impressed across a very high resistance so that the current drawn is very low, on the order of 5×10^{-11} A. Since the resistance of the cell may be as large as 10^8 ohms, this means a voltage drop of 0.005 V:

$$E = I \times R = 5 \times 10^{-11} \times 10^8 = 0.005$$

or an error of 0.5% at 1.000 V.

The direct-reading instruments are the most popular today, particularly since they can be line-operated. Several commercial models are available from such companies as Beckman, Leeds and Northrup, Coleman, Fisher, and Corning. The operating instructions are either printed directly on the instrument or furnished in a pamphlet by the manufacturer.

Experiment 47. *Acid-Base Titrations*

The following experiments are chosen to illustrate the types of titration curves obtained with a strong acid, a weak acid, and a polyprotic acid. The data can be used for standardizing a solution, analyzing an unknown, or determining the dissociation constant of a weak acid.

A typical experimental setup is shown in Fig. 22.4. The instructor will explain to you the operation of the pH meter and the precautions you should observe.

Procedure (a) *Strong Acid–Strong Base*. Prepare solutions of about 0.1 N hydrochloric acid and sodium hydroxide as directed in Experiment 1, page 611. Then pipet a 25.00-mL aliquot of the acid into a 250-mL beaker and dilute the solution to about 100 mL with distilled water. Insert the electrodes into the solution, being sure that they dip about $\frac{1}{2}$ in. below the surface. Adjust the mechanical or magnetic stirrer and set up a buret containing the base solution as shown in Fig. 22.4. Measure and record the pH of the solution before the addition of any titrant. Then add from the buret about 5 mL of the base solution and again measure the pH. Record this value as well as the buret reading at this point.

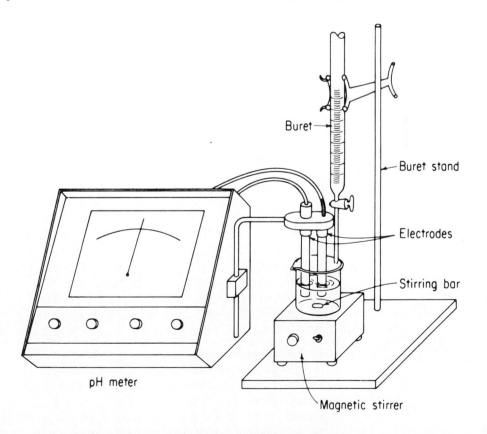

Figure 22.4 Potentiometric titration using pH meter.

Proceed in this manner to record the pH and buret readings after the addition of about 10, 15, and 20 mL of titrant. Add the titrant in about 1-mL intervals until the equivalence point is almost reached. (It may save time to run rapidly through the first titration to approximately locate the equivalence point and then to do the second titration carefully.) Continue to add the titrant in 0.1-mL intervals until the equivalence point is passed. It will be evident when this point is reached because of the large change in pH that occurs. Finally, record two additional readings at about 5 and 10 mL of excess titrant.

Make the following plots of the data: (1) pH vs. milliliters of NaOH; (2) $\Delta pH/\Delta V$ vs. milliliters of NaOH; and (3) $\Delta^2 pH/\Delta V^2$ vs. milliliters of NaOH. Determine the volume of base required by the acid from each plot and also calculate the volume by the analytical method described in Chapter 12. Calculate the volume ratio of the two solutions.

Repeat the titration on two additional aliquots of acid (Note), adding to the first two drops of methyl red indicator and to the second two drops of phenolphthalein. Note the pH values at which these indicators change color and compare these with the data given in Table 6.3, page 146. Average the values obtained for the volume ratios and calculate the precision of the measurement.

Note

If standard solutions are already at hand, one titration is sufficient to illustrate the titration curves.

(b) Weak Acid–Strong Base. Weigh a sample of about 0.7 to 0.9 g of pure, dry potassium acid phthalate (Note) on the analytical balance. Dissolve the sample in about 100 mL of distilled water and titrate with the sodium hydroxide. Measure the pH at different increments of titrant as in part (a).

Plot the titration data in the same manner as above and compare the curves obtained for the weak acid with those obtained for the strong acid. Justify the selection of the indicator used in Experiment 3, page 613. Calculate the normality of the base and acid solutions. If further use is to be made of these solutions, the standardization should be repeated (consult the instructor.)

Note

If the normalities of the solutions are already known, a sample of unknown purity can be titrated.

(c) pK of an Unknown Acid. Dissolve a sample of an unknown acid (consult the instructor) in about 100 mL of distilled water in a 250-mL beaker and titrate the solution with standard sodium hydroxide as described in part (a). Plot the data as pH vs. milliliters of NaOH and determine the equivalence volume. Read from the curve the pH at half the volume required to reach the equivalence point. At this halfway point $pH = pK_a$, and the acid constant is thus determined. Report this value to your instructor and repeat the titration if she wishes this done. She may also wish to know the concentration of the unknown acid, if this is a solution, or the equivalent weight of the unknown if is a solid. If the sample is a solid, it should be weighed on an analytical balance before it is dissolved.

(d) Titration of Phosphoric Acid. Pipet 25.00 mL of a phosphoric acid solution of unknown concentration into a 250-mL beaker. Dilute the solution to about 100 mL, insert the electrodes, and titrate with standard sodium hydroxide as directed in part (a). You should observe two breaks in the titration curve, one around pH 4 to 5 and the other around pH 9 to 10. Plot the titration curve as pH vs. milliliters of NaOH. Determine the following from the curve and report to the

instructor: (a) the molarity of the acid solution and (b) the values of pK_{a_1} and pK_{a_2} for the first two dissociation constants of phosphoric acid (Note).

Note

For acids whose dissociation constants are greater than about 10^{-3} to 10^{-2}, an appreciable error is made by using the expression $pH = pK_a$ halfway to the equivalence point. The following expression can be used for such stronger acids:

$$pH = pK_a - \log \frac{C - [H^+]}{C + [H^+]}$$

The volume to the equivalence point, V_e, is first determined; then the pH at $\frac{1}{2} V_e$ is read and the value of $[H^+]$ calculated. The term C is the concentration of the anion halfway to the equivalence point and is given by

$$C = \frac{\frac{1}{2} V_e \times N}{\text{total volume}}$$

where the total volume is $\frac{1}{2} V_e$ plus the starting volume. The latter volume should be known to ± 10 mL.

Experiment 48. *Redox Titrations*

There are many redox titrations that can be used to illustrate the potentiometric technique. In this experiment the titration of iron(II) with dichromate or cerium(IV) solution is carried out. The indicator electrode is platinum, and the reference is a saturated calomel electrode.

Procedure *(a) Titration of Iron with Dichromate.* Prepare a standard solution of potassium dichromate by weighing accurately about 1.25 g of the pure, dry salt. Dissolve the salt in water, transfer the solution to a 250-mL volumetric flask, and dilute the solution to the mark. Fill a buret with the solution.

Weigh accurately a sample of about 1.3 to 1.5 g of pure iron(II) ammonium sulfate. Dissolve the salt in about 100 mL of distilled water and add about 10 mL of concentrated sulfuric acid.

Insert a platinum and a saturated calomel electrode into the solution and adjust the stirrer if one is to be used. If a pH meter is to be employed, set the instrument to measure potential rather than pH. The platinum wire is the positive electrode. Add about 5 mL of the dichromate solution from the buret and then measure the potential. Continue the titration in the usual manner, recording the potential and volume of dichromate solution, until about 10 mL of excess titrant is added.

Make the following plots of the data: (1) potential vs. milliliters of titrant, (2) $\Delta E/\Delta V$ vs. milliliters of titrant, and (3) $\Delta^2 E/\Delta V^2$ vs. milliliters of titrant. Determine the volume of dichromate used, and from this volume and the weight of iron(II) ammonium sulfate, calculate the normality of the dichromate solution. Compare this value with the normality calculated from the weight of dichromate dissolved in 250 mL of solution. Repeat the titration if desired (consult the instructor).

(b) Titration of Iron with Cerium(IV). Prepare 250 mL of a standard cerium(IV) solution from cerium(IV) ammonium nitrate as directed on page 636. Dissolve a sample of iron ore and reduce it with tin(II) chloride as directed in Experiment 26, page 634. Stop the procedure after adding excess tin(II) chloride; that is, do

not remove the excess tin(II) ion with mercury(II) chloride. The titration curve will show two breaks: the tin reacting first, followed by the iron. The difference in volume to the two end points gives the volume of titrant used by the iron sample.

Insert a platinum and a saturated calomel electrode into the iron solution in a 250-mL beaker. Adjust the volume to about 100 mL and set the pH meter to measure potential rather than pH. The platinum wire is the positive electrode. Add about 5 mL of the cerium(IV) solution from a buret and measure the potential. Continue the titration in the usual manner, recording the potential and volume of cerium(IV) solution, until about 10 mL of excess titrant is added.

Plot the potential vs. milliliters of titrant and measure the difference in volume between the first and second breaks. From this volume and the weight of the sample, calculate the percentage of iron in the sample. Report this value to the instructor and repeat the determination if necessary.

Also plot $\Delta E/\Delta V$ vs. milliliters of titrant (for the titration of iron) and $\Delta^2 E/\Delta V^2$ vs. milliliters of titrant and compare these with the plots obtained in Experiment 47(a).

Experiment 49. *Precipitation Titrations*

A piece of polished silver wire serves as the indicator electrode for titrations involving silver ions. This electrode is available commercially for use with pH meters. A calomel electrode cannot dip directly into the solution being titrated, since silver chloride would precipitate; instead, this electrode is placed in a saturated potassium nitrate solution in a separate vessel, and the two solutions are joined by a salt bridge containing potassium nitrate.[13] Directions are given for the titration of chloride with silver nitrate, and for the titration of a chloride-iodide mixture.

Procedure *(a) Titration of Chloride.* Prepare a standard solution of silver nitrate by weighing accurately about 4.5 g of the pure salt. Dissolve the salt in water, transfer the solution to 250-mL volumetric flask, and dilute the solution to the mark.

Accurately weigh a dried chloride sample of unknown purity (consult the instructor as to the weight required) and dissolve the salt in about 100 mL of distilled water. Insert the silver electrode and the potassium nitrate salt bridge into the solution and adjust the stirrer. Set the pH meter to read potential. The silver wire is the positive electrode. Add about 5 mL of the silver nitrate solution from the buret and then measure the potential. Continue the titration in the usual manner, recording the potential and volume of silver nitrate, until about 10 mL of excess titrant is added.

Make the following plots of the data: (1) potential vs. milliliters of silver nitrate, (2) $\Delta E/\Delta V$ vs. milliliters of silver nitrate, and (3) $\Delta^2 E/\Delta V^2$ vs. milliliters of silver nitrate. Determine the volume of titrant, and from this volume and the molarity calculate the percentage of chloride in the sample. Repeat the titration if desired (consult the instructor).

(b) Titration of a Chloride-Iodide Mixture. The standard solution of silver nitrate is prepared as in part (a). Secure from the instructor a solution containing both chloride and iodide ions of unknown concentration. Pipet a 25-mL aliquot into a 250-mL beaker and dilute the solution to about 100 mL with water. Insert the sil-

[13] Since the pH of the solution changes little during the titration, it is possible to use a glass electrode in place of a calomel one as the reference electrode, thereby eliminating the salt bridge.

ver electrode and potassium nitrate salt bridge and adjust the stirrer. Set the *p*H meter to read potential. Titrate with silver nitrate in the usual manner, recording the potential and volume of titrant, noting carefully the volumes near the two equivalence points.

Plot the potential vs. milliliters of titrant and determine the volumes of titrant used by iodide and chloride. Calculate the molarities of these two ions and report the results to the instructor.

22.7

ELECTROLYSIS

The principles of electrolysis are discussed in Chapter 13. Directions are given here for the determination of copper in a solution free of interfering metals and for the separation of copper and nickel by electrolysis.

Experiment 50. *Electrolytic Determination of Copper*

Apparatus A schematic diagram of the necessary apparatus is given in Fig. 13.2. The source of direct current is often a 6-V rectifier rather than a storage battery, and an ordinary rheostat of suitable size serves as the adjustable resistance. An inexpensive unit that can be easily assembled has found wide usage.[14] It is desirable to provide mechanical stirring if the equipment is available. Commercial "electroanalyzers" which provide for rotation of one electrode to effect stirring can be purchased.

The electrodes are usually platinum, although copper gauze can be employed as the cathode for the deposition of copper. The cathode is usually a cylinder of platinum gauze, and the anode is a spiral of platinum wire or a small cylinder if the electrode is rotated.

Procedure Clean the platinum electrodes by immersing them in warm 1:3 nitric acid for about 5 min. Then rinse them well with tap water and distilled water. Place the gauze cathode on a watch glass and dry it in an oven at about 105°C. Cool the cathode in a desiccator and then weigh it on the analytical balance. Avoid touching the gauze with the fingers, as this may leave grease on the surface and prevent copper from adhering.

The solution to be electrolyzed should contain about 2 mL of sulfuric acid and 1 mL of nitric acid (Note 1) per 100 mL. Obtain this solution or a solid unknown from the instructor (Note 2). Connect the electrodes properly to the apparatus, placing the spiral anode inside the gauze cylinder. Make sure that the electrodes do not touch. Raise the beaker (a tall-form one is convenient) around the electrodes and adjust the height so that the lower edge of the cathode almost touches the bottom of the beaker. About $\frac{1}{4}$ in. of the top of the cathode should not be covered by the solution. The solution can be diluted if necessary with distilled water. Cover the beaker with a split watch glass and adjust the rheostat to give its full resistance.

Turn on the stirrer and close the circuit to start the electrolysis. Only a small current should flow, since the resistance is high. Gradually lower the resistance until the current is 2 to 4 A and the voltage is below 4 V (Note 3). Electrolyze at this voltage until the blue color of copper disappears (usually about 30 to 45 min). Add 0.5 g of urea (Note 4), continue the electrolysis for 15 min longer, and then add sufficient water to cover completely the top of the cathode (Note 5). Continue

---

[14] P. J. Elving, J. R. Hayes, and M. G. Mellon, *J. Chem. Ed.*, **30**, 254 (1953).

the electrolysis for an additional 15 min using 0.5-A current, and if no copper is deposited on the fresh cathode surface, the deposition is complete.

To stop the electrolysis, turn off the stirrer but do not interrupt the current at this time. Remove the support under the breaker and slowly lower the beaker with one hand while washing the exposed portion of the cathode with a stream of water from the wash bottle. As soon as the cathode is completely out of the solution, cut off the current and raise a beaker of distilled water to cover the electrodes. Wash the electrodes with a second portion of distilled water and then disconnect the cathode. Dip the cathode into a beaker of acetone or alcohol and place it on a watch glass in the oven at about 10°C for 5 min (Note 6). Cool the electrode to room temperature and then weigh it accurately.

If a prepared solution was used, report the total weight of copper in your solution. If copper ore was employed, calculate the percentage of the metal in the sample. Duplicate samples should be run; consult the instructor for instructions.

The copper can be removed from the electrode by placing it in warm $1:3$ nitric acid for a few minutes. Then rinse the electrode well with tap water and distilled water.

Notes

1. If the acid has a yellow tinge, indicating the presence of oxides of nitrogen, boil the solution until the oxides are expelled. Nitric acid improves the nature of the copper deposit by preventing the evolution of hydrogen at the cathode.
2. This solution can be prepared by dissolving pure copper foil in nitric acid according to procedure (b) of Experiment 36. It is then diluted in a volumetric flask, and aliquot portions are given to the students. Alternatively, commercial unknowns of copper oxide which are readily dissolved in sulfuric acid can be used. If a brass sample is being analyzed, the filtrate from the lead determination by the sulfate method can be electrolyzed after first adjusting the volume to about 100 mL and adding 1 mL of concentrated nitric acid.
3. If a mechanical or magnetic stirrer is not available, the electrolysis can be carried out without stirring by using a current of about 0.5 A. The electrolysis should then be allowed to run overnight.
4. Nitrite prevents complete depositon of copper and is removed by urea according to the equation

$$2NO_2^- + 2H^+ + (NH_2)_2CO \rightleftharpoons CO_2 + 2N_2 + 3H_2O$$

The nitrite is formed by the reaction

$$2H^+ + NO_3^- + 2e \rightleftharpoons H_2O + NO_2^-$$

5. If the solution is not to be used for further analysis, a few drops can be removed with a pipet and tested with concentrated ammonia. The deep blue of the copper-ammonia complex indicates the incomplete deposition of copper.
6. Do not heat the electrode any longer than this, since the surface of the copper becomes oxidized easily.

Experiment 51.   *GLC Analysis of Benzene in Cyclohexane*[15]

This experiment involves the determination of benzene in cyclohexane using gas-liquid chromatography. This topic is discussed in detail in Chapter 17.

Procedure   Secure a stock solution of 10.0 v/v% benzene in cyclohexane from the instructor. With this solution prepare six standards in the 0.2 to 1.0% range. Use a 10-mL

[15] This experiment was designed by Professor Hubert L. Youmans of Western Carolina University, who has kindly given us permission to use it here.

buret with a Teflon stopcock to measure the stock solution into 25-mL volumetric flasks. Dilute to volume with spectrographic-grade cyclohexane. The standards used in Experiments 43 and 44 can be used if available. Take a 25-mL volumetric flask to the instructor for an unknown sample and dilute it to volume with cyclohexane.

The column you will use is 10 ft long and 5 mm i.d. It is packed with 20% Carbowax 20M on 80- to 100-mesh Chromosorb W. Carbowax 20M is poly(ethylene glycol) of average molecular weight 20,000. Chromosorb W is made by calcining diatomite with sodium carbonate at about 900°C. The fluxing agent causes fusion of the smaller particles to form larger microamorphous silica aggregates, some of which contain cristobalite crystals.

To fill a syringe, first pump the plunger with the needle tip below the surface of the liquid. When all air is expelled, draw more than the required amount of liquid into the syringe. Then with the needle pointing upward, adjust the plunger to the desired mark.

Insert the needle into the septum to the limit; keep the needle straight while inserting. Do not bend it. Push the plunger in, then withdraw the needle. The three steps should be a smooth, continuous action.

To obtain retention times, stop the chart, mark it at the pen, and then inject the sample and start the chart simultaneously. The retention time, t_R, is the distance from the starting point to the maximum of a peak multiplied by the chart speed. In calculating n, the number of theoretical plates, chart distances may be used because chart speed is a constant that occurs in both numerator and denominator.

To obtain measured volumetric flow rares, F_{meas}, measure the time in seconds for a soap film in the flowmeter to go from 0 to 10 mL. Divide the number of seconds into 600. This gives F_{meas} in milliliters per minute.

Operate the column at 90°C with a helium flow rate of 60 mL/min. Run chromatograms of the standards and unknown. Plot peak area vs. concentration of benzene for the standards. From this calibration curve, determine the concentration of benzene in your unknown.

I

Tables of Equilibrium Constants and Standard Potentials

ACID	FORMULA	K_a	pK_a	CONJUGATE BASE	K_b	pK_b
Acetic	CH_3COOH	1.8×10^{-5}	4.74	Acetate ion	5.6×10^{-10}	9.26
Ammonium ion	NH_4^+	5.6×10^{-10}	9.26	Ammonia	1.8×10^{-5}	4.74
Anilinium ion	$C_6H_5NH_3^+$	2.2×10^{-5}	4.66	Aniline	4.6×10^{-10}	9.34
Arsenic	H_3AsO_4	6.0×10^{-3}	2.22	Dihydrogen arsenate ion	1.7×10^{-12}	11.78
Dihydrogen arsenate ion	$H_2AsO_4^-$	1.1×10^{-7}	6.96	Monohydrogen arsenate ion	9.1×10^{-8}	7.04
Monohydrogen arsenate ion	$HAsO_4^{2-}$	3×10^{-12}	11.5	Arsenate ion	3×10^{-3}	2.5
Arsenious	H_3AsO_3	6×10^{-10}	9.2	Dihydrogen arsenite ion	1.7×10^{-5}	4.8
Dihydrogen arsenite ion	$H_2AsO_3^-$	3×10^{-14}	13.5	Monohydrogen arsenite ion	3×10^{-1}	0.5
Benzoic	$HC_7H_5O_2$	6.6×10^{-5}	4.18	Benzoate ion	1.5×10^{-10}	9.82
Boric	H_3BO_3	5.8×10^{-10}	9.24	Dihydrogen borate ion	1.7×10^{-5}	4.76
Carbonic	H_2CO_3	4.6×10^{-7}	6.34	Bicarbonate ion	2.2×10^{-8}	7.66
Bicarbonate ion	HCO_3^-	4.4×10^{-11}	10.36	Carbonate ion	2.3×10^{-4}	3.64
Chloroacetic, mono	$CH_2ClCOOH$	1.4×10^{-3}	2.85	Monochloroacetate ion	7.1×10^{-12}	11.15
Chloroacetic, di	$CHCl_2COOH$	5×10^{-2}	1.3	Dichloroacetate ion	2×10^{-13}	12.7
Chloroacetic, tri	CCl_3COOH	2×10^{-1}	0.7	Trichloroacetate ion	5×10^{-14}	13.3
Chromic	H_2CrO_4	1.8×10^{-1}	0.74	Bichromate ion	5.6×10^{-14}	13.26
Bichromate ion	$HCrO_4^-$	3.2×10^{-7}	6.49	Chromate ion	3.1×10^{-8}	7.51
Citric	$H_3C_6H_5O_7$	8.4×10^{-4}	3.08	Dihydrogen citrate ion	1.2×10^{-11}	10.92
Dihydrogen citrate ion	$H_2C_6H_5O_7^-$	1.8×10^{-5}	4.74	Monohydrogen citrate ion	5.6×10^{-10}	9.26
Monohydrogen citrate ion	$HC_6H_5O_7^{2-}$	4.0×10^{-6}	5.40	Citrate ion	2.5×10^{-9}	8.60
Cyanic	$HOCN$	2×10^{-4}	3.7	Cyanate ion	5×10^{-11}	10.3
Diethylammonium ion	$C_4H_{10}NH_2^+$	7.7×10^{-12}	11.11	Diethylamine	1.3×10^{-3}	2.89

TABLE A-1 Dissociation Constants of Weak Acids and Bases (25°C)

ACID	FORMULA	K_a	pK_a	CONJUGATE BASE	K_b	pK_b
Dimethylammonium ion	$C_2H_6NH_2^+$	1.9×10^{-11}	10.72	Dimethylamine	5.2×10^{-4}	3.28
Ethylammonium ion	$C_2H_5NH_3^+$	1.8×10^{-11}	10.75	Ethylamine	5.6×10^{-4}	3.28
Formic	$HCOOH$	1.8×10^{-4}	3.74	Formate ion	5.6×10^{-11}	10.26
Hydrazinium ion	$N_2H_5^+$	7.7×10^{-9}	8.11	Hydrazine	1.3×10^{-6}	5.89
Hydrocyanic	HCN	7.2×10^{-10}	9.14	Cyanide ion	1.4×10^{-5}	4.86
Hydrofluoric	HF	6×10^{-4}	3.22	Fluoride ion	1.7×10^{-11}	10.78
Hydrogen sulfide	H_2S	1×10^{-7}	7.0	Bisulfide ion	1.1×10^{-7}	7.0
Bisulfide ion	HS^-	1×10^{-15}	15.0	Sulfide ion	10	-1.0
Hypochlorous	$HOCl$	3×10^{-8}	7.5	Hypochlorite ion	3×10^{-7}	6.5
Hydroxylammonium ion	NH_3OH^+	9.1×10^{-7}	6.04	Hydroxylamine	1.1×10^{-8}	7.96
Methylammonium ion	$CH_3NH_3^+$	2×10^{-11}	10.7	Methylamine	5×10^{-4}	3.3
Nitrous	HNO_2	4.5×10^{-4}	3.35	Nitrite ion	2.2×10^{-11}	10.65
Oxalic	$H_2C_2O_4$	5.6×10^{-2}	1.25	Bioxalate ion	1.8×10^{-13}	12.75
Bioxalate ion	$HC_2O_4^-$	5.4×10^{-5}	4.27	Oxalate ion	1.9×10^{-10}	9.73
Phenol	C_6H_5OH	1.3×10^{-10}	9.89	Phenolate ion	7.7×10^{-5}	4.11
Phosphoric	H_3PO_4	7.5×10^{-3}	2.12	Dihydrogen phosphate ion	1.3×10^{-12}	11.88
Dihydrogen phosphate ion	$H_2PO_4^-$	6.2×10^{-8}	7.21	Monohydrogen phosphate ion	1.6×10^{-7}	6.79
Monohydrogen phosphate ion	HPO_4^{2-}	4.8×10^{-13}	12.32	Phosphate ion	2.1×10^{-2}	1.68
Phthalic	$C_6H_4(COOH)_2$	1.1×10^{-3}	2.96	Biphthalate ion	9.1×10^{-12}	11.04
Biphthalate ion	$C_6H_4C_2O_2H^-$	3.9×10^{-6}	5.41	Phthalate ion	2.6×10^{-9}	8.59
Pyridinium ion	$C_5H_5NH^+$	7.1×10^{-6}	5.15	Pyridine	1.4×10^{-9}	8.85
Sulfuric	H_2SO_4	Strong	—	Bisulfate ion	Weak	—
Bisulfate ion	HSO_4^-	1.2×10^{-2}	1.92	Sulfate ion	8.3×10^{-13}	12.08
Sulfurous	H_2SO_3	1.7×10^{-2}	1.77	Bisulfite ion	5.9×10^{-13}	12.23
Bisulfite ion	HSO_3^-	6.2×10^{-8}	7.21	Sulfite ion	1.6×10^{-7}	6.79
Tartaric	$H_2C_4H_4O_6$	9.4×10^{-4}	3.03	Bitartrate ion	1.1×10^{-11}	10.97
Bitartrate ion	$HC_4H_4O_6^-$	2.9×10^{-5}	4.54	Tartrate ion	3.4×10^{-10}	9.46

TABLE A-2 Stepwise Formation Constants of Complex Ions

LIGAND	CATION	IONIC STRENGTH	TEMP., °C	LOGARITHM OF EQUILIBRIUM CONSTANT			
				K_1	K_2	K_3	K_4
Ammonia	Ag^+	1.0	25	3.37	3.78		
	Cd^{2+}	2.1	25	2.74	2.21	1.37	1.13
	Cu^{2+}	1.0	25	4.27	3.55	2.90	2.18
	Ni^{2+}	1.0	25	2.36	1.90	1.55	1.23
	Zn^{2+}	2.0	30	2.37	2.44	2.50	2.15
Chloride	Ag^+	0.2	25	2.85	1.87	0.32	0.86
	Fe^{3+}	1.0	25	0.62	0.11	−1.40	−1.92
	Hg^{2+}	0.5	25	6.74	6.48	0.85	1.00
	Pb^{2+}	1.0	25	0.88	0.61	−0.40	−0.15
Cyanide	Cd^{2+}	3.0	25	5.48	5.14	4.56	3.58
	Hg^{2+}	0.1	20	18.00	16.70	3.83	2.98
EDTA	Ag^+	0.1	20	7.32			
	Al^{3+}	0.1	20	16.13			
	Ba^{2+}	0.1	20	7.76			
	Ca^{2+}	0.1	20	10.70			
	Cd^{2+}	0.1	20	16.46			
	Co^{2+}	0.1	20	16.31			
	Cu^{2+}	0.1	20	18.80			
	Fe^{2+}	0.1	20	14.33			
	Fe^{3+}	0.1	20	25.1			
	Hg^{2+}	0.1	20	21.80			
	Mg^{2+}	0.1	20	8.69			
	Mn^{2+}	0.1	20	13.79			
	Ni^{2+}	0.1	20	18.62			
	Pb^{2+}	0.1	20	18.04			
	Sr^{2+}	0.1	20	8.63			
	Th^{4+}	0.1	20	23.2			
	TiO^{2+}	0.1	—	17.3			
	VO^{2+}	0.1	20	18.77			
	Zn^{2+}	0.1	20	16.50			
Thiocyanate	Ag^+	4.0	25	4.59	3.70	1.77	1.20
	Fe^{3+}	1.8	18	1.96	2.02	<−0.41	>−0.14
	Ni^{2+}	1.0	20	1.18	0.46	0.17	
Thiosulfate	Ag^+	0.2	20	10.00	3.36		

Note: Stepwise constants are defined as follows:

$$M^{4+} + X^- \rightleftharpoons MX^{3+} \qquad K_1 = \frac{[MX^{3+}]}{[M^{4+}][X^-]}$$

$$MX^{3+} + X^- \rightleftharpoons MX_2^{2+} \qquad K_2 = \frac{[MX_2^{2+}]}{[MX^{3+}][X^-]} \qquad \text{etc.}$$

TABLE A-3 Solubility Product Constants

COMPOUND	FORMULA	SOLUBILITY PRODUCT CONSTANT, K_{sp}	COMPOUND	FORMULA	SOLUBILITY PRODUCT CONSTANT, K_{sp}
Aluminum hydroxide	$Al(OH)_3$	5×10^{-33}	Magnesium ammonium phosphate	$MgNH_4PO_4$	3×10^{-13}
Barium carbonate	$BaCO_3$	5×10^{-9}	Magnesium carbonate	$MgCO_3$	3×10^{-5}
Barium chromate	$BaCrO_4$	2×10^{-10}	Magnesium fluoride	MgF_2	7×10^{-9}
Barium fluoride	BaF_2	3×10^{-6}	Magnesium hydroxide	$Mg(OH)_2$	1×10^{-11}
Barium iodate	$Ba(IO_3)_2$	2×10^{-9}	Magnesium oxalate	MgC_2O_4	9×10^{-5}
Barium oxalate	BaC_2O_4	2×10^{-7}			
Barium sulfate	$BaSO_4$	1×10^{-10}	Manganese(II) hydroxide	$Mn(OH)_2$	2×10^{-13}
			Manganese(II) sulfide	MnS	1×10^{-16}
Cadmium carbonate	$CdCO_3$	3×10^{-14}	Mercury(II) sulfide	HgS	3×10^{-52}
Cadmium oxalate	CdC_2O_4	1×10^{-8}	Mercury(I) bromide	Hg_2Br_2	3×10^{-23}
Cadmium sulfide	CdS	5×10^{-27}	Mercury(I) chloride	Hg_2Cl_2	6×10^{-19}
			Mercury(I) iodide	Hg_2I_2	7×10^{-29}
Calcium carbonate	$CaCO_3$	5×10^{-9}			
Calcium fluoride	CaF_2	4×10^{-11}	Nickel sulfide	NiS	1×10^{-25}
Calcium oxalate	CaC_2O_4	2×10^{-9}			
Calcium sulfate	$CaSO_4$	3×10^{-5}	Silver arsenate	Ag_3AsO_4	1×10^{-22}
			Silver bromate	$AgBrO_3$	6×10^{-5}
Copper(II) hydroxide	$Cu(OH)_2$	2×10^{-19}	Silver bromide	$AgBr$	4×10^{-13}
Copper(II) iodate	$Cu(IO_3)_2$	1×10^{-7}	Silver carbonate	Ag_2CO_3	8×10^{-12}
Copper(II) oxalate	CuC_2O_4	3×10^{-8}	Silver chloride	$AgCl$	1×10^{-12}
Copper(II) sulfide	CuS	4×10^{-38}	Silver chromate	Ag_2CrO_4	2×10^{-12}
			Silver cyanide	$Ag[Ag(CN)_2]$	2×10^{-12}
Copper(I) bromide	$CuBr$	5×10^{-9}	Silver hydroxide	$AgOH$	2×10^{-8}
Copper(I) chloride	$CuCl$	3×10^{-7}	Silver iodate	$AgIO_3$	3×10^{-8}
Copper(I) iodide	CuI	1×10^{-12}	Silver iodide	AgI	1×10^{-16}
Copper(I) thiocyanate	$CuSCN$	4×10^{-14}	Silver oxalate	$Ag_2C_2O_4$	5×10^{-12}
			Silver sulfide	Ag_2S	1×10^{-48}
			Silver thiocyanate	$AgSCN$	1×10^{-12}
Iron(III) hydroxide	$Fe(OH)_3$	1×10^{-36}			
Iron(II) hydroxide	$Fe(OH)_2$	8×10^{-16}	Strontium carbonate	$SrCO_3$	2×10^{-9}
Iron(II) oxalate	FeC_2O_4	2×10^{-7}	Strontium fluoride	SrF_2	3×10^{-9}
Iron(II) sulfide	FeS	4×10^{-19}	Strontium oxalate	SrC_2O_4	6×10^{-8}
			Strontium sulfate	$SrSO_4$	3×10^{-7}
Lead carbonate	$PbCO_3$	2×10^{-13}			
Lead chloride	$PbCl_2$	2×10^{-5}	Thallium(I) chloride	$TlCl$	2×10^{-4}
Lead chromate	$PbCrO_4$	2×10^{-14}	Thallium(I) sulfide	Tl_2S	1×10^{-22}
Lead fluoride	PbF_2	5×10^{-8}			
Lead hydroxide	$Pb(OH)_2$	3×10^{-16}	Zinc carbonate	$ZnCO_3$	3×10^{-8}
Lead iodate	$Pb(IO_3)_2$	3×10^{-13}	Zinc hydroxide	$Zn(OH)_2$	2×10^{-14}
Lead sulfate	$PbSO_4$	2×10^{-8}	Zinc oxalate	ZnC_2O_4	3×10^{-9}
Lead sulfide	PbS	3×10^{-28}	Zinc sulfide	ZnS	1×10^{-24}

TABLE A-4 Standard Potentials

REDOX COUPLE	$E°$	REDOX COUPLE	$E°$
$F_2 + 2H^+ + 2e \rightleftharpoons 2HF(aq)$	3.06	$2H_2SO_3 + 2H^+ + 4e \rightleftharpoons S_2O_3^{2-} + 3H_2O$	0.40
$F_2 + 2e \rightleftharpoons 2F^-$	2.87	$Fe(CN)_6^{3-} + e \rightleftharpoons Fe(CN)_6^{4-}$	0.36
$O_3 + 2H^+ + 2e \rightleftharpoons O_2 + H_2O$	2.07	$VO^{2+} + 2H^+ + e \rightleftharpoons V^{3+} + H_2O$	0.36
$S_2O_8^{2-} + 2e \rightleftharpoons 2SO_4^{2-}$	2.01	$Cu^{2+} + 2e \rightleftharpoons Cu$	0.34
$Co^{3+} + e \rightleftharpoons Co^{2+}$	1.82	$Hg_2Cl_2 + 2e \rightleftharpoons 2Hg + 2Cl^-$	0.28
$H_2O_2 + 2H^+ + 2e \rightleftharpoons 2H_2O$	1.77	$IO_3^- + 3H_2O + 6e \rightleftharpoons I^- + 6OH^-$	0.26
$MnO_4^- + 4H^+ + 3e \rightleftharpoons MnO_2 + 2H_2O$	1.70	$AgCl + e \rightleftharpoons Ag + Cl^-$	0.22
$PbO_2 + SO_4^{2-} + 4H^+ + 2e \rightleftharpoons PbSO_4 + 2H_2O$	1.69	$HgBr_4^{2-} + 2e \rightleftharpoons Hg + 4Br^-$	0.21
$Au^+ + e \rightleftharpoons Au$	1.68	$Cu^{2+} + e \rightleftharpoons Cu^+$	0.15
$HClO_2 + 2H^+ + 2e \rightleftharpoons HClO + H_2O$	1.64	$Sn^{4+} + 2e \rightleftharpoons Sn^{2+}$	0.15
$HClO + H^+ + e \rightleftharpoons \frac{1}{2}Cl_2 + H_2O$	1.63	$S + 2H^+ + 2e \rightleftharpoons H_2S$	0.14
$Ce^{4+} + e \rightleftharpoons Ce^{3+}$	1.61	$CuCl + e \rightleftharpoons Cu + Cl^-$	0.14
$Bi_2O_4 + 4H^+ + 2e \rightleftharpoons 2BiO^+ + 2H_2O$	1.59	$AgBr + e \rightleftharpoons Ag + Br^-$	0.10
$BrO_3^- + 6H^+ + 5e \rightleftharpoons \frac{1}{2}Br_2 + 3H_2O$	1.52	$S_4O_6^{2-} + 2e \rightleftharpoons 2S_2O_3^{2-}$	0.08
$MnO_4^- + 8H^+ + 5e \rightleftharpoons Mn^{2+} + 4H_2O$	1.51	$CuBr + e \rightleftharpoons Cu + Br^-$	0.03
$PbO_2 + 4H^+ + 2e \rightleftharpoons Pb^{2+} + 2H_2O$	1.46	$2H^+ + 2e \rightleftharpoons H_2$	0.00
$Cl_2 + 2e \rightleftharpoons 2Cl^-$	1.36	$HgI_4^{2-} + 2e \rightleftharpoons Hg + 4I^-$	−0.04
$Cr_2O_7^{2-} + 14H^+ + 6e \rightleftharpoons 2Cr^{3+} + 7H_2O$	1.33	$Pb^{2+} + 2e \rightleftharpoons Pb$	−0.13
$MnO_2 + 4H^+ + 2e \rightleftharpoons Mn^{2+} + 2H_2O$	1.23	$CrO_4^{2-} + 4H_2O + 3e \rightleftharpoons Cr(OH)_3 + 5OH^-$	−0.13
$O_2 + 4H^+ + 4e \rightleftharpoons 2H_2O$	1.23	$Sn^{2+} + 2e \rightleftharpoons Sn$	−0.14
$IO_3^- + 6H^+ + 5e \rightleftharpoons \frac{1}{2}I_2 + 3H_2O$	1.20	$AgI + e \rightleftharpoons Ag + I^-$	−0.15
$ClO_4^- + 2H^+ + 2e \rightleftharpoons ClO_3^- + H_2O$	1.19	$CuI + e \rightleftharpoons Cu + I^-$	−0.19
$Br_2(aq) + 2e \rightleftharpoons 2Br^-$	1.09	$Ni^{2+} + 2e \rightleftharpoons Ni$	−0.25
$Br_2(liq) + 2e \rightleftharpoons 2Br^-$	1.07	$V^{3+} + e \rightleftharpoons V^{2+}$	−0.26
$Br_3^- + 2e \rightleftharpoons 3Br^-$	1.05	$PbCl_2 + 2e \rightleftharpoons Pb + 2Cl^-$	−0.27
$VO_2^+ + 2H^+ + e \rightleftharpoons VO^{2+} + H_2O$	1.00	$Co^{2+} + 2e \rightleftharpoons Co$	−0.28
$AuCl_4^- + 3e \rightleftharpoons Au + 4Cl^-$	1.00	$PbBr_2 + 2e \rightleftharpoons Pb + 2Br^-$	−0.28
$NO_3^- + 4H^+ + 3e \rightleftharpoons NO + 2H_2O$	0.96	$PbSO_4 + 2e \rightleftharpoons Pb + SO_4^{2-}$	−0.36
$NO_3^- + 3H^+ + 2e \rightleftharpoons HNO_2 + H_2O$	0.94	$PbI_2 + 2e \rightleftharpoons Pb + 2I^-$	−0.37
$2Hg^{2+} + 2e \rightleftharpoons Hg_2^{2+}$	0.92	$Cd^{2+} + 2e \rightleftharpoons Cd$	−0.40
$AuBr_4^- + 3e \rightleftharpoons Au + 4Br^-$	0.87	$Cr^{3+} + e \rightleftharpoons Cr^{2+}$	−0.41
$Cu^{2+} + I^- + e \rightleftharpoons CuI$	0.86	$Fe^{2+} + 2e \rightleftharpoons Fe$	−0.44
$Hg^{2+} + 2e \rightleftharpoons Hg$	0.85	$2CO_2(g) + 2H^+ + 2e \rightleftharpoons H_2C_2O_4(aq)$	−0.49
$Ag^+ + e \rightleftharpoons Ag$	0.80	$Cr^{3+} + 3e \rightleftharpoons Cr$	−0.74
$Hg_2^{2+} + 2e \rightleftharpoons 2Hg$	0.79	$Zn^{2+} + 2e \rightleftharpoons Zn$	−0.76
$Fe^{3+} + e \rightleftharpoons Fe^{2+}$	0.77	$H_2O + e \rightleftharpoons \frac{1}{2}H_2 + OH^-$	−0.83
$PtCl_4^{2-} + 2e \rightleftharpoons Pt + 4Cl^-$	0.73	$Cr^{2+} + 2c \rightleftharpoons Cr$	−0.91
$Q + 2H^+ + 2e \rightleftharpoons H_2Q$	0.70	$Mn^{2+} + 2e \rightleftharpoons Mn$	−1.18
$O_2 + 2H^+ + 2c \rightleftharpoons H_2O_2$	0.68	$Al^{3+} + 3e \rightleftharpoons Al$	−1.66
$PtBr_4^{2-} + 2e \rightleftharpoons Pt + 4Br^-$	0.58	$Mg^{2+} + 2e \rightleftharpoons Mg$	−2.37
$MnO_4^- + e \rightleftharpoons MnO_4^{2-}$	0.56	$Na^+ + e \rightleftharpoons Na$	−2.71
$H_3AsO_4 + 2H^+ + 2e \rightleftharpoons HAsO_2 + 2H_2O$	0.56	$Ca^{2+} + 2e \rightleftharpoons Ca$	−2.87
$I_3^- + 2e \rightleftharpoons 3I^-$	0.54	$Sr^{2+} + 2e \rightleftharpoons Sr$	−2.89
$I_2(s) + 2e \rightleftharpoons 2I^-$	0.54	$Ba^{2+} + 2e \rightleftharpoons Ba$	−2.90
$Cu^+ + e \rightleftharpoons Cu$	0.52	$K^+ + e \rightleftharpoons K$	−2.93
$4H_2SO_3 + 4H^+ + 6e \rightleftharpoons S_4O_6^{2-} + 6H_2O$	0.51	$Li^+ + e \rightleftharpoons Li$	−3.05

Source: From W. M. Latimer, *Oxidation Potentials*, 2nd ed., Prentice-Hall, Inc., Englewood Cliffs, N.J., 1952.

TABLE A-5 Some Formal Potentials

REDOX SYSTEMS	STANDARD POTENTIAL	FORMAL POTENTIAL	SOLUTION
$Ce^{4+} + e \rightleftharpoons Ce^{3+}$	—	1.23	1 M HCl
		1.44	1 M H$_2$SO$_4$
		1.61	1 M HNO$_3$
		1.7	1 M HClO$_4$
$Fe^{3+} + e \rightleftharpoons Fe^{2+}$	+0.771	0.68	1 M H$_2$SO$_4$
		0.700	1 M HCl
		0.732	1 M HClO$_4$
$Cr_2O_7^{2-} + 14H^+ + 6e \rightleftharpoons 2Cr^{3+} + 7H_2O$	+1.33	1.00	1 M HCl
		1.05	2 M HCl
		1.08	3 M HCl
		1.08	0.5 M H$_2$SO$_4$
		1.15	4 M H$_2$SO$_4$
		1.03	1 M HClO$_4$
$Fe(CN)_6^{3+} + e \rightleftharpoons Fe(CN)_6^{4-}$	+0.356	0.48	0.01 M HCl
		0.56	0.1 M HCl
		0.71	1 M HCl
		0.72	1 M H$_2$SO$_4$
		0.72	1 M HClO$_4$
$H_3AsO_4 + 2H^+ + 2e \rightleftharpoons H_3AsO_3 + H_2O$	+0.559	0.557	1 M HCl
		0.557	1 M HClO$_4$
$TiO^{2+} + 2H^+ + e \rightleftharpoons Ti^{3+} + H_2O$	+0.1	0.04	1 M H$_2$SO$_4$
$Pb^{2+} + 2e \rightleftharpoons Pb$	−0.126	−0.14	1 M HClO$_4$
$Sn^{2+} + 2e \rightleftharpoons Sn$	−0.136	−0.16	1 M HClO$_4$
$V^{3+} + e \rightleftharpoons V^{2+}$	−0.255	−0.21	1 M HClO$_4$

II

Balancing Oxidation-Reduction Equations

We shall describe briefly the oxidation number and the half-reaction methods for balancing redox equations. Both methods are presented in most general chemistry texts today.

Oxidation Number Method

In this method oxidation numbers are assigned to atoms in the reactants and products, and any changes in these numbers are attributed to a loss or gain of electrons. The number of electrons gained by the oxidizing agent is made equal to the number lost by the reducing agent by selecting appropriate coefficients for these two reactants.

Although the assignment of oxidation numbers can be made on an arbitrary basis, it is convenient to select the numbers as follows:

1. In ionic compounds the oxidation number is taken to be the same as the number of electrons gained or lost by the elements in forming the ion. For example, in sodium chloride, Na^+Cl^-, the oxidation number (also called the oxidation state) of sodium is $+1$, that of chlorine -1. In zinc oxide, ZnO, zinc is taken as $+2$, oxygen as -2.
2. In covalent compounds of known structure, where electrons are shared by two atoms, the electrons are counted as belonging to the more electronegative of the two atoms. In HCl, for example, chlorine is assigned a number of -1, hydrogen a number of $+1$. In a covalent bond between like atoms, such as Cl_2,

one electron is assigned to each atom, making the oxidation number of each Cl atom zero. The oxidation number of any elemental substance, such as Na, H_2, O_3, P_4, etc., is zero.

In assigning oxidation numbers in compounds containing hydrogen, it is customary to start with hydrogen as $+1$. The only exception occurs in ionic hydrides, such as Na^+H^-, where it is taken as -1. Oxygen is next assigned a value of -2 except in peroxides, such as H_2O_2, where the value must be -1 if hydrogen is $+1$.

It should be noted that the algebraic sum of oxidation numbers in a neutral molecule is zero. Similarly, in a complex ion the sum of the oxidation numbers of the atoms must equal the charge of the ion. For example, in the ion HPO_4^{2-}, the oxidation numbers are $H = +1$, $O = -2$, and $P = +5$, giving the sum $+1 + 5 + 4(-2) = -2$.

The following examples illustrate the use of the oxidation number method to balance equations. Since the reactions occur in aqueous solution, we may need to add H^+, OH^-, or H_2O to balance oxygen and hydrogen atoms.

EXAMPLE APP.1   Potassium permanganate, $KMnO_4$, oxidizes oxalic acid, $H_2C_2O_4$, in acid solution to form manganous ion, Mn^{2+}, and carbon dioxide, CO_2:

$$MnO_4^- + H_2C_2O_4 \longrightarrow Mn^{2+} + CO_2$$

Balance the equation, adding H^+ and H_2O as needed.

Note that the oxidation number of manganese changes from $+7$ in MnO_4^- to $+2$ in Mn^{2+}. Hence each MnO_4^- ion gains five electrons. The oxidation number of carbon changes from $+3$ in $H_2C_2O_4$ to $+4$ in CO_2, a loss of one electron per carbon atom. Since each $H_2C_2O_4$ molecule contains two carbon atoms, the molecule loses two electrons. We can equate the electrons gained and lost by taking $2MnO_4^-$ ions for each $5H_2C_2O_4$ molecules. Hence we write

$$2MnO_4^- + 5H_2C_2O_4 \longrightarrow 2Mn^{2+} + 10CO_2 \tag{1}$$

Next we balance the ionic charge by placing H^+ ions where needed. We need six plus charges on the left to make the charge $+4$ on each side of the arrow. The process is completed by adding eight molecules of water to the right side to balance the oxygen and hydrogen atoms.

$$2MnO_4^- + 5H_2C_2O_4 + 6H^+ \longrightarrow 2Mn^{2+} + 10CO_2 + 8H_2O \tag{2}$$

□

EXAMPLE APP.2   Chromate ion, CrO_4^{2-}, oxidizes sulfite ion, SO_3^{2-}, to sulfate, SO_4^{2-}, and is reduced to chromite, CrO_2^-, in basic solution:

$$CrO_4^{2-} + SO_3^{2-} \longrightarrow CrO_2^- + SO_4^{2-}$$

Balance the equation, adding water and OH^- as needed.

Note that the oxidation number of chromium changes from $+6$ to $+3$, a gain of three electrons. Sulfur changes from $+4$ to $+6$, a loss of two electrons. Hence we write

$$2CrO_4^{2-} + 3SO_3^{2-} \longrightarrow 2CrO_2^- + 3SO_4^{2-}$$

Next we balance the ionic charge by placing two hydroxide ions on the right side of the equation. The addition of one molecule of water to the left completes the balancing.

$$2CrO_4^{2-} + 3SO_3^{2-} + H_2O \longrightarrow 2CrO_2^- + 3SO_4^{2-} + 2OH^- \qquad \square$$

Half-Reaction Method

A redox equation can be balanced by writing separate reactions for the oxidizing and reducing agents. These half-reactions are balanced by adding electrons where needed. The number of electrons gained and lost in the two reactions are made the same, and the two equations are added to give the desired result.

EXAMPLE APP.3   Balance the following equation for the reaction which occurs in aqueous acid.

$$MnO_4^- + C_2O_4^{2-} \longrightarrow Mn^{2+} + CO_2$$

Add H_2O and H^+ where needed.
First treat the oxidizing agent:

$$MnO_4^- \longrightarrow Mn^{2+}$$

Add water to the right to balance the oxygen:

$$MnO_4^- \longrightarrow Mn^{2+} + 4H_2O$$

Add H^+ to the left to balance hydrogen:

$$MnO_4^- + 8H^+ \longrightarrow Mn^{2+} + 4H_2O$$

Balance the charge by placing five electrons on the left:

$$MnO_4^- + 8H^+ + 5e \longrightarrow Mn^{2+} + 4H_2O \qquad (1)$$

Second, treat the reducing agent:

$$C_2O_4^{2-} \longrightarrow 2CO_2$$

This is balanced except for the charge. Add two electrons to the right:

$$C_2O_4^{2-} \longrightarrow 2CO_2 + 2e \qquad (2)$$

Multiply Eq. (1) by 2 and Eq. (2) by 5, and add to give the final balanced equation:

$$2MnO_4^- + 5C_2O_4^{2-} + 16H^+ \longrightarrow 2Mn^{2+} + 10CO_2 + 8H_2O \qquad \square$$

EXAMPLE APP.4   Balance the following reaction in basic solution:

$$Cr(OH)_3 + IO_3^- \longrightarrow I^- + CrO_4^{2-}$$

First treat the reducing agent:

$$Cr(OH)_3 \longrightarrow CrO_4^{2-}$$

Add H_2O to the left and H^+ to the right:

$$Cr(OH)_3 + H_2O \longrightarrow CrO_4^{2-} + 5H^+$$

Since we wish the equation to show OH^- ions rather than H^+ ions, we can add $5OH^-$ to both sides. This yields

$$Cr(OH)_3 + H_2O + 5OH^- \longrightarrow CrO_4^{2-} + 5H_2O$$

or

$$Cr(OH)_3 + 5OH^- \longrightarrow CrO_4^{2-} + 4H_2O$$

Now balance the charge by adding three electrons to the right:

$$Cr(OH)_3 + 5OH^- \longrightarrow CrO_4^{2-} + 4H_2O + 3e \qquad (1)$$

Next treat the oxidizing agent similarly, giving

$$IO_3^- + 6H^+ \longrightarrow I^- + 3H_2O$$

$$IO_3^- + 6H_2O \longrightarrow I^- + 3H_2O + 6OH^-$$

$$IO_3^- + 3H_2O + 6e \longrightarrow I^- + 6OH^- \qquad (2)$$

Multiply Eq. (1) by 2 and add it to Eq. (2), giving

$$2Cr(OH)_3 + IO_3^- + 4OH^- \longrightarrow 2CrO_4^{2-} + I^- + 5H_2O \qquad \square$$

EXERCISES

1. Balance the following half-reactions in aqueous acid:
   (a) $MnO_4^- \longrightarrow Mn^{3+}$
   (b) $Cr_2O_7^{2-} \longrightarrow Cr^{3+}$
   (c) $H_3AsO_3 \longrightarrow H_3AsO_4$
   (d) $PH_3 \longrightarrow HPO_3^{2-}$
   (e) $S_2O_3^{2-} \longrightarrow SO_4^{2-}$

2. Balance the following half-reactions in aqueous base:
   (a) $MnO_4^- \longrightarrow MnO_2$
   (b) $O_2 \longrightarrow OH^-$
   (c) $Cu_2O \longrightarrow Cu(OH)_2$
   (d) $CrO_4^{2-} \longrightarrow Cr(OH)_3$
   (e) $Sn^{2+} \longrightarrow SnO_3^{2-}$

3. Balance the following equations, adding H_2O and H^+ where needed. Where basic medium is indicated, show OH^- ions rather than H^+ ions. Use either the oxidation number of half-reaction method, as you prefer.
   (a) $H_2O_2 + MnO_4^- \longrightarrow Mn^{2+} + O_2$
   (b) $SO_3^{2-} + Br_2 \longrightarrow SO_4^{2-} + Br^-$
   (c) $I_2 \longrightarrow IO_3^- + I^-$ (basic)
   (d) $Mn^{2+} + BiO_3^- \longrightarrow MnO_4^- + Bi^{3+}$
   (e) $HO_2^- + CrO_2^- \longrightarrow CrO_4^{2-}$ (basic)
   (f) $Al + NO_3^- \longrightarrow AlO_2^- + NH_3$ (basic)
   (g) $MnO_4^- + VO^{2+} \longrightarrow VO_3^- + Mn^{2+}$
   (h) $C_7H_8O + Cr_2O_7^{2-} \longrightarrow$
       $C_7H_8O_2 + Cr^{3+}$
   (i) $SbH_3 + Cl_2O \longrightarrow H_4Sb_2O_7 + Cl^-$
   (j) $FeS + NO_3^- \longrightarrow Fe^{3+} + NO + S$
   (k) $MnO_4^- + CN^- \longrightarrow MnO_2 + CNO^-$
       (basic)
   (l) $UO_5^{2-} \longrightarrow UO_2^{2+} + O_2$
   (m) $Zn + NO_3^- \longrightarrow Zn^{2+} + NH_4^+$
   (n) $Fe + ClO_4^- \longrightarrow Fe^{3+} + Cl_2$
   (o) $Pt + NO_3^- + Cl^- \longrightarrow$
       $PtCl_6^{2-} + NO_2$
   (p) $VO_2^+ + V^{2+} \longrightarrow VO^{2+}$
   (q) $S_2O_3^{2-} + I_2 \longrightarrow S_4O_6^{2-} + I^-$
   (r) $I^- + Cr_2O_7^{2-} \longrightarrow I_2 + Cr^{3+}$
   (s) $Cl^- + MnO_4^- \longrightarrow Cl_2 + Mn^{2+}$
   (t) $MnO_4^- + Mn^{2+} \longrightarrow MnO_2$ (basic)

III

Answers to Odd-Numbered Problems

CHAPTER 2.   Errors and the Treatment of Analytical Data

1. (a) Oxygen;
(b) H, 0.07 ppt; 0, 0.02 ppt; Fe, 0.05 ppt
3. (a) $\bar{x}$ = 16.64; M = 16.64; R = 0.12; $\bar{d}$ = 0.04;
$\bar{d}$ (rel.) = 2.4 ppt; s = 0.046; v = 0.28%;
(b) 16.64 ± 0.05 from s and R
5. (a) A: 0.09; 5.4 ppt; B: 0.11; 6.6 ppt.
(b) The work of A is more accurate and precise.
7. 0.0514; 7.8 ppt
9. 10 g
11. (a) 0.001 ppt; (b) 0.1
13. (a) 0.4 mg, 2 ppt; (b) 1.2 mg, 2 ppt; (c) 2 mg, 2 ppt
15. (a) 25; (b) 2; (c) 1.7; (d) 1; (e) 1
17. (a) No; (b) (i) Yes; (ii) Yes; (iii) No
19. (a) No; (b) (i) Yes; (ii) No; (iii) No
21. No
23. No

25. (a) No; (b) 243
27. (a) 10.69; (b) 10.06
29. Yes; smaller s based on larger n
31. (a) 12.35 ± 0.13; (b) 12.35 ± 0.09
33. (a) 4; (b) 9
35. (a) 14.6; 0.1; (b) 2.083; 0.004
37. (a) 3; (b) 0.2; (c) 1; (d) 1; (e) 0.01; (f) 5
39. (a) 10.31; (b) 2.06 × 10^{-3}; (c) 17.939
41. (a) 4.683; (b) 0.4190 − 5 or −4.581; (c) 2.24 × 10^4; (d) 1.16
43. Absolute −0.06; relative 2.5 ppt
45. 0.3 g
47. (a) y = −17.6x + 0.832; (b) 0.02; (c) 0.7;
(d) −17.6 ± 1.6; (e) 0.023;
(f) (i) −0.001; (ii) −0.0007

CHAPTER 3.   Titrimetric Methods of Analysis

1. (a) 0.637; (b) 0.0100; (c) 0.500
3. (a) 8.00; (b) 25.0; (c) 170
5. (a) 4.20; (b) 14
7. (a) 0.102; (b) 0.012; (c) 0.090
9. (a) 3.24; (b) 26

11. (a) 0.00800; (b) 750
13. 3.2 liters
15. Acidic; 0.0050 M
17. (a) (i) 0.949; (ii) 1.054; (b) 0.1048
19. (i) $CaO + 2HCl \longrightarrow CaCl_2 + H_2O$

(ii) $2MnO_4^- + 3CN^- + H_2O \longrightarrow 2MnO_2 + 3CNO^- + 2OH^-$

(iii) $Ag^+ + SCN^- \longrightarrow AgSCN$

(iv) $Hg^{2+} + 2Cl^- \longrightarrow HgCl_2$

(a) 28.038; (b) 36.461; (c) 52.6780; (d) 32.558; (e) 169.873; (f) 97.182; (g) 162.30; (h) 74.551

21. (a) 0.100; (b) 0.0640; (c) 0.200

23. 0.1231

25. 0.09790

27. NaOH, 0.09890; HCl, 0.09658

29. (a) 9.289; (b) 13.01

31. (a) 20.9; (b) 27.6

33. 1.25

35. 0.410

37. 39.27

39. 20.07

41. $T = a/t \times M \times MW$; $T = N \times EW$

43. (a) 0.82; (b) 3.3; (c) 0.85

45. 31.93

47. 10.7

CHAPTER 4.   Gravimetric Methods of Analysis

1. (a) 13.83%; (b) 22.79

3. 23.00%

5. 2.47366

7. (a) 0.5343; (b) 0.25704; (c) 0.962000

9. 29 ml

11. 2.15 g

13. 2.00

15. 22.991

17. 55.8

19. 120

21. 11.17 NaCl; 30.26% NaBr

23. (a) 0.4929 g; (b) 0.2235 g

CHAPTER 5.   Review of Chemical Equilibrium

1. (a) 0.16; (b) 0.80; (c) 0.30

3. (a) 0.0756; (b) 0.373; (c) 0.487

5. (a) 2.2×10^{-8}; (b) 1×10^{-5}

7. (a) 1.10; (b) 0.40; (c) -0.70; (d) 9.40

9. (a) 1.78; (b) 4.0×10^{-15}; (c) 5.0×10^{-10}; (d) 1.8×10^{-6}

11. (a) 11.40; (b) 1.60; (c) 5.30; (d) 4.74

13. (a) 5.0×10^{-6}; (b) 0.71%; (c) 0.20 M

15. (a) 11.36; (b) 2.24; (c) 2.55; (d) 11.18

17. (a) 2.9; (b) 8.95; (c) 13.00

19. (a) 8.61; (b) 4.96; (c) 7.00

21. (a) 8.1; (b) 8.77; (c) 8.0

23. 2.5×10^{-4}

25. (a) 1.30; (b) 2.30; (c) 7.00; (d) 12.65; (e) 7.00

27. (a) 0.05, 0.03, 67%; (b) 0.024, 0.019, 26% (c) 0.0087, 0.0080, 8.8%; (d) 0.0030, 0.0029, 3.4%; (e) 9.5×10^{-4}, 9.5×10^{-4}, 0%

29. $s = 0.0095\ M$; $[Br^-] = [Ag(NH_3)_2^+] = 0.0095\ M$; $[Ag^+] = 4.2 \times 10^{-11}\ M$; $[AgNH_3^+] = 4.0 \times 10^{-7}$; error negligible

31. (a) 0.04; (b) 0.02; (c) 3×10^{-3}

33. (a) No; (b) Yes

35. (a) 2×10^{-4}; (b) 0.3

37. 0.02

39. 5×10^{-9}

41. 2.0×10^{-17}; (b) 1.0×10^{-4}

43. (a) 100; (b) 120

CHAPTER 6.   Acid-Base Titrations

1. (a) 1.36; (b) 1.51; (c) 1.74; (d) 2.54; (e) 4.46; (f) 7.00; (g) 9.54; (h) 11.60. Bromcresol purple, bromthymol blue, or neutral red.

3. (a) 5.25; (b) 8.61; (c) 9.14; (d) 10.14; (e) 10.81; (f) 10.81; (g) 10.84; (h) 11.60. Titration not feasible.

5. (a) 10.90; (b) 9.68; (c) 9.26; (d) 8.56; (e) 6.40; (f) 5.42; (g) 4.43; (h) 2.54. Neutral red or p-nitrophenol.

7. (a) 3.26; (b) 3.43; (c) 3.74; (d) 4.22; (e) 5.74; (f) 6.74; (g) 7.74

9. (a) 55; (b) 95; (c) 99.5; (d) 100; (e) 100; (f) 100

11. (a) 3.71; (b) 3.56; (c) 4.04

13. 10.45

15. 8.04

17. (a) 2.4%; (b) 24%

19. (a) 0.0075; (b) 1.2; (c) 99

21. 19.32

23. (a) 2.14; (b) 3.5

25. 53 ml HCl; 47 ml NH$_3$

27. (a) I, 0.15; II, 0.30; (b) I, 0.18; II, 0.40

29. I, 0.40; II, 0.33; III, 0.30

31. A, 9.26; B, 8.65

33. 4.59

35. (a) 9.70; (b) 9.70; (c) 9.70. Titration not feasible; approximations not valid.

37. (a) 8.00, 8.80, 9.60; (b) 8.00, 8.85, 9.70; (c) 8.00, 8.91, 9.82

39. 7.9×10^3

41. 2×10^{-4}

CHAPTER 7. Acid-Base Equilibria in Complex Systems

1.

	Equation (11)	Equation (12)
(a)	5.00	5.00
(b)	5.15	5.00
(c)	7.00	7.00
(d)	7.00	7.00
(e)	7.98	8.50
(f)	7.15	8.50

3.

	$[H_2L^+]$	$[HL]$	$[L^-]$
(a)	0.037	0.013	1.8×10^{-10}
(b)	2.1×10^{-12}	0.0017	0.048

5. (a) 7.21; (b) 9.77; (c) 7.21; (d) 7.21
7. (a) 5.41; (b) 11.84; (c) 4.67; (d) 8.35
9. (a) 5.70, 5.81; (b) 5.84, 5.92
11. 125
13. 60
15. (a) 7.03; (b) 7.39

17. (a) 0.019; (b) 7.24
19.

pH	H_3C	H_2C^-	HC^{2-}	C^{3-}
1.0	1.00			
2.0	0.92	0.08		
3.0	0.54	0.45	0.008	3×10^{-5}
4.0	0.091	0.766	0.137	0.006
5.0	0.003	0.285	0.508	0.203
6.0		0.11	0.20	0.79
7.0			0.024	0.976
8.0			0.002	0.998

21. (a) 40, 15; (b) 20, 30; (c) 30, 40; (d) 18, 18;
(e) 30, 0; (f) 0, 25
23. 19.56% Na_2CO_3; 10.11% NaOH
25. (a) 40.91; (b) 40.00
27. (a) 4.18; (b) (i) 60.2; (ii) 13.1; (iii) 1.5; (iv) 0.15;
(v) 0.015

CHAPTER 8. Complex Formation Titrations

1. (a) 0.01214; (b) 2.259
3. 205 ppm Ca; 136 ppm Mg
5. 9.94
7. Bi, 29.04; Pb, 16.19
9. 42.17
11. 1.4×10^{-12}; 3.5×10^{-7}; 0.35
13. 5×10^{-5}
15. 10.46

17. (a) 7.92; (b) 11.17; (c) 15.23
19. (a) 1.70; (b) 2.17; (c) 4.70; (d) 7.49; (e) 10.30;
(f) 12.00
21. (a) 1.0×10^{10}; (b) 1.0×10^9
23. (a) 0.48; (b) 0.84; (c) 0.99; (d) 100
25. (a) (i) 7.08; (ii) 7.44; (iii) 11.30; (iv) 14.52
(b) (i) 3.44; (ii) 3.82; (iii) 9.48; (iv) 14.52
27. (a) 1.6×10^{-2}; (b) 1.0×10^{19}; (c) 63

CHAPTER 9. Solubility Equilibria and Precipitation Titrations

1. (a) 3.95×10^{-13}; (b) 1.99×10^{-19}; (c) 2.00×10^{-12}
3. (a) 7×10^{-9}; (b) 2×10^{-6}; (c) 5×10^{-6}
5. (a) 1×10^{-19}; (b) 1×10^{-30}; (c) 4×10^{-9}
7. 0.0007 M
9. 1.7 M
11. (a) 1×10^{-4}; (b) 2×10^{-22}; (c) 1×10^{-13};
(d) 2×10^{-3}
13. (a) $p\,Cl = 2.0$; $p\,Ag = 8.0$; (b) $p\,Cl = 8.5$;
$p\,Ag = 1.5$
15. (a) 1.00; (b) 1.48; (c) 4.00; (d) 8.0; (e) 12.0;
(f) 14.0

17. (a) 4.30; 4.28; (b) 4.30; 3.89; (c) 4.30; 2.98
19. 2.5×10^{13}; 4×10^{-14}
21. (a) 8.45, 8.80, 9.80, 10.30; (b) 7.04, 7.38, 8.38,
8.89
23. (a) -1.00; (b) 1.35
25. (a) -1.26; (b) 0.74; (c) 4.50; (d) 6.50; yes
27. (a) 1×10^4; (b) 1×10^8
29. 50%
31. 6×10^{-11}
33. 0.04 M

CHAPTER 10. Oxidation-Reduction Equilibria

1. (a) -0.04, left, Co negative; (b) $+0.03$, right, Co positive; (c) -0.44, left, Pt negative; (d) $+1.48$, right, Pt(Ce) positive
3. (a) -0.83 V; (b) 0.00 V; (c) 0.82 V; (d) 0.00 V

5. (a) 3.8×10^{154}; (b) $[Al^{3+}] = 0.0067$, $[Sn^{2+}] = 2.0 \times 10^{-53}$; (c) -1.69 V
7. (a) 50; (b) Right; (c) -1×10^4 joules; (d) 1×10^4 joules, left

9. Derivation
11. (a) 1.39 V; (b) 0.13 V; (c) 0.74 V; (d) −0.34 V
13. (a) −0.30 V; (b) −0.28 V; (c) −0.26 V;
    (d) −0.22 V; (e) −0.14 V; (f) −0.08 V
15. 0.196
17. (a) 5.2×10^{62}; (b) 2.8×10^{-10}
19. (a) −0.37 V; (b) +0.87 V
21. (a) 1.0×10^{8}; (b) 0.24 V

23. (a) $E = 0.25 + 0.059\ pH$; (b) 0.37 V; (c) 1.00 V
25. (a) 0.78 V; (b) 0.54 V
27. (a) 4.1×10^{-84}; (b) 1.8×10^{-5}; (c) 9.6×10^{-42};
    (d) 6.3×10^{47}
29. (a) 0.14 V; (b) −0.21 V
31. (a) 0.79 V; (b) 0.94 V
33. 74
35. (a) 3.6×10^{47}; (b) 6.07

CHAPTER 11.   Applications of Oxidation-Reduction Titrations

1. (a) $2Mn^{2+} + 5BiO_3^- + 14H^+ \longrightarrow$
       $2MnO_4^- + 5Bi^{3+} + 7H_2O$
   (b) $2Cr^{3+} + 3H_2O_2 + 10OH^- \longrightarrow$
       $2CrO_4^{2-} + 8H_2O$
   (c) $2Cr^{3+} + 3S_2O_8^{2-} + 7H_2O \longrightarrow$
       $Cr_2O_7^{2-} + 6SO_4^{2-} + 14H^+$
   (d) $2V^{5+} + SO_2 + 2H_2O \longrightarrow$
       $2V^{4+} + SO_4^{2-} + 4H^+$
   (e) $Cu^{2+} + Ag + Cl^- \longrightarrow Cu^+ + AgCl$
3. (a) $C_2H_6O_2 + 2MnO_4^- + 6H^+ \longrightarrow$
       $2CO_2 + 2Mn^{2+} + 6H_2O$; EW = 6.2068
   (b) $C_2H_6O_2 + 6Ce^{4+} + 2H_2O \longrightarrow$
       $2H_2CO_2 + 6Ce^{3+} + 6H^+$; EW = 10.345
   (c) $C_2H_6O_2 + HIO_4 \longrightarrow$
       $2H_2CO + IO_3^- + H_2O + H^+$; EW = 31.034
5. (a) 21.91; (b) 31.32

7. 5.52 ml
9. 38.0 ml
11. 65.66% KHC_2O_4; 34.34% $Na_2C_2O_4$
13. 26.76
15. 22.96
17. 0.236
19. 291
21. 2.455
23. Fe, 0.05852; Cr, 0.08688
25. (a) $3CHO_2^- + 2MnO_4^- + H_2O \longrightarrow$
       $2MnO_2 + 3CO_2 + 5OH^-$; (b) 27.43
27. 40.0
29. (a) $IO_3^- + 5I^- + 6H^+ \longrightarrow$
       $3I_2 + 3H_2O$; 0.02000 M; (b) 40.00
31. (a) $K = 6.3 \times 10^{47}$; (b) $Q = 1 \times 10^{16}$, right

CHAPTER 12.   Potentiometric Methods of Analysis

1. (a and b)

Volume	pH	E, V
0.00	1.30	0.323
10.00	1.44	0.331
20.00	1.60	0.341
30.00	1.81	0.353
40.00	2.15	0.373
49.00	3.17	0.434
49.50	3.48	0.452
49.60	3.57	0.457
49.70	3.70	0.465
49.80	3.87	0.475
49.90	4.18	0.494
50.00	7.00	0.660
50.10	9.82	0.827
50.20	10.12	0.845
50.30	10.30	0.856
50.40	10.42	0.863
50.50	10.52	0.869
52.00	11.12	0.904
55.00	11.51	0.927

(c) The curve should resemble Figure 12.3 (a).
3. The curves should resemble Figures 12.3 (b) and
   (c) in text.
5. 39.20 mL to 2 decimal places
7. 0.8%
9. $E_{cell} = 0.45 − 0592\ pH$
11. 4.27×10^{30}
13. (a) 0.11 V (b) 0.25 V (c) 0.47 V

CHAPTER 13. Other Electroanalytical Methods

1. (a) 1.23 V (b) 1.23 V (c) 1.23 V (d) 1.23 V
3. 0.95 V
5. (a) 1.17×10^{-12} M (b) 1.41×10^{-39} M
7. 919.7 C
9. (a) H_2 (b) Zn

11. 7.77
13. 48.8
15. 319.9 μg
17. (a) -0.50 V (b) n = 1
19. 8.58×10^{-6} cm^2/sec

CHAPTER 14. Spectrophotometry

1. infinity and zero, respectively
3. (a) 0.046 (b) 0.155 (c) 0.347 (d) 1.155 (e) 2.000
5. (a) 95.6% (b) 40.4% (c) 1.07%
7. 0.776
9. 2.06 μg/mL
11. (a) 138 (b) yes
13. 0.22%
15. (a) 0.500, 0.750, and 1.000
    (b) 56.2, 31.6, 17.8, and 10.0%
        differences: 24.6, 13.8, and 7.8
    (c) absorbance values: 0.250, 0.500, 0.750
        100, 56.2, 31.6, 17.8 % T
        differences: 43.8, 24.6, 13.8
17. 79 mg/100 mL
19. 0.155 mg/100 mL
21. (a) MX
    (b) appreciable dissociation of MX where there is

no excess of either M or X to force the complexation reaction to completion

23. Values to plot:

mL	A
0.20	0.279
0.50	0.697
0.80	1.116
1.10	1.535
1.50	2.093
2.20	2.790
2.50	2.790
3.00	2.790

Lower A-values can be obtained by changing to a wavelength were ε for BiY$^-$ is smaller.

25. 3.97 ppm

CHAPTER 16. Solvent Extraction

1. (a) 89.9
   (b) As I_2 in the aqueous phase reacts, more moves from the CCl$_4$ to satisfy K_D, and eventually all the I_2 in the system has reacted with thiosulfate
3. (a) 49.5 (b) 5.82 mg
5. 10.7
7. 8.8×10^{-5}
9. (a) equation given in text
   (b) 2.5, 3.26, 3.50, 3.74, 5.00

11. n = 4; $K_{ex} = 1$
13.

tube #	f
0	9.5×10^{-11}
1	4.8×10^{-8}
2	9.5×10^{-6}
3	9.5×10^{-4}
4	4.8×10^{-2}
5	9.5×10^{-1}

CHAPTER 18. Liquid Chromatography

1. (a) 13.78% (b) 3.20 meqt/g
3. 106
5. 622.5

Index

Single-pan balance, 605 *(figure)*
Size exclusion chromatography, 544
Soda ash, determination of alkalinity of, 617
Sodium bismuthate, as oxidant, 293
Sodium carbonate, as primary standard, 614
Sodium hydroxide:
  composition of concentrated solution, 584 *(table)*
  preparation of solution, 612
  standardization of solution, 613
Sodium oxalate:
  as primary standard, 296, 631
  determination of, 633
Sodium thiosulfate, preparation and standardization of solutions of, 303, 640
Solubility:
  equilibria, 111
  factors affecting, 234
  thermodynamics of, 235 *(box)*
Solubility product constant, 111 123, 234, 665 *(table)*
  calculation of, 267
Solvents, for spectrophotometery:
  UV-VIS region, 421 *(table)*
  IR region, 421
Sources, in spectrophotometers, 403
Specific gravity, 47
Specific indicator, 283, 287
Spectrophotometers, 402
  double-beam, 412
  single-beam, 403
Stability constant, 113, 123, 189, 664 *(table)*
Standard addition, method of, 375
Standard deviation, 14, 17, 38
Standardization, 43, 50, 63
Standard potential, 258, 287, 666 *(table)*
Standard reference materials, 580 *(box)*
Standard solution, 43, 63
Standard states, 100, 123
Starch indicator, 301
  preparation of, 639 *(note)*
Steps, in an analysis, 3
Stepwise formation constants, 199
Stirring rods, 585
Stoichiometry, 47
Strong electrolyte, 94
Student's *t*, 18, 19 *(table)*, 38
Substitution method of weighing, 605
Successive approximations, method of, 107
Sulfamic acid, 158
Sulfate, gravimetric determination of, 86, 626
Sulfosalicylic acid, 159
Sulfur, gravimetric determination of, 626
Supersaturation, 74
Suspensoid, 73, 89
Systematic equilibrium calculation, 116
  in acid-base systems, 139
Systematic errors, 8

T

Taring devices, 606
Technical-grade chemicals, 583
THAM, 159, 615
Theoretical plate, 495
Thermal conductivity detector, 517
Thermal excitation, in emission spectroscopy, 452
Thermobalance, 80, 89
Thermocouple, 409
Thermogravimetric analysis, 80
Thin-layer chromatography (TLC), 555
Thiocyanate ion, as indicator, 232
Tin(II) chloride, as reductant, 293, 634
Titanium(III), as reducing agent, 309
Titer, 67 *(problem)*
Titrant, 43
  in nonaqueous solvents, 165
Titration curves, 120, 123
  acid-base, 131
  amperometric, 379
  complex formation, 205
  photometric, 423
  precipitation, 226
  redox, 275
Titrations, 44, 63
  amperometric, 379
  of a mixture of two acids, 188
  in nonaqueous media, 166, 617
  photometric, 422
  potentiometric, 331
Titrimetric analysis, 6, 44
Tolerances, for volumetric glassware, 599 *(table)*
Top-loading balance, 607 *(figure)*
Trace constituent, 2
Transition potentials, 284 *(table)*
Transmittance, 400
Tributyl phosphate, as extractant for U and Pu, 473 *(box)*
Trien, 200
TRIS, 159, 615
t-test, 21, 38
Two-pan balance, 604

U

Ultramicro analysis, 2
Ultraviolet determination of benzene in cyclohexane, 649
Ultraviolet-visible (UV-VIS) spectrophotometry, 394
Unidentate ligand, 198, 223
  titrations involving, 220
U.S.P. reagents, 583

V

Van Deemter equation, 502
Variability, 15
Variance, 17, 38
Variance-ratio test, 22, 38
Vibrational energy, 391
Vibrational relaxation in excited molecules, 444
Vinegar, acetic acid content of, 616
Vitamin C, determination of, 639
Volhard method, 231, 248, 619
Voltage, 254, 287
  in electrolysis, 345
Volumetric flasks, 590, 591
  (*figure*)
Von Weimarn's theory, 74

W

Wash bottle, 585 (*figure*)
Washing precipitates, 77, 593
Wavelength, 389
Wave number, 389
Weak electrolyte, 94
Weight, 603
Weight percent, 49, 63
Winkler method, 304, 309

Z

Zimmermann-Reinhardt solution, 296, 309
Zipax, 557
Zorbax, 558
Zwitterion, 181, 193